Arthropods of Tropical Forests

Spatio-temporal Dynamics and Resource Use in the Canopy

Arthropods are the most diverse group of organisms on our planet and the tropical rainforests represent the most biologically diverse of all ecosystems. This book, written by 79 authors contributing to 35 chapters, aims to provide an overview of data collected during recent studies in Australia, Africa, Asia and South America. The book focusses on the distribution of arthropods and their use of resources in the rainforest canopies, providing a basis for comparison between the forest ecosystems of the main biogeographical regions. Topics covered include the distribution of arthropods along vertical gradients and the relationship between the soil/litter habitat and the forest canopy. The temporal dynamics of arthropod communities, habitats and food selection are examined within and among tropical tree crowns, as are the effects of forest disturbance.

This important book is a valuable addition to the literature used by community ecologists, conservation biologists, entomologists, botanists and forestry experts.

Yves Basset is a Tupper fellow at the Smithsonian Tropical Research Institute, Panama. Vojtech Novotny is based in the Czech Academy of Sciences, Ceske Budejovice, Czech Republic. Scott Miller is Chairman of the Department of Systematic Biology, National Museum of Natural History, Smithsonian Institution in Washington DC, USA, and Roger Kitching holds the Chair of Ecology at Griffith University in Brisbane, Australia.

ARTHROPODS
OF TROPICAL FORESTS

Spatio-temporal Dynamics and Resource Use in the Canopy

Edited by:

Yves Basset
Smithsonian Tropical Research Institute, Panama City, Republic of Panama

Vojtech Novotny
Czech Academy of Sciences and University of South Bohemia, Ceske Budejovice, Czech Republic

Scott E. Miller
National Museum of Natural History, Smithsonian Institution, Washington DC, USA

and

Roger L. Kitching
Griffith University, Brisbane, Australia

CAMBRIDGE
UNIVERSITY PRESS

PUBLISHED BY THE PRESS SYNDICATE OF THE UNIVERSITY OF CAMBRIDGE
The Pitt Building, Trumpington Street, Cambridge, United Kingdom

CAMBRIDGE UNIVERSITY PRESS
The Edinburgh Building, Cambridge CB2 2RU, UK
40 West 20th Street, New York, NY 10011-4211, USA
477 Williamstown Road, Port Melbourne, VIC 3207, Australia
Ruiz de Alarcón 13, 28014 Madrid, Spain
Dock House, The Waterfront, Cape Town 8001, South Africa

http://www.cambridge.org

First published 2003

Printed in the United Kingdom at the University Press, Cambridge

Typeface Ehrhardt MT 9.5/12 pt *System* LaTeX 2_ε [TB]

A catalogue record for this book is available from the British Library

Library of Congress Cataloguing in Publication data

Arthropods of tropical forests : spatio-temporal dynamics and resource use in the canopy /
edited by Yves Basset . . . [et al.].
p. cm.
Includes bibliographical references (p.).
ISBN 0 521 82000 6 (hb)
1. Arthropoda – Ecology – Tropics. 2. Forest canopy ecology – Tropics. I. Basset, Yves, 1960–
QL434.6 .A78 2003
595′.1734–dc21 2002067235

ISBN 0 521 82000 6 hardback

Contents

List of contributors — *page* viii
Foreword — xiii
THOMAS E. LOVEJOY
Preface — xv

PART I Arthropods of tropical canopies: current themes of research

Introduction — 3

1 Canopy entomology, an expanding field of natural science — 4
YVES BASSET, VOJTECH NOVOTNY, SCOTT E. MILLER AND ROGER L. KITCHING

2 Methodological advances and limitations in canopy entomology — 7
YVES BASSET, VOJTECH NOVOTNY, SCOTT E. MILLER AND ROGER L. KITCHING

3 Vertical stratification of arthropod assemblages — 17
YVES BASSET, PETER M. HAMMOND, HÉCTOR BARRIOS, JEREMY D. HOLLOWAY AND SCOTT E. MILLER

4 Determinants of temporal variation in community structure — 28
RAPHAEL K. DIDHAM AND NEIL D. SPRINGATE

5 Herbivore assemblages and their food resources — 40
VOJTECH NOVOTNY, YVES BASSET AND ROGER L. KITCHING

PART II Vertical stratification in tropical forests

Introduction — 57

6 Distribution of ants and bark-beetles in crowns of tropical oaks — 59
ULRICH SIMON, MARTIN GOSSNER AND K. EDUARD LINSENMAIR

7 Vertical and temporal diversity of a species-rich moth taxon in Borneo — 69
CHRISTIAN H. SCHULZE AND KONRAD FIEDLER

8 Canopy foliage structure and flight density of butterflies and birds in Sarawak — 86
FUMITO KOIKE AND TERUYOSHI NAGAMITSU

9 Stratification of the spider fauna in a Tanzanian forest 92
LINE L. SØRENSEN

10 Fauna of suspended soils in an *Ongokea gore* tree in Gabon 102
NEVILLE N. WINCHESTER AND VALERIE BEHAN-PELLETIER

11 Vertical stratification of flying insects in a Surinam lowland rainforest 110
BART P. E. DE DIJN

PART III Temporal patterns in tropical canopies

Introduction 125

12 Insect responses to general flowering in Sarawak 126
TAKAO ITIOKA, MAKOTO KATO, HET KALIANG, MAHAMUD BEN MERDECK, TERUYOSHI NAGAMITSU, SHOKO SAKAI, SARKAWI UMAH MOHAMAD, SEIKI YAMANE, ABANG ABDUL HAMID AND TAMIJI INOUE

13 Arthropod assemblages across a long chronosequence in the Hawaiian Islands 135
DANIEL S. GRUNER AND DAN A. POLHEMUS

14 Seasonality of canopy beetles in Uganda 146
THOMAS WAGNER

15 Seasonality and community composition of springtails in Mexican forests 159
JOSÉ G. PALACIOS-VARGAS AND GABRIELA CASTAÑO-MENESES

16 Seasonal variation of canopy arthropods in Central Amazon 170
JOSÉ CAMILO HURTADO GUERRERO, CLÁUDIO RUY VASCONCELOS DA FONSECA, PETER M. HAMMOND AND NIGEL E. STORK

17 Arthropod seasonality in tree crowns with different epiphyte loads 176
SABINE STUNTZ, ULRICH SIMON AND GERHARD ZOTZ

PART IV Resource use and host specificity in tropical canopies

Introduction 189

18 How do beetle assemblages respond to anthropogenic disturbance? 190
ANDREAS FLOREN AND K. EDUARD LINSENMAIR

19 Organization of arthropod assemblages in individual African savanna trees 198
KARSTEN MODY, HENRYK A. BARDORZ AND K. EDUARD LINSENMAIR

20 Flower ecology in the neotropics: a flower–ant love–hate relationship 213
KLAUS JAFFÉ, JOSÉ VICENTE HERNANDEZ, WILLIAM GOITÍA, ANAÍS OSIO, FRANCES OSBORN, HUGO CERDA, ALBERTO ARAB, JOHANA RINCONES, ROXANA GAJARDO, LEONARDO CARABALLO, CARMEN ANDARA AND HENDER LOPEZ

21 Taxonomic composition and host specificity of phytophagous beetles in a dry forest in Panama 220
FRODE ØDEGAARD

22 Microhabitat distribution of forest grasshoppers in the Amazon 237
CHRISTIANE AMÉDÉGNATO

23 Flowering events and beetle diversity in Venezuela 256
SUSAN KIRMSE, JOACHIM ADIS AND WILFRIED MORAWETZ

PART V Synthesis: spatio-temporal dynamics and resource use in tropical canopies

Introduction 269

24 Habitat use and stratification of Collembola and oribatid mites 271
ANDREAS PRINZING AND STEFFEN WOAS

25 Insect herbivores feeding on conspecific seedlings and trees 282
HÉCTOR BARRIOS

26 Hallowed hideaways: basal mites in tree hollows and allied habitats 291
MATTHEW D. SHAW AND DAVID E. WALTER

27 Arthropod diel activity and stratification 304
YVES BASSET, HENRI-PIERRE ABERLENC, HÉCTOR BARRIOS AND GIANFRANCO CURLETTI

28 Diel, seasonal and disturbance-induced variation in invertebrate assemblages 315
TIMOTHY D. SCHOWALTER AND LISA M. GANIO

29 Tree relatedness and the similarity of insect assemblages: pushing the limits? 329
ROGER L. KITCHING, KAREN L. HURLEY AND LUKMAN THALIB

30 A review of mosaics of dominant ants in rainforests and plantations 341
ALAIN DEJEAN AND BRUNO CORBARA

31 Insect herbivores in the canopies of savannas and rainforests 348
SÉRVIO P. RIBEIRO

32 Canopy flowers and certainty: loose niches revisited 360
DAVID W. ROUBIK, SHOKO SAKAI AND FRANCESCO GATTESCO

33 How polyphagous are Costa Rican dry forest saturniid caterpillars? 369
DANIEL H. JANZEN

34 Influences of forest management on insects 380
MARTIN R. SPEIGHT, JURIE INTACHAT, CHEY VUN KHEN AND ARTHUR Y. C. CHUNG

35 Conclusion: arthropods, canopies and interpretable patterns 394
YVES BASSET, VOJTECH NOVOTNY, SCOTT E. MILLER AND ROGER L. KITCHING

References 407
Index 469

Contributors

Henri-Pierre Aberlenc
Centre de Coopération Internationale en Recherche Agronomique pour le Développement, TA 40/L, Campus International de Baillarguet – CSIRO, 34398 Montpellier Cedex 5, France; email: henri-pierre.aberlenc@cirad.fr

Joachim Adis
Tropical Ecology Working Group, Max-Planck-Institute for Limnology, Postfach 165, 24302 Plön, Germany; email: adis@mpil-ploen.mpg.de

Christiane Amédégnato
Centre National de la Recherche Scientifique, UPRESA 8079 ESE, Laboratoire d'Entomologie, Muséum National d'Histoire Naturelle, 45 rue de Buffon, 75005 Paris, France; email: amedeg@mnhn.fr

Carmen Andara
Departamento de Biología de Organismos, Universidad Simón Bolívar, Apartado 89000, Caracas 1080A, Venezuela

Alberto Arab
Departamento de Biología de Organismos, Universidad Simón Bolívar, Apartado 89000, Caracas 1080A, Venezuela

Henryk A. Bardorz
Lehrstuhl für Tierökologie und Tropenbiologie, Universität Würzburg, Biozentrum, Am Hubland, D-97074 Würzburg, Germany

Héctor Barrios
Programa de Maestría en Entomología, Universidad de Panamá, Panamá; email: hbarrios@ancon.up.ac.pa

Yves Basset
Smithsonian Tropical Research Institute, Apartado 2072, Balboa, Ancon, Panamá, and Programa de Maestría en Entomología, Universidad de Panamá; email: bassety@tivoli.si.edu

Valerie M. Behan-Pelletier
Biodiversity Program, Research Branch, Agriculture and Agri-Food Canada, K. W. Neatby Building, 960 Carling Avenue, Ottawa, Ontario, Canada K1A 0C6; email: behanpv@EM.AGR.CA

Leonardo Caraballo
Departamento de Biología de Organismos, Universidad Simón Bolívar, Apartado 89000, Caracas 1080A, Venezuela

Gabriela Castaño-Meneses
Laboratorio de Ecología y Sistemática de Microartrópodos, Depto. Biología, Fac. Ciencias, Universidad Nacional Autónoma de México, 04510 México; email: gcm@hp.fciencias.unam.mx

Hugo Cerda
Centro de Estudios de Agroecología Tropical, Universidad Nacional Experimental Simón Rodríguez, Apartado 47925, Caracas 1041, Venezuela

Chey Vun Khen
Forest Research Centre, Forestry Department, PO Box 1407, 90715 Sandakan, Sabah, Malaysia; email: cheyvk@tm.net.my

Arthur Y. C. Chung
Forest Research Centre, Forestry Department, PO Box 1407, 90715 Sandakan, Sabah, Malaysia; email: artycc@hotmail.com

Bruno Corbara
Laboratoire de Psychologie Sociale de la Cognition, Université Blaise Pascal, 63037 Clermont-Ferrand, France; email: corbara@srvpsy.univ-bpclermont.fr

Gianfranco Curletti
Museo Civico di Storia Naturale, CascinaVigna, Cas. Post. 89, 10022 Carmagnola (TO), Italy; email: musnat@mail.comune.carmagnola.to.it

Alain Dejean
Laboratoire d'Ecologie Terrestre, Université Toulouse III, 31062 Toulouse, France; email: dejean@cict.fr

Bart P. E. De Dijn
National Zoological Collection of Suriname, University of Suriname, Leysweg Campus, PO Box 9212, Paramaribo, Suriname
Present address: STINASU, C. Jongbawstr. 14, PO Box 12252, Paramaribo, Suriname; email: research@stinasu.sr

Raphael K. Didham
Zoology Department, University of Canterbury, Private Bag 4800, Christchurch, New Zealand; email: r.didham@zool.cantebrury.ac.nz

Konrad Fiedler
Department of Animal Ecology I, University of Bayreuth, D-95440 Bayreuth, Germany; email: Konrad.fiedler@uni-bayreuth.de

Andreas Floren
Lehrstuhl für Tierökologie und Tropenbiologie, Universität Würzburg, Biozentrum, Am Hubland, D-97074 Würzburg, Germany; email: floren@biozentrum.uni-wuerzburg.de

Cláudio Ruy Vasconcelos da Fonseca
Coordenação em Entomologia, Instituto Nacional de Pesquisas da Amazônia, Alameda Cosme Ferreira 1756, Caixa Postal 478, CEP 69011-970, Manaus, Amazonas, Brasil; email: rclaudio@inpa.gov.br

Roxana Gajardo
Departamento de Biología de Organismos, Universidad Simón Bolívar, Apartado 89000, Caracas 1080A, Venezuela

Lisa M. Ganio
Department of Forest Science, Oregon State University, Corvallis, Oregon 97331-2907, USA; email: schowalt@bcc.orst.edu

Francesco Gattesco
Università degli Studi di Milano, Dipartimento di Biologia, Sezione Ecologia, via Celoria 26, 20133 Milano, Italy; email: fgattesco@hotmail.com

William Goitía
Centro de Estudios de Agroecología Tropical, Universidad Nacional Experimental Simón Rodriguez, Apartado 47925, Caracas 1041, Venezuela

Martin Gossner
Lehrstuhl für Landnutzungsplanung und Naturschutz, Technische Universität München-Weihenstephan, Am Hochanger 13, 85354 Freising, Germany; email: Martin.Gossner@lrz.tu-muenchen.de

Daniel S. Gruner
Department of Zoology and EECB Program, University of Hawai'i at Manoa, 2538 The Mall/Edmondson Hall, Honolulu, HI 96822, USA; email: dgruner@hawaii.edu

Abang Abdul Hamid
Forest Research Centre, Forest Department, Pusat Penyelidikan Hutan, Batu 6 Jalan Penrissen, 93250 Kuching, Sarawak, Malaysia

Peter M. Hammond
Department of Entomology, The Natural History Museum, Cromwell Road, London SW7 5BD, UK; email: p.hammond@nhm.ac.uk

José Vicente Hernandez
Departamento de Biología de Organismos, Universidad Simón Bolívar, Apartado 89000, Caracas 1080A, Venezuela

Jeremy D. Holloway
Department of Entomology, The Natural History Museum, Cromwell Road, London SW7 5BD, UK; email: jdh@nhm.ac.uk

Karen L. Hurley
Cooperative Research Centre for Tropical Rainforest Ecology and Management, Australian School of Environmental Studies, Griffith University, Brisbane QLD 4111, Australia; email: K.Hurley@mailbox.gu.edu.au

José Camilo Hurtado Guerrero
Coordenação em Entomologia, Instituto Nacional de Pesquisas da Amazônia, Alameda Cosme Ferreira 1756, C.P. 478, CEP 69011-970, Manaus, Amazonas, Brasil; email: camilo@inpa.gov.br

The late Tamiji Inoue
Formerly of Center for Ecological Research, Kyoto University, 520-2113 Otsu, Japan

Jurie Intachat
Owley Farm, South Brent, Devon TQ10 9HN, UK

Takao Itioka
Graduate School of Bioagricultural Sciences, Nagoya University, Nagoya 464-8601, Japan; email: b51552b@cc.nagoya-u.ac.jp

Klaus Jaffé
Departamento de Biología de Organismos, Universidad Simón Bolívar, Apartado 89000, Caracas 1080A, Venezuela; email: Kjaffe@usb.ve

Daniel H. Janzen
Department of Biology, University of Pennsylvania, Philadelphia, PA 19104, USA; email: Djanzen@sas.upenn.edu

Het Kaliang
Forest Research Centre, Forest Department, Pusat Penyelidikan Hutan, Batu 6 Jalan Penrissen, 93250 Kuching, Sarawak, Malaysia; email: hetl@tm.net.my

Makoto Kato
Graduate School of Human and Environmental Studies, Kyoto University, Sakyo-ku, Kyoto 606-8501, Japan; email: Kato@bio.h.kyoto-u.ac.jp

Susan Kirmse
Institute of Botany, University of Leipzig, Johannisallee 21-23, D-04103 Leipzig, Germany
Present address: Am Hopfenberg 19, D-99096 Erfurt, Germany; email: Susan.Kirmse@freenet.de

Roger L. Kitching
Cooperative Research Centre for Tropical Rainforest Ecology and Management, Australian School of Environmental Studies, Griffith University, Brisbane QLD 4111, Australia; email: r.kitching@mailbox.gu.edu.au

Fumito Koike
Department of Environmental and Natural Sciences, Graduate School of Environmental and Information Sciences, Yokohama National University, 79-7 Tokiwadai, Hodogaya-ku, Yokohama 240-8501 Japan; email: koikef@ynu.ac.jp

K. Eduard Linsenmair
Lehrstuhl für Tierökologie und Tropenbiologie, Universität Würzburg, Biozentrum, Am Hubland, D-97074 Würzburg, Germany; email: ke-lins@biozentrum.uni-wuerzburg.de

Thomas E. Lovejoy
Heinz Center for Science, Economics and the Environment, 1001 Pennsylvania Ave NW, Washington DC, USA; email: lovejoy@heinzctr.org

Hender Lopez
Departamento de Biología de Organismos, Universidad Simón Bolívar, Apartado 89000, Caracas 1080A, Venezuela

Mahamud Ben Merdeck
Forest Research Centre, Forest Department, Pusat Penyelidikan Hutan, Batu 6 Jalan Penrissen, 93250 Kuching, Sarawak, Malaysia

Scott E. Miller
Department of Systematic Biology, National Museum of Natural History, Smithsonian Institution, Washington DC 20560-0105, USA; email: miller.scott@nmnh.si.edu

Karsten Mody
Lehrstuhl für Tierökologie und Tropenbiologie, Universität Würzburg, Biozentrum, Am Hubland, D-97074 Würzburg, Germany; email: mody@biozentrum.uni-wuerzburg.de

Sarkawi Umah Mohamad
Forest Research Centre, Forest Department, Pusat Penyelidikan Hutan, Batu 6 Jalan Penrissen, 93250 Kuching, Sarawak, Malaysia

Wilfried Morawetz
Institute of Botany, University of Leipzig, Johannisallee 21-23, D-04103 Leipzig, Germany; email: morawetz@rz.uni-leipzig.de

Teruyoshi Nagamitsu
Hokkaido Research Centre, Forest and Forest Products Research Institute, Hitsujigaoka-7, Toyohira, Sapporo Hokkaido 062-8516, Japan; email: nagamit@ffpri-hkd.affrc.go.jp

Vojtech Novotny
Institute of Entomology, Czech Academy of Sciences and Biological Faculty, University of South Bohemia, Branisovska 31, 370 05 Ceske Budejovice, Czech Republic; email: novotny@entu.cas.cz

Frode Ødegaard
Norwegian Institute for Nature Research, Tungasletta 2, NO-7845 Trondheim, Norway; email: frode.odegaard@nina.no

Frances Osborn
Instituto de Biomedicina y Ciencias Aplicadas, Universidad de Oriente, Cumana, Venezuela

Anaís Osio
Departamento de Biología de Organismos, Universidad Simón Bolívar, Apartado 89000, Caracas 1080A, Venezuela

José G. Palacios-Vargas
Laboratorio de Ecología y Sistemática de Microartrópodos, Depto. Biología, Fac. Ciencias, Universidad Nacional Autónoma de México, 04510 México, D. F.; email: jgpv@hp.fciencias.unam.mx

Dan A. Polhemus
Department of Systematic Biology, National Museum of Natural History, Smithsonian Institution, Washington DC 20560-0105, USA; email: polhemus@nmnh.si.edu

Andreas Prinzing
Institute of Zoology, Dept V (Ecology), University of Mainz, Becherweg 13, D-55099 Mainz, Germany; email: prinzing@oekologie.biologie.uni-mainz.de

Sérvio P. Ribeiro
Departmento de Ciências Biológicas, Instituto de Ciências Exatas e Biológicas, Universidade Federal de Ouro Preto, 35.400-000 Ouro Preto, MG, Brazil; email: spribeiro@iceb.ufop.br

Johana Rincones
Departamento de Biología de Organismos, Universidad Simón Bolívar, Apartado 89000, Caracas 1080A, Venezuela

David W. Roubik
Smithsonian Tropical Research Institute, Apartado 2072, Balboa, Ancon, Panamá; email: roubikd@tivoli.si.edu

Shoko Sakai
Graduate School of Human and Environmental Studies, Kyoto University, Yoshida-Nihonmatsu-cho, Sanyo-ku, Kyoto 606-8501, Japan; email: sakai@bio.h.kyoto-u.ac.jp

Timothy D. Schowalter
Department of Entomology, Oregon State University, Corvallis, Oregon 97331-2907, USA; email: Schowalt@bcc.orst.edu

Christian H. Schulze
Department of Animal Ecology I, University of Bayreuth, D-95440 Bayreuth, Germany
Present address: Institute of Agroecology, Waldweg 26, Georg-August-University, D-37073 Göttingen, Germany; email: c.schulze@uaoe.gwdg.de

Matthew D. Shaw
Department of Zoology and Entomology, University of Queensland, St Lucia, QLD 4072, Australia; email: mshaw@zen.uq.edu.au

Ulrich Simon
Lehrstuhl für Landnutzungsplanung und Naturschutz, Technische Universität München-Weihenstephan, Am Hochanger 13, 85354 Freising, Germany; email: Ulrich.Simon@lrz.tu-muenchen.de

Line L. Sørensen
Zoological Museum, University of Copenhagen, Universitetsparken 15, DK-2100 Copenhagen Ø, Denmark; email: llsorensen@zmuc.ku.dk

Martin R. Speight
University of Oxford, Department of Zoology, South Parks Road, Oxford OX1 3PS, UK; email: martin.speight@zoology.ox.ac.uk

Neil D. Springate
Department of Entomology, The Natural History Museum, Cromwell Road, London SW7 5BD, UK; email: nds@nhm.ac.uk

Nigel E. Stork
Cooperative Research Centre for Tropical Rainforest Ecology and Management, James Cook University, Cairns Campus, PO Box 6811, Cairns QLD 4870, Australia; email: nigel.stork@jcu.edu.au

Sabine Stuntz
Lehrstuhl für Botanik II der Universität Würzburg, Julius-von-Sachs-Platz 3, D-97082 Würzburg, Germany; email: Sabine@stuntz.com

Lukman Thalib
Cooperative Research Centre for Tropical Rainforest Ecology and Management, Australian School of Environmental Studies, Griffith University, Brisbane QLD 4111, Australia

Thomas Wagner
Universität Koblenz-Landau, Institut für Biologie, Rheinau 1, 56075 Koblenz, Germany; email: thwagner@uni-koblenz.de

David E. Walter
Department of Zoology and Entomology, University of Queensland, St Lucia, QLD 4072, Australia; email: d.walter@mailbox.uq.edu.au

Neville N. Winchester
Department of Biology, University of Victoria, PO Box 3020, Victoria, British Columbia V8W 3N5, Canada; email: tundrast@uvvm.uvic.ca

Steffen Woas
State Museum of Natural History, Department of Zoology, Pf 11 13 64, 76063 Karlsruhe, Germany

Seiki Yamane
Department of Biology, Faculty of Science, Kagoshima University, Kagoshima 890-0065, Japan; email: sky@sci.kagoshima-u.ac.jp

Gerhard Zotz
Botanisches Institut der Universität Basel, Schönbeinstrasse 6, CH-4056 Basel, Switzerland and the Smithsonian Tropical Research Institute, Apartado 2072, Balboa, Ancon, Panamá; email: gerhard.zotz@unibas.ch

Foreword

Life in the tropical forest canopy is vividly different. That was instantly apparent – and must be to anyone – in the first moments I spent atop one of the very first towers built in 1965 in the forests on the outskirts of the main Amazon port city of Belem. Bright sunlight contrasted with the gloom below. So too did the dramatic daily fluctuations in temperature with the relatively constant conditions of the understorey. There was a very different fauna: brilliantly coloured birds, a plethora of arthropods and a squirrel monkey, which showed no fear of this larger primate newly arrived in its canopy world.

Exciting as it was, the tower 'technology' was frustrating in its limitations. So much lay simply beyond reach, sight or measurement – not very different from William Beebe's first descent into the ocean depths in a bathyscape. It was obvious there was an exciting world to be studied in the treetops and that the tropical forests could never be properly understood while the treetops remained tantalizingly beyond most reach.

The tropical forest canopy remains one of the great frontiers of biological exploration, but development of new means of access has stimulated a flurry of investigation. That this volume goes to press but 5 years after publication of *Canopy Arthropods* is testament to the activity. The astonishing diversity of the arthropods is exciting to investigate and document, yet that is but the first step in understanding the upper reaches of what Darwin described as 'one great wild tropical hothouse made by nature for herself', as well as understanding the articulation with the shaded world below.

Already in this volume there are questions of evolution and ecology, patterns of life histories, grist for theory and hints of the role of canopy arthropods in ecosystem function. One can but dimly perceive, yet keenly anticipate, what another 5 years will yield. Fortunately, the adventure involved in canopy research is of a sort that readily captures the public imagination, which I fervently hope can be harnessed for conservation of these forests that are so rapidly in retreat. Edward O. Wilson had cogently noted the great irony of organic evolution, namely just at the moment it has reached a point where it can understand itself, it is destroying the very living library upon which such understanding must be based. Nowhere is that more true than in tropical forests and their canopies.

Thomas E. Lovejoy
President, Heinz Center for Science, Economics and the Environment
Washington, DC, USA

Preface

The idea of bringing together an updated account of worldwide research on arthropods of tropical forest canopies originated during the preparation of the symposium *Host and Microhabitat Use by Arthropods in Tropical Forest Canopies* for the *XXI International Congress of Entomology*, held in Brazil in August, 2000 (Gazzoni, 2000). At that time, we realized that since the publication of *Canopy Arthropods* (Stork *et al.*, 1997), the information available on assemblages of canopy arthropods had improved significantly. Many research groups had new data that better explained arthropod distributions in forest canopies and their use of food resources. We invited active research groups to report on original, previously unpublished, results from a major, current research project addressing the ecological dynamics of arthropod communities in the canopy. Contributions were selected to cover diverse themes, arthropod taxa, forest types and sampling methods at various tropical locations.

These chapters are organized into five sections: an introduction, three sections of case studies and a synthesis. The first section includes a general introduction, a review of recent techniques of canopy access used by entomologists and three broad reviews of the state-of-the-art of the main current research themes in canopy entomology. Each of these reviews introduces a particular section of case studies. These sections represent the main factors of variance in the distribution of arthropod communities in the canopy and, at the same time, the main themes of current research: (i) patterns of spatial distribution emphasizing vertical stratification and relationships with the soil/litter habitat; (ii) patterns of temporal distribution including annual variability, seasonality and diel activity; and (iii) patterns of resource use and host specificity. The synthesis section includes research topics for which more information is available and which are particularly amenable to discussion. The concluding chapter summarizes interpretable patterns in the distribution of arthropod assemblages in tropical canopies.

This volume aims to provide a balanced overview of an active and burgeoning area of research, targeting the most diverse group of organisms – the arthropods – in the most diverse terrestrial ecosystems – the tropical forests. We believe that the 35 contributions, written by 79 authors and peer-reviewed by renowned experts, fulfil this aim. We have attempted to provide a balanced coverage for each of the main biogeographical regions in the tropics, the Indo-Australasian, African and neotropical regions. One recurrent problem in canopy science is high pseudoreplication, caused by physical constraints on sampling schemes. Comparing datasets from different biogeographical regions may improve replication and may help to explore whether community patterns hold at different scales and geographic locations. This strategy also helps to highlight lacunae in our knowledge, such as the lack of data originating from Africa, and from dry tropical forests in general.

Peer-review of this diverse and multiauthored volume was not an easy task. We would like to thank the following reviewers for their help: R. Beaver, V. Behan-Pelletier, P. Borges, J. Bridle, J. Burns, C. Carbonell, J. Coddington, C. R. da Fonseca, A. Davis, A. Dejean, R. Didham, T. Erwin, K. Fiedler, A. Floren, L. Fox, E. Guilbert, D. Gruner, J. Hill, H. Höfer, J. Holloway, J. Intachat, K. Jaffé, F. Koike, S. Leather, M. Leponce, R. Leschen, E. Linsenmair, J. Majer, G. McGavin, R. Norton, M. Nummelin, F. Ødegaard, B. Orr, C. Ozanne, J. Palacios-Vargas, M. Paoletti, A. Prinzing, B. Reynolds, S. Ribeiro, D. Roubik, T. Schowalter, T. Schultz, U. Simon, L. Sørensen, M. Toda, T. Wagner and N. Winchester.

The production of this volume benefited greatly from the assistance of Karolyn Darrow at the Smithsonian Institution. We are also grateful to our editors at Cambridge University Press, Ward Cooper, Jillian Culnane and Jane Ward. Henri-Pierre Aberlenc, Hirochika Setsumasa and Michel Boulard discussed the book cover and provided pictures for it.

The preparation of this book also owns much to discussions with many colleagues and assistants, who helped the four editors in various ways in their different research projects in tropical rainforests. These frequent companions of canopy and fogging adventures are too numerous to acknowledge individually here but special thanks to our associates with the canopy raft and canopy crane, and to the parataxonomists who worked with us in Papua New Guinea, Guyana and Gabon. The US National Science Foundation has generously supported the collaboration of three of us (Y. B., V. N. and S. M.) since 1994, contributing to the development of many of the ideas presented here.

Yves Basset, Vojtech Novotny, Scott Miller
and Roger Kitching
Panama, Budejovice, Washington & Brisbane
January 2002

REFERENCES

Gazzoni, D. L. (ed.) (2000). *XXI International Congress of Entomology*. Londrina, Brazil: Embrapa Soja.

Stork, N. E., Adis J. A., & Didham, R. K. (eds.) (1997). *Canopy Arthropods*. London: Chapman & Hall.

PART I

Arthropods of tropical canopies: current themes of research

Introduction

The aim of these preliminary chapters is to introduce the reader to tropical forest canopies, their characteristics and how they affect their arthropod inhabitants, and to the main problems that entomologists must face when implementing their protocols or experiments in the canopy. These chapters will be of interest to both the tropical and temperate entomologist/ecologist, as they often stress differences between tropical and temperate forests, and the resources that they provide to arthropods. There has sometimes been controversy, mostly semantic, as to what constitutes a 'canopy', especially in tropical rainforests. A brief chapter by the editors clarifies this issue and stresses the significance of tropical forest canopies in entomological and ecological research, and in biological sciences in general. Readers that may not be familiar with methods of accessing the canopy and conducting entomological projects there will be particularly interested to read the second chapter, also written by the editors. It reviews the latest methods providing access to the canopy with examples of entomological studies, and it compares the different advantages and inconveniences of each method with particular reference to entomological work. It also points out the major problems that must confront canopy entomologists and provides some hints as how to overcome them. As far as methods of collecting arthropods in forest canopies are concerned, we leave the reader to refer to the volume of Stork *et al.* (1997a), in which several contributions deal with various methodological aspects of collecting arthropods in the canopy.

One of the main differences between tropical and temperate forests is the much sharper abiotic and biotic gradients from the floor to the upper canopy in the former, especially in closed wet tropical forests. In Ch. 3, Basset *et al.* contrast arthropod stratification in tropical and temperate forests. They emphasize the distinctness of the upper canopy in terms of abiotic and biotic factors and, possibly, of the associated fauna. Arthropod stratification in tropical forests is examined with particular reference to the various habitats and resources available above the forest floor. In particular, the authors conclude that arthropod assemblages in the soil and canopy cannot be studied in isolation, an opinion shared by several contributors to this volume.

For many arthropods in temperate forests, food resources are in a sense largely predictable temporally. The annual cycle is initiated with bud break in the spring and closes with leaf fall and dormancy in autumn and winter. From this viewpoint, patterns are much more complex in tropical forests, as plants there exhibit a variety of leaf, flowering and fruiting phenologies, with concomitant temporal changes in the abundance of insect herbivores and their enemies. In Ch. 4, Didham and Springate review the factors responsible for long- and short-term temporal changes in the diversity and abundance of arthropod communities in tropical canopies. They adopt a most interesting approach by considering first the effects of anthropogenic disturbance, tree growth rates and senescence patterns. They then ask to what extent does tree phylogeny and/or successional status mediate the expression of temporal trends in arthropod communities in tree canopies.

Rainforest canopies contain an exceptionally rich array of organic resources for an equally diverse number of consumers; it is, therefore, very tempting as well as difficult to study. To illustrate this point and close this section, Novotny *et al.* discuss more particularly in Ch. 5 the different food resources available to insect herbivores in the canopy of tropical forests. They show that complex patterns of plant use by herbivores result both from fluctuating resource availability in space and time and from interactions between herbivores and their predators and parasitoids. In doing so, they review the main methodological approaches, results and remaining challenges in this field of study.

1

Canopy entomology, an expanding field of natural science

Yves Basset, Vojtech Novotny, Scott E. Miller and Roger L. Kitching

In 1929, O. W. Richards and his colleagues hoisted light traps into the canopy of a Guyana forest and became the first to collect arthropods quantitatively from the canopy of any tropical rainforest (Hingston, 1930, 1932). Today, canopy science has become a novel, burgeoning and exciting field in the natural sciences, as evidenced by the ever increasing number of publications focussing on this habitat (Nadkarni & Parker, 1994; Nadkarni *et al.*, 1996). The vitality of canopy science can be traced back to a series of studies about the canopy flora and fauna in tropical forests performed about 20 years ago, in which entomology figured prominently (e.g. Hallé *et al.*, 1978; Perry, 1978; Wolda, 1979; Erwin & Scott, 1980; Nadkarni, 1981; Lowman & Box, 1983; Adis *et al.*, 1984; Stork, 1987a). This sparked a lively and continuing scientific interest in tropical forest canopies and their inhabitants (e.g. Nadkarni & Parker, 1994; Stork & Best, 1994; Lowman *et al.*, 1995; Stork *et al.*, 1997a,c).

The term 'canopy' has been used by different authors to mean rather different things. A definition gaining acceptance is the aggregate of every tree crown in the forest, including foliage, twigs, fine branches and epiphytes (Nadkarni, 1995). In a recent review of definitions, Moffett (2000) promotes the view that 'canopy' should be regarded as all elements of the vegetation above the ground and urges us to distance ourselves from the anthropocentric view that only high tree canopies deserve that designation. He supports this view by reference to agricultural science, in which even meadows are regarded as having canopies (e.g. Monteith, 1965). He reminds us of the tacit assumption underlying much of ecological science that processes and dynamics can be scaled up across many orders of magnitude. Of course, if this highly inclusive view of 'canopy' is adopted, then the science needs a careful set of definitions for the height subdivisions to which we must refer for descriptive clarity.

The term 'understorey' may be defined as the vegetation immediately above the forest floor and reachable by the observer or, if such measurements are available, the zone with less than 10% light transmittance (Parker & Brown, 2000). The French word *canopée* denotes the interface between the uppermost layer of leaves and the atmosphere (e.g. Hallé & Blanc, 1990). It has been translated as 'canopy surface' (e.g. Bell *et al.*, 1999) or 'outer canopy' (Moffett, 2000). Further, 'upper canopy' refers to the canopy surface and the volume immediately below (a few metres) that may be occupied by arthropods foraging specifically in the upper region of the canopy. This zone may be distinct only in tall, wet and closed tropical forests and its occurrence is discussed by several contributions in this volume (Chs. 22 and 27). Emergent trees and the air above the canopy may be termed the 'overstorey' as this layer is important for the dispersal of several arthropod groups (Amédégnato, 1997; Compton *et al.*, 2000). Note that use of the above terms does not necessarily mean that the forest is stratified: they are convenient to describe the origin of the material collected. Forest strata are best described in terms of segments of gradients, rather than height above the ground (Parker & Brown, 2000). For cogent descriptions of the canopy structure and canopy surface, see Bongers (2001) and Birnbaum (2001), respectively.

The expression 'canopy arthropods' also needs a brief explanation. Although some species spend their entire life cycle in the canopy ('canopy residents' *sensu* Moffett, 2000), others may live in the soil/litter habitat as immature stages and may move later into the canopy to feed on other resources as adults (Ch. 3). We use the term 'canopy arthropods' to refer to all those species that are dependent in any way on the canopy at least at some stage of their life cycle.

The role of forest canopies in key ecosystem processes within the biosphere, such as energy flows, biogeochemical cycling and the dynamics of regional and global climates, cannot be understated. The forest canopy is the principal site of energy assimilation in primary production, with ensuing intense interchange of oxygen, water vapour and carbon dioxide. Most photosynthetic activity in the biosphere occurs in the canopy, and forest canopies account for almost half of the carbon stored in terrestrial vegetation (e.g. Lowman & Nadkarni, 1995; Wright & Colley, 1996; Malhi & Grace, 2000).

In addition, scientists and the media have been captivated even more by the near countless species of animals and plants sustained by tropical forest canopies. The majority of these organisms is still unknown or undescribed. Erwin (1983a) termed the canopy of tropical forests 'the last biotic frontier', referring to the vast, but poorly studied, richness of organisms, particularly arthropods, resident in the canopy. This epithet and the underlying idea was seized upon and expanded by authors including Wilson (1992) and Moffett (1993). This volume focusses on arthropods of tropical canopies. Yet, canopies of all types, including temperate and tropical forests, play a crucial role in the maintenance of ecological processes and biodiversity. Since different forces may structure tropical and temperate systems (e.g. Turner *et al.*, 1996), comparative research is vital. We hope that this book will stimulate similar parallel efforts to bring together most of the information on arthropods of temperate canopies.

Tropical forest canopies represent fascinating environments for entomologists and ecologists for many reasons. First, with the exceptional diversity of their arthropod communities, tropical forest canopies may be the most species-rich habitat on Earth (e.g. Erwin, 1983a), although the soil/litter of tropical forests is another strong contender (e.g. André *et al.*, 1994; Hammond, 1994; Stork, 1988). Most of the contributions in this volume confirm the impressive diversity of canopy arthropods in the tropics.

Second, despite increasing interest in their study, the fauna of tropical canopies remains largely unknown and has been the subject of much controversy among entomologists and ecologists. Even the simplest questions remain unanswered, such as: how many arthropod species live in the canopy of various tree species and forest types, what is their resource base, or how have their ecological niches evolved (e.g. May, 1994). For example, the deceptively simple question as to whether tropical canopy herbivores are more or less specialized than their temperate counterparts, or than counterparts foraging in the understorey, is far from answered (e.g. Basset, 1992a; Gaston, 1993). Several contributions in this volume address these issues (Chs. 5, 21, 22 and 29).

Third, tropical canopies represent a key arena in which biologists can study the interactions of multiple species within communities and test hypotheses on evolution and coevolution. Studies of tropical environments and their organisms have contributed significantly to the advancement of modern ecology (Chazdon & Whitmore, 2002). Canopy ecology is a young, 'frontier' science, exploring a largely unknown environment, establishing basic patterns and developing nascent theories. We expect that synecological studies of canopy organisms, documenting the rich variety of their interactions in this complex habitat, will boost our ability to generate and test evolutionary scenarios. Because of the high diversity of canopy arthropods, they also represent model organisms for studying macroecology and discovering general patterns and rules in nature (Lawton, 1999). The synecology of arthropod assemblages in the canopy is a central theme of several contributions in this volume (Chs. 19, 20, 26, 29, 30 and 33).

Last, rapid habitat loss in tropical forests makes their canopy inhabitants particularly vulnerable to endangerment or even extinction. Dissemination of scientific information to foster scientific interest in canopy arthropod communities is crucial for the survival of canopy communities, as well as for those who study them. For example, sound estimates of species loss cannot be inferred from ground-based studies alone; data on the distribution and ecology of canopy arthropods are essential to predict the effects of forest disturbance and fragmentation on species loss (e.g. Willott, 1999; Kitching *et al.*, 2000). These issues are discussed in Chs. 18, 28 and 34.

Most of the biological activity in tropical rainforests is concentrated in the upper canopy rather than the understorey (e.g. Hallé & Blanc, 1990; Parker, 1995). Many abiotic and biotic characteristics of the canopy are different from those in the underlying understorey. The higher illumination levels in the canopy promote rapid rates of photosynthesis, which, in turn, promotes high plant production, thereby sustaining a more abundant and diverse community of animals than in the understorey (Wright & Colley, 1996). The vertical stratification of resources and organisms is one of the key characteristics of tropical forests, particularly of wet forests. Many contributions in this volume explore how different and unique canopy arthropods are compared with their counterparts in the understorey (Chs. 6–8, 11, 25 and 27), or in the soil/litter habitat (Chs. 9, 10, 24 and 26).

Other characteristics promoting a distinctive canopy fauna relate to the discrete distribution of resources in the canopy. The high rate of plant production in the canopy is often not continuous (e.g. van Schaik *et al.*, 1993), thus shaping specific strategies and life histories in insect herbivores and their associated assemblages of predators and parasitoids. This important theme is addressed by several contributions (Chs. 4, 12, 14–16). Others emphasize the temporal scale at which these complex interactions occur, from diel activity patterns (Chs. 23, 27 and 28) to transannual effects (Ch. 13).

Arthropods and their food resources may experience different microclimates within the heterogeneous environment of tropical forests (e.g. Lowman, 1995; DeVries *et al.*, 1999a). This is most apparent along the vertical gradients of closed wet tropical forests, but it also occurs horizontally when natural tree falls modify the structure of the canopy. Several contributions examine the spatial distribution of particular resources and its consequences for canopy arthropods: the physical qualities of leaves (Ch. 31), the accumulation of litter (Ch. 10) and the presence of epiphytes (Ch. 17), vines (Ch. 31) or flowers (Chs. 20 and 23).

Many of the earlier studies of canopy arthropods took primarily a faunistical approach (reviews in Erwin, 1989, 1995; Stork & Hammond, 1997; Basset 2001b). They often relied upon indirect sampling methods, such as light trapping or pyrethrum knockdown (canopy fogging). Workers were limited to ground level and arthropod life histories and population dynamics could only be inferred indirectly from their data. In addition to improvements in fogging techniques, recent methodological developments in canopy access (some reviewed in Moffett & Lowman, 1995) allow the observation of canopy arthropods *in situ* and their live collection.

Today, methods of canopy access that are favoured by entomologists include construction cranes, canopy towers, canopy rafts, aerial sledges, aerial walkways or single rope techniques (see Ch. 2). Collecting methods are accordingly more diverse and reflect the increasing complexity of the questions that are pursued by canopy biologists and entomologists. The contributions in the present volume truly reflect this revolution in canopy access, sampling methodologies and the consequent maturation of research.

2

Methodological advances and limitations in canopy entomology

Yves Basset, Vojtech Novotny, Scott E. Miller and Roger L. Kitching

ABSTRACT

Many of the earlier studies of canopy arthropods relied upon indirect sampling methods, such as light trapping or pyrethrum knockdown (canopy fogging). Usually, observers were limited to ground level, with few opportunities to study canopy organisms directly. Life histories and population dynamics could only be inferred from such data. In addition to improvements in fogging techniques, recent methodological developments (such as construction cranes, canopy towers, canopy rafts, aerial sledges, aerial walkways or single rope techniques) have broadened our ability to access the canopy and allow the observation and collection of canopy arthropods *in situ*. Collecting methods are accordingly much more diverse and reflect the increasing complexity of the questions that are pursued by canopy biologists. In this chapter, we review past and recent methods of canopy access that have allowed entomologists to sample arthropods in tropical forest canopies. In doing so, we stress the advantages and limitations of each method, from an entomological viewpoint. We further review key problems in tropical canopy entomology and discuss possible remedies.

INTRODUCTION

Logistical problems in tropical canopy entomology are dictated by the physical environment and canopy access, as well as by the formidable biodiversity present in the canopy. How to sample efficiently the canopy habitat, how to document adequately the life history of its inhabitants, how to perform manipulative experiments there and how to archive the ensuing data and collections efficiently are recurrent problems. It is not an exaggeration to state that the resolution of these interrelated problems would take us a large step closer to understanding the diversity and distribution of life on Earth.

Surveys of arthropods in the canopy of tropical rainforests are a recent field of investigation. The oldest attempt to collect quantitative information on the invertebrates in the canopy of tropical rainforests appears to be the pioneering efforts led by O. W. Richards, who hoisted light traps up in the canopy in 1929 during the Oxford University expedition in Guyana (Hingston, 1930, 1932; Sutton, 2001). With the exception of the erection of towers and occasional insect collection from these in the 1950s (Haddow *et al.*, 1961; Haddow & Ssenkubuge, 1965), the mass collection of arthropods from the canopy did not progress notably until the development of ground-based fogging and light-trapping techniques in the 1970s (fogging: Roberts, 1973; Erwin & Scott, 1980; Gagné & Howarth, 1981; light trapping: Sutton, 1979; Sutton & Hudson, 1980; Holloway, 1984a). From then, methods adopted by entomologists to access, collect and experiment in the canopy greatly diversified. Canopy access and canopy entomology in the tropics are reviewed historically in Mitchell (1982), Erwin (1989, 1995), Moffett (1993), Moffett and Lowman (1995), Lowman and Wittman (1996) and Sutton (2001), among others.

This contribution examines recent advances and the most significant remaining limitations in canopy entomology. Improved canopy access, which in recent years has represented the most significant development in canopy entomology and science (Barker & Pinard, 2001), is discussed at more length.

ADVANCES IN CANOPY ENTOMOLOGY

Canopy access

Recent progress in canopy access has allowed entomologists to sample arthropods *in situ*, in better conditions and for longer periods of time than previously, and to increase notably the number of spatial replicates. A brief review of the main methods of canopy access favoured by entomologists follows. These methods are further compared in terms of spatial and temporal replicates, time investment in the field and productivity of arthropod material (Table 2.1) and some are illustrated in Fig. 2.1.

Ground-based techniques and their evolution

Ground-based techniques that provide indirect access to the canopy have always been popular with entomologists. For example, out of 17 tropical studies reported in Stork *et al.* (1997a), 12 were based on the ground and did not include direct access to the canopy and sampling or observation *in situ* of its arthropod inhabitants. An increasing number of studies are concerned with sampling arthropods *in situ*, as the present volume attests (about half of contributions relied on some form of canopy access, the other half were ground based).

The favourite technique, insecticide knockdown (known as canopy fogging or canopy misting), includes hoisting in the canopy a radio-controlled fogging or misting machine that dispenses insecticide in different directions. The dying arthropods fall on collecting trays located just above ground level. Alternatively, one may fog or spray from the ground. Other popular ground-based methods include hoisting light traps (or other types of trap) in the canopy using pulley systems, which allows convenient surveys of the traps. For protocols, advantages and limitations of these and other techniques used in canopy entomology, see Erwin (1983b), Stork and Hammond (1997), Basset *et al.* (1997b), Adis *et al.* (1998b) and Kitching *et al.* (2000).

The first attempts to use insecticide knockdown in the tropics often targeted plantations or relatively open and low vegetation (e.g. Gibbs *et al.*, 1968; Roberts, 1973; Gagné & Howarth, 1981; Room, 1975). Other studies performed in primary and tall rainforests followed and sparked a vigorous interest in both canopy arthropods and the techniques of insecticide knockdown itself (e.g. Erwin & Scott, 1980; Adis *et al.*, 1984, 1997; Stork, 1987b; Paarman & Kerck, 1997). The main advantages of this technique are the quick implementation of a systematic and productive protocol (Table 2.1) that produces reasonably clean samples and that it is ideal for general surveys of forest tracts and large-scale taxonomic work, as well as comparative studies. Apart from some technical limitations (e.g. dependence upon weather conditions, sampling often limited to daybreak, etc.), the main disadvantages include the fact that the specimens collected are dead or moribund, the difficulty in tracing the precise origin of the specimens to a specific habitat within the tree fogged, the often low number of spatial replicates available, and the likely escape of larger and more robust individuals. In addition, selective sampling of the upper canopy in tall forests is difficult, as the method yields a mixture of specimens originating from different forest strata.

In recent years insecticide knockdown has been improved significantly by several techniques.

- The observer controls directly the action of the fogging machine by climbing the tree with the single-rope technique (see below); this improves the efficiency of fogging, particularly within tall trees (e.g. Erwin, 1989; Basset, 1995; Floren & Linsenmair, 1997b).
- The collecting trays are set up immediately below the foliage or tree parts fogged (e.g. Ellwood & Foster, 2000; Ch. 34). This prevents the drift of small specimens away from trays located at ground level.
- Particular tree species are selectively fogged by stretching a cotton roof, preventing collection of arthropods from neighbouring trees (Floren & Linsenmair, 1997b).
- Reducing the insecticide concentrations enables live insects to be collected for rearing and observation (e.g. Adis *et al.*, 1997; Paarmann & Kerck, 1997).

Fogging methods and results are discussed further in Chs. 13–16, 18, 19, 29 and 34.

The interest in traps, especially light traps, for sampling the canopy fauna originated from several studies, particularly those of Wolda (1979) and Smythe (1982) on Barro Colorado Island, Panama, of Sutton (1979) in Brunei, and of Sutton and Hudson (1980) in Zaire. These traps are highly productive for nocturnal insects, particularly moths and beetles (Table 2.1). The size and composition of catches in such light traps, however, is highly dependent on environmental factors such as temperature, relative humidity, other ambient light sources and air movement. Correction factors can be computed but are likely to be region or even site specific (e.g. Bowden, 1982). The range of attraction to traps depends on taxa and is often difficult to estimate. The usual response to these various problems has been to

Table 2.1. *Comparison of canopy access methods with regard to spatial and temporal replicates, duration of field work and arthropod material produced. Representative references have been selected to cover different forest types. Note that sites may have been defined differently among studies, and that the productivity of studies depend on many parameters, including collecting method, target group and forest type*

Method/reference	Tree species	Tree individuals	Sites	Duration	Individuals collected	Collecting method	Target group
Ground-based techniques							
Erwin, 1983b	?	10	4	2 months	24 350	Pyrethrum knockdown	Arthropods
Stork, 1987b	5	10	1	12 days	23 874	Pyrethrum knockdown	Arthropods
Kitching *et al.*, 1993	10 × 10 m plots	Various	16	3 years	32 951	Pyrethrum knockdown	Arthropods
Allison *et al.*, 1997	6	51	>3	>2 weeks	45 464	Pyrethrum knockdown	Arthropods
Floren & Linsenmair, 1997b	3	10	>2	?	155 000	Pyrethrum knockdown	Arthropods
Wagner, 1998	4	64	3	?	3952	Pyrethrum knockdown	Chrysomelidae
Missa, 2000	17	38	1	13 months	7899	Pyrethrum knockdown	Curculionidae
Wolda, 1982	1	1	1	4 years	87 547	Light trap	Homoptera
Sutton *et al.*, 1983a	4	4	4	38 days	98 569	Light trap	Arthropods
Single rope technique							
Basset, 1996	10	>100	2	1 year	52 858	Hand-collecting/beating	Insect herbivores
Platforms, towers and walkways							
Corbet, 1961b	>2?	>4?	1	3 months	1951	Light trap	Culicidae
Kato *et al.*, 1995	1?	1?	1	1 year	1 023 008	Light trap	Arthropods
Canopy crane							
Blüthgen *et al.*, 2000b	27	66	1	3 months	?	Hand-collecting/baiting	Ants and homopterans
Ødegaard, 2000a	24	37	1	2 years	35 479	Hand-collecting/beating	Phytophagous beetles
Y. Basset & H. Barrios, unpublished data	>80	>100	1	1 year	>2000	Hand-collecting/beating	Curculionoidea
Canopy raft							
Basset *et al.*, 1992	15	>15	4	4 days	2271	Beating	Insect herbivores
Dejean *et al.*, 1999, 2000e	*c.* 30	34	3	Several days	>10 000	Hand collecting	Ants
Basset *et al.*, 2001	>10	>10	3	15 days	9026	Beating/different traps	Insect herbivores
Canopy sledge							
Dejean *et al.*, 1992b, 1999	62	167	Many	*c.* 6 hours	>50 000	Hand collecting	Ants
Dejean *et al.*, 1998, 1999	*c.* 60	71	Many	*c.* 3 hours	>20 000	Hand collecting	Ants
Dejean *et al.*, 2000b	106	220	Many	Several days	>60 000	Hand collecting	Ants
Basset *et al.*, 2001a	56	78	Many	*c.* 16 hours	626	Beating	Insect herbivores
Treetop bubble							
Basset *et al.*, 2001a	>30[a]	>30[a]	2	4 days	4003	Sticky traps	Insect herbivores

[a]Depends on the length of the transect

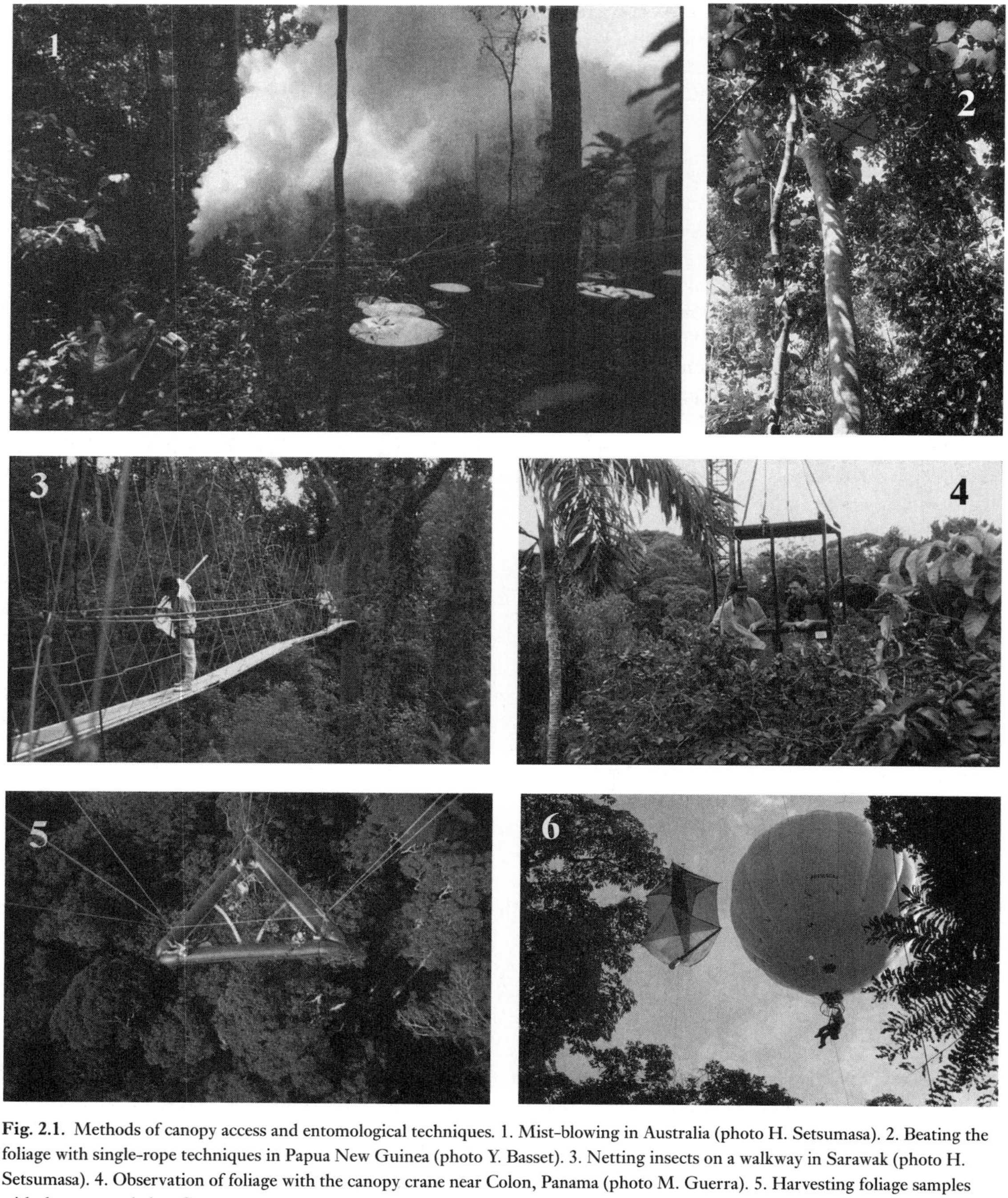

Fig. 2.1. Methods of canopy access and entomological techniques. 1. Mist-blowing in Australia (photo H. Setsumasa). 2. Beating the foliage with single-rope techniques in Papua New Guinea (photo Y. Basset). 3. Netting insects on a walkway in Sarawak (photo H. Setsumasa). 4. Observation of foliage with the canopy crane near Colon, Panama (photo M. Guerra). 5. Harvesting foliage samples with the canopy sledge, Cameroon (photo H. Setsumasa). 6. Surveying interception-flight traps with the treetop bubble, Gabon (photo H. Setsumasa).

use more traps more often. Recent studies have used light traps on towers erected in rainforests (e.g. Kato *et al.*, 1995; Willott, 1999; Chs. 7 and 12) or suspended from ropes in the high canopy (Kitching *et al.*, 2000).

Single-rope technique

Perry (1978) appears to have been the first scientist to modify the single-rope technique used in caving to climb tall rainforest trees. The observer climbs with harnesses and jumars on a static rope anchored at ground level. For later developments and safety hints, see Perry and Williams (1981), Whitacre (1981), Landsberg and Gillieson (1982), Dial and Tobin (1994), Laman (1995), Barker (1997) and Barker and Sutton (1997). The equipment is inexpensive and allows a protocol to be developed based on spatial replication (Table 2.1). However, in addition to safety and liability concerns, the mobility of the climber is often restricted (but see description of the more unrestrained arborist method in Dial & Tobin (1994)) and the upper canopy or the crown periphery are often out of reach, unless climbing from emergent trees. In addition, the availability of suitable branches able to bear the weight of climber and equipment necessarily place constraints on precisely which locations can be accessed.

Entomologists have used this technique to sample the fauna of epiphytes (e.g. Nadkarni & Longino, 1990; Rodgers & Kitching, 1998; Ch. 10), to study herbivory (e.g. Lowman, 1985; Sterck *et al.*, 1992; Barone, 2000), to set up and survey various traps (e.g. Basset, 1991a; Ch. 6) or to collect live arthropods *in situ* (e.g. Basset, 1996; Longino & Colwell, 1997). The technique is also increasingly used to test or improve the efficiency of insecticide knockdown (e.g. Ellwood & Foster, 2000; Ch. 18). Regular improvements in speleological equipment (such as the improved ascender developed by the PETZL™ company, which may save up to 30% in energy during the climb) promise that entomologists will still be using this method in the future, particularly to reach the lower parts of the canopy and as a complement to the more intensive kinds of access provided by other methods discussed below.

Platforms and towers

Medical entomologists often set up platforms and towers in rainforests to study insect vectors, as did Bates (1944) in Columbia, Galindo *et al.* (1956) in Panama, and Haddow *et al.* (1961) and Haddow and Ssenkubuge (1965) in East Africa. Paulian (1947) used a sophisticated system of platforms and lifts to collect various taxa in the canopy of a lowland rainforest in Ivory Coast. Le Moult (1955) collected butterflies from platforms in French Guiana, and Cachan (1964) studied the seasonality and vertical stratification of Scolytinae from a tower located in the Ivory Coast. McClure (1966) collected various insect taxa from his platform in Malaysia. These structures tend to be relatively inexpensive and may also be replaced by cheaper scaffoldings (e.g. Jackson, 1996). However, their fixed access cannot be chosen randomly: appropriate clearings, adjacent trees or other constraints associated with tower construction impose limitations. In addition, foliage, flowers or fruits may be difficult to reach by the observer. Using a different approach, some workers have established small individual platforms in trees (Nadkarni, 1988). A new generation of larger tree platforms shaped like icosahedrons, which can be conveniently set up within tree crowns and moved elsewhere, seem promising for many sampling purposes including the light trapping of insects (Ebersolt, 2000; Hallé *et al.*, 2000). In the present volume, several authors relied on towers for canopy access (Chs. 7 and 34). A different technique, the canopy boom, is mentioned here in the interests of completeness but did not generate wide enthusiasm among entomologists following the initial pollination studies performed with it in Malaysia (e.g. Ashton *et al.*, 1995).

Walkways

Canopy walkways to conduct scientific research were first built in Malaysia to study ecto- and endoparasites of mammals (Muul & Lim, 1970; Muul, 1999). Other walkways were constructed in Panama, Papua New Guinea and Sulawesi (Sugden, 1985) and, among other work, were used to study herbivory (Wint, 1983). More recently, walkways have been in use in Australia, Belize and Peru to study herbivory and to survey arboreal mites (Walter & O'Dowd, 1995; Lowman, 1997; Walter *et al.*, 1998). These structures may well be affordable by research institutions and are safe (e.g. Lowman & Bouricius, 1993). They expand canopy access for sampling from points to transects, in contrast with platforms, towers and single-rope access (Muul & Lim, 1970). Access to the upper canopy, however, is difficult. A recent trend has been to combine platforms and walkways, such as at Blue Creek, Belize (Lowman & Bouricius, 1995), or towers and walkways as in the

Canopy Observation System in Lambir Hills, Sarawak (Inoue *et al.*, 1995; Yumoto *et al.*, 1996). Numerous entomological contributions resulted from the latter effort, with a special emphasis on insect pollination and seasonality (e.g. Kato *et al.*, 1995; Momose *et al.*, 1996; Sakai *et al.*, 1999a). In Ch. 12, Itioka *et al.* used walkways to gather entomological data.

Canopy cranes

Several construction tower cranes have been erected in tropical rainforests, such as in Panama (two cranes), Venezuela, Australia and Sarawak (Parker *et al.*, 1992; Wright, 1995; Wright & Colley, 1996; Stork *et al.*, 1997c). A crane operator controls the position of the crane gondola, from which observers can perform a variety of tasks. Many entomological studies have been performed or initiated with tropical canopy cranes, using different sampling methods (e.g. Roubik, 1993; Ødegaard, 2000a; Ødegaard *et al.*, 2000; Blüthgen *et al.*, 2000b; Basset, 2001a). The main advantages are the safety and excellent access within much of the canopy (less so in the lower part), and the possibility of obtaining many temporal replicates (Table 2.1). This is particularly useful for behavioural and life-history studies. Problems related to pseudoreplication at the meso-scale are acute within the relatively small and fixed crane perimeter (seldom exceeding 1 ha in area), and the costs of purchasing, erecting and maintaining a crane are expensive, particularly in remote locations. Crane use may be restricted during stormy or windy weather. One exciting development of canopy cranes is the Canopy Operation Permanent Access System (COPAS), currently being developed in French Guiana (Lohr, 2000). This system is similar to that using canopy cranes but the gondola is supported by a helium balloon and moves across a triangular line supported by three masts. Masts could be moved or added after the triangular area has been well studied, thus providing improved spatial replication. Several contributions in this volume present data that have been obtained using canopy cranes: Chs. 20, 21, 23, 25, 28 and 32.

Canopy raft and sledge

The canopy raft ('*Radeau des Cimes*') is a 580 m^2 platform of hexagonal shape, consisting of air-inflated beams and Aramide^{TM} (polyvinyl chloride) netting. An air-inflated dirigible of 7500 m^3 raises the raft and sets it upon the canopy. The raft is positioned on particular sites upon the canopy and moved every 2 weeks by the dirigible. Access to the raft is provided by single-rope techniques (Hallé & Blanc 1990; Ebersolt, 1990). The sledge ('*Luge des Cimes*') is a triangular platform of about 16 m^2 that is suspended below the dirigible and 'glides' over the canopy at low speed (Ebersolt, 1990; Lowman *et al.*, 1993a). Several entomological teams have worked with either the canopy raft or sledge, using a variety of collecting methods (e.g. Basset *et al.*, 1992, 1997b, 2001a; Dejean *et al.*, 1992b, 1998, 1999, 2000b,e; Sterck *et al.*, 1992; Lowman, 1997). The mobility of the raft, and particularly of the sledge, is ideal to obtain spatial replicates (Table 2.1). The infrastructure needed is expensive, however, and long-term temporal replicates are difficult to obtain. Access to the foliage is mainly restricted to the periphery of the raft. Flights with the sledge are restricted to the early mornings and times of good weather. In this volume, Chs. 10, 27 and 30 present data obtained with either the canopy raft or sledge.

Treetop bubble

The Treetop bubble ('*Bulle des Cimes*') is an individual 180 m^3 helium balloon of 6 m in diameter that runs along a fixed line set up in the upper canopy (Hallé *et al.*, 2000; Cleyet-Marrel, 2000). The system is independent from the canopy raft and sledge although the dirigible used to move the canopy raft is used to install the transect line. The observer is seated in a harness suspended below the balloon. He or she moves along the line with jumars. Different transects of several hundred metres have been set up, but longer transects of several kilometres are planned. So far, the bubble has been used to set up and survey different traps in the upper canopy (Basset *et al.*, 2001a), but other entomological applications are certainly possible. The equipment needed is relatively inexpensive and spatial replicates along line transects can be easily obtained (Table 2.1). Long-term temporal replicates along these transects could also be achieved. Possible limitations may be the relative instability of the observer because of the buoyancy of the balloon, and the difficulty in accessing the lower canopy. One promising development may be to elaborate protocols that would allow setting up the line with professional tree-climbers, instead of relying on the dirigible. This would make the method affordable to many research institutions. In Ch. 27, Basset *et al.* report data obtained with the treetop bubble.

Sampling and performing experiments in the canopy

With enhanced canopy access, entomologists can now focus their attention on the one hand in expanding their spatial and temporal replicates and, on the other, carrying out detailed process studies at selected 'super-access' locations. Basset *et al.* (2001a), for example, reported on one of the first attempts to factor out the effects of site, stratum and time in the distribution of arthropods in a tropical rainforest, using different methods of canopy access and sampling. They found that the effects of stratum were most important, representing between 40 and 70% of the explained variance in arthropod distribution, depending on the collecting method used. Site effects represented between 20 and 40% of the variance, whereas time effects (diel activity) explained a much lower percentage of variance (6–9%). These data stress the importance of replication among canopy sites and the appreciably different arthropod fauna that forages in the understorey compared with the upper canopy, where microclimatic conditions appear to be very different.

Enhanced access to the canopy also means that entomologists can now perform extensive and selective sampling with sufficient replicates in key canopy habitats, as opposed to more systematic sampling of individual trees. Targeting key habitats such as the upper canopy layer, blooming trees, lianas, epiphytes or dead suspended wood may be one strategy to refine relevant hypotheses on habitat and resource use in the canopy. This line of research is evident in recent studies (e.g. Nadkarni & Longino, 1990; Berkov & Tavakilian, 1999; Compton *et al.*, 2000; Ellwood & Foster, 2000; Ødegaard, 2000a; Basset *et al.*, 2001a), and in many contributions to this volume.

Manipulative experiments *in situ* are also needed to improve our understanding of arthropod distribution in the canopy. To date, few such examples exist, but they are bound to increase in the future. For example, Dial and Roughgarden (1995) removed *Anolis* lizards from tree crowns in Puerto Rico and monitored the resulting changes in the foodweb, particularly within arthropod groups. V. Novotny (unpublished data) has likewise performed multiple-choice feeding experiments *in situ* in the canopy in Panama, using canopy cranes. Mark-recapture experiments would also help greatly to study arthropod dispersal in the canopy. However, these experiments are likely to be challenging for some time (T. Roslin, personal communication), given the relatively low arthropod densities in the canopy (Basset, 2001b), their high spatial aggregation (Novotny & Leps, 1997; Novotny & Basset, 2000) and the difficulties in accessing multiple sites in the canopy.

The collection and rearing of live specimens from the canopy also represents another recent trend in the study of canopy arthropods that is promising (e.g. Paarmann & Paarmann, 1997; Novotny *et al.*, 1999b; Ch. 25). Rearing juvenile specimens provides adult specimens tractable for taxonomic studies, and the behaviour of live specimens can be studied either *in situ* or in the laboratory. Specimens collected alive can also be used subsequently in a variety of experiments, investigating, for example, resource use.

Archiving of data and collections

Recently, several projects have documented the rich tropical insect fauna by training local people ('parataxonomists') in the basics of insect collecting, mounting and sorting morphospecies; digital photography; and the use of simple, yet powerful computer databases (e.g. Janzen *et al.*, 1993; Longino & Colwell, 1997; Novotny *et al.*, 1997; Basset *et al.*, 2000). To date, none of these efforts has targeted specifically the canopy habitat, but it is only a question of time before entomologists train efficient parataxonomists on a large enough scale to cope with the enormous arthropod diversity in the canopy. These strategies can yield high-quality insect material and data, which are also available for subsequent taxonomic studies, within a relatively short time.

The identification of specimens collected in the canopy and their permanent storage are other problems that entomologists must face (see below). We expect some improvements in the routine identification of specimens that belong to known, and named, species. Extended computer hardware and software now allows the routine inclusion of digital pictures of specimens and characters in sophisticated databases, and this information can be circulated readily among colleagues over the internet and worldwide web. Large public databases, such as Ecoport (www.ecoport.org) and taxonomic tools are beginning to be available widely on the internet. As access to the internet, worldwide web and the nodes of expert taxonomists and their taxonomic tools from tropical countries improves, identification of specimens belonging to described species should be facilitated, and the ecological information linked to these species should expand. For a discussion on networking, entomological databases and the worldwide web, see Miller (1994).

CAVEATS AND LIMITATIONS IN CANOPY ENTOMOLOGY

The caveats and limitations in canopy entomology can be grouped in four categories: sampling limitations, taxonomic limitations, interpretation of ecological data, and conservation threats.

Sampling limitations

One recurrent problem in canopy entomology is low sample size, not so much in terms of number of specimens collected but rather in terms of number of replicates available for statistical analysis. This often results from difficult, expensive, partial or constrained canopy access. It often leads to pseudoreplication within the sampling universe and to disturbance and possible interference with the object being studied (Barker & Pinard, 2001). To what extent this problem may be serious may be dependent upon both taxa and scale. Data about larger, more active, stronger flying, better dispersing arthropod taxa are more likely to suffer from pseudoreplication than those related to more sedentary and physically smaller taxa, as sampling units are more likely to be independent in the latter. For similar taxa, the scale at which their distribution is analysed (e.g. microsites on a leaf, leaves, branches, crown segments, trees or forest plots) is also crucial.

For example, the study of insect host specificity in tropical rainforests appears to be constrained by at least three critical issues: sample size, number of singleton and rare species, and aggregation patterns of arthropods. Since the vegetation is highly diverse in rainforests, the sample size needed to estimate the true range of a species of herbivorous insect must be high, although no guidelines exist at the moment. Sufficient spatial and temporal replicates need to be combined with natural history data. Insufficient sampling and the mass effect described by Shmida and Wilson (1985) partly explain why so many species are represented by singletons in canopy samples (e.g. Morse *et al.*, 1988; Allison *et al.*, 1997). In tropical rainforests, the distribution of many insect herbivores is aggregated on the foliage, even for generalist species (Basset, 2000). This is reflected in their apparent high host specificity and rarity at low sample size (Novotny & Basset, 2000). This issue is also discussed in Ch. 29.

The positioning of access systems in the canopy is almost always nonrandom and opportunistic, particularly for fixed structures. An associated potential limitation is that samples obtained may not be representative of the fauna of the wider but less-accessible canopy (Barker, 1997).

Once sampling methods have been selected appropriately to investigate particular hypotheses, protocols should, as far as possible, be standardized (e.g. Adis *et al.*, 1998b, Kitching *et al.*, 2001, for insecticide knockdown). This will ensure subsequent comparison of valuable data across studies (Erwin, 1995; Stork *et al.*, 1997b). In practice, this comparability has seldom been sought.

Taxonomic limitations

Sampling techniques have greatly influenced present knowledge of canopy invertebrates. Invertebrates other than arthropods, although often abundant in epiphytic habitats, phytotelmata and perched litter, are little studied. The abundance of several arthropod groups, such as Acari, Collembola and Isoptera, is almost certainly seriously underestimated. The meagre taxonomic information available is usually focussed on a few relatively better known groups, such as Coleoptera and Lepidoptera (Basset, 2001b).

In addition to this conspicuous lack of information at the higher taxonomic level, many challenges are inherent in dealing with more detailed analyses of tropical arthropods. Taxonomic sufficiency, or using the level of identification appropriate to the study question, is important (Pik *et al.*, 1999; Slotow & Hamer, 2000): some studies may be accomplished with resolution at the level of order, genus or functional group, whereas other studies require species or morphospecies. Identification to species involves three steps: (i) sorting specimens into similar groups based on external characters; (ii) refining these groups based on detailed examination of accepted taxonomic characters (often including genitalic dissection) or external knowledge of patterns of polymorphism, sexual dimorphism, etc.; and (iii) associating these groups of specimens with formal names, based on literature, reference collections and comparisons with type specimens. Many early ecological studies on tropical arthropods only went to the first step, resulting in many errors. The present standard of practice is the second step, which we refer to as a morphospecies (although some authors use the same term for the first step). This can effectively be done with parataxonomists, if sufficient training, identification aids and quality control are provided (Cranston & Hillman, 1992; Basset *et al.*, 2000). The third level often requires collaboration with professional taxonomists, as well as access to type specimens scattered in European museums. Because of the cost in money and time of accomplishing the third step, many studies have assigned numbers and voucher specimens to their morphospecies, thus providing an 'interim taxonomy' (Erwin, 1991b) that allows the species to be referenced and its identity to be verified in the future.

Of all the species collected in canopy arthropod studies, few are identified as described species (e.g. Erwin, 1995; Ch. 21). Further, in most studies only very narrow taxonomic groups are actually taken fully through the third step above by exhaustive comparisons with type specimens (e.g. Roberts, 1993; Curletti, 2000). Our own work in Papua New Guinea has found, however, that a surprisingly high portion of morphospecies can be linked to names if sufficient effort is made (S. E. Miller, unpublished data). With the continuing crisis in systematics (e.g. Miller, 1991, 2000), description of new species collected in the canopy is going to be an increasingly difficult task (e.g. Kitching, 1993). Although coded morphospecies resolve many of the taxonomic problems in local studies, examination of ecological information associated with these unnamed species across multiple localities or studies is difficult. One solution will be to deposit and link information on the morphospecies (e.g. digital pictures, genitalia drawings, ecological data, etc.) into large public databases, such as Ecoport (www.ecoport.org), but this not an ideal solution without the features of a full taxonomic framework. In addition, the permanent storage of the material collected is also problematic (Stork & Gaston, 1990). Means of streamlining taxonomic practices have been suggested (e.g. Erwin & Johnson, 2000), but until images of old type specimens are routinely available in searchable databases, associating names with tropical arthropods will remain challenging.

Several major taxonomic initiatives have been proposed (such as the Global Taxonomy Initiative (GTI), BioNet International, Systematics Agenda 2000, All Taxa Biodiversity Inventories (ATBI), Global Biodiversity Information Facility (GBIF), All Species Inventory), which, *inter alia*, would build the capacity for taxonomic understanding of canopy arthropods (Cracraft, 2000). These approaches vary in their feasibility but, at this point, they remain largely unfunded. There has been widespread recognition of the crisis in staffing and available expertise in arthropod taxonomy since the 1980s (Wilson, 1985; Hawksworth & Ritchie, 1993; Miller & Rogo, 2002) but, in general, there has been little response from government agencies and other funding bodies. More progress is being made in databasing of existing formal collections, in cataloguing already named species and in documenting available expertise. Nevertheless, the parataxonomic approaches already described will likely remain the methodology of choice for canopy ecologists for some time to come.

For studies dealing with the biology of particular species, it is especially important that voucher specimens of both insects and host plants be placed in appropriate repositories for future reference (Huber, 1998; Ruedas *et al.*, 2000).

Interpretation of ecological data

To date, most of the information on canopy arthropods results from surveys of the canopy habitat, isolated from other forest habitats (but see Stork & Brendell, 1993; Kitching *et al.*, 2001; Chs. 9 and 26). Whether the canopy should be studied on its own or jointly with other forest habitats, such as soil and litter, is debatable. Many insect herbivores, such as some chrysomelids and curculionids, feed on roots as larvae and later migrate into the canopy to feed as adults on leaves. Although it is relatively easy to report differences in the occurrence of particular species of beetles in the adult stage either in the soil or in the canopy, our understanding of the relationships between the canopy and soil should also proceed by assessing how many insect species depend on the soil/litter habitat during their juvenile stages and on the canopy during their adult phase. Understanding the distribution of adult insects in the canopy may require solid data on their distribution as larvae in the soil (Basset & Samuelson, 1996). Further, comparison between the litter and canopy faunas may emphasize specific adaptations of arboreal invertebrates that may be important from a conservation viewpoint. Nevertheless, multimethod, multihabitat studies are essential if statements are to be made about the overall arthropod diversity of the forest: the assumption, tacit since Erwin & Scott's (1980) article, that the species richness of the forest is totally canopy dominated is certainly not true.

Rightly, Stork *et al.* (1997b) advised that entomological studies in the canopy should be integrated with other groups of organisms and studies of ecosystem processes. A sound understanding of biotic relationships in the canopy may require baseline knowledge of the entire rainforest ecosystem, an additional challenge in itself.

Conservation threats

The habitats that are the objects of study of canopy entomologists are disappearing fast (e.g. Bowles *et al.*, 1998), and with them an unknown but presumably vast number of arthropod species (e.g. Lawton & May, 1995). Canopy entomologists can play an active role in forest conservation by focussing their technical research on topics directly relevant to nature conservation issues (see the concluding chapter of this volume). In particular, they should attract the attention of the media, public and policy-makers by disseminating popular accounts of the tropical canopy and its arthropod inhabitants, aimed at both developing and developed countries (e.g. Basset & Springate, 1993; Floren & Linsenmair, 2000b; Novotny, 2000; Ødegaard, 2000b).

CONCLUSION

The limitations of each method of canopy access are obvious, as are those of each collecting method. There is no doubt that the choice of access technology and sampling methods must be tailored to the particular scientific questions being posed. For example, with regard to invertebrate samples obtained *in situ*, seasonal aggregation may be better studied with construction cranes, whereas spatial aggregation may be better studied with mobile devices such as the canopy raft or the canopy sledge (Table 2.1). This multifaceted approach calls for increasing collaborative effort (e.g. Nadkarni & Parker, 1994; Stork & Best, 1994; Erwin, 1995), involving not only researchers but also parataxonomists, and the use of multiple and complimentary techniques to create, for example, a 'canopy station' (e.g. A. W. Mitchell (cited in Lowman *et al.*, 1995); Hallé *et al.*, 2000; Mitchell, 2001). In particular, the more 'mobile' methods (single-rope technique, raft, sledge, treetop bubble) could be used to assess the representativeness of samples and observations obtained with 'fixed' methods (towers, walkways, cranes). One can also imagine merging different methods of canopy access and sampling, such as, for example, performing insecticide knockdown with the canopy sledge.

Not only do canopy entomologists need to expand greatly their sampling universe in the canopy, they also need to study the distribution of canopy arthropods at a much finer scale than done previously, by accurately tracking arthropod resources in space and time. Ideally, manipulative experiments should be performed at these meso- and microscales. One way to succeed in tackling these various problems would be to develop local inventories near or at canopy stations, as suggested by several authors (e.g. Janzen, 1993a; Stork, 1994; Godfray *et al.*, 1999). Significant progress in understanding arthropod distribution in the canopy requires us to solve these different challenges.

3

Vertical stratification of arthropod assemblages

Yves Basset, Peter M. Hammond, Héctor Barrios, Jeremy D. Holloway and Scott E. Miller

ABSTRACT

We review the information available on the vertical distribution of arthropods in tropical forests, especially rainforests. In these forests, faunal boundaries are likely to occur between the soil/litter layers, between these and the canopy, and in some instances between the upper canopy and lower levels. The major determinants of arthropod vertical distribution can be grouped in four categories: abiotic factors, forest physiognomy and tree architecture, resource availability, and arthropod behaviour *per se*. Many arthropod species are likely to forage at preferred levels within the rainforest canopy, to locate their preferred food resources, for example. Strict stratification in the canopy of closed and wet tropical forests has been reported for certain scavengers and fungal feeders, herbivores and ants but is less evident for generalist predators and biting flies. With respect to stratification of arthropods, the most evident and probably key distinction between temperate and tropical wet forests lies in the lack of pronounced vertical gradients (in microclimate and biotic factors) in the former. In particular, the presence of an upper canopy layer in closed tropical rainforests that is well delineated in terms of physiognomy and microclimate provides the most obvious explanation for the occurrence of richer and more distinctive arthropod assemblages in the uppermost parts of these forests compared with temperate forests. Overall, unambiguous data on the differential occupancy of vertical space by tropical forest arthropods remain few. With respect to the task of assembling and interpreting additional data, we stress the importance of appropriate methodology, of utilizing data available for temperate forests, of obtaining natural history information on the arthropods studied, and of placing investigations of canopy arthropods firmly in the context of the forest system as a whole.

INTRODUCTION

Vertical stratification (as opposed to altitudinal stratification) represents the distribution of organisms along the vertical plane (e.g. Dajoz, 1982) and is more or less well-marked depending on study systems (plant species in forests, plankton in lakes, microarthropods in soils, etc.). This chapter discusses the extent of vertical stratification of arthropods in tropical forests. For the sake of consistency, and as generally convenient locators of samples or observations, we generally employ the terms soil, litter, understorey, upper canopy, overstorey and canopy (the latter encompassing the three previous levels), as defined in Ch. 1. However, this is not intended to imply that these layers are necessarily well demarcated or that their fauna are necessarily distinct.

First, some necessary words of caution. One must agree with Smith (1973), for a similar argument about vegetational strata, that for rigorous demonstration of arthropod stratification, data must be collected at numerous, randomly located sampling points along vertical, ground-to-canopy transects. Because of the difficulties of canopy access, such data are virtually lacking. Data are often collected nonrandomly in the litter, understorey and upper canopy, more rarely in the midcanopy. For the few datasets available (see below), this gives an impression of faunal discontinuity among different putative strata, which is almost certainly an artefact. However, as discussed in the next section, faunal discontinuities may be real at the boundaries between the litter and understorey, and perhaps also between the canopy surface and the lower parts of the canopy.

Few resident arthropod species are to be found evenly distributed through any given tropical forest at any one time. In addition, different life stages may occupy different parts of a forest, and individuals too may move in response to temporal changes (daily rhythms, weather, season) in their environment, or to fulfil varying needs (dispersal, mate-finding, etc.). Against this dynamic background, it is not surprising that there is inconsistency in just what is meant by 'stratification' as applied to arthropod assemblages. Following Intachat and Holloway (2000), we discriminate between (i) preferences in the vertical distribution of organisms from ground to the overstorey (aggregated as opposed to uniform or random distribution), and (ii) strong clumping of these preferences in true 'strata' within the vertical column, resulting in clear 'faunal boundaries' and distinct arthropod assemblages. We consider the latter to be strict stratification. We recognize, however, that assemblages may be time limited (e.g. evident only at certain seasons), and that for some species assemblage membership may be confined to a particular life stage. The distinction between (i) and (ii) above may require subtle statistical analysis (e.g. Rodgers & Kitching, 1998) and may also depend on the forest layers being compared (e.g. understorey versus midcanopy, understorey versus upper canopy, etc.).

Many entomologists have taken advantage of recent advances in canopy access to study canopy arthropods (see Ch. 2). Much of this recent ecological literature would leave the reader with the impression that the study of vertical stratification in tropical forests had not begun before the 1970s. On the contrary, there is a diverse literature, much of it in medical entomology journals, on much earlier efforts to study vertical stratification of biting flies and other economically important insects (e.g. Bates, 1944; Mattingly, 1949). Further, many observations on the vertical distribution of arthropods in tropical forests, made in the course of largely ground-based studies, stem from even earlier times. Many of the findings of the great Victorian naturalists such as Alfred Russell Wallace and Henry Bates (see Elton, 1973), along with those of their successors, remain instructive but represent a seemingly underappreciated resource. For many of the early investigators, the simple expedient of felling selected trees (or seeking out freshly fallen ones) provided the opportunity to collect many canopy arthropods (e.g. Bryant, 1919). Early efforts to study stratification outside of medical entomology included Allee (1926) on Barro Colorado Island, Panama and Hingston (1930, 1932) and associates in Guyana. The methods may have been primitive by today's standards, but the scale of samples taken on various kinds of platform suspended in trees remains impressive.

A particularly ground-breaking effort was the series of studies of Haddow *et al.* (1961), who used a 40 m tower, originally built at Mpanga, Uganda, in 1958 for mosquito studies. Haddow's team found that some climatic parameters showed little variation vertically (temperature and saturation deficiency, a measure of humidity), whereas others did (wind and light). Vertical stratification was observed in breeding sites of mosquitoes, but patterns of stratification varied among different insect groups. In some groups, males and females exhibited different patterns. In addition, because of the cycles involved in flight patterns, some kinds of trap gave biased views of overall population activities. Haddow's group also published some of the first detailed observations of insect behaviour above the canopy, noting especially the swarming activities of mosquitoes. The data were used in broader discussions of interactions between endogenous (genetic) and exogenous (environmental) components in determining patterns of insect behaviour (Corbet, 1966).

In this review, we first discuss the compartmentalization of tropical forests and the extent of arthropod stratification there. We then examine the information available on gradients of species richness, discuss determinants of arthropod stratification in tropical forests and comment upon the extent of stratification across different arthropod guilds.

COMPARTMENTALIZATION AND THE EXTENT OF ARTHROPOD STRATIFICATION IN TROPICAL FORESTS

There is a large body of literature originating from the study of forest entomology in temperate areas showing that (i) the soil fauna is stratified among different soil layers (e.g. Gisin, 1943; Stebayeva, 1975); (ii) the soil/litter fauna is in large measure distinct from that of the forest above (e.g. Luczak, 1966; Cherrill & Sanderson, 1994; Osler & Beattie, 2001); (iii) many arthropods, especially herbivores, show vertical preferences in their distribution within the canopy (e.g. Morris, 1963; Nielsen, 1978; Gross & Fritz, 1982; Philipson & Thompson, 1983; Bogacheva, 1984); however (iv) distinct herbivore or other arthropod assemblages are not generally recognizable at different canopy levels, even, in the case of leaf-feeders, between the foliage of seedlings and that of conspecific mature trees (e.g. Fowler, 1985; Godfray, 1985; Schowalter & Ganio, 1998; Le Corff & Marquis,

1999). Item (iii) is well known to forest entomologists and often results in stratified sampling of pest populations (e.g. Morris, 1960).

Are these findings equally applicable to tropical forests? Available evidence suggests this is generally so for conclusions (i) and (ii) (e.g. Duviard & Pollet, 1973; Schal & Bell, 1986; Adis *et al.*, 1989; Hammond, 1990; Longino & Nadkarni, 1990; Longino & Colwell, 1997; Brühl *et al.*, 1998; Rodgers & Kitching, 1998). Items (iii) and (iv) are more difficult to evaluate, and discussion of them forms the substance of the present review. Before examining vertical gradients of species richness within tropical forests, we must first comment on the compartmentalization of these forests and related topics in general. Differences in the vertical distribution of arthropods in different forest types (e.g. montane, dry, lowland) are discussed in the concluding chapter of this book.

What is immediately clear is that variability is not expressed in most instances entirely or even mainly in terms of vertical strata. Indeed, it is arguable that, once away from the forest floor, strictly vertical stratification is not the norm. Nevertheless, in the canopy, using the term in its broadest sense, some stratification is evident. For example, the extremes in physical conditions experienced at the interface between the forest and the free air above justify the recognition of this uppermost part of a forest as a distinct stratum. Many abiotic and biotic characteristics of the upper canopy of closed tropical rainforests are different from forest layers below, especially from the understorey. For example, in a rainforest in Cameroon, the characteristics of the canopy surface are more akin to chaparral shrub vegetation than to familiar rainforest understorey vegetation (Bell *et al.*, 1999). Whereas the upper canopy receives close to 100% of the solar energy, less than 1% of this energy reaches the understorey (Parker, 1995). Average light availability decreases up to two orders of magnitude over short distances from the external surface to a few centimetres inside the canopy (e.g. Mulkey *et al.*, 1996). Levels of ultraviolet, fluctuation of relative humidity and air temperature, and wind speed are notably higher in the upper canopy than in the understorey (e.g. Blanc, 1990; Parker, 1995; Barker, 1996). Water condensation at night is frequent within the upper canopy but absent in the understorey (e.g. Blanc, 1990). The leaf area density and the abundance of young leaves, flowers and seeds are also usually higher in the upper canopy than beneath (Parker, 1995; Hallé, 1998). Leaf turnover and nitrogen translocation, upon which many sap-sucking insects depend, are well marked in the upper canopy (Basset, 1991e; 2001a). The leaf buds of the upper canopy appear to be extremely well protected against desiccation and herbivory (Bell *et al.*, 1999). Further, levels of secondary metabolites that are biologically active within individual trees are much higher in leaves of the upper canopy than in leaves situated at the base of the crown (Hallé, 1998; Downum *et al.*, 2001).

Beneath the upper canopy, vertical strata are much less clearly demarcated in terms of the physical and biotic features most likely to determine arthropod distributions. There are obvious exceptions: for example, individual tree cavities may exhibit clear internal vertical stratification. Are there alternative or better ways of talking about a forest in terms of 'naming of the parts'? This is not a trivial matter as the answer may have a direct bearing on how we go about studying and reporting on within-forest distribution. In much of the space between the ground surface and the upper canopy, a useful approach may be to deal with 'compartments', for example, the main trunk area of trees, 'free space', other spatially defined blocks of the canopy, and the habitats/microhabitats such as fungus fruiting bodies, carrion, inflorescences and so forth, distributed within them.

Although a growing body of literature on 'stratification' in tropical forests exists mainly from the 1970s, few datasets on the composition of arthropod assemblages in tree crowns are extensive enough to evaluate whether these are truly stratified or compartmentalized (but see the contributions in Part II). The main shortcomings responsible for this situation are the absence of any reference to the composition of assemblages found at lower levels; the comparison of the distribution of various taxa that differ significantly in biology; and a lack of uniformity in the methods used to sample or appraise different assemblages, sometimes even among different putative strata.

As discussed in the next section, most arthropod species in most forests are effectively confined to the lowest layers (soil and soil surface habitats). Nevertheless, many species associated with soil and soil surface habitats are represented in samples taken from compartments above the forest floor, where they are often best regarded as vagrants, tourists or, at most, short-term visitors (e.g. Adis, 1984b; Hammond, 1990); others are more appropriately regarded as 'stratum generalists' (Hammond *et al.*, 1997). Knowledge of the assemblages tied to these lowest forest strata is, therefore, essential (e.g. Haddow, 1961) if sample data from the canopy are to be used effectively in the characterization of the various assemblages occupying the upper levels of a forest.

Many studies have found, in general, significant differences in composition and abundance of arthropods at different vertical levels in the trees, but questions of what is actually being sampled and the lack of understanding of the biology of the organisms have limited the conclusions that can be drawn. Some studies find greater abundance at higher levels (e.g. Basset *et al.*, 1992), whereas others find greater abundance at lower levels (e.g. Wolda *et al.*, 1998) or no significant differences are observed (e.g. Intachat & Holloway, 2000). Further, a large body of literature focusses on arthropod samples obtained from the forest 'canopy', usually referring to samples obtained 15 m or more above the ground, by various methods (reviewed in Basset, 2001b). Most studies with insecticidal fogging (e.g. Erwin, 1995), light traps (e.g. Sutton, 1983; Wolda *et al.*, 1998) or by felling trees (e.g. Amédégnato, 1997; Basset *et al.*, 1999) do not sample the upper canopy selectively and efficiently. Improved canopy access (Ch. 2) has allowed entomologists to refine their sampling protocols, in order to obtain replicated samples of the upper canopy and question whether stratification is maintained both during day and at night (Basset *et al.*, 2001a), or during seasonal events (Ch. 7).

Based on a review of recent literature, we argue in the following sections that vertical stratification may be more distinct in tropical than temperate forests but may only concern certain taxa during certain life stages.

VERTICAL GRADIENTS OF SPECIES RICHNESS: SOIL VERSUS CANOPY FAUNA

Globally, are arthropod faunas more species rich in the canopy than in the soil of tropical rainforests? This question (although not the only one of relevance: see Basset *et al.*, 1996a) has been central to global estimates of arthropod species richness derived from surveys of arboreal arthropods on particular host trees (e.g. Erwin, 1982; May, 1990). Erwin (1982) contended that the canopy fauna was the most species rich but the subsequent evidence seems contrary (e.g. Hammond, 1990, 1995; André *et al.*, 1992; Hammond *et al.*, 1997; Walter *et al.*, 1998).

It is worth reiterating that the problem of generalizing for all taxa and forest types remains. There are four additional issues, both for and against Erwin's contention, that are relevant to this particular debate. First, it is difficult to compare soil and canopy faunas, since they need to be surveyed by different sampling methods (see Ch. 9). For example, extremely high densities of springtails occur in the canopy of certain dry forests in Mexico (Palacios-Vargas *et al.*, 1998) but how do they compare with springtail densities in the soil and litter? Sample size is different and not directly comparable. Furthermore, the number of individuals collected is not a valid criterion in this context, since it is highly dependent on the amount of habitat sampled. The volume of habitat sampled may be a better descriptor of sample size, but it is difficult to estimate for the canopy habitat.

Second, Acari are often dominant but underestimated in arboreal habitats (e.g. Walter, 1995), whereas they are relatively well sampled in soil and litter habitats. Since this taxon is dominant in the soil of rainforests (e.g. Stork, 1988), comparison between the faunas of soil and canopy must ensure that Acari have been well sampled in the latter. To date, no tropical study has had sufficient scope to survey representatively *all* arthropod taxa within a vertical transect of forest.

Third, assessing the diversity of soil versus canopy biota evidently also depends on patterns of β-diversity. If faunal turnover is rather high in the canopy (because of the relatively high specialization of insect herbivores and associated specific predators and parasitoids on particular host-tree species, see Ch. 5), compared with that in the soil, it may be inappropriate to compare the diversity of equivalent projected areas of canopy and soil. Monodominant stands aside, the β-diversity of canopy communities may be much higher than that of soil communities in rainforests. For example, the β-diversity (and 'host specificity') of soil mites is very low in Australia (Osler & Beattie, 2001). Note, however, that at the appropriate scale a correlation between below-ground and above-ground biodiversity may exist (Hooper *et al.*, 2000). In particular, plant diversity, because of the production of diverse root exudates, can lead to increased diversity of mutualistic soil microflora, the first link of a cascade of effects resulting in increased diversity of other soil animals (Lavelle *et al.*, 1995).

Last, as emphasized repeatedly (e.g. Hammond, 1990; Basset & Samuelson, 1996; Chs. 2 and 5), faunal comparisons rely on the taxonomic study of adult specimens, and juveniles are rarely accounted for, be they spiders or beetles. The most likely situation is for juveniles to develop in the soil, to move up into the canopy as adults and to feed and disperse from there (e.g. Hammond, 1990; Basset & Samuelson, 1996). In addition, one must factor in seasonal migrations upward into the canopy, especially during flooding (e.g. Adis, 1981, 1997a; Erwin & Adis, 1982). Do we always study the soil and canopy fauna separately in these conditions?

DETERMINANTS OF ARTHROPOD STRATIFICATION IN TROPICAL FORESTS

This and the following section are more relevant to the canopy fauna *per se*. Several popular hypotheses involving concepts such as tree architecture (Lawton, 1983), resource concentration (Root, 1973) or resource base (Price, 1992) could explain vertical gradients of insect diversity in tropical rainforests. These explanations are not mutually exclusive and can account only partially for the observed gradients. There may be many, most likely interrelated, factors that may induce arthropod stratification in the canopy of tropical rainforests. Depending on vegetation type, latitude and so on, some factors may be locally more significant than others. We discuss four categories of determinants with reference to how they may tune arthropod behaviour, from a coarse to a finer scale of behaviour: (i) abiotic factors; (ii) forest physiognomy and tree architecture; (iii) resource availability; and (iv) arthropod behaviour *per se*, including search for enemy-free space and dispersal. Each of these categories of determinants influences the lower one in the hierarchy but, for the sake of simplicity, we will discuss them separately.

Abiotic factors

Abiotic factors such as light, levels of ultraviolet, air temperature, relative humidity, wind and water condensation, to cite but a few, may have direct as well as indirect effects (i.e. through their strong influence upon other determinants) on arthropod stratification. Their significance should not be underestimated. Bates (1944), for example, recorded the flight of different species of mosquitoes at various heights in the canopy in Columbia. He noted that it is easy to find sections of the forest in which light is greater at ground level than in the canopy, but the gradients of humidity and temperature seem never to be reversed by local conditions in the canopy. He also observed that mosquitoes react primarily to the humidity gradient, not to the light gradient. Accordingly, arthropod stratification could be maintained readily in tropical forests by strong gradients of abiotic factors alone. Similarly, the stratification of bark beetles in a lowland rainforest in the Ivory Coast is maintained by differences in relative humidity along the vertical transect (Cachan, 1974). The vertical distribution of Diptera in lowland dipterocarp forests in Malaysia is significantly affected by wind speed and minimum air temperature (Ng & Lee, 1980).

The quantity and quality of sunlight may also influence strongly the photosynthetic process at different levels in the forest, both in terms of primary and secondary metabolites. This, in turn, is likely to influence the quantity and quality of resources available to insect herbivores and their vertical distribution (see further discussion in Ch. 5). For example, most insect herbivores feeding on the Australian rainforest tree *Argyrodendron actinophyllum* Edlin respond primarily to the availability of young foliage, which depends directly on the local light regime. In this case, stratification is not well marked as it depends on local differences in illumination within tree crowns, which can vary substantially among individual trees (Basset, 1992c).

Wind speed within different forest layers may influence significantly insect flight in temperate forests (e.g. Nielsen, 1987), but this has not been well studied in tropical forests (Ng & Lee, 1980). Schal (1982) reported that, in Costa Rica, cockroaches stratify vertically both inter- and intraspecifically along micrometeorological gradients. This observation relates to the ascent of warm air and pheromone dispersion at night and represents a mate-finding strategy. Perching of dung beetles on the foliage also perhaps represents a form of behavioural thermoregulation and/or a strategy to maximize the detection of scents in the air and a consequent resource partitioning (e.g. Young, 1984; Davis, 1999b). Other factors, such as dust accumulation on the foliage, with effects that have been well studied for rainforest vertebrates (e.g. Ungar *et al.*, 1995), could also influence the vertical distribution of arthropods.

Forest physiognomy and tree architecture

The forest physiognomy, including features such as the height of the canopy, the disposition of large tree trunks, the leaf area index, the occurrence of free space (openness) and so forth, as well as the tree architecture (Lawton, 1983: height, biomass, size and abundance of leaves, flowers, seeds, etc.), represents resources in their own right. However, these features also have a bearing on the way that the 'primary' resources are distributed and may influence arthropod foraging activity. For example, the localization of flight height in Lepidoptera is not so developed in the more open forests of lesser stature such as lowland forests of Sulawesi and Seram, compared with that in tall Bornean forests (J. D. Holloway, personal observation). In addition, the extent of faunal stratification may depend on the slope of the terrain (Sutton, 1983) and on local flooding regimes (e.g. Adis, 1997a).

Tree architecture, including the varying biomass of conspecific seedlings, saplings and trees, is a significant determinant of the richness of associated insect herbivores in tropical trees (e.g. Basset *et al.*, 1999; Basset, 2001a; Caraglio *et al.*, 2001; Chs. 5 and 25). Differences in the volume of habitat available often correlate with resource availability (e.g. higher occurrence of young foliage, flowers and seeds in mature trees than in seedlings or saplings).

Studies of vegetational strata in temperate and tropical forests suggest two possible further hypotheses for the maintenance of arthropod stratification in tropical forests (Smith, 1973).

1. Plant stratification, by providing clear 'flight paths' for insects, birds and bats above and below each stratum, may increase the probability of pollination or seed dispersal. In other words, production of open areas in the canopy through stratification may have selective value.
2. Canopy-level predation on tree flowers, fruits, buds and leaves may select for aggregation of the foliage of different plant species into one or more common strata. If many tree species produce mature foliage at a common level (as opposed to each species producing mature foliage at its own characteristic level above the ground), any herbivore specializing in a particular species would need to spend more time and energy searching for its host. So stratification might confer protection against herbivory.

Despite the proven existence of clear flight paths for arthropods in rainforests (e.g. Shelly, 1988; Ch. 8), it is difficult to comment on the validity of the first hypothesis without further data. Recent and comprehensive data about the host specificity of tropical insect herbivores, including several tens of replicates of both individuals and tree species, suggest that many insect species specialize at the generic or familial, rather than specific, plant level (Novotny *et al.*, 1999b, 2002a). For mixed and botanically diverse forests, such data would not appear to support the second hypothesis.

Resource availability

Resource availability and its use by insect herbivores in the canopy are discussed in more details in Ch. 5. As already emphasized, the quantity and quality of resources for herbivores (young foliage, flowers, fruits, seeds, etc.) differ between the understorey and upper canopy and globally this should result in higher abundance/diversity of herbivores in the upper canopy, as well as the occurrence of strata specialists (e.g. Basset *et al.*, 1992, 2001a; Basset, 2001a; Chs. 5 and 25). The major volume of tropical forests is in the canopy, and a wide range of habitats are scattered or non-existent in the understorey. For example, the greater part of production and structural diversity of lianas occur in the mid- or upper canopy (Hegarty & Caballé, 1991) and many herbivores specialize on them (e.g. Stork, 1987a; Ødegaard, 2000a). Low values of leaf area index in the understorey, as compared with that in the upper canopy, are likely to affect not only the resources available to insect herbivores but also how they can escape their potential enemies.

The quality of resources may also represent a significant factor. Hallé (1998) has argued that the exposure of canopies should result in high concentration and diversities of compounds, either developmentally controlled or induced by light, wind, desiccation and/or exposure to herbivores and pathogens. Yet, the evidence for increases in compounds relative to the understorey is meagre and primarily from colorimetric assays for tannins and total phenols (Coley & Barone, 1996). Recently, Downum *et al.* (2001) showed that the crowns of rainforest trees produce significantly more secondary compounds and at higher concentrations than do understorey saplings. Some of the compounds are biologically active and could help to reduce damage from herbivory and disease. The canopy samples from each species showed dramatic increases (by more than four times) for the number of compounds and their relative concentrations. The greatest number of compounds was produced from tree crowns: those exclusively from the crowns were half or more of the total number of compounds detected although a few compounds were produced in the understorey alone. These differences may result in discrete habitats, depending on the age and physiognomy of the forest stand but also on the ecological characteristics of the host plant (taxonomic isolation, height, crown volume, growth patterns and phenology). In turn, this would select for rather specialized herbivores either in the upper canopy or understorey. Accordingly, plant phylogeny may well influence both temporal patterns (Ch. 4) and the vertical stratification of arthropod assemblages in rainforests.

Certain resources are evidently more abundant in the understorey and near the ground, such as dead wood, litter, dung, fallen fruits, carrion, the availability of specific prey, and so on. These resources may attract different assemblages specific to the understorey (e.g. Davis *et al.*, 1997; DeVries & Walla, 2001; Schulze *et al.*, 2001). For example, most dung beetles are more species rich near the ground but certain species specialize on perched dung (Davis *et al.*, 1997). These canopy species rarely forage near the ground in primary forests but may sometimes be present at ground level in logged forests, tracking their preferred resources (Davis & Sutton, 1998).

Many adults of herminiine Noctuidae fly at low levels, where the larvae feed on litter and detritus (Holloway, 1984b; as Hypeninae). This poses the question 'to what extent is there resource fidelity?' For holometabolous flying insects, larval densities, pupation sites and adult flight levels may (e.g. Beccaloni, 1997; Willmott *et al.*, 2001) or may not (e.g. Van Klinken & Walter, 2001) be correlated. Is a larval/adult correlation to be deemed stronger if the intervening pupation site level is different (canopy larvae dropping to pupate in litter)? Is such a correlation stronger in taxa where the adults do not feed so resources are not 'pulling' in different directions (foliage versus nectar/fruit/carrion/salts)? Once again, these questions highlight the difficulty of interpreting distributional data and putative stratification without considering the entire life cycle of the species studied.

Arthropod behaviour

Specific patterns of arthropod behaviour may generate preferred distribution in the vertical plane and, perhaps more rarely, strict stratification. One example of the latter could be mimicry rings, where groups of species become locked into interdependence in some way at a certain level in the forest (e.g. Papageorgis, 1975; Mallet & Joron, 1999). For example, the mimicry rings of ithomiines (Nymphalidae) show some patterns in flight height (e.g. Medina *et al.*, Beccaloni, 1997; DeVries *et al.*, 1999b; 1996) that may be tuned to local predator knowledge and could be a response to different guilds of predators that forage in different habitats (Beccaloni, 1997).

As for preferred vertical distributions, it is well known that sexual differences in adult butterfly behaviour may be resource based and may lead to observations of flight at different forest levels (e.g. Holloway, 1984c). For example, the males of papilionids and pierids are more prone to disperse and fly more in the open, whereas females are less often observed and fly more in the forest interior, searching for oviposition sites. Phenomena such as hill-topping and migration (and the response to light) are usually dominated by male butterflies. In addition, certain groups, such as Nymphalidae or Lycaenidae, may be more or less territorial, with males favouring sunlight gaps for perching (e.g. the Amathusiinae) or occupying specific areas of the forest floor (Holloway, 1984c; Novotny *et al.*, 1991).

Enemy-free space (e.g. Schal & Bell (1986) for cockroaches), competition (Enders (1974) for spiders) and aggregation of conspecifics are other factors that may induce preferences for particular forest levels, and perhaps stratification in tropical forests. Adult Japanese beetles, *Popillia japonica* Newman (Scarabaeidae), aggregate and feed most heavily in the upper canopy of their host plants. However, they begin to feed in the upper canopy for reasons unrelated to host nutritional variation (e.g. behavioural thermoregulation, visual orientation to the host silhouette), and top-down defoliation follows as additional beetles are attracted to feeding-induced volatiles acting as aggregation kairomones (Rowe & Potter, 1996).

Finally, arthropod dispersal may or may not promote strict stratification. Fig wasps in Borneo, for example, disperse mostly in the overstorey even where species are associated with host trees that do not fruit in the canopy. In this case, once the fig wasps detect the species-specific volatiles released by their host figs, they then may fly down into the canopy, where the lower wind speeds allow them to fly actively upwind to their hosts

(Compton *et al.*, 2000). In Uganda, many mosquitoes that feed almost exclusively at ground level may rise to the canopy after sunset and form substantial swarms in the overstorey (Haddow, 1961; Haddow & Corbet, 1961a; Haddow & Ssenkubuge, 1965).

With the above list of determinants in mind (abiotic factors, forest structure, disposition of habitats/ resources and arthropod behaviour *per se*), the explanation for observed stratification of arthropods in forest may be sought through consideration of these factors. Unfortunately, our current knowledge of the autoecology of canopy arthropods in tropical forests is in most cases crude. For now, we will discuss resource availability and its influence on the vertical distribution of arthropods by contrasting representative taxa from different feeding guilds.

VERTICAL GRADIENTS AND ARTHROPOD GUILDS

We review briefly and separately the extent of stratification for the following guilds: biting flies; scavengers and fungal-feeders, including dead wood eaters; herbivores, including pollinators; predators and parasitoids; and ants.

Biting flies

There is a large body of information on both the vertical and horizontal distribution of representatives of the biting flies in tropical rainforests, originating from studies in medical entomology. The food resources of the guild do not appear to be well segregated along the vertical plane. However, foraging of biting flies is often influenced by abiotic factors such as humidity and light (e.g. Bates, 1944) and this may induce strong preferences for particular forest levels, depending on the structure of the forest. Indeed, stratification of adult biting flies does not appear to be well marked, although most species do show height preferences and, collectively, are often more abundant in the understorey (e.g. Ceratopogonidae: Arias & Freitas, 1982; Aguiar *et al.*, 1985; Azevedo *et al.*, 1993; Veras & Castellon, 1998; Culicidae: Bates, 1944; Mattingly, 1949; Murillo *et al.*, 1988). This pattern appears to be similar for larvae in their breeding sites (e.g. Galindo *et al.*, 1956; Corbet, 1961a; Lounibos, 1981; Lopes *et al.*, 1983; Murillo *et al.*, 1988). Certain species of Ceratopogonidae are well known to prefer foraging at canopy level (e.g. Arias & Freitas, 1982; Aguiar *et al.*, 1985), and some Culicidae move to feed from canopy to ground and vice versa during daily vertical migrations (e.g. Haddow, 1961; Haddow & Ssenkubuge, 1965; Deane *et al.*, 1984). The latter phenomenon is of considerable medical importance, as it provides a link between the fauna of the forest canopy and that at ground level, including humans (Haddow & Ssenkubuge, 1965). Phytotelmata, particularly those provided by Bromeliaceae, are favourable breeding sites for many culicid vectors of human malaria and filariasis (e.g. Pittendrigh, 1948; Zavortink, 1973; Lounibos, 1981).

Scavengers, fungal-feeders and dead wood eaters

As well as the information presented in this volume (see notably Chs. 10 and 24), several studies more specifically targeting the scavenging and fungal-feeding fauna report on upward migrations in the canopy (e.g. Adis, 1984b), distinct faunas in the litter and canopy and associated flight preferences (e.g. Cachan, 1964; Schal, 1982; Walter, 1983; Young, 1983; Schal & Bell, 1986; Hammond, 1990; Rodgers & Kitching, 1998; Yanoviak, 1999; De Abreu *et al.*, 2001; Van Klinken & Walter, 2001), or the higher species richness and/or abundance in the litter compared with the canopy (e.g. Nadkarni & Longino, 1990; Paoletti *et al.*, 1991; Basset *et al.*, 1992, 2001a; Davis *et al.*, 1997; Hammond *et al.*, 1997; Walter *et al.*, 1998; Basset, 2001a).

Habitats for this guild, such as dead wood and associated fungi, appear to be relatively discontinuous and discrete along a vertical transect of rainforest (Ch. 25). This may limit the dispersal of some forest floor species higher up in the canopy and maintain distinct assemblages at different levels (Rodgers & Kitching, 1998). Although the amount of dead wood and suspended soil may not be negligible in the canopy, their abundance is highest at ground level (Nadkarni & Longino, 1990; Martius & Bandeira, 1998). In addition, the low relative humidity in the upper canopy may hinder fungal growth there. Consequently, we would expect rather different assemblages of scavengers and fungal-feeders at different heights in rainforests, perhaps with a specialized fauna able to cope with the harsh environmental conditions of the upper canopy. These conditions include greater illumination, more wind and frequent cycles of wetting/drying (Dajoz, 2000). Strict stratification could, therefore, occur for this guild. To date, selective data relevant to the upper canopy and comparison with the lower levels of the forest are rare but indicate a clear stratification for representatives of Collembola, Acari and Buprestidae of the genus *Agrilus* (Rodgers & Kitching, 1998; Walter *et al.*, 1998; Curletti, 2000; Basset *et al.*, 2001a). However, the abundance and diversity of this guild should be highest near the ground (Hammond, 1990).

Herbivores

Given that the food resources for many herbivores, such as leaves, flowers and fruits, are more abundant in the upper canopy than in the understorey of wet rainforests (e.g. Hallé, 1998; Ch. 5), the abundance and diversity of many herbivorous taxa should be higher in the former strata. This appears to be the case for homopterans, herbivorous beetles, flower-visiting butterflies, caterpillars, fig wasps and certain euglossine bees (e.g. Wolda, 1979; Erwin, 1982; Sutton *et al.*, 1983a; Basset *et al.*, 1992, 1999, 2001a; Spitzer *et al.*, 1993; Kato *et al.*, 1995; De Oliveira & Campos, 1996; Compton *et al.*, 2000; Basset, 2001a; Schulze *et al.*, 2001; E. Charles, personal communication). Differences in foliage quality between the upper canopy and understorey (e.g. Downum *et al.*, 2001; see above and Ch. 5) may induce a clear stratification of herbivores, as reported in several studies (e.g. Amédégnato, 1997; Basset, 2001a; Basset *et al.*, 2001a; Ch. 25), particularly when taxa have a narrow host range, such as is the case for many gall-makers and leaf-miners (Medianero, 1999; Valderrama, 1999: faunal overlap between the upper canopy and understorey <1%). For many herbivorous beetles, this stratification may be complex, with juvenile stages feeding in the soil on roots and the adults feeding at different levels in the canopy (Basset & Samuelson, 1996; E. Charles, personal communication).

Exceptions to these 'general' patterns are also common. Fruit-feeding nymphalid butterflies and geometrid moths appear to be more active, abundant or species rich in the understorey than in the upper canopy (e.g. DeVries, 1987b; DeVries *et al.*, 1997; DeVries & Walla, 2001; Schulze & Fiedler, 1998; Intachat & Holloway, 2000; Schulze *et al.*, 2001). In general, bees do not forage consistently by strata. However, some understorey specialists are known, and species that prefer to forage in the upper canopy may have specific physiological traits, such as capacity for heat loss during flight, or may be nocturnal (Roubik, 1993; Roubik *et al.*, 1995).

This lack of concordance may be a result of differences in resource use and preferences across life stage, since larval distribution and adult flight behaviour may not necessarily be correlated. In some cases, juvenile stages may be feeding actively in the canopy, but adults may be dispersing in more open parts of the forests, such as in the understorey (in pristine forests) or overstorey. Indeed, sometimes they may not feed at all.

A few words should be said here about the Janzen-Connell model. In brief, this model states that patterns of herbivore attack below the parent tree are density dependent and decrease with increasing distance from the parent tree (Janzen, 1970; Connell, 1971). This process could promote botanical diversity by prohibiting establishment of young trees near conspecific parents (e.g. Janzen, 1970; Connell, 1971). One of the implicit assumptions in this model is that most insect herbivores that feed on seedlings are specialists that originate from and feed on the parent tree (Leigh, 1994). This assumption has repeatedly proven false when examining the distribution of insect herbivores on conspecific seedlings and mature trees in different tropical wet forests (Basset *et al.*, 1999; Basset, 2001a; Willmott *et al.*, 2001; Ch. 25); data considering leaf damage, such as Barone (2000) are unconvincing for reasons discussed in Basset & Höft (1994) and Hadwen *et al.* (1998). Janzen (1970) formulated the model when studying an insect–plant system in a dry forest in Costa Rica. This suggests that the model and its assumptions may be valid in situations where few barriers to the dispersal or maintenance of insect herbivores exist between mature trees and their conspecific seedlings. Alternatively, the model may apply only to postdispersal attack of seeds and may be irrelevant to attack upon seedlings or saplings near the parent tree.

Predators and parasitoids

The stratification and migrations of carabid beetles have been well studied in the Amazon (Adis, 1982; Erwin & Adis, 1982). Nevertheless, in general, few data exist on specific groups of predators and parasitoids in tropical canopies (Godfray *et al.*, 1999). Most data refer to levels of abundance and do not detail distribution patterns for specific species. For example, in Panama, the abundance of insect predators and parasitoids, but not that of spiders, was higher on the foliage of mature *Pourouma bicolor* Martius than on conspecific saplings (Basset,

2001a). In Gabon, parasitoids, particularly Scelionidae, were more abundant and active in the upper canopy than in the understorey (Basset *et al.*, 2001a). In Borneo, Encyrtidae are more abundant in the overstorey and upper canopy than in the understorey, but Mymaridae show the reverse trend (Compton *et al.*, 2000).

The extent of stratification of these groups probably depends on whether they specialize on certain prey/hosts or not. Generalist predators may not be strata specific, depending on whether they can tolerate environmental differences among different strata. Specialist predators and parasitoids are more likely to forage within the discrete habitats of their prey, as is known to occur for certain temperate species (e.g. Hollier & Belshaw, 1993; Redborg & Redborg, 2000) and are, therefore, more likely to be restricted to certain forest strata. Note that tropical parasitoids are not necessarily host specific, as many target egg masses of different arthropod hosts (Noyes, 1989b).

Ants

Depending on their nesting ecology, ant assemblages may often be distinct along vertical transects of rainforests (e.g. Wilson, 1959; Itino & Yamane, 1995; Brühl *et al.*, 1998; Yanoviak & Kaspari, 2000). Strict stratification has been reported in several studies (e.g. Longino & Nadkarni, 1990; Brühl *et al.*, 1998). One extreme specialization is represented by the famous ant gardens occurring in the canopy of neotropical forests (e.g. Wheeler, 1942; Davidson, 1988; Cedeño *et al.*, 1999). Although many studies have reported a high abundance or diversity of ants in the canopy of tropical rainforests (e.g. Erwin, 1983b; Stork, 1987b; Wilson, 1987; Tobin, 1991; Basset *et al.*, 1992; Dejean *et al.*, 1999), this is not a general rule. Often, it reflects but a few ant species able to feed on plant and homopteran exudates in the canopy, but which also may prey on other arthropods (e.g. Tobin, 1991; Itino & Yamane, 1995; Davidson, 1997; Kaspari & Yanoviak, 2001; Ch. 30). The choice of support trees by arboreal ants is not random (Ch. 30) and is perhaps related to the abundance and fitness of ant-attended homopterans on putative host trees.

One example is particularly eloquent. In one lowland forest in southern Cameroon, ants were significantly more abundant in the upper canopy than in the understorey, and many were attending Coccoidea in the upper canopy (Basset *et al.*, 1992; Dejean *et al.*, 2000e). Not far from there, in a similar lowland rainforest in central Gabon, the reverse trend occurred. In this case, Psylloidea were much more abundant than Coccoidea and few homopterans were attended by ants in the upper canopy (Dejean *et al.*, 2000b; Basset *et al.*, 2001a).

In short, many arthropod species are likely to forage at preferred levels within the rainforest canopy, in order to locate their preferred food resources. Reinforcing this tendency, as well as acting in their own right, are the additional determinants of stratification discussed in the previous section. Strict stratification in closed and wet tropical forests has been reported for certain scavengers and fungal-feeders, herbivores and ants, but it appears less likely for generalist predators and biting flies.

CONCLUSIONS

One essential difference between temperate and tropical wet forests may well prove to be the lack of pronounced vertical gradients in the former because of the less drastic vertical changes in microclimate and biotic factors (see Ch. 24). Although the upper canopy of wet tropical rainforests is often structurally and environmentally distinct from lower forest levels (e.g. Bell *et al.*, 1999), this stratum (or the extent of its distinctiveness), seems to be lacking in temperate forests. Lowman *et al.* (1993b) have suggested that the upper canopy of a temperate forest has proportionally fewer niches available to organisms compared with tropical forests. This, and the discontinuity of available habitats, could explain the stronger stratification observed in tropical forests and their richer fauna compared with temperate forests. Vertical gradients of species richness within wet tropical rainforests may be akin to gradients of latitudinal richness, when considered as a result of the control exerted by solar energy over organic diversity in conditions of unlimited water resources (Turner *et al.*, 1987).

One goal in planning new research on the spatial distribution of arthropods in tropical forest canopies should be to understand how these arthropods are distributed through the forest as a whole. Without this context, and especially when attempting to interpret the results of mass sampling, the composition of assemblages will be difficult to establish and the unique features of assemblages hard to determine. Assessing under what conditions a distinct upper canopy layer and its more or less specific fauna is maintained is also a research priority.

Much remains to be gained by examining what is known of stratification and compartmentalization of forest arthropod assemblages where these are best documented: in parts of the northern temperate zone (e.g. Dajoz, 1980; Strong *et al.*, 1984; Barbosa & Wagner, 1989; Schowalter, 2000). Of course, the extent to which the situation in a moist temperate forest may be extrapolated to the tropics is uncertain. Still, our understanding of arthropod distribution (vertical or horizontal) will progress considerably following appropriate comparisons of temperate and tropical systems.

To improve protocols and satisfy the rigour demanded by Smith (1973), sampling artefacts and biases need to be factored out, particularly where a bait such as light or fruit is being used, as it may pull an insect away from its usual 'cruising level' (see for example Byers *et al.* (1989) for calibration of flight height distribution in bark beetles and DeVries & Walla (2001) for mark–recapture studies). The simplest solution to these problems is to avoid any method of sampling that involves attractants, especially those, such as light, that are responded to from some distance. For valid comparisons, a prerequisite is for samples to be of the same type. However, even when the same method of sampling is employed, results obtained may be strongly affected by variables such as the mobility/activity levels of the arthropods themselves (e.g. pitfall traps, Malaise traps) or trap 'apparency' (e.g. pan traps), which, in turn, may be largely a function of structural differences (such as 'openness') in the immediate environment. Further, a minimum number of replicates is obviously needed to factor out site effects (Basset *et al.*, 2001a; DeVries & Walla, 2001) and it would be preferable to obtain selective samples along the whole vertical transect, not just from the understorey and upper canopy. There may be preference for traps that sample passively (suction, Malaise, intercept), though these rarely provide useful samples of groups like Lepidoptera.

It is difficult to draw general conclusions about the vertical distribution of insects in tropical forests. Long-term studies, such as those of Roubik (1993) and DeVries & Walla (2001), suggest that temporal movements up and down in response to changes in the environment (food and nectar sources, as well as microclimate) and in the insect populations (mating behaviours, for example) are of critical importance and can only be understood with long-term observations. With respect to vertical stratification, data concerning samples of insects taken while in flight will always present a particular problem. Appropriate understanding of the biology of the individual species involved will, however, help to determine whether such data have any bearing on vertical gradients or stratification of some persistence, or merely catalogue the numbers of short-term or occasional visitors to a particular canopy level. Last, an improved framework for describing the within-forest distributions of arthropod species is badly needed. Ideally, this should reflect, at least in some measure, the ways in which arthropods themselves experience the heterogeneity of their setting.

In the context of tropical forests, the most evident explanations for a pronounced stratification of arthropod assemblages are marked vertical gradients of environmental factors and discontinuities in the occurrence of available habitats. These characteristics are most strikingly exhibited by tall, closed, mixed and wet rainforests growing on flat terrain. Since these same forests are those under the greatest threat, particularly from logging, the study of arthropod vertical distribution in situations where stratification is most evident may, besides its taxonomic, ecological and evolutionary interests, also have important implications in conservation biology.

ACKNOWLEDGEMENTS

This review has benefited from many useful exchanges of information and opinion with co-workers, too numerous to list individually, in the field of forest entomology. The second author thanks those that have organized, facilitated or helped with his access to and use of various canopy facilities, or have contributed to field investigations that have furnished data most pertinent to the topic of arthropod stratification in tropical forests. In most instances, the latter have included colleagues at the Natural History Museum, London, notably Martin Brendell, Stuart Hine and Nigel Stork. To the last-mentioned a particular debt is owed; without his drive and enthusiasm for studies of treetop insects, the second author's opportunities to investigate canopy insect assemblages first-hand would have been much curtailed. The studies of arthropod vertical distribution in tropical rainforests of the first and third authors are supported in part by the Smithsonian Tropical Research Institute and the University of Panama.

4

Determinants of temporal variation in community structure

Raphael K. Didham and Neil D. Springate

ABSTRACT

There is pronounced temporal variation in the diversity and abundance of canopy arthropods, even in supposedly 'aseasonal' tropical forests. Various combinations of biotic and abiotic factors have been put forward to explain these trends, but no framework exists within which to interpret patterns. Temporal variation in community dynamics is still largely thought of as a climatic phenomenon. The main reason for this misconception is a lack of distinction between the proximate cues and the driving mechanisms of temporal periodicity. We review major abiotic (abiotic regulation and abiotic resource tracking) and biotic (biotic resource tracking, resource competition and predation) determinants of temporal periodicity in arthropod assemblages and distinguish between these direct mechanistic processes and at least two indirect processes (apparent regulation and apparent competition) that influence temporal periodicity. These processes are important local determinants of temporal variation in community structure, but they cannot be considered independently of large-scale, regional determinants of temporal variation in community structure (e.g. history, large-scale periodic events and phylogeny). The relative importance of regional processes is not widely acknowledged because of a lack of appreciation of the full time-scale of temporal periodicity in biotic communities. We review temporal periodicity at the species and community levels, from cycles at subannual frequencies to frequencies of the order of tens of thousands of years. Regional processes play a major role in determining variation in community structure. In particular, host-plant phylogenetic traits may have an important bearing on arthropod temporal dynamics. Since host-tree phylogeny is expressed in traits such as tree phenology (flower timing, leaf flush and seed-set), growth rate and senescence patterns, and since many arthropod species track these resources through time, we conclude that there is a strong effect of host-tree phylogeny on temporal trends in associated arthropod species. These findings have implications for the effects of human disturbance on canopy arthropod assemblages, as changes in host-tree dynamics (e.g. species composition) may lead to previously unsuspected changes in short-term and long-term temporal variation in community structure.

INTRODUCTION

Temperate forests are defined and structured by strong seasonal trends in the abundance, diversity and composition of terrestrial communities (Fretwell, 1972; Tauber *et al.*, 1986). Surprisingly, temporal variation in arthropod assemblages can be just as pronounced in supposedly aseasonal tropical climates (Bigger, 1976; Wolda, 1978a, 1988, 1992; Denlinger, 1986; Boinski & Fowler, 1989). In this review, we highlight the nature and magnitude of temporal periodicity in canopy arthropod assemblages and provide a much-needed general framework within which empirical findings can be interpreted. We focus primarily on the mechanistic determinants of temporal periodicity, and on some of the broader problems and challenges faced in the study of temporal canopy ecology.

As a relatively young field of study, canopy ecology has barely progressed beyond small-scale descriptive studies of primarily spatial phenomena (Lowman & Moffet, 1993; Stork *et al.*, 1997b; Moffett, 2000). Not surprisingly, relatively few studies have focussed on temporal periodicity in assemblages of tropical canopy arthropods, and most of these have been short term (Erwin & Scott, 1980; Lowman, 1982b, 1985;

Hijii, 1983; Adis, 1984a,b; Paarmann & Stork, 1987a,b; Basset, 1988, 1991c; Stork & Brendell, 1990; Kato *et al.*, 1995; Recher *et al.*, 1996a; Smythe, 1996; Heatwole *et al.*, 1997; Wolda *et al.*, 1998; Palacios-Vargas *et al.*, 1999). Furthermore, fruitful comparisons among arthropod taxa, between sites or across tree species are largely negated by differences in the trap types and methodologies employed in different studies (e.g. Basset *et al.*, 1997b). These problems present practical difficulties for the study of temporal ecology, but more importantly there are major conceptual limitations evident in the interpretation of temporal patterns. There is still the popular misconception that temporal variation is of relatively minor importance in structuring tropical arthropod assemblages and is more of a confounding variable to take into consideration when studying the spatial ecology and diversity of canopy communities. We argue that the scale of temporal variation in biotic communities has not been fully appreciated, and that it is at least as significant a factor structuring tropical assemblages as is spatial variation.

We structure our review around three major conceptual problems in temporal canopy ecology. First, many ecologists still believe that abiotic factors alone are the main driving forces behind temporal periodicity. We emphasize that this is not the case for most organisms, and that this misconception stems largely from a lack of distinction between the proximate cues for change and the actual determinants of temporal periodicity. Clarification of the distinction between temporal cues and driving mechanisms of periodicity is fundamental to the extension of descriptive observations into future hypothesis testing.

Second, many canopy studies lack a context of the scale within which their temporal results should be interpreted, simply because there is a lack of long-term data from forest canopies. We quantify the approximate time-scale of known temporal periodicity in canopy and noncanopy communities, from subannual frequencies to frequencies of the order of tens of thousands of years.

Third, we address the relative importance of local versus regional processes in determining temporal variation in biotic assemblages. In particular, it is not well recognized that host-tree phylogeny has a major effect on temporal, as well as spatial, variation in canopy arthropod assemblage structure.

THE CUES AND DRIVERS OF TEMPORAL VARIATION

Most arthropod species are highly sensitive to variation in environmental conditions (Andrewatha & Birch, 1954; Milne, 1957; Tauber & Tauber, 1976; Southwood, 1968; Willmer, 1982; Wigglesworth, 1984; Price, 1997). In particular, many temperate and tropical arthropods use specific temperature, moisture or light conditions as triggers to initiate life-history development (Bradshaw, 1974; Tauber & Tauber, 1976; Denlinger, 1986; Tauber *et al.*, 1986; Caceres, 1997). For example, rainfall triggers emergence from dry season diapause in the stem-boring larvae of *Busseola fusca* (Fuller) (Lepidoptera: Noctuidae) (Usua, 1973), some Tabanidae (Diptera) species in Africa (Bowden, 1976) and in numerous other tropical insects (Denlinger, 1986). However, subsequent fluctuations in the population densities of these species may be determined only partially by moisture availability or perhaps may not be related to increasing rainfall at all (but see Kaspari *et al.*, 2001). In other words, although rainfall might be the *cue* for life-history development, other factors such as temperature or resource availability may actually *drive* variability in growth, survival, reproduction and mortality rates, which ultimately control population dynamics. A clear distinction between the cues and drivers of temporal periodicity is crucial for interpreting trends in population dynamics and for gaining any sort of mechanistic understanding of the major processes controlling temporal variation in community structure.

The problem is that the cues for temporal periodicity are almost always strongly correlated with the mechanistic determinants of population change, making it difficult to distinguish one from the other. The reason for this is the strong selective advantage to the detection of cues to the future availability of limiting resources, and thus the maximization of access to these resources before they are consumed or disappear. Resources, in this context, can be anything from food, water, mates or territory, to windows of 'enemy-free space' and time. The situation is further confounded because there are probably as many cues and drivers of temporal periodicity as there are different life-history strategies, and the temporal cue for one species may be the driver for another. For example, abiotic factors such as temperature or rainfall drive the phenology of leaf flush in many species of tropical plants (e.g. Frankie *et al.*, 1974; van Schaik

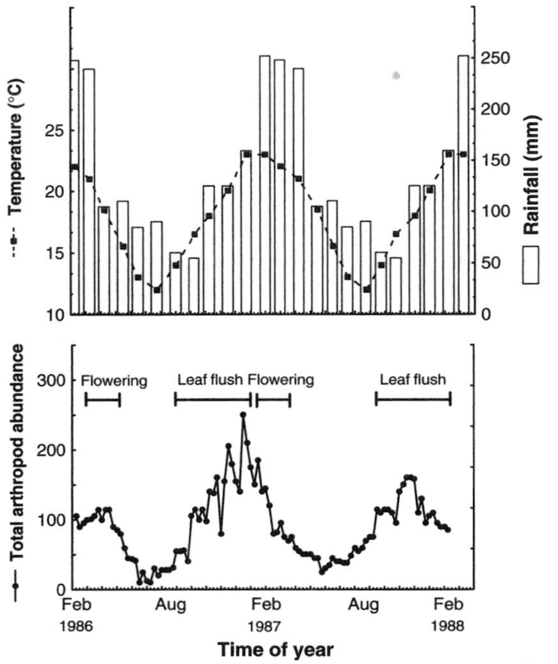

Fig. 4.1. The relationship between canopy arthropod abundance in the crowns of *Argyrodendron actinophyllum* Edlin (Sterculiaceae) trees near Brisbane, Australia, and variation in biotic and abiotic seasonality (after Basset, 1991c). Temporal variation in canopy arthropod density is strongly correlated with fluctuations in temperature, rainfall and tree phenology, making it impossible to resolve the factor(s) driving arthropod community dynamics.

et al., 1993; Reich, 1995; Williams *et al.*, 1997; Borchert, 1998) (Fig. 4.1). These variables, at the same time, cue the emergence of herbivore species that specialize on newly emerged leaves within a very narrow window of opportunity. Temperature and rainfall, while highly correlated with herbivore abundance (Fig. 4.1), do not (necessarily) drive herbivore dynamics, they merely act as a reliable indicator of resource availability: hence, the problem of correlation versus causation that arises when trying to assess the major determinants of temporal periodicity (Fig. 4.1) (e.g. Basset 1999a).

The degree to which species rely on correlated cues to resource availability depends on the reliability of the indicator and the degree of spatio-temporal heterogeneity in resource availability. Conditions under which cues accurately and reliably forecast resource availability, and conditions under which resource availability varies predictably through time, both result in the strongest coupling of life-history responses with cues. In situations where species have evolved finely tuned responses to environmental triggers, the fitness costs of responding to false cues can be devastating. For example, Foster (1974) recorded uncharacteristically heavy rainfall during the dry season in Panama in 1970, which elicited early reproduction and the subsequent fruit failure of many plant species during the latter half of the rainy season. Response to the false cue resulted in massive resource shortages for canopy frugivores, increased animal mortality and increased leaf herbivory and tree damage (Foster, 1974).

Cues for temporal periodicity are either abiotic or biotic. In the majority of studied cases, exogenous *abiotic* factors cue changes in life-history stage, or phenophase (Bradshaw, 1974). Often this involves initiation or termination of diapause (Caceres, 1997), even in tropical insects (Denlinger, 1986). The most common of these factors are temperature, moisture and photoperiod (reviews in Tauber & Tauber, 1976; Denlinger, 1986; Wolda, 1988; Caceres, 1997). It is not so widely appreciated that some species instead use *biotic* cues to temporal periodicity in resource availability. Rarely, biotic triggers are evoked endogenously without direct feedback from the environment (such as in obligate diapausing species: Bradshaw, 1974). More commonly, however, other species respond to exogenous biotic cues. For example, the seed bugs *Jadera obscura* (Westwood) and *Jadera aeola* (Dallas) (Hemiptera: Rhopalidae) in Panama apparently break diapause when they detect chemical cues related to the availability of palatable seeds of Sapindaceae (Wolda, 1988), and the jewel beetle, *Melanophila acuminata* DeGeer (Coleoptera: Buprestidae), in Germany congregates to reproduce at a food source by detecting volatile chemicals from burning trees over distances of up to 50 km (Schütz *et al.*, 1999).

Although largely unstudied, it is likely that biotic cues may be more commonly used by endophytic arthropods buffered from direct abiotic conditions (e.g. leaf-miners or stem-borers) than by free-living, exophytic organisms on canopy surfaces. Examples of such biotic cues for endophytes might be translocation of nutrients within plant tissues during the growing season, or seasonal changes in the allocation of plant defensive chemicals (e.g. Feeny, 1970; Potter & Kimmerer, 1986), which may even affect the temporal dynamics and diversity of endoparasitoids feeding on herbivores within plant tissues (Gauld *et al.*, 1992).

It is important to reiterate that, although suboptimal responses to biotic or abiotic cues can have important fitness consequences for individuals, temporal cues are distinct from the driving mechanisms behind temporal variation in population density.

DRIVING MECHANISMS OF TEMPORAL PERIODICITY: A FRAMEWORK

There have been major advances in our understanding of spatial ecological processes and the factors that influence the distribution and abundance of organisms since the 1970s (Kareiva, 1990, 1994b; Pickett & Cadenasso, 1995; Hanski & Gyllenberg, 1997; Ranta *et al.*, 1997; Gaston, 2000). Much of the current framework of spatial ecology is founded on the explicit contrast of local versus regional determinants of community structure (Cornell & Lawton, 1992; Ricklefs & Schluter, 1993; Caley & Schluter, 1997; Srivastava, 1999; Loreau, 2000). Although temporal variation in the population dynamics and community structure of organisms has long been recognized, it is often considered less important than spatial variation and no theoretical framework is available to evaluate the relative roles of different processes. In particular, the relative importance of local versus regional processes controlling temporal periodicity has not been explicitly considered. Explanations for temporal trends in arthropod assemblage dynamics have largely centred around small-scale local determinants of periodicity, such as climatic seasonality, with abiotic factors almost invariably considered to be the dominant driving mechanisms. We emphasize that this is not the case for all organisms. Temporal variation in local population dynamics is driven by both abiotic *and* biotic factors.

Initially, we focus here on establishing a framework for local determinants of temporal variation in community structure and return later to local versus regional processes. We discuss seven possible mechanisms controlling variation in species density through time (i.e. population dynamics), and which may extend to the control of temporal variation in community structure (to the extent that the summed variation in population densities of many species equates to changes in community assembly over time). These mechanisms can be categorized as biotic versus abiotic, and direct versus indirect (Fig. 4.2).

	Direct	Indirect
Abiotic	Abiotic regulation Abiotic resource tracking	Apparent regulation
Biotic	Biotic resource tracking Resource competition Predation	Apparent competition

Fig. 4.2. A framework for the mechanisms controlling temporal periodicity in population abundance.

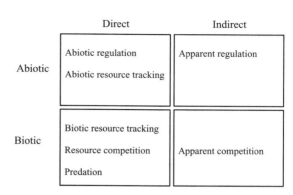

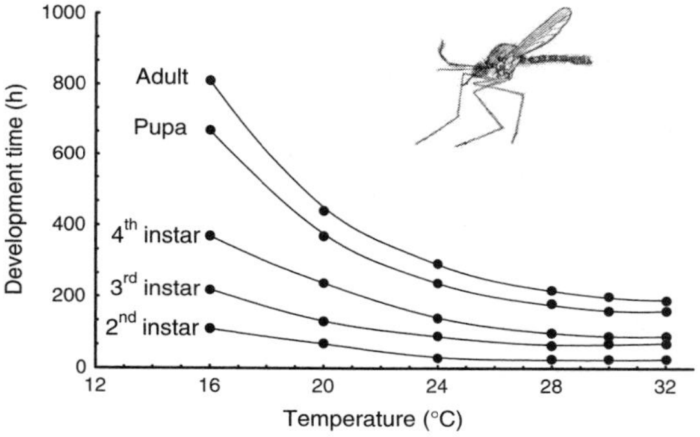

Fig. 4.3. Abiotic regulation. Temperature controls the rate of development of the mosquito *Aedes aegypti* (L.) (Diptera: Culicidae). (Redrawn, with permission, from Gullan & Cranston (2000). *The insects: An outline of Entomology*, 2nd edn. Copyright 2000 Blackwell Science Ltd.)

Direct mechanisms

Abiotic regulation

Most arthropods are poikilothermic, receiving the heat required for growth and development directly from the environment (Willmer, 1982; Wigglesworth, 1984). Even small variations in temperature, within the limits of physiological tolerance, can have marked effects on arthropod growth rates (Wigglesworth, 1984; Gullan & Cranston, 2000) (Fig. 4.3). For this reason, temperature can have a direct regulatory effect on arthropod population dynamics. At the local scale, the most obvious effect of temperature-controlled variation in growth and development is seasonal fluctuation in arthropod density (Wolda, 1978a, 1988).

Although seasonal variation in temperature is of critical importance for many temperate forest species, tropical forests generally have a more equitable temperature regime, and most days of the year exceed the developmental threshold for most arthropod species (Bigger, 1976; Denlinger, 1986; Wolda, 1989). More frequently, moisture is a seasonally limiting factor in many tropical forests, and temporal variation in available water (rainfall, humidity, moisture content) drives arthropod population dynamics (Wolda, 1988, 1989; Intachat *et al.*, 2001). However, even in the tropics, temperature and rainfall covary to such a degree that without properly controlled experimentation it is impossible to discern which factor drives population dynamics for which species (Basset, 1988, 1991c; Wolda, 1989) (e.g. Fig. 4.1). Although such experiments have not been performed for canopy arthropod species, Wright (1992) shows how experimental watering of *Psychotria* shrubs (Rubiaceae) in Panama disrupted the temporal synchrony of leaf production and flowering that was normally constrained by dry season moisture deficit (Wright & Cornejo, 1990; Wright, 1992).

Although many organisms are limited by moisture availability, active small-bodied taxa, such as arthropods, with high surface area to volume ratios and high metabolic rates are particularly sensitive to desiccation (Willmer, 1982; Wigglesworth, 1984). Within tropical forests, daily and seasonal fluctuations in temperature and moisture deficit are most extreme in the forest canopy (e.g. Haddow & Corbet, 1961b; Rees, 1983; Parker, 1997; Parker & Brown, 2000). In many ways, the upper canopy is more similar to an arid desert environment than a moist rainforest (Bell *et al.*, 1999). Where moisture availability drives population dynamics, it is likely that arthropod species utilizing the forest canopy for major stages of growth and development will exhibit greater temporal periodicity in population density than comparable ground-dwelling species.

Other abiotic factors that correlate strongly with arthropod growth and development, such as photoperiod, solar radiation and wind, are best thought of either as cues for temporal periodicity or as indirect factors operating through altered temperature and moisture regimes. For example, photoperiod itself does not drive increases or decreases in abundance, but it does cue initiation or termination of diapause in order to avoid seasonal fluctuations in limiting conditions (see references above). The amount of insolation can have an important bearing on the densities of surface-active arthropods, but this factor operates indirectly through increased desiccation. Similarly, wind may alter local population density by increasing rates of desiccation. There are situations, however, where high wind speed can drive spatial patterning in population density by mechanical abrasion of arthropods active on leaf surfaces (Claridge & Wilson, 1981; Moran & Southwood, 1982; Southwood *et al.*, 1982a,b; Didham, 1997), and where wind causes local changes in abundance through enhanced dispersal rates (den Boer, 1990), but there are no data to suggest that such effects operate regularly through time to induce temporal periodicity in population dynamics.

Abiotic resource tracking

One interesting subset of abiotic effects on arthropod population dynamics is the situation where an abiotic regulating mechanism is also a resource that is utilized for growth and development. In the broadest sense, moisture acts as both a regulator and a resource for all organisms, whereas temperature, for example, is solely a regulator of abundance. More specifically, some organisms require water-filled tree holes or pools in which to breed (e.g. Paradise & Dunson, 1997; Kitching, 2000; Yanoviak, 2001), and the availability of this resource drives temporal variation in population abundance (Macia & Bradshaw, 2000). Note that pool availability is strongly correlated with, but not solely dependent on, rainfall. In this context, physical habitat space is an abiotic resource that varies through time, thus driving temporal periodicity.

Biotic resource tracking

Perhaps the most important direct biotic mechanism driving temporal variation in species abundance is resource tracking. For example, herbivores are more abundant when their leaf resources are more abundant, and predator populations increase and decrease with fluctuations in prey numbers. Once again, resources can be interpreted very broadly to include food, mates, habitat availability or other biotic factors. For canopy arthropods, variation in food availability is determined to a large extent by host-tree phenology; seasonal variation in leaf production, flowering and fruiting (Faeth *et al.*, 1981; van Schaik *et al.*, 1993; Heatwole *et al.*, 1997; Novotny & Basset, 1998). While the interplay of abiotic regulation and biotic resource tracking is far from clear (e.g. Basset, 1999a; Intachat *et al.*, 2001) (Fig. 4.1), it is obvious that resource limitation is a key factor driving population dynamics for many species (Nakamura *et al.*, 1990; Matson & Hunter, 1992; van Schaik *et al.*, 1993).

Temporal synchronicity in the mast-seeding of south-east Asian Dipterocarpaceae provides an example of the importance of resource availability in structuring canopy arthropod assemblages. Total seed production by Dipterocarpaceae is highly synchronous among tree species, across large spatial scales, and at multiannual time intervals (Medway, 1972; Janzen, 1974; Ashton *et al.*, 1988; Yap & Chan, 1990; Curran *et al.*, 1999; Curran & Leighton, 2000; Curran & Webb, 2000; Nakagawa *et al.*, 2000; Wich & van Schaik, 2000). Temporal peaks in the abundances of insect herbivores and insectivorous vertebrates and invertebrates coincide with mast-seeding events in dipterocarp forests (McClure, 1966; Janzen, 1974; Whitmore, 1975; Toy *et al.*, 1992; Kato *et al.*, 1995; Curran & Leighton, 2000). Examples include increases in the abundances of pollinating *Thrips* L. (Thysanoptera: Thripidae) and '*Lemurothrips*' (Ashton *et al.*, 1988) (although '*Lemurothrips*' is a *nomen nudum*, possibly intended to be *Megalurothrips* Bagnall, but there is some doubt as to the identification of the species; L. A. Mound, personal communication) and weevil seed predators (Coleoptera: Curculionidae) (Toy *et al.*, 1992), both exhibiting resource tracking through time. Since dipterocarps dominate south-east Asian forests, patterns in tree phenology and resource availability for associated arthropod species in these forests differ markedly from those in neotropical and African rainforests, where trees typically exhibit annual or biennial periodicity (Richards, 1952; Foster, 1974; Gentry, 1974; Gentry & Emmons, 1987; Roth, 1987; Schatz, 1990; Knowles & Parrotta, 1997; Tutin & White 1998).

In dipterocarp forests, extreme resource fluctuation not only drives small-scale temporal periodicity but also the evolution of unusual life-history strategies of some arthropod species (Roubik, 1989). Social apine and xylocopine bees (Hymenoptera: Apidae) follow the boom and bust cycle of dipterocarp flowering by being very long lived (surviving up to 3–4 years) and passing the long interval between flowering events by entering diapause, both of which are extremely rare traits for bees. When the host trees do flower, the bees produce from one to four rapid generations within the flowering year and most die after reproducing (Roubik, 1989). This reproductive strategy is in stark contrast to the relatively stable temporal dynamics of related neotropical bees (Proctor *et al.*, 1996).

Resource competition

Because of strong competition for limited resources, species often partition resource use through time and space (Pontin, 1982; Grover, 1997). A good example is the temporal sequence of flowering of dipterocarp trees. There is strong competition among dipterocarp trees flowering at multiannual intervals for access to a limited available pool of pollinating thrips (Ashton *et al.*, 1988). Access to flower resources drives an increase in thrip abundance (resource tracking), but initial thrip densities during the intermast interval would not be high enough, and would not increase rapidly enough, to pollinate effectively all dipterocarp species flowering synchronously. Consequently, dipterocarp species have evolved a fine temporal partitioning of flowering to ensure effective pollination and reduce the likelihood of cross-species pollen transfer (Fig. 4.4) (Ashton *et al.*, 1988). The interesting thing is that different dipterocarp species have quite different flower timing despite the fact that all species use the same temporal cue to initiate flowering: the El Niño Southern Oscillation (Ashton *et al.*, 1988; Curran *et al.*, 1999). Moreover, despite variable flower timing, dipterocarps exhibit highly synchronous mast-seeding (indicating compensatory seed development times across species). The reason for this is that seed phenology in dipterocarps is driven by an entirely different mechanism (predation) than is flower phenology (resource competition) (Toy *et al.*, 1992; Curran & Leighton, 2000; Curran & Webb, 2000).

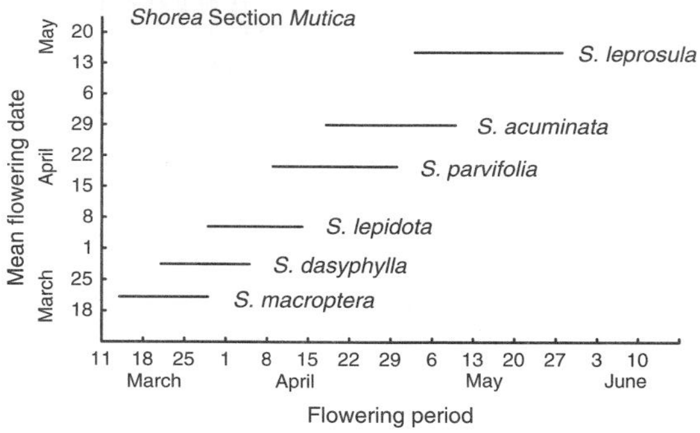

Fig. 4.4. Resource competition. Staggered flowering among species of *Shorea* (Section *Mutica*) trees (Dipterocarpaceae) in south-east Asia in response to strong competition for a limited pool of available *Thrips* and '*Lemurothrips*' (Thysanoptera) pollinators (Redrawn, with permission, from Ashton *et al.*, 1988. Copyright 1988 University of Chicago Press.)

In addition to the partitioning of temporal access to pollinators by plants, arthropods competing for access to flower resources also show complementary temporal trends in abundance owing to resource competition. For example, long-tongued and social host plant-specific bees are favoured towards the end of the wet season in the neotropics because there are fewer plants flowering at this time and they tend to have relatively few, large flowers that stay open for a long time (Proctor *et al.*, 1996). Earlier in the wet season, these bee species cannot compete with other generalist flower foragers.

Predation and parasitism

The third major biotic determinant of temporal variation in species abundance is predation, often more generally referred to as top-down or recipient control. Although consumption by predators and parasitoids may impose direct mortality on organisms, it does not necessarily drive population dynamics (Price, 1987; Matson & Hunter, 1992; Polis & Strong, 1996). Predator density may track prey resources through time, producing highly correlated predator–prey cycles, but prey dynamics may actually be driven by resource availability or other factors (Krebs *et al.*, 1995). The degree of top-down control on population density can also vary through time and space (Hunter & Price, 1992). Nevertheless, for some species, over a range of conditions, predation is an important determinant of temporal variation in population density.

One example is the role of predation as the driving mechanism behind synchronous mast-seeding in southeast Asian dipterocarp trees (Janzen, 1974; Curran & Leighton, 2000). Intraspecific and interspecific synchronization of seed production means that a greater proportion of seeds are likely to survive vertebrate and invertebrate predation to germinate. Curran & Leighton (2000) found that seedling establishment only occurred in Borneo during major masting years, when some seeds were able to escape predation. Their findings conform to Janzen's (1974) predator satiation hypothesis, except that 'satiation' is operating at the landscape level rather than at the local level (Curran *et al.*, 1999). Regional escape from predation, resulting from the inability of predators to respond numerically to the rapid increase in available resources, appears to be the mechanism driving temporal synchronicity in seed production (Curran & Leighton, 2000; Curran & Webb, 2000). Similar evidence exists for seed predation driving seed phenology in other systems (Albrectsen, 2000), and for the role of herbivory in controlling leaf-flush phenology (Aide, 1988, 1992; Yukawa, 2000).

Excellent experimental evidence for predator-driven population cycles in canopy arthropods comes from Turchin and co-workers (1999), studying the southern pine beetle, *Dendroctonus frontalis* Zimmermann (Coleoptera: Scolytinae), in the USA. Population density of a predatory beetle, *Thanasimus dubius* (F.) (Coleoptera: Cleridae), cycles in phase with *D. frontalis* (Fig. 4.5**a**) showing classic resource tracking. The difficulty is inferring a causal link in the dynamic relationship between the predator and prey species from time series data alone. The beauty of the study by Turchin *et al.* (1999), and similar recent work (e.g. Hudson *et al.*, 1998), is the combination of time-series data with field experiments and mechanistic theoretical models. Using simple predator exclusion, Turchin *et al.* (1999) showed that survival rates of *D. frontalis* were significantly greater when protected from predators than when exposed, driving a (time-lagged) density-dependent increase in mortality rates (Fig. 4.5**b**).

Indirect effects

In addition to the direct mechanistic determinants of temporal dynamics, there are many situations in which there may be strong correlation between biotic or abiotic variables and population density, but the cause and effect relationship is mediated by indirect effects through a third variable. Indirect effects and higher-order interactions in ecology (Strauss, 1991; Billick & Case, 1994; Wootton, 1994) are perhaps the most difficult to detect and interpret, but they occur ubiquitously in diverse, multispecies communities (Kareiva, 1994a). Two examples, apparent regulation and apparent competition, serve to illustrate the processes involved.

Apparent regulation

In the case of parasites, pathogens or concealed herbivores within a host organism, the internal parasite is effectively buffered from external conditions, particularly in the case of parasites within homeothermic hosts. However, because the population density of host organisms may respond strongly to abiotic conditions, there will also be a strong correlation between abiotic variation and parasite population density (in the landscape). The proximate cause of temporal variation in parasite density in this hypothetical example is actually resource (host) availability, but ultimately host density itself is driven by abiotic regulation. Accordingly, there is a strong indirect effect of abiotic regulation on parasite density, which we will term *apparent regulation*.

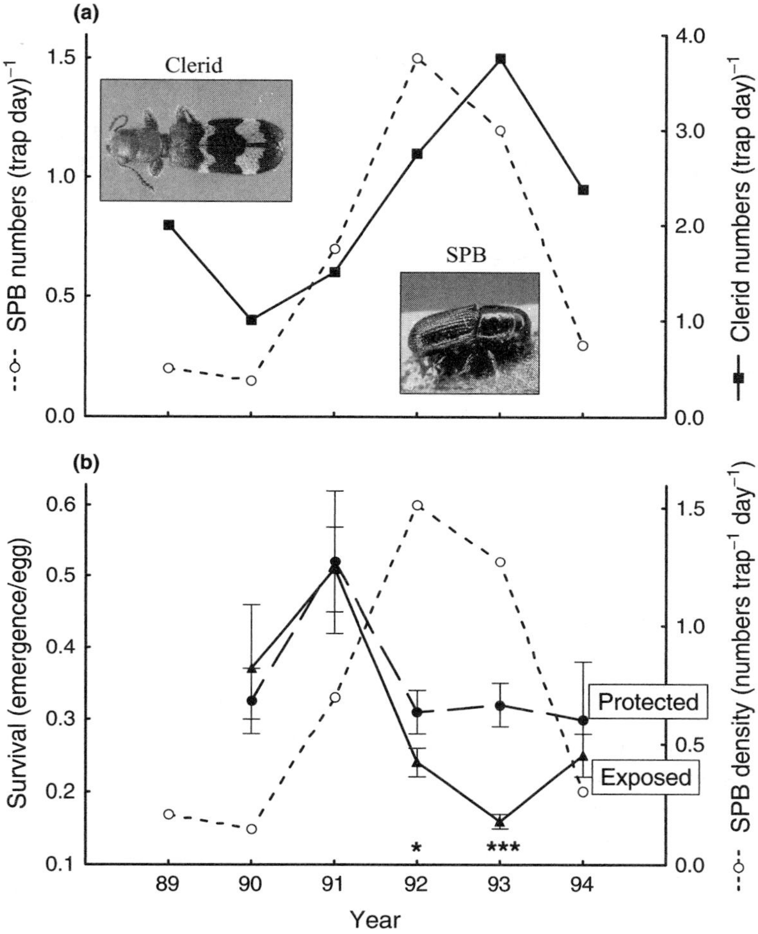

Fig. 4.5. Predation. (**a**) Population density of the predator *Thanasimus dubius* (F.) (Coleoptera: Cleridae) tracks the abundance of its prey, the southern pine beetle (SPB), *Dendroctonus frontalis* Zimmermann (Coleoptera: Scolytidae), through time. (**b**) Using simple predator exclosures, survival (emergence/egg) of SPB is significantly higher when protected from predators, providing experimental evidence that predation drives population dynamics in this system. $^{*}p < 0.05$; $^{***}p < 0.001$. (Reprinted, with permission, from Turchin *et al.*, (1999). Dynamic roles of predators in population cycles of a forest insect: an experimental test. *Science*, **285**, 1068–1070. Copyright 1999 American Association for the Advancement of Science.

Apparent competition

Resource competition causes the population densities of competing organisms to cycle out of phase with one another (Grover, 1997). However, it is widely recognized that the same patterns can be observed in species that do not interact directly with each other, if both species are subject to mortality by the same predator exhibiting density-dependent prey-switching (e.g. Lawton & Strong, 1981). Population density of either species is directly dependent on predation but is also indirectly dependent on the population density of the apparent competitor.

A similar phenomenon occurs where temporally separated herbivores apparently compete by influencing host plant chemistry and hence food quality (Faeth, 1986). In this case, population density of either herbivore is directly dependent on food quality but is also indirectly dependent on the feeding activity and population density of the apparent competitor (Masters & Brown, 1992; Masters, 1995).

A network of direct and indirect links within reticulate food-webs influences population fluctuations for any given species (Polis & Strong, 1996). The relative strengths of species interactions and the relative importance of different driving mechanisms probably vary through time and space to a considerable degree (Hunter & Price, 1992; Berlow *et al.*, 1999), making it extremely difficult to define the mechanisms controlling temporal dynamics. Only by linking time-series observations with experimentation and dynamic modelling can these controlling mechanisms be resolved (Turchin *et al.*, 1999).

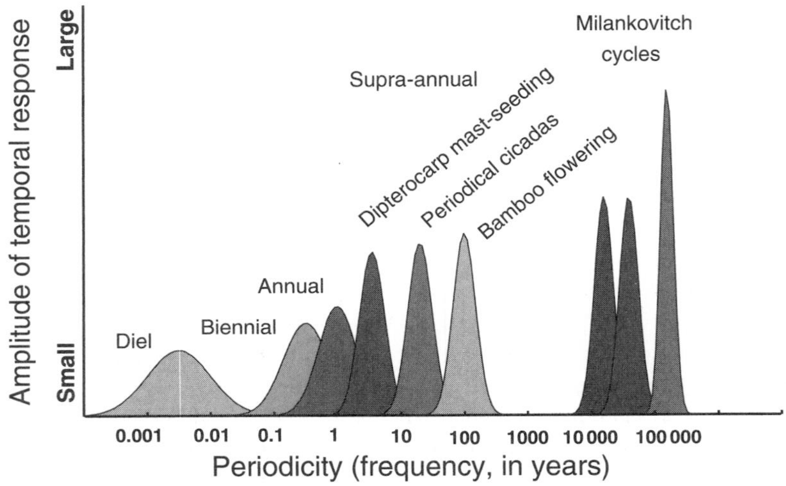

Fig. 4.6. The approximate time-scale of known temporal periodicity in biotic communities. Note that diel periodicity results from the spatial movement of organisms between different habitats; therefore, it does not strictly belong in the framework of temporal variation in population dynamics. See text for explanation of periodical cycles.

SMALL-SCALE VERSUS LARGE-SCALE TEMPORAL VARIATION

Population oscillations observed at local or even regional scales today are rarely stable through time and do not occur against a static backdrop of spatial or temporal processes (Wolda, 1992). Both the spatial and temporal histories of sites (and species) have a major bearing on current population dynamics (Cappuccino & Price, 1995). Many large-scale (or 'regional') processes effectively define the conditions under which small-scale local determinants of temporal variation operate. Important regional processes include the spatial history of the site (stochastic variation in biogeography, topography and disturbance), temporal history (frequency and magnitude of disturbance events, stochastic fluctuations in biotic and abiotic processes and periodic fluctuations in climate) and phylogenetic history. Spatial ecology has wrestled with the relative importance of local versus regional processes as determinants of community structure since the 1970s (Cornell & Lawton, 1992). The emerging consensus is that local communities are almost never saturated with species, such that local diversity is determined almost entirely by regional diversity and the processes that generated it (Cornell & Lawton, 1992; but see Srivastava, 1999; Loreau, 2000). Of course there are situations where extreme local environmental or competitive adversity do control community assembly (Drake, 1990; Weiher & Keddy, 1995). There is no reason to expect that the same will not hold true for temporal variation in community structure as well.

Temporal ecology has been slow to recognize the importance of regional processes controlling the population dynamics of species, perhaps because of the short-term nature of most temporal studies or because of an inadequate appreciation of the full time-scale of periodicity in biotic communities. We find that periodicity in species abundance and composition occurs at much larger time-scales than most biologists realize; across time-scales ranging from subannual frequencies (seasonality) to frequencies on the order of tens of thousands of years. Here we review the time-scale of known periodicity in biotic communities (Fig. 4.6).

A number of canopy studies have focussed on very high-frequency temporal variation in community composition between day and night, known as diel periodicity (Springate & Basset, 1996). More correctly, however, diel periodicity stems from either spatial shifts in the habitat utilization of species or temporal variation in activity rates, but not from temporal variation in population density; consequently, it does not genuinely belong in the present framework. The same is probably true of the effects of lunar cycles on the distribution and abundance of night-flying insects (Brown & Taylor, 1971; Perfect & Cook, 1982).

For tropical organisms, subannual, particularly biennial, periodicity is much more common than is annual periodicity (Bigger, 1976; Roth, 1987; Boinski & Fowler, 1989). Most temporal trends identified in canopy arthropod studies fall into the biennial and annual periodicity categories (e.g. Hijii, 1983; Adis, 1984a; Paarmann & Stork, 1987b; Basset, 1988, 1991c; Nakamura *et al.*, 1990; Kato *et al.*, 1995; Smythe, 1996; Heatwole *et al.*, 1997; Palacios-Vargas *et al.*, 1999). Some organisms, of course, cycle at supra-annual frequencies, such as the 3–5 year cycles of dipterocarp trees and their associated insect species (see above), 13- and 17-year periodical cicadas (Karban, 1997; Cox & Carlton, 1998), and some bamboo species that synchronize flowering at intervals of 76 to 120 years and then die after reproducing only once (Keeley & Bond, 1999) (Fig. 4.6). The extent of supra-annual periodicity in arthropod assemblages has almost certainly been underestimated (e.g. Turchin *et al.*, 1999).

Less well appreciated is low-frequency periodicity in biotic communities associated with periodic variation in the amount of insolation reaching the Earth (Fig. 4.6). Regular variation in solar orbital cycles, known as Milankovitch cycles, periodically bring the sun much closer to the Earth, or much further away from the Earth, than at other times. Orbital forcing is known to drive changes in glacial ice sheets at frequencies of about 19 000 years (precession) and 41 000 years (obliquity) (Laskar *et al.*, 1993). In addition, Willis *et al.* (1999) have further described 124 000 year periodicity in terrestrial vegetation change in Pliocene Europe (3.0–2.6 million years ago). Using stratigraphic pollen samples, they show regular shifts in tree dominance from subtropical to boreal canopies at a period of 124 000 years that appears to be an internally driven, nonlinear response of the Earth's climate to solar cycles (Willis *et al.*, 1999).

At these large temporal scales, we are talking about huge changes in canopy communities: major shifts in the composition of dominant tree species, biogeographical changes and species divergence and extinction on evolutionary time-scales. Although local biotic and abiotic determinants of temporal periodicity are important in structuring small-scale temporal variation in community structure, whole ecosystems can clearly undergo periodic climate forcing that has a much greater effect on the long-term temporal dynamics of associated arthropod species.

REGIONAL DETERMINANTS OF TEMPORAL PERIODICITY

Most regional processes that have a bearing on the temporal dynamics of populations or communities are likely to be stochastic historical accidents, rather than large-scale periodic events (such as Milankovitch cycles). It is difficult, then, to envisage how random stochastic events such as biogeographic accidents, large-scale disturbance or phylogenetic history can have a bearing on deterministic trends in current population dynamics. That is, unless we consider that past history is important in setting up local conditions that favour a particular form of periodic cycling; for example, if stochastic plate tectonics and mountain building in the past have led to regional atmospheric or oceanic circulation patterns (the Gulf Stream, El Niño, the Atlantic hurricane path) that enhance seasonal climatic variation. As a consequence, current temporal periodicity cannot be interpreted without knowledge of past spatial and temporal history.

For tropical canopy arthropods, one particularly important regional determinant of species distribution patterns is their relationship with host-tree species and the influence of plant community composition on arthropod assemblage structure. We know that the history of vegetation composition, as expressed in the phylogenetic relationships among present-day plant species, plays a very important role in the spatial structuring of canopy arthropods (Mitter *et al.*, 1991; Armbruster, 1992; Basset, 1992a) – but what about temporal structuring? We ask the question, 'Does host-tree phylogenetic status mediate the expression of temporal trends in arthropod community structure?'

Phylogenetics and the comparative method are important in community ecology because related species share many ecological traits through common ancestry (Harvey & Pagel, 1991; Miles & Dunham, 1993). But do related species share common temporal dynamics? Many closely related arthropod species certainly do, by virtue of sharing phylogenetically conservative life-history strategies (Eggleton & Vane-Wright, 1994). In addition, quite unrelated species of canopy arthropod may share similar temporal dynamics if the driving force behind temporal periodicity is imposed by traits of a shared host tree. For canopy arthropods on tropical trees, we would argue that if temporal variation in host-tree phenology is phylogenetically conservative, and if canopy arthropods track phenologically generated host resources through time, then plant phylogeny may heavily influence temporal variation in associated arthropod abundance.

Much hinges on the phylogenetic basis of host-plant phenology as this is the resource link for

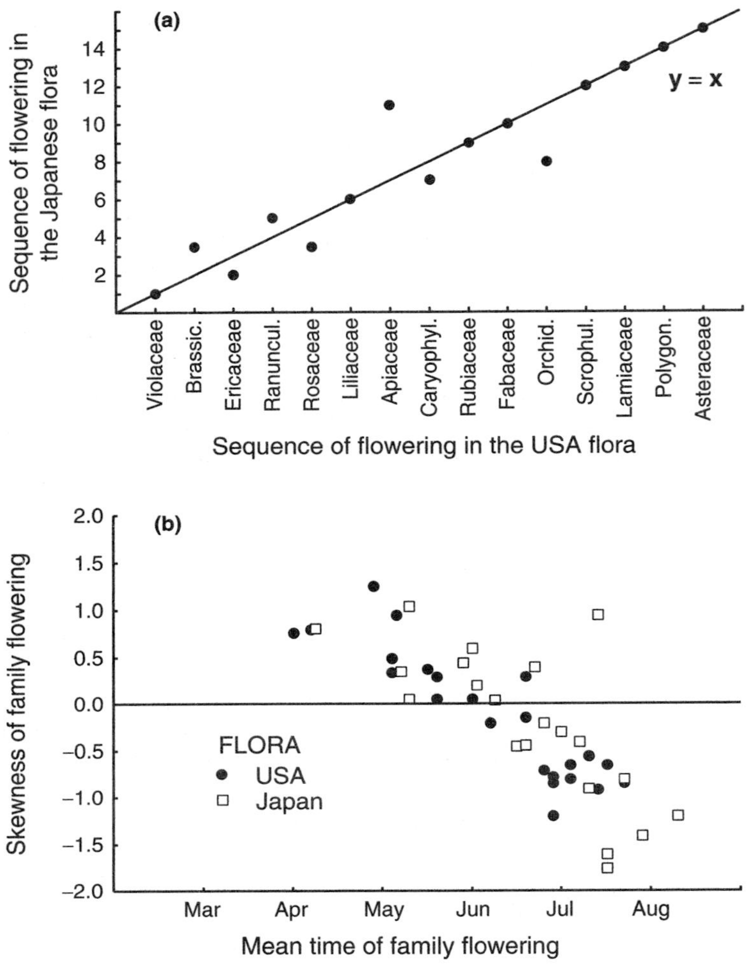

Fig. 4.7. The phylogenetic basis of host-plant phenology. (**a**) The sequential timing of flowering among plant families. (**b**) The skewness of the distribution of flowering of individual plants within families. Both have a strong phylogenetic component and are highly conservative through evolutionary time. See text for full family names. (Redrawn, with permission, from Kochmer & Handel, 1986. Copyright 1986 Ecological Society of America).

arthropods. The best example of phylogenetically conservative host-plant phenology is given by Kochmer & Handel (1986), who compared the timing of flowering across plant families in Japan versus the USA. Despite having completely different species assemblages in the two countries, early-season flowering families in the USA, such as Ericaceae, flowered early in Japan, and late-season flowering families in the USA, such as Asteraceae, flowered late in Japan. There was a near-perfect match in the sequence of flowering in the two floras (Fig. 4.7a) and, incredibly, a near-perfect match in the skewness of flower timing as well (Fig. 4.7b), suggesting that flowering phenology is phylogenetically highly conservative (Kochmer & Handel, 1986).

As with flower phenology, it is likely that many other plant phenological traits, such as leaf-flush and fruiting phenology, also have a strong phylogenetic component. As canopy arthropod density is strongly affected by host-plant phenology (see references above), these findings imply that the temporal dynamics of canopy arthropod assemblages may depend to a large degree on plant species composition and the phylogenetic makeup of forest canopies. Take the extreme example of south-east Asian canopies dominated by Dipterocarpaceae exhibiting supra-annual periodicity of flowering and fruiting. Although not all insect species in these forests are directly associated with dipterocarps (e.g. Intachat *et al.*, 2001), the reproductive ecology of Dipterocarpaceae nevertheless imposes extreme fluctuations in resource availability for many species in south-east Asia and drives, both directly and indirectly, supra-annual periodicity in canopy insect abundance. In contrast, other tropical regions are dominated by plant families with biennial or subannual reproductive periodicity, such as Lecythidaceae or Bignoniaceae in the neotropics (Foster, 1974; Gentry, 1974; Roth, 1987) and, in these forests, we typically find insects with subannual temporal variability. This difference may be the result, in large part, of the influence of host-tree phylogeny.

It is important to recognize that the phylogenetic history of a flora is only one of many large-scale regional determinants of community structure. No single factor accounts for the distribution and abundance of all species, either spatially or temporally. As in spatial ecology, there are almost certainly situations in which extreme local fluctuations in biotic and abiotic factors override broad, regional determinants of temporal variation. Although the traits of species such as phenology might be phylogenetically conservative over evolutionary time, there is often pronounced small-scale intraspecific genetic variation in tropical tree species (Thomson *et al.*, 1991; Chase *et al.*, 1996). Such genetic variability may lead to marked local variability in host phenology and, hence, resource availability for herbivorous insects (Basset 1991d).

A first step, then, in trying to compare temporal variation in canopy insects between different tree species, different habitats or different regions must be to control for host-tree phylogeny.

CONSERVATION IMPLICATIONS

Recognizing the important phylogenetic component in plant phenology, and the influence of host-tree phenology on the temporal dynamics of canopy arthropod species, we realize that any change in tree species (phylogenetic) composition must affect temporal oscillations in arthropod populations. Scientists and conservation managers are far more conscious of the spatial habitat and host affinities of canopy arthropods, and of the devastating spatial restructuring of arthropod assemblages that occurs following tropical forest logging and fragmentation (Aizen & Feinsinger, 1994a,b; Laurance *et al.*, 1997, Stork *et al.*, 1997b; Didham *et al.*, 1998; Curran *et al.*, 1999; Nepstad *et al.*, 1999). It is also important to appreciate that tropical forest disturbance may also trigger previously unsuspected changes in the temporal dynamics of arthropod species. Forest disturbance, whether by clear-felling, burning, or selective logging, alters the composition of tree species (Johns, 1992; Laurance *et al.*, 1997, 1999; Nepstad *et al.*, 1999). The pioneer trees that recolonize these disturbed sites typically have a different species composition from those in undisturbed forest, with a different phylogenetic makeup and different phenologies (Hubbell *et al.*, 1999; Webb 2000). As a consequence, altered tree species composition will almost certainly cause changes in the spatial and temporal dynamics of insects associated with disturbed tropical forest edges (Cates & Orians, 1975; Reader & Southwood, 1981). The exact nature of such changes remains uncertain, but it seems likely that any change in the temporal dynamics of canopy arthropod species (including increased frequencies of pest outbreaks) may have long-term implications for ecosystem stability and the conservation of biodiversity (Roland, 1993; Curran *et al.*, 1999).

CONCLUSIONS

In conclusion, we suggest there needs to be clearer distinction between the cues and drivers of temporal periodicity, and that the relative roles of local versus regional determinants of temporal variation should be more explicitly resolved. In particular, host-plant phylogeny should be controlled for in studies of both the spatial *and* temporal ecology of canopy communities. The strong phylogenetic component to host-tree phenology implies that any change in tree species composition caused by logging or fragmentation of tropical forests will have a major impact on temporal variation in resource availability, and hence on arthropod assemblage dynamics.

ACKNOWLEDGEMENTS

We would like to thank the editors of *Arthropods of Tropical Forests* for their kind invitation to present our work at the *XXI International Congress of Entomology*, Foz do Iguaçu, Brazil, 2000. Thanks to Rob Ewers, Jeremy Holloway and an anonymous reviewer for useful comments on earlier drafts. We would like to thank Blackwell Science Ltd for permission to reprint Fig. 4.3, to the University of Chicago Press for permission to reprint Fig. 4.4 (© 1988 by The University of Chicago Press. 0003-0147/88/3201-0007#02.00. All rights reserved), to the American Association for the Advancement of Science for permission to reprint Fig. 4.5, and to the Ecological Society of America for permission to reprint Fig. 4.7.

5

Herbivore assemblages and their food resources

Vojtech Novotny, Yves Basset and Roger L. Kitching

ABSTRACT

Ultimately, all communities are limited by the resources available to them, but the process of matching the supply and demand sides in rainforest ecosystems is rarely straightforward. When live plants or animals are exploited as resources, the pattern of their use is a particularly complex result of the interplay between adaptations on the part of predators or parasites and the counter-adaptations of their potential prey or host species. This phylogenetical component is further modified by often equally complex patterns of resource availability in space and time, resulting from contemporaneous ecological processes. Rainforest canopies contain an exceptionally rich array of organic resources for an equally diverse number of consumers, thereby constituting a very attractive as well as difficult field for study. We review briefly the main methodological approaches, results and remaining challenges in this field of study, focussing primarily on ecological aspects of the plant–herbivore, host–parasitoid and prey–predator interactions and their importance in shaping invertebrate assemblages in the rainforest canopies of three continents.

INTRODUCTION

The canopies of tropical rainforests provide a vast array of different food resources to arthropods. Since forest canopies are the principal site of energy assimilation by primary production (e.g. Lowman & Nadkarni, 1995), they are particularly attractive to phytophagous arthropods (mostly insects and mites). One can further argue that plant resources and the arthropods that feed on them in the canopy can be more conveniently studied than any other arthropod-resource system in this habitat, perhaps with the exception of fungi and fungal-feeders. For example, leaf-chewing species can be collected alive and tested on a variety of potential food plants (e.g. Basset & Novotny, 1999; Novotny *et al.*, 1999b). Without doubts, there are problems associated with determining the diet of larvae and adults, respectively, but these problems (see below) are common to most arthropod taxa.

This chapter discusses the plant resources available to insect herbivores in the canopy of tropical forests. As the authors are most familiar with this category of herbivores, we will refrain from referring explicitly to phytophagous mites, although some of the items discussed here may also apply to this group. Ecological information about phytophagous mites in tropical canopies is rather scarce and mostly originates from Australia (e.g. Walter *et al.*, 1994, 1998; Walter & O'Dowd, 1995; Walter & Behan-Pelletier, 1999).

We will first concentrate on the different types of canopy resource available and their characteristics. We discuss briefly resource patchiness by considering its spatial and temporal availability within the canopy. A number of problems are associated with the study of insect herbivores in tropical canopies. For example, our knowledge of insect/plant use in the canopy is greatly constrained by two main factors: plant phylogeny and the spatial distribution of the herbivores themselves. We will then discuss more particularly the constraints on the ecological specialization of insect herbivores and review patterns of host specificity, as they are currently understood. We conclude by providing some suggestions on how our current knowledge of insect–plant interactions in the canopy of tropical forests may be improved.

TYPE AND AVAILABILITY OF RESOURCES IN THE CANOPY

The main plant resources used by insect herbivores in canopies and their characteristics are summarized in Table 5.1. These may be classified according to the plant parts (tissues) and plant life-forms available to insects in the canopy. Sap is understood here as a liquid carried by vascular tissues, such as in the phloem and xylem vessels. The sap in mesophyll tissues represents the contents

Table 5.1. *The main plant resources used by insect herbivores and their availability in the canopy of tropical forests*

Plant part	Spatial availability	Temporal availability	Insect taxa[a]	References
Leaves				
Trees	High but depends on the relative palatability and toughness of mature and young foliage	Restricted for young foliage; depends on tree phenology and successional status	Some Orthoptera, Lepidoptera, Phasmida, Chrysomelidae some Curculionidae, *Atta*, etc.	Coley, 1983; Basset, 1996; Basset & Novotny, 1999; Ødegaard, 2000a
Lianas	High, same as above	As above but higher than for trees because of often rapid growth	As above	Benson *et al.*, 1976; Ødegaard, 2000a
Vascular epiphytes	Patchy, depends on support trees	Probably restricted as mature leaves of many species are tough	Lepidoptera, Curculionidae, etc.	Brévignon, 1990, 1992; Frank, 1999
Nonvascular plants	As above?	Probably not as restricted as above	Lithosiinae, Tingidae	Drake & Ruhoff, 1965
Sap[b]				
Trees	High	High, but may vary in quality because of translocation events	Homoptera, some Heteroptera Thysanoptera[c], *Atta*, etc.	Novotny, 1993; Novotny & Leps, 1997; Basset, 1991d
Lianas	High	High, may be of high quality compared with that of trees	As above, in particular Membracidae	Blüthgen *et al.*, 2000a
Vascular epiphytes	Patchy	Relatively high?	As above	Blüthgen *et al.*, 2000a
Flowers				
Trees	Patchy	Patchy	Apidae, Agaonidae Curculionidae, Chrysomelidae, Lepidoptera	Roubik, 1989; Compton & Hawkins, 1992 Basset & Samuelson, 1996
Lianas	Patchy	Patchy	As above	Ødegaard, 2000a
Vascular epiphytes	Probably low?	Probably low?	As above	Singer & Cocucci, 1999
Seeds/fruits				
Trees	Patchy	Patchy to low, depends on supra-annual phenology	Lepidoptera, Bruchinae, Curculionidae, some Diptera, etc.	Janzen, 1971a; Hopkins, 1983; DeVries *et al.*, 1997
Lianas	Patchy	As above?	As above	Ødegaard, 2000a
Vascular epiphytes	Probably low?	Probably low?	As above	–
Stems[d]				
Trees	High for dead wood of small branches	May depend on stochastic events such as treefall, wind break, etc.	Cerambycidae, Curculionidae, Scolytinae, etc.	Tavakilian *et al.*, 1997; Jordal & Kirkendall, 1998
Lianas	High for soft tendrils	High for soft tendrils	As above, plus Baridinae and Chlamisinae for tendrils	Ødegaard, 2000a
Vascular epiphytes	Low	Low	As for trees, above	–

[a]Refers to known examples of taxa feeding on these resources, with particular reference to canopy studies in the tropics.
[b]Phloem and xylem sap (see text).
[c]Most feed on mesophyll tissues (see text).
[d]Including live and dead bark, cambium and inner wood.

of mesophyll cells, which are exploited by various taxa (most Thysanoptera and Typhlocybinae: Novotny & Wilson, 1997). In essence, it is identical to the leaf resources exploited by leaf-chewing species, with the exception of the selective avoidance of certain parts of the leaf (e.g. lactifers). Indeed, factors affecting the local species richness of leaf chewers and mesophyll-feeders appear similar on certain host plants, such as *Ficus* spp. (Basset & Novotny, 1999). For the sake of simplicity, we also consider the 'stem' resource as including bark, cambium and heartwood.

The main differences among the plant resources provided by temperate and tropical forests are the much higher occurrence of lianas and vascular epiphytes in the latter (e.g. Gentry & Dobson, 1987; Nieder *et al.*, 2001). This has important consequences for many arthropod species, including insect herbivores, leading to an enhanced local species richness (Stuntz *et al.*, 1999; Ødegaard, 2000a; Chs. 17 and 21). Liana abundance and biomass in tropical forests increase on nutrient-rich soils and with forest disturbance (Laurance *et al.*, 2001). Thus, the magnitude of these effects may be optimal in dry forests exposed to disturbance, as within the crane perimeter studied by Ødegaard (2000a) in Panama. In addition, it is very possible that the proportion of dead suspended wood in the canopy of tropical forests may also be higher than that in temperate forests. For example, in Central Amazonian forests, the biomass of dead wood on the forest floor varies from 1.8 to 33.0 ton (dry weight) ha^{-1}. There are a further 1.1 to 7.6 ton ha^{-1} represented by standing dead trees and 1.1 to 45.6 ton ha^{-1} represented by dead suspended wood (Martius & Bandeira, 1998). Connell *et al.* (1997) have also argued that 'subcanopy' (midcanopy) gaps, created by dead standing trees in the subcanopy, are commonplace in rainforests. Hence, the amount of dead suspended or standing wood temporarily available to xylophagous insects is likely to be considerable in such forests. In particular, this resource may increase the availability of habitats for wood-eating taxa able to develop in branches of small diameter (e.g. *Agrilus* spp., Curletti, 2000). Parenthetically, we note that the few remaining 'old-growth' temperate forests may also have substantial amounts of suspended dead and dying wood in their canopies.

The spatial distribution of these various plant resources is heterogeneous in the canopy (Table 5.1). Although leaves are plentiful in the canopy, most of them may be unsuitable, or at least suboptimal, for insect herbivores, as they may be too tough or unpalatable. Leaf availability may depend on the relative differences in palatability to insects between mature and young leaves. Many insect herbivores strongly prefer young over mature foliage, resulting in higher damage on younger than more mature leaves (e.g. Coley, 1983) and/or an aggregated insect distribution on young foliage (e.g. Basset, 1991d, 1996). Even, for example, on *Pourouma bicolor* Martius (Cecropiaceae) in Panama, a pioneer tree supposedly palatable (Coley, 1983), 95% of its associated leaf-chewing species are only able to attack young foliage (Basset, 2001a). There are, of course, many exceptions to this pattern. First, many leaf-mining and galling species are able to develop in mature leaves, sustaining a low growth rate but to some extent avoiding predators and parasitoids (e.g. Faeth, 1980). Second, large leaf-chewing insects with strong mandibles, such as some Orthoptera, Phasmida or Lepidoptera, may be able to cut through relatively tough mature foliage, (e.g. Saturniidae: Bernays & Janzen, 1988). In the same manner, some sap-sucking insects may be able to insert their stylets and feed on relatively tough tissues, such as phloem and xylem vessels (Pollard, 1968). Last, aggregation of conspecifics and collective feeding may overcome the initial toughness of leaves (e.g. Morris *et al.*, 1996).

The diversity and abundance of epiphytes in the canopy can sometimes be very high. For example, 109 species of epiphytes were growing on 20 m^2 of branch surface of several neighbouring trees in Ecuador (Nowicki, 1998). However, the ratio of epiphyte species to the total number of vascular plants, the epiphytic index, varies greatly among biogeographical regions. The epiphytic index, for example, is usually much higher in the neotropics than in Africa (Niedler *et al.*, 2001). The local occurrence of epiphytes also depends greatly on a variety of factors, such as rainfall, cloud cover, microclimatic gradients and host-tree structure, and epiphytes are particularly abundant in montane tropical forests (Niedler *et al.*, 2001). Nevertheless, their interactions with arthropods, and specifically with insect herbivores, have rarely been studied explicitly (e.g. Benzing, 1983; Lowman *et al.*, 1996, 1999; Schmidt & Zotz, 2000).

Live wood is rarely attacked by insects and where this does occur it usually involves the action of gall-making insects or bark beetles (e.g. Dajoz, 1980).

However, meristem feeders are more common and many bore into flower stalks, buds or young shoots. Likewise, the stems and tendrils of lianas represent important resources for tropical insects in the canopy (Ødegaard, 2000a). Dead suspended wood is exploited by many taxa, mostly beetles. Newly dead wood has a short persistence, and saproxylic beetles associated with this habitat may be more prone to disperse than counterparts inhabiting more decomposed wood (such as in the hollow trunks of living trees), which may persist for a longer period (Ranius & Hedin, 2001; see also Ch. 26).

The availability in time of these various resources to insect herbivores also differs (Table 5.1). The production of young foliage on trees greatly depends on their successional status, phenology and on rainfall (e.g. Coley, 1983). Rapid growth and high productivity in lianas, on the one hand, ensures that their leaves, stems and tendrils are available to insect herbivores most of the year (Hegarty & Caballé, 1991). On the other hand, the mature leaves of many vascular epiphytes are relatively tough and thick, and their leaf production is rather low (Zotz, 1995), so that they may be far less attractive to insect herbivores than are lianas. High-quality sap may also be difficult to track for sap-sucking insects and may be related to translocation of nitrogen, in leaves, flowers or seeds. Basset (1991d), for example, indicated that specialist psyllids attack different tissues of high nitrogen content, such as young foliages and flowers, on one species of rainforest tree in Australia. The occurrence of flowers, and particularly of fruits and seeds, in the canopy may represent the most difficult resource to track for insect herbivores. Setting aside specialized, presumably coevolved, pollinators, many flower visitors are not highly specific (e.g. Ødegaard, 2000a; Ch. 23). Accordingly, at least for generalist herbivores feeding on flowers (many Membracidae, Chrysomelidae, Curculionidae, etc.), it should not be too difficult to locate suitable resources, especially if the herbivores are reasonably good fliers (Chrysomelidae) or are dispersed and attended by ants (Membracidae). Locating adequate seeds or fruits, especially if their production is not annual, appears to be harder. Complex diapauses dependent on host phenology may be required to exploit these resources, as is known to occur in some temperate insect herbivores (e.g. Roques, 1988). To date, very little is known about such nonannual resources in the canopy of tropical rainforests and the insects that may thrive on them – with the possible exception of mass flowering of dipterocarps pollinated by thrips in south-east Asia (e.g. Appanah, 1993; see also Ch. 4).

Perhaps the most obvious difference in terms of temporal availability of resources between the canopies of tropical and temperate forests is the higher predictability of these resources in the temperate forests. Consider tree leaves: temperate forests show a predictable succession of phenological events broadly synchronous among species, starting with bud break in spring and ending with senescence and leaf fall in autumn. This leads to a distinct succession of suites of insect herbivores during the course of the year (e.g. Dixon, 1976). In tropical forests, the diversity of plant species, their asynchronous phenology and their responses to climatic factors are such that such distinct patterns are unlikely to occur. However, patterns of leaf damage and herbivore activity may be more predictable in locations with well-marked dry seasons than in perhumid locations: compare Panama (Aide, 1992; Ch. 25), New Guinea (Novotny & Basset, 1998) and Guyana (Basset, 2000).

In addition to the higher specific variety of leaves and the coexistence of young and mature leaves for longer periods in the canopies of tropical forests, there is another important distinction differentiating them from temperate canopies. This relates to vertical structuring. Vertical complexity and stratification is higher in tropical forests than in temperate forests (Smith, 1973; Terborgh, 1985); in particular, vertical gradients in microclimatic and biotic gradients are much steeper in tropical rainforests than in temperate forests (e.g. Parker, 1995; Hallé, 1998). The quality of sunny and shaded leaves in tropical rainforests is, therefore, likely to differ substantially, even across conspecific plants. In wet tropical rainforests, this results not only in distinct faunas of herbivores foraging in the upper canopy and understorey, respectively (e.g. Amédégnato, 1997; Basset *et al.*, 2001a), but also in different herbivore assemblages feeding on conspecific seedlings/saplings and mature trees (Basset *et al.*, 1999; Basset, 2001a; Ch. 25). Note that studies relying largely on the examination of leaf damage are unable to demonstrate these faunal differences (e.g. Barone, 2000). Extreme heterogeneity in vertical abiotic and biotic gradients, such as in tall, closed, wet and tropical rainforests, may lead to a specialized herbivore fauna being restricted to the upper canopy layer (Basset *et al.*, 2001a; Ch. 3). Much of this paragraph has concerned leaves but similar arguments may also apply to the quality of sap in the canopy of tropical forests.

To close this section, it is tempting to classify food resources according to their variability in carrying capacity over time or space. On the one hand, a large variability in carrying capacity over time (e.g. in the mass production of flowers or seeds) increases the benefit of dispersal and would select for a higher degree of mobility in insect consumers. On the other hand, a large variability in carrying capacity over space (e.g. in the occurrence of epiphytes or dead wood) would rather select for more resident individuals as, on average, dispersal would result in an individual reaching an environment worse than that it originated from (Ranius & Hedin, 2001). In turn, a large variability in carrying capacity over space could in some instances select for host-specificity in insects, while a large variability in carrying capacity over time could select for less-specialized species. However, since food resources in tropical canopies vary notoriously both in time and space (Table 5.1), the significance of this variability for insect herbivores remains difficult to evaluate. For example, one may predict that herbivores able to feed on mature leaves of epiphytes should be rather host specific, but data are presently lacking to evaluate this proposition.

ARE THERE KEYSTONE RESOURCES FOR INSECT HERBIVORES IN THE CANOPY?

Before attempting to answer this question, we need to discuss what might be the most commonly used plant resources in the canopy. It is difficult to answer this directly, but we can consider, as a guide, the number of species of insect herbivores restricted to particular resources (guilds), and the number of such species represented in large-scale collections from rainforest canopies (see Basset, 2001b for a list of such datasets). This compilation indicates that, among herbivores and in order of decreasing importance, Curculionidae, Lepidoptera, Chrysomelidae, Thysanoptera and Homoptera are often species rich in the canopy. The order of this list, of course, depends greatly on the aims of each study and the availability of taxonomic expertise to sort the material. However, if it has at least some relevance to the biological reality, then it suggests that leaves, sap and perhaps flowers (upon which many Curculionidae and Chrysomelidae feed) represent the most common resources used by insect herbivores in the canopy. Most likely, the young leaves of trees may be the most commonly used resource, but they could be locally supplanted by young leaves of lianas, which may be available for longer periods in the canopy. This generalization is complicated by issues related to leaf palatability and the local occurrence of confamilial or congeneric hosts (Novotny *et al.*, 2002a).

In considering the likely occurrence of keystone resources and the plant species that could provide them, we note that similar resources may be keystone in certain locations and not in others. For example, *Ficus* spp. are often considered as a keystone resource for canopy animals in the neotropics (Terborgh, 1986) and Malaysia (Lambert & Marshall, 1991), but not in Africa (Gautier-Hion & Michaloud, 1989) or in New Guinea (Basset *et al.*, 1997a). In the last study, the overlap of herbivores across *Ficus* spp. was high and precluded considering the leaves or figs of any particular species as a keystone resource. Despite many species of *Cecropia* being inhabited by ants, this genus has been deemed 'the most hospitable tree of the tropics' (Skutch, 1945). Many animal species feed on its leaves, nectar and inflorescences, food bodies and on the ants themselves. Despite being pioneer trees, the leaves of *Cecropia* ssp. are not highly palatable to insect herbivores (Y. Basset & H. Barrios, unpublished data) and these species do not support an unusually high species richness of folivorous insects (Y. Basset, personal observation). Examples such as these could be multiplied. We conclude that if keystone resources for insect herbivores do exist in the canopy of tropical forests, they are likely to be related to leaf resources and, most likely, will be of only local importance.

METHODOLOGICAL PROBLEMS IN THE STUDY OF RESOURCE USE BY HERBIVORES

Herbivore 'communities' as they are usually studied represent ensembles *sensu* Fauth *et al.* (1996): that is, sets of species delimited by their common phylogeny (they are insects), use of resources (feeding on plants) and physical location (in a forest canopy). These ensembles are intersections of three principally different species groups: taxa (groups of species of common descent), guilds (groups of species that exploit the same class of environmental resource in a similar way) and communities (groups of species occurring in the same place at the same time). These ensembles invite further dissection with regard to their phylogeny, resource use and spatio-temporal distribution.

Herbivore phylogeny and studies of resource use

Phylogeny shapes not only the composition of herbivore assemblages but also the way they are studied. Even primarily ecological studies are often limited in their scope to certain taxa, rather than guilds, usually for methodological rather than logical reasons. For instance, 23 of the 30 contributions amenable to comparison in this volume focussed primarily on taxa, rather than guilds. This is perhaps inevitable, but we should be at least aware of the fact that the taxonomic tradition shapes our knowledge of rainforest ecosystems as it makes some ecological interactions more likely and others less likely to be studied – which may bear little relationship to their 'real' importance.

Ecological relationships across taxa of herbivores, where each requires a different method of study, are particularly poorly known. For instance, lepidopteran caterpillars and leaf-cutter ants are perhaps the two most important taxa of insect herbivores (as measured by their leaf damage) in Neotropical rainforests (Barone, 1998; Leigh, 1999); however, as far as we are aware, there is no study addressing their mutual interactions in a rainforest ecosystem. Similarly poorly explored are interactions between insect and vertebrate herbivores competing for leaf resources in rainforests (however, see Estrada & Coates-Estrada, 1986). Seed predation by vertebrates and invertebrates is perhaps a well-known exception to this pattern (e.g. Janzen, 1971a; Lamprey *et al.*, 1974), originating from long-term interests in rainforest regeneration.

In contrast, many studies focus on groups of species that do not represent any coherent ecological or evolutionary units. For instance, modern analyses have demonstrated that the Auchenorrhyncha, a favourite group for ecological studies (e.g. Wolda, 1979; Novotny, 1993), is not monophyletic (Sorensen *et al.*, 1995). It includes species from three different guilds, phloem-, xylem- and mesophyll-feeders (Novotny & Wilson, 1997), and some of these guilds are of multiple evolutionary origin, either within the Auchenorrhyncha (xylem-feeders) or elsewhere (the two other guilds). A similarly questionable unit for ecological analysis is the microlepidoptera. Ecological analyses of such groups are difficult to interpret, as they are neither taxa nor guilds. Ecological patterns, such as body size distribution, can often be analysed only when these groups are further divided into separate taxa or guilds (Novotny & Basset, 1999; Hodkinson & Casson, 2000).

There is a great imbalance in our level of knowledge among taxonomic groups in the canopy. Coleoptera is by far the most often studied arthropod order (Basset, 2001b). Likewise, 8 out of 12 contributions dealing with specific herbivore taxa in this volume focus on beetles. We can only guess at our overall level of ignorance of insect–plant interactions in the canopy, given such an overwhelming reliance on studies of Coleoptera. Consider, for example, the full tally of species likely to be herbivorous within a large and relatively well-known area such as Australia (which extends well into the tropical zone). We summed 14 300 species of beetles likely to be herbivores and 21 400 species of non-beetle herbivores (from the Lepidoptera, Hemiptera, Orthoptera, etc.) in Waterhouse (1991). The actual numbers may be influenced by a range of factors, not all of which are biological, that are themselves unimportant. The main point is that there is a far from negligible fraction of herbivores in the canopy that is not beetles, and we should also attempt to study this fraction. How many studies to date, for example, have reported on canopy tortricids or cicadellids in the tropics?

Resource use and spatial distribution of herbivores

Many studies of resource use by herbivores are unduly restricted to a particular habitat, even where the herbivores they study use resources from several habitats. This is particularly frequent in habitats that require specific methods of study or access, including the canopies of rainforests. This is an important bias in the study of canopy herbivores, which are often dependent on extracanopy resources or environments during a part of their life cycle.

As already emphasized, beetles rank as one of the most popular foci in the studies of rainforest insects, particularly by canopy fogging techniques (e.g. 9 from 14 assemblage studies in Stork *et al.* (1997a) were restricted to beetles). Numerous studies addressed their local species richness (Erwin & Scott, 1980; Basset *et al.*, 1996a; Ødegaard *et al.*, 2000; Missa, 2000), β-diversity (Allison *et al.*, 1997), host specificity (Mawdsley & Stork, 1997; Ødegaard *et al.*, 2000,

Ch. 29), colonization dynamics (Floren & Linsenmair, 1998a) or patterns of body size (Morse *et al.*, 1988). All this effort, however, has been concentrated almost exclusively on the adults, which are often the only life stage present in the canopy (Basset & Novotny, 1999; Novotny *et al.*, 1999b). Resource utilization by adults of herbivore beetles in the canopy can be of rather marginal importance compared with that by their larvae, which develop in wood, flowers, seeds or roots. Some of these larval resources may be present in the canopy but an appreciable proportion of beetle larvae must originate from dead wood and roots in the understorey and soil. Further, there is an indication that some adult beetles may not feed at all in the canopy (Y. Basset & H. Barrios, unpublished data). Therefore, larvae, as the more host-specific stage, are conceivably the key stage to be studied if we want to understand the composition and dynamics of canopy beetles, despite the fact that they only rarely reside in the canopy. Unfortunately, larval biology is particularly poorly known for beetle herbivores in rainforests (Jolivet & Hawkeswood, 1995; Tavakilian *et al.*, 1997). A redirection of effort may be indicated, for example to lepidopteran herbivores where the larvae are clearly visible, can be readily collected and can be bred through to produce identifiable adults.

In the tropics, congeneric species of herbivores may sometimes have quite different life histories and/or forage in different habitats (e.g. Basset *et al.*, 2001b). Accordingly, it is difficult to provide a list of the most common higher taxa that are most likely to be restricted to the canopy in both their larval and adult stages. Many (but not all) Eumolpinae, Galerucinae, Alticinae, and Entiminae feed on roots as larvae and are, therefore, unlikely to fall into this category. Most Cerambycidae, Anthribidae, Brentidae, Cossoninae, Molytinae, Zygopinae, Scolytinae and Platypodinae feed on wood and/or fungi. They may breed either in the dead wood of the understorey or litter or in the dead suspended wood of the canopy. The final category of taxa is most likely to have representatives associated with the canopy throughout their life cycle: the Hispinae (leaf-rollers/miners), Attelabidae (leaf-rollers), Chlamisinae (stem feeders), Baridinae (stem and flower feeders), Anthonominae, Tychiinae (flower feeders), Bruchinae and Curculioninae (seed eaters). Note that a speciose taxon such as the Cryptorhynchinae defies such crude classification, its members feeding on a variety of resources (sources for larval food resources: Anderson, 1995; Jolivet & Hawkeswood, 1995). Turning from the Coleoptera, the Derbidae, one of the most species-rich hemipteran families on rainforest vegetation (Casson & Hodkinson, 1991), can be found in the canopy only as adults while their immatures, suspected to be fungal-feeders (O'Brien & Wilson, 1985), are virtually unknown. Similarly, the nymphs of the most abundant species of sap-sucking insect present in the canopy of mature *Pourouma bicolor* trees in Panama, *Bebaiotes* sp. (Achilixiidae), are also unknown and may well live in the understorey or litter (Basset, 2001a).

It is clear that canopy assemblages of herbivorous insects cannot be studied in isolation from those of other parts of the rainforest ecosystem. It is unfortunate and ironical that most canopy studies target Coleoptera, a taxon including many representatives not restricted to the canopy habitat throughout their life cycle. Other taxa of insect herbivores may either feed during their whole life cycle in the canopy (presumably most Thysanoptera, Hemiptera and Orthoptera) or feed as larvae in the canopy with subsequent small intake of food as adults in the canopy also (presumably most Lepidoptera and Diptera).

Numerous studies infer resource use by canopy insects from their spatial distribution in the rainforest, rather than from direct observation of their feeding habits. This approach is often chosen more for convenience than any other reason. However, the diverse rainforest vegetation is particularly unsuitable for such inferences concerning usually highly mobile insect herbivores. For example, capture–recapture studies of butterflies yielded average dispersal distances from one to several hundreds of metres (Scott, 1975; Mallet, 1986; Novotny *et al.*, 1991) and these are probably underestimates. Fig wasps, thought to be poor fliers, may disperse distances up to 10 km (Nason *et al.*, 1996). Colonization by herbivores of isolated *Passiflora* plants located 50 m away from conspecifics was as rapid as those not so isolated (Thomas, 1990b). Although data on herbivore dispersal in rainforests are scarce, it can be safely assumed that each rainforest tree is within dispersal distance of a herbivore fauna from plants growing at least in the surrounding hectare, but most likely from a larger area of the forest. Dispersal distances of only 400 m correspond to a 50 ha area becoming a source of herbivore colonizers as well as of transient 'tourist' species. Three plots of this size from humid rainforests included respectively 305, 817 and 1171 species of plants with diameter at breast height > 1 cm (Plotkin *et al.*, 2000), not to mention smaller lianas, epiphytes and nonvascular plants. Dispersal of herbivores from such a large number of plant species represents a significant source of noise impeding inference on resource use from spatial distribution of herbivores among plants.

Resource use by herbivores in an ecological context

Most ecological data are contextual: that is, they make sense only in connection with other data. Resource use can be most profitably studied either as the use of the various types of resource (leaves, wood, flowers, seeds, etc.) provided by a single tree (or a population of a single tree species) or as the use of a particular resource across the whole plant community. Ideally, both approaches should be combined, however, in practice, no study has been fully successful in implementing either of these two protocols. Basset (1993) studied various resources in the canopy of a single tree species whereas Janzen (1988a and subsequent unpublished studies) focussed on folivorous caterpillars but included all plant species in a tropical forest. Several other studies of herbivores have selected a subset of local rainforest vegetation, including either closely related species (flower eaters: Compton & Hawkins, 1992; seed eaters: Hopkins, 1983; leaf chewers: Marquis, 1991), distantly related species (seed eaters: Janzen, 1980a; leaf chewers: Basset, 1996; wood borers: Tavakilian *et al.*, 1997), or a mixture of closely, distantly and locally abundant species (leaf chewers: Barone, 1998; Novotny *et al.*, 1999b, 2002a; Ødegaard, 2000a).

The task of studying resource use across many classes of resource on many plant species by many species of herbivore is daunting but is probably the only approach that will bring substantial progress in our understanding of herbivore assemblages in rainforests. Most of the studies to date take a narrow focus targeting specific interactions among a tiny subset of species from the whole ecosystem, which may be interesting in its own right but does not contribute much to the understanding of the whole ecosystem (Godfray *et al.*, 1999). This point is illustrated by recent progress in the study of rainforest vegetation, which has benefited from detailed studies of a large proportion of plant species across relatively large (50 ha) areas (Condit, 1997).

Long-term studies on a relatively large scale are also desirable inasmuch as they are likely to be the best approach to understanding how insect herbivores exploit highly fluctuating resources through time. This subject has been little studied in the canopy and in the tropics in general. Insects specializing on intermittently available resources may finely tune their activities to the phenology of the host, by diapausing if necessary (e.g. Janzen, 1971b; Rockwood, 1974; Denlinger, 1986); switch among other resources within the same host plant, such as young leaves and flowers (e.g. Basset, 1991d); or switch to secondary hosts (cf. Basset, 2000 for seedling insects).

ECOLOGICAL SPECIALIZATION AND RESOURCE USE

Spatial and temporal dynamics of plant resources

The plant's availability, nutritional quality, antiherbivore defences and the associated predators, parasites and competitors are the principal factors circumscribing the plant–herbivore interaction from the viewpoint of the plant (Michaud, 1990). Plant traits must be evaluated in the context of the herbivore's response to them – its efficiency in locating the plant, utilizing the plant's nutrients and its ability to counteract, tolerate or circumvent the plant's defences and the associated parasites, predators and competitors (Novotny, 1994).

The plant's availability to herbivores is determined by its spatial distribution and the temporal variability of its populations. Short-lived plants with unpredictable population dynamics, particularly pioneers in early stages of secondary succession, are colonized, in temperate forests at least, predominately by generalist herbivores (Jaenike, 1990). The temporal variability of plant populations may be less constraining in rainforests, where ecological succession often starts with pioneer trees rather than with annual herbs. Woody pioneers, although short lived compared with other trees, nevertheless represent a sufficiently predictable and 'permanent' resource to be colonized by specialized herbivores (Leps *et al.*, 2001).

Plant rarity may be a more important factor constraining plant use by specialists. Dixon *et al.* (1987) suggested that low species diversity of aphids in the tropics is a result of poor colonization efficiency and high host specificity of aphids combined with low population density of many plant species in rainforests. An interesting, but unexplored, question is whether this lack of aphids is compensated by higher species richness in other phloem feeders.

Unfortunately, information on herbivore communities on rare plants is limited, as they are particularly difficult to study in rainforests. Most of the studies (e.g. Basset, 1996; Barone, 1998; Basset & Novotny, 1999) are, for practical reasons, focussed on common plant species. Rainforest flora includes many species of epiphytes such as orchids (Gentry, 1990), which are often rare, small in size and sometimes relatively short lived. These species would be prime candidates for the study of the effects of plant abundance on their use by herbivores in the canopy. We await such studies.

Individual plants can escape from specialized herbivores when they grow far from their conspecifics, but such escape may require extremely low population densities for woody plants, which are generally long lived and large (Thomas, 1990b). In contrast, seedlings grow often <100 m from the canopy of their parent tree, but as small resource units separated from the canopy by a steep vertical gradient of microclimatic conditions they host mostly generalist herbivores (Basset, 1999b) at lower densities than in the canopy (Basset *et al.*, 1999). Seeds, which are small and ephemeral resources, can escape from their herbivores more easily when they are isolated; that is, when they disperse far from their parent tree. Seed infestation by a bruchid beetle on *Scheelea* palm seeds, for instance, declined between 16 and 100 m away from the parent tree (Wright, 1983). Mast fruiting represents an efficient means of escape in time by seeds from their predators (Toy *et al.*, 1992) but is relatively rare in rainforests.

Unlike whole plants, specific plant resources exploited by herbivores in the rainforests are often short lived and unpredictable. Young leaves are a high-quality, rare and short-lived resource for herbivores (Scriber & Slansky, 1981; Coley & Aide, 1991) and their availability is an important determinant of species diversity in leaf-chewing communities (Basset, 1996; Basset & Novotny, 1999). The rate of damage to young leaves by herbivores is several times higher than for mature leaves (Coley & Barone, 1996) and is higher in rainforests than in any other forest type (Coley & Aide, 1991). Some rainforest species developed costly strategies, such as delayed greening of expanding leaves, in order to limit their potential for damage by herbivores (Kursar & Coley, 1992). Plants with unpredictable, synchronous production of rapidly expanding and maturing leaves would be expected to be colonized by more generalist communities of herbivores than those with young leaves continuously available, but this proposition remains untested. In support of this, the proportion of specialist herbivores on distantly related species of New Guinean trees depended in part on the availability of young foliage all year round (Basset, 1996).

Plant defences and enemies of herbivores

Tropical vegetation is renowned for its high diversity and incidence of alkaloids (Levin, 1976), latex and other secondary metabolites (Coley & Barone, 1996), but also for a diversity of counter-adaptations by herbivores. For instance, plants with a novel antiherbivore defence – latex and resin canals – were probably temporarily freed from herbivore pressure and diversified more than their sister lineages not possessing this defence (Farrell *et al.*, 1991). A variety of counter-measures, including vein cutting and leaf trenching, which interrupts latex or resin flow, are, however, common among contemporaneous herbivores (Dussourd & Eisner, 1987; Becerra, 1994), and lactiferous rainforest trees, such as *Ficus*, support herbivore communities as diverse as those on nonlactiferous plants (Basset & Novotny, 1999; Novotny *et al.*, 1999b).

Tropical rainforest, and the canopy in particular, is an environment with a high risk of predation, generated mostly by ants (Jeanne, 1979; Olson, 1992; Novotny *et al.*, 1999a). A large proportion of rainforest plants attract ants using extrafloral nectaries (Pemberton, 1998) and obligate ant–plant mutualisms are also common (Jolivet, 1996a). Insect herbivores, that is potential prey, can respond using a variety of defence strategies – the avoidance of enemy-filled space – or succumb to predators. A wide range of these responses are probably common, but we lack a broader picture of how intense pressure from predators shapes herbivore assemblages. Intense foraging by ants can be responsible for a low abundance of externally feeding herbivores (particularly of those that are poorly defended) in rainforest canopies. Externally feeding larvae of beetles are conspicuously missing from rainforest assemblages. Among leaf-beetles, the scarcity of freely feeding larvae in rainforest canopies results in the prevalence of Eumolpinae, Galerucinae and Alticinae in most assemblages (Stork, 1987a; Farrell & Erwin, 1988; Basset & Samuelson, 1996; Wagner, 1997; Novotny *et al.*, 1999b) since all species of Eumolpinae and many Galerucinae and Alticinae have subterranean, root-feeding larvae (Jolivet & Hawkeswood, 1995). In contrast, approximately 70% of the 560 chrysomelid species from Central Europe have leaf-feeding larvae (Warchalowski, 1985; J. Bezdek, personal communication).

Feeding on roots as larvae may be one strategy developed by beetles to specialize on and stay tuned to the host phenology by using translocation clues in the root tissues, thus predicting the optimum timing for adult emergence and feeding on young leaves in the canopy. At the same time, they may avoid potential predators such as ants. Another strategy is to develop relatively mobile nymphs that feed freely in the canopy and that can either tolerate/escape ants (Orthoptera, some Hemiptera) or are tended by them (Coccoidea, Membracidae, Aphididae, etc.). Larvae of Lepidoptera, which are relatively species rich in the canopy, specialize on concealed resources (leaf rolls, leaf-mines and galls, stems, flowers and seeds), appear to be highly toxic (as are the few beetle larvae free-feeding on the foliage: e.g. Jolivet, 1987) or themselves associate with ants (e.g. Lycaenidae: Kitching, 1987; Fiedler, 1995).

A high incidence and wide variety of chemical defences against predators and parasitoids among tropical herbivores (Sime & Brower, 1998) have been proposed as explanations of the low species diversity of certain host-specific parasitoid taxa in the tropics (Gauld *et al.*, 1992). Since these defences can also be effective against predators, we speculate that high predation pressure by ants indirectly reduces the resource base and hence the species richness of parasitoids of herbivorous insects. High predation risk can influence also the host specificity of herbivores, as diet specialists seem to be better protected from predators than are generalists (Dyer, 1995).

One example: *Coelomera* attacking *Pourouma* in Panama

Consider a specific example that illustrates the constraints acting on the distribution of specific herbivores in rainforests and the wide range and nature of data required for identifying and interpreting such causal effects. Basset (2001a) studied the herbivorous fauna feeding on saplings and mature trees of *Pourouma bicolor* (Cecropiaceae), a pioneer tree, in a wet forest in Panama. There was very little faunal overlap between saplings and mature trees. In particular, the most abundant leaf-chewing species on mature trees was *Coelomera* sp. (Chrysomelidae: Galerucinae). The 35 known species of this Neotropical genus feed on Cecropiaceae and appear to be quite specialized (e.g. Jolivet, 1987). Both the adults and the free-living larvae feed on young leaves and shelter within the large stipules of trees. If disturbed, both adults and larvae exude a nauseating fluid, which may act as chemical defence. Adults, larvae and their characteristic leaf damage were not observed on saplings of *P. bicolor* in the 6 ha plot studied.

From the viewpoint of this specialized species, trees may be more hospitable than saplings for at least five possible reasons, all subject to debate.

1. Production of young foliage at the study site was three times higher, and more predictable, in trees than in saplings.
2. The presence of more abundant, larger and stronger stipules in trees than in saplings may protect *Coelomera* sp. from desiccation and predation.
3. Leaf palatability may be higher on trees than on saplings; the main factors affecting leaf palatability may be leaf toughness and chemistry, since leaf pubescence is superficially similar on both host stages. However, tree leaves are tougher and, when damaged, exude more latex than those of the saplings. If overall leaf palatability is higher on trees than on saplings, perhaps permitting higher loads of chewers on trees, then the foliage of saplings must be extremely well defended, as has been found in some other tropical trees (e.g. Langenheim & Stubblebine, 1983). In fact, 89% of larvae tested in multiple choice feeding experiments on *P. bicolor* in captivity preferred young leaves from trees over young leaves from saplings ($n = 18$), but adults fed on both ($n = 3$). Consequently, this issue cannot be resolved without a thorough chemical analysis of the leaf material.
4. Microclimatic effects must be of consequence and may represent a behavioural barrier for many insects dispersing either in the sunny upper canopy or in the shady understorey (e.g. Moore *et al.*, 1988). These

effects may be significant even when the host is a pioneer species and grows in relatively more open forest, such as in the present case.

5. Enemy-free space may also account for the patterns of distribution of *Coleomera* sp. The abundance of insect predators and parasitoids was higher on trees than on saplings; ants showed the reverse trend. Other predators (mostly spiders) had similar levels of abundance on both host stages. Accordingly, it is difficult to comment on the relative importance of enemy-free space on saplings and trees for *Coelomera* sp. It may further depend on the success of its defences against particular enemies, including vertebrate predators, and would require behavioural studies.

The conclusion of this short example is that, although it is difficult enough to outline the distribution of herbivores in the rainforest, it is far more complex to establish what might be the causal factors behind the observed distributions.

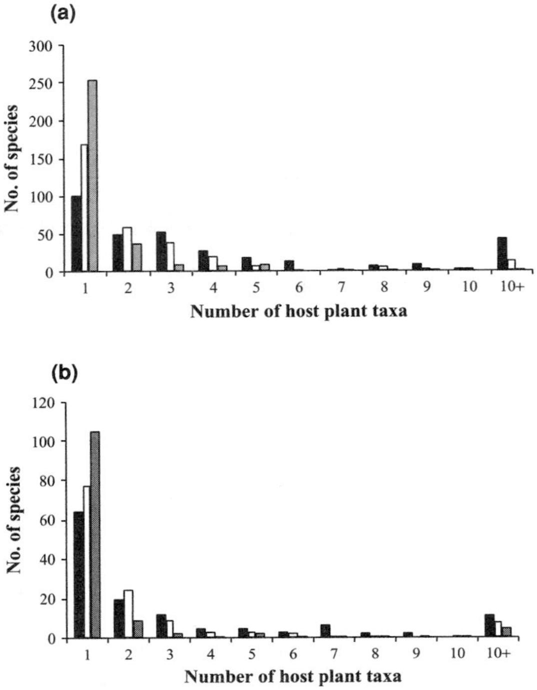

Fig. 5.1. Host plant taxa used as food. (a) The number of species of butterflies of the Australian continent feeding on 1, 2, . . . , 10 + plant taxa (extracted from Braby, 2000). (b) The number of species of the tephritid fruit-flies of south-east Asia feeding on 1, 2, . . . , 10 + plant taxa (extracted from Allwood *et al.*, 1999). Closed bars, plant species; open bars, plant genera; grey bars, plant families.

PATTERNS OF HOST SPECIFICITY

There are sound optimality arguments in support of the evolution of both narrow and broad host-plant ranges in rainforest insects (Basset, 1992a). In short, high diversity of chemical defences displayed by rainforest plants and high predation pressure by generalist predators have been suggested as agents selecting for restricted host range in herbivores, while high spatial variability, scarcity and unpredictability of resources, especially of young foliage, are suggested as factors promoting the evolution of polyphagy. Nevertheless, the predominance of specialists (at plant-specific, generic or familial levels) found in many taxa of herbivorous insect cannot be fully explained by optimality arguments and may well reflect an elevated speciation/extinction ratio in specialized versus polyphagous lineages (Jermy, 1993; Menken, 1996).

Monophagy, oligophagy and polyphagy

There are many species of insect that are restricted to single species of tree (Connor *et al.*, 1980; Claridge & Wilson, 1981). Where we have sufficiently complete information, as is the case with, say, butterflies and fruit-flies, then characteristic distributions of host specificity emerge. Figure 5.1 presents data on the known food plants of the entire butterfly fauna of the Australian continent and on the tephritid fruit-flies of south-east Asia (Allwood *et al.*, 1999; Braby, 2000). In fact, these data show that for both groups functional monophagy is the commonest feeding pattern across all species. Relative to the entire fauna, this monophagy becomes even more marked when the analysis is done at the level of the plant genus and greater still at the the family level. That having been said, there are many herbivore species showing all grades of nonmonophagy from the exploitation of two taxa of plant through to *Bactrocera papayae* Drew & Hancock (Tephritidae), which is recorded as feeding on an astounding 193 species of plants in 114 genera in 50 families! Figure 5.2 is redrawn from Novotny *et al.* (2002a) and shows the host-plant ranges at a variety of taxonomic levels for three groups of herbivores in lowland forest in Papua New Guinea. The figure underlines the

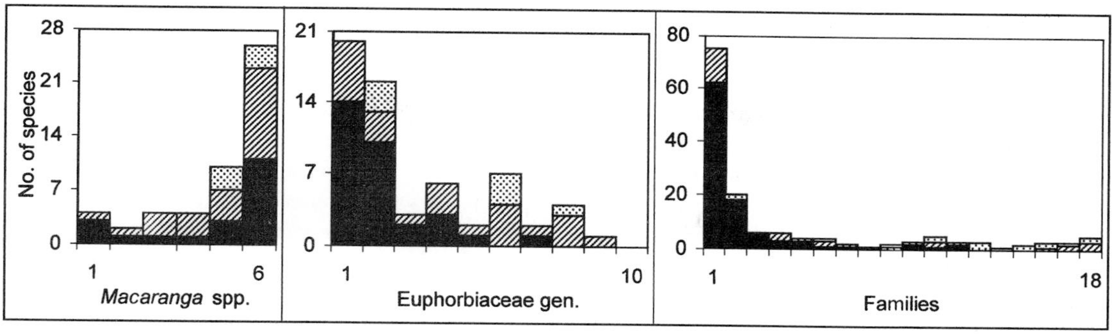

Fig. 5.2. Host-plant range of leaf-chewing insects on congeneric *Macaranga* species, on hosts from different genera of Euphorbiaceae and on hosts from different families of flowering plants. Black bars, Lepidoptera; hatched bars, Coleoptera; stippled bars, orthopteroids.

fact that within a fauna different groups of herbivores may show dramatically different degrees of host-plant fidelity. In the herbivorous Lepidoptera, Coleoptera and orthopteroids of a New Guinea forest, the Lepidoptera showed the greatest degree of host specificity (although within the plants examined this was most apparent at the generic level), followed by the Coleoptera. The orthopteroids were in general the most catholic feeders of the groups being compared. Even simple comparisons of the kind presented here can lead to oversimplistic interpretations. Janzen (Ch. 33) discusses at length the various sorts of polyphagy exhibited within the Costa Rican Saturniidae. He notes cases where the species may show polyphagy but individuals do not; the contrasting situations where even single individuals may be polyphagous; the extraordinary increase in food-plant range that may occur during an outbreak population; and, finally, the restraints on potential polyphagy by limits in the diversity of the local flora. All of these considerations lead us, further, to caution that even the best of datasets should not be assumed to be final and complete. Even in the butterflies, there are many familiar species where the full range of food-plants is poorly understood, particularly in the case of monocot feeders such as the Satyrinae and Hesperioidea.

Highly specific phytophages are generally presumed to have coevolved to varying degrees (but note the important comments in Ch. 33). Sometimes this level of coevolution is extreme. Gall-formers (Askew, 1980), some leaf-miners (Common, 1990), specialized seed predators (Janzen, 1970; Oberprieler, 2003) and so forth are cases in point. Often even insects that we regard as 'specialists' – such as the large assemblage of oak associates in the cool temperate regions (Kennedy & Southwood, 1984; see also the general treatment of Lawton & Schröder, 1977) – are, in fact, oligophages, specializing not in a single species but in a particular genus or group of related genera. Further, even within that set of host species, not all will be equally selected – and the preferred set may change from place to place within the range of the herbivore. The degree to which such oligophagous specialization is obligate or facultative will reflect the chemical characteristics of the plants involved. Examples of sets of herbivore getting ecologically 'marooned' so to speak by the chemical peculiarities of their host plants, however, are widespread within both the gymnosperms and the angiosperms. Examples that have featured in recent important studies include the coniferophagous weevil family Nemonychidae (Farrell, 1998) and the assemblage of specialized herbivores on the angiospermatous Asclepiadaceae, Moraceae and Apocynaceae (Futuyma & Moreno, 1988; Dobler & Farrell, 1999). Of course, many insects are genuinely polyphagous and are able to feed on a very wide range of host plants. Generally, such polyphages will still avoid some of the more chemically challenging groups of plants but nevertheless remain remarkably catholic.

Considerations of this kind have lead to the redefinition of host-plant specificity as a statistical concept better represented by distributions of levels of exploitation of a range of host species and the degree of exploitation of each by particular herbivores. Diserud and Ødegaard (2000) use a beta binomial model for this purpose while Mawdsley and Stork (1997) follow May

(1990) in calculating a weighted sum reflecting the number of co-hosts for each insect species encountered on a focal tree. Both of these approaches ideally require complete species lists for each tree species: a requirement that is rarely met even for temperate forests (but see Ward & Spalding, 1993) and that remains a distant dream for tropical systems.

A diversity of herbivores

Differing degrees of specialization among different major groups of herbivore restrict the degree to which conclusions about one group may be generalized to all of the Insecta. Novotny and his co-workers (2002a) have demonstrated recently that, even though a simple power curve readily represents the relationship between insect herbivores and tree species within a tropical rainforest in northern New Guinea, different orders of insects have quite different exponent values, reflecting the degree of specialization characteristic of each. Of the three major herbivorous orders, the Lepidoptera showed the greatest degree of specialization followed by the Coleoptera and, finally, the Orthoptera.

Insect–plant associations often exhibit considerable evolutionary conservatism and are also a product of chance historical events. In consequence, we can expect phylogenetic relationships among host plants to account for a significant part of the similarities observed in the contemporary assemblages of herbivores, these similarities being only partially attributable to causal effects of local and/or regional factors (Farrell & Mitter, 1993; Losos, 1996). There is no theoretical framework for predicting whether host specificity and species richness of insects feeding locally on rainforest tree species can be attributed largely to local processes, to regional processes or, being a result of idiosyncrasies of evolutionary history of plant–herbivore interactions, to neither of them. Empirical studies taking into consideration the whole range of local and regional variables, as well as the phylogeny of host plants, are, therefore, much needed, especially in the tropics.

CONCLUSIONS

The study of insect herbivores and their resource use in the canopy of tropical rainforests represents a formidable challenge. We need sampling efforts, sampling universes, observations and results of experiments that are commensurate with the size and scope of the study system, possibly one of the most species-rich on Earth. To date, no study has been able to provide us with data that may be useful to extrapolate patterns at the forest level. Studies are too short in duration, too restricted in terms of insect or plant taxa, too restricted in terms of spatial or temporal replicates originating from the study hosts, or they do not provide direct observations of insect–plant interactions and feeding records. No matter what mathematics is applied to a poor, unrepresentative dataset, the result is going to be equivocal.

To remedy this, we need to increase our knowledge by:

- increasing the focus on taxa other than beetles and comparing patterns of plant use by these various taxa
- increasing plant replicates so as to include a representative subset of the forest stand, in terms of phylogeny, abundance and biomass
- increasing spatial replicates so as to include replicated blocks (sites) within the forest and sufficient replicates of conspecific plants: the number of replicates needed most likely will depend on plant characteristics, such as phylogeny, abundance, ecological plasticity, phenology, and so on (although, for these reasons, it is difficult to recommend a figure, as many as 50 or 100 replicates are indicated (*cf.* Basset & Novotny, 1999; Novotny *et al.*, 1999b, 2002a)
- increasing temporal replicates so as to include supra-annual events
- studying the relationships between the canopy and soil/litter faunas.

This list may look rather depressing, especially when considering the difficulties of canopy access and the associated physical and financial constraints. Where do we start? A 'canopy station' (see Ch. 2) would greatly help to get representative and sufficient replicates both spatially and temporally, and to develop long-term experiments. Numerous replicates and experiments could be obtained or performed by local parataxonomists (e.g. Novotny *et al.*, 1997; Basset *et al.*, 2000) and they could be productively trained by experienced and professional tree-climbers to access many parts of the canopy. Perhaps the biggest challenge of all may be to develop the framework within which canopy ecologists and taxonomists can (and will) work together to achieve these goals.

ACKNOWLEDGEMENTS

Our current studies of insect–plant interactions in the tropics are funded by the US National Science Foundation (DEB-94-07297, DEB-96-28840 and DEB-97-07928), the Czech Academy of Sciences (A6007106, Z 5007907), the Czech Ministry of Education (ES 041), the Czech Grant Agency (206/99/1115), the Darwin Initiative, the Smithsonian Tropical Research Institute (Tupper fellowship) and the Cooperative Research Centre for Tropical Rainforest Ecology and Management.

PART II

Vertical stratification in tropical forests

Introduction

Many earlier studies of canopy arthropods in the tropics commented on the apparent richness of the fauna high in the canopy, as opposed to that on the lower vegetation. However, with rare exceptions such as medical studies of mosquito vectors, few studies actually quantified faunal turnover occurring from the forest floor to the overstorey. Many earlier studies relied on fogging or mist-blowing, which have methodology that make it difficult to study specifically arthropod stratification. Assessing whether arthropod stratification exists in tropical forests, and to what extent, is a relatively recent trend in canopy entomology, resulting mainly from improved canopy access.

The case studies included in this section attempt specifically to quantify arthropod stratification in tropical forests. The authors target different taxa, with various collecting methods and in diverse locations. The first three contributions are based in Australasia, the next two originate from Africa and the last one belongs to the neotropics. The reader particularly interested in the topic of arthropod stratification is also invited to examine Chs. 24–26.

Simon *et al.* in Ch. 6 consider two different taxa, Scolytinae and Formicidae, and examine their fine distribution within the crowns of tropical oaks in Borneo. Whereas arthropod stratification has often been examined from forest floor to the upper canopy, few studies examined the extent of stratification within individual tree crowns. This chapter is also interesting in that it contrasts patterns within the same sampling universe for arthropod taxa associated in varying degrees to the host plant and with very different foraging behaviour.

In Ch. 7, Schulze and Fiedler discuss the results of light trapping at different heights within a forest in Borneo and concentrate on a species-rich moth taxon not well-studied in the tropics: Pyraloidea. The interesting new feature of this study is the comparison of temporal patterns and change in faunal similarity between the understorey and upper canopy. Such an approach, contrasting both temporal and spatial determinants of arthropod distribution, appears particularly well suited to the study of heterogeneous and complex habitats such as tropical rainforests.

Koike and Nagamitsu in Ch. 8 examine the vertical distribution of flight patterns of butterfly and birds in Sarawak. Their approach involves assembling vertical quadrats within the canopy, from which observations can be easily quantified. This approach is commendable since it allows relatively precise quantification of other variables that could influence arthropod distribution patterns, such as leaf area index. Recording such variables is rather challenging in the complex three-dimensional habitat provided by tropical rainforests.

Sørensen (Ch. 9) focusses on spiders and studies their vertical distribution from the forest floor to the canopy in a Tanzanian forest. The author uses different collecting techniques, each calibrated and targeting spiders in a particular habitat along a vertical transect within the forest. One of the merits of the study is to compare the fauna of the litter and forest floor with that foraging on the vegetation and also to consider a taxon relatively unspecialized with regard to association with different plant species. Such studies are needed for comparison with the distribution patterns of other taxa that may be more intimately associated with plant species and that may respond primarily to the physiological and ecological changes of their host mediated by vertical gradients.

Suspended soils in the tropics have long fascinated scientists. However, quantitative studies of their microarthropod faunas and how they are distributed vertically are still rare. Winchester and Behan-Pelletier provide an interesting analysis of such habitats within an *Ongokea* tree in a Gabonese forest (Ch. 10). With a simple, easily repeatable methodology, they investigate the distribution of various invertebrate taxa at different heights within the tree trunk and also on branches at various replicated distances from the tree trunk. Their study stresses again the complex three-dimensional systems that entomologists must face in tropical forests.

Eventually, De Dijn explores vertical gradients of insect abundance and diversity in forest plots affected by varying degrees of disturbance in Suriname (Ch. 11). Contrasting patterns of arthropod stratification in pristine versus disturbed forests should figure prominently on the research agenda since it could help us to predict (and perhaps avoid) changes in arthropod distribution resulting from anthropogenic disturbance. The author uses an inexpensive methodology that may be attractive to colleagues restricted by limited funding and infrastructure.

6

Distribution of ants and bark-beetles in crowns of tropical oaks

Ulrich Simon, Martin Gossner and K. Eduard Linsenmair

ABSTRACT

The activity of arthropods in the uppermost and lower crowns of nine tropical oaks was studied from May to October 1997 in a montane forest of the Kinabalu National Park, Sabah, Borneo, Malaysia. The sampling covered a period from the end of the wet season to the beginning of the next wet season. Worker ants (Hymenoptera: Formicidae) and bark-beetles (Coleoptera: Scolytinae) were collected by arboreal pitfall traps and flight-interception traps, respectively. The number of individuals and species of both taxa did not differ significantly between the two canopy layers. Although bark-beetles showed no determinable differences in their occurrence across the strata studied, β-diversity of ants among traps set in the upper part of the trees was significantly lower than in the lower traps, and ant subfamilies were not distributed evenly between the crown layers.

INTRODUCTION

Forests are layered or stratified ecosystems. This stratification results from both a non-uniform distribution of plant parts and from microclimate (Chazdon & Fetcher, 1984; Parker, 1995). Many studies have dealt with the vertical organization of plant assemblages (e.g. Terborgh, 1985; Oliver & Larson, 1990; Hallé, 1995). Animal assemblages are influenced, in turn, by this stratification, which results partly from drastic differences between the upper and lower habitats in both temperate (Simon, 1995; Schubert, 1998) and tropical forests (Smith, 1973; Longino & Nadkarni, 1990; Nadkarni & Longino, 1990; Paoletti *et al.*, 1991; Kato *et al.*, 1994; Brühl *et al.*, 1998). More precisely, there have been several studies comparing the arthropod fauna among forest layers (Sutton *et al.*, 1983a; Basset, 1992c), but only a few comparing the arthropods along a vertical gradient of the resources with which they are associated (Basset *et al.*, 1999; Ch. 25). Only a few publications analyse the fine distribution of arthropods along these gradients (Davis & Sutton, 1998; Simon & Linsenmair, 2001).

Traps are advantageous for studying the spatial distributions of arthropods, since they survey arthropod assemblages 'on the spot'. Repeated trapping allows an assessment of the assemblage composition at a defined site, and the impact on assemblage structure can be assumed to be low. Such an approach has seldom been used for studying the distribution patterns of arthropods within tree crowns in a tropical forest (Compton *et al.*, 2000; Simon & Linsenmair, 2001).

Setting up traps in different layers of the canopy allowed us to assess differences in the assemblages of ants and bark-beetles that may exist within tropical tree crowns. Ants are one of the most important insect groups in tropical rainforests both in terms of their number and their impact on the associated flora and fauna (Fittkau & Klinge, 1973; Hölldobler & Wilson, 1990a; Tobin, 1995). There are studies that indicate differences in the occurrence of ants in different layers of tropical forests (Longino & Nadkarni, 1990; Brühl *et al.*, 1998). The first question to answer was whether there is a stratification of the ants that are active on the bark surface on a smaller scale within tree crowns. Beetles are highly diverse in the tropics (Erwin, 1982; Stork *et al.*, 1997a). Beetle assemblages in tree crowns reflect forest types (Wagner, 2000), are said to be more or less tree specific (Stork & Brendell, 1990; Wagner, 2000) and offer a new look on mechanisms of community assemblage in tree crowns (Floren *et al.*, 1998; Floren & Linsenmair, 1998a). Ødegaard *et al.* (2000) found that previous estimates of host specificity of beetles, and hence of species richness, were too high. Scolytinae, in particular, are very species rich in the tropics (Schedl, 1981; Kirkendall, 1993). Ecological requirements of the majority of tropical species are, in contrast to temperate forests (Grüne, 1979; Szujecki, 1987), rather broad (Kirkendall, 1993). There are indications that there is a stratified occurrence of species of bark-beetles along the vertical extent of trees as well as within the crowns of forest trees in tree species of temperate forests (Capecki, 1969; Safranyik *et al.*, 2000). This leads to the question whether bark-beetles also show differences in occurrence *within* tree crowns, that is, on a finer scale.

METHODS

Sampling site and study trees

The study was carried out in the Mount Kinabalu National Park, Malaysian Federal State of Sabah, in the north-western part of Borneo. The park covers an area of approximately 75370 ha (Ismail & Din, 1995). Emergent oak trees of the genus *Quercus* occur only around Sayap Ranger Station at the western boundary of the park (6° 10′ N, 116° 35′ E, altitude *c.* 1000 m above sea level). The forest in this area is characterized as a 'lower montane forest' by Kitayama (1992) or as a 'hill dipterocarp forest' by Rahman *et al.* (1995).

Nine trees of the genus *Quercus* were investigated. They grew along the valley of the river Wariu, near Sayap Ranger Station (Fig. 6.1). All of them were tall trees (Table 6.1), which emerged 10 to 20 m above the main canopy layer. The first branching of the study trees occurred within the upper closed layer of tree crowns in the forest, between 17 and 28 m above the ground.

All these trees had leaves with a sunken midrib and 5–10 secondary nerves, a leaf margin minutely toothed in the upper half and acorns of the same shape. According to Cockburn (1972), these traits are typical for *Quercus subcericea* A. Camus (1933), a species that occurs in lowland and lower montane forest from 300 to 2600 m asl but is supposed to reach a height of 20 m. Accordingly, an unequivocal identification of the trees at the species level has proved impossible, as inflorescences could not be obtained. Voucher specimens of leaves and acorns have been collected and are deposited in the collection of one of us (U. S.).

Most tree sites were situated on the flat banks of Kemantis Creek or the Wariu river (Trees 1, 2, 6–9). Tree 3 grew on the flat top of a small watershed between

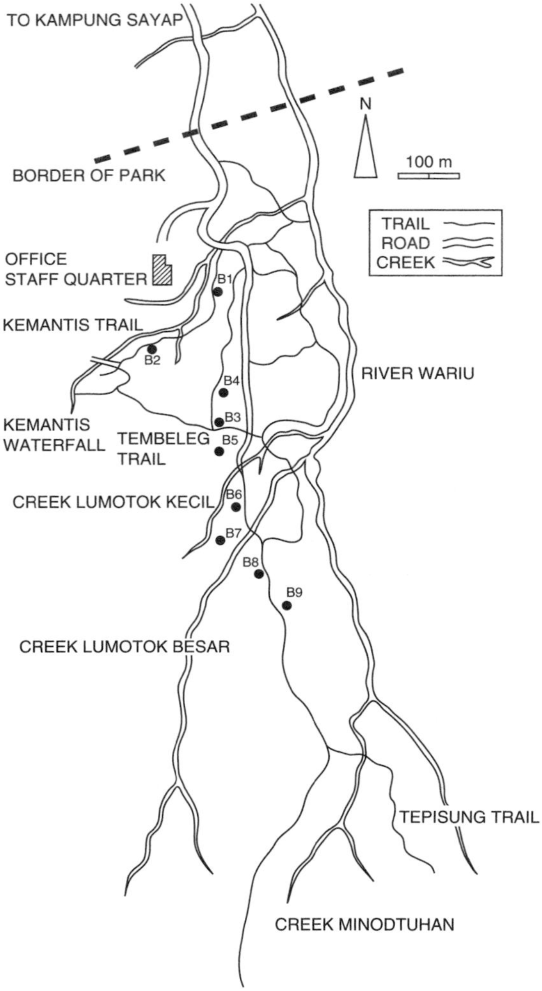

Fig. 6.1. Situation map of the study site, redrawn from a local hiking map, with indication of the study trees (B1 to B9).

Kemantis Creek and the Wariu river (Fig. 6.1) trees 4 and 5 grew on more or less steep slopes (30° inclination) on the east-faced sides of the Wariu river bed.

Insect collecting and processing

Two different types of trap were used in this study (see also Simon & Linsenmair, 2001). First, flight-interception traps made of gauze with funnels of cotton attached above and below (modified after Basset *et al.*, 1997b) were used. The gauze was 1 m long and 50 cm wide. Two intersecting rectangles of gauze were crossed rectangularly to make up the central barriers. The upper

Table 6.1. *Characteristics of study trees and height of the two types of trap set up in the crown of study trees*

Tree	Tree height (m)	DBH (m)	FIT-U (m)	FIT-L (m)	APT-U (m)	APT-L(m)
B1	44.0	2.1	35.0	27.0	37.0	29.0
B2	38.5	1.4	27.0	22.5	27.5	21.5
B3	35.0	1.5	26.5	16.5	27.0	17.0
B4	35.0	1.3	23.5	17.0	24.5	19.5
B5	36.0	1.8	28.5	21.0	29.0	20.0
B6	34.0	1.5	28.0	20.5	29.0	22.0
B7	36.0	2.0	27.5	19.5	30.0	21.0
B8	45.0	2.2	37.0	26.5	38.5	27.5
B9	35.0	1.4	24.0	14.5	25.0	17.0

DBH, diameter at breast height; FIT, flight-interception trap; APT, arboreal pitfall trap; U, upper crown; L, lower crown.

funnel was attached to a box with a transparent lid via a plastic tube in order to trap arthropods trying to escape by flying upwards. At the base of the bottom funnel was a sample jar with a hole in its cover. This type of trap combines the advantages of a window trap and a Malaise trap. It samples mainly flying insects and is one of the most efficient methods for sampling beetles (Schubert, 1998; Simon & Linsenmair, 2001). Arboreal pitfall traps were also used, modified from the design of Weiss (1995), who used them on spruce trunks in a temperate forest in Germany. Small plastic buckets were cut in half longitudinally, each half supporting a small plastic bag filled with killing agent. The half-bucket and plastic bag were pressed to the surface of the bark by an elastic band, ensuring a tight fit of the rim of the plastic bag to the bark, and smooth movement of bark-dwelling arthropods into the trap. The width of the pitfall trap was about 12 cm.

All sampling jars were filled with an aqueous solution of 1% copper sulphate, combined with a small amount of detergent. This agent kills arthropods very quickly and, although an irritant, is nontoxic to adult humans.

The disposition of the traps in the crowns was always similar: one flight-interception trap and one arboreal pitfall trap in the uppermost reachable parts of the tree crowns, and a similar pair of traps at the level of the first branching of the trees. The absolute height of the traps above ground varied with the height of the oak trees studied and the distance to the first branch (Table 6.1).

We climbed the oaks using single-rope techniques (Perry, 1978). Traps were surveyed twice a month during the period from 4 April 1997 to 15 September 1997. Damaged traps were repaired during sampling and eventually replaced.

Insects were sorted to order, and other arthropods to various higher taxa (see Simon & Linsenmair, 2001). We present data on ants only from the arboreal pitfall traps; ants were sorted to morphospecies, and identification to genus level was made using Bolton (1994). Data for the subfamily Scolytinae in this paper were obtained only from flight-interception trap catches, which were sorted to morphospecies.

Data analyses

We used the Jaccard index for the estimation of β-diversity: the lower the Jaccard value, the higher the β-diversity (see Magurran, 1988). We used this index to compare ant and scolytine assemblages between trees, and within trees between the two studied layers.

For statistical analyses we used the program STATISTICA 5.1 (Statsoft, 1997), except for the Wilcoxon–Wilcox test (see below). We used the Wilcoxon matched-pairs test to compare insect abundance in the upper and lower parts of the trees. Comparisons of Wilcoxon-tests with t-tests (asymptotic efficiency) showed that the loss of efficiency is low, as the tests have 95.5% of the power of the comparable parametric test (Weber, 1986). To compare the differences in species composition between the upper and the lower parts of the tree, we calculated a sign-test based on

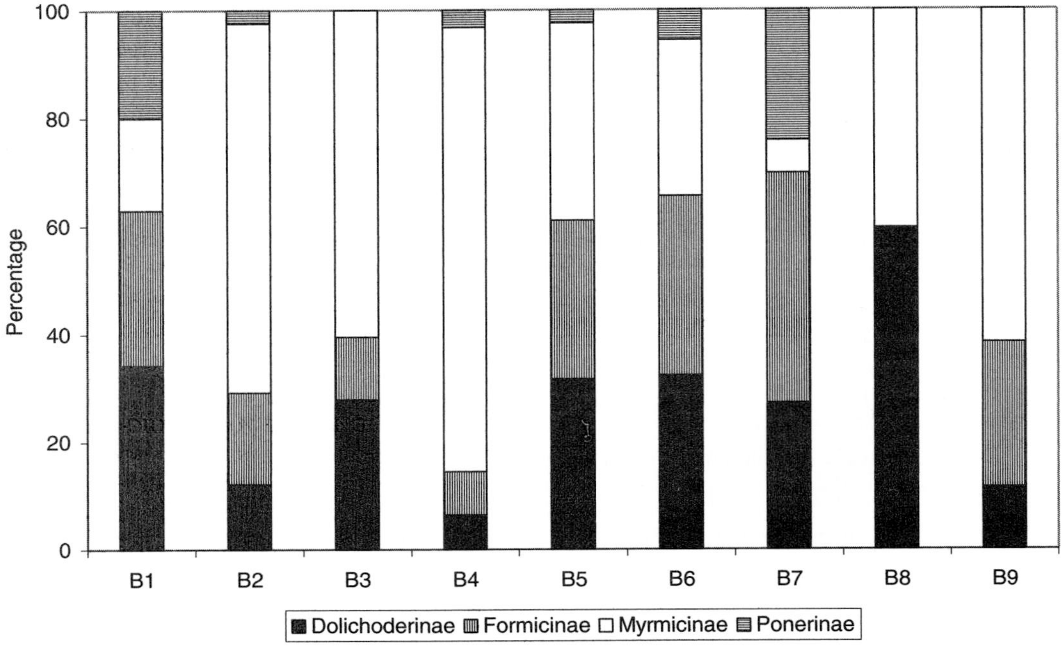

Fig. 6.2. Proportion of ant individuals collected on each study tree, detailed for each subfamily.

presence/absence data, in which species were regarded as replicates. To detect differences in the similarity values (Jaccard index) between crown parts, we performed a Friedman test and, as a *post-hoc* test, the Wilcoxon–Wilcox test (Köhler *et al.*, 1995).

RESULTS

Ants

In total, 494 ant specimens representing 71 morphospecies were collected with the arboreal pitfall traps. Twenty-three species (32.4%) were singletons, and another 23 species occurred only as pairs. Accordingly, one third of all ant morphospecies was represented by three or more individuals.

Four subfamilies of Formicidae were collected, with Myrmicinae dominating the number of individuals collected (43%), followed by Dolichoderinae (28%), Formicinae (23%) and Ponerinae (6%). The distribution of species richness was slightly different. Although Myrmicinae still dominated (41%), Formicinae were better represented (35%). The Dolichoderinae represented 17% of species and the Ponerinae 7%.

The distribution of ant individuals and species varied among the trees. In five trees, Myrmicinae were the most abundant group (trees B2, B3, B4, B5, B9), whilst Dolichoderinae (trees B1 and B8) and Formicinae (tree B7) were most abundant in two and one tree, respectively. In tree B6, a more or less equal number of Dolichoderinae, Formicinae and Myrmicinae occurred (Fig. 6.2). Myrmicinae were the most species-rich subfamily on six trees, whereas Formicinae dominated in the other study trees (Table 6.2).

Numbers of species of ants did not differ significantly between the two strata (Table 6.2; Wilcoxon matched-pairs test, two-tailed, $z = 0.948$, $p = 0.343$). As shown in Fig. 6.3, there is a slight, but statistically nonsignificant, decrease in species number in the lower traps. Within each of the subfamilies only the species numbers of the Formicinae differed significantly (Table 6.2; Wilcoxon matched-pairs test; $z = 1.960$; $p = 0.049$), showing a preference for the upper area of the canopy. Neither the number of individuals of Formicinae nor the species and individual numbers of the other subfamilies differed statistically (all with $p > 0.05$) across layers.

Table 6.2. *Number of ant species collected in arboreal pitfall traps in different trees and heights, detailed by subfamilies*

Study tree	Dolichoderinae	Formicinae	Myrmicinae	Ponerinae
Upper crown				
B1	3	5	2	1
B2	1	4	4	0
B3	3	3	6	0
B4	1	3	4	1
B5	3	7	4	0
B6	2	5	5	2
B7	4	8	1	1
B8	2	1	3	0
B9	1	1	4	0
Lower crown				
B1	0	1	0	1
B2	4	1	6	1
B3	1	2	6	0
B4	1	3	5	1
B5	2	2	5	1
B6	3	3	6	0
B7	3	6	1	1
B8	3	2	6	0
B9	0	2	3	0

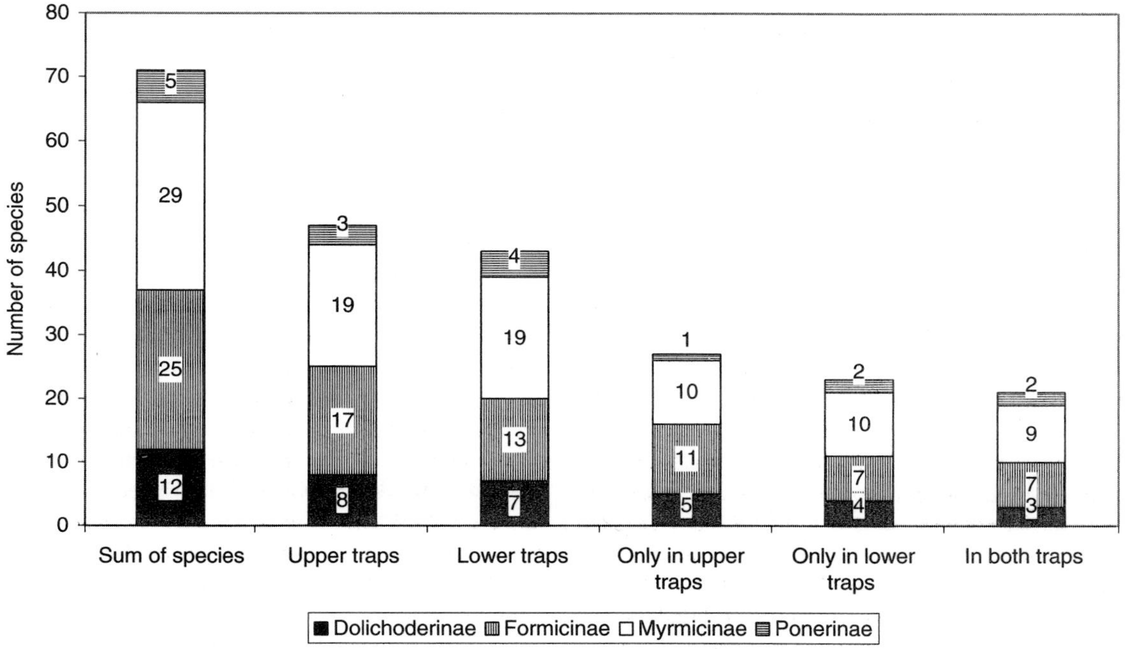

Fig. 6.3. Number of ant species collected, detailed by subfamilies and different sampling situations.

Table 6.3. *Number of individuals caught in arboreal pitfall traps for the five most abundant ant species, detailed by study tree and sampling height*

Study tree	*Technomyrmex* sp. 1		*Technomyrmex sp.* 4		*Pheidole* sp.1		*Camponotus* sp.1		*Vollenhovia* sp. 6	
	U	L	U	L	U	L	U	L	U	L
B1	4	0	7	0	0	0	4	0	4	0
B2	1	1	0	0	7	0	0	1	0	0
B3	1	0	1	9	1	2	0	3	0	0
B4	0	1	0	3	3	16	0	1	0	0
B5	0	6	4	0	1	2	3	0	4	1
B6	2	38	0	0	4	5	2	3	17	1
B7	1	0	1	1	0	0	4	2	0	0
B8	11	5	1	3	0	1	2	1	4	2
B9	3	0	0	0	3	0	1	1	5	1
p value[a]	0.834		0.893		0.612		0.575		Not testable	

U, upper crown; L, lower crown.
[a] Wilcoxon matched-pairs test.

Only four species were frequent enough (i.e. occurring on six trees or more) to be tested with a matched-pairs test (Table 6.3). None of these species showed significant preferences for one of the two sampling strata. Since the Myrmicinae *Vollenhovia* sp. 6 occurred only on five trees, the matched-pairs test was not applied. However, more individuals of this species were collected in the upper layer, suggesting a preference for this part of the crown.

We tested possible ant preferences for height in a single tree with a sign-test using species as replicates. A significant difference ($p < 0.05$) existed only in one tree (B5), with more individuals found in the upper traps.

A Friedman test of Jaccard indices calculated for data between the two layers and within each study layer showed one of the groups to be significantly different ($p = 0.020$). It appeared that the indices for the samples from arboreal pitfall traps of the upper layer (APT-U) against the lower layer (APT-L), and the indices among trees in the lower layer (APT-L), were not statistically different from each other (Wilcoxon–Wilcox test: $W_{(3; 36; 0.05)} < 19.0$; $p > 0.05$). The indices calculated among trees in the upper layer (APT-U versus APT-U) were significantly different from both of the other sets of comparisons (Fig. 6.4; $W_{(3; 36; 0.05)} > 19$; $p < 0.05$).

Bark-beetles

The flight-interception trap catches included 420 specimens of Scolytinae, of which 414 were assigned to 164 morphospecies (six specimens were too damaged). The majority (72.5%) of morphospecies were represented by only one or two specimens (91 singletons, 28 pairs).

The number of morphospecies among study trees was not evenly distributed ($\chi^2 = 36.81$; $df = 8$; $p < 0.001$). Tree B3 had more morphospecies than average, and trees B6, B7 and B9 had fewer morphospecies than average (Fig. 6.5).

The number of morphospecies of bark-beetle did not differ significantly between the two sampling heights (Table 6.4; Wilcoxon matched-pairs test, $z = 0.118$; $p = 0.910$). The most common bark-beetles ($n > 10$) did not show significant preferences for either of the two sampling heights, as was the case with the ants. A Friedman test of Jaccard indices calculated for within-tree samples and between-tree samples showed no significant differences ($p > 0.05$; Fig. 6.6).

In a last analysis, we performed a sign test to estimate the preference of the species assemblage of bark-beetles for sampling heights, at the single tree level. In two trees, the species assemblage was significantly different between the upper and the lower crown. In tree B4, more species were collected in the lower

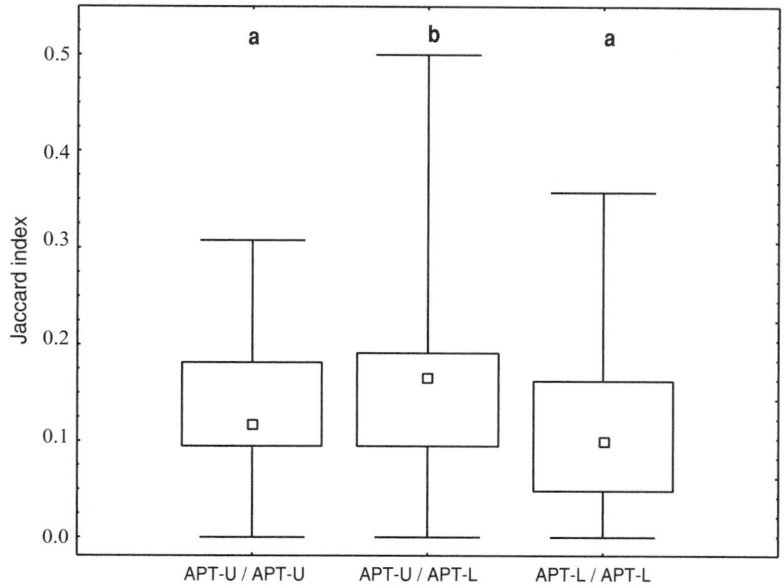

Fig. 6.4. Comparison of the Jaccard indices of the ants within heights and between heights. The small box indicates the median value; the large box is the 25–75% confidence interval; a and b indicate significant differences (Friedman test $p < 0.02$; Wilcoxon–Wilcox text $p < 0.05$); APT, arboreal pitfall trap; U, upper crown; L, lower crown.

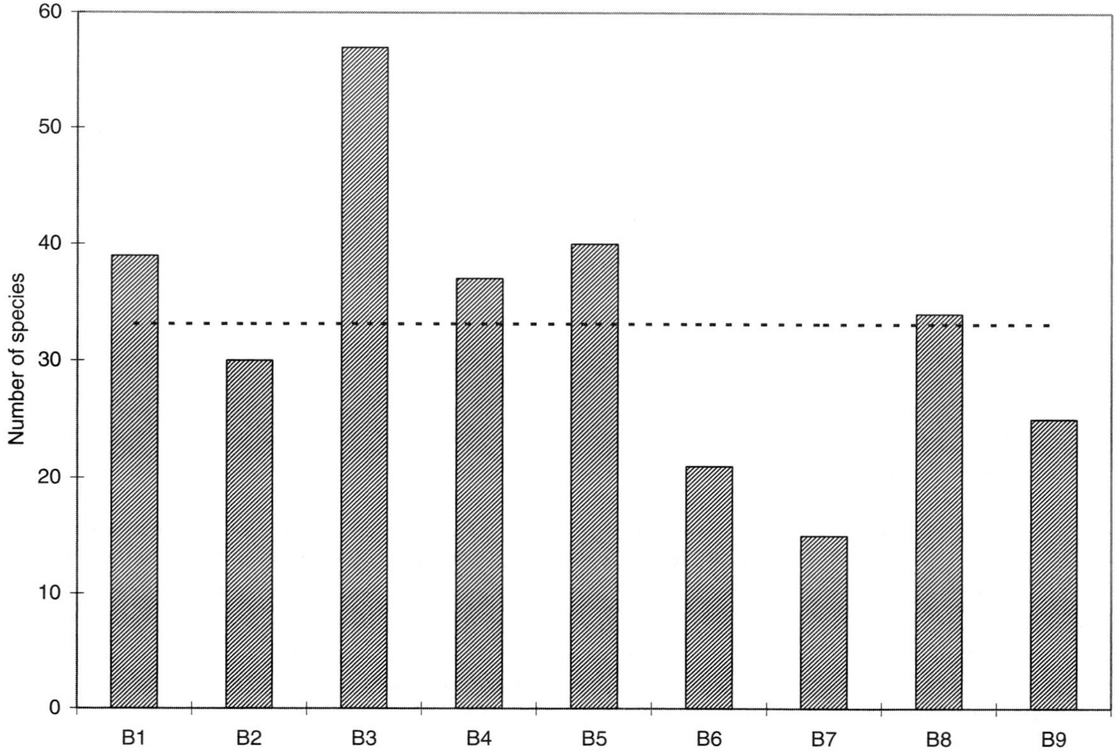

Fig. 6.5. Number of species of bark-beetles collected per tree for trees B1 to B9. The mean number of species per tree is indicated by a dotted line.

Table 6.4. *Number of species of bark-beetle collected by flight-interception traps in study trees, detailed by sampling height*

Study tree	Upper crown	Lower crown
B1	18	26
B2	13	21
B3	30	31
B4	10	29
B5	21	23
B6	17	7
B7	10	6
B8	22	13
B9	14	12

The p-value of a Wilcoxon matched–pairs test was 0.910.

traps ($n = 35$; $z = 3.381$; $p < 0.001$), and in tree B6, species preferred the upper crown ($n = 19$; $z = 2.294$; $p = 0.022$). For the three other trees p values were not significant.

DISCUSSION

Sampling of bark-dwelling arboreal ants

The number of ants collected was low compared with the number of ants sampled by insecticide fogging, particularly when entire nests are sampled. Floren & Linsenmair (1997b) presented data on 500 to 5000 individuals of ants from a single *Aporusa* tree crown (Euphorbiaceae) in south-east Asia, and Adis *et al.* (1998a) collected 948 specimens of ants in a single crown of *Goupia glabra* Aubl. (Celastraceae) in an Amazonian upland forest. In our studies, however, the number of ants was higher than in studies using baits (Itino & Yamane, 1994; U. Simon & R. Bertele, unpublished data). Arboreal pitfall traps generate data on the intensity of activity of ants. This intensity of activity was very low at both sampling heights. An observational study of the activity of ants on similar trees also indicated low activity. Baits exposed on these trees (tuna, sugar) resulted in a dramatic increase of ant activity within a short time-span (around 10 min; U. Simon & R. Bertele, unpublished data). Under undisturbed circumstances, the density of active ants in oak tree crowns may be relatively low; high densities of ants on the study oak trees occurred only where there was disturbance or during recruitment after detection of prey.

Ant distribution within tropical oak tree crowns

The composition of the ant subfamilies in our samples was similar to the results for arboreal ants presented by Brühl *et al.* (1998) for another study site (Poring Hot Spring) at Mount Kinabalu National Park. Formicinae

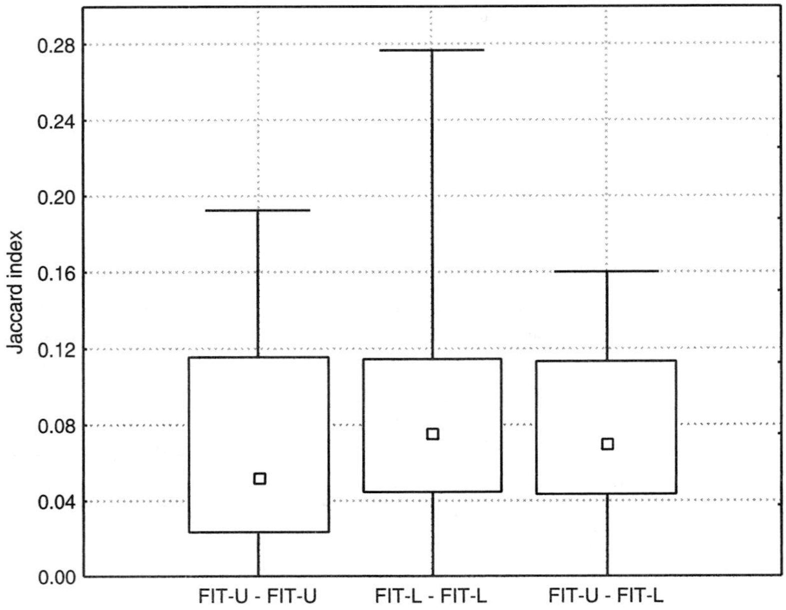

Fig. 6.6. Comparison of the Jaccard indices for bark-beetles within heights and between heights. The small box indicates the median value; the large box is the 25–75% confidence interval; FIT, flight-interception trap; U, upper crown; L, lower crown. There were no significant differences ($p > 0.05$, Friedman test).

and Myrmicinae were species rich, and Ponerinae, in contrast to samples obtained from the soil and litter layer, were species poor. Surprisingly, no species of Pseudomyrmicinae was sampled, even though species of the genus *Tetraponera* are typical inhabitants of lower vegetation, either of tree crowns (Brühl *et al.*, 1998) or in the lower canopy (Itino & Yamane, 1994). However, typical genera of tree-inhabiting groups of ants were collected. Götzke (1993), for example, found Ponerinae and species of *Pachycondyla* and *Ponera* to be tree-dwelling just as we did. Floren (1995) reported the genus *Diacamma* from tree crowns.

No statistical difference either in species number or in individual number of ants could be detected between the two sampling heights. Only one species, *Vollenhovia* sp. 6, showed a tendency to prefer the upper crown.

We conclude that worker ants of species dwelling in tree crowns apparently use the whole tree crown (and perhaps partly or completely the trunk) without any discrimination based on height or exposure. These results were corroborated for worker ants in a companion study (U. Simon and R. Bertele, unpublished data). Ants on the study oaks apparently exhibit a generalist foraging strategy. In contrast, alates may need particular structural and microclimatic features for their establishment. The spatial distribution of alates will be discussed elsewhere.

The β-diversity of ants in the upper crowns of study trees was significantly lower than in the lower crowns. Although the total number of species in upper traps was higher than in lower traps (48 to 44) there were many lower traps that shared no common species among themselves. Whereas nesting is restricted mainly to the lower parts of the trees (U. Simon & R. Bertele, unpublished data), foraging activity of worker ants was not significantly different between the two study heights. The reduced β-diversity in the upper crown may indicate a microclimatic limitation of ant occurrence in these emergent trees.

Sampling bark-beetles in tree crowns

Scolytinae are often neglected in studies on beetle diversity. This is because of the difficult taxonomy of the taxon, and the comparatively low proportions of the subfamily obtained in insecticide-fogging samples (e.g. Davies *et al.*, 1997; Wagner, 1997). In our study, bark-beetles presented at least a tenth of all beetle specimens collected by the flight-interception traps. We think that this high proportion reflects the sampling method: flight-interception traps in tree crowns catch bark-beetles arriving for reproduction as well as Scolytinae leaving after hatching. We cannot exclude the possibility that a proportion of the bark-beetles caught are just passing by on dispersal. How large a proportion these 'tourists' represent could only be estimated by studying the bark-beetles in the wood of the study trees itself (see Büchs (1988) 'bark emergence eclector').

Distribution of bark-beetles within tree crowns

In our study, we expected that there would be different microclimatic conditions within the lower part of the closed canopy compared with those in the exposed upper parts of these emergent trees. Although no measurements of climatic data were available, the distribution of mosses, lichens and vascular epiphytes was different across study heights. There were more mosses and vascular epiphytes in the lower parts, and more crustaceous lichens in the upper crowns, indicating, at least, differences in the air and/or the bark moisture levels (U. Simon & S. Unsicker, unpublished data). Accordingly, we expected differences in species composition, or at least in the individual numbers of Scolytinae caught within the two contrasting layers. Overall, no such pattern was detected. This indicates that the exposed, and most likely heated, outer parts of the *Quercus* crowns that we studied were not avoided by bark-beetles. This is reflected in the similar values for β-diversity that we obtained for within-tree and between-tree assemblages of Scolytinae. In the upper parts of the crowns, bark-beetles must be adapted to or at least be tolerant of dry and hot conditions.

Conclusions: tree crowns as single units?

With regard to information on the trees and insect taxa presented in the present contribution, we might conclude that tree crowns represent a single unit used by arthropods without discrimination by height or exposure. However, several pieces of evidence are counter-indicative.

- The β-diversity of ants is lower in the upper crown, indicating differences in the composition of the faunal assemblages.
- The mechanism of colonization and recolonization of tree crowns is mainly by chance (Floren & Linsenmair, 1997b; Floren *et al.*, 1998). However,

within single trees the species assemblages of bark-beetle in the tree crown was demonstrably different. Consequently, the random (re-)colonization of tropical tree crowns often results in a (random?) subdivision of the assemblage.

- Basset (1991b, 1992c) reported a nonuniform distribution of herbivory and herbivores in *Argyrodendron actinophyllum* F. Muell., an Australian rainforest tree, and Lowman (1985, 1995) found conspicuous differences in the spatial distribution of herbivory in rainforest tree crowns.
- Simon & Linsenmair (2001) showed that, at the ordinal level, the arthropod assemblages of the same tree species were different. They also found significant differences in the occurrence of ants in the upper and lower crown of the same study trees, mainly revealed by samples from flight-interception traps. Since in this study, ant workers, collected by pitfall traps, showed no statistical difference in activity between upper and lower crowns, the reason for the difference observed by Simon & Linsenmair (2001) may well be that alates prefer the lower parts of the tree crowns, probably because of the increased amount of accumulated litter and the more equable microclimate.

Ricklefs (2000, p. 84) stated that for the study of tropical plant assemblages: 'when the abundance of many species averages one individual per hectare, local communities of species within hectare plots have little meaning'. In our study, from two thirds to three quarters of the morphospecies of ant and bark-beetle were singletons. Ricklefs (2000, p. 84) concluded that 'old concepts of ecological communities and the sampling strategies accompanied them, should be discarded'. If his statement is taken to include insect trapping among the 'old . . . strategies' then we must beg to disagree. We do agree that more satisfying data could be obtained with a more extended sampling protocol including more study sites on a larger scale, as reported by Pitman *et al.* (1999) for Amazonian tree species. In our opinion, a trapping protocol as discussed above *does* provide relevant insights into the composition of canopy arthropod assemblages, the processes occurring within them and their dynamics.

We do not know if the bark-beetles we collected have their breeding habitats at the crown heights that we studied or whether the nests of the ants encountered were close to the trap positions. An unknown number of individuals of both taxa are also likely to be 'tourists'. For this study, the issue was of secondary importance. The conspicuous differences in faunal composition at the ordinal level between the upper and the lower crown of the study trees found by Simon & Linsenmair (2001) was not reflected within these two numerically and ecologically important taxa.

ACKNOWLEDGEMENTS

We are grateful to the Director of Sabah Parks, Datuk Ali Lamri, for permission to work in Kinabalu Park, and to Mr Francis Liew and Mr Wong for their support. Ansow Gunsalam of the Botanical Department of Kinabalu Park helped us to find the study trees around Sayap Ranger Station. Peninsus Guliong and his family provided great hospitality at Sayap Ranger Station. Many thanks to our Malaysian worker Soimin Magindol, who surveyed and maintained the traps over a period of more than 4 months with great reliability. Furthermore, three of our students, Heinrich Bardorz, Reinhard Bertele and Sibylle Unsicker, together with Jan Pfütze helped us to start the project and with surveying and dismantling the traps. The fieldwork was supported by travel grants of the Tropical Canopy Programme (1994–1998) of the European Science Foundation.

7

Vertical and temporal diversity of a species-rich moth taxon in Borneo

Christian H. Schulze and Konrad Fiedler

ABSTRACT

Using the very species-rich moth superfamily Pyraloidea, we have studied diversity and temporal variation of a moth assemblage in the understorey and canopy layer of a Bornean hill dipterocarp forest. A total of 5043 specimens representing 575 species was collected using a 15 W blacklight for 13 nights each in the canopy (2831 specimens, 405 species) and understorey (2212 specimens, 399 species). Alpha-diversity (expressed as rarefied species number per nightly catch or Fisher's α) was temporally stable and similar in both strata. When based on cumulative rather than nightly samples, species diversity was significantly higher in the understorey than in the canopy. Size and variation of total catch, as well as temporal fluctuations of relative abundances of the commonest species, did not differ between the forest strata. Between-sample similarity diminished with the time intervening between sampling nights, indicating substantial dynamics in ensemble composition despite temporally stable α-diversity. Similarity matrices showed high correlations between understorey and canopy ensembles, indicating largely parallel temporal dynamics. Nonlinear multidimensional scaling based on NESS (normalized estimate of shared species) index values as similarity measures partitioned temporal (collecting periods) and spatial β-diversity (stratification) in a highly significant fashion. Results remained unchanged when rare species ($\leq$5 individuals) were excluded, demonstrating the robustness of these patterns against chance sampling effects. The distinction between canopy and understorey pyraloid faunas was largely a consequence of differences in species abundances – few species qualified as stratum specialists. Temporal dynamics of diversity, or of species composition, did not differ between strata. Accordingly, for this large group of insect herbivores, processes regulating assemblage dynamics may be inferred to be surprisingly similar in different forest strata.

INTRODUCTION

Patterns in temporal variation of population abundances and the underlying processes are of central interest to ecologists (Link & Nichols, 1994; Stewart-Oaten *et al.*, 1995). The analysis of temporal variation of abundance or density of selected species has been the main aim of a number of studies (e.g. Berryman, 1996; Hunter & Price, 1998). However, apart from studies focussing on succession, recolonization or food-web theory (e.g. Polis *et al.*, 1997), temporal variation in the structure and diversity of whole terrestrial communities has been less frequently addressed (for Lepidoptera, e.g. Barlow & Woiwod, 1989; Wolda *et al.*, 1994; Orr & Häuser, 1996; Intachat & Holloway, 2000). Such studies are particularly needed for less-accessible, yet extremely species-rich ensembles such as those of arthropods in tropical forest canopies. For example, competing concepts about the mechanisms responsible for maintaining high tropical biodiversity (Huston, 1994; Rosenzweig, 1995; Gaston, 2000) make different predictions about the 'stability' of diverse communities. There is a shortage of empirical data that would allow for a critical appraisal of such hypotheses.

Although a growing number of studies have addressed temporal dynamics of tropical insect ensembles (e.g. Wolda, 1978a, 1992; Basset, 1991c; Novotny & Basset, 1998), there are still few attempts to follow their dynamics with taxonomic resolution down to species level in both the understorey and canopy (e.g. DeVries *et al.*, 1997, 1999a; DeVries & Walla, 2001). Insecticidal fogging, now popular for the study of canopy arthropods (e.g. Floren & Linsenmair, 1997b, 1998a), disrupts 'natural' dynamics and implies substantial disturbance of the area around the target tree, which inevitably drives off many of the more mobile organisms. Lepidoptera, for example, though abundant in all tropical forests, are usually rare in fog samples (Floren & Linsenmair, 1997b, 1999; Chey *et al.*, 1998). Moreover, it is difficult to obtain adequate samples from understorey layers, where the structure of the vegetation usually does not allow the execution of fogs that may be compared with those in the upper canopy (Barker & Pinard, 2001). Less-invasive methods of monitoring faunal dynamics, such as the use of bait or light traps, do not present this problem.

We address the temporal variation in diversity and species composition of a species-rich ensemble (we use the term ensemble here for that fraction of a guild which also belongs to a single monophyletic taxon, in accordance with Fauth *et al.*, 1996) of highly mobile herbivorous insects in two strata of perhumid hill dipterocarp forest in Borneo. Samples were obtained by attracting nocturnal insects to light (hereafter termed 'light-trapping' for brevity, although we did not use automatic light traps). Since only a rather small number of individuals per night is removed from the natural community, sampling should have minimal impact on natural dynamics. Moreover, light trapping can be performed using identical methods in the canopy and understorey. The moth superfamily Pyraloidea (*sensu* Munroe & Solis, 1999) was used as the target group since pyraloid moths are usually the single most abundant lepidopteran taxon represented in light-trap samples from south-east Asia, frequently accounting for 40–50% of the nightly catch (Robinson *et al.*, 1994). Therefore, each nightly catch generally provides a sufficiently large sample with regard to the number of individuals and species collected to allow for statistical analyses. The Pyraloidea is also very rich in species and contains almost all larval feeding habits to be found among the Lepidoptera as a whole (Robinson *et al.*, 1994; Munroe & Solis, 1999). Accordingly, light trapping of abundant and diverse pyraloid moths promises to be an excellent mode of assembling empirical datasets for assessing and comparing community patterns across forest strata.

We address the following five issues. (i) How variable are the communities over time with regard to α-diversity? Do forest strata differ in α-diversity? (ii) How variable is species composition over time (temporal β-diversity)? If there is a difference between canopy and understorey fauna (i.e. spatial β-diversity), is this stratification temporally stable? (iii) Do patterns of spatial and temporal β-diversity change if rare species are excluded from analysis? In most tropical insect samples, the majority of species are represented by few individuals only, either since they are truly rare or just occur as 'tourists'. Hence, exclusion of such rare species reveals whether the observed patterns are robust against chance sampling effects. (iv) Do temporal dynamics of community composition differ between forest strata? Such differences would be expected if the composition of the canopy community was more strongly governed by stochastic factors compared with the understorey species assemblages. (v) Do abundances of dominant species differ in temporal variability between forest strata? If canopy faunas were subject to more intense stochastic fluctuations, one would expect higher variation in abundances relative to the understorey fauna.

METHODS

Study area

Moths were collected at the south-eastern margin of Mount Kinabalu National Park (Sabah, Malaysia) near Poring Hot Springs (6° 2.9′ N, 116° 42.0′ E, altitude 580 m above sea level (asl)). Forests at this altitude are termed hill forest by Kitayama (1992) and are dominated by trees of the Dipterocarpaceae. There is a well-developed, multilayered forest canopy with the largest trees reaching far more than 50 mm height. Kitayama *et al.* (1999) presented a detailed description of the climate of Mount Kinabalu. The climate is generally perhumid with annual rainfall at Poring ranging between 2000 and 3000 mm and monthly rainfall falling below 50 mm only during El Niño events. There were two sampling seasons, from early March to mid-May 1997, and from late August to late September 1997. We sampled moths at two sites, one in the canopy and one in the understorey. A canopy walkway facilitated access to the upper vegetation layers of the forest. The canopy light trap was placed 45 m above ground level, on a platform attached to a >60 m high emergent tree. Because of the terrain – on a steep slope – the light trap there largely sampled moths flying in the upper and middle canopy of this multilayered forest. The understorey light trap was run in close proximity to the trees supporting the canopy walkway. The understorey site was characterized by an almost complete multilayered canopy cover, with rather sparse understorey vegetation (mostly dipterocarp seedlings), as usual for unlogged hill dipterocarp forest. For further description of the sites see Schulze (2000) and Beck *et al.* (2002).

Light trapping

Moths were attracted to an ultraviolet light tube (Sylvania™ blacklight-blue, F 15W/BLB-TB), which was placed in a white, reflective gauze cylinder (height 1.60 m, diameter 0.80 m). The use of the accumulator-driven, low-energy light tube was considered sufficiently local to minimize attraction of moths from neighbouring vegetation layers. Even far stronger light sources usually attract only moths within a radius of less than 30 m (e.g. Butler & Kondo, 1991; Muirhead-Thomson, 1991). The tree-crown site was situated immediately within the upper canopy layer above the corresponding understorey site. Moths attracted to the light and settling on the gauze were caught by hand using cyanide-charged killing jars. Sampling was undertaken from 18:30 to 21:00 local time for 13 nights in each stratum between March and May (eight temporal replicates per stratum) and in August and September (five temporal replicates), resulting in a total of 26 samples. Restriction of light trapping to 2.5 h was considered sufficient as we observed that catch dropped considerably during the last 30 min of each sampling session. However, certain species that fly only late at night or towards early morning may have been missed.

Pyraloid moths were identified using available literature as well as reference and type materials in museum collections and by consultation with taxonomists (species lists: Schulze, 2000). Specimens with distinctive external and/or genitalia features, which could not be found in the literature consulted, were sorted to morphospecies level. Although morphospecies delimitations may not be free of errors in every single case (e.g. in taxonomically underworked taxa such as the subfamily Phycitinae), such errors should be equally distributed across samples and thus not affect our analyses. Among the Pyraloidea larvae of the monophyletic subfamily, Acentropinae (= Nymphulinae: Speidel, 1998) have highly aberrant life histories: their aquatic larvae feed on algae or aquatic vascular plants (Stoops *et al.*, 1998; Munroe & Solis, 1999). Since it may be misleading for diversity assessments to combine organisms from very different guilds, we excluded Acentropinae moths from all analyses presented below. Available information on larval food requirements (Munroe & Solis, 1999) suggests that all pyraloids represented in our samples can be considered as terrestrial herbivores or, rarely, detritivores, although expected host plants range from mosses to herbs and trees. Throughout this chapter, the term 'Pyraloidea' therefore refers precisely to all families and subfamilies of Pyraloidea except Acentropinae (see Table 7.2, below).

Measuring α-diversity

The selection of diversity measures always depends on the scale and aims of a particular study (e.g. Purvis & Hector, 2000). Numbers of species are unsuitable for comparisons when (i) the numbers of recorded individuals vary considerably between samples, but this variation is beyond the control of the observer; or (ii) if communities are incompletely sampled. Therefore, measures of α-diversity that control for sampling effort are required. The α-parameter of the log-series distribution (Fisher's α hereafter) is frequently used for the study of tropical moth communities (Barlow & Woiwod, 1989; Robinson & Tuck, 1993; Willott, 1999; Intachat & Holloway, 2000). The log-series model has many advantages for application to real datasets (even if these do not follow a log-series distribution precisely; Hayek & Buzas, 1997). Gamma type species abundance distributions (of which the log-series is a special case) emerge under a range of biologically realistic conditions (Engen & Lande, 1996). However, estimates of Fisher's α from real samples may be strongly influenced by sample size in species-rich tropical insect assemblages with a high ratio of species number to the number of individuals (Robinson, 1998; see below). Therefore, we used, in addition, Hurlbert's rarefaction procedure (Hurlbert, 1971) to quantify α-diversity for the largest shared number of specimens (e.g. Achtziger *et al.*, 1992). Fisher's α, Hurlbert's rarefaction and their variance estimates were calculated using programs contained in Kenney & Krebs (2000). The fit of data to log-series and log-normal distributions was assessed using the octave method as implemented in the software package by Henderson & Seaby (1998). The same program was used to compare estimates of Fisher's α between samples using a permutation test (with 10000 permutations) originally proposed by Solow (1993).

Measuring β-diversity

Most measures of between-habitat diversity are sensitive to size, α-diversity and incompleteness of samples, leading to overestimates of dissimilarity whenever incomplete samples are compared (Wolda, 1981; Lande, 1996). Morisita's index largely avoids these problems (Wolda, 1981; Robinson & Tuck, 1996), yet species with low or medium abundances (the majority in tropical communities) have little influence on index values. The NESS index (Grassle & Smith, 1976) provides an opportunity to control for such effects. NESS values indicate normalized estimates of shared species if samples of the size m were drawn from two communities. For $m = 1$, the NESS index is equivalent to Morisita's index, while the contribution of rare species increases with rising m. Since samples drawn from tropical moth ensembles usually contain many rare species and are almost never 'complete' (in the sense that all species available at a site would be covered by sampling), we here use the NESS index as a measure of β-diversity.

We present NESS values for the highest possible value of m as well as for $m = 1$. NESS values for all possible pairs of light-trapping nights were calculated with a computer program provided by S. Messner (personal communication). Based on these index values, we conducted a two-dimensional ordination of all samples using nonlinear multidimensional scaling (MDS). MDS has two advantages over other ordination techniques such as correspondence analysis (CA) or principal component analysis (PCA). First, it can be performed with any similarity measures, thus allowing the use of NESS index values that are particularly suitable to express similarity among rich, incomplete samples with many rare species (Süßenbach *et al.*, 2001). Second, MDS is independent of linearity assumptions and does not require transformation of data. Therefore, MDS ordinations are particularly robust and lack certain problems frequently encountered with CA or PCA (see Minchin (1987) and Clarke (1993) for details and references). We did not employ canonical correspondence analysis (CCA: Palmer, 1993), since application of such constrained ordination techniques (Økland, 1996) is only useful for testing of specific hypotheses about extrinsic factors that govern species distributions. In the present study, in contrast, we were primarily interested in uncovering (the previously undescribed) patterns in the animal ensembles and generating hypotheses as to their causal background. The use of two-dimensional MDS plots was considered sufficient, since raw stress (which is a measure of the goodness-of-fit of the representation of the true similarity matrix values by the ordination: Clarke, 1993) ranged from 0.14 to 0.16. Moreover, visualization in Shepard diagrams (see Clarke, 1993; StatSoft, 1999) revealed a good data fit (plots not shown). The two-dimensional scores of samples were used to compare strata and seasons using a standard two-way multiple analysis of variance (MANOVA). Three-dimensional MDS ordinations were also calculated and showed a better fit (raw stress 0.09–0.11; data not shown), but subsequent MANOVA on the dimensional scores showed that the third dimension never contributed significantly to a segregation of samples according to stratum or collection period.

Tropical insect ensembles usually contain a very high proportion of rare species (e.g. Robinson & Tuck, 1996; Floren & Linsenmair, 1998a). Rare species may pose a problem in assessing differences between communities, because (i) it cannot readily be proven whether they really have a functional connection with the community in question (the 'tourist species' problem) and (ii) random sampling effects should be particularly strong with rare species (which will be absent in many samples just by chance, owing to their scarcity). Accordingly, we performed the same analyses of β-diversity (calculations of NESS indices and subsequent MDS ordination) on a restricted dataset from which we excluded all species that were represented by five or fewer individuals in the total cumulative collection material (i.e. whose individual abundances accounted for less than 1% of the entire dataset).

One would expect samples drawn from a community to become increasingly dissimilar with increasing time between light-trapping nights. To test for the existence of such a relationship, the generalized regression approach for matrix correlations introduced by Mantel (1967) was used. This procedure overcomes the nonindependence of data points (here: pairwise similarity values). We calculated the standardized Mantel test statistic r, which is effectively equivalent to Pearson's correlation coefficient (Cavalcanti, 2001). The significance of r was tested by a modified t-test, comparing the observed matrix with 10 000 random permutations. Both r (together with its standard deviation (SD)) and t values are presented. Calculations were performed with the software published by Cavalcanti (2001).

Quantifying variation in the relative abundance of individual species

The SD of log-transformed abundances is most commonly used as measure of variability (e.g. Wolda, 1992). However, McArdle *et al.* (1990) suggested alternative measures such as the coefficient of variation (CV) to overcome the problem of zeros, a common feature of real sample data. We adopted the CV in the form corrected for bias owing to sample size, together with its variance estimate, as proposed by Sokal & Braumann (1980):

$$\mathrm{CV} = \left(1 + \frac{1}{4n}\right)\frac{\mathrm{SD}(N)}{\overline{N}} \quad (7.1)$$

$$\mathrm{SD_{CV}} = \left[\frac{\mathrm{CV}^2}{2n}\left(\frac{n}{n-1} + 2\mathrm{CV}^2\right)\left(1 + \frac{1}{4n}\right)^2\right]^{1/2} \quad (7.2)$$

where SD(N) is the standard deviation of abundances, $\overline{N}$ is the mean abundance and n the number of samples. Differences between CV values were tested for significance using the modified t-test introduced by Sokal & Braumann (1980). Since most tropical insect species tend to occur in low densities most of the time (Wolda, 1992), we restricted quantitative comparisons of abundances to the 10 most abundant species in either forest stratum. Mean abundances (±1SD) of these 10 commonest species ranged from 4.23 ± 2.12 to 8.84 ± 9.19 individuals per night in the canopy, and from 2.38 ± 3.54 to 16.54 ± 16.26 individuals per night in the understorey.

Light traps do not provide real 'abundance' data, since the capturing success of a light trap depends on a range of external factors that vary between sampling sites and nights (background light intensity owing to lunar cycle, temperature, rainfall: Muirhead-Thomson, 1991; Yela & Holyoak, 1997). We instead used relative abundances, that is, numbers of individuals of one species relative to total nightly catch. Temporal variability of the relative abundance of a species was then expressed as the CV of all 13 nightly values and compared statistically between forest strata. In addition, we applied a repeated measures analysis of variance (ANOVA) on the relative abundances of the 10 commonest species of either stratum to explore whether any temporal influence could be found. Different moth species do not respond uniformly to light sources. However, by restricting our analyses to only those species that were common in our samples (i.e. which responded readily to the artificial light source), differential catchability should not affect our data and conclusions.

All standard statistical analyses, as well as MDS and tests for temporal autocorrelations, were performed using the package STATISTICA™ 5.5 (StatSoft, 1999). Unless stated otherwise, means are given ±1SD.

RESULTS

Alpha-diversity: stratification and temporal dynamics

A total of 5043 specimens representing 575 species was sampled during 26 nights in 1997 in the canopy and understorey layer (Table 7.1). In all, representatives of two families and thirteen subfamilies of Pyraloidea were encountered (Table 7.2). The number of specimens caught per night did not differ significantly between canopy (217.8 ± 119.3) and understorey (170.2 ± 79.7) (t-test, $t_{24} = 1.197$, $p > 0.24$). The number of species recorded per night was also similar (canopy: 94.2 ± 29.1; understorey: 85.7 ± 28.4; t-test, $t_{24} = 0.75$, $p > 0.45$).

Evaluated for the total catch per stratum (i.e. summed over the 13 nightly samples each), the gross taxonomic composition of pyraloid faunas was similar in the canopy and understorey (Table 7.2). Neither with regard to individuals ($\chi^2_{5\mathrm{df}} = 10.23$, $p = 0.069$; for statistical comparisons the subfamilies Galleriinae, Scopariinae, Heliothelinae, Crambinae, Schoenobiinae, Cybalomiinae, Musotiminae and Evergestiinae were combined to avoid too small expected values per cell in the contingency table) nor species ($\chi^2_{5\mathrm{df}} = 6.11$, $p = 0.30$) was a significant difference in subfamilial composition apparent. The cumulative samples of pyraloid moths from both strata were similar to each other in rank-abundance relationships. However, none showed a good fit to a log-series distribution (canopy: Fisher's α (±1SE) = 129.4 ± 4.5, $\chi^2_{5\mathrm{df}} = 39.3$, $p < 0.0001$; understorey: Fisher's α = 141.9 ± 5.7, $\chi^2_{5\mathrm{df}} = 19.5$, $p < 0.0016$). Both contained more rare species than expected under that model, but the proportions of species represented by only one (singletons), or two individuals (pairs), respectively, were similar between strata (canopy: 41.5% of total species were singletons, 16.5% pairs; understorey: 39.6% singletons, 18.0% pairs). Cumulative samples showed a better fit to a log-normal abundance distribution, but

Table 7.1. *Numbers of Pyraloidea individuals and species recorded in 13 sampling nights at one canopy and one understorey site in a mature hill dipterocarp forest near Poring (Mt Kinabalu National Park)*

Canopy			Understorey		
Date	Individuals	Species	Date	Individuals	Species
3 March	126	66	11 March	121	71
18 March	100	69	19 March	76	51
24 March	98	56	25 March	107	52
29 March	349	120	28 March	168	91
3 April	164	88	5 April	144	77
9 April	207	103	17 April	107	78
20 April	203	99	3 May	191	94
10 May	263	112	9 May	182	101
30 August	313	120	31 August	350	147
6 September	512	159	7 September	246	97
11 September	175	81	10 September	232	108
15 September	99	61	16 September	65	42
22 September	222	91	21 September	223	105
Total	2831	405	Total	2212	399

Table 7.2. *Gross taxonomic composition (at the subfamilial level) of pyraloid samples collected at Poring*

Family/ subfamily	Canopy		Understorey	
	Species	Individuals	Species	Individuals
Pyralidae				
Pyralinae	24	67	26	69
Galleriinae	1	1	1	2
Epipaschiinae	38	119	30	79
Phycitinae	36	64	20	36
Crambidae				
Scopariinae	1	2	1	1
Heliothelinae	0	0	2	3
Crambinae	4	18	4	9
Schoenobiinae	1	4	0	0
Cybalomiinae	1	1	0	0
Musotiminae	3	4	6	10
Odontiinae	11	46	13	22
Evergestiinae	2	6	1	3
Pyraustinae	283	2499	295	1978

in the canopy a marginally significant deviation from the model, caused by an excess of rare species, was observed (understorey: $\chi^2_{5df} = 6.25$, $p = 0.28$; canopy: $\chi^2_{6df} = 13.23$, $p < 0.04$).

Regardless of the measure chosen, the cumulated pyraloid ensemble was more diverse in the understorey than in the canopy (Fisher's α, Solow test: $p < 0.001$; expected species number $\pm$1SE of canopy if rarefied to size of understorey sample: 354.2 ± 5.9 spp., understorey: 399 spp., $p < 0.001$).

Despite prominent fluctuations of moth abundance (fivefold) and species number (more than threefold) recorded in individual nights, α-diversity was relatively constant over time in both vegetation layers. Rarefied expected species numbers for a minimum common sample size of 65 individuals are shown in Fig. 7.1. Mean expected species numbers per night, averaged over the 13 samples each, were almost identical in the canopy (43.9 ± 2.5) and the understorey (43.5 ± 4.2; $t_{24} = 0.36$, $p > 0.7$). When rarefaction values were calculated over the entire cumulative samples rather than for individual nights, only slightly (and not significantly) higher diversity estimates were apparent (understorey: 47.4 ± 3.2; canopy: 46.7 ± 3.3). There was no

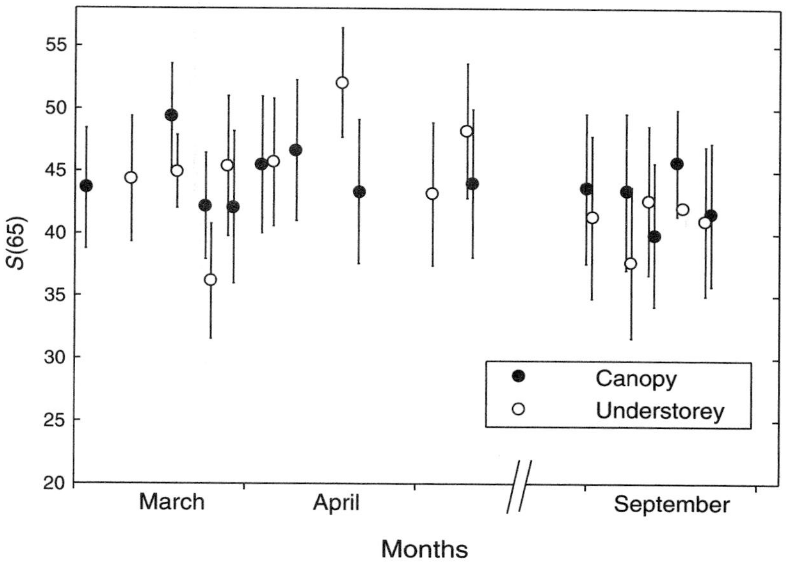

Fig. 7.1. Temporal variation of Pyraloidea diversity in understorey and canopy layer. Diversity is measured as the expected numbers of species $S(q)$ [(calculated by Hurlbert rarefaction for largest shared number of specimens ($q = 65$)]. The bar is the 95% confidence limits for 13 light-trapping nights in each vegetation layer.

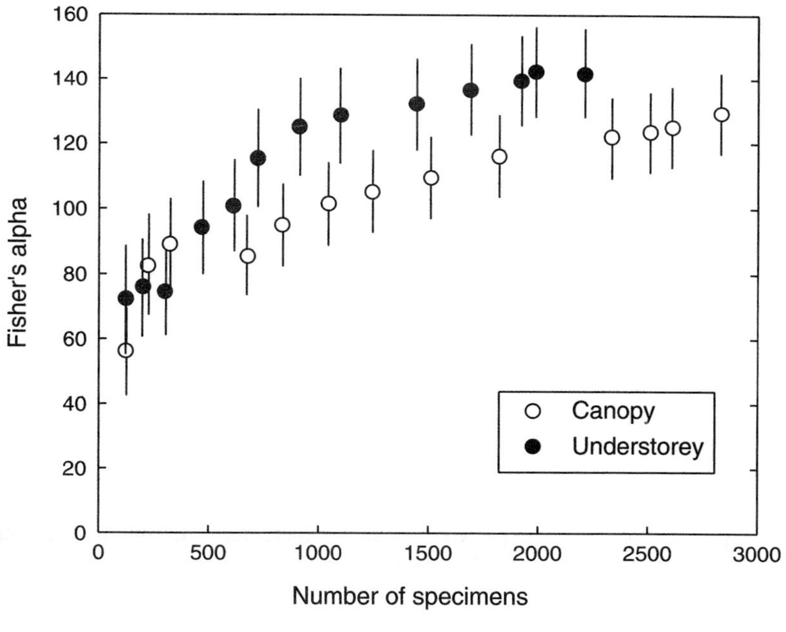

Fig. 7.2. Estimates of Fisher's α of the logarithmic series (95% confidence limits) as a function of cumulative sample size for Pyraloidea. In both strata, Fisher's α increases logarithmically with sample size ($y = a \log(x + b)$; canopy: $a = 15.42 \pm 0.26$ (SE), $b = -84.90 \pm 15.94$, $r^2 = 0.925$; understorey: $a = 17.84 \pm 0.44$, $b = -83.78 \pm 16.97$, $r^2 = 0.877$).

significant temporal autocorrelation in the time series of rarefaction diversity estimates for either of the two forest layers ($p > 0.3$ for first and all higher-order time lags). Fluctuations of diversity between sampling nights were similar in the forest understorey (range, 36.7–52.1), and canopy (range, 39.9–49.4). Coefficients of variation were 0.467 ± 0.007 for the canopy and 0.479 ± 0.012 for the forest understorey (t-test: $t_{24} = 0.838$, $p > 0.3$).

If Fisher's α was used as the diversity measure, largely the same patterns emerged. Mean α-values per

Table 7.3. *Values of the normalized estimate of shared species (NESS) index for a sample size parameter of* m = *1*

	3 Mar	18 Mar	24 Mar	29 Mar	3 Apr	9 Apr	20 Apr	10 May	30 Aug	6 Sep	11 Sep	15 Sep	22 Sep
3 Mar		0.873	0.784	0.652	0.614	0.859	0.416	0.560	0.521	0.525	0.502	0.604	0.522
18 Mar	0.875		1.000	0.678	0.894	0.849	0.303	0.443	0.804	0.690	0.548	0.763	0.826
24 Mar	0.722	1.000		0.563	0.534	0.864	0.395	0.477	0.639	0.509	0.554	0.647	0.620
29 Mar	0.752	0.925	0.848		0.907	0.678	0.343	0.761	0.589	0.600	0.492	0.459	0.579
3 Apr	0.706	0.974	0.809	0.895		0.725	0.316	0.561	0.621	0.621	0.475	0.510	0.740
9 Apr	0.761	0.865	0.908	0.829	0.855		0.404	0.672	0.697	0.603	0.588	0.685	0.675
20 Apr	0.722	0.783	0.865	0.769	0.765	0.890		0.329	0.312	0.291	0.304	0.381	0.241
10 May	0.627	0.764	0.718	0.793	0.737	0.805	0.726		0.551	0.573	0.389	0.388	0.448
30 Aug	0.557	0.861	0.816	0.703	0.620	0.749	0.674	0.712		0.903	0.764	0.711	0.858
6 Sep	0.600	0.795	0.702	0.699	0.648	0.726	0.659	0.742	0.927		0.794	0.877	0.881
11 Sep	0.626	0.779	0.767	0.671	0.673	0.742	0.704	0.602	0.831	0.881		0.856	0.736
15 Sep	0.586	0.836	0.780	0.606	0.566	0.717	0.771	0.571	0.793	0.878	0.819		0.711
22 Sep	0.610	0.886	0.797	0.656	0.736	0.775	0.574	0.619	0.891	0.890	0.897	0.777	
11 Mar	0.369	0.387	0.533	0.447	0.426	0.494	0.558	0.592	0.588	0.531	0.532	0.505	0.496
19 Mar	0.359	0.295	0.408	0.555	0.532	0.519	0.557	0.645	0.560	0.552	0.516	0.420	0.574
25 Mar	0.449	0.514	0.665	0.580	0.524	0.625	0.636	0.651	0.639	0.597	0.614	0.488	0.590
28 Mar	0.219	0.319	0.440	0.335	0.299	0.350	0.493	0.443	0.487	0.417	0.452	0.504	0.414
5 Apr	0.319	0.310	0.508	0.364	0.321	0.380	0.392	0.473	0.505	0.454	0.458	0.408	0.470
17 Apr	0.553	0.527	0.682	0.586	0.501	0.634	0.625	0.669	0.572	0.535	0.542	0.491	0.571
3 May	0.492	0.557	0.599	0.489	0.469	0.571	0.584	0.530	0.544	0.504	0.468	0.543	0.546
9 May	0.740	0.600	0.859	0.757	0.691	0.761	0.804	0.711	0.790	0.719	0.745	0.748	0.829
31 Aug	0.610	0.399	0.602	0.543	0.498	0.547	0.665	0.570	0.409	0.295	0.297	0.320	0.377
7 Sep	0.604	0.440	0.612	0.546	0.457	0.616	0.674	0.532	0.447	0.306	0.323	0.400	0.375
10 Sep	0.635	0.528	0.726	0.651	0.550	0.680	0.703	0.683	0.583	0.452	0.435	0.402	0.496
16 Sep	0.767	0.506	0.864	0.647	0.571	0.774	0.831	0.789	0.681	0.595	0.730	0.685	0.581
21 Sep	0.554	0.482	0.634	0.570	0.491	0.636	0.663	0.541	0.451	0.329	0.401	0.340	0.412

night were 70.4 ± 12.6 for the canopy and 74.1 ± 18.2 for the understorey ($t_{24} = 0.61$, $p > 0.54$). There was again no temporal autocorrelation ($p > 0.3$ in both strata). However, estimated α-values were less stable from night to night than rarefied expected species numbers. Their CV values were higher than the respective figures for rarefied species numbers in both forest strata (t-tests; canopy: $t_{24} = 4.704$, $p < 0.0001$; understorey: $t_{24} = 3.760$, $p < 0.001$). Moreover, in the understorey Fisher's α had larger ranges (canopy: 56.0–81.4; understorey: 39.9–107.0) as well as slightly higher CV values (canopy: 0.507 ± 0.004; understorey: 0.540 ± 0.011; t-test: $t_{24} = 2.784$, $p < 0.02$). Estimates of Fisher's α strongly depended on sample size and reached relatively stable values only at sample sizes far higher than 1000 individuals (Fig. 7.2).

Stratification and temporal β-diversity

Two-dimensional scaling based on NESS indices for all possible pairwise combinations of samples (Table 7.3) showed a clear separation of both vegetation layers as well as between the two light-trapping periods (Fig. 7.3). For the highest possible value of the parameter m ($m = 32$ in our case), this separation was similar to the ordination obtained with $m = 1$.

A MANOVA on the scores of each sample for dimensions 1 and 2 showed that the segregations apparent in Fig. 7.3 were all highly significant statistically (Table 7.4), but the interpretation of dimensions slightly differed depending on the parameter value of m (Table 7.5). For $m = 32$, dimension 1 effectively separated ensembles of the two forest strata, while dimension 2 was responsible for a highly significant *(above diagonal) and* m = *32 (below diagonal)*

11 Mar	19 Mar	25 Mar	28 Mar	5 Apr	17 Apr	3 May	9 May	31 Aug	7 Sep	10 Sep	16 Sep	21 Sep
0.551	0.577	0.438	0.747	0.690	0.393	0.499	0.193	0.199	0.098	0.165	0.135	0.226
0.398	0.481	0.187	0.522	0.523	0.273	0.255	0.293	0.295	0.134	0.200	0.420	0.225
0.372	0.388	0.129	0.366	0.355	0.172	0.238	0.078	0.242	0.088	0.097	0.087	0.160
0.389	0.283	0.164	0.526	0.511	0.268	0.415	0.153	0.237	0.135	0.148	0.307	0.230
0.420	0.303	0.204	0.592	0.589	0.334	0.395	0.232	0.286	0.178	0.227	0.470	0.267
0.413	0.304	0.140	0.582	0.586	0.264	0.327	0.180	0.248	0.096	0.133	0.121	0.155
0.175	0.141	0.075	0.228	0.306	0.130	0.174	0.072	0.097	0.056	0.076	0.066	0.095
0.268	0.149	0.094	0.401	0.387	0.196	0.371	0.163	0.230	0.181	0.158	0.157	0.250
0.319	0.315	0.134	0.446	0.508	0.319	0.361	0.157	0.505	0.252	0.287	0.387	0.366
0.482	0.378	0.147	0.584	0.576	0.380	0.441	0.230	0.722	0.521	0.602	0.606	0.666
0.339	0.368	0.151	0.437	0.489	0.372	0.531	0.178	0.552	0.314	0.436	0.157	0.429
0.483	0.445	0.182	0.492	0.640	0.518	0.504	0.220	0.565	0.402	0.569	0.355	0.562
0.483	0.272	0.180	0.544	0.466	0.316	0.375	0.121	0.612	0.482	0.518	0.657	0.519
	0.883	0.713	0.983	0.737	0.663	0.698	0.701	0.580	0.496	0.664	0.617	0.638
1.000		0.839	0.975	0.792	0.514	0.577	0.368	0.365	0.188	0.428	0.257	0.480
1.000	0.996		0.630	0.544	0.438	0.629	0.292	0.211	0.160	0.320	0.218	0.312
0.973	1.000	0.948		0.998	0.629	0.624	0.748	0.597	0.409	0.597	0.542	0.616
0.847	0.824	0.878	0.961		0.607	0.624	0.551	0.557	0.220	0.411	0.312	0.484
0.814	0.710	0.773	0.730	0.707		0.799	0.700	0.407	0.209	0.397	0.348	0.441
0.813	0.696	0.788	0.691	0.704	0.874		0.756	0.498	0.295	0.446	0.335	0.543
0.735	0.541	0.694	0.791	0.627	0.786	0.874		0.633	0.449	0.575	0.539	0.651
0.613	0.557	0.583	0.684	0.680	0.689	0.693	0.795		0.813	0.876	0.706	0.936
0.549	0.493	0.521	0.657	0.504	0.549	0.626	0.647	0.874		0.940	0.765	0.840
0.752	0.724	0.735	0.748	0.669	0.791	0.709	0.679	0.834	0.868		0.802	0.922
0.581	0.476	0.563	0.608	0.510	0.538	0.519	0.504	0.571	0.590	0.651		0.761
0.697	0.664	0.737	0.727	0.624	0.761	0.760	0.820	0.968	0.885	0.878	0.594	

stratum/period interaction. With $m = 1$, dimension 1 again separated the two forest strata, but dimension 2 now differentiated sampling periods and only a marginally significant interaction between both factors was found.

A significant negative correlation existed between pairwise similarity values (NESS, $m = 32$) of individual samples and the time-span between sampling dates. This correlation was equally strong for the canopy (Mantel test: $r = -0.626 \pm 0.109$, $t = 5.773$, $p < 0.0001$) and understorey samples ($r = -0.614 \pm 0.108$, $t = 5.711$, $p < 0.0001$). The relationship is shown graphically in Fig. 7.4. In both forest layers, similarity between samples taken on individual nights tended to degrade quite quickly depending on the time intervening between two sampling events. However, with increasing time-span between two sampling nights, degradation of similarity attenuated asymptotically. Similarities remained relatively high even when the second sample of a pair was taken almost 200 days after the first. Although the data points are not independent and little weight should be placed on the curve-fitting statistics, the graphical representation of data points suggests a roughly exponential relationship (Fig. 7.4), with no clear difference in the asymptote value between forest canopy and understorey. The similarity matrices for canopy and understorey pyraloid ensembles were highly correlated with each other (Mantel test: $r = 0.451 \pm 0.140$, $t = 3.253$, $p = 0.0008$).

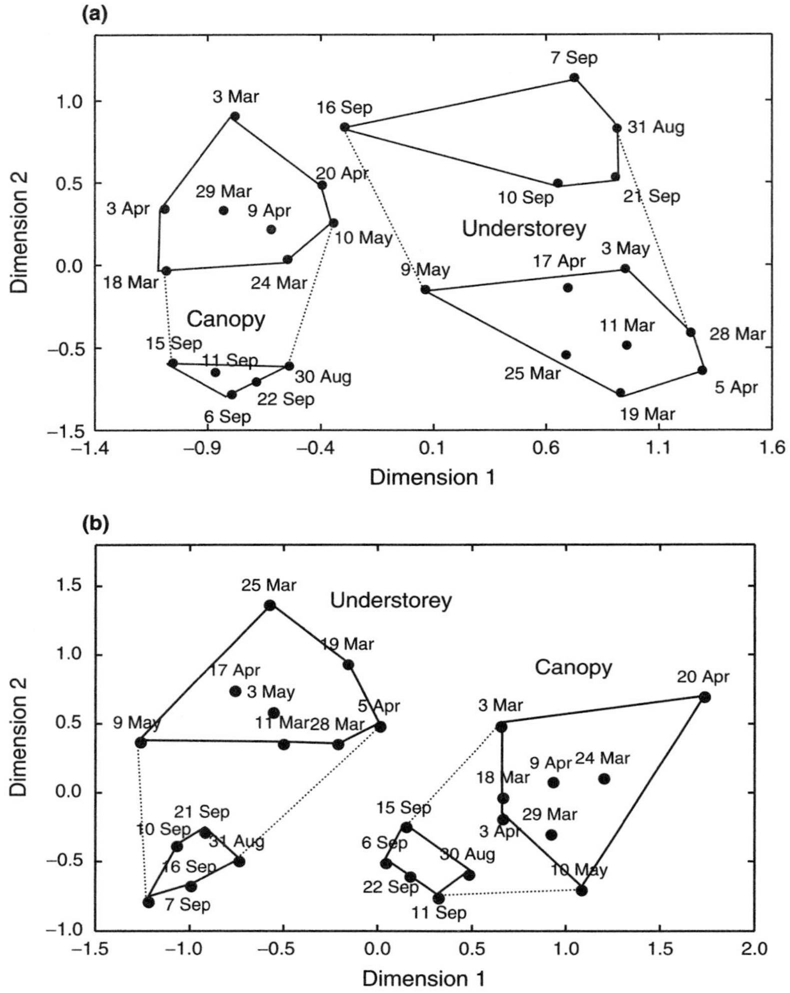

Fig. 7.3. Ordination by nonlinear two-dimensional scaling of nightly pyraloid moth samples in understorey and canopy according to NESS (normalized estimate of shared species) similarity. (**a**) $m = 32$, stress $= 0.154$; (**b**) $m = 1$, stress$= 0.149$. Within canopy and understorey, respectively, samples from the same period are connected by thick solid lines, and sampling periods (March to April; August to September) are connected by thin broken lines. Partitioning of vertical and temporal β-diversity was highly significant (see Tables 7.3 and 7.4).

Stratification and temporal dynamics when rare species are neglected

Exclusion of all rare species of Pyraloidea (i.e. those represented in our total catch by five or fewer individuals) did not change the patterns of spatial and temporal β-diversity, although this resulted in a very much contracted database (4167 individuals belonging to 169 species, i.e. almost 83% of specimens but less than 30% of the species total). Two-dimensional ordination based on NESS values with $m = 23$ (the maximum parameter value possible in this reduced dataset; plots not shown) revealed very distinct segregations between canopy and understorey faunas (MANOVA: Wilk's $\lambda = 0.134$, Rao's $R_{2;21} = 67.9$, $p < 0.0001$), as well as between the samples from the two field periods (Wilk's $\lambda = 0.176$, Rao's $R_{2;21} = 49.1$, $p < 0.0001$). Strata were strongly distinguished by the first dimension ($F_{1;22} = 118.2$, $p < 0.0001$), and sampling periods segregated along the second dimension ($F_{1;22} = 77.7$, $p < 0.0001$). There was no significant stratum/period interaction (Wilk's $\lambda = 0.962$, Rao's $R_{2;21} = 0.42$, $p > 0.6$).

Similar results recurred for the two-dimensional ordination based on NESS values with $m = 1$. Here, forest strata (MANOVA: Wilk's $\lambda = 0.175$, Rao's $R_{2;21} = 49.4$, $p < 0.0001$) as well as sampling periods (Wilk's $\lambda = 0.236$, Rao's $R_{2;21} = 34.1$, $p < 0.0001$) segregated clearly, and only a marginally significant

Table 7.4. *Results of a multiple analysis of variance on the scores of dimensions 1 and 2 extracted by multidimensional scaling of normalized estimate of shared species (NESS) similarity values*

Effect	Wilks' λ	Rao's R (df = 1;21)	p value
$m = 32$			
Stratum	**0.139**	**64.59**	**<0.00001**
Period	0.919	0.926	0.41
Stratum × period	**0.163**	**53.94**	**<0.00001**
$m = 1$			
Stratum	**0.146**	**61.15**	**<0.00001**
Period	**0.295**	**25.06**	**<0.00001**
Stratum × period	0.754	3.432	0.051

m, sample size.
Bold figures show significant effects ($p < 0.05$).

interaction term was found (Wilk's $\lambda = 0.761$, Rao's $R_{2;21} = 3.30$, $p = 0.057$).

Matrix correlations with the rare species excluded were very similar to those obtained with the full dataset. Similarity between samples degraded significantly with the time intervening between sampling nights (Mantel test; canopy: $r = -0.619 \pm 0.108$, $t = 5.66$, $p < 0.0001$; understorey: $r = -0.465 \pm 0.104$, $t = 4.49$, $p = 0.0002$). Correspondence between both strata was even higher when only the more common species were included ($r = 0.536 \pm 0.147$, $t = 3.67$, $p = 0.0002$).

Fluctuations in abundance

Overall, abundance of pyraloid moths (i.e. totals captured per night) showed very similar temporal variation in the canopy (CV$= 0.456 \pm 0.094$) and understorey (CV$= 0.440 \pm 0.103$; t-test: $t_{24} = 0.112$, $p > 0.50$). Taking values of temporally closest samples (presented in the same lines in Table 7.1) as data pairs, total nightly catch was also closely related across the two strata (individuals: Pearson's $r = 0.655$, $p = 0.0076$; species: $r = 0.647$, $p = 0.0084$). There was no obvious difference with respect to the fluctuations in relative abundance of the 10 most common species of either stratum: mean CV values (canopy: 0.692 ± 0.508; understorey: 0.959 ± 0.398) did not differ between strata (Mann–Whitney U-test: $z = 1.295$, $p > 0.19$; Fig. 7.5). Relative abundances of the 10 most common species of either stratum did not show any consistent patterns of variation according to stratum or time (repeated measures ANOVA; stratum: $F_{1;18} = 0.123$, $p > 0.72$; time: $F_{12;216} = 1.06$, $p > 0.39$; stratum/time interaction: $F_{12;216} = 0.805$, $p > 0.64$; homogeneity of variances approved with Levene's test). Three pyraloid

Table 7.5. *Results of two-way analysis of variance on the scores of dimensions 1 and 2 extracted by multidimensional scaling of normalized estimates of shared species (NESS) similarity values, partitioned according to the main factors (stratum, period, stratum × period) and scaling dimensions*

Effect	NESS $m = 1$		NESS $m = 32$	
	F(df = 1; 22)	p value	F(df = 1; 22)	p value
Stratum				
DIM1	**104.8**	**<0.00001**	**107.5**	**<0.00001**
DIM2	**12.63**	**0.002**	**5.57**	**0.028**
Period				
DIM1	1.47	0.24	**22.52**	**<0.0001**
DIM2	0.80	0.38	**39.94**	**<0.0001**
Stratum × period				
DIM1	0.46	0.50	0.99	0.33
DIM2	**111.4**	**<0.00001**	**5.00**	**0.0 36**

DIM, dimension; m, sample size.
Significant effects ($p < 0.05$) are in bold face.

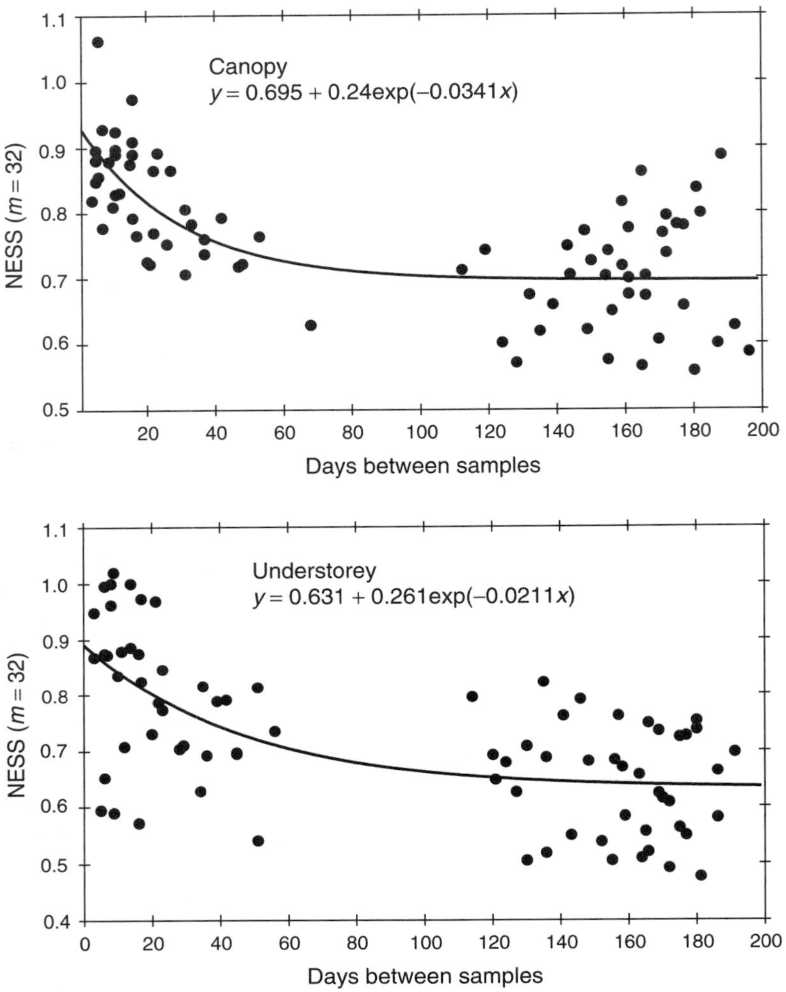

Fig. 7.4. Degradation of between-sample similarity (expressed as the normalized estimate of shared species (NESS) index for $m = 32$, where m is sample size) as a function of time intervening between two sampling events. In both forest strata, similarity degrades approximately in exponential fashion ($y = a + b\ \exp(cx)$; canopy: $a = 0.695 \pm 0.013$ (SE), $b = 0.240 \pm 0.036$, $c = -0.034 \pm 0.011$, $r = 0.699$; understorey: $a = 0.631 \pm 0.028$, $b = 0.261 \pm 0.038$, $c = -0.021 \pm 0.012$, $r = 0.638$).

species scored among the top 10 species in the canopy as well as in the understorey.

DISCUSSION

Light-trap data and biodiversity assessment

Light-trap data are frequently used in biodiversity research (Intachat & Woiwod, 1999). Yet, controversy continues as to whether reliable inferences on community structure and diversity can be obtained from such data. Moreover, the physiological mechanisms underlying attraction of moths to light sources are far from being settled (Spencer *et al.*, 1997). Three specific criticisms have been put forward. First, light-traps sample selectively rather than randomly from populations. Second, artificial light sources attract moths from a distance; consequently samples may be diluted, to an unknown degree, by specimens that do not belong to the sampled community. Third, the effective attraction radius may depend on habitat structure (e.g. density of vegetation). The first criticism is evidently true (not all moth species are equally responsive to any given light source: Butler *et al.*, 1999), yet the same applies to any other field method in animal ecology (Southwood & Henderson, 2000). For example, in knockdown samples obtained by insecticidal fogging, Lepidoptera are systematically underrepresented (Chey *et al.*, 1998). Recent work with low-power light tubes (Butler *et al.*, 1999) has shown

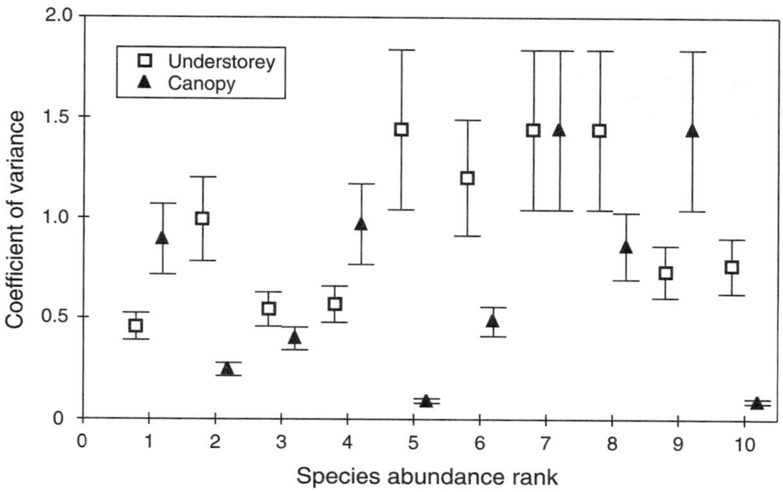

Fig. 7.5. Coefficients of variation (95% confidence limits) of relative abundance for the 10 commonest pyraloid species in canopy and understorey, arranged according to abundance rank of species (1, most abundant).

that differences between light-trap and Malaise-trap (a putatively 'passive' and thus random-sampling method) catches are far less pronounced than previously thought (Taylor & Carter, 1961). Generally, interpretation of any such dataset must keep in mind that statements are only possible for those organisms which are sensitive to the sampling method used.

The second criticism is potentially more serious. However, even with far stronger light sources (usually 125 or 250 W bulbs) than employed in our study (15 W tube), the effective sampling radius is less than 30 m (Muirhead-Thomson, 1991). In our dataset, the very fact that we observed a clear segregation of canopy and understorey samples in both sampling periods and including, as well as excluding, rare species indicates that attraction of moths from the 'wrong' habitat was obviously not a serious problem. If we had attracted substantial fractions of the canopy fauna down to the understorey light source, and vice versa, we would expect a much lower differentiation between the samples. Some cross-attraction between habitats may have occurred, but certainly not to a degree that would undermine the validity of our results.

The third criticism again could have serious implications since in habitats where catch efficiency is higher one would expect much larger sample sizes, with all the consequences for biodiversity estimates. However, in our dataset, the canopy light source (where one might have believed the effective catch radius to be larger because of the more open vegetation structure in the upper canopy) did not attract a larger number of moths than the understorey light. The well-known adverse effects of moonlight on catch efficiency of light traps (Muirhead-Thomson, 1991; Yela & Holyoak, 1997) also recurred at our sampling sites, but we did not notice any differences in this regard between canopy and understorey (C. Schulze & K. Fiedler, unpublished data). We generally avoided light trapping around full moon and visited sampling sites in random order to exclude any systematic bias of moonlight. Moreover, restriction of sampling periods to the first 2.5 hours of darkness further diminished potential effects of moonlight, as the moon was not high in the sky at that time.

Pyraloid moths as model organisms

In south-east Asia, more than 1500 pyraloid species occur (Robinson *et al.*, 1994). From Mt Kinabalu Park alone, approximately 850 species have been recorded so far (Schulze, 2000). Out of this large species pool, 575 species were observed at the canopy and understorey sites analysed here in more detail. Extrapolations from these samples (Colwell & Coddington, 1994) indicated that one should expect about 550–700 species in the forest canopy and in the understorey (Schulze, 2000). Therefore, with 405 and 399 species, respectively, neither ensemble was near being sampled exhaustively after 13 light-trapping nights.

Despite incomplete coverage, pyraloid moths proved to be an excellent target group in which to study ensemble diversity and, in particular, temporal aspects of ensemble composition. Although the numbers of individuals and species recorded per night varied conspicuously (a common phenomenon: Barlow & Woiwod, 1989; Intachat & Woiwod, 1999), nightly samples were always sufficiently large to allow for comparisons between individual collecting nights. The most important reasons for such variation in the number of species and specimens attracted by light are background illumination by the moon and weather (e.g. Yela & Holyoak, 1997; Butler *et al.*, 1999). Therefore, meaningful assessment of pyraloid diversity requires the use of measures that control for effects of sample size. In this respect, Hurlbert's rarefaction method yielded slightly more stable results than Fisher's α with the relatively small nightly sample sizes. Even when based on large cumulative samples, values of Fisher's α for pyraloids at Poring still tended to increase. For Bornean microlepidoptera (Robinson & Tuck, 1996) as well as for Malaysian Geometridae (Intachat & Holloway, 2000), asymptotic behaviour of Fisher's α has been reported with much smaller samples (see also Barlow & Woiwod, 1990). Apparently, the pyraloid fauna at Poring is so diverse that only at very large sample sizes will assessments based on Fisher's α become reliable. Values of Fisher's α for the understorey ($\alpha = 142$), canopy ($\alpha = 129$) and for the two strata combined ($\alpha = 167$) fell within the range reported for pyraloid ensembles of dipterocarp forests at low to moderate elevations elsewhere in south-east Asia (α = 91–218: Barlow & Woiwod, 1989; Robinson & Tuck, 1993; Robinson *et al.*, 1995).

Assessment of β-diversity in such rich and incompletely sampled communities calls for the application of similarity measures that allow control for sampling effects. Here, the NESS index performed well (Wolda, 1996; Willott, 1999) and produced a more effective ordination of Bornean pyraloid communities relative to ecological gradients than the binary Sørensen index (Schulze, 2000). In combination with MDS, this measure allowed a perfect partitioning of between-sample differences into a spatial (stratification) and a temporal (time of sampling) component.

Temporal dynamics of moth communities

In our Bornean samples, the α-diversity of pyraloids fluctuated moderately but irregularly, and this occurred to the same extent in both the forest canopy and understorey. Abundances of the more common species as well as total catch also showed no consistent temporal trends. In addition, a matrix correlation test indicated that temporal changes in ensemble composition were highly correlated across the two forest strata. These results suggest that temporal dynamics were similar in pyraloid ensembles from the canopy and understorey.

MDS ordinations effectively separated samples from the two collecting periods, showing that pyraloid ensembles from Borneo did change over time. Samples taken at the same site tended to be increasingly dissimilar as the period between times of sampling increased. Degradation of between-sample similarity occurred at roughly the same rate in both forest strata, with highest degradation occurring in the first few days or weeks between samples. With increasing time lag between samples, however, similarity tended to converge to an asymptotic value (NESS index values in the range 0.6–0.7) and never fell below 0.45. Hence, a substantial fraction of the pyraloid ensembles was sufficiently stable to indicate some minimum level of similarity even between samples taken as much as 200 days apart. As a consequence, despite temporal fluctuations in community composition, similarity remained higher between samples taken from the same stratum at different times than in samples taken in the same period but in different strata.

The reasons for the observed temporal dynamics of pyraloid ensembles remain speculative. Larvae of almost all pyraloid moths are herbivores (Munroe & Solis, 1999). One would expect their abundance and diversity to be regulated by the availability of suitable food resources. Many pyraloid larvae do not feed on mature foliage but rather focus on nutrient-rich young leaves, shoots, flowers or fruits. This is also the reason for the status of many south-east Asian pyraloid moths as economic pests (Kalshoven, 1981; Common, 1990; Khoo *et al.*, 1991). Since, in Malesian rainforests, leaf flushing, flowering and fruiting of many plants seldom occurs continuously (Medway, 1972; Appanah, 1985; Sakai *et al.*, 1999b; Wich & van Schaik, 2000; but see Putz, 1979; Reich, 1995), it is plausible that the phenology of host plants at least partly accounts for the temporal dynamics of pyraloid ensembles (for Geometroidea see Intachat *et al.*, 2001). However, nothing is known at present about the larval biology of almost any of pyraloid species covered by our samples so this question cannot be addressed explicitly.

Even in perhumid rainforests, where a true dry season does not occur regularly, pronounced fluctuations in the abundance of individual species or entire guilds are frequent (e.g. Kato *et al.*, 1995; Novotny & Basset, 1998; Intachat & Holloway, 2000; Intachat *et al.*, 2001), although consistent temporal trends are not always apparent (Orr & Häuser, 1996; Basset, 2000). At least some south-east Asian moth ensembles appear to be largely aseasonal. Barlow & Woiwod (1989) ran light traps for a year in Peninsular Malaysia and did not observe any regular cyclic patterns in the fluctuation of abundance, number of recorded species and diversity (measured as Fisher's α). Similarly, little evidence of seasonality was found in moth samples from Sulawesi (Barlow & Woiwod, 1990). Despite pronounced temporal fluctuations in total catch, geometrid diversity (Fisher's α) showed no clear cyclic trends either (Intachat & Holloway, 2000). Rather, geometroid diversity responded to weather conditions (Intachat *et al.*, 2001). These observations together with the data reported here suggest that α-diversity of moth ensembles from perhumid south-east Asian forests remains rather stable throughout the year, whereas abundances as well as abundance relationships may change profoundly, depending on the site and taxa under consideration.

In all published studies of which we are aware where temporal trends in tropical moth faunas have been investigated, the reported data concern either single species or refer only to numbers of species, individuals or some kind of α-diversity statistics. No quantitative data on species turnover and/or fluctuations in abundance among entire ensembles are given. Our study seems to be the first to confirm that assemblage composition of moths in tropical rainforests may be highly dynamic even if α-diversity remains stable and high. It will be rewarding to see whether similar time series for other taxa of moths or for Pyraloidea in other forest types show assemblage dynamics of the same magnitude.

How different are canopy and understorey moth communities?

Since Erwin's (1982) remarkable results on beetle diversity in Panamanian forest canopies, the crowns of tropical forests have gained the reputation of having the greater part of terrestrial biodiversity. Although a hot debate over Erwin's extrapolations has ensued (e.g. Ødegaard 2000c; Basset, 2001b), there is general consensus that tropical forest canopies are extremely rich in species. As an implicit corollary, canopy faunas are frequently assumed to be richer or more diverse than corresponding habitats in the forest understorey. The high structural complexity and microclimatic diversity of forest canopies (Parker, 1995) underpins this notion of canopies as habitats that provide manifold niches and are subject to intense fluctuations. Therefore, both deterministic and stochastic models of biodiversity regulation (Huston, 1994; Rosenzweig, 1995) predict diversity will be highest in the canopy stratum, at least with regard to those animals that have direct trophic links with the vegetation (note that the highest photosynthetically active plant biomass is found in the canopy).

Our data on pyraloid ensembles give insight into possible differences between canopy and understorey layers with regard to α-diversity, species composition and temporal dynamics. Pyraloid ensembles at Poring were very similar in temporal dynamics in the two strata. We failed to observe consistent stratum differences in variation of total catch, rarefied species numbers or relative abundances of the commonest species. In contrast, similarity matrices for nightly catches were highly correlated between both strata.

As far as α-diversity was concerned, pyraloid ensembles in the canopy and understorey were almost identical at our study site. Analysis of the entire catch by rarefaction as well as using Fisher's α even showed that the understorey ensemble was significantly more diverse than the canopy ensemble. The same was also true when comparing a neighbouring understorey site from a slightly disturbed forest with the canopy site (Schulze, 2000). Similar results have been obtained for Malaysian Geometridae, where understorey samples are consistently more diverse than those drawn from the canopy (Intachat & Holloway, 2000; Beck *et al.*, 2002). Willott (1999) reported that at Danum Valley (Borneo) the diversity of the entire 'macro-moth' species assemblage in the understorey was equal to or even slightly higher (depending on the measure used) than in the canopy. Hence, canopy ensembles of Lepidoptera are by no means universally more diverse than the corresponding understorey ensembles (see Schulze *et al.*, 2001). Although nectarivorous groups (for example, most Sphingidae, many butterflies) are distinctly more diverse in the canopy, others (for example, rotten-fruit-feeding Nymphalidae, the nonfeeding fraction of Arctiinae) have similar or even higher diversity in dark forest undergrowth.

Despite similarities in α-diversity and temporal dynamics, we did observe a clear distinction between the canopy and understorey fauna. This segregation was not just a product of the many rare species in our samples, since ordination results remained robust when species represented by five or less individuals were excluded. Hence, among the Pyraloidea there is pronounced vertical stratification (see Willott, 1999, for all 'macro-moths'; Schulze *et al.*, 2001; but see Intachat & Holloway, 2000; Beck *et al.*, 2002 for Malaysian Geometridae). Few pyraloid species qualified as true canopy specialists. Of the 405 species recorded in our canopy samples, 285 (70.4%) appeared also in some samples taken at ground level. When rare species (as defined above) are omitted, the proportion of exclusive canopy species further diminishes to less than 10% (Schulze, 2000). Thus, stratification in our samples was not primarily caused by the existence of stratum specialists but rather by variation in stratum-specific patterns in abundances. This is reflected in a comparison of recorded abundances per species per stratum. Among the 59 species represented by a total of 20 or more individuals in our samples, the null hypothesis of equal abundances in canopy and understorey was upheld for only 13 species. Thirty-three species were significantly more abundant in the canopy, 13 more abundant in the understorey: one of the common species with uneven stratification of abundances appeared to be exclusive to the canopy.

The mechanisms responsible for stratification of Bornean ensembles of Pyraloidea remain speculative. In contrast to many tropical forest butterflies, where vertical movements are well documented (e.g. feeding, mating and egg laying may occur in different strata), such behavioural data are nonexistent for most moths. Given their relatively small body size, one might expect pyraloid moths not to fly long distances when searching for mates or oviposition substrates. However, we frequently sampled small-bodied Acentropinae (with aquatic larvae) in the canopy, suggesting that pyraloid moths are mobile enough to fly dozens to hundreds of metres per night (C. Schulze & K. Fiedler, unpublished data). Nevertheless the very fact that we did observe substantial stratification of pyraloids (despite their mobility) suggests that habitat or resource preferences exist to account for this spatial pattern.

Whether this stratification is more related to adult feeding (almost all Pyraloidea have a well-developed proboscis as adults (Common, 1990; Munroe & Solis, 1999), suggesting that they do visit nectar sources) than to larval host–plant relationships remains to be elucidated. In general, given the contrasting resource requirements of holometabolous Lepidoptera during their larval and adult stages, stratification of ensembles of adult Lepidoptera may reflect needs of the adult butterflies or moths, as well as the availability of larval resources (Schulze *et al.*, 2001). In addition, predation pressure on adult Lepidoptera by aerial-hawking birds (Schulze *et al.*, 2001) and bats (Rydell, 1995; Kalko, 1998) is higher in the canopy than the understorey. The canopy, in spite of its higher foliage biomass, presents strongly fluctuating microclimatic conditions, which may not present a favourable environment for desiccation-prone soft-bodied larvae. The high levels of natural enemies such as birds and ants (Davidson, 1997) also diminishes the suitability of the canopy as moth habitat. In line with these expectations, caterpillars are consistently rare in south-east Asian canopy samples obtained by insecticidal fogging (Floren & Linsenmair, 1997b, 1999; Chey *et al.*, 1998). Hence, taxon-specific patterns of resource use combined with abiotic and biotic constraints could well explain why, among south-east Asian Lepidoptera (one of the most speciose taxa of insect herbivores), maximum diversity is not necessarily reached in the canopy of the forest.

CONCLUSIONS

Pyraloid moth assemblages are very diverse in tropical rainforests. Pyraloidea are unique among south-east Asian moths in that sufficiently large samples can be collected during each individual light-trapping night to warrant quantitative evaluation. Therefore, pyraloid ensembles provide insight into temporal aspects of diversity and community composition at a level of resolution that cannot be achieved with other taxa, such as Geometridae. Our samples clearly show that canopy and understorey ensembles are distinct. The understorey fauna was, however, slightly more diverse than the canopy ensemble. This is in line with similar findings from south-east Asia on Geometridae and some other macro-moths. Pyraloid assemblage dynamics were similar in both strata. How far these spatio-temporal patterns can be generalized to other herbivore taxa or forest types needs to be tested with similar time series of data.

ACKNOWLEDGEMENTS

We thank the directors of Sabah Parks, Datuk Lamri Ali and Francis Liew, for granting permission to work at Kinabalu Park, and the very helpful staff for their hospitality. In particular we thank Prof. Maryati Mohamed (Universiti Malaysia Sabah, Kota Kinabalu) for her great hospitality and support, and the Economic Planning Unit at the Prime Minister's Department of Malaysia for granting a research permit. We thank Jutta Klein, Edda Roddewig and Eva and Roswitha Mühlenberg for their assistance with data collection and setting moths, and Stefan Messner for providing the computer program to calculate NESS index values. Dr Wolfgang Speidel (Zoological Museum A. Koenig, Bonn), Dr Matthias Nuß (Zoological Museum, Dresden), and Dr Christoph L. Häuser (State Museum of Natural History, Stuttgart) assisted with moth identification, morphospecies delimitations and provision of taxonomic literature. We are particularly thankful to Dr Bert Orr and Dr Yves Basset for their critical comments on an earlier manuscript draft. The study was supported by grants from the Deutsche Forschungsgemeinschaft (Fi 547/4-1 and 4-2).

8

Canopy foliage structure and flight density of butterflies and birds in Sarawak

Fumito Koike and Teruyoshi Nagamitsu

ABSTRACT

We examined the effects of canopy foliage structure on the flight density distribution of butterflies (Papilionoidea) and birds (Aves) in a vertical section of a lowland mixed dipterocarp forest in Lambir Hills National Park, Sarawak. A rope system installed between a tree tower and an emergent tree gave a vertical section 80 m wide by 40 m high. The distribution of foliage density, a measure of canopy structure, was quantified by canopy tomography using hemispherical photographs taken with the help of the rope system. Because trapping of such flying animals was difficult, flight density was assessed by direct observation. The rope system was used as a series of vertical quadrats that allowed us to determine flight position. Canopy structure consisted of a stratified lower layer with dense continuous foliage below 20 m height and an unstratified upper layer above this. Butterfly flights were not randomly distributed and were most frequently recorded in the lower part (25 to 30 m height) of the gaps in the upper layer, where no foliage existed above the gaps but was present at lower levels and in the surroundings. Butterfly flights were never recorded far above the canopy and rarely occurred through dense foliage. Most (69%) bird flights occurred above 40 m height; more than 72% of above canopy flights were by swallows (Apodidae). The high bird activity above the canopy may depress butterfly flights there through predation pressure. The study demonstrates that the rope system is effective for detecting animal distributions in relation to the vertical and horizontal structure of canopy foliage in tropical rainforests.

INTRODUCTION

The structure of canopy foliage is an important factor affecting the spatial distribution of animals in tropical rainforests (Sutton *et al.*, 1983b). Foliage intercepts strong light and wind, facilitating animal activity (Ross, 1981; Raupach, 1989), and impedes flight and vision. The canopy provides foods for various animals (Toda, 1977; Yumoto, 1987; Lowman, 1995; Momose *et al.*, 1998) through flowering, fruiting and new leaf production, and it offers shelter from predators (Chai & Srygley, 1990; Srygley & Chai, 1990; Ohsaki, 1995). Consequently, canopy foliage structure is crucial for analysing patterns of animal distribution in the forest (Koike *et al.*, 1998).

Tropical rainforests are characterized by complex canopy structure (Richards, 1952; Whitmore, 1984; Popma *et al.*, 1988). Sparse distribution of emergent trees results in a canopy structure with an unstratified upper layer and a dense lower layer (Ashton & Hall, 1992; Koike & Syahbuddin, 1993). Such sparse emergent trees cause horizontal heterogeneity of foliage density in the upper layer, which in turn causes horizontal variation in animal distribution in the forest canopy (Koike *et al.*, 1998). The quantitative measurement of foliage canopy structure and its relationship to the activity of arboreal animals has rarely been examined, with the exception of our previous study in south-east Asia (Koike *et al.*, 1998) and a study of drosophilid fly distribution in a cool-temperate forest by Toda (1992). Toda (1992) studied the three-dimensional distribution of drosophilid fly assemblages in secondary forest in an artificial landscape and found canopy layer species were also present at a low height above canopy gaps and at the forest edge. Several grassland species were also present at the forest edge facing grassland.

Traps, insecticide fogging and branch clipping techniques have been used to study insect fauna in canopies (reviewed by Basset *et al.*, 1997b). However, it is difficult to use traps for larger animals such as birds and butterflies. Mist nets may be set for collecting birds but are unlikely to be practicable in canopy studies owing to the denseness of foliage and branches (Stokes *et al.*, 2000). Bait traps specific to some butterflies have been used (Daily & Ehrlich, 1995; DeVries *et al.*, 1997), and an ultrasonic sensor has been used to determine the distribution of bats in flight (Crome & Richards, 1988). However, these methods are specific to the target animals, and it is difficult to catch a wide range of animals by such single methods. Direct observation is effective for assessing the flight behaviour of various large animals. This entails recording the precise spatial position of the animals. No suitable method of determining spatial position in forest canopies has been proposed yet and, accordingly, quantitative observations are scarce.

In this study, canopy structure is described in terms of the distribution of foliage density within a vertical section of a lowland mixed dipterocarp forest. A rope system was used to define vertical quadrats to assess the flight positions of butterflies (Papilionoidea) and birds (Aves) by direct observation. To examine the impact of horizontal and vertical canopy structure on the spatial distribution of these animals, we examined the relationship between the distributions of foliage density and flight density of these animals.

METHODS

The study was conducted in the Canopy Biology Plot in Lambir Hills National Park, Sarawak, Malaysia (4° 20′ N, 113° 50′ E, altitude 150–250 m). The climate, topography and flora of the study site have been described by Inoue *et al.* (1994) and Nagamasu *et al.* (1994). The plot is situated in a lowland mixed dipterocarp forest on clay soil. Common tree families in the emergent layer are Dipterocarpaceae (*Shorea*, *Dipterocarpus*, *Dryobalanops*), Leguminosae (*Koompasia*, *Sindora*, *Dialium*) and Sapotaceae (*Madhuca*) (Nagamasu *et al.*, 1994). Momose *et al.* (1998) studied plant–pollinator relationships in the forest. Forest floor shrubs of *Ixora* (Rubiaceae) and *Clerodendron* (Verbenaceae) and lianas of *Bauhinia* (Leguminosae) were pollinated exclusively by butterflies. However, butterflies visit various flower species even though they do not contribute to pollination, and the spatial distribution of flower resources for butterflies is not known. Kato *et al.* (1995) documented the vertical distribution of nocturnal insects captured by ultraviolet light traps in the forest. Individuals of Hymenoptera (including fig wasps), Coleoptera, Hemiptera, Diptera and Lepidoptera were abundant. Body length distribution had a peak at 0.8–2.2 mm.

The study plot includes two tree towers and nine walkways of 298.3 m total length (Inoue *et al.*, 1994). Although cranes give observers three-dimensional access to a cylindrical space of canopy (Schowalter & Ganio, 1998), a rope system, allowing two-dimensional observation, is cheaper and easier to install. A rope system 80 m wide was set up between a tree tower and an emergent tree at a height of 40 m (Fig. 8.1). Eight vertical ropes were installed at horizontal intervals of 10 m. This system remains available for use in the Canopy Biology Plot of Lambir Hills National Park and details are presented in Koike (1994).

Hemispherical photographs were taken upward from the rope system at intervals of 10 m horizontally and 2.5 m vertically (Fig. 8.1). A camera with a hemispherical lens was maintained pointing upwards using a giro-horizon. Height was adjusted using the vertical lines of the rope system, and the shutter was released electronically. Relative light intensity was determined from the hemispherical photographs using the method of Anderson (1964). The two-dimensional distribution of foliage density on the vertical canopy section was determined by canopy tomography from the hemispherical photographs. The canopy section was divided into subareas 9.5 m wide by 7.2 m high. The foliage density of each subarea was checked against the image data from the hemispherical photographs using an overlay-tracing model employing a least-squares fitting method. The principle of this method is the same as is used in medical computed tomography. The foliage density is the expected number of leaves intercepted by a 1 m column (unit m^{-1}). Koike (1985) described the details and limitations of the method.

The rope system created vertical quadrats that we used to observe the flight behaviour of animals. Conspicuous tags were fixed every 5 m on each vertical line to form a grid of rectangles 10 m horizontally by 5 m vertically. The horizontal borders of the grid were parallel to the slope of the ground. The flights of butterflies and birds were observed during the day (07:30 to 14:30) at the end of the dry season in 1993 from a walkway 30 m above the ground and 20 m from the canopy section. During the observation period, there was no prominent flowering of canopy trees. Two people observed with the

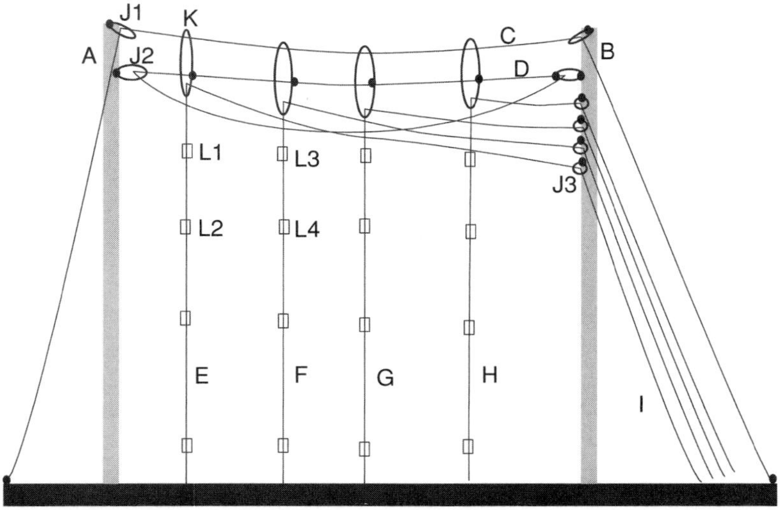

Fig. 8.1. Diagram of the rope system. **A** and **B** are free-standing structures. **A** is an emergent tree and **B** is a tree tower in the system of Lambir Hills National Park. **C** is the main rope supporting the weight, and **D** is a ring rope to adjust the horizontal positions. **J1**, **J2** and **J3** are metal rings (carabiners). **K** is a curtain ring. **E–H** are vertical ropes used to pull up the camera and to fix the markers. These lines are adjusted from position **I**. Small filled circles show the points at which ropes and carabiners are fixed. The area surrounded by markers **L1–L4** is one of the vertical quadrats. To set up the apparatus, first we climbed up **B** and shot a single leader line to **A** by cross-bow; we then fixed the main rope and ring rope to the leader line. Next we climbed up **A**, pulled across the leader line and set up **J1** and **J2**. Finally, we climbed up **B** again and installed the main rope and ring rope. The curtain rings **K** were fixed to the ring rope. The horizontal positions of the vertical ropes were adjusted by the ring rope, and vertical ropes were held down by weights. The ring rope was fixed when the system was in use.

naked eye, each covering a 40 m width of canopy. Some portions of the canopy section below 20 m height could not be seen because of the dense foliage there. Observation periods totalled 6.5 hours. Because precise identification was difficult in this method, butterflies and birds were identified at least to class, and sometimes to family.

The flight density per square metre of canopy section per hour was calculated for each quadrat. The hypothesis that flight density is even throughout the canopy was evaluated using a chi-square test. The region above the canopy was excluded because its height is not limited and therefore, flight density could not be calculated. Because some grids had no observations, the Monte-Carlo chi-square test was used instead of the more usual parametric version.

Environmental factors that may determine flight density were analysed by linear regression using a stepwise variable selection procedure. The dependent variable was flight density; the independent variables were relative light intensity, foliage density and height above the ground.

RESULTS

The canopy was heterogeneous vertically and horizontally (Fig. 8.2). There was a high foliage density layer below 20 m and scattered foliage of tall emergent crowns from 20 to 60 m.

In all, we counted 74 butterfly and 29 bird flights. Butterfly taxa recorded were Lycaenidae, Papilionidae, Pieridae and Nymphalidae (including Danainae). The distribution of butterfly flights was uneven in the canopy section ($p < 0.001$; Fig. 8.3). Most flights of butterflies occurred through the 'valleys' of the canopy surface at 25 to 30 m height, where there was no overarching foliage, foliage density being high only in the lower layer and surrounding parts of the canopy. No flights were observed above the canopy and only a few flights occurred through dense foliage. Butterflies preferred open spaces and lower heights in the canopy (Table 8.1).

Bird flights above the canopy represented 69% of those observed (Fig. 8.4). More than 72% of bird flights above the canopy were of swifts (Apodidae). The

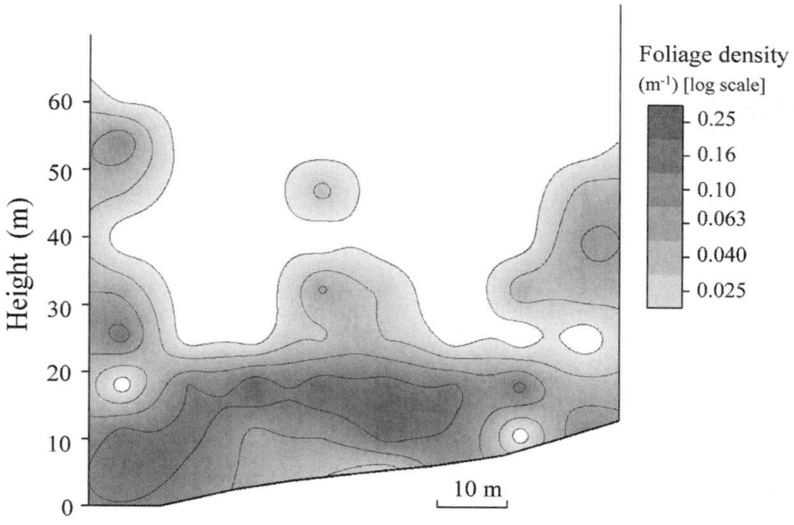

Fig. 8.2. Distribution of foliage density in a vertical canopy section. Data are presented at a higher spatial resolution (9.5 m horizontal × 7.2 m vertical) than previously (Koike, 1996).

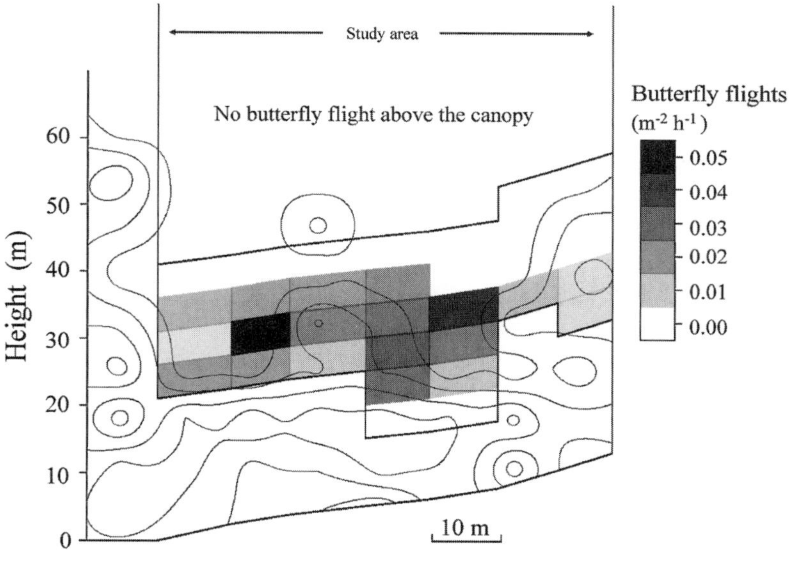

Fig. 8.3. Flight density of butterflies in the canopy section. Flights were counted in the area bounded by thick lines and above the canopy. The background contour lines indicate foliage density, as in Fig. 8.2.

distribution of bird flights was not random ($p < 0.05$), as some duplication of observations occurred for some locations. Preferences were difficult to determine (Table 8.1).

DISCUSSION

Koike *et al.* (1990) and Koike and Hotta (1996) compared two-dimensional distributions of foliage density among climax forest types ranging from an alpine tree limit to a tropical rainforest in humid eastern Asia and identified characteristics of canopy structure in each. Tropical rainforests are characterized by a dense lower layer and an unstratified upper layer (Koike & Syahbuddin, 1993; Koike *et al.*, 1998), whereas in subtropical and warm-temperate forests, upper layers are dense and continuous and lower layers sparse (Koike *et al.*, 1990; Koike & Syahbuddin, 1993; Koike & Hotta, 1996). Both upper and lower layers are stratified in cool-temperate and subalpine forests. The forest in the Canopy Biology Plot has

Table 8.1. *Environmental factors determining the flight density of butterflies and birds*

Dependent variable	Model with one independent variable[a]	Model with two independent variables[a]
Butterfly flight density	−Height above ground*	−Height above ground** −Foliage density*
Bird flight density	NS	NS

[a]The independent variables are relative light intensity at the grid position, foliage density at the grid position, and grid height above the ground. In results of stepwise regression, '−' indicates a negative contribution.
$^*p < 0.05$; $^{**}p < 0.01$; NS: not significant.

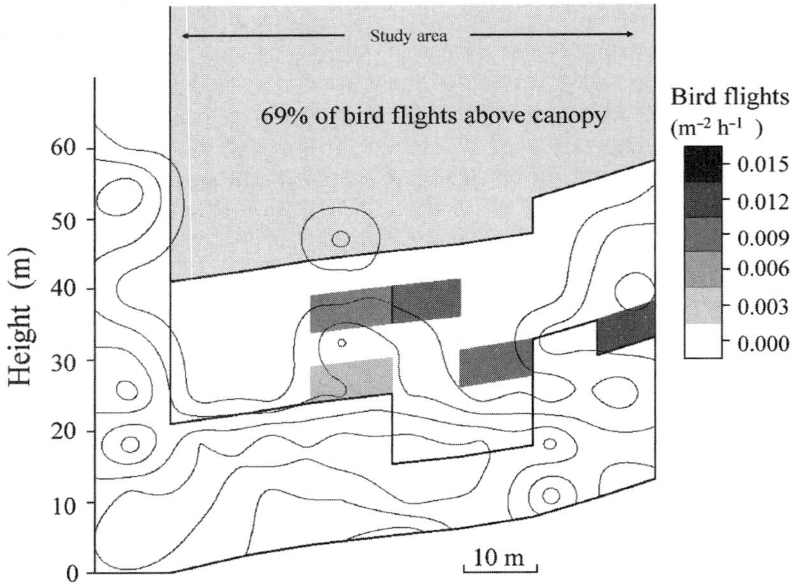

Fig. 8.4. Flight density of birds in the canopy section. Flights were counted in the area bounded by thick lines, and above the canopy. The background contour lines indicate foliage density, as in Fig. 8.2.

the typical canopy structure of the tropical rainforest. Such structure has been suggested also by tree size distribution and profile diagrams in south-east Asia and Mexico (Richards, 1952; Popma *et al.*, 1988; Ashton & Hall, 1992). The dense lower layer causes vertical heterogeneity and the unstratified upper layer causes horizontal heterogeneity. Consequently, the canopy structure of tropical rainforests is heterogeneous in both vertical and horizontal dimensions.

In horizontally even forests, height relative to the upper forest layer expresses spatial position in the canopy structure. Insect flight distribution has been investigated through the use of vertically arranged traps and has been discussed in terms of differences between canopy and understorey (Toda, 1977). Such a dichotomy between the two strata may also be relevant to biological interactions such as pollination (Yumoto, 1987).

Insect flight distribution in most Asian tropical rainforests should be considered not only in the vertical dimension but also in the horizontal dimension, because the upper layers of the forests are not stratified. The distribution of insects along single vertical axes in tropical rainforests has been studied (e.g. Kato *et al.*, 1995; Compton *et al.*, 2000), although the results are not sufficient to show the overall spatial variation in insect

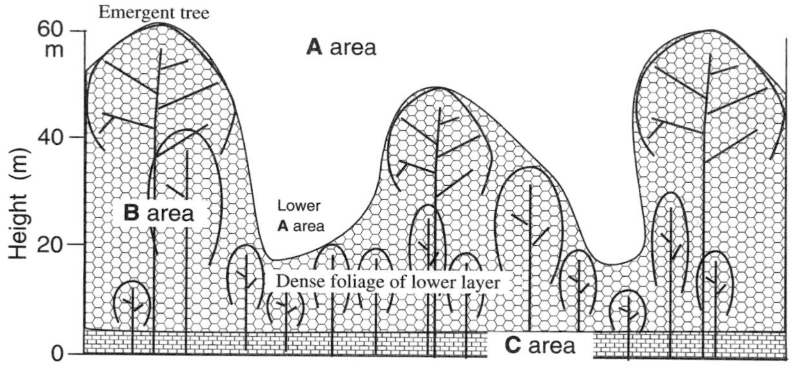

Fig. 8.5. Diagram of canopy structure and classification of parts of the canopy by insect assemblage types in a hypothetical tropical rainforest in south-east Asia. The positions where insect assemblage types occur (A to C) follow Koike *et al.* (1998).

distributions. Koike *et al.* (1998) investigated the spatial distributions of foliage density and insect assemblages in a vertical section of a tropical rainforest and showed that tropical rainforest canopy was divided into three major parts on the basis of the associated insect assemblages: upper open space (**A**), where fig wasps and thrips were abundant; inside the canopy (**B**), with intermediate composition; and forest floor (**C**), where Coleoptera were abundant (Fig. 8.5). The **A** area penetrates down into the forest coincident with canopy gaps. The area examined in the present study included the **A** area and the upper parts of the **B** area. Butterfly flights were frequent in the lower part of the **A** area, especially in tree gaps, but were rare in the **B** area and the upper part of the **A** area (Fig. 8.3).

A similar pattern of vertical distribution of lepidopteran flights was observed in nocturnal hawk moths (Sphingidae) trapped by ultraviolet light on another tree tower about 160 m from the present study site: the abundance of these moths peaked at 35 m height and decreased at both above 45 m and below 25 m (Kato *et al.*, 1995). The sphingids also seemed to show a height preference for the layer just above the dense foliage of the lower layer. In contrast, the abundance of other nocturnal lepidopterans tended to be higher in the understorey (Kato *et al.*, 1995).

Intachat and Holloway (2000), for peninsular Malaysia, reported that the abundance of light-trapped moths of the superfamily Geometroidea tended to be lower at 30 m height than at 1 and 15 m. DeVries *et al.* (1997), using bait traps in Ecuador, showed that the abundance of fruit-feeding nymphalid butterflies was lower in the canopy than in the understorey. Although quantitative information about foliage distribution is lacking, these lines of evidence suggest that some butterflies and moths fly at lower levels. The forest floor space (**C** area) may be another active area, in addition to the space just above the dense foliage of the lower layer (lower **A** area; Fig. 8.5).

Environmental conditions and predation pressure seem to be responsible for the lower frequency of flights in the open space over the upper canopy. As distance above the foliage surface increases, daytime dryness and wind velocity increase, possibly rendering flight difficult even for large lepidopterans (Taylor, 1974; Boiteau *et al.*, 1999). Most fast-flying large lepidopterans, except for species having defensive chemicals, are palatable and are frequently attacked by birds (Srygley & Chai, 1990; Chai & Srygley, 1990; Ohsaki, 1995). Bats also eat sphingid moths at night (Jones, 1990; Rydell & Lancaster, 2000). The high swift activity observed during daytime in this study showed that such birds tend to prefer open space over the canopy (Fig. 8.4). These microclimate and predation factors may prevent butterflies from flying in the open space over the canopy and may lead to their preference for the lower parts of the canopy, as in 'valleys' surrounded by canopy foliage. In contrast to the large active fliers, which are attacked by vertebrates, small passive fliers free from vertebrate predators, such as fig wasps and thrips, travel over the canopy (Koike *et al.*, 1998; Compton *et al.*, 2000).

ACKNOWLEDGEMENTS

We thank Yves Basset for his helpful comments and suggestions.

9

Stratification of the spider fauna in a Tanzanian forest

Line L. Sørensen

ABSTRACT

The vertical distribution of arthropods in the vegetation of tropical forests in Africa has rarely been examined. In the present study, the stratification of true spiders (Araneae) was investigated in a montane forest in Tanzania. Spiders were chosen because they are generally not host specific. Their distribution relies mainly on the structure of the habitat and on prey availability (only a few species are prey specific). Sampling was carried out in a 1 ha plot in a mature and relatively homogeneous forest at the end of the rainy season. Three strata were recognized and sampled: the forest floor (the forest litter and vegetation below 50 cm), the understorey (vegetation between 50 cm and 3 m) and the canopy (vegetation taller than 3 m). The spider assemblage of the forest floor was most diverse (Simpson's index), although the canopy and the understorey were more species rich. Spider composition of the understorey was more similar to that of the canopy than to that of the forest floor. Of the 175 species collected in total, 27 species were confined to the canopy (15%), whereas 32 species (18%) were restricted to the forest floor. Even given that different sampling methods do not sample all taxa with equal efficiency, specialization of spiders from certain families within a particular strata was evident. This was even true for species that are regarded as being relatively mobile, like linyphiids. The present study indicates that the canopy should be included if a reasonably complete inventory of spiders is to be obtained.

INTRODUCTION

Tropical rainforests are reputed to support most of the world's biodiversity (Wilson, 1988a). One possible reason for this high diversity may be the richness of vertically distributed microhabitats in these forests, although it is disputed which strata may be the most important in this regard. Based on Panamanian beetle data, Erwin (1982) assumed that as many as 66% of the beetles were confined to the forest canopy, 20% of which were host-specific species. Stork (1988), by comparison, estimated that no more than 20% of the beetle species within his plot on Borneo were unique to the canopy. Similarly, Brühl *et al.* (1998) in their study on Borneo found that 25% of the ants were unique to the canopy, and Hammond (1990) found only 10% of beetles to be unique to the canopy in Sulawesi. This lack of consensus, which precludes a precise estimate of global arthropod species richness (1–80 million species), indicates the need for studies of the distribution of taxa in stratified forests and of differences among forest types. The resulting understanding of stratification will be important for management and conservation of the arthropod diversity of tropical forests.

Only a few tropical forest surveys have assessed the relative importance of the canopy and ground strata for the distribution of species within a particular area. These studies have focussed on four groups of invertebrates:

- highly mobile communities of insects (e.g. Sutton & Hudson, 1980; Sutton, 1983; Sutton *et al.*, 1983a; Toda, 1992; DeVries *et al.*, 1997; Intachat & Holloway, 2000) sampled by various kinds of trap, attractive or nonattractive, and targeting specific taxa: baited traps have been used for beetles, butterflies and *Drosophila* (e.g. DeVries *et al.*, 1997) and unbaited traps, such as light traps and sticky traps (e.g. Sutton, 1983; Sutton *et al.*, 1983a; Intachat & Holloway, 2000), for these and other taxa
- herbivorous insects in the upper canopy layer (e.g. Basset *et al.*, 1992)

- all arthropods (e.g. Stork 1988; Hammond, 1990; Hammond *et al.*, 1997): sampled using techniques such as flight-interception traps, litter samples, pitfall traps and yellow pan traps (Hammond, 1990), Malaise traps (Hammond, 1990; Hammond *et al.*, 1997) and canopy fogging (Stork, 1988; Hammond, 1990)
- ground-living and canopy-dwelling arthropods (e.g. arthropods associated with ground and arboreal 'soil' in the canopy) (Longino & Nadkarni, 1990; Nadkarni & Longino, 1990; Paoletti *et al.*, 1991; Brühl *et al.*, 1998; Rodgers & Kitching 1998; Walter *et al.*, 1998).

Despite differences in sampling methodology and analytical approaches, the studies above show that faunal stratification occurs in tropical forests and that there is a gradual turnover of faunal composition from the ground to the canopy. Further, some studies reported a strong shift in faunal composition (e.g. Brühl *et al.*, 1998). There is, however, no consensus as to the relative importance of the canopy for arthropods. There is a weak tendency for the canopy, broadly defined, to be richer in arthropod species, whereas the abundance of individuals is generally higher at ground level. Exceptions are the studies of butterflies by DeVries *et al.* (1997) and by Intachat and Holloway (2000), of macro- and microarthropods by Longino and Nadkarni (1990) and of ants by Brühl *et al.* (1998).

The present study focusses on spiders, a cosmopolitan group, locally common and diverse. The group is not host specific, their distribution depending rather on the physical structure of the environment and on the availability of prey (Greenstone, 1982; Uetz, 1991; Halaj *et al.*, 2000).

The objectives of the present study were to determine how the richness and the composition of the spider fauna change vertically between the forest floor and the canopy within 1 ha of montane forest in Tanzania. The study aimed to answer the following questions:

1. Are spider assemblages different among different microhabitats and strata surveyed?
2. Does spider abundance differ among strata?
3. Is spider species richness in the canopy different from that in the lower strata of the forest (ground level and understorey)?

METHODS

Study area

Spiders were collected in a montane forest of the Uzungwa Scarp Forest Reserve, Uzungwa Mountains, Tanzania (campsite 08°′ 22′05″ S, 35°′ 58′ 42″ E) during a 2-week period in May 1997 at the end of the rainy season. Sampling was performed on the forest floor, in the understorey and in the canopy layer within a 1 ha plot on a ridge (1800 to 1900 m above sea level) next to the campsite. The Uzungwa Mountains are part of the Eastern Arc Mountains, which are known for their high degree of floral and faunal endemism (Lovett & Wasser, 1993; Fjeldså, 1999; see these sources for full biological descriptions of the Uzungwa Mountains). The forest is primary and mature mixed forest, in which the common trees are *Allanblackia* spp., *Albizia gummifera* (Gmelin) Smith, *Parinari excelsa* Sabine, *Agauria salicifolia* (Lam.) Hook and *Aphloia theifomis* (Vahl) Benett. In the sampling area, the forest was undisturbed by humans and was homogeneous. The understorey was fairly open with some *Tabernaemontana* sp., and the canopy did not exceed 30 m in height. Mosses and lichens were common on stems and branches in the canopy. The structure of the forest appeared rather similar to a mature European beech forest, with emergent trees of rather similar size and very few smaller trees (these smaller trees were avoided in the canopy sampling). The rainfall is approximately 2000 mm per year.

Sampling methods

Spiders were collected by a team of 10 people within an 11-day period. Two collectors mainly sampled from the canopy while the others collected from the lower strata of the forest. The team was a mix of experienced and less-experienced collectors (three professional arachnologists, six students and one parataxonomist). The forest was divided into three strata: the *forest floor* (including ground, litter, low shrubs and herbs below 0.5 m); a higher herb, bush and shrub layer from 0.5 m to 3 m (*understorey*); and an upper, closed *canopy* stratum (including all foliage above 3 m, which was in practice almost exclusively above 15 m).

It was not feasible to access the canopy using cranes or other technical devices. The canopy was, therefore, sampled using insecticidal knockdown sampling (canopy fogging) using a PulsFog^{TM} K-10 Standard Thermal fogger (PulsFog, Germany) releasing a solution of 0.8% natural pyrethrum diluted in water. Sampling took place early in the morning or late in the afternoon, when there was no wind. The sampling targeted vegetation higher than that which could be reached by hand, beating or sweep net. There were few small trees and the sampling was, therefore, focussed on the closed canopy (and the tree trunks). It was not possible to ensure that the uppermost canopy was well sampled, because both rising fog and falling arthropods may have become trapped by the lower portion of the canopy. On 4 consecutive days, four areas of canopy were sampled within the 1 ha plot. Each sample consisted of 90 subsamples of 1 m^2. The insecticide was applied from the ground and each application (2 litres) took 30 minutes. A 2-hour drop-time was allowed after the application of the insecticide.

Various semiquantitative and quantitative sampling methods (Coddington *et al.*, 1991, 1996) were used to sample the lower strata: forest floor and understorey. The forest floor was sampled by searching for spiders living in 'cryptic' habitats (litter, logs and mosses) and manual collecting where spiders were collected from the ground and vegetation up to knee-level ('ground'). Because the herb layer was relatively thin in the study area it was given less attention than other habitats. It was sampled mainly by sweep netting ('sweeping'). Samples obtained by sweeping were included as part of the forest floor sample in the analyses of differences across strata.

The understorey stratum (shrub layer) was sampled by beating the vegetation and collecting the spiders from a tray ('beating') and by manual collection of spiders from shrubs and trunks from knee-level to as high as one can reach ('aerial' *sensu* Coddington *et al.*, 1996). Each sample of the methods – cryptic, ground, sweeping, beating and aerial – was timed, with only active sampling time being measured. One hour of active sampling represented one sample. Overall sampling totalled 200 hours, approximately one third during the day and two thirds at night. In order to avoid bias by collector, the eight collectors sampling forest floor and understorey applied all methods both at night and during the day. Summary statistics of the inventory by method are given in Table 9.1.

Collected material was stored in 70% alcohol and the spiders were subsequently sorted to morphospecies (distinctive, recognizable morphological units, hereafter 'species') because knowledge of the spider fauna remains incomplete. Juvenile spiders (representing ~60% of the specimens) were excluded because they could not be identified to species level.

Statistical methods

Species-accumulation curves (200 random resamplings of all samples for each method and for each stratum) were generated using the software EstimateS 5.0.1 (Colwell, 1997b; Longino, 2000). To compare species diversity among strata, Simpson's indices (Magurran, 1988) were also calculated.

Complementarity values between pairs of strata were calculated as the number of species unique to each stratum divided by the combined number of species of the two strata (Colwell & Coddington, 1994). Analyses were performed among all three combinations of strata: canopy, understorey and forest floor.

To investigate further the differences in spider assemblages among strata, a chi-square test (adopting the null hypothesis of no difference in overall distribution of individuals within families among strata) was performed on the number of individuals within the 13 most abundant families. This included the families Thomisidae and Salticidae, often abundant families in lowland tropical forest (Russell-Smith & Stork, 1994, 1995; Silva, 1996), which were only the 12th and 13th most abundant families here. These 13 families represented 94% of the total specimens collected.

The number of spider species for each stratum and for the total plot was estimated with the Chao 1 estimator using the software EstimateS 5.0.1 (Colwell, 1997b). Chao 1 (Chao, 1984) was selected rather than other estimators because it can be used to estimate species richness based on the number of observed species, number of singletons (species represented only by one specimen) and doubletons (species represented by two specimens) in a single sample and is, therefore, widely applicable. The Chao 1 estimates gave intermediate species-richness estimates when compared with the other estimators included in the programme EstimateS (results not shown).

The completeness of sampling by method and by stratum was calculated by dividing the species-richness estimate (Chao 1) by the observed number of species (performance in Table 9.1). As the species-accumulation curves and curves of the species-richness estimator did not tend to an asymptote, species richness was estimated based on 200 resamplings of 500 adult specimens for comparison among sampling methods

Table 9.1. *Summary of variables (±standard deviation) detailed by sampling methods and strata surveyed*

	Beating	Cryptic	Ground	Aerial	Sweeping	Forest floor	Understorey	Canopy (fogging)	Total
Samples	49	45	41	48	17	103	97	325	525
Species	92	62	66	72	42	96	114	95	175
Species in 500 individuals	55.8 ± 5.7	59.1 ± 1.3	60.9 ± 2.6	53.0 ± 3.4	39.7 ± 1.3	63.44 ± 4.1	57.9 ± 5.6	62.5 ± 3.6	78.1 ± 6.7
Adult spiders	1783	558	642	1141	584	1784	2924	1608	6316
Singletons	27	22	19	25	13	27	36	26	40
Doubletons	16	11	7	10	7	14	16	19	20
Unique	14	11	5	6	1	32	21	27	–
Simpson index	8.13	9.53	18.83	6.60	4.62	14.66	7.97	9.21	14.40
Chao 1	114.8 ± 11.5	84 ± 12.4	91.8 ± 16.2	103.3 ± 16.9	54.1 ± 8.8	122.0 ± 13.2	154.5 ± 18.0	112.8 ± 9.1	215 ± 16.7
Performance	78.1 ± 5.4	68.4 ± 9.8	76.8 ± 6.8	71.2 ± 7.5	72.7 ± 7.8	74.6 ± 7.2	74.6 ± 6.4	80.6 ± 4.9	81.1 ± 5.3
Chao 1 for 500 individuals	91.9 ± 20.2	83.0 ± 13.7	87.5 ± 17.6	81.6 ± 19.1	52.9 ± 9.5	89.0 ± 16.7	95.2 ± 21.5	98.5 ± 19.1	118.2 ± 19.4

Chao 1, estimated number of species using the Chao 1 estimator (see text); singletons, single representatives of a species; doubletons, two representatives of a species; unique, species unique to sampling method or stratum; performance, Chao 1 divided by the observed number of species.

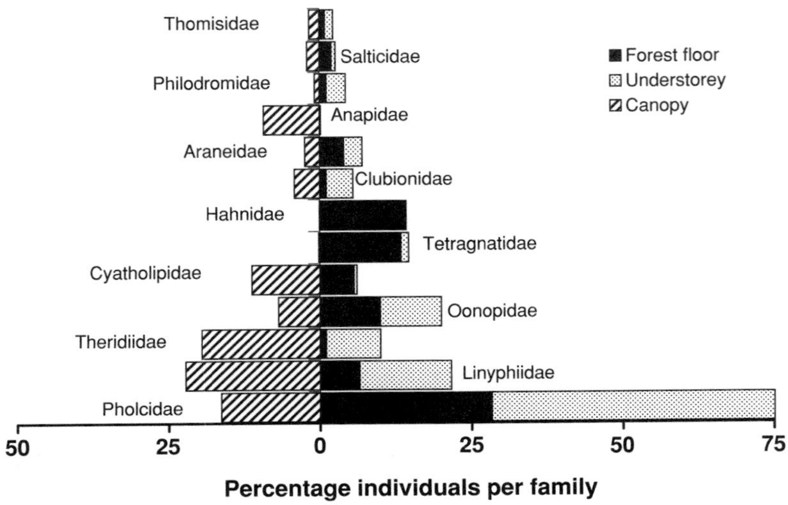

Fig. 9.1. The relative abundance of the most abundant families of spiders in the three defined strata: canopy, understorey and forest floor.

and strata. The number of species for 500 specimens and estimated richness are given in Table 9.1.

RESULTS

In total, the three strata sampled included 6316 adult specimens representing 175 species in 33 families. Forty (22%) of the species collected were represented by only one specimen (singletons) and ~80% were species new to science. Overall, 1784 adult spiders were collected on the forest floor, representing 96 species in 28 families. Thirty-one of these species were collected only on the forest floor and 11 (11%) of them were singletons. In the understorey, 2924 adult spiders of 114 species representing 29 families were collected, including 23 species unique to this stratum (21%). Fourteen (13%) of these were singletons. In the canopy, 1608 adults belonging to 95 species (including 24 families) were collected. Of these, 27 (29%) species were collected only in the canopy, including 15 (16%) singletons (Table 9.1 summarises the data by method and strata). The data at family level for the three strata within the plot are summarised in Table 9.2.

Dissimilarities between strata

The spider communities on the forest floor and in the canopy were the most different (complementarity 79.2) whereas the fauna in the understorey and in the canopy shared more species (complementarity 53.8). The fauna of the forest floor shared more species with the understorey (complementarity 58.1) than with the canopy.

The distribution of individuals in the 13 most abundant families also showed considerable differences among the three strata (Table 9.2, Fig. 9.1), as demonstrated by a chi-square test (chi-square$_{3,24}$ = 5557, $p < 0.0001$). The same pattern was reflected at species level (Fig. 9.2). Pholcidae were common in all strata, most noticeably in the understorey. The pholcid species collected from the forest floor were different from those found in the higher strata. The family Linyphiidae was very common in the canopy, along with the family Anapidae, which is normally considered to be associated with the litter and low vegetation (Dippenaar-Schoeman & Jocqué, 1997). The families Tetragnathidae and Hahnidae were sampled almost exclusively on the forest floor, whereas the family Cyatholipidae was most common on the high vegetation (Fig. 9.1). One species that might be referred to the family Oonopidae was found primarily in the canopy, and although other species of Oonopidae were found on the ground, very few individuals were collected in the understorey. This family is normally considered to consist of ground dwellers (Dippenaar-Schoeman & Jocqué, 1997).

Further examples of the stratification of species within families are given in Fig. 9.2. Species of Linyphiidae, such as *Mecynidis* spp. and *Callitrichia sellafrontis* Scharff 1990 were collected primarily in the understorey and low vegetation whereas *Callitrichia* sp. n. was

Table 9.2. *Numbers of adult spiders, species, singletons and pairs for each spider family collected in the 1 ha plot, detailed by forest strata (see methods)*

Family	Total data				Forest floor				Understorey				Canopy			
	Ind	Sp	1s	2s	Ind	Sp	1s	2s	Ind	Sp	1s	2s	Ind	Sp	1s	2s
Agelenidae	38	2	–	1	19	1	–	–	19	2	–	1	–	–	–	–
Amaurobiidae	4	1	–	–	–	–	–	–	2	1	–	1	2	1	–	1
Anapidae	152	3	1	–	2	2	2	–	2	2	2	–	148	2	–	–
Araneidae	201	12	3	–	71	3	1	–	91	10	4	–	39	10	3	2
Barychelidae	6	3	1	1	4	2	1	–	2	1	–	1	–	–	–	–
Clubionidae	217	6	1	–	19	2	1	–	131	5	1	–	67	6	1	1
Corinnidae	32	6	1	1	7	2	–	1	12	3	–	1	13	4	1	–
Ctenidae	45	3	–	–	42	3	–	–	3	1	–	–	–	–	–	–
Cyatholipidae	596	3	–	–	18	3	–	–	264	3	–	1	314	2	–	–
Dictynidae	19	1	–	–	1	1	1	–	10	1	–	–	8	1	–	–
Gnaphosidae	3	1	–	–	3	1	–	–	–	–	–	–	–	–	–	–
Hahnidae	257	4	–	–	252	4	–	1	4	2	–	2	1	1	1	–
Heteropodidae	5	2	–	1	–	–	–	–	3	2	1	1	2	1	–	1
Linyphiidae	915	16	2	3	117	9	1	2	441	7	2	1	357	12	3	4
Liocranidae	32	6	3	–	29	5	2	1	3	2	1	1	–	–	–	–
Lycosidae	52	1	–	–	51	1	–	–	1	1	1	–	–	–	–	–
Mimetidae	47	6	1	1	6	5	4	1	38	5	1	–	3	3	3	–
Mysmenidae	3	2	1	1	2	1	–	1	–	–	–	–	1	1	1	–
Oonopidae	296	5	1	–	101	4	–	1	15	2	1	–	180	3	1	–
Palpimanidae	1	1	1	–	–	–	–	–	–	–	–	–	1	1	1	–
Philodromidae	128	2	–	–	20	2	–	–	95	2	–	1	13	1	–	–
Pholcidae	2134	7	2	–	506	5	1	–	1366	5	1	–	262	4	–	–
Salticidae	88	20	9	2	35	7	4	1	20	10	6	1	33	9	3	2
Scytodidae	14	1	–	–	–	–	–	–	3	1	–	–	11	1	–	–
Segestriidae	10	3	–	2	8	3	–	2	2	1	–	1	–	–	–	–
Selenopidae	15	1	–	–	1	1	1	–	6	1	–	–	8	1	–	–
Symphytognathidae	3	1	–	–	–	–	–	–	–	–	–	–	3	1	–	–
Tetragnathidae	279	6	–	1	239	6	1	–	37	5	2	–	3	2	1	1
Theridiidae	581	37	9	4	178	14	3	2	293	30	9	3	110	23	6	6
Thomisidae	83	8	3	1	16	4	2	–	40	5	2	–	27	4	1	–
Theridiosomatidae	38	2	–	1	28	2	1	–	10	2	1	–	–	–	–	–
Uloboridae	14	1	–	–	2	1	–	1	10	1	–	–	2	1	–	1
Zodariidae	8	2	1	–	7	2	1	–	1	1	1	–	–	–	–	–
Total	6316	175	40	20	1784	96	27	14	2924	114	36	16	1608	95	26	19

Ind, adult spiders; Sp, species; 1s, singletons; 2s, doubletons.

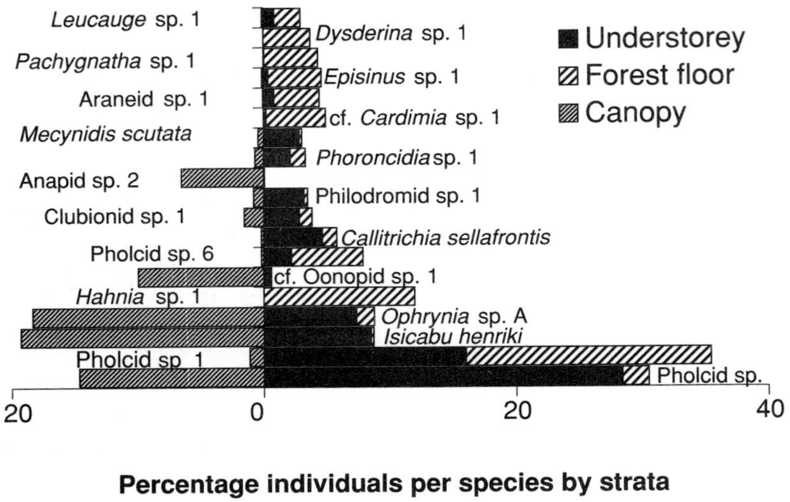

Fig. 9.2. The relative abundance of the 19 most abundant of the 170 species of spiders in the three defined strata: canopy, understorey and forest floor.

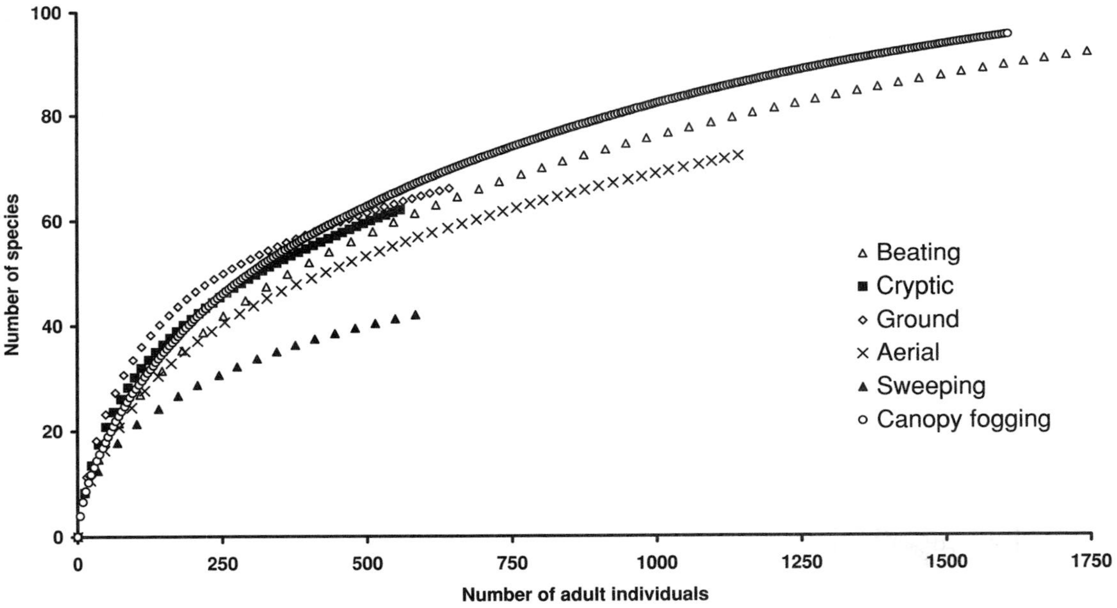

Fig. 9.3. Species-accumulation curves for spiders collected with different sampling methods within different habitats.

collected in the canopy. Species of *Ophrynia* were observed mainly in the understorey and canopy, and *Ophrynia* sp. A (8% of the total number of adults in the plot) was very common in both layers. *Lepthyphantes* spp. were found only on the forest floor.

Species richness

The species-accumulation curves for the different sampling methods (Fig. 9.3) show that species saturation was not reached using any of the methods. Sweeping samples yielded the lowest number of species, and the shallow slope of the sweeping curve indicates a low within-sample diversity (Simpson's index 4.62). In contrast, ground sampling (in particular Simpson's index 18.83) showed a steep slope, indicating that the method detects many species in low numbers (high within-sample diversity). The canopy fogging seems to detect more species overall compared with beating and ground samples. Although the same microhabitats were surveyed by vegetation beating and by aerial manual collecting in the understorey, the observed species richness was lower using the latter method. Manual collecting of the understorey may miss some of the cursorial hunting spiders and rare spiders that are collected with beating, and sampling by beating is also likely to access more habitat volume than manual collecting of the different microhabitats. All methods contributed unique species within each stratum, resulting in an increase in species richness when samples were combined.

The results of the calculation of Simpson's indices for each stratum (Table 9.1) support the results of the species-accumulation curves (Fig. 9.4). In particular, the forest floor displays higher within-sample diversity (Simpson index 14.66) than both the understorey and the canopy, but the fact that its species-accumulation curve intersects with the curves for both other strata (Fig. 9.4) indicates that its overall assemblage richness is lower. The within-sample diversity is higher for the canopy but close to that of the understorey; it is not clear whether their curves would intersect upon extrapolation.

Overall, the total estimated species richness within the 1 ha plot was 215 ± 17. The sample of the understorey seemed the richest in species whereas the canopy sample gave the lowest estimated richness. Comparable subsamples showed that the estimated richness of the canopy, in fact, was highest, but that the understorey was almost as rich.

DISCUSSION

Spiders are adapted to a wide range of habitats and it is difficult to use comparably efficient collecting methods

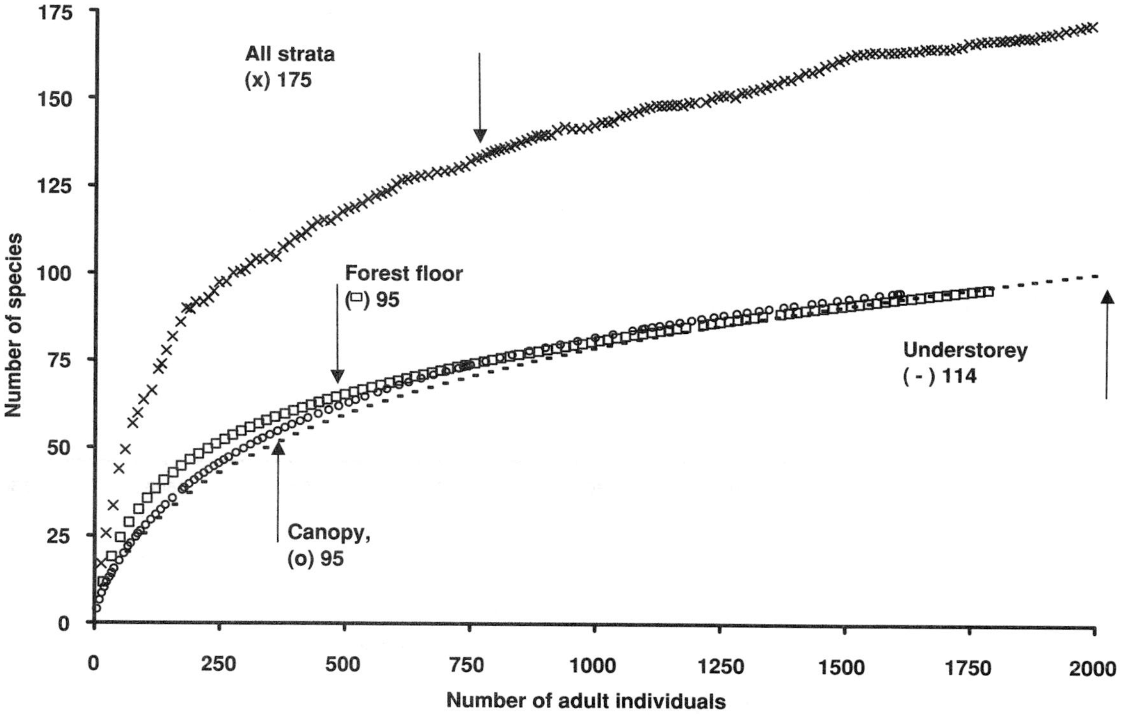

Fig. 9.4. Species-accumulation curve for spiders collected from the forest floor, understorey and canopy, and for all strata combined. The total number of species observed is given for each area.

to cover all the different microhabitats in which they live (Turnbull, 1973). Because the original purpose of the inventory was in part to obtain data for the study of species richness and to assess the importance of the canopy, several methods were used, including manual collecting methods. Longino and Colwell (1997) have stressed the importance of using methods in 'strict inventories' that sample complementary sets of species. In many tropical rainforests, arthropod species richness is so high and many species are so rare (Novotny & Basset, 2000) that it is nearly impossible to obtain sufficient complete data to do actual comparisons across different strata based on species lists or for species richness (Longino, 2000). This incomplete sampling is reflected in the accumulation curves (Figs. 9.3 and 9.4) and emphasizes the difficulties of obtaining sufficient samples for reliable estimates of species richness.

A high number of rare species (singletons and pairs) were observed in the Tanzanian samples when considering either sampling methods or the strata sampled (Table 9.2). Differences in patterns of diurnal activity were accounted for by sampling both at night and during the day. However, the apparent rarity of some species might be explained by seasonality. The present knowledge of the biology of the Tanzanian spider fauna is limited and it is, therefore, not possible to assess the extent to which phenology can be used to explain the occurrence of rare species. In the present samples, 80% of the species were undescribed taxa. There is clearly a need for more studies on tropical invertebrates both from the forest floor and the canopy.

This is one of the first studies that focusses specifically on the vertical distribution of spiders in tropical forests. This vertical distribution indicated complex assemblages with a strong stratification pattern across the defined strata (forest floor, understorey and canopy), even though the understorey was more intensely sampled than the other strata.

The canopy and understorey showed much similarity in species composition whereas the canopy shared fewer species with the forest floor. Since the understorey included low vegetation (above 50 cm) it was expected that this stratum would have shown the highest species diversity, with an influx of forest floor and canopy fauna. However, the understorey appears to have a distinctive fauna, with more species shared with the canopy than with the forest floor. The forest floor, furthermore, has a number of distinctive species confined to it.

A larger sample size would undoubtedly have yielded additional species in all strata and also would have included species that were previously considered 'unique' in other strata. Nevertheless, the data still reflect a high degree of habitat specificity. The linyphiids, of which few species were unique to single strata, are a good example. They showed a high degree of specialization, with the majority of individuals of most species collected in a single stratum, even though juvenile linyphiids are known to disperse by ballooning. Millidge and Russell-Smith (1992) found that linyphiids collected in the canopy of Brunei and Sulawesi rainforests were clearly distinct species from those of the lower strata.

The uniqueness of the canopy fauna (approximately 20% unique species) was close to the data reported by Stork (1988: 17%) and Brühl *et al.* (1998: 25%). Although considering the 19 most common species revealed no unique species to the canopy, distinct differences in species abundance were apparent (Fig. 9.2). Many species appeared to have their highest populations in the canopy, suggesting that the canopy supports a distinctive fauna.

Further studies are needed in order to clarify whether there is an exchange of species between the canopy and the ground or whether species occurring in high numbers in the canopy but collected in low numbers at ground level should be considered as transient at the lower levels. Because of their predatory nature, spiders cannot be classified as 'tourists' (Moran & Southwood, 1982) even though their biology indicates that they do have certain habitat preferences. Species from the canopy are more likely to occur as 'transient' in lower strata than vice versa.

Studies in temperate climates have shown that the composition of spider assemblages are determined by the structure of the habitat and its microclimate (Duffey, 1966; Scheidler, 1990; Halaj *et al.*, 2000). Vertical stratification has previously been shown for spiders in temperate deciduous forests (e.g. Elliott, 1930; Gibson, 1947; Turnbull, 1960) and in pine forests (Docherty & Leather, 1997). Despite the fact that many temperate species overwinter in the ground and migrate in the spring to higher strata, stratification is still apparent although the phenomenon of species tending to extend their ranges into adjacent strata has been observed (Turnbull, 1960). Seasonal variation in the composition of the spider assemblage in the temperate forests is high (Elliott, 1930; Gibson, 1947; Turnbull, 1960).

Turnbull (1960) found that the abundance of spiders is much higher in the ground stratum (forest floor) than in any of the other strata, and that the field layer (understorey) was richest in species. He attributed these observations mainly to the large open non-inhabitable spaces in the canopy. In lowland Cameroon, Basset *et al.* (1992) also reported that, when beating the vegetation, the understorey was richer in number of families of spiders than the upper canopy, though no significant difference could be detected in number of individuals.

In the present study, the forest floor turned out to be the most diverse stratum though the canopy was more species rich. In the canopy and understorey, many species occurred in low abundances, and a few species of pholcids and a single species of linyphiid dominated in terms of adult numbers (Sørensen, 2000). This might indicate that the fauna is less dense at ground level, perhaps because of fewer available microhabitats. However, this could also be explained by the choice of sampling methods. Canopy fogging, despite the method's apparent inability to access cryptic living species, might sample more broadly than manual collecting methods, where the collector may be biased against less-abundant species. This could be the reason why more species are represented in low numbers in the canopy compared with those collected manually from the ground and understorey.

Canopy fogging, as applied in the present study, will not allow the same degree of resolution of animal distribution across microhabitats as the methods applied on the forest floor and in the understorey. Since the method was applied from the ground, the sampling probably focussed on the fauna of the lower parts of the canopy, which may be more comparable with the understorey with regard to microclimate. It is likely that the exposed upper canopy would reveal a fauna that differs from that of the lower canopy; for instance, it is likely that active hunters will be more abundant in the former location. The high species richness in the canopy might be explained by a high density and diversity of prey (Sutton & Hudson, 1980; Sutton *et al.*, 1983a; Sutton, 1983; Nadkarni & Longino, 1990; DeVries *et al.*, 1997). Spiders, then, differ from less-mobile organisms such as mites, which exhibit greater diversity at the forest floor.

The stratification pattern of the spiders and other invertebrate groups (Stork, 1988; Brühl *et al.*, 1998) seems similar. If correct, this implies that a relatively high proportion of the fauna is confined to the canopy. The results of the present study indicate that for reasonably complete inventories of the spider fauna the rich canopy fauna must be included. Nevertheless, a survey of the understorey in the present study encountered a high proportion of the species occurring in the canopy, albeit in lower numbers.

ACKNOWLEDGEMENTS

I especially thank Nikolaj Scharff and Per de Place Bjørn (ZMUC), Johanna Heinonen (University of Helsinki), Jonathan Coddington, Scott Larcher and Jeremy Zujko-Miller (USNM, Washington DC), Elia Mulungu (Parabiologist, Iringa), Bruno Nyundo (University of Dar es Salaam) and Innocent Zilihona (Tanzania Forestry Research Institute) for their invaluable help in the field, and Renner Baptisita (USNM, Washington DC), Sidsel Larsen, Søren Jensen and Aslak Jørgensen (ZMUC) for their help with sorting. I thank the Commission of Science and Technology (COSTECH) for kindly granting research permits, the Tanzania Catchment Forestry Office, Iringa, for allowing the expedition to work in forests under their jurisdiction, and the Division of Wildlife for granting CITES permits. I acknowledge Dr Felista M. Urassa, Dr Jacob G. Yarro and Dr Kim M. Howell (University of Dar es Salaam) and the Enreca programme (Danish International Development Assistance, Danida) for support to Tanzanian students, and the North/South Priority Research Area (University of Copenhagen) for funding. Finally I thank the Danish Centre for Tropical Biodiversity (CTB) and David Moyer (WCS, Tanzania) for logistical support; and Robert Colwell (University of Connecticut), Peter M. Hammond (The Natural History Museum), Pernille Thorbek (DMU), Keith Sunderland (Horticulture Research International), Jette Andersen, Henrik Enghoff, Nikolaj Scharff and Christian Frimodt-Møller (University of Copenhagen) for comments on the manuscript.

10

Fauna of suspended soils in an *Ongokea gore* tree in Gabon

Neville N. Winchester and Valerie Behan-Pelletier

ABSTRACT

The invertebrate species inhabiting suspended soils in the crown of a mature *Ongokea gore* tree in Gabon, West Africa, were assessed at two heights and four distances from the trunk. Access to the canopy was by single-rope techniques, with logistic support from 'Océan Vert', and a hand-held corer was used to collect random, replicated samples from branches. This sampling methodology is a model for rapid assessment of canopy habitats in rainforests of the world. Height from the ground affected significantly the abundance of most invertebrates, with Thysanoptera and Formicidae more abundant at 32 m and Nematoda, Collembola and Coleoptera more abundant at 42 m. However, height was not significant for Acari, the dominant component of the invertebrate assemblage in all samples. Among mites, Oribatida were the dominant group in these suspended soils. Their mean number and maximum abundance on each branch at each distance were consistently higher than those of all other invertebrates. Of the 25 oribatid species collected from these suspended soils, similar numbers of species (14–15) were found over the first three distances from the trunk (0.0–2.0 m) but declined towards the branch tip (10 spp.). Species shared between each distance ranged from a low of 7 to a high of 11. *Tectocepheus velatus* and undescribed species of *Chaunoproctus, Scheloribates* and *Unguizetes* were the most abundant species found at all branch distances. In contrast, there were 13 species occurring at a single distance only, of which the most common were *Trimalaconothrus* sp., *Clavazetes* sp. and *Benoibates* sp. Results from this study support previous conclusions that arboreal habitats are reservoirs for arthropod biodiversity, and that oribatid mites are among the dominant and most species-rich arthropods in the suspended soil habitat.

INTRODUCTION

Canopies of natural rainforests in tropical regions contain largely undescribed and little understood assemblages of arthropods that have greatly expanded estimates of the total number of species present (Erwin, 1983a; Stork *et al.*, 1997a; Basset & Kitching, 1991; Basset, 1997; Didham, 1997; Davis *et al.*, 1997; Hammond *et al.*, 1997; Kitching *et al.*, 1997). However, it is less widely appreciated that Acari are a dominant component of this fauna (Walter & Behan-Pelletier, 1999). Acari are among the most speciose and abundant arachnid group in arboreal microhabitats (Koponen, *et al.*, 1997; Winchester,1997a; Schowalter & Ganio, 1998) and are often the most abundant arthropods (Nadkarni & Longino, 1990). Because most canopy sampling methods used to estimate total diversity (e.g. insecticide fogging) are not efficient for collecting mites (Walter, 1995; Walter *et al.*, 1998) and the range of microhabitats available for colonization is large (Winchester, 1997a), current figures for the diversity of mites living in rainforest canopies are underestimates (Walter & O'Dowd, 1995). Within the Acari, Oribatida are often the dominant mite group and the most diverse taxonomically in forest canopies (Walter & Behan-Pelletier, 1999).

Traditionally, the forest canopy has been viewed as a substrate source for decomposition processes that occur after litter falls to the forest floor (Dickinson & Pugh, 1974). However, decomposition processes also occur within the canopy and provide a substrate that supports a diverse invertebrate community (Delamare-Deboutteville, 1951; Nadkarni & Longino, 1990; Winchester, 1997a,b). These rich accumulations of debris and litter, or 'suspended soils' (*sensu* Delamare-Deboutteville, 1951), provide the habitat for the high abundance and diversity of Acari and, in particular, of oribatid mites (Nadkarni & Longino, 1990; Winchester *et al.*, 1999).

Many Oribatida associated with decomposition substrates show little evidence of trophic specialization, yet species ostensibly coexist (Anderson, 1978b). Microhabitat specialization has been proposed to explain oribatid coexistence (Aoki, 1967; Anderson, 1978b) and, subsequently, in several studies of oribatid faunas in forests, has been used to explain how potentially competitive species reduce interactions by becoming spatially separated (Hammer, 1972; Fujikawa, 1974). Specific oribatid species, for example, are associated with lichens (Seyd & Seaward, 1984; Prinzing & Wirtz, 1997), moss (Seniczak & Plichta, 1978), corticolous habitats (André, 1984, 1985; Nicolai, 1989; Wunderle, 1992a,b), cortical cankers (Nannelli *et al.*, 1998) and suspended soils in temperate rainforests (Carroll, 1980; Winchester *et al.*, 1999).

Information on the diversity and abundance of suspended soil-associated invertebrates in the tropical forests of West Africa is almost totally lacking (but see Delamare-Deboutteville, 1951), as is information on species assemblages, community structure and ecology. This paucity of data is far from unique: data are lacking for microarthropod diversity in the canopies of most tree species in most temperate and tropical rainforests. In turn, this leads to difficulty in assessing these often remote forests, and there is no agreement on suitable, repeatable collecting methods for rapid evaluation of canopy microarthropod diversity. In this chapter, we describe the first documentation of invertebrates that inhabit suspended debris accumulations in the crown of a rainforest tree in Gabon, using a simple, repeatable sampling method that is geared towards collection of microarthropods. In particular we describe the composition of the arboreal oribatid mite assemblage associated with these suspended soils and discuss whether or not height from the ground and distance from the trunk affect species richness and evenness.

METHODS

Study area

The study area is located in a lowland tropical forest (Forêt des Abeilles, La Makandé) in Gabon, West Africa (0° 40′ 39″ S, 11° 54′ 35″ E, 200–700 m above sea level). Precipitation may vary, but the mean annual precipitation is in excess of 1600 mm and mean annual temperature is 24 °C (Fréty and Dewynter, 1998). Upper canopy height is variable and large emergent trees are uncommon. Typically, canopy height varies from 35 to 45 m. The physical aspects of this forest are described by Fréty and Dewynter (1998) and Hallé (2000).

Canopy access and tree selection

Canopy access was made possible using the logistics of Océan Vert at La Makandé during mid-January to mid-March 1999 (Hallé, 2000; Basset *et al.*, 2001a). Our sampling programme included the use of the *radeau des cimes* (canopy raft) and the *luge* (sledge) for selection of a suitable tree to sample. (Details of this equipment and its use are described in http://www.radeau-des-cimes.com/publi.htm. An individual tree of *O. gore* (Angeuk) was selected and canopy access was gained by using single-rope techniques (Mitchell, 1982).

O. gore trees are the largest of the Olacaceae, often reaching 50 m in height, and are characteristic of old, stable mature forests (White & Abernethy, 1997). The long straight trunk without buttresses has a grey-coloured thin bark that flakes off easily. The tree crown contains deep pockets of suspended soils that support several species of ferns and lichens and associated invertebrates.

Suspended soil sampling, sorting and identification

The sampling programme for canopy arthropods takes into account recommendations made from the senior author's 1992–1999 canopy studies. The suspended soil was sampled in the crown (>30 m) of a mature *O. gore* tree. From each of five branches, three to five replicate samples were collected using a hand-held moss/soil corer (3 cm × 5 cm). Three branches were sampled at 32 m and two branches were sampled at 42 m. A total of 23 cores was collected during the first week in March, 1999. Arthropods were removed from each core by washing. Samples were washed through a 1 cm sieve into a saturated saline solution in plastic bags and submersed in this solution for 48 hours. Contents of the saturated saline solution were then sieved through a nest of two sieves: an upper one with 120 μm mesh and a lower one with 45 μm mesh, emptied into a whirl-pak™ bag and preserved in 75% ethanol. Height from the ground and distance from the trunk were recorded for each sample.

All invertebrates were counted, and Acari separated. Representative oribatid specimens were slide mounted using the procedures outlined by Krantz (1978). Specimens were identified to genus and species where possible and deposited in the Canadian National Collection of Insects and Arachnids, Agriculture and AgriFood Canada, Ottawa, and at the Pacific Forestry Centre, Victoria, British Columbia.

Data analyses

Total abundance of invertebrates collected from each branch was tabulated. Summary statistics for the Acari and Oribatida for each branch and distance along the branch were calculated using the statistical package SPSS 6.1 for Windows (SPSS, 1994). Oribatid species were arranged in tabular form and divided into generalist and specialist categories based on their occurrence over four branch distances. The proportional representation of Acari, oribatid mites and all other identified (i.e. placed in a taxonomic category) invertebrates sampled was calculated as the ratio of taxon abundance to total invertebrate abundance. The data were not normally distributed; therefore, they were transformed using an arcsine square root procedure before plotting means and standard errors (Sokal & Rohlf, 1995). Faunal composition based on ordinal classification best fitted a negative binomial distribution and the original data were log-transformed (ln+1) before plotting means and standard errors of abundance and richness of oribatid mites as a function of distance from the trunk. Distances were divided into categories: 1, 0.0–0.5 m; 2, 0.5–0.75 m; 3, 0.75–2.0 m; and 4, > 2.0 m. Oribatid species composition for all distances were tabulated and expressed in a stylized ball-and-stick model (similarity diagram).

A chi-squared contingency analysis was used to test for independence between height and ordinal composition. Single-factor analyses of variance were used to test for the effects of distance on oribatid mite abundance and richness. All tests were performed using SPLUS 4.5 for Windows (MathSoft, 1996) at a significance level of 0.05.

Table 10.1. *Total abundance of canopy invertebrates (adults and immatures) extracted from 23 suspended soil cores collected in an Ongokea gore tree crown*

	Abundance branch[a]				
	1	2	3	4	5
Height from ground (metres)	32	32	32	42	42
Phylum Arthropoda					
Class Insecta					
Order Blattaria	2	1	0	7	9
Order Coleoptera	10	15	4	40	26
Family Staphylinidae	0	1	4	24	40
Order Diptera	18	26	4	46	30
Order Homoptera	1	1	5	8	0
Family Reduvidae	3	1	0	0	0
Family Coccidae	2	2	8	0	0
Order Hymenoptera	9	2	3	9	10
Family Formicidae	3	8	4	4	2
Order Lepidoptera	1	0	0	5	0
Order Psocoptera	5	2	3	4	1
Order Thysanoptera	12	13	20	7	4
Others (e.g. larvae, pupae, mostly insects)	9	14	38	20	14
Class Collembola	4	4	4	18	8
Class Arachnida					
Order Araneae (includes Pseudoscorpiones)	17	2	6	8	8
Order Acari	352	232	268	172	439
Class Crustacea					
Subclass Copepoda	0	5	3	3	0
Phylum Nematoda	8	3	0	52	1
Total	456	332	374	427	592

[a] Distances from the tree trunk are pooled for each branch.

RESULTS

The suspended soil community

A total of 2174 invertebrates was collected from the 23 suspended soil cores from the *O. gore* tree crown (Table 10.1). The Arthropoda was the dominant invertebrate phylum (95.6%). Numerically dominant arthropod classes were the Insecta (32.2%) and Arachnida (69.2%). Most notably, the subclass Acari constituted (67.8%) of the total invertebrate fauna. Copepoda (0.5%) and Nematoda (2.9%) were also found in the canopy suspended soils. Faunal composition at the ordinal level varied significantly at the $p < 0.0001$ level ($\chi^2_{0.05,13} = 95.79$) between heights, largely because of a higher abundance of Coleoptera, Blattaria and Nematoda (treated as a single group of orders in the analysis) at the upper height (42 m from the ground),

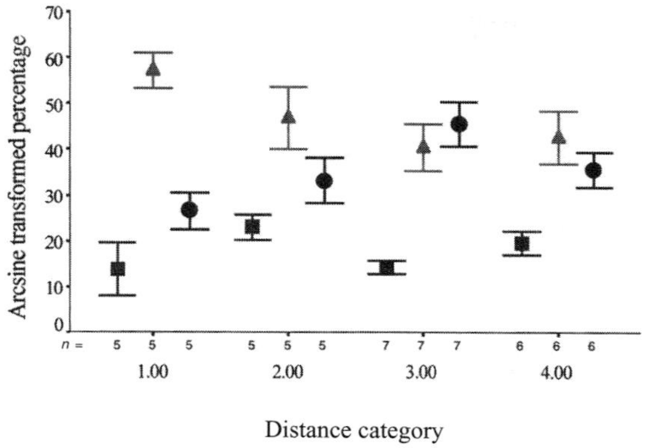

Fig. 10.1. Mean (±1SE) proportional representation of Acari (■), Oribatida (▲) and all other invertebrates (●) as a function of increasing distance from the trunk in an *Ongokea gore* tree crown, Gabon, West Africa. Distance categories 1, 0.0–0.5 m; 2 0.5–0.75 m; 3, 0.75–2.0 m; 4, >2.0 m. Data are presented as arcsine squared root transformed proportions (mean ± SE values in degrees).

and Diptera, Hymenoptera and Thysanoptera at the lower height (32 m from the ground). Proportional representation of all invertebrates other than Acari gradually increased with distance from the trunk and was highest at distance 2 from the trunk (Fig. 10.1).

Canopy oribatida

Oribatida dominated the acarine suspended soil assemblage, representing 64.25% of total mite abundance.

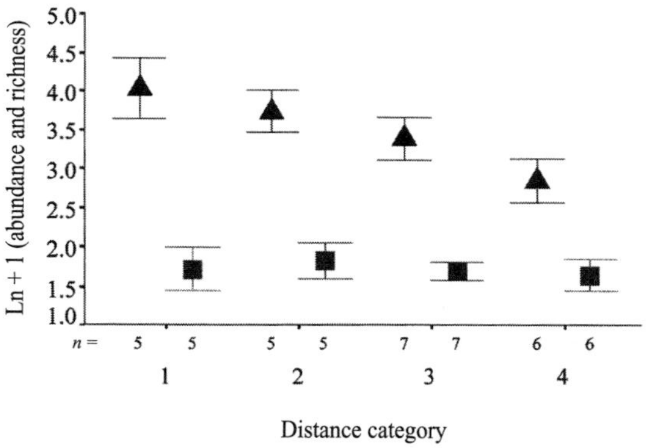

Fig. 10.2. Mean (±1SE) abundance (▲) and mean (±1SE) species richness (■) of Oribatida as a function of distance along the branch from the trunk in an *Ongokea gore* tree crown. Distance categories 1, 0.0–0.5 m; 2, 0.5–0.75 m; 3, 0.75–2.0 m; 4, >2.0 m.

Table 10.2. *Comparison of summary statistics for abundance of Acari and Oribatida from each branch (distances pooled) and for each distance (branches pooled)*

Category[a]	Mean	SE	Minimum	Maximum	*n*
Acari					
B1	14.60	7.86	0.00	42.00	5
B2	9.00	3.75	0.00	25.00	6
B3	12.17	4.87	1.00	31.00	6
B4	6.67	0.33	6.00	7.00	3
B5	2.33	0.60	1.00	3.00	3
Oribatida					
B1	63.80	8.48	45.00	86.00	5
B2	30.17	10.02	5.00	71.00	6
B3	38.83	18.88	13.00	131.00	6
B4	50.33	20.10	12.00	80.00	3
B5	15.67	3.84	10.00	23.00	3
Acari					
D1	11.20	5.03	0.00	23.00	5
D2	17.80	7.10	3.00	42.00	5
D3	5.14	0.96	1.00	7.00	7
D4	7.67	4.70	1.00	31.00	6
Oribatida					
D1	70.00	20.38	13.00	131.00	5
D2	45.80	9.73	14.00	71.00	5
D3	35.29	10.16	12.00	80.00	7
D4	17.50	5.10	5.00	40.00	6

[a]B, branch (1–3 at 32 m; 4 and 5 at 42 m from the ground); D, distance from trunk (1, 0.0–0.5 m; 2, 0.5–0.75 m; 3, 0.75–2.0 m; 4, >2 m).

The mean number of oribatid mites and their maximum abundance on each branch at each distance were consistently higher than all other Acari combined (Table 10.2). Oribatida did not exhibit a significant decrease in mean abundance (Fig. 10.2) as distance from the tree trunk increased ($F_{0.05,3,19} = 2.73$, $0.10 > p > 0.20$). In addition, oribatid species richness was not significantly different ($F_{0.05,3,19} = 0.15$, $p > 0.50$) as distance from the tree trunk increased (Fig. 10.2). The proportional representation of Acari (excluding Oribatida) remained constant over all distances (Fig. 10.1). Oribatid mites represented the largest proportional component at all distances, with the largest representation closest to the trunk and gradually declining as distance increased (Fig. 10.1).

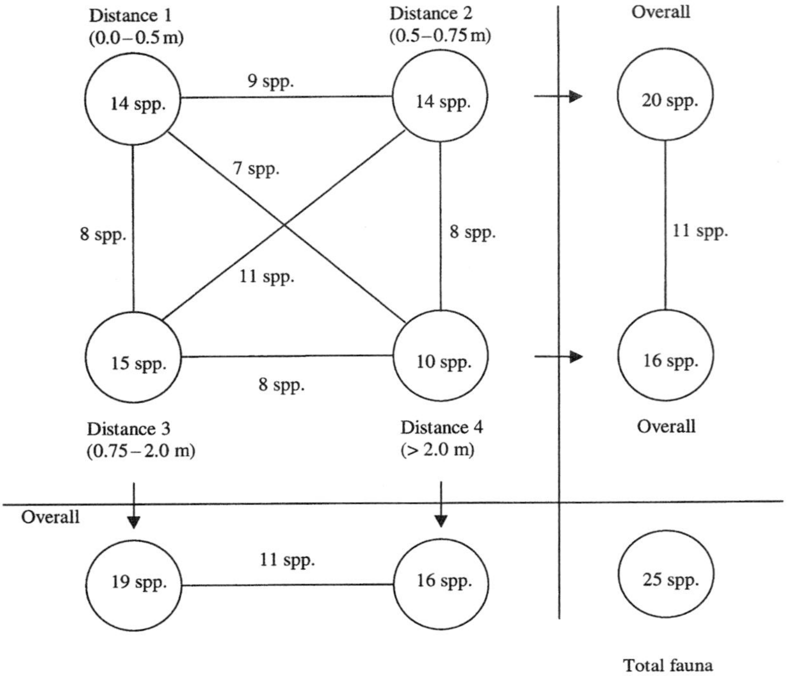

Fig. 10.3. Stylized ball-and-stick model depicting the number of oribatid species shared between distances from trunk for all branches. Numbers within the circles represent the total number of species collected at each distance. Numbers along the lines represent the number of species shared between distances.

A total of 25 oribatid species representing 15 families was collected from all suspended soils (Table 10.3, Fig. 10.3). The number of species was similar at each of the four distances from the tree trunk and ranged from 15 species recorded at distance 3 to 10 species recorded at distance 4 (Fig. 10.3). Twenty species occurred when the two closest distances (1 and 2) were pooled compared with 16 species in the two farthest distances (3 and 4; Fig. 10.3). Species shared between each distance ranged from a high of 11 (distances 2 and 3) to a low of seven (distances 1 and 4) (Fig. 10.3). *Tectocepheus velatus* (Michael) ($n = 246$), *Chaunoproctus sp.* 1 ($n = 89$), and *Scheloribates sp.*1 ($n = 87$) were the most abundant of the seven 'generalist species'; that is, those found at all branch distances (Table 10.3). *Notophthiracarus* sp. ($n = 18$) and *Scapheremaeus* sp. 1 ($n = 87$) were the only species found at all three distances. Three species occurred at two distances, the most abundant of which was *Phauloppia* sp.1 (Table 10.3). In contrast, there were 13 specialist species, occurring at but a single distance, the most common of which were *Trimalaconothrus* ($n = 13$), *Clavazetes* sp. 1 ($n = 7$) and *Benoibates* sp. 1 ($n = 8$).

DISCUSSION

Faunal composition

O. gore tree crowns are one of the only tree species in the Forêt des Abeilles that support accumulations of suspended soils. This suspended soil habitat is located in a continuous manner along branches located between 32 and 46 m above ground in the canopy and is discontinuous between branches and from the ground, the tree trunk being devoid of epiphytes and debris accumulations. There is a diverse community of invertebrates associated with these soils, primarily composed of Arthropoda, with the Acari numerically dominant and, mainly, represented by Oribatida. Overall faunal composition was similar to that found in other tropical studies of suspended soil microarthropod communities (Nadkarni & Longino, 1990; Wunderle, 1992a) but, as Walter & Behan-Pelletier (1999) noted, data are limited since sampling of suspended soils is commonly neglected in tropical canopy research. However, the widespread pattern of faunal composition with Oribatida the most abundant group of arthropods is similar to findings for many other tropical

Table 10.3. *Number of individuals of each species collected from suspended soil cores at each branch distance*

Distribution of taxa[b]	Distance[a]				Distribution of taxa[b]	Distance[a]			
	1	2	3	4		1	2	3	4
Generalists					**Specialists** *cont.*				
Tectocepheidae					Malaconothridae				
Tectocepheus velatus	111	123	11	1	*Trimalaconothrus* sp. 1	–	–	–	13
Tegeozetes sp. 1	1	39	9	13	Carabodidae				
Oppiidae					*Carabodes* sp. 1	–	1	–	–
Amerioppia sp. 2	9	3	2	3	Otocepheidae				
Scheloribatidae					*Clavazetes* sp. 1	7	–	–	–
Scheloribates sp. 1	19	37	18	13	Oppiidae				
Scheloribates sp. 3	7	22	2	1	*Amerioppia* sp. 1	–	–	2	–
Mochlozetidae					Cymbaeremaeidae				
Unguizetes sp. 1	6	66	15	6	*Scapheremaeus* sp. 2	1	–	–	–
Chaunoproctidae					Licneremaeidae				
Chaunoproctus sp. 1	18	43	14	14	*Licneremaeus* sp. 1	1	–	–	–
Generalist–specialists					*Licneremaeus* sp. 2	–	–	3	–
Phthiracaridae					Micreremidae				
Notophthiracarus sp.	–	8	8	2	*Micreremus* sp. 1	–	–	1	–
Cymbaeremaeidae					Scheloribatidae				
Scapheremaeus sp. 1	84	1	2	–	*Scheloribates* sp. 2	–	–	–	2
Specialists–generalists					Oripodidae				
Oripodidae					*Benoibates* sp. 1	8	–	–	–
Exoripoda sp. 1	–	7	1	–	Oribatulidae				
Protoripoda sp. 1	–	1	1	–	*Phauloppia* sp. 2	1	–	–	–
Oribatulidae					Galumnidae				
Phauloppia sp. 1	26	1	–	–	*Notogalumna* sp. 1	1	–	–	–
Specialists					**Totals**				
Phthiracaridae					Total number of families	13	11	11	8
Steganacarus sp.	–	3	–	–	Total number of species	14	14	15	10
					Total number of individuals	350	297	190	105

[a]Generalist species occur at all four distances; generalist–specialist species occur at three of four distances; specialist–generalist species occur at two of four distances; specialist species occur at a single distance.
[b]Data pooled for branches and heights (distance categories as in Table 10.2).

(Norton & Palacios-Vargas, 1987; Wunderle, 1992a; Behan-Pelletier *et al.*, 1993; Walter, 1995; Franklin *et al.*, 1998) and temperate (Carroll, 1980; André, 1984, 1985; Behan-Pelletier & Winchester, 1998; Schowalter & Ganio, 1998; Winchester *et al.*,1999; Fagan, 1999) tree species and canopy habitats.

In contrast, Delamare-Deboutteville (1951), using Berlese extractors, found Formicidae and Collembola to be the most abundant arthropods in suspended soils at 25 m and 45 m, respectively, in the canopy of a *Parinarium* sp. in the Ivory Coast. Collembola were more abundant than Acari in a number of canopy studies using fogging techniques, reviewed by Watanabe (1997). Similarly, Koponen *et al.* (1997), using arboreal pitfall traps, found Formicidae were more abundant than Acari in a Finnish forest. Factors accounting for the widespread dominance of Oribatida appear directly related to tree architecture that allows for debris accumulation and subsequent suspended soil formation over time. Accumulations of litter in the crown region of trees, on branches and in the forks of branches vary among tree species and rainforest types, yet oribatid mites are usually the dominant arthropods in all these accumulations of organic matter. Individual mites are almost certainly utilizing these debris accumulations as food resources and suitable living space, providing protection from changing microclimatic conditions and from predation by other invertebrates.

Faunal composition and height

Faunal composition differed significantly with increasing height, as Coleoptera, Blattaria and Nematoda were more abundant at 42 m and Diptera, Hymenoptera and Thysanoptera were more abundant at 32 m. Vertical stratification has been recorded for Coleoptera (Erwin & Scott, 1980; Hammond *et al.*, 1997), Diptera (Didham, 1997) and for several other orders (e.g. Stork, 1987b; Basset, 1991a; Kitching *et al.*, 1997). This may reflect the differential ability of organisms to track the availability and suitability of suspended soils. As a habitat resource, suspended soils differ with height. For example, as height increases debris accumulations become patchy and usually decrease in thickness. Most of the invertebrates we collected fed on decomposing material and associated bacterial and fungal growth, which varies with debris accumulation, changing phenological patterns, variable microclimatic conditions and increased moisture retention. In contrast, oribatid mites were not segregated by height, suggesting little variation between microclimate conditions of importance to them in suspended soils at both heights.

Oribatid assemblage structure and habitat use

The 25 oribatid species recorded in this study represent the only records from suspended soils in a lowland African rainforest. Although our data are preliminary and use a rapid access and sampling methodology, they support the thesis that suspended soils contain an abundant and diverse assemblage of oribatid mites. This fauna is dominated by members of the oribatid lineage, Brachypylina (22 species, 88%), a proportion common to all canopy habitats. In contrast, Brachypylina constitute only about 74–77% of the soil and litter oribatid fauna (Behan-Pelletier & Walter, 2000). Among Brachypylina, many families such as the Carabodidae, Cymbaeremaeidae, Licneremaeidae, Scheloribatidae, Oripodidae, Mochlozetidae and Oribatulidae are known for their arboreal habits (Walter & Behan-Pelletier, 1999). Some genera in these families, such as *Scapheremaeus*, *Licneremaeus*, *Exoripoda*, *Protoripoda*, *Benoibates*, and *Phauloppia*, are restricted to canopy habitats, and species have diversified within these habitats. Little is known of the habitat associations of species of *Unguizetes* and *Chaunoproctus*, but a species of *Clavazetes* was recorded from the phylloplane in subtropical forests in Australia (Walter & Behan-Pelletier, 1993) and *Scheloribates* is known from suspended soils in temperate rainforests (Winchester *et al.*, 1999) and from trunk microhabitats in tropical rainforests (Franklin *et al.*, 1998). Other than *Tectocepheus velatus*, a cosmopolitan species found in soil, litter and arboreal habitats, all species in Table 10.3 share the short, clubbed-to-globose shape of the bothridial setae (sensillus) common to most canopy Oribatida (Aoki, 1967; Walter & Behan-Pelletier, 1999).

Oribatid density and species richness were not strongly associated with distance from the trunk. However, species composition changed over distance. The six 'specialist species' represented by a single individual suggest that a more intense sampling effort may be required to determine the full complement of mites inhabiting suspended soils along the length of the branch. Furthermore, it is possible that species composition changes over time with the development of suspended debris patches, similar to the succession of species found in litter bag studies either on the ground (Walter, 1985) or in the canopy (Fagan, 1999). Since we only sampled at a single time interval, it is likely that our sample design was adequate to describe the more common oribatids present, but not the complete species composition or, obviously, changes in abundance that occur over time.

The number of species shared among the four branch distances was low (11). As branch distance from the trunk increased, diameter decreased and suspended soil depth and moisture content also decreased. Among the seven 'generalist species', *Tectocepheus velatus* is a known thelytokous species (Norton *et al.*, 1993). It is possible that *Unguizetes* sp. 1, *Amerioppia* sp. and *Tegeozetes* sp. 1 are also thelytokous as no males have been recorded, though this may reflect the single time interval we sampled. Furthermore, adult *Unguizetes* sp. 1 carried up to 20 eggs at a time, an unusually high number for iteroparous oribatid mites, though higher egg numbers have been recorded for the thelytokous *Archegozetes longisetosus* Aoki (Honciuc, 1996).

The 13 species (*Steganacarus* sp., *Trimalaconothrus* sp. 1,*Carabodes* sp.1,*Clavazetes* sp. 1, *Amerioppia* sp.1, *Scapheremaeus* sp.2, *Licneremaeus* sp.1, *Licneremaeus* sp.2, *Micreremus* sp.1, *Scheloribates* sp.2, *Benoibates* sp. 1, *Phauloppia* sp.2 and *Notogalumna* sp.1) that appear to be restricted to a particular distance may lend support to the hypothesis that a component of the oribatid species assemblage is effected by subtle changes in suspended soils associated with distance from the trunk, such as depth of soil, moisture content or resource availability and quality. Certainly, species assemblages associated with decomposition events impose a series of successional episodes on decomposer ecosystems following the breakdown of their own resources (Wallwork, 1983). It is unlikely that distance from a source of colonization is itself a factor, since the suspended soil in our study was located in a continuous manner along branches. Overall, our results indicate that the oribatid fauna associated with these suspended soils is diverse and parallels trends in oribatid composition and distribution from different tree species and canopy habitats across geographically distinct forest ecosystems.

CONCLUSIONS

The canopy of a mature *O. gore* tree in the tropical rainforest of Gabon contained a rich assemblage of invertebrates inhabiting suspended soils. Oribatida dominated this fauna with a number of ubiquitous arboreal soil species (e.g. *Tectocepheus velatus*, *Chaunoproctus* sp., *Scheloribates* sp. and *Unguizetes* sp.). Factors accounting for the widespread dominance of Oribatida appear directly related to a tree architecture that allows for debris accumulation and subsequent suspended soil formation over time. Species may be distributed spatially the way they are as a result of factors such as feeding specificity, microclimatic preferences and structural characteristics that provide protection from predation by other invertebrates. We found that distance from the trunk rather than height from the ground was an important factor affecting mite species distribution in the *O. gore* tree crown. Oribatid species composition changed along the length of the branch, with several species only occurring at one distance (e.g. *Trimalaconothrus* sp., *Clavazetes* sp., and *Benoibates* sp.). Overall faunal composition was similar to that found in other tropical studies of suspended soils (Nadkarni & Longino, 1990; Wunderle, 1992a) but differences reported in other studies (e.g. Delamare-Deboutteville, 1951; Watanabe, 1997) clearly point out the need to harmonize sampling techniques that look specifically at resident suspended soil microarthropods when making comparisons. Our data support the thesis that suspended soils in geographically separated rainforests contain an abundant and diverse assemblage of oribatid mites and that basic physical characteristics of the tree (e.g. distance along branch) are among the many variables that contribute to this diversity.

Future direction

This analysis of suspended soil arthropod diversity in Gabon uses a rapid, repeatable sampling methodology and is the first in a series of proposed projects to analyse canopy microarthropod communities in ancient rainforests of the world. The ultimate goal of this research is to document species richness and examine common factors that shape the distribution, diversity, abundance and community composition of resident arthropods in high-canopy ecosystems. Not only is this information essential for assessing global biodiversity but it will also provide linkages among study sites that have direct relevance to global forest biodiversity issues. Ultimately this research should move us beyond descriptions of pattern, such as that presented in this paper, towards a more complete understanding of the functional role of canopy arthropods in ecosystem processes.

ACKNOWLEDGEMENTS

The study of arthropod biodiversity in ancient rainforests incorporates several ideas and talents. In particular, we are grateful to Prof. Francis Hallé and the staff of the Makandé 99 Scientific Mission to Gabon for making this research possible. In particular, we would like to thank Mr Kevin Jordan for his expertise in gaining safe access to the canopy and to Drs Richard Ring and Yves Basset for their valuable comments on this manuscript. This project is promoted by DIVERSITAS-IBOY (International Biodiversity Observation Year) as an example of a global biodiversity project that gathers and disseminates new information about biodiversity. Further information on IBOY and our scientific research programme conducted in Gabon can be viewed on the webpage: http://www.icsu.org/DIVERSITAS/Iboy/index.html

11

Vertical stratification of flying insects in a Surinam lowland rainforest

Bart P. E. De Dijn

ABSTRACT

From two 16 ha study plots in lowland, mixed rainforest at Akintosula (Surinam), samples of the understorey insect fauna were obtained with yellow pot traps placed on the ground; the canopy fauna was sampled with yellow pots suspended in tree crowns. Insect abundance and species richness per trap were compared between the two strata (understorey and canopy), across three different sampling periods and between the crowns of two common tree taxa, *Inga* (Mimosoidea) and *Tetragastris* (Burseraceae). Analyses were based on counts of the number of adult individuals of winged insects (abundance) and morphospecies (species richness) per trap for Coleoptera, Phoridae, non-phorid Diptera, non-ant Hymenoptera and Homoptera. Based on repeated measure analyses of variance for each of the insect groups investigated except Homoptera, the abundance per trap was significantly different between strata in at least one of the two study plots; where significant, abundance in the canopy was less than in the understorey. Based on Wilcoxon's test, Homoptera abundance per trap always differed significantly between strata, and in all these cases abundance was higher in the canopy. For insect groups such as Coleoptera and Diptera, there were significant differences in abundance between different sampling periods, which was interpreted as evidence of seasonality. The results based on the species-richness data were largely similar, but because the morphospecies were not cross-compared between sampling units, these results must be interpreted with caution. The species richness of Diptera in canopy traps in *Inga* was significantly different and higher than that in traps in *Tetragastris*. Overall evidence of between-tree-taxon differences was, however, weak, presumably because of limited sample size. The assumption that the canopy holds the bulk of the rainforest insects and arthropods, for example in terms of number of species, is discussed in the light of results in published studies as well as the present one. It is concluded that the general pattern may be the opposite, except for sap-sucking insects and possibly their allies and some other herbivorous groups. It is suggested that the construction of exhaustive inventories of canopy arthropods may be less important than detailed spatio-temporal studies of a variety of arthropod groups, including aspects of their behaviour.

INTRODUCTION

Research interest in the vertical distribution of insects in tropical forests heightened after Erwin & Scott (1980) presented a first analysis of data on beetles obtained by canopy fogging in Panama. There, the application of insecticides with fogging equipment in the crowns of a single tree species yielded an unprecedented diversity of species, most of which were assumed to be restricted to the forest canopy. The initial results on beetles inspired Erwin (1982) to extrapolate and estimate global land arthropod species richness at several tens of millions. This estimate was, amongst other things, based on the assumptions that beetles are a representative arthropod taxon, that most of the beetles present in the canopy are 'canopy specialists' that do not occur in the forest understorey and that the canopy has a much richer fauna than the understorey. Erwin's assumptions and estimate have been the subject of several critical reviews (e.g. May, 1990; Stork, 1993). Based on a recent study undertaken in an Indonesian rainforest, Hammond *et al.* (1997) concluded that less than 15% of the beetle species may be restricted to the canopy as adults, a larger percentage is restricted to the understorey and the rest presumably are stratum generalists. If we were to generalize this result, then an arthropod sample taken in the rainforest canopy would tend to be less species rich than a comparable sample taken in the understorey. This is quite contrary to the assumptions of Erwin (1982), and to the general conclusions drawn by Sutton (1989) in relation to flying insects.

Following Erwin's (1982) extrapolation, the discussion and research focus has been very much on arthropod species richness in rainforest canopies, and much less on arthropod abundance in different strata. When Basset *et al.* (1992) published a study on foliage arthropods in a Cameroon rainforest, they stressed that there were still few published studies on arthropod densities in forest canopies and even fewer documenting vertical stratification. Recent publications provide new data on arthropod stratification in tropical forests, for example on taxa such as fig wasps (Compton *et al.*, 2000), butterflies (DeVries *et al.*, 1997, 1999a), moths (Intachat & Holloway, 2000), Collembola (Rodgers & Kitching, 1998), mites (Walter *et al.*, 1998) and herbivorous insects (Basset *et al.*, 1999; Barone, 2000). The conclusions reached in these and other studies (e.g. Wolda & Wong, 1988; Hammond *et al.*, 1997) are varied: canopy faunas can be seen as divergent and often much richer than understorey faunas, as equally rich and largely similar to understorey faunas or are not interpreted in terms of richer or poorer but in terms of hosting certain arthropod taxa for different reasons at different times (e.g. for dispersal versus reproduction; in the dry versus the rainy season). This suggests that the research questions in relation to stratification need not be solely about species richness or diversity but also may concern the temporal aspects of stratification, and the behavioural and ecological meaning of the stratifications that are observed.

A variety of methods can be used to sample canopy arthropods, and some are more useful than others for the investigation of stratification (Basset *et al.*, 1997b). Canopy fogging, despite its high profile and frequent application since the early 1980s (cf. literature cited in Hammond *et al.*, 1997), has the distinct disadvantage that it does nor permit clear distinction between the actual canopy fauna and the fauna inhabiting lower strata (Basset *et al.*, 1997b). There appears to be no equivalent technique available to fog the lowest or intermediate forest strata. Sampling by sweeping with a net or by bagging branches is in principle more than adequate for assessing the densities of arthropods in different strata, but it requires excellent canopy access, which is simply not available at many research sites. In the absence of infrastructure allowing researchers to actually enter the canopy, it is nevertheless possible to suspend traps in the canopy, such as yellow pot traps (cf. Methods). When installed at different heights in the rainforest, traps can obtain comparable arthropod samples from different strata (Sutton, 1989; Casson & Hodkinson, 1991; Compton *et al.*, 2000).

A problem with traps is that they generally capture only a fraction – usually an unknown fraction – of the resident fauna. Consequently, they tend to be less appropriate for estimating absolute population densities. Trap contents will reflect the relative density (abundance) of those species active near the traps or attracted to them. Each kind of trap and sampling method in general has its particular biases and shortcomings, as reviewed by Southwood (1978). Comparative studies undertaken in an Indonesian rainforest, involving different insect sampling methods, indicate that Hymenoptera (Noyes, 1989a) and Hemiptera (Casson & Hodkinson, 1991) can be sampled successfully by means of commonly used insect traps, such as yellow pan traps. Nevertheless, each sampling method, including canopy fogging, was biased towards capturing certain Hymenoptera and Hemiptera subtaxa more than others. Wolda (1978a, 1979) and Wolda & Wong (1988), who investigated rainforest Homoptera in Panama, reached the same conclusion in relation to this taxon and, in addition, pointed out that the number of individuals trapped may reflect the level of activity rather than population density. This is obvious in the case of Homoptera, which diapause during the dry season and, accordingly, cannot be trapped in large numbers at that time. The fact that traps tend to reflect activity and may not capture all members of larger target taxon is an important point. The abundance of insects collected by means of a trap must be regarded as a function of their level of activity and their proneness to be intercepted by the kind of trap used. The catches may not reflect reliably the relative abundance of target taxa in the habitat where the trap was placed. A strict interpretation of trap contents as reflecting relative abundance of the taxon in the habitat or ecosystem (as in Wolda, 1978a,b) – that is, without considering the effect on trapping of situation-dependent differences in level of activity – can lead to erroneous conclusions.

Despite such complications with interpretation, trapping remains useful to obtain baseline data on flying insects (Wolda & Wong, 1988; Sutton, 1989) because using traps may represent one of the few practical options to obtain samples from habitats that are hard to access directly (by an observer or collector), such as the canopy. Hammond (1994) concludes that samples of modest size obtained by means of activity-based traps are very useful for comparative purposes. Such samples contain few species that do not 'belong' at the site where the traps are set up; they are biased towards common species and against rare ones, most of which can be regarded as vagrants ('tourists') anyway. This may imply that samples obtained by, say, pitfall traps or yellow pans may be more useful than fogging or Malaise samples. They certainly are easier to process because the sample contents are considerably less bulky. This matter is discussed by Sutton (1989), who suggests that some kinds of insect trap may even need 'a modicum of inefficiency' in order to be practical.

Data on numbers of individuals and species of insects trapped in the canopy and understorey at Akintosula, a Guiana Shield rainforest site, are presented below. The data will be used, principally, to test a null hypothesis of equal canopy and understorey insect abundance and species richness. They will be used, also, to touch briefly on the subject of differences in insect abundance and species richness between the crowns or canopies of different tree genera (also relevant to the evaluation of the Erwin (1982) extrapolation). Seasonality is considered also in our analyses of the Akintosula data. Contrary to popular (mis)conception, neotropical forests are not aseasonal (Janzen, 1983c). Insect populations in rainforests undergo considerable seasonal fluctuations (e.g. Wolda, 1978b; Wolda & Flowers, 1985; Pearson & Derr, 1986; Wolda & Wong, 1988; Novotny & Basset, 1998). The vegetation most certainly undergoes distinct seasonal changes in Surinam, where there are distinct wet and dry seasons (Schultz, 1960). The subject of insect seasonality is touched upon only briefly here as sampling was replicated in three periods, each in a different season.

To summarize, the following alternative hypotheses are addressed:

H1. abundance and species richness of flying insects trapped differs between canopy and understorey
H2. abundance and species richness of flying insects trapped in tree crowns differs between tree taxa
H3. abundance and species richness of flying insects trapped changes seasonally.

METHODS

General

The study reported here was subject to severe logistic, financial and staffing constraints typical of many small 'Third World' research organizations. Limited funds and field infrastructure were available and, in consequence, there was no direct canopy access. Yellow pot trapping is a technique that suggested itself because it was cheap and would yield – even in the absence of direct canopy access – comparable, modest-sized canopy and understorey samples of speciose insect groups. Other 'typical' constraints were the absence of taxonomic expertise beyond the order/family level, and the lack of adequate equipment to create useful reference collections of morphospecies. Accordingly, cross-matching of morphospecies between different individual sampling units (trap contents) was out of the question. Consequently, no figures could be generated that estimate overall species richness, that is, richness at scales beyond that of individual traps.

The original purpose of the trapping of insects at Akintosula was to assess the long-term impact of sylvicultural treatments on the fauna in general. It was part of a preliminary environmental impact assessment. The hypotheses tested below using the Akintosula insect data were formulated *a posteriori*, and the original study was not designed to address them specifically. The analyses presented should be regarded as exploratory (i.e. as 'data analysis' *sensu* Sokal & Rohlf, 1995). They involve the use of (small) subsets of data filtered out of a larger body of original data, something that may limit the power of the statistical tests.

Study area

The study was performed at Akintosula, in the hilly lowlands of the interior of Surinam, about 90 km inland from the Atlantic Ocean. The area has mixed, mesophytic rainforest on loamy to gravelly soil, which extends to the south and east. Just a few kilometers to the north and west, the soil becomes sandy and is covered with more xerophytic forest. The Mapane creek, a tributary of the Commewijne river, drains the area. It is a clearwater creek, along which are pockets of marsh forest; that is, seasonally flooded, palm-dominated forest. The mixed rainforest in the general area had a closed canopy some 20–35 m above soil level, with some emergent trees reaching up to 40–55 m. *Inga* spp. (Mimosoidea) and *Tetragastris altissima* (Aublet) Swart

(Burseraceae) were common tree species in this forest (Schultz, 1960; B. P. E. De Dijn, personal observation).

Insect sampling took place at two 16 ha rainforest plots: the first was a less disturbed plot south of the Mapane creek, and the second was an experimentally treated plot at the other side of the creek. The plots were located some 600 m from each other. Both plots fringed the creek at one side and reached the top of a low ridge at the other side. Although most of the terrain was well (>1 m) above the rainy season level of the creek, some 5–10% of the forest flooded occasionally at the peak of the long rainy season (May–July). The treated plot had been logged and sylviculturally treated some 20 years earlier. This treatment involved the poisoning of all nontimber trees larger than 20 cm DBH (diameter at breast height) and cutting of larger lianas and palms (details in De Graaf, 1986). The less disturbed plot had been lightly logged several times, but not poisoned. Part of the forest in this plot is 'liana forest'; that is, a generally low forest (canopy at 10–20 m) with dense liana mats in the canopy that occasionally descend to the ground (in treefall gaps). Both forest plots had a closed canopy except where there had been recent treefalls. In contrast to the less-disturbed plot, the treated plot had no emergent trees but many early-secondary forest trees of the genera *Cecropia* and *Pourouma* (Cecropiaceae).

Insect sampling and processing

Four parallel north–south transects of 400 m length each were made, 100 m distance from each others, through each of the two study plots. Each transect was subdivided into four 100 m segments, the midpoint of which was marked in the field with a PVC pole. Each of these poles marked a sampling point that may be regarded as the centre of an imaginary 100 m × 100 m (1 ha) nonoverlapping quadrat (resulting in a 4 × 4 quadrat pattern in each plot).

Yellow pot traps were used, a trap type that primarily collects insects which are flying near the trap and are attracted to it because of its colour, yellow being attractive to a host of phytophagous insects and their predators (Southwood, 1978; Noyes, 1989a). Yellow plastic pots 8.5 cm in height and 15.0 cm in diameter were half-filled with a diluted formaldehyde solution (2–4%; with some liquid soap). Around each of the PVC marker poles, three yellow pots for sampling understorey insects were set up at ground level, about 10 m from each other (in a triangular pattern). Three pots were placed at each sampling point in an attempt to ensure the successful retrieval of the contents of at least one trap. Only one of the trap contents retrieved (randomly selected) per sampling point was considered for data analysis. Yellow pots for sampling canopy insects were suspended by means of thick (about 1 mm) nylon threads and pulled up into the crown of canopy trees, some 0–15 m from the nearest PVC marker pole. The exact position of each trap in the canopy was variable, but traps were always placed in the lower half of the closed upper canopy (never in the upper half of the canopy and not in emergent trees), which meant that most were at 15 to 24 m above the ground (one trap was at 12 m near a clearing in high forest, and three traps were between 10 and 12 m in liana forest). For trap placement, an ultrathin nylon thread was first shot over a canopy branch with a sling-shot; this was used to put in place the thick nylon thread from which the canopy pot was suspended. As the installation of the canopy traps proved to be time and labour intensive, only one trap was installed at each sampling point. All trap pots were filled with formaldehyde solution within 2 days (half of the pots at each plot on day 1 and the rest on day 2) and were emptied in the same sequence 3 days later. Insects were recovered from the traps by filtering the trap fluid, using coffee filters placed in percolators. The tree species in which canopy traps were placed were identified with the help of a field technician able to recognize the local tree species.

The understorey and canopy pots were serviced three times in the course of 7 months: August 1994 (long wet season), October 1994 (long dry season) and January 1995 (short dry season). In this fashion, three understorey and three canopy samples were retrieved by means of the same trap type at the same sampling points. Although traps at all 16 sampling points per plot were serviced during sampling periods 1 and 2, only 12 traps per plot were serviced during sampling period 3. This was because the working conditions in the field had deteriorated considerably by January 1995 (much treefall, increased human disturbance of the plots and the sampling set up). The contents of the pots in the understorey could often not be recovered because of miscellaneous accidents (e.g. disturbance by large animals or humans), and canopy pots could often not be serviced properly because the suspension threads got stuck (this problem got worse over time, possibly because of wear of the threads).

The focus of this study was on flying adult insects, not on immatures, flightless or typically soil-inhabiting arthropods. This was commensurate with the use of the yellow pot traps. The insects recovered from each sampling unit (i.e. each individual yellow pot) were sorted into the following groups: Coleoptera, Phoridae, non-phorid Diptera, non-ant Hymenoptera and Homoptera (the last were not sorted out of the January 1995 understorey samples). For each group, the number of adult (winged) individuals and morphospecies per sampling unit were counted. These counts were interpreted as measures of the abundance of the insect groups in the immediate vicinity of the traps. The insect species trapped were not identified; instead, morphospecies were simply distinguished on the basis of large differences in body length (×2 or more) or differences in colour pattern, integument texture or shape of the appendages. This was done by two parataxonomists, who were trained and assisted by an experienced entomological technician. For each insect group, the number of (adult) morphospecies per sampling unit was counted. These counts were interpreted as measures of species richness. The orders Hymenoptera, Coleoptera and Diptera were selected because they were the most speciose groups of flying insect usually captured by means of yellow pot traps in Surinam (e.g. University of Suriname, 1996). Homoptera was selected because it was observed in the field at Akintosula that substantial numbers of this order were present in the canopy samples (cf. Results). Collembola were often caught in large numbers in the traps (B. P. E. De Dijn, personal observation) but were not counted because they are flightless and presumably representative of the soil fauna rather than that of the understorey. Ants, although occasionally captured in substantial numbers in yellow pot traps in Surinam (B. P. E. De Dijn, unpublished results), were excluded from the Hymenoptera counts. This was not just because most ants are flightless and many are soil dwellers but also because the ants constitute a taxon that is behaviourally very different from most other Hymenoptera (cf. Davidson & Patrell-Kim, 1996). Phoridae were sorted out separately because they had been observed to dominate strongly yellow pot samples taken elsewhere in Surinam (University of Suriname, 1996) and, in addition, were easy to distinguish.

Data analysis

To test the hypothesis of differences in abundance and species richness between canopy and understorey (H1), a repeated measure general linear model (GLM) analysis of variance (ANOVA) model was used, with the following *a priori* factors (sources of within-subject variability): stratum (canopy and understorey), and time (sampling periods 1 to 3). A repeated measure analysis was appropriate because the traps placed at the same sampling point (near the same PVC marker pole) represented repeated measurements in space and time of the same subject: the subject being the rainforest in the immediate vicinity of a marker pole. Given its design, the repeated measure analysis also allowed for the assessment of the effect of seasonality on insect abundance and species diversity (H3). The data from each plot were analysed separately.

The hypothesis on insect abundance and species richness in canopies of different tree genera (H2) was tested using the data on the two most frequently sampled tree genera, *Inga* and *Tetragastris*. The data used were derived from trapping that happened to have taken place in individual trees belonging to these two common tree genera. Data from both plots were pooled to increase sample size and thus the power of the analyses. Student t-tests were performed separately for each sampling period. Series of simple paired analyses were preferred because using a repeated measure ANOVA design would mean that the many subjects (i.e. individual trees) for which the dataset was incomplete would be excluded from the analysis, which would result in a serious decrease in sample size.

Figures for number of individuals and morphospecies trapped were log transformed ($\ln[x + 1]$) before hypothesis testing, subjected to runs tests, and also tested for normality (Kolmogorov–Smirnov test) before and after transformation. All data passed the runs tests ($p > 0.05$). All untransformed and transformed data passed the normality tests ($p > 0.05$) except for the Homoptera understorey data, and the Coleoptera canopy data of the first sampling period. The latter was a minor problem, but the former was important enough to warrant using nonparametric methods for the Homoptera data. Wilcoxon's tests were, therefore, used as alternatives to investigate H1 (separate test for each sampling period; H2 not investigated nonparametrically); Mann–Whitney tests were used as an alternative to investigate H3. For the sake of uniformity and comparability, not only the Homoptera data but also that on the other insect groups were analysed using these nonparametric methods. All the above mentioned tests were performed using SPSS^{TM} Ver. 8.0 software for Windows^{TM}, in agreement with Sokal and Rohlf (1995).

The number of significant test results observed was compared with the number of significant results that could be expected on the basis of chance alone if the null hypothesis were correct (at $p < 0.05$). This was done by comparing the frequency of observed significant and nonsignificant outcomes with frequencies calculated on the basis of a binomial distribution, given $p = 0.05$ and $q = 0.95$ (calculations based on Sokal & Rohlf (1995) and performed using Quattro Pro^{TM} software).

RESULTS

General

In the course of the three sampling periods, the contents of a total of 157 yellow pots were retrieved successfully and considered for analysis (19 were not). A total of 5019 adult insects were sorted out of these trap contents (2567 from the sylviculturally treated, 2452 from the less-disturbed plot) comprising 675 Coleoptera, 2272 non-phorid Diptera, 334 Phoridae, 1011 non-ant Hymenoptera, and 727 Homoptera. Summary statistics in relation to the various samples obtained are presented in Table 11.1.

Vertical stratification

The GLM repeated measure ANOVAS with stratum and sampling period as *a priori* factors produced the most comprehensive evidence of vertical stratification of flying insects at the study site. The results of these analyses are presented in Table 11.2. Stratum proved to be a significant factor ($p < 0.05$) determining Coleoptera abundance in the traps at both plots investigated. As far as the Coleoptera species richness (i.e. number of Coleoptera morphospecies) per trap is concerned, the trend was the same. However, the results of the analyses indicate that in this case stratum was only a significant factor at the sylviculturally treated plot, not at the less-disturbed plot. Data on other insect groups led to comparable results. At the sylviculturally treated plot, stratum was a significant factor determining non-phorid Diptera abundance in the traps, but not their species richness. At the less-disturbed plot, stratum was not a significant factor for non-phorid Diptera, but it was a significant one for Phoridae and non-ant Hymenoptera. At this plot there were, in general, more individuals and morphospecies of phorids and non-ant hymenopterans per trap in the understorey than in the canopy. At the sylviculturally treated plot, stratum was not a significant factor in relation to these two insect groups.

The numbers of significant outcomes (i.e. number of rejections of the null hypothesis at $p < 0.05$) obtained here were compared with the corresponding binomial probabilities of getting, by pure chance, at least as many such outcomes if the null hypothesis were actually correct. In general, for $n =$ total number of outcomes, and $a =$ number of significant outcomes: if $n = 4$, then $p(a \geq 1) = 0.185$, $p(a \geq 2) = 0.014$, and $p(a \geq 3) < 0.001$. The observed outcomes in relation to insect stratification can be summarized as follows: for $n = 4$ insect groups tested, $a = 2$ at the sylviculturally treated plot, and $a = 3$ at the less disturbed plot. Given that both $p(a \geq 2) < 0.05$ and $p(a \geq 3) < 0.05$, the conclusion can only be that the number of rejections of the null hypothesis observed at each plot was higher than the number expected in the case where the null hypothesis was correct. In relation to species richness, the number of observed significant outcomes was more than expected for the less disturbed plot ($a = 2$), but not for the sylviculturally treated plot ($a = 1$; $p(a \geq 1) > 0.05$). Combining all the outcomes in relation to both plots investigated ($n = 8$), the conclusion would be that, overall, more significant differences in abundance and species richness were detected than would be expected in a situation where the null hypothesis were correct (given that here $p(a \geq 3) = 0.006$).

Where tests indicated that differences in abundance or species richness per trap between the two strata were significant (cf. Table 11.2 and above), it is informative to look at the approximate magnitude of the differences by comparing the corresponding averages in Table 11.1. The average abundance of Coleoptera per canopy trap was up to twice that in understorey traps, or exponentially higher (10 times; first sampling period, sylviculturally treated plot). The latter outcome is based, however, on data that are less suitable for such a comparison given that they are not normally distributed (Kolmogorov–Smirnov test: $p < 0.05$); the former (second sampling period, less-disturbed plot) is based on more suitable, approximately normally distributed data (Kolmogorov–Smirnov test: $p > 0.05$). The average species richness of Coleoptera per canopy trap was up to 1.5 times that per understorey trap (second sampling period), in one case approximately twice (first sampling period; less-suitable data). For non-phorid Diptera, the average abundance per understorey trap at the sylviculturally treated plot was

Table 11.1. *Summary statistics on adult insects captured by means of yellow pot traps in the rain forest canopy and understorey of a sylviculturally treated plot (T) and a less-disturbed plot (L) at Akintosula (Surinam), over three sampling periods*

	Period 1[a]		Period 2[a]		Period 3[a]	
	Canopy	Understorey	Canopy	Understorey	Canopy	Understorey
Sampling units[b]						
T	15	16	15	15	8	11
L	16	16	10	12	11	12
Adult insect individuals per trap[c]						
Coleoptera						
T	0.9 ± 1.3	3.1 ± 1.6	6.9 ± 2.8	7.6 ± 2.8	5.0 ± 1.6	5.3 ± 2.2
L	0.3 ± 0.4	3.2 ± 1.2	2.9 ± 2.2	6.9 ± 4.0	4.7 ± 1.6	6.3 ± 2.4
Non-phorid Diptera						
T	8.7 ± 4.6	38.9 ± 7.4	7.4 ± 3.3	13.7 ± 5.7	3.6 ± 2.3	11.1 ± 5.9
L	9.3 ± 2.9	32.1 ± 8.9	4.7 ± 3.0	13.0 ± 4.4	8.5 ± 6.2	7.7 ± 5.4
Phoridae						
T	0.9 ± 0.5	2.6 ± 1.5	1.9 ± 1.2	3.1 ± 2.0	1.3 ± 1.7	2.5 ± 1.5
L	1.4 ± 0.5	2.1 ± 0.8	1.3 ± 0.8	4.9 ± 3.9	1.4 ± 1.1	1.8 ± 1.4
Non-ant Hymenoptera						
T	7.3 ± 2.7	8.0 ± 2.5	4.8 ± 2.0	5.9 ± 2.1	3.0 ± 1.3	4.9 ± 2.1
L	8.8 ± 3.5	8.8 ± 2.3	2.9 ± 1.4	9.1 ± 5.2	2.5 ± 1.4	7.4 ± 3.8
Homoptera						
T	6.3 ± 4.9	0.0 ± 0.0	9.3 ± 3.3	2.2 ± 1.0	7.0 ± 3.3	–
L	6.8 ± 4.3	0.3 ± 0.4	13.4 ± 8.0	2.7 ± 1.0	11.5 ± 5.1	–
Insect morphospecies per trap[d]						
Coleoptera						
T	0.4 ± 0.4	1.4 ± 0.4	4.7 ± 1.7	3.9 ± 0.7	5.0 ± 1.6	3.4 ± 1.0
L	0.3 ± 0.4	1.4 ± 0.5	2.1 ± 1.0	3.0 ± 1.3	4.2 ± 1.5	4.0 ± 1.0
Non-phorid Diptera						
T	4.9 ± 1.7	10.1 ± 1.7	5.0 ± 2.0	6.7 ± 2.1	2.9 ± 1.7	6.1 ± 2.6
L	6.4 ± 1.6	7.7 ± 2.1	3.5 ± 1.8	6.8 ± 2.1	4.0 ± 1.9	4.0 ± 2.3
Phoridae						
T	0.7 ± 0.4	1.6 ± 0.6	1.0 ± 0.4	1.5 ± 0.5	0.5 ± 0.4	1.5 ± 0.7
L	1.1 ± 0.3	1.6 ± 0.3	0.8 ± 0.5	1.5 ± 0.7	0.8 ± 0.5	1.1 ± 0.6
Non-ant Hymenoptera						
T	5.9 ± 1.9	6.6 ± 2.1	3.4 ± 1.1	4.8 ± 1.6	2.6 ± 1.3	4.6 ± 2.1
L	7.1 ± 2.4	7.0 ± 1.6	2.6 ± 1.3	5.5 ± 1.1	2.3 ± 1.2	6.3 ± 2.7
Homoptera						
T	1.5 ± 0.5	0.0 ± 0.0	7.1 ± 2.6	1.9 ± 1.0	5.4 ± 2.4	–
L	1.6 ± 0.6	0.3 ± 0.4	7.8 ± 3.9	2.2 ± 0.7	8.3 ± 2.1	–

[a]Periods: 1, August 1994 (long wet season); 2, October 1994 (long dry season); 3, January 1995 (short wet season).

[b]No. trap contents successfully retrieved and considered for analysis.

[c]Mean ± 2SE (where SE is standard error of the mean: approximate amplitude of 95% confidence interval for the mean); equivalent to abundance per trap.

[d]Mean ± 2SE (where SE is standard error of the mean: approximate amplitude of 95% confidence interval for the mean); equivalent to species richness per trap.

Table 11.2. *Significance of* a priori *factors, stratum and sampling period, in general linear model repeated measure ANOVAS of abundance and species richness for insects trapped with yellow pots at two rainforest plots, one sylviculturally treated (T) and the other less-disturbed (L), in Akintosula*

Abundance (A) and species richness (S) for taxa	Factor stratum		Factor sampling period		Interaction between factors	
	T ($n = 8$)	L ($n = 7$)	T ($n = 8$)	L ($n = 7$)	T ($n = 8$)	L ($n = 7$)
Coleoptera						
A	0.005*	0.045*	<0.001*	0.004*	<0.001*	0.716
S	0.042*	0.207	<0.001*	0.001*	<0.001*	0.791
Non-phorid Diptera						
A	0.007*	0.084	0.002*	0.003*	0.085	0.055
S	0.056	0.625	0.051	0.021*	0.380	0.290
Phoridae						
A	0.211	0.013*	0.452	0.299	0.905	0.065
S	0.095	0.025*	0.450	0.270	0.610	0.534
Non-ant Hymenoptera						
A	0.270	0.026*	0.479	0.500	0.154	0.774
S	0.382	0.023*	0.523	0.538	0.239	0.793

*$p < 0.05$.

up to approximately four times that per canopy trap and average species richness was up to approximately twice (first sampling period). For Phoridae and non-ant Hymenoptera, the average abundance per understorey trap at the less-disturbed plot was up to approximately three times that per canopy trap and average species richness was up to approximately twice.

The above analyses do not relate to abundance and species richness of Homoptera. Differences in numbers of individuals or species of Homoptera per trap between canopy and understorey traps were investigated using Wilcoxon tests (pairwise tests for each sampling period separately). The results are presented in Table 11.3, together with the results of the same type of test applied to the data on the insect groups already analysed above using the GLM ANOVAS. The abundance and species richness of Homoptera per trap differed significantly between canopy and understorey (in all cases: $p < 0.05$). Table 11.1 indicates that, on average, both abundance and species richness per trap were much lower in the understorey than in the canopy, at both plots and for both sampling periods for which there were data available. For Homoptera, obviously, the number of significant outcomes obtained in relation to abundance as well as species richness was much more than could be expected if the null hypothesis were correct: for $n = 4$ tests on data of two periods and two plots, and $a = 4$ (all observed outcomes significant), $p(a \geq 4) < 0.001$ (based on binomial probabilities).

For Homoptera, the average abundance and species richness in the canopy traps was approximately three to four times that in the understorey ones, at least during the second sampling period (cf. Table 11.1). During the first sampling period, no or hardly any Homoptera were caught in the understorey traps, and the difference between the two strata in average numbers of Homoptera caught was exponential.

The results of analogous Wilcoxon tests applied to data on the other insect groups show that there were many cases of significant differences in abundance (10 out of 24 cases) and species richness (9 out of 24 cases) between canopy and understorey traps (Table 11.3). For each of these insect groups (Coleoptera, non-phorid Diptera, Diptera, and non-ant Hymenoptera), the observed number of significant outcomes ($n = 6$; $a = 2$ to 4) was higher than expected on the basis of chance alone if the null hypothesis were correct (here, for example, $p(a \geq 2) = 0.033$).

Table 11.3. *Significance of differences in abundance and species richness, based on Wilcoxon's test, between insects trapped in the rainforest canopy and understorey by means of yellow pots at a sylviculturally treated plot (T) and a less-disturbed plot (L) in Akintosula, over three sampling periods*

Abundance (A) and species richness (S) for taxa	Period 1[a]		Period 2[a]		Period 3[a]	
	T ($n = 15$)	L ($n = 16$)	T ($n = 15$)	L ($n = 10$)	T ($n = 8$)	L ($n = 11$)
Coleoptera						
A	0.016*	0.004*	0.363	0.262	0.345	0.423
S	0.009*	0.017*	0.623	0.372	0.042*	0.824
Non-phorid Diptera						
A	0.001*	0.001*	0.031*	0.005*	0.161	0.594
S	0.002*	0.443	0.162	0.028*	0.208	0.878
Phoridae						
A	0.032*	0.235	0.421	0.035*	0.291	0.591
S	0.049*	0.285	0.142	0.039*	0.086	0.404
Non-ant Hymenoptera						
A	0.955	0.469	0.636	0.007*	0.106	0.036*
S	0.802	0.507	0.414	0.009*	0.092	0.049*
Homoptera						
A	0.001*	0.002*	0.002*	0.008*	–	–
S	0.001*	0.007*	0.002*	0.005*	–	–

[a]Periods: 1, August 1994 (long wet season); 2, October 1994 (long dry season); 3, January 1995 (short dry season).
*$p < 0.05$.

Where Wilcoxon tests yielded significant outcomes, abundance or species richness per trap were, on average, higher in the understorey than in the canopy, with just one exception where the situation was the reverse. This was for species richness of Coleoptera during sampling period 3 at the sylviculturally treated plot (Table 11.1).

Sampling period

Sampling period proved to be a significant factor determining the abundance and species richness of Coleoptera at both plots, for determining non-phorid Diptera abundance at both plots, and Diptera species richness at the less-disturbed plot only. At the less-disturbed plot, the general trend over time for Coleoptera was one of increasing average abundance and species richness per trap (based on averages in Table 11.1), whereas the trend for non-phorid Diptera at the same plot was a declining one. At the sylviculturally treated plot, the trends were parallel, but only from sampling period 1 to 2 (the increasing/decreasing trend levelled off towards sampling period 3). Sampling period was not a significant factor for Phoridae or non-ant Hymenoptera.

The observed number of significant outcomes ($n =$ 4 insect groups tested at each plot) was higher than expected if the null hypothesis was true for insect abundance ($a = 2$ at both plots; cf. binomial probabilities cited earlier). For insect species richness it was only so at the less-disturbed plot ($a = 2$), but not at the sylviculturally treated plot ($a = 1$). Combining the outcomes in relation to both plots ($n = 8$), the conclusion is that, overall, more significant differences in abundance ($a = 4$) and species richness ($a = 3$) were detected than expected if the null hypothesis were correct ($p(a \geq 3) = 0.006$).

Table 11.4. *Abundance and species richness per trap of adult insects captured by means of yellow pot traps in tree crowns of* Inga *and* Tetragastris *at Akintosula, over three sampling periods*

Abundance (A) and species richness (S) of taxa[a]	Period 1[b]		Period 2[b]		Period 3[b]	
	Inga ($n = 6$)	*Tetragastris* ($n = 8$)	*Inga* ($n = 5$)	*Tetragastris* ($n = 7$)	*Inga* ($n = 4$)	*Tetragastris* ($n = 6$)
Coleoptera						
A	0.3 ± 0.4	0.5 ± 0.8	7.6 ± 6.6	2.6 ± 2.0	5.3 ± 4.0	4.0 ± 1.4
S	0.3 ± 0.4	0.5 ± 0.8	5.6 ± 4.4	1.9 ± 1.2	3.8 ± 3.5	4.0 ± 1.4
Non-phorid Diptera						
A	14.0 ± 7.9	8.5 ± 7.6	13.0 ± 7.6	4.3 ± 1.4	14.0 ± 15.0	2.7 ± 1.8
S	8.5 ± 4.1	3.4 ± 1.8	8.0 ± 4.5	3.1 ± 1.5	4.8 ± 3.8	2.3 ± 1.4
Phoridae						
A	1.3 ± 1.0	0.9 ± 0.7	0.6 ± 0.8	2.1 ± 2.1	0.5 ± 1.0	1.7 ± 2.2
S	1.2 ± 0.8	0.6 ± 0.4	0.4 ± 0.5	1.0 ± 0.6	0.5 ± 1.0	0.7 ± 0.7
Non-ant Hymenoptera						
A	12.0 ± 8.1	6.0 ± 3.0	3.4 ± 2.6	5.0 ± 4.0	3.8 ± 3.0	3.2 ± 1.8
S	7.3 ± 5.2	5.4 ± 2.6	2.8 ± 1.9	3.9 ± 2.1	3.3 ± 2.6	3.0 ± 1.7
Homoptera						
A	4.8 ± 5.4	11.0 ± 9.9	19.0 ± 13.0	8.1 ± 5.3	14.0 ± 13.0	7.5 ± 2.8
S	0.8 ± 0.6	1.9 ± 0.8	11.0 ± 5.6	5.3 ± 3.3	8.5 ± 3.7	6.5 ± 3.1

[a]Average ± standard error of mean (approximate amplitude of 95% confidence interval for the mean).
[b]Periods: 1, August 1994 (long wet season); 2, October 1994 (long dry season); 3, January 1995 (short dry season).

The average abundance and species richness of Coleoptera per trap changed over time by a factor of approximately 1.5 (second to third sampling period) or exponentially (first to second or third sampling period; cf. Table 11.1). For non-phorid Diptera, the average abundance per trap changed by a factor of up to approximately three; average species richness per trap changed by a factor of up to approximately two.

Tree taxon

In the course of the three sampling periods, a total of 36 yellow pot trap contents were successfully retrieved from *Inga* and *Tetragastris* tree crowns (six were not). A total of 1040 adult insects were sorted out of these trap contents: 107 Coleoptera, 312 non-phorid Diptera, 45 Phoridae, 207 non-ant Hymenoptera and 369 Homoptera (Table 11.4).

Student's t-test and Mann–Whitney tests of the null hypothesis showed that there are no differences in average abundance and diversity of insects per trap between *Inga* and *Tetragastris* (Table 11.5). The null hypothesis was only rejected for the Student's t-test results for non-phorid Diptera: twice ($a = 2$) on the basis of species richness data and just once on the basis of abundance data ($a = 1$). Based on Mann–Whitney tests, the null hypothesis was rejected for only one out of three sampling periods ($a = 1$), on the basis of species richness data (for abundance data: $a = 0$). Given that $n = 3$ (three sampling periods), two significant outcomes ($a = 2$) is more than would be expected if the null hypothesis were indeed correct ($p(a \geq 2) = 0.007$), meaning that the observed difference in average number of non-phorid Diptera trapped is indeed significant. When test results on all insect groups (five) and all sampling periods (three) are combined, the number of significant outcomes obtained does not differ from that expected if the null hypothesis were correct (for $n = 15$; $p(a \geq 2) = 0.171$). This suggests that in

Table 11.5. *Significance of differences in abundance and species richness, based on Mann–Whitney and Student's* t*-tests, between insects trapped in the crowns of* Inga *and* Tetragastris *spp. by means of yellow pots at Akintosula, over three sampling periods*

Abundance (A) and species richness (S) for taxa	Period 1[a]		Period 2[a]		Period 3[a]	
	M-W test	T test	M-W test	T test	M-W test	T test
Coleoptera						
A	0.871	0.909	0.141	0.103	0.591	0.952
S	0.871	0.909	0.097	0.058	0.827	0.522
Non-phorid Diptera						
A	0.291	0.175	0.057	0.023*	0.165	0.203
S	0.058	0.031*	0.049*	0.028*	0.238	0.455
Phoridae						
A	0.456	0.518	0.234	0.234	0.469	0.462
S	0.282	0.352	0.268	0.211	0.762	0.681
Non-ant Hymenoptera						
A	0.271	0.455	0.622	0.543	0.831	0.726
S	0.649	0.803	0.459	0.462	0.975	0.841
Homoptera						
A	0.265	0.353	0.146	0.107	0.741	0.277
S	0.056	0.076	0.101	0.082	0.386	0.433

M-W, Mann–Whitney; T test, Student's *t*-test.

[a]Periods: 1, August 1994 (long wet season) for six *Inga* and eight *Tetragastris* crowns; 2, October 1994 (long dry season) for five *Inga* and seven *Tetragastris* crowns; 3, January 1995 (short dry season) for four *Inga* and six *Tetragastris* crowns.

*$p < 0.05$.

general – taking into account all insect groups – the null hypothesis cannot be rejected.

The average abundance of non-phorid Diptera per trap in *Inga* crowns was up to three times that in *Tetragastris* and the average non-phorid Diptera species richness per trap in *Inga* crowns was up to two times that in *Tetragstris* (cf. Table 11.4).

DISCUSSION

The methodology used at Akintosula was biased towards abundant and actively flying insects, because the trapping relied on the attraction of passing insects to a short-range visual cue. The trapping was severely biased towards insects attracted to yellow. Accordingly, such abundant taxa as Lepidoptera will be sampled hardly at all, and most probably various subgroups of Hymenoptera and Homoptera were sampled to a much lesser extent than other subgroups (cf. Noyes, 1989a; Casson & Hodkinson, 1991). This was possibly also the case for Coleoptera and Diptera. The fact that some subgroups are well sampled in the Akintosula data should not be taken as evidence that the larger taxa to which they belong are equally well sampled. The data may not even be representative of the entirety of the focal groups themselves. Nevertheless, our approach has been a purely comparative one, and there is no evidence to suggest that any biases were different between samples. It has been suggested that differences in trap visibility (as affected, say, by surrounding vegetation) may result in biases (Noyes, 1989a). Additional data we have collected at Kabo (Surinam) suggest that such bias is unlikely. At Kabo, yellow pots were placed on the ground in virgin and disturbed forest and measurements were made within a radius of 10 m around the traps of the density of vegetation from ground level to about 1 m above the ground (B. P. E. De Dijn, unpublished data). There was no significant correlation (Kendall's tau; $p > 0.05$) between abundance or species richness of insects trapped at Kabo and the density of vegetation at forest floor level (which varied considerably).

Whether the numbers of insects captured at Akintosula reflect levels of insect activity rather than insect abundance is another issue. There are no additional data available from Surinam to address this problem. Given observations of activity shifts among Panamanian Homoptera (Wolda, 1978b, 1979; Wolda & Wong, 1988), one cannot exclude the idea that the extremely poor catches of Homoptera in the canopy at Akintosula during the first sampling period may have been the result of low levels of activity. The very poor catches of Coleoptera during that same period may also point to low levels of activity rather than low abundance. Only by detailed studies involving direct sampling (e.g. with a sweep net) and observations would one be able to address this issue adequately (see also discussion in Sutton, 1989).

Given that trap contents were not cross-compared, the Akintosula data reflect the abundance and species richness of active insects in the vicinity of individual trapping units, that is, at smaller spatial scales than is the case in studies that involve the extrapolation of species richness on the basis of cross-comparable data from arrays of sampling units (such as Erwin and Scott, 1980). This makes the Akintosula data less comparable with data from such studies. The large standard errors in Tables 11.1 and 11.4 indicate that numbers of insect individuals and species trapped were highly variable, which is not surprising given that samples were taken in a tropical rainforest – an ecosystem that is well known for its great heterogeneity – and given that the traps essentially captured only part of the fauna active in the microhabitat where the trap happened to be placed. In line with Basset *et al.* (1992), we note that the sampling of microhabitats is not inherently less valid than that of sampling macrohabitats, especially when the goal is to study spatio-temporal patterns in a comparative manner (as opposed to undertaking inventories).

The results of the Akintosula study indicate that in most cases insect abundance and species richness per trap were either significantly lower in the canopy than in the understorey or not significantly different between canopy and understorey (at least for such major taxa as Diptera, Hymenoptera, and Coleoptera). Only for Homoptera, were abundance and species richness per trap significantly higher in the canopy than in the understorey. This is in agreement with data reviewed by Sutton (1989) and recent data where felled, adult trees in a Guyanese rainforest had more species of herbivores than conspecific seedlings (Basset *et al.*, 1999). Relatively lower or about equal species richness of Coleoptera in the canopy compared with the understorey is contrary to the assumption of Erwin (1982), but in line with results of Hammond *et al.* (1997, cf. Introduction). Intachat & Holloway (2000) observed that species richness and abundance of geometrid moths in the canopy was similar to that in the lower strata of Malaysian rainforest. May (1990) and Stork (1993) reviewed published evidence and concluded that the canopy probably holds substantially less arthropod species than the remainder of the forest (i.e. the understorey *and* the soil). The Akintosula results would seem be in agreement with this. Basset *et al.* (1992) did conclude, however, that density and species richness of foliage arthropods was higher in the canopy of a Cameroon rainforest. In line with this, Barone (2000) concluded that herbivore densities in the canopy are higher in Panama. The methods applied in Cameroon were, however, strongly biased towards less active insects present during the day only. This probably also applies to the Panamanian studies. Studies of two taxa of flightless arthropods in Australia, mites (Walter *et al.*, 1998) and Collembola (Rodgers & Kitching, 1998), led to the conclusion that the canopy had a distinct fauna (with many canopy specialists), which was, however, not richer than the soil or understorey fauna. The general pattern may be that sap-sucking insects, as well as the ants tending them (cf. discussion in Basset *et al.*, 1992) and other herbivorous arthropods, are more abundant and species rich in the rainforest canopy, in contrast to the majority of flying insects and flightless arthropod groups. The published results are not always easily comparable, for a variety of reasons: in some studies both immatures and inactive adults have been recorded; in others only data on active adults have been used (as in the Akintosula study). Given the complexity of insect life cycles and behaviour, it is not surprising that patterns of stratification are fundamentally different for different developmental or life stages.

As the focus of the analysis of the Akintosula data was on stratification, the results on seasonality and between-tree differences in abundance and species richness of insects trapped will not be discussed at length. The observed seasonal changes in abundance and species richness of some insect groups at Akintosula is unsurprising given the very obvious seasonality of the climate in Surinam (Schultz, 1960) and is in general agreement with data obtained elsewhere on, for example, rainforest Homoptera (Wolda, 1978b, 1979), Coleoptera (Erwin & Scott, 1980) and moths (Intachat & Holloway, 2000). Differences in arthropod abundance and species richness between tree species have been observed by Basset (1999b) in Guyana and Barone (2000) in Panama. The failure to detect, on the basis of the Akintosula data, convincing between-tree taxon differences in insect abundance or species richness may have simply been a consequence of the very small sample sizes.

The Akintosula study and some other studies discussed above do seem to indicate that for several speciose insect groups the rainforest understorey may be equally or more hospitable than the canopy. Although comprehensive data are lacking on most tropical insects, it would appear that some of them modify their level of activity in the course of the year and because of this may appear to be less abundant when activity-based sampling techniques are used. When data on tropical Homoptera (Wolda & Wong, 1988) and fig wasps (Compton *et al.*, 2000) are considered, then an important aspect is that the stratification of tropical forest insects is not necessarily stable: these insects can shift from one forest stratum to the other as a function of season, or opportunistically (i.e. situation dependent) in order to switch from one type of activity to another (e.g. from dispersing to reproducing). In the light of these conclusions, efforts to inventory the 'canopy' insect fauna of rainforests may not be very useful. Canopy and understorey faunas may exist as distinct entities only at certain times and places (patches within the forest). It is important to investigate what insects are actually doing in the various strata of tropical rainforest patches, and to what extent and for what reason individuals move between strata.

What does all this mean for researchers wanting to undertaken forest canopy or stratification research in less-developed countries, which typically have very limited taxonomic expertise, only basic laboratory infrastructure and museum facilities, as well as no canopy access infrastructure? Baseline data of limited quality, such as presented above for Akintosula, can of course be generated almost anywhere, irrespective of the constraints mentioned. It is obvious, though, that if the quality of the research on stratification – as well as on spatial and temporal patterns in general – in rainforest environments is to improve, direct canopy access is an important prerequisite, as is some level of enhancement of taxonomic expertise and improvement of physical infrastructure. An important improvement on the Akintosula methods would have involved the cross-comparison of samples. Cross-comparing requires much time and labour for sorting, pinning and the actual comparison of specimens from different samples. This might be feasible only after a considerable narrowing of the study, in terms of its taxonomic focus. This would mean, however, that the study would largely lose its appeal as a baseline study on rainforest insects or arthropods *in general*. The choice may simply be determined by such mundane matters as whether results are needed for an impact study (as part of an Environmental Impact Assessment) or whether the goal is a rigorous investigation of scientific hypotheses on stratification.

ACKNOWLEDGEMENTS

Thanks to Antonio Brack-Egg, Kenneth Tjon and Eddy Stuger for facilitating the research; special thanks to Anil Gangadin, Judy Cramer and Danielle Bruining for field and laboratory assistance, and to Rudy Armaketo and John Pinas for identifying trees. My thanks also to all others who helped with the field work, including people from the villages of Pierrekondre and Powaka. Financial support for the research was obtained from the UNDP via Project RLA/92/G32, and from the VVOB, a Belgian development cooperation organization; the Dutch Tropenbos Foundation supported a very informative visit to its research site in Guyana.

PART III

Temporal patterns in tropical canopies

Introduction

Many entomological studies in the tropics have investigated arthropod seasonality, and how it compared with that in temperate forests. The myth of a stable environment in the tropics has been criticized numerous times. From the often descriptive emphasis on peaks of arthropod activity, recent studies have shifted more towards understanding how fluctuating resources affect the temporal variation of arthropod species and the structure of communities. The case studies in this section reflect this trend and include two contributions from Australasia, one from Africa and three from the neotropics. The reader particularly keen in studying temporal patterns may also consider the two chapters that are related to this topic in the synthesis section.

Itioka *et al.* (Ch. 12) analyse the effects of pluriannual mass and general flowering in dipterocarp forests of Sarawak on the seasonality of various insect taxa. Their approach considers both antophilous and non-antophilous insects, and possible indirect effects of general flowering on the latter. One of the values of the study is to compare temporal trends at different heights within the canopy, in a fashion similar to Schulze and Fiedler in Ch. 7.

In their analysis of arthropod assemblages associated with *Metrosideros* trees in Hawaii (Ch. 13), Gruner and Polhemus look at a very different time scale, of several million years. In doing so, they touch an important issue in ecology: the relative influence of local versus regional factors in determining local species richness. Their contribution also provides an excellent example of island ecology and reminds us that canopy habitats are as varied as forest types are.

In Ch. 14, Wagner examines beetle seasonality with canopy fogging in forest plots of varying degrees of disturbance in Uganda. The approach is thus similar to that of De Dijn in Ch. 11. In addition, the author also focusses on beetles associated with a particular tree species, *Rinorea*, growing in different forest habitats experiencing different regimes of relative humidity. This allows him to partition conveniently the effects of relative humidity/rainfall, habitat structure and host-tree phylogeny on beetle seasonality.

Palacios-Vargas and Castaño-Meneses contrast the seasonality of springtails in different forest types experiencing different climates (Ch. 15). They survey different habitats to check for migration of springtail species up in the canopy. Their study argues once again against studying the canopy fauna in isolation, because of the risk of avoiding important spatio-temporal patterns or patterns of resource use.

Faunal turnover during the wet and dry seasons in tropical forests is an important topic of research. It may help to understand the maintenance of high local diversity in tropical forests, as opposed to temperate forests. This theme is tackled in Ch. 16 by Hurtado Guerrero *et al.*, who use canopy fogging to discuss such seasonal turnovers in an Amazon forest.

The last contribution in this section, Ch. 17, asks whether the habitat structure or carrying capacity of the environment may influence arthropod seasonal patterns. The study system of Stuntz *et al.* consists of a host tree, *Annona*, with differing epiphyte loads in Panama. The authors examine the relative effects of the phenology of the host and of that of epiphytes on the seasonality of arthropod foraging within *Annona*.

12

Insect responses to general flowering in Sarawak

Takao Itioka, Makoto Kato, Het Kaliang, Mahamud Ben Merdeck, Teruyoshi Nagamitsu, Shoko Sakai, Sarkawi Umah Mohamad, Seiki Yamane, Abang Abdul Hamid and Tamiji Inoue

ABSTRACT

The temporal trends in abundance of five insect orders or suborders (Blattaria, Coleoptera, Heteroptera, Homoptera and Orthoptera) and two species of predatory wasp (*Provespa anomala* and *P. nocturna*) were monitored by monthly light trapping from January 1994 to August 1997 in a tropical lowland dipterocarp forest in Sarawak, Malaysia. We examined whether the abundance in terms of individuals of the target insect groups increased in response to two multispecies mass flowering (general flowering) events, which occurred from March to November in 1996 and from March to June in 1997. It is well known that some anthophilous insects show an increase in population in response to general flowering. In this contribution, we examined whether non-anthophilous insects also increase in response to general flowering. The abundance of Blattaria, Orthoptera and Homoptera, all of which included a few anthophilous species, was not significantly higher during general flowering periods than during other periods. Both *P. anomala* and *P. nocturna* did not show any increase concomitant with general flowering. However, Coleoptera and Heteroptera, which include some anthophilous insects, showed significant increase in abundance in response to general flowering. These results suggest that populations of insects feeding on flowers or food resources provided by flowers increase in response to general flowering, but that generalist insect predators and phytophagous insects other than anthophilous ones do not respond either to general flowering or the population increase of anthophilous insects during such flowering. The patterns of changes in abundance in relation to the occurrence of general flowering were different across different above-ground strata. In Blatteria and Heteroptera, the abundance was highest during postflowering periods and lowest during nonflowering periods at the canopy (35 m) level, and it was lowest during post-flowering periods and highest during flowering periods at the under-canopy (17 m) and ground level (1 m). The abundance of Coleoptera was highest during postflowering periods and lowest during nonflowering periods at the canopy level, whereas it was lowest during non-flowering periods and highest during flowering periods at under-canopy and ground levels. Therefore, it is suggested that increase of food resources derived from increased flower availability during general flowering periods does not cause a clear increase in abundance of non-anthophilous insects, but that space and food utilization in relation to the vertical position in the forest canopy is affected by the occurrence of general flowering.

INTRODUCTION

The forest canopy is an important centre for the cycling of various nutrients through food-webs in the terrestrial forest ecosystem. To understand the structure and functions of the food-web, the dynamic aspects of interactions between various organisms need to be known. Most of the organisms concerned are insects. Insects are not only dominant in species number but are also important in the roles they play in food-webs (Wilson, 1992). Insects constitute a large part of the herbivore and pollinator assemblages, and important fractions of predator and decomposer assemblages in terrestrial ecosystems. Consequently, to understand the food-web structures and their functions within whole forests and ecosystems, the dynamics and functions of insect assemblages in the forest canopy need to be well studied (Lowman & Nadkarni, 1995; Stork *et al.*, 1997a).

The vegetation of the moist humid lowlands of south-east Asia is characterized by dipterocarp trees with a height of more than 60 m (Whitmore, 1990), high species richness (Whitmore, 1984; Ashton, 1988) and general flowering – a unique mass-flowering phenomenon in which multiple species of canopy trees flower together. General flowering and simple mass-flowering differ in terms of the participation of many or only single tree species. Most tree species of the canopy mass-flower over several months, during which tree species flower sequentially yet do so only at intervals of every 4 to 5 years (Yap & Chan, 1990; Ashton, 1991; Appanah, 1993; Sakai *et al.*, 1999b). Because flower and fruit resources in the forest canopy increase remarkably during general flowering periods and decrease thereafter, they may affect the population and community dynamics of insects associated with flowers or fruits of these trees. How insect populations and communities respond to general flowering remains to be studied.

We have collected insects in the forest canopy by monthly light trapping since August 1992 in a lowland mixed dipterocarp forest in Borneo, in order to monitor temporal changes in insect abundance. The study period covered two general flowering events (Momose *et al.*, 1998; Sakai *et al.*, 1999b). To date, based on light-trapping data, Kato *et al.* (2000) have shown that several scarabaeid species of different food habits exhibit a range of temporal fluctuations in population size in relation to general flowering events. One species of *Anomala*, an anthophilous generalist, showed clear supra-annual patterns in response to general flowering. However, three chafer species belonging to the Melolonthini, showed a clear pattern of seasonal fluctuation. *Parastasia bimaculata* Guérin, a specific pollinator of a plant species that flowers constantly, showed hardly any temporal fluctuation. Itioka *et al.* (2001) reported that the abundance of the giant honeybee, *Apis dorsata* F., which is a dominant flower visitor, increased rapidly, perhaps because of long-distance movement in response to general flowering. Accordingly, it has been suggested that populations of flower-utilizing insects fluctuate in response to general flowering. Although the study of Kato *et al.* (2000) illustrated the fluctuation patterns of populations of some non-anthophilous scarabaeids, the population-level responses of insects other than flower-associated ones to general flowering events have not been investigated.

This paper examines whether the abundance of insects, especially non-anthophilous ones, in the canopy increases in response to general flowering events, based on the data from monthly light-trapping surveys. We compare the abundance of several taxonomic groups of light-attracted insect taxa during flowering, post-flowering and nonflowering periods in order to answer the following questions.

1. Do non-anthophilous insect populations or assemblages in the forest canopy increase in response to general flowering?
2. If so, by how much do they increase?
3. Does the degree to which they increase in response to general flowering depend on their food habits and/or on the height of their habitats?

Specifically, we have focussed on several groups of light-attracted insect taxa that have different food habits: two species of nocturnal predacious *Provespa* wasps, blattarians, coleopterans, heteropterans, homopterans and orthopterans. Because of the difficulty in sorting and identifying a large quantity of insect specimens at the species level, we treated light-trapped insects as ordinal (or subordinal) groups as above (five in all), except for the two *Provespa* species.

METHODS

The study site was Lambir Hills National Park, Sarawak, Malaysia (4° 2′ N, 113° 50′ E, 120–250 m above sea level). The park is located about 10 km inland from the coast in the northern part of Sarawak. The climate is humid tropical with a weak seasonal change in rainfall (Kato *et al.*, 2000). Most of the park is covered with primary evergreen forests, much of which is classified as lowland mixed dipterocarp forest (Ashton, 1991). The forest canopy is approximately 35–40 m high and emergent trees penetrate the canopy layer, attaining heights of more than 70 m (Inoue & Hamid, 1994; Inoue *et al.*, 1995).

Flowering and fruiting phenology of approximately 580 individual plants representing 305 species in 56 families were monitored twice a month within the Canopy Biology Plot (8 ha, 200 m × 400 m) and a belt transect along the Waterfall Trail (5 ha, 1 km × 50 m) (Sakai *et al.*, 1999b).

Canopy insects were sampled by light trapping at a site in the Canopy Biology Plot. The sampling site was a 35 m tall tree tower for canopy observation (Inoue *et al.*, 1995). Light-trap collection was done at 1, 17, and 35 m up the tree tower. We used modified Pennsylvania ultraviolet light traps consisting of a 20 W fluorescent lamp tube driven by on-line power, two crossed transparent boards for intercepting the flight path of light-attracted flying insects, and a bucket filled with 50% ethanol to receive the falling insects. To avoid the overlap of light emanation at the three levels, each light trap was shaded by the platform of the tree tower. The light traps were operated throughout the night over four consecutive nights around the period of new moon. The insect catches during the four consecutive nights of one trapping period constituted the 'monthly' catch.

All the insects trapped in the buckets were recovered every morning. Larger insects (>7 mm) and all coleopterans, heteropterans and homopterans were mounted. All mounted insect specimens were recorded. The material was deposited at the Forest Research Centre in Kuching, Sarawak.

Six target insect groups were selected for the analysis, as follows.

1. All species of Blattaria: most of these are decomposers that feed on humus, litter and humic soils.
2. All species of Orthoptera: these include both omnivores and leaf-chewing herbivores.
3. All species of Homoptera: most of these are sap-sucking herbivores.
4. All species of Heteroptera: the majority of these are sap-sucking herbivores but others are more or less polyphagous predators.
5. All species of Coleoptera. Coleoptera biomass was the largest of all insect orders trapped at light at the study site (Kato *et al.*, 1995). Their individual numbers were also much higher than those of the other five insect groups analysed here.
6. Two species of *Provespa* wasps (*P. anomala* Saussure and *P. nocturna* van der Vecht). These are nocturnal polyphagous predators, and they are easily identified at species level. We treated them separately in the analyses.

General flowering occurred at the study site March–August 1992, March–November 1996, March–June 1997 and February–July 1998 (Momose *et al.*, 1998; Sakai *et al.*, 1999b; Kato *et al.*, 2000; Fig. 12.1). With reference to flowering intensity and context of general flowering event, we classified data from the monthly

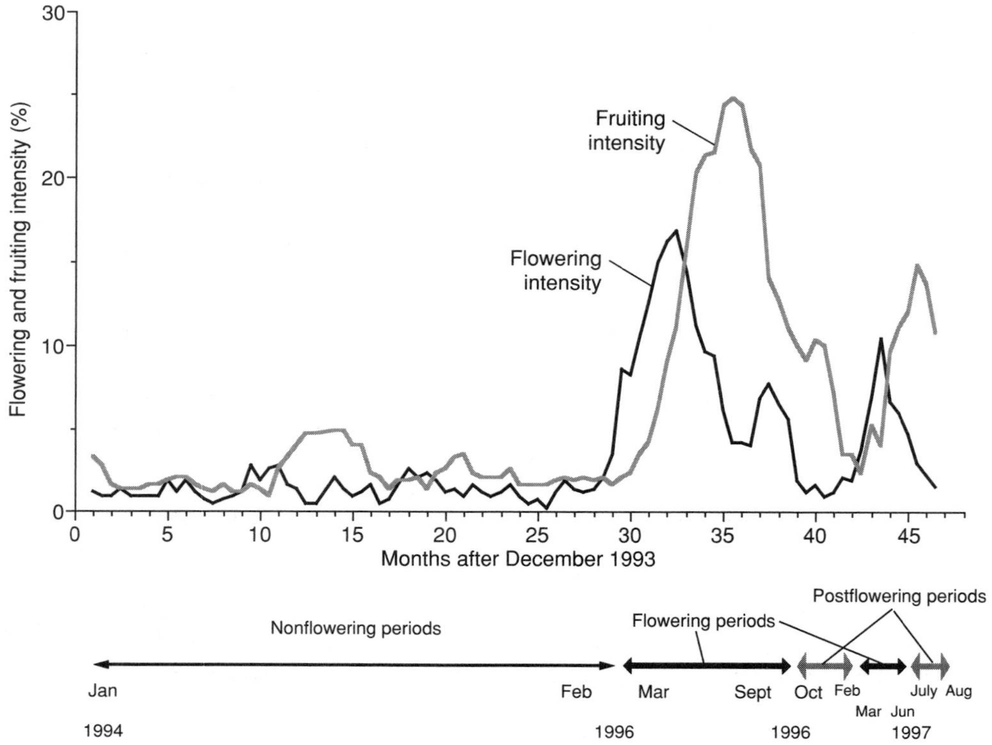

Fig. 12.1. Changes in flowering and fruiting intensities at Lambir Hills National Park, Sarawak, Malaysia (after Kato *et al.*, 2000, upper part), and classification of the period with reference to flowering intensity and the context of general flowering events (lower part).

catches into three periods: (i) the monthly catches from January 1994 to February 1996 as the nonflowering period, (ii) those from March 1996 to September 1996 and from March 1997 to June 1997 as flowering periods, and (iii) those from October 1996 to February 1997 and from July 1997 to August 1997 as postflowering periods (Fig. 12.1). During the postflowering period, flowering intensity (Sakai *et al.*, 1999b, Kato *et al.*, 2000) was similar to that during the nonflowering period but fruiting intensity was much higher than during the nonflowering period (Sakai *et al.*, 1999b; Kato *et al.*, 2000). Although light trapping was carried out from August 1992, monthly catches before 1994 were omitted from the analysis in order to avoid the delayed effects of the preceding general flowering event observed in 1992.

We analysed the effects of flowering status (as differences in the three periods categorized with reference to flowering intensity and the context of a general flowering event) and height (three above-ground levels: ground level at 1 m, under-canopy level up to 17 m, and canopy level up to 35 m) on the abundance of each insect group using two-way analyses of variance together with an analysis of the interaction effects. For the analysis, we used the general linear model procedure in the SAS statistical package version 6.12 (SAS, 1995).

RESULTS

For the five target higher taxa of insects, the abundance in terms of individual numbers did not show clear patterns (Fig. 12.2). In addition, the abundance of Blattaria, Orthoptera and Homoptera did not appear to be consistently higher during general flowering periods than in the other two periods across the three heights (Fig. 12.3 and Table 12.1). For Coleoptera and Heteroptera, however, the average abundance during the nonflowering period was significantly lower than during the other two periods (Fig. 12.3 and Table 12.1).

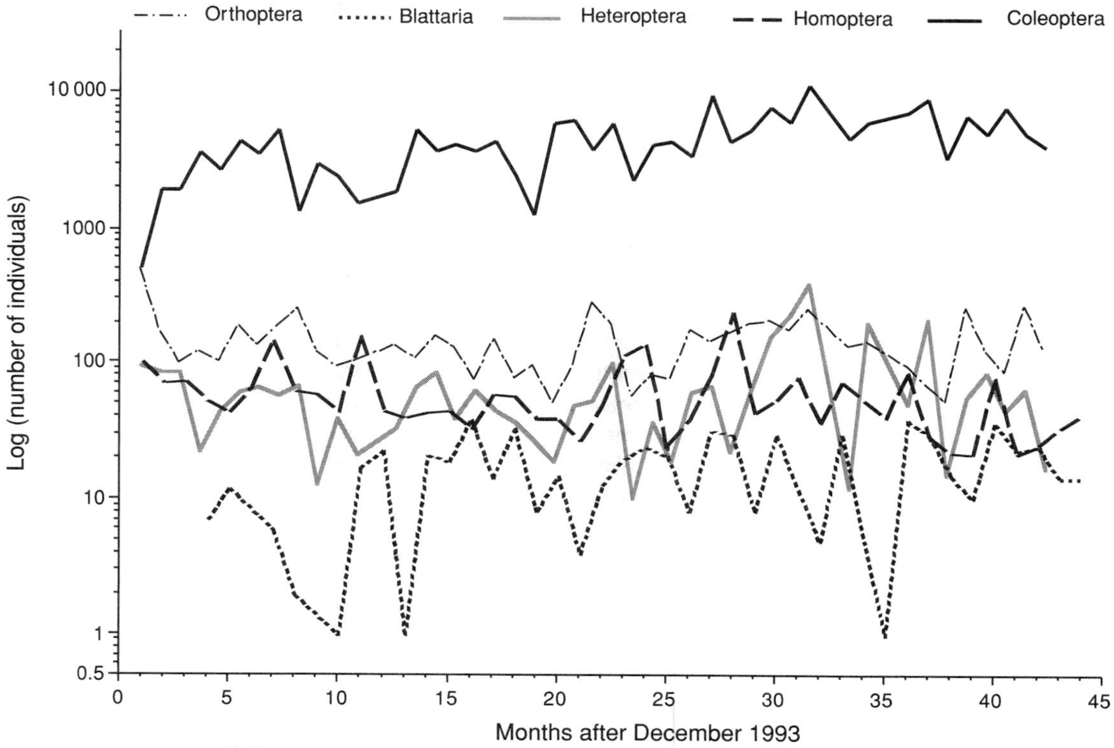

Fig. 12.2. Temporal trends in monthly light-trap catches of five insect taxa – Orthoptera, Blattaria, Heteroptera, Homoptera and Coleoptera – from January 1994 to August 1997.

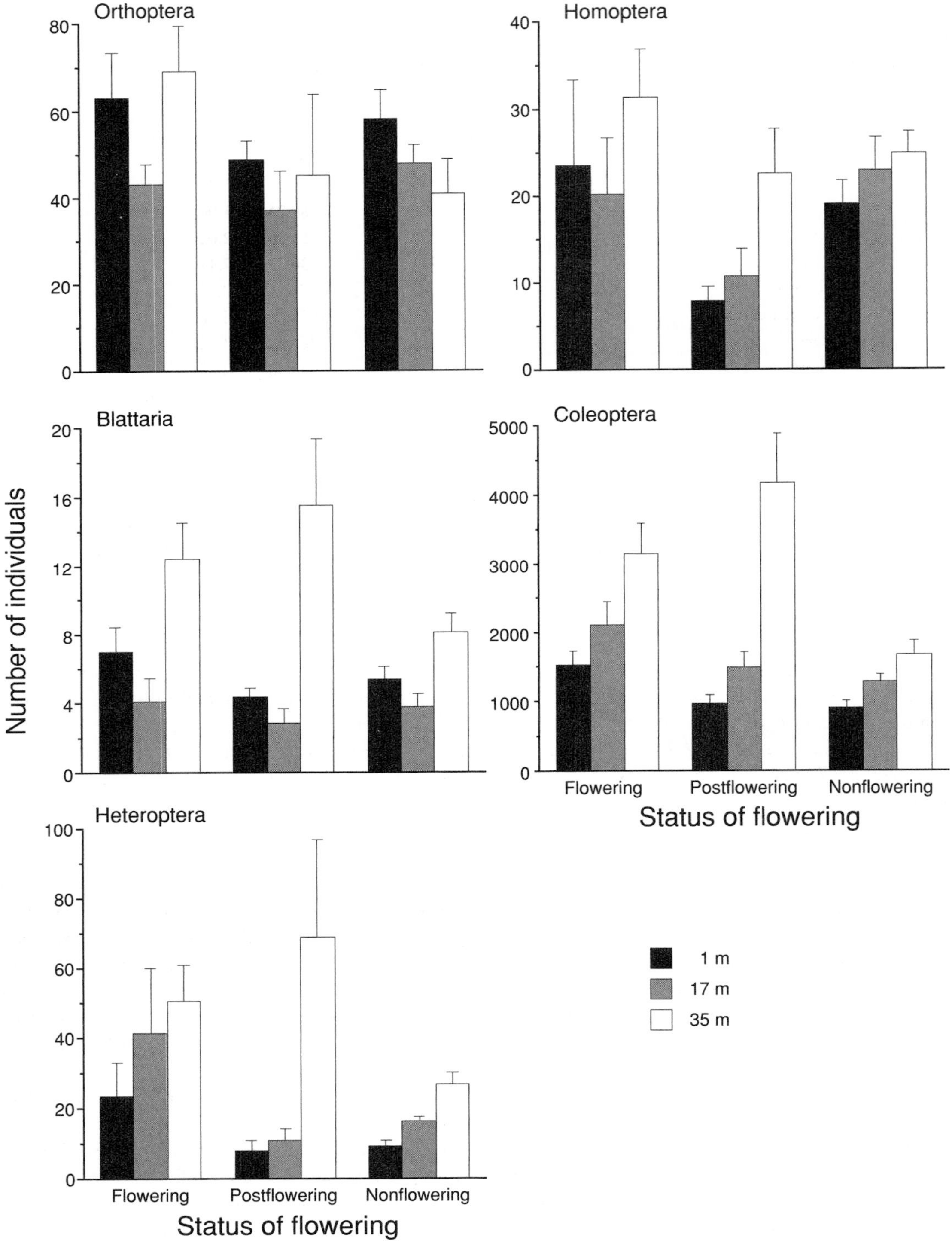

Fig. 12.3. Average monthly numbers of light-trapped insects of five taxa at 1, 17 and 35 m levels during three trapping periods with different flowering status. Error bars are ± 1SE.

Table 12.1. *Two-way analysis of variance (ANOVA) of the effects of flowering status, height and the interaction effect on the abundance of light-trapped insects in five orders*

Order	Source	Degrees of freedom	SS	MS	*F* value	*p* value
Orthoptera	Flowering period	2	3 111.57	1 555.79	1.52	0.2237
	Height	2	3 241.79	1 620.90	1.58	0.2103
	Flowering status × height	4	4 356.97	1 089.24	1.06	0.3786
Blattaria	Flowering status	2	102.59	51.30	2.62	0.0780
	Height	2	1 072.71	536.35	27.43	0.0001
	Flowering status × height	4	228.10	57.02	2.92	0.0255
Heteroptera	Flowering period	2	10 117.70	5 058.85	6.36	0.0024
	Height	2	20 561.16	10 280.58	12.92	0.0001
	Flowering status × height	4	8 197.90	2 049.48	2.57	0.0411
Homoptera	Flowering period	2	1 697.33	848.66	2.89	0.0592
	Height	2	1 692.19	846.09	2.89	0.0597
	Flowering status × height	4	581.38	145.35	0.50	0.7388
Coleoptera	Flowering period	2	26 932 138.05	13 466 069.03	16.40	0.0001
	Height	2	56 836 164.13	28 418 082.07	34.60	0.0001
	Flowering status × height	4	19 731 141.14	4 932 785.29	6.01	0.0002

SS, sum of squares; MS, mean square.

When we separated the effect of height in the canopy on the abundance of insect groups, the increase of individual numbers during the flowering period was apparent in some cases (Fig. 12.3 and Table 12.1). In Coleoptera and Heteroptera, the effects on abundance of flowering status, of height and the two-factor interaction effect were all significant. In both orders, the abundance at 35 m was higher during the postflowering period than during the other two periods, whereas, at 1 and 17 m, the abundance was higher during the flowering period than during the other periods. Similar patterns were observed in Blattaria, although the effect of flowering status was only marginally significant. In the three taxa, the abundance at 35 m was higher than those at 1 and 17 m throughout the three periods.

In Homoptera, the effect of height was close to being significant. Further, at each height, the abundance of Homoptera was lowest during the postflowering period, and the difference in abundance between the flowering and nonflowering periods was small in comparison with the differences between the postflowering and the other two periods. For the Orthoptera, the effects of flowering status, height and the two-factor interaction were all nonsignificant. Differences in abundance were clear neither across the three heights nor among the flowering periods. It was notable that the relative abundance at 1 m in comparison with those at the other two heights appeared to be higher than for the other orders.

The temporal trends of abundance in the two *Provespa* spp. did not show clear patterns and no increase in response to general flowering was apparent (Fig. 12.4). In contrast, the abundance of *P. nocturna* decreased during the flowering period (Fig. 12.5), and the effect of flowering status on its abundance was significant (Table 12.2). The effect of flowering status was not significant at all for *P. anomala* (Table 12.2), although the abundance at 1 m was higher during the flowering period compared with the nonflowering period (Fig. 12.5). The effect of height on the abundance was significant in both species. The abundance of each species was notably lower at 35 m than at the other heights. For *P. anomala*, the difference in abundance across heights appeared to increase during the flowering period.

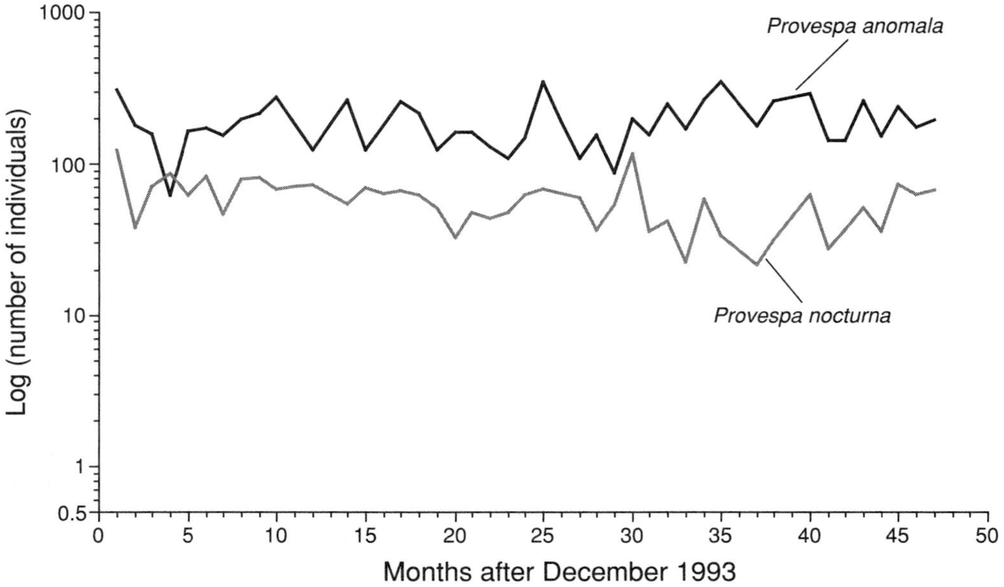

Fig. 12.4. Temporal trends in monthly light-trap catches of workers of two nocturnal wasp species, *Provespa anomala* and *P. nocturna*, from January 1994 to August 1997.

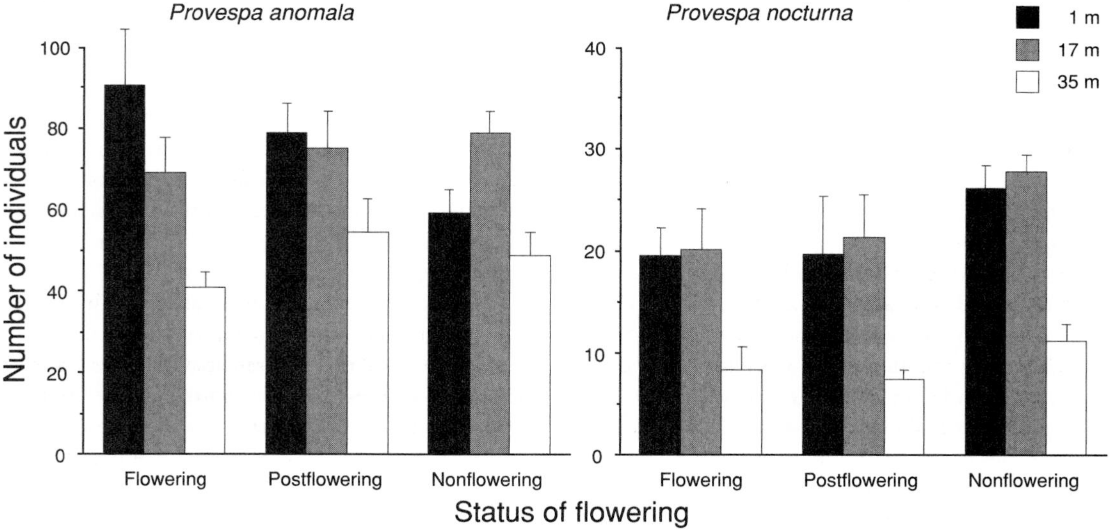

Fig. 12.5. Average monthly numbers of workers of two nocturnal wasp species, *Provespa anomala* and *P. nocturna*, at 1, 17 and 35 m levels during three trapping periods with different flowering status. Error bars are ± 1SE.

DISCUSSION

It is reasonable to predict that the abundance of anthophilous insects, especially generalist flower-eating insects, will increase in response to a general flowering event because their food resources become abundant then. This prediction has been supported by the observed increase of some insect species in response to general flowering (Kato *et al.*, 2000; Itioka *et al.*, 2001). However, it is not well understood what are the

Table 12.2. *Two-way analysis of variance (ANOVA) of the effects of flowering status, height and the interaction effect on the abundance of light-trapped workers of two* Provespa *spp.*

Species	Source	Degrees of freedom	SS	MS	*F* value	*p* value
Provespa anomala	Flowering period	2	1 099.10	549.55	0.71	0.4947
	Height	2	15 613.82	7 806.91	10.06	0.0001
	Flowering status × height	4	8 296.52	2 074.13	2.67	0.0353
Provespa nocturna	Flowering period	2	978.91	489.45	5.10	0.0075
	Height	2	3 864.02	1 932.01	20.13	0.0001
	Flowering status × height	4	92.74	23.19	0.24	0.9143

responses of insects to general flowering events in terms of their food habits. The present study provided some answers to, and a base for the further discussion of, such problems.

The two species of *Provespa* wasp do not utilize flowers as their main food but are thought to be generalist predators (Maschwitz & Hänel, 1988; Martin, 1995). Their abundance increased neither in response to general flowering nor during the postflowering period. These results suggest that generalist predatory species hardly show any numerical response to general flowering, and that they do not respond to increases in abundance of the few associated anthophilous insects. In addition, our results indicate that they prefer ground or under-canopy levels as foraging habitat rather than canopy level. This habitat preference may be one of the reason why they do not respond to general flowering, as flower resources increase in the forest canopy rather than in lower strata (Sakai *et al.*, 1999b).

The abundances of Blattaria and Orthoptera did not show clear increases in response to general flowering. Most species of these two orders do not utilize flowers as their main food. Most cockroach species are scavengers and most Orthopterans are folivorous scavengers, predators or omnivores. It is most likely that this explains their lack of response to general flowering.

The abundance of Coleoptera and Heteroptera was clearly higher during the flowering period. It is likely that the increase in abundance of the two taxa during the flowering period may be attributed to their relatively high proportion of anthophilous species. Some species of Heteroptera are known to prefer inflorescences, flower and flower bud as feeding sites, where they are able to get nutrients of high quality. Moreover, many anthophilous species belong in the Coleoptera, for example, anthophilous species in Scarabaeidae, Nitidulidae, Mordellidae, Cerambycidae and Chrysomelidae. Accordingly, we may hypothesize that the increase of heteropterans and coleopterans during general flowering may be a consequence of an increase of anthophilous species within the two taxa in response to the increase in flower resource during general flowering. The hypothesis should be examined by further analysis at the species level using the data from species whose food and spatial habits are well known. Homopterans did show an increase in response to general flowering even though the numbers were slightly less than significant. This suggests that some homopteran species get higher growth and reproductive performance when they suck sap on and near inflorescences or single flowers.

Because the Coleoptera and Heteroptera include many frugivorous species, and because fruiting intensity is much higher than usual during the postflowering period, we infer that the remarkable increases in the abundance of these taxa at canopy level (35 m) during the postflowering period may well be a reflection of increases in some frugivorous species. However, it is possible that the increase of Blattaria at the canopy level during the postflowering period might be caused by an increase in humus in the canopy. A large amount of humus, derived from abundant fruits and seeds, may be generated in the forest canopy during the postflowering period. Attention must be paid to the delayed effects upon insect abundance resulting from the increase of flower resources during the preceding period.

This study suggests that there is correlation between food habit and patterns of population fluctuation in relation to general flowering, and that the effects of general flowering on insect abundance are influenced by environmental conditions – such as height – in the forest structure. The abundance of light-trapped insects was high in the canopy, and it is reasonable to infer that the numerical responses to flowering are different between ground and canopy level. This indicates the necessity for investigating the dynamics of the component species of the insect assemblage in the forest canopy. In this study, except for two *Provespa* species, the analyses were performed on the data grouped at ordinal or subordinal level. We propose that future studies should be carried out on many individual species to show up differences in population responses, such as those identified in our studies of the two species of *Provespa*.

Several studies have shown that the abundances of various insect guilds in the forest canopy show clear seasonal fluctuations (Wolda, 1978b; Lowman, 1985; Basset, 1991c,d; Roubik, 1993). However, our results provided no evidence that the abundances of dominant generalist pollinators, sap-sucking herbivores or generalist predators showed seasonal fluctuation. This discrepancy is caused by differences in the nature of seasonality of climates across study sites. In our case, the seasonal patterns are weak, with monthly precipitation always greater than 140 mm and the air temperature always over 18 °C (Momose *et al.*, 1994) with remarkable supra-annual climatic changes (Kato *et al.*, 2000). Previous studies have been carried out in subtropical areas or in the tropical areas with clear seasonal climatic changes of alternating dry and rainy seasons. The seasonal increases of leaf production and flower/fruit production, which follow the seasonal increase of precipitation and the change from dry to wet seasons, respectively, were tracked by sap-sucking herbivores (Wolda, 1978b; Lowman, 1985; Basset, 1991c,d) and flower-associated insects (Roubik, 1993). These phenomena were hardly apparent in our systems. In our study sites, the 'vague' seasonality in terms of climatic conditions and consequent plant phenological changes seems to result in weak seasonality in the temporal patterns of insect abundance and the large impact of general flowering events on the abundance of particular insects. No clear seasonality was observed even in shoot production of the canopy trees in our study site (S. Sakai, unpublished data; T. Ichie, unpublished data).

ACKNOWLEDGEMENTS

We are indebted to H.S. Lee (Forest Department of Sarawak) and K. Ogino (the University of Shiga Prefecture) for their support and organization of this study. We thank B. Rapi and R. Johan (Lambir Hills National Park, Sarawak, Malaysia) and the laboratory technicians of the Entomological Section, Forest Research Centre, Forest Department, Sarawak, for servicing the ligh-traps and for mounting and labelling insect specimens. This study was partly supported by a Grant-in-Aid for Scientific Research (No. 04041067, 06041013, and 09NP1501) from the Ministry of Education, Science and Culture, Japan. It was approved by the State Secretary, Sarawak, and Director of Forests, Sarawak, under the reference number 80/PKM/1335/5/79 on 6 October 1992, as The Long-Term Forest Ecology Research Project at Lambir Hills National Park.

13

Arthropod assemblages across a long chronosequence in the Hawaiian Islands

Daniel S. Gruner and Dan A. Polhemus

ABSTRACT

The Hawaiian archipelago forms a linear time series of replicated communities, with ages of volcanic origin varying over seven orders of magnitude. Important ecosystem attributes, such as parent substrate, climate and elevation, may be held constant at selected sites along the series. Previous studies along a montane mesic chronosequence (300 years to 4.1 million years) have shown that rates of many ecosystem processes peak at sites of intermediate age of volcanic substrate (20 000 years to 1.4 million years). At all these sites, the canopy is dominated by a single tree species, *Metrosideros polymorpha* (Myrtaceae), and arthropod lineages are largely conserved. These features make the islands an ideal natural laboratory to test the relative importance of ecological versus historical and evolutionary hypotheses of community organization. In 1997, we used pyrethrum canopy fogging to sample free-living arthropods on 44 *M. polymorpha* trees at five sites on three islands. Total arthropod densities were highest at more productive sites of intermediate age. Arthropod abundance tracks the unimodal pattern of resource availability, and the concentration of foliar nitrogen was highly correlated with mean arthropod density among sites. However, overall species diversity of true bugs (Heteroptera) and carabid beetles (Coleoptera: Carabidae) increased as a function of geological age. Moreover, one of three indigenous tribes of Carabidae (Psydrini) was completely absent from, and presumably has not colonized, the oldest site. Although contemporary ecological processes appear to determine arthropod abundance in local communities, regional species pools may determine local richness and historical processes or phylogenetic accidents may place constraints on composition.

INTRODUCTION

As this volume attests, canopy arthropod biology has surged in popularity during the 1990s. This growing interest follows a high-profile debate concerning the number of species found in tropical tree canopies and, by extrapolation, on Earth (e.g. Erwin, 1982; Stork, 1988; Adis, 1990; Erwin, 1991a; Gaston, 1991a,b; Ødegaard *et al.*, 2000). As with any scientific field in its infancy, the work that followed was largely descriptive and exploratory. As forest area dwindles globally, there is an urgency to sample as many regions and tree species as possible in a short time. We are learning more about the basic biology of canopy denizens and the essential taxonomic foundation is being laid. This provides an increasing opportunity to use arthropod communities to test important ecological hypotheses. Indeed, canopy arthropod communities are thought by many practitioners to be realistic models for the examination of community structure and dynamics (Nadkarni, 1994; Stork *et al.*, 1997a). Tree arthropod communities are spatially discrete and are easier to sample more completely, widely and accurately, and with greater replication than more complex model communities such as coral reefs or islands (Moran & Southwood, 1982).

Ecologists have been forced to acknowledge that the temporal and spatial scale of observation influences variability and emergent patterns of any system (Levin, 1992). Community ecologists have relied heavily on theory regarding processes that limit the diversity and structure of local communities (e.g. May & MacArthur, 1972; Schoener, 1974), while neglecting macroevolutionary and biogeographical processes such as speciation, colonization and chance historical events that add species over larger temporal and spatial scales. There is growing realization that the availability of regional diversity pools may be a major constraint on local species diversity and community structure (Cornell, 1985; Ricklefs, 1987; Ricklefs & Schluter, 1993; Colwell & Lees, 2000). The repeated observation of high proportions of 'singletons' and otherwise rare species in tree arthropod communities draws particular attention to these considerations (Allison *et al.*, 1997; Novotny & Basset, 2000). Moreover, short-term colonization dynamics can be highly unpredictable, seemingly without any structure (Floren & Linsenmair, 1997b, 1999).

Is there recognizable, predictable structure in tree arthropod communities, or are these communities simply stochastic assemblages drawn from the regional species pool? We use a well-characterized space-for-time chronosequence in the Hawaiian Islands to examine the relative importance of local ecological variables and geological age in the structure of canopy arthropod communities. Older islands in the archipelago are known to harbour more diverse species pools of arthropods, but there have been no controlled comparative investigations of local community structure. Using these tightly constrained sites across 4 million years of ecosystem development, we seek to understand the relative importance of local (e.g. biological interactions, resources) and regional (e.g. history, phylogeny) processes in community structure. If local resource quantity or quality is important, then arthropod variables (e.g. density or species diversity) may correlate with higher nutrient levels and rates of cycling at sites of intermediate age. Alternatively, if accumulated biogeographical and historical effects determine community structure, then arthropod variables will be an increasing function of geological age.

METHODS

Study system

The Hawaiian Archipelago is an excellent location for ecological studies (Vitousek, 1995). The volcanic archipelago has been generated as the north-westerly moving Pacific plate passes over a stationary 'hot-spot' of upwelling basaltic lava (Clague & Dalrymple, 1989). Accordingly, the main high islands represent a time series from northernmost Kauai to the southernmost, volcanically active, Hawaii. At carefully selected sites along the island chain, ecosystem state factors can be held remarkably constant while others are allowed to vary in wide, but well-defined and constrained ways. Along the gradient of substrate age (Chadwick *et al.*, 1999; Fig. 13.1), elevation, annual average temperature and precipitation, topographic position, land use history and the dominant tree species are all held constant. All soils on the chronosequence are derived from the same tephritic parent material of the original volcanic shield surface (Lockwood *et al.*, 1988; Wolfe & Morris, 1996) but ecosystems span over seven orders of magnitude in developmental age (Table 13.1). Along this gradient, soil and foliar nutrients and rates of many ecosystem processes peak at sites of intermediate age (Crews *et al.*, 1995; Vitousek *et al.*, 1995; Herbert & Fownes, 1999; Table 13.1). These features make the Hawaiian Islands uniquely suited for disentangling the relative contributions of ecological and evolutionary factors for community structure.

The myrtaceous tree *Metrosideros polymorpha* Gaudichaud-Beaupré (Myrtaceae) is the dominant tree of the wet and more mesic forests on all major islands. Subtle intraspecific genetic differences with associated distinct morphologies do exist among populations of *M. polymorpha* (Joel *et al.*, 1994; Cordell *et al.*, 1998). These represent, however, minor variations compared with the species and family level differences complicating studies elsewhere (Kelly & Southwood, 1999). Because *M. polymorpha* exists in monodominant stands through much of its range and supports relatively few epiphytes, the arboreal fauna is more homogeneous than in most tropical systems. The fauna of *M. polymorpha* is composed primarily of birds and arthropods: reptiles, mammals (except for one species of bat) and amphibians are absent from the native fauna. Although particular species of bird or arthropod may be restricted to a single island or habitat, species in different areas are often closely related members of the same lineages (Swezey, 1954). Accordingly, the arthropod communities on *M. polymorpha* are eminently well suited to robust and replicable comparisons across a variety of ecological regimes.

M. polymorpha in Hawaii was the target of the first ever pyrethrum-fogging studies of canopy arthropods (Gagné, 1979, 1981). Moreover, the canopy may be one of best remaining habitats to study relatively undisturbed ecosystem and community pattern and process in Hawaii. Beginning with Polynesian colonization (Kirch, 1982), most native habitat below 1000 m has been cleared for agriculture or other human uses, resulting in widespread extinction and a degraded lowland

Table 13.1. *Site characteristics of the Hawaiian mesic chronosequence. All sites are on tephra substrate, at similar elevation (1200 m), approximate mean annual temperature (16 °C), and approximate mean annual precipitation (2500 mm)*

Site (code)	Parent material age ($\times 10^3$ year)	Elevation (m)	Community basal area (m^2 ha^{-1})/% *Metrosideros*	Net primary productivity (kg m^{-2} yr^{-1} ± SE)	Foliar nitrogen (% dry mass ± SE)	Foliar phosphorus (% dry mass ± SE)
Volcano (VO)	0.3	1175	35.8/81	1576.9 ± 127.25	0.87 ± 0.04	0.060 ± 0.006
Laupahoehoe (LA)[a]	20	1170	33.6/83	1604.4 ± 86.868	1.42 ± 0.05	0.101 ± 0.006
Kohala (KH)	150	1122	35.7/83	1781.6 ± 81.160	1.14 ± 0.10	0.113 ± 0.013
Molokai (MO)	1400	1210	37.3/86	1962.9 ± 98.906	1.06 ± 0.06	0.085 ± 0.004
Kauai (KA)	4100	1134	38.0/88	1429.6 ± 148.74	0.86 ± 0.04	0.061 ± 0.002

Adapted from Crews *et al.* (1995); Vitousek *et al.* (1995); Herbert & Fownes (1999).

[a]Data for 'Laupahoehoe ash' is given, although arthropods were sampled at 'Laupahoehoe flow', which is dated at 5000 BP. Although community basal area and NPP data are unavailable for Laupahoehoe ash, the foliar nitrogen and phosphorus levels are nearly identical to Laupahoehoe flow (Vitousek *et al.*, 1995).

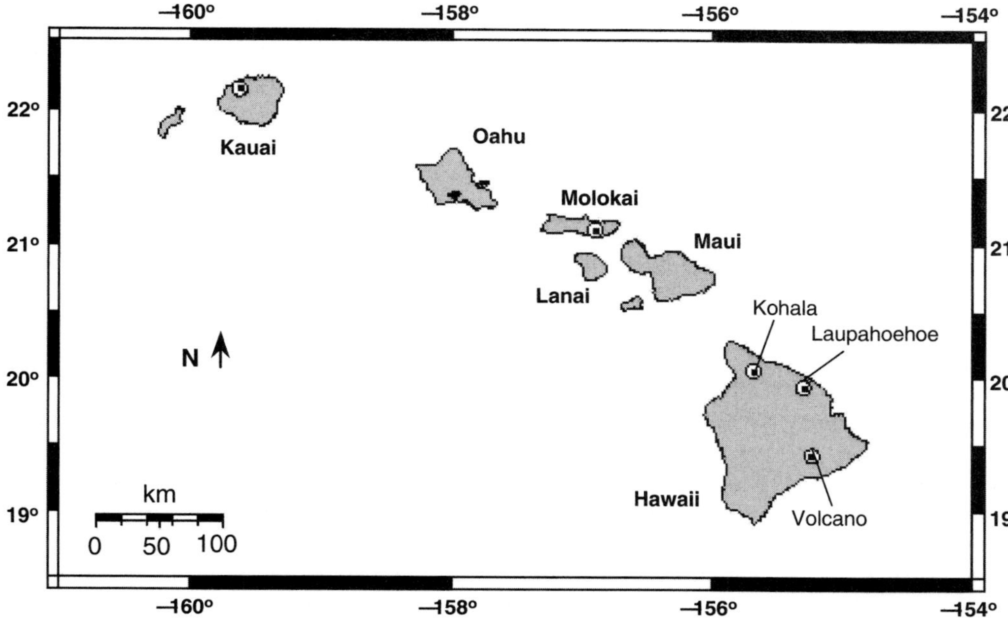

Fig. 13.1. Location of the five sites in the Hawaiian mesic chronosequence that were sampled for canopy arthropods in October and November of 1997. From youngest to oldest, the sites are Volcano (300 years; Hawaii Volcanoes National Park, Hawaii Island), Laupahoehoe flow (5000 years; Laupahoehoe Forest Reserve, Hawaii Island), Kohala (150 000 years; Kohala Forest Reserve, Hawaii Island), Molokai (1.4 million years; Nature Conservancy Kamakou Preserve, Molokai Island) and Kauai (4.1 million years; Napali-Kona Forest Reserve, Kauai Island). Sites are denoted by half-filled circles.

arthropod fauna (Roderick & Gillespie, 1998). Montane mesic and wet forests still persist in wide swathes at elevations above 1000 m, but invasions of exotic ungulates within the last 200 years have led to devastation of many native understorey plant and animal communities even in these areas (Vitousek *et al.*, 1987).

Arthropod sampling

We sampled arthropods from *M. polymorpha* at five sites (Fig. 13.1) in October and November of 1997 by pyrethrum knockdown following canopy fogging. We used the pyrethrum formulation 'pyrenone 100' (1% pyrethrins, 5% piperonyl butoxide and 94% isoparafinnic petroleum). Stork and Hammond (1997) write: 'fogging probably has its greatest potential use in ecological studies where the reliability and comparability of samples is of paramount importance'. This statement underscores our conviction that the specimens and data obtained will be most useful in community level comparisons within this study. We compare conspecific trees of a low, relatively homogeneous canopy. There remains great difficulty in comparing results with other studies by fogging because of large interstudy variation in the chemical compounds used, their concentrations, drop times and other factors.

We sampled during the same year, season and local weather conditions, but sites could not be sampled simultaneously because of logistical constraints (Table 13.2). Trees were fogged during the period from sunrise to early evening in good weather conditions only (without precipitation or wind), and individual trees were fogged until the entire crown was saturated with fog for at least 10 seconds. On average, we obtained samples from 10 *M. polymorpha* trees at each of four sites (Table 13.2). The fifth site, Laupahoehoe flow (see Vitousek *et al.*, 1995), is excluded from some analyses because poor weather limited the sample size and data for certain variables (e.g. foliar nitrogen is lacking; Table 13.2). We avoided flowering trees.

Arthropods dropped for 1 hour onto a strategically placed array of three to six collection trays each of 1.5 m^2. The trays were constructed on a frame of collapsible PVC piping, which suspended white sheets 0.75 m above the forest floor. Because our goal was to collect samples from at least 10 trees at five sites within a short seasonal duration, we chose larger trays rather than using the conventional 1 m^2 standard size. This allowed a greater coverage per unit time and effort and allowed more trees to be sampled. The total area of trays used for each tree was roughly proportional to the canopy spread, and trays were arranged for sampling from all horizontal segments of the canopy. All arthropods were collected into 70% ethanol as they dropped, sorted into gross taxonomic categories in the laboratory and entered into the Biota^{TM} database program (Colwell, 1997a). In this contribution, we report only results on density for the total community, and diversity patterns only for Carabidae (Coleoptera), Heteroptera and Cicadellidae (Homoptera). Because their ecological habits are similar to many phytophagous Heteroptera, Cicadellidae are analysed and reported together with the Heteroptera.

Ecological variables

In addition to the collection of arthropods, we recorded a number of ecological variables from each tree. Foliar morphology was classified as either glabrous or pubescent. Tree height was estimated using a clinometer. We recorded diameter at breast height for each tree but converted it to total stem basal area (πr^2) because some trees had multiple major stems. Canopy volume was estimated using measures of canopy breadth in each cardinal direction and the difference between the minimum foliage height and total tree height ($(4\pi r^3)/3$). Canopy cover was estimated using a spherical densiometer. Densiometer measurements were taken at four distances (0, 3, 6 and 12 m) from each tree bole, in all four cardinal directions. At each measure, we counted the number of quarter-squares of cover and multiplied this by 1.04 to get percentage cover. An average of the 12 measures was used for each tree. Site means of foliar phosphorus were obtained from the literature (Vitousek *et al.*, 1995).

In January and February of 1998, Gruner clipped several independent terminal branches from each cardinal direction of each tree at four sites (except Laupahoehoe). Accordingly, for each tree, there were four subsamples, except in several cases where foliage did not extend beyond the bole in a particular cardinal direction. From sampled branches, only the most recently expanded, mature cohorts were selected for nitrogen analyses. Individual, discrete leaf flushes on *Metrosideros* spp. are readily identifiable by searching for scars and leaf size attenuation between leaf clumps (Porter, 1972). Good indicators of this also include well-developed apical or lateral buds or additional but

Table 13.2. *Sample size, sampling dates and means of local ecological variables measured at the five sites of the Hawaiian mesic chronosequence*

Site (code)	Number of trees	Dates fogged	Height (m ± SE)	Basal area (cm^2 ± SE)	Canopy volume (m^3 ± SE)	Canopy cover (% ± SE)	Foliar [N] (% d.m. ± SE)
Volcano (VO)	9	4–6 Oct 97	13.03 ± 0.49	675.89 ± 96.27	121.01 ±19.42	89.22 ± 1.16	0.96 ± 0.026
Laupahoehoe (LA)	4	6 Nov 97	11.75 ± 3.28	5207.29 ± 2946.97	450.21 ± 262.96	n/a	n/a
Kohala (KH)	11	12, 14 Oct 97	10.13 ± 0.45	676.91 ± 111.62	43.13 ± 7.09	83.45 ± 2.15	1.44 ± 0.042
Molokai (MO)	10	23–24 Oct 97	8.46 ± 0.28	1302.00 ± 213.96	57.23 ± 7.80	90.6 ± 1.29	1.13 ± 0.047
Kauai (KA)	10	29–30 Oct 97	9.63 ± 0.49	712.50 ± 169.95	34.30 ± 5.46	77.5 ± 4.45	1.16 ± 0.034

n/a, not available.

obviously recent leaf whorls or flower expansions beyond the target cohort. In addition to standardizing the unit of measure, these cohorts are most likely to have been young leaves during the arthropod sampling 4 months earlier. Herbivores grazing on *M. polymorpha* use young, expanding buds and leaves preferentially (Nishida *et al.*, 1980; D. Gruner, personal observation).

Foliar samples were dried to constant mass at 65 °C and consigned for nitrogen analysis to the Agricultural Diagnostic Service Center, Department of Agronomy and Soil Science, University of Hawaii. Their technique uses automated Kjeldahl digestion measuring the percentage total nitrogen per dry weight of leaf tissue (Schuman *et al.*, 1973; Isaac & Johnson, 1976). Based on data from 10 uniform standards (foliar nitrogen ($\pm$ standard error): $0.918 \pm 0.009\%$), the analysis was judged sufficiently precise and reliable for comparisons among sites. Although many insects cannot feed on lignin-rich vascular tissues such as the leaf midrib, most studies have been concerned only with total leaf attributes. We evaluated this assumption by comparing leaf plugs to whole leaves (minus petioles) from the same branches of 12 trees (three per site). Leaf plugs were taken using a cork hole-borer of known cross-sectional area. Nitrogen was analysed on a per tree basis, but an equal area of plugs was included in the sample for each cardinal direction of a tree. Leaves and leaf plugs were dried in a 65 °C oven to constant mass for at least 48 hours and ground using a Wyley mill. Paired t-test analysis detected no difference between nitrogen concentrations from leaf plugs and whole leaves ($t = -0.27$, $p > 0.5$, $n = 12$). We report only the percentage dry weight nitrogen from leaf plugs.

Statistical analyses

Arthropod density for each tree was calculated from the total catch (or the total of a particular taxon or taxon group) on all trays for each tree divided by the total area of those trays. Densities were logarithmically (log) transformed to meet assumptions for normality, and ecological variables and substrate age were log-transformed as necessary. In cases where zero values were present (e.g. carabid beetles were not collected from every tree) we used the transformation $\log(x + 1)$ for that data set. We used the arcsin-square root transformation to normalize percentage canopy cover.

Exploratory regression analysis was used to determine the predictive value of ecological variables and substrate age for density and species-richness response variables, on a per tree basis within sites. The 'best' linear model was selected using the 'all-possible-regression' procedure with maximum adjusted r^2 criterion (Neter *et al.*, 1990): the 'best-subsets' procedure in the statistical package Minitab™ (Minitab, 1998). All first-order independent variables are fed into the procedure, and output is ranked by highest adjusted r^2 in a hierarchy of the number of independent variables. The selected model had the highest adjusted r^2 and a Cp value approximating the number of independent variables (Neter *et al.*, 1990). The Cp criterion is concerned with the total mean squared error of the fitted values for each subset of the regression model; a value approximating the number of independent variables is considered the least biased (Neter *et al.*, 1990). Simpler models (fewer predictor variables) were preferred over complex, and the 'best' models were investigated further using one-way analysis of variance (ANOVA). We used the residuals of these ANOVA models to test for terms of secondary importance in the best subsets procedure, and plotting techniques to look for two-way interactions or quadratic terms. Patterns within plots of residuals against the log of substrate age for several regressions suggested that additional relationships were present. Accordingly, we removed the log-transformation of substrate age, rescaled (million years, instead of years) and centred the variable by subtracting the mean, and repeated the 'all-possible-regression' procedure with the centred age variable and its quadratic term.

To examine among-site differences in arthropod densities, we used one-way ANOVA with post-hoc comparisons (Tukey's, family error = 0.05) and simple correlations among site means. The fifth site, Laupahoehoe, was included where data were available. In addition, we tested for correlations of arthropod density with site averages of net primary productivity and foliar phosphorus using data from the literature (Vitousek *et al.*, 1995; Herbert & Fownes, 1999; Table 13.1).

Fisher's α (the diversity index) was calculated using the EstimateS™ software package (Colwell, 1997b). These values are obtained through an iterative process that cannot produce independent, discrete values and reasonable estimates of variance on a per tree basis with our levels of replication (three to six trays per tree). Site indices and error terms can be generated and compared graphically, but this estimate is based on the pool of all subsamples from that site. Diversity calculated in this manner cannot be compared statistically within or among sites because there are no estimates of the index for the replicates. True bugs (Heteroptera) and leafhoppers (Homoptera: Cicadellidae) are treated as a unit and analysed separately and together with ground beetles (Coleoptera: Carabidae). These are first steps towards a forthcoming analysis of total arthropod diversity at these sites.

RESULTS

Arthropod density

The pattern of arthropod density is consistent with the unimodal pattern of ecosystem processes along the mesic chronosequence that peak at sites of intermediate age (20 000–1 400 000 years Fig. 13.2). Site was identified as the most important term in exploratory regression analysis. The relationship of site identity to log arthropod density was highly significant ($F_{3,36} = 20.61$, $p < 0.001$). All within-site variables were not significant when regressed against the residuals of this ANOVA. We repeated this procedure for density of Heteroptera and Carabidae separately and obtained the same result for Heteroptera. However, additional variation in Carabidae density was explained by a combination of leaf morphology ($F = 4.69$, $p = 0.037$), tree height ($F = 7.93$, $p = 0.008$), canopy cover ($F = 1.99$, $p = 0.107$), basal area ($F = 5.77$, $p = 0.022$) and a

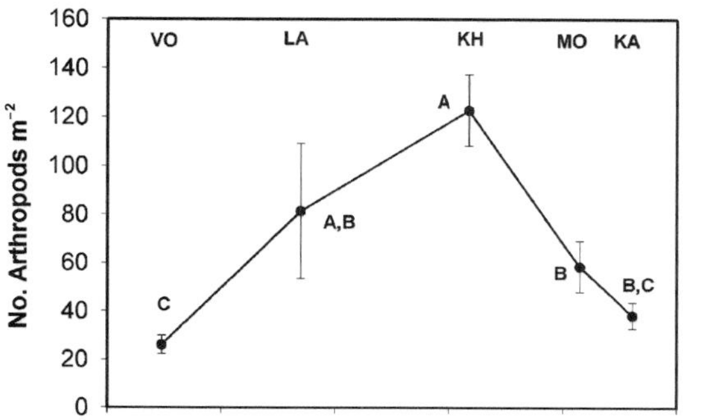

Fig. 13.2. Total arthropod density, expressed as a function of tray capture area, at sites along the chronosequence (see Fig. 13.1 and Table 13.1 for description of sites). Site codes are listed at top. Unique letters indicate significant difference (Tukey's tests). Values are site means ± 1SE.

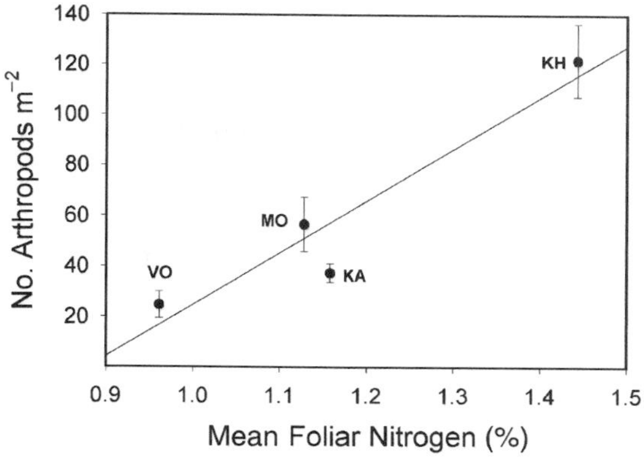

Fig. 13.3. Relationship of foliar nitrogen with mean canopy arthropod densities per site (Pearson $r = 0.952$, $p < 0.05$). Densities are expressed as a function of capture tray area for individual trees. Laupahoehoe is excluded because foliar nitrogen data were not collected. See Table 13.1 for site codes. Values are site means ± 1SE.

negative relationship with canopy volume ($F = 4.25$, $p = 0.047$) (total model $r^2 = 0.28$, $p < 0.05$).

In post-hoc (Tukey's) comparisons of site densities, the densities at intermediate-aged sites were higher than at the ends of the gradient (Fig. 13.2). Kohala had significantly higher densities than all other sites except Laupahoehoe, which had a high variance as a result of the low sample size. Laupahoehoe did not differ significantly from Molokai and Kauai, the oldest site, but the youngest site (Volcano) differed from all others except the oldest site.

Most correlations with arthropod density among the means of the five sites were not significant. Of the data collected for this study, only foliar nitrogen was correlated significantly with mean density ($r = 0.952$, $n = 4$, $p < 0.05$; Fig. 13.3). This is a particularly strong relationship considering the low site replication. If site identity is removed from the preliminary regression analysis on all fogged trees ($n = 40$), arthropod density is strongly related to foliar nitrogen ($r^2 = 0.42$). This circumstantial evidence lends support to the importance of resource variables for predicting arthropod density. However, much of the variation of foliar nitrogen and arthropod density is among sites, so foliar nitrogen cannot be distinguished from site identity. Of other published site data, foliar phosphorus concentration (Vitousek *et al.*, 1995) was also strongly correlated to mean arthropod density among sites ($r = 0.965$, $p < 0.01$).

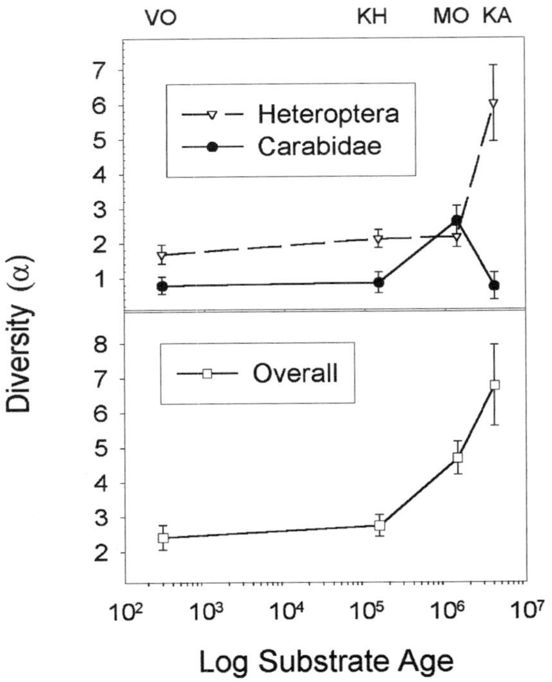

Fig. 13.4. Species diversity (measured by Fisher's α) as a function of geological age. Site codes are listed at top (see Fig. 13.1 and Table 13.1 for description of sites). Heteroptera is inclusive of leafhoppers (Homoptera: Cicadellidae) because identifications were available, and diversity is inclusive of native and adventive species. Overall diversity steadily rises, although major groups analysed separately show individualistic trends owing, in part, to historical factors. Values are site means ± 1SE.

Arthropod species diversity

Fisher's α revealed idiosyncratic patterns for the taxon groups individually, but the overall diversity (of included groups) increased with geological age (Fig. 13.4). No model effectively predicted Heteroptera species richness, but substrate age and its quadratic term were important to total richness and to carabid richness, along with tree basal area (total richness $= 8.80 + 1.18\text{age} - 1.13(\text{age*age})$; $r^2 = 0.495$, $p < 0.001$; Carabidae richness $= -2.87 + 0.999\text{age} - 0.874(\text{age*age}) + 1.05\log(\text{BA})$; $r^2 = 0.725$, $p < 0.001$).

DISCUSSION

Local influences on community structure

Early in ecosystem development in Hawaii, nitrogen levels are low, while phosphorus is increasingly available from weathering of parent rock. Sites of intermediate age have more available nitrogen and phosphorus, but phosphorus becomes occluded in recalcitrant forms in older soils (Crews *et al.*, 1995). Net primary productivity and foliar nitrogen and phosphorus levels reflect this gradient of soil fertility, reaching their highest values at intermediate-aged sites along the mesic chronosequence (Vitousek *et al.*, 1995). The peak rate of cycling of nitrogen and phosphorus is at Kohala (150 000 years old; Herbert & Fownes, 1999), tapering off towards both ends of the age gradient. Kohala is also the site with the highest densities of canopy arthropods (Figs. 13.2 and 13.3). These correlations are consistent with dynamic equilibrium models (e.g. DeAngelis, 1980) that predict faster growth rates, and thus greater organismal abundance and diversity, at higher levels of resource flux. Preliminary studies from North Queensland, Australia reveal differences in canopy arthropod densities between adjacent areas underlaid by basaltic versus granite parent rock (Kitching *et al.*, 1997), although their relative fertilities were not reported.

It was surprising that all within-site ecological variables failed to account for variation in arthropod densities. Heteroptera, which are predominantly phytophagous or else feed on plant products *in Metrosideros* sp. (many lygaeids utilize seeds and fruits), and the predatory ground beetles were strongly influenced by site identity. Within-habitat variables, with the sole exception of foliar nitrogen, explained additional variation beyond site identity for ground beetles, but not unambiguously. Circumstantial evidence implicates the importance of foliar nitrogen as a predictor of total arthropod density or diversity (e.g. Recher *et al.*, 1996b). Many animals seem limited by organic nitrogen in the form of specific proteins or amino acids, and nitrogen has been suggested as the primary limiting resource for arthropods and other herbivores (White, 1978, 1984; Mattson, 1980). Although plant nitrogen content ordinarily is in the range 1–7%, animals are 7–14% nitrogen by weight (Mattson, 1980). Much of the variation in foliar nitrogen that we observed was among sites. This may indicate that arthropods on *M. polymorpha* respond numerically to resource availability, but only on a large scale and not on a tree-by-tree or patch basis. The high correlation of arthropod density with mean foliar phosphorus concentrations supports this notion.

Foliar phenological variables, such as leaf production and development, were not measured in this study, despite their clear importance to arthropod community structure in other canopy studies (e.g. Basset, 1991d; Basset & Novotny, 1999). Phytophages on *M. polymorpha* are dependent on young foliar tissues, such as expanding buds or newly flushed leaves (Nishida *et al.*, 1980; D. Gruner, personal observation). Phenological data with *M. polymorpha* on the scale of whole trees would be difficult to obtain and too crude to be useful. Vegetative flushing occurs at all times of year, usually asynchronously within single trees, and there is no distinct beginning or end to an annual phenological cycle (Gerrish, 1989). *M. polymorpha* is an evergreen tree with leaf life-spans of 2 years or more (Cordell, *et al.* 2001). Buds, expanding buds/flushing leaves, mature leaves and senescent leaves ordinarily are all present at any given time on a mature (not senescent) tree. These processes appear to vary without discernible pattern among different morphological varieties, age classes, altitudes and climates (Porter, 1972). A complementary study of the mechanisms of *Metrosideros* leaf expansion and herbivore attack will be reported elsewhere.

A major problem with interpreting densities of arthropods sampled by fogging is that it can be difficult to quantify the volume of canopy habitat contributing to each tray (Basset *et al.*, 1997b). Larger canopy dimensions may contribute more individuals to a sample but also may provide more interception substrate, removing individuals from a sample. In this study, all trays were placed directly beneath 100% canopy cover – if this was not possible for a given tray, we used fewer trays. Then we calculated canopy volume using the horizontal and vertical canopy dimensions. We concede this solution is just a first step towards dealing with the problem. The foliage of the canopy tends to be concentrated towards the edges, with larger twigs and branches towards the bole. Different trophic guilds can be expected to respond differently to these different substrates. As a next step towards dealing with this, we will analyse data at the guild level as a function of tray distance to the bole. With existing data and analyses, however, canopy volume was unimportant in all regression models.

Regional and historical influences on community structure

There are few studies that attempt to examine community structure or species diversity as a function of geological age (references in Borges & Brown, 1999). The Hawaiian Islands are a well-constrained natural laboratory for such questions. In studies of floristic diversity at the same sites on the mesic chronosequence, Kitayama and Mueller-Dombois (1995) concluded that quantitative species changes were effected by differences in soil fertility among sites. However, these local processes did not explain all the variation, and a simple unimodal pattern was not observed for plant richness or diversity. Richness and diversity were also influenced by differing regional floristic richness in combination with forest structure (Kitayama & Mueller-Dombois, 1995). Older islands tend to have greater regional richness, but also higher local endemism and reduced geographic ranges. Cluster analyses group the plant communities at intermediate-aged sites (Molokai and Kohala) as most similar in composition – even more similar than sites on the same island. These vegetation data from the mesic chronosequence 'support the idea that changes in community composition are driven by soil development as well as by geographic proximity of the sites' (Crews *et al.*, 1995, p. 1420).

In similar fashion, our preliminary canopy arthropod data appear to show an accumulating diversity of arthropods with increasing geological age. These results should be viewed with considerable caution until results from more taxonomic groups are available. The contrasting idiosyncratic patterns exhibited by Heteroptera and Carabidae serve to emphasize this point. With the Carabidae, haphazard phylogenetic accidents confound a pattern otherwise consistent with geological history. The ground beetles increase in richness and diversity along the chronosequence until Kauai, where there is a dramatic drop (Fig. 13.4). This drop is traceable to the fact that one lineage which is important in tree faunas (tribe Psydrini), especially in the richest site on Molokai, is entirely absent from Kauai and apparently never colonized the island (Liebherr & Zimmerman, 2000). In a synthesis using checklists from several island groups, Becker (1992) concluded that Coleoptera are inferior colonizers of remote locations relative to Heteroptera. Geological evidence suggests that all the main Hawaiian Islands, with the exception of Kauai, were at one time united above sea level with the next youngest volcano (Carson & Clague, 1995). The Kauai/Oahu strait apparently was the largest barrier to gene flow in the genus *Ptycta* (Psocoptera: Psocidae) as two diverse complexes are restricted only to Kauai (Thornton, 1984).

Graphical techniques can help to resolve the relative importance of local and regional processes in determining community structure. Local richness for a number of regions is plotted against the regional pool of potential colonists (i.e. α- compared with γ-diversity (Whittaker, 1975)). Most studies have found a linear relationship between regional and local richness (e.g. Cornell, 1985; Compton *et al.*, 1989; Cornell & Lawton, 1992; Ricklefs & Schluter, 1993; Griffiths, 1997; but see Terborgh & Faaborg, 1980; Tonn *et al.*, 1990). This pattern of 'proportional sampling' suggests that local diversity is merely a function of the regional pool of species, and local interactions are not sufficient to limit local diversity. In a saturated or interactive community, by contrast, the number of local species would approach an asymptote with regional richness. The analysis does not end there, as one cannot infer whether interactions occur simply from the shape of the local–regional richness relation (Cornell and Lawton, 1992). Moreover, there are serious problems with these techniques (e.g. Hartley, 1998; Srivastava, 1999). It is clear that other data and analyses are needed to resolve this important question (e.g. Basset, 1996; Basset & Novotny, 1999). In order to determine if species diversity or richness is a function of the regional pool of species, we must begin to understand the regional pool and how it varies with geological age. In many tropical regions, this is an impossible task because the regional faunas are so poorly known, especially for the arboreal taxa (Basset & Novotny, 1999). In the Hawaiian Islands, this question can be addressed only for a few restricted taxa. For many groups for which adequate information is available, species ranges are smallest and number of species is highest on the older islands, Oahu and Kauai (Gillespie *et al.*, 1997; Roderick & Gillespie, 1998).

Peck *et al.* (1999) used an extensive checklist for Hawaiian terrestrial arthropods (Nishida, 1997) and linear regression procedures to explore relationships of the fauna with general abiotic variables, such as area, elevation and island age. They found area to be the best general predictor of arthropod species and generic diversity, and of the species to genus ratio. Area and elevation, which are themselves positively correlated, were both important for the number of endemic species. However, Peck *et al.* (1999) include coral atolls, the older northwestern Hawaiian Islands, which represent highly influential leverage points in the regressions. These islands are two to five orders of magnitude smaller in area and elevation, and they are as much as 29 million years older. Some of the islands have no basaltic substrate remaining and most are represented by little or no native habitat or habitat diversity. As they state, 'an analysis of the data as two separate and independent subsets (the young, large main islands and the old, small northwestern islands) did not produce significant results' (Peck *et al.*, 1999, p. 532). At these scales, geological age has an immense impact on arthropod species diversity, although it is a negative one that is tightly interwoven with local habitat characteristics. Another problem, moreover, with using a large checklist for analyses of this type is accounting for biases associated with unequal collecting effort and changes in taxonomic concepts.

Cowie (1995) used path analysis to investigate similar variables for a catalogue of land snails in the Hawaiian Islands. Cowie (1995) was able to quantify variable sampling effort and found it to be a very important factor in the analysis. Physical variables also proved important, including habitat diversity, which was estimated by the number of plant community types present on an island. Island age, *per se*, was not a good predictor of the number of species. Cowie (1995) found a clumped pattern of lineages consistent with a 'colonization and radiation' model of diversification, and this was not strictly concordant with age. However, the youngest islands had fewest species and lineages, suggesting that these islands have not had adequate time for radiations of hyperdiverse lineages.

It has been suggested that *Metrosideros* is a geologically recent colonist of the islands (Wright *et al.*, 2000). This 'clock' argument dates the colonization to within the Pleistocene, but Kauai and Molokai were formed earlier. Pollen data from the Salt Lake tuff on Oahu places *Metrosideros* spp. in the islands for at least 100 000 years. An alternative interpretation of the molecular data is that the lack of differentiation between Hawaiian and New Zealand species is maintained by constant gene flow owing to the high vagility of the seeds (S. Grose, personal communication). Moreover, mounting evidence based on pollen records, fossilized isotope ratios and other data suggests that these chronosequence sites differed under different climate regimes. However, the climatic changes probably led to gradual altitudinal migration, rather than elimination, of ecotypes within each region (Hotchkiss *et al.*, 2000).

If *Metrosideros* is a recent arrival, then contemporary arthropod communities are a subset of the regional species pool that has shifted hosts within a short evolutionary time frame. *Metrosideros* is the dominant tree in a wide variety of ecosystems in Hawaii, from dry to wet, from sea level to tree line, from new lava flows to old growth forest (Dawson & Stemmerman, 1990). Moreover, the Hawaiian *Metrosideros* has a well-developed radiation of gall-forming insects (Homoptera: Psyllidae) that has apparently developed this complex habit *in situ*. These host shifts would have occurred in ecological resource regimes not appreciably different from those at present, at least for the older sites. Radiations onto *Metrosideros* should be a stochastic reflection of the regional species pool unless constrained by local deterministic processes. If the older islands are richer, as is suggested by widespread and pervasive anecdote, then diversity and richness of the canopy fauna should be an increasing function of age if unconstrained by local processes. A more recent colonization of the islands by *Metrosideros* than indicated by the oldest substrates poses no special problems for the interpretation of these arthropod data along the chronosequence.

These preliminary data appear to support contemporary ecological factors as important to arthropod densities on an ecosystem scale, but not on the level of individual trees. Overall diversity increases with geological age, but these patterns must be viewed with caution until more groups are added to the analysis. Replicate sampling on short temporal scales (e.g. diel, seasonal) may allow improved examinations of the importance of local factors to faunas of individual trees, but larger temporal processes and historical accidents may constrain diversity of local canopy arthropod communities. More comprehensive future analyses will integrate guild structures and species diversity of additional lineages into this framework.

ACKNOWLEDGEMENTS

We thank the Division of Forestry and Wildlife of the State of Hawaii, Hawaii Volcanoes National Park, the Joesph Souza Center at Kokee State Park, Parker Ranch and the Nature Conservancy of Hawaii for access to field sites; C. P. Ewing and J. K. Liebherr ('BugStrafe 97'), and K. Magnacca, S. Fretz and L. Santiago for field assistance; J. K. Liebherr for taxonomic assistance with the Carabidae; K. Heckmann for laboratory assistance; D. Foote and the research staff at Hawaii Volcanoes National Park for logistical assistance; and A. Taylor and R. Colwell for miscellaneous discussions leading to improvement in the study or the manuscript. Research was supported by NSF DGE-9355055 & DUE-9979656 to the Ecology, Evolution and Conservation Biology programme of the University of Hawaii at Manoa, the John D. and Catherine T. MacArthur Foundation, the Science to Achieve Results (STAR) programme of the US Environmental Protection Agency, and by the Drake Fund of the Smithsonian Institution.

14

Seasonality of canopy beetles in Uganda

Thomas Wagner

ABSTRACT

Beetles were collected by insecticidal fogging of two tree species, *Rinorea beniensis* (Violaceae) and *Cynometra alexandri* (Caesalpiniaceae), in Budongo Forest, a seasonal rainforest in Uganda that is characterized by a pronounced dry season in January/February. Eight trees of each species were fogged in adjacent plots of primary and selectively logged forest; *R. beniensis* was also sampled in a swamp forest situated between the other forest types. Field study was performed in June/July 1995 during the wet season and in the same plots in January 1997 during the dry season. During the wet season, a total of 19 675 beetles, assigned to 966 morphotypes were collected, whereas 16 583 individuals (779 morphotypes) were collected during the dry season. In comparison with other arthropod groups collected, Coleoptera were more abundant during the wet season (18.3–27.6% of all arthropods), and less abundant during the dry season (14.1–17.6%). In particular, the mould-feeding Latridiidae and the predominantly mycetophagous Alleculinae (Tenebrionidae) showed a marked decrease moving from wet to dry season. The phytophagous Alticinae (Chrysomelidae), Apionidae and Curculionidae showed a great increase in abundance from wet to dry seasons. Seasonal change was lowest in the swamp forest, but greater in forest plots with increasing aridity. Most of those species were very small (2–3 mm in length) and it is probable that they had no food-plant relationship to the tree species fogged but accumulated in the dense, relatively humid tree crowns during the dry season. This resulted in a high beetle faunal overlap across different tree species within one season, but much less overlap between conspecific trees between seasons. This indicated great influence of seasonality and habitat structure on beetle distribution, but low influence of tree species, even with regard to phytophagous beetles.

INTRODUCTION

A comprehensive overview of a local canopy fauna is the most rigorous way to establish a better understanding of the mechanisms that generate and maintain the enormous diversity of canopy arthropod assemblages in wet tropical forests (Godfray *et al.*, 1999). In an ongoing project on the influence of the tree species, management effects and forest type on canopy arthropods in central and east African forests, this research concentrated on Budongo Forest, a seasonal rainforest in western Uganda. Seasonal change in climatic conditions, particularly the availability of water, has considerable effect on the vegetation and may be crucial in determining the dynamics of the arthropod assemblages.

Long-term studies on seasonal fluctuations of abundance in tropical forest arthropods have been made on litter arthropods (Levings & Windsor, 1985), on Homoptera captured at lights (Wolda, 1979, 1992), on other insects (Ricklefs, 1975), on Coleoptera fogged in Central America (Erwin & Scott, 1980) and on canopy arthropods in Sulawesi (Stork & Brendell, 1990) and in Queensland (Frith & Frith, 1985; Basset, 1991c; Kitching & Arthur, 1993). Only a few studies have concentrated on the seasonal change between distinct seasons, while others were long-term studies including, but not concentrating on, seasonal change.

In all studies on canopy arthropods, beetles have been identified as one of the most species-rich and abundant arthropod groups. Beetles show a high ecological diversity, since they include predatory, mycetophagous, saprophagous and phytophagous species. However, this diversity and the species richness of Coleoptera cause major problems, since the identification to species is impossible for most of the taxa. Assigning sampled material to morphotypes can be an effective method and has been carried out by our working group and numerous cooperating taxonomists for several years (Wagner, 1996a).

In this study, the seasonal change in the fauna of canopy beetles of two tree species has been compared between wet and dry seasons in adjacent plots of primary, selectively logged and swamp forest. Data are analysed with respect to the influence of forest structure and selective logging on the canopy beetle fauna, and the study focusses on distribution patterns and seasonal change, especially of phytophagous beetles.

METHODS

Study site

Research was carried out in the Budongo Forest Reserve in western Uganda around the field station of the Budongo Forest Project (1° 45′ N, 31° 35′ E, 1200 m above sea level). This forest consists of selectively logged compartments next to old primary forest areas and has a well-documented history (Eggeling, 1947; Plumptre & Reynolds, 1994; Plumptre, 1996). Dense forest covers an area of 428 km^2, which is divided into compartments for logging purposes. Selective logging started before 1915, increased between 1935 and 1960, when mechanical logging operated in the forest, and was reduced, subsequently, in the early 1970s. No recent logging has been or is being carried out in the compartments investigated. The dominant forest type in Budongo is *Cynometra* forest, where *C. alexandri* C. H. Wright (Caesalpiniaceae) forms up to 70% of the upper canopy trees, and mixed forest dominated by some *Celtis* spp. (Ulmaceae) and the mahoganies *Khaya anthotheca* (Welwitsch) C. DeCandolle and *Entandrophragma* spp. (Meliaceae; Eggeling, 1947).

Around the field station, an exceptionally large amount of timber (80 $m^3\ ha^{-1}$) was selectively logged between 1947 and 1952 (Plumptre & Reynolds, 1994). This compartment is now similar to a secondary forest, with great gaps in the upper canopy, no distinct canopy layering and a dense cover of young shrubs and trees in the understorey. Primary forest occurs in compartment N15, which has been protected as a Nature Reserve since the 1930s. It is a typical mature rainforest, with distinct canopy layers. The ironwood *C. alexandri*, and the mahoganies *K. anthotheca* and *Entandrophragma utile* (Dawe & Sprague) are the dominant species of the upper canopy at between 30 and 40 m. The vegetation of the lowest storey, between 10 and 12 m, is dense, with *Rinorea beniensis* (Welwitsch ex. Olivier) (Violaceae) as one of the dominant tree species. Because of the dense canopy, the ground is heavily shaded and lacks a shrub layer. The selectively logged forest and the primary forest are separated by a small river. The vegetation on the poorly drained soils next to the river is a distinct swamp forest with the palm *Raphia farinifera* (Gaertner) Hylander (Arecaceae) as the characteristic plant taxon. The two-storied canopy is much lower and naturally more open and less homogeneous than in the primary forest on higher elevations (Eggeling, 1947). The plant associations and vegetation structure of primary and selectively logged forest outside the swamp are more similar to each other than either is to the swamp forest.

The mean annual rainfall in Budongo Forest from 1992 to 1995 was about 1600 mm (A. Plumptre, personal communication), with a relatively constant precipitation from March to November, and a dry season from late December to February. During this short but pronounced dry season, about 25% of the tree species loose their leaves (Synnott, 1985), which leads to a more intensive irradiation of the understorey and ground vegetation and the drying of the soils, particularly in the selectively logged forest compartments.

Data collecting and analysis

R. beniensis and *C. alexandri* are abundant species in this forest and were selected for the fogging experiments. Hereafter, the two tree species are designated by their generic names. With a maximum height of 14 m, the former is the dominant tree species of the first canopy layer, whilst the latter grows up to 50 m in height and dominates the upper canopy. In this study only a few large trees of *Cynometra* were fogged: most of them were young trees between 12 and 20 m in height. Trees were fogged during the wet season in June/July 1995 and during the dry season in January 1997. With the exception of two short rain showers, there was no precipitation for 30–40 days before fogging the trees during the dry season, whilst there was rain nearly every day during the wet season. For both seasons, eight trees of *Rinorea* were fogged in the adjacent compartments of primary, selectively logged and swamp forest; eight trees of *Cynometra* were fogged in the primary and selectively logged forest, but not in the swamp forest where this species is rare. Different individual trees were fogged in the same compartments at each season. A group of eight conspecific trees per forest type and season is designated as a 'collecting unit'. The maximum distance between individually fogged trees in the primary and selectively logged forest compartments is about 4 km, the swamp forest being situated between both other forest types. Fogged trees were neither flowering nor fruiting. Canopies of trees selected for fogging were isolated from those of other trees as far as was possible, i.e., vertical distance to the canopy of the next tree was at least 5 m horizontally, without overlap to other trees.

Before fogging, 16 funnel-shaped nylon sheets, 1 m^2 each, were fixed with ropes under the canopy. Trees were fogged for approximately 4 minutes with a Swingfog SN-50 and a solution of natural pyrethrum with 1% active ingredient, and arthropods were collected for 90 minutes after fogging. The total height of fogged trees and the height of the lowest branch carrying dense leaves were measured. In the selectively logged forest, *Rinorea* are densely covered with leaves even on the lowest branches, whilst in the primary and swamp forest lowest branches sometimes have sparse leaf cover; consequently these branches were excluded when taking these measurements. The difference between the total height of tree and lowest branch with denser leaf cover, multiplied by 16 (area of collecting sheets), gave a measurement of the exploited canopy volume. This standardized sampling regime, especially the fogging of many trees of one species only at one locality, ensures a good comparison of the arthropod assemblage between forest types, tree species and seasons.

Material was sorted to major taxa groups (usually on order level) and the numbers of individuals were counted. Numbers of individuals of distinct beetle taxa were partly transformed to percentage data for each tree. Percentages of the collecting units between seasons and forest types were compared using one-way analysis of variance (ANOVA). Faunal similarities and overlap of several beetle groups between seasons, forest types and tree species were calculated by Morisita–Horn indices (Krebs, 1989; Wolda, 1981). These values were calculated for all combinations in pairs (64×64) then arithmetic means for each collecting unit were derived. Means of the Morisita–Horn indices between trees of each particular collecting unit were also calculated. However, Morisita–Horn indices were detailed only for the most abundant beetle groups Chrysomelidae, Curculionidae and the predatory Staphylinidae (Aleocharinae, Paederinae, Pselaphinae, Steninae, Staphylininae, Tachyporinae and Xantholininae).

RESULTS

Abundance of Coleoptera and correlations with canopy volume

On all 48 trees of *Rinorea* and 32 trees of *Cynometra* fogged, a total canopy volume of 8064 m^3 was surveyed and 36 256 beetles were collected (Table 14.1). The percentage of beetles within arthropod canopy fauna varied between 18.3 and 27.6% during the wet season but showed less variation (14.1–17.6%) during the dry season (Table 14.1).

Generally, Coleoptera and Formicidae were the most abundant arthropod groups. Coleoptera as a percentage of the entire arthropod fauna decreased significantly from wet to dry season, on both tree species in the primary and selectively logged forest (one-way ANOVA: primary forest: $F = 31.6$, $p < 0.001$; selectively logged forest: $F = 3.87$, $p < 0.05$), whereas there was no significant difference between individuals of *Rinorea* in the swamp forest ($F = 0.63$). During the wet season, Coleoptera and Formicidae each represented, on average, about 25% of the arthropod fauna, but during the dry season, Formicidae were much more abundant (32%) than Coleoptera (15%) (Wagner, 2001).

The number of beetles fogged per tree was correlated strongly with canopy volume (Fig. 14.1). Regressions slopes were much steeper in *Rinorea*, where also a much higher number of beetles per cubic metre exploited canopy could be collected (Table 14.1). While the percentage of beetles decreased from wet to dry season, the density of beetles on the foliage (beetle m^{-3} canopy) was nearly constant. There was no significant change in Coleoptera per cubic metre on *Rinorea*, but data for *Cynometra* revealed a significant higher density of beetles during the dry season. This probably reflected an effect of different size between the trees fogged during both seasons (see below).

The mean beetle number per tree for different collecting units varied between 250 (*Cynometra*, primary forest, dry season) and 664 (*Rinorea*, selectively logged forest, wet season) (Table 14.1, Fig. 14.2). Beetle species per collecting unit varied between 241 species (*Cynometra*, primary forest, dry season) and 420 species (*Rinorea*, swamp forest, wet season) (Appendix Table 14A.1). From 40 trees fogged during the wet season,

Table 14.1. *Tree height and canopy volume, arthropod numbers and percentage and density of Coleoptera for each collecting unit*

Collecting unit	Tree height (m)	Canopy volume (m^3)	Arthropods/tree (SD)	Coleoptera percentage (SD)	Coleoptera per m^3 (SD)
Wet season					
Cynometra					
Primary forest	13–33	1696	2120 (215)	18.3 (3.05)	1.84 (0.33)
Selectively logged forest	10.5–30	1376	2214 (154)	21.8 (2.87)	1.61 (0.15)
Rinorea					
Primary forest	9.5–12	560	1835 (84)	22.8 (2.42)	5.98 (1.41)
Selectively logged forest	7.5–10	768	2404 (123)	27.6 (1.83)	6.92 (0.87)
Swamp forest	7.5–11	568	2411 (141)	20.6 (3.12)	6.99 (1.15)
Dry season					
Cynometra					
Primary forest	7.5–13	672	1643 (132)	15.5 (3.54)	2.98 (0.75)
Selectively logged forest	9–12.5	592	2212 (179)	16.9 (2.13)	4.85 (1.02)
Rinorea					
Primary forest	8.5–13.5	672	2845 (98)	16.9 (1.39)	5.74 (0.42)
Selectively logged forest	7–11.5	624	3997 (145)	14.1 (1.12)	7.21 (0.94)
Swamp forest	7.5–12	536	2384 (132)	17.6 (1.87)	6.25 (1.18)

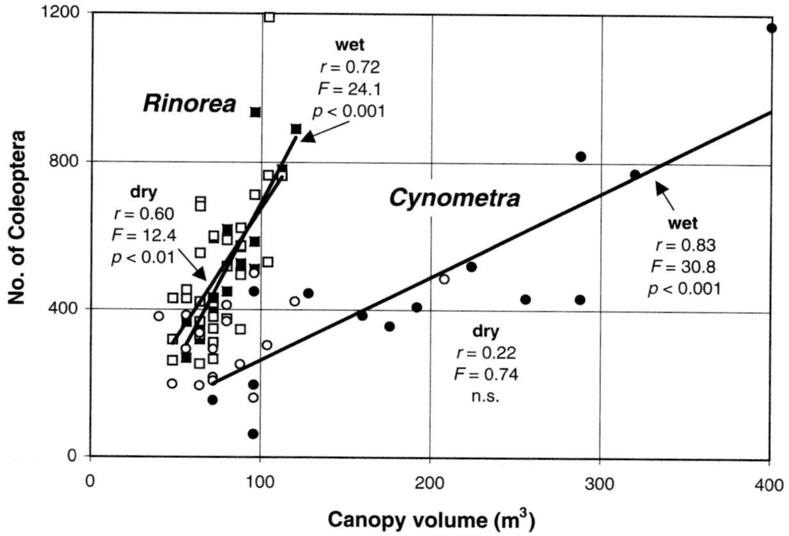

Fig. 14.1. Correlation between the volume of canopy and beetle abundance for study trees in different collecting units by linear regression. *Cynometra*, wet season (●); *Cynometra*, dry season (○); *Rinorea*, wet season (■); *Rinorea*, dry season (□).

19 675 beetles were collected and assigned to 966 morphotypes. During the dry season, 16 583 beetles (779 morphotypes) were collected. Despite this high number of specimens, most species were extremely rare. During the wet season, 41.4% of species were singletons, and 84.4% were collected with less than 10 individuals. During the dry season, 37.5% of species were singletons, and 78.4% were collected with less than 10 individuals.

The most abundant beetle species during the wet season was a *Melanophthalma* sp. (Latridiidae; 10.3% of all beetle individuals collected), followed by a *Bruchidius*

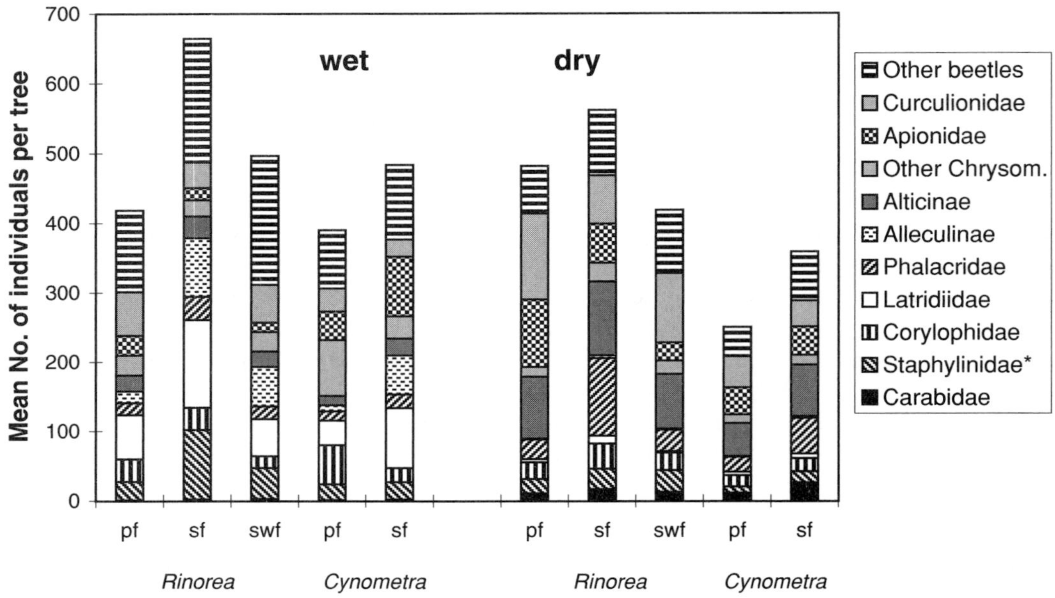

Fig. 14.2. Mean number of individuals per tree ($n = 8$ conspecific trees) for the major beetle taxa (*Staphylinidae without nonpredatory Oxytelinae and Scaphidiinae), per tree species, forest type and season. pf, primary forest; sf, selectively logged forest; swf, swamp forest.

sp. (Bruchinae; 7.9%), an Alleculinae (5.7%) and a *Cypha* sp. (Aleocharinae, Staphylinidae; 5.5%). During the dry season other species dominated: a Notarinae (Curculionidae; 11.3%), an Apionidae (6.2%), and a *Longitarsus* sp. (Alticinae, Chrysomelidae; 4.1%).

Seasonal change in beetle abundance

Seasonal change in faunal composition was similar among the major beetle groups on different tree species and forest types (Table 14.2). The strongest decrease in abundance from wet season (9.1 to 19.2% of beetle specimens for different collecting units) to dry season (<1%) was observed in the mould-feeding Latridiidae (Fig. 14.3, Table 14.2). Similarly, the mostly mycetophagous Alleculinae showed a significant decrease from wet to dry season for all collecting units, while another group of mycetophagous beetles, the Corylophidae, showed no such strong reduction in abundance (Fig. 14.3, Table 14.2).

The most abundant predatory beetle taxa, Carabidae and Staphylinidae, also showed a similar pattern for all collecting units. Carabidae increased significantly in abundance from dry to wet season, the principal cause of this being a single species of Bembidiinae, which was very abundant in the entire forest during the dry season. Generally, predatory Staphylinidae decreased from wet to dry season, particularly on *Rinorea* in the selectively logged forest, where a few species were very abundant during the wet season and absent during the dry season.

The most common groups of phytophagous beetle increased in abundance from the wet to the dry season. This was most significant for Curculionidae, Apionidae (with the exception of *Cynometra*: wet season, selectively logged forest) and Alticinae, while other chrysomelid taxa together showed no significant change between seasons (Fig. 14.3, Table 14.2). One exception was *Cynometra* in the primary forest, where very high numbers of Eumolpinae were collected on a single, large tree. Curculionidae, Apionidae and Alticinae together were also abundant during the wet season (on average 20.3% of all beetle individuals) but much more abundant during the dry season (50.4%).

Seasonal change in faunal composition

The faunal composition also changed between seasons, particularly in phytophagous beetles (Curculionidae (Fig. 14.4) and Chrysomelidae (Fig. 14.5)). Faunal overlap for Curculionidae within one season was of similar degree even between different tree species. Overlap

Table 14.2. *Results of one-way analysis of variance (ANOVA) considering the percentages of beetle individuals per tree between seasons*[a]

	Rinorea beniensis						*Cynometra alexandri*			
	Primary		Selectively logged		Swamp		Primary		Selectively logged	
Taxa	$p <$	F	$p <$	F	$p <$	F	$p <$	F	$p <$	F
Carabidae	0.05	4.92	0.05	7.32	0.05	4.32	0.01	12.3	0.05	8.37
Staphylinidae	0.05	5.34	**0.001**	32.8	NS	1.27	0.05	7.18	NS	1.23
Corylophidae	0.05	7.02	NS	2.39	NS	3.17	**0.001**	41.3	NS	0.32
Latridiidae	**0.001**	56.4	**0.001**	65.1	**0.001**	34.7	**0.001**	31.6	**0.001**	48.6
Phalacridae	NS	1.50	**0.001**	33.7	0.05	5.76	0.01	13.2	**0.001**	28.3
Alleculinae	0.05	7.21	**0.001**	44.1	0.01	17.3	NS	2.18	0.01	20.3
Alticinae	**0.001**	76.1	**0.001**	54.8	**0.001**	45.9	**0.001**	81.9	**0.001**	101
Other Chrysomelidae	0.05	8.58	NS	3.10	NS	2.18	**0.001**	39.5	0.05	6.78
Apionidae	**0.001**	30.8	0.01	12.4	0.05	8.65	0.05	7.80	0.01	16.3
Curculionidae	**0.001**	46.4	0.01	22.3	**0.001**	38.2	0.01	16.3	0.01	12.2

NS, not significant.

[a]Degrees of freedom 15; $p < 0.001$ indicated in bold.

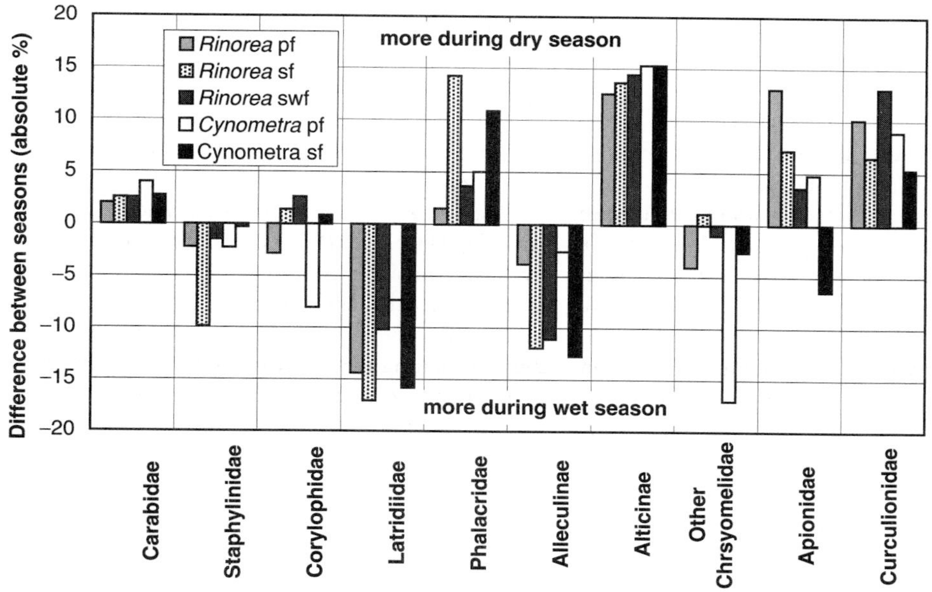

Fig. 14.3. Seasonal changes in the major beetle taxa, as indicated by the mean percentage difference in the contribution to the whole of beetle assemblages between seasons. Abbreviations as in Fig. 14.2.

Curculionidae 147 sp. 4650 ind.	Season	Forest type	*Cynometra* wet pf	*Cynometra* wet sf	*Cynometra* dry pf	*Cynometra* dry sf	*Rinorea* wet pf	*Rinorea* wet sf	*Rinorea* wet swf	*Rinorea* dry pf	*Rinorea* dry sf	*Rinorea* dry swf
Cynometra	wet	pf	**0.81**									
		sf	0.51	**0.44**								
	dry	pf	0.07	0.08	**0.94**							
		sf	0.15	0.19	0.59	**0.59**						
Rinorea	wet	pf	0.81	0.53	0.06	0.17	**0.81**					
		sf	0.26	0.31	0.03	0.16	0.29	**0.50**				
		swf	0.59	0.47	0.06	0.19	0.63	0.37	**0.56**			
	dry	pf	0.17	0.16	0.92	0.63	0.17	0.09	0.14	**0.92**		
		sf	0.07	0.17	0.50	0.51	0.10	0.32	0.19	0.56	**0.59**	
		swf	0.06	0.08	0.95	0.55	0.06	0.06	0.06	0.91	0.47	**0.97**

Shading key
< 25
< 40
< 55
< 70
> 70

Fig. 14.4. Morisita–Horn indices calculated for Curculionidae across tree species, forest types and seasons. Numbers in bold indicate means of trees compared within each collecting unit; others are means of trees compared between collecting units. Abbreviations as in Fig. 14.2.

Chrysomelidae 193 sp. 7634 ind.	Season	Forest type	*Cynometra* wet pf	*Cynometra* wet sf	*Cynometra* dry pf	*Cynometra* dry sf	*Rinorea* wet pf	*Rinorea* wet sf	*Rinorea* wet swf	*Rinorea* dry pf	*Rinorea* dry sf	*Rinorea* dry swf
Cynometra	wet	pf	**0.47**									
		sf	0.35	**0.45**								
	dry	pf	0.23	0.32	**0.46**							
		sf	0.19	0.27	0.48	**0.49**						
Rinorea	wet	pf	0.44	0.49	0.41	0.32	**0.73**					
		sf	0.37	0.53	0.41	0.35	0.68	**0.70**				
		swf	0.20	0.19	0.21	0.19	0.17	0.20	**0.35**			
	dry	pf	0.22	0.28	0.52	0.49	0.41	0.39	0.22	**0.60**		
		sf	0.23	0.31	0.44	0.46	0.34	0.37	0.24	0.54	**0.52**	
		swf	0.20	0.25	0.47	0.45	0.34	0.33	0.19	0.58	0.45	**0.54**

Shading key
< 25
< 40
< 55
< 70
> 70

Fig. 14.5. Morisita–Horn indices calculated for Chrysomelidae (including Bruchinae) among tree species, forest types and seasons. Presentation follows Fig. 14.4.

during the dry season between *Rinorea* and *Cynometra* in different forest types varied between 51 and 95% without stronger differences between tree species and forest types. However, overlap between seasons on conspecific trees was much lower and varied between 6 and 19%. Only the faunal overlap of curculionids on *Rinorea* in the selectively logged forest attained a relative similarity between seasons of 32% (Fig. 14.4).

In general, Chrysomelidae showed a similar pattern, but differences between seasons were less pronounced than for Curculionidae. Furthermore, specific patterns relevant to forest types could also be derived from the Morisita–Horn matrix (Fig. 14.5). Overlap of the chrysomelid fauna on *Rinorea* during the wet season in the swamp forest was very different from all other collecting units (Wagner, 1999). Also the chrysomelid fauna of *Cynometra* during the wet season in the primary forest showed some peculiarities (see Discussion).

Faunal overlap for Staphylinidae showed a low influence of seasonal effects. Overlap was very low between different collecting units (most values were less than 20%; Fig. 14.6), whereas the similarity of the staphylinid fauna between the eight trees within one collecting unit was much higher, reaching values up to 84% on *Rinorea* during the wet season in the selectively logged forest.

DISCUSSION

Relative abundance of Coleoptera

Data on seasonality of tropical arthropods reveal that rainfall and humidity are important factors affecting the abundance of arthropods in the tropics. Studies on the seasonal fluctuations of arthropods in savannas (Denlinger, 1980) and in tropical forests have shown a great decrease in arthropod abundance during the dry season (Janzen & Schoener, 1968; Frith & Frith, 1985; Kitching & Arthur, 1993). In Budongo Forest, arthropod density (i.e. numbers of arthropods per tree) between seasons was similar in the swamp forest but showed a significant increase from the wet to the dry season in the primary and selectively logged forest (Wagner, 2001). Ensifera, Lepidoptera (larvae), Heteroptera and Coleoptera were less abundant during the dry season, and Psocoptera, Cicadina, parasitoid Hymenoptera and Formicidae in particular increased significantly from the

Staphylinidae 189 sp. 2509 ind.	Season	Forest type	*Cynometra* wet pf	*Cynometra* wet sf	*Cynometra* dry pf	*Cynometra* dry sf	*Rinorea* wet pf	*Rinorea* wet sf	*Rinorea* wet swf	*Rinorea* dry pf	*Rinorea* dry sf	*Rinorea* dry swf
Cynometra	wet	pf	**0.35**									
		sf	0.28	**0.33**								
	dry	pf	0.24	0.21	**0.32**							
		sf	0.09	0.07	0.21	**0.33**						
Rinorea	wet	pf	0.16	0.18	0.10	0.06	**0.23**					
		sf	0.04	0.15	0.03	0.03	0.22	**0.84**				
		swf	0.06	0.11	0.02	0.02	0.17	0.42	**0.29**			
	dry	pf	0.17	0.10	0.14	0.14	0.12	0.02	0.02	**0.33**		
		sf	0.15	0.14	0.18	0.39	0.17	0.06	0.06	0.21	**0.49**	
		swf	0.14	0.13	0.16	0.40	0.15	0.04	0.04	0.21	0.48	**0.48**

Shading key: < 25; < 40; < 55; < 70; > 70

Fig. 14.6. Morisita–Horn indices calculated for Staphylinidae (including Pselaphinae, excluding Oxytelinae, Scaphidiinae) between tree species, forest types and seasons. Presentation follows Fig. 14.4.

wet to the dry season, resulting in this overall increase in arthropod density (Wagner, 2001).

A general decrease of Coleoptera from wet to dry season was noted comparing the percentages of Coleoptera to the entire arthropod fauna (Table 14.1). The mean percentage during the dry season decreased from 3% on *Cynometra* in the primary forest, and *Rinorea* in the swamp forest to 14.5% on *Rinorea* in the selectively logged forest for pair-wise comparisons of collecting units. Conversely, the absolute number of beetles collected per cubic metre canopy foliage was approximately similar for both seasons on *Rinorea*. The significant increase in beetle number per canopy volume on *Cynometra* from wet to dry season was caused by the sampling protocol. During the wet season, the eight trees of *Cynometra* fogged were between 20 and 33 m tall, while the maximum tree height of all *Cynometra* trees fogged during the dry season was less than 13 m. Since the effectiveness of the method decreased strongly with tree height when the trees were fogged from the ground, it is highly probable that only a smaller portion of the upper canopy arthropod fauna could be collected from higher canopies, and, consequently, the number of arthropods collected per canopy volume of large trees was lower.

Seasonal changes in the major beetle groups

The greatest decrease in abundance from wet to dry season for all collecting units was recorded for mycetophagous beetles (Fig. 14.3, Table 14.2). Mould-feeding Latridiidae decreased markedly from wet to dry season (see also Erwin & Scott, 1980), when it is probable that their preferred food was rare. Corylophidae using the same resource may be better adapted to drier conditions, since they showed only a significant decrease in the primary forest. Predatory Staphylinidae were represented by smaller percentages during the dry season in all collecting units, with the smallest value observed on *Rinorea* in the selectively logged forest. Since this observation depended on a single species of *Cypha*, which was extremely abundant in the selectively logged forest during the wet season, this change should not be overstated.

Chrysomelidae, with the exception of Alticinae, showed a decrease in abundance from wet to dry season, which was exceptionally pronounced on *Cynometra* in the primary forest. This resulted mainly from the high abundance of some species of Eumolpinae, which were predominantly collected from large *Cynometra* trees fogged in the primary forest. Differences in the faunal composition between understorey and upper canopy in this group may be crucial, because faunal change in stratification has been found in several other studies on tropical forest arthropods (e.g. Sutton & Hudson, 1980; Hammond, 1990; Basset *et al.* 1992). Conversely, the lower abundance of Chrysomelidae (with the exception of Alticinae) or other phytophagous beetles like Lagriinae during the dry season may be explained by insufficient food resources (Wagner, 1999).

There is some evidence that the abundance of phytophagous insects peaks during leaf-flush periods (Basset, 1991c; Aide, 1993) because of their preference for young leaves (Price *et al.*, 1980; Raupp, 1985; Coley & Aide, 1991). Unfortunately, nothing is known about the feeding habits and host–plant relationships of the phytophagous beetles collected in Budongo Forest, nor of most other afrotropical species. It is not clear whether these species do feed on the tree species fogged. Most of the Alticinae and Apionidae collected obviously had no direct food–plant relationship to *Rinorea* and *Cynometra*, since dissections of many specimens of these taxa revealed empty mid- and hindguts (Wagner, 1999). Increase in the abundances of these taxa must result from other factors. During the dry season, the canopies of trees with dense foliage like *Rinorea* are often the most humid habitats, and small and soft-bodied insects are known to accumulate along a humidity gradient. Most of the Curculionidae collected during the dry season (and only rarely during the wet season) were also of small body size, within the range of flea beetles and apionids. This aggregation, sometimes combined with dormancy in comparatively humid sites during the dry season, is known for many insect groups (Denlinger, 1986). Such aggregations can also originate from the distribution patterns of understorey arthropods along a humidity gradient in adjacent forest types (Janzen & Schoener, 1968).

Seasonal changes in different forest types

With the exception of Curculionidae and, to a lesser extent, Corylophidae, seasonal changes in beetle composition were lowest on *Rinorea* in the swamp forest. Absence of rain for about 2 months had no obvious negative influence overall on canopy arthropods in the swamp forest. Despite the reduction in the water level of the stream during the dry season, soils surrounding the stream were constantly moist. A permanent water supply is one important factor for the development and maintenance of the distinct swamp vegetation, characterized by low leaf abscission during the dry season and high density of epiphytic mosses and lichens on trees. This also maintains a high structural diversity in plant architecture and diversity of habitats (see Lawton, 1983; Williamson & Lawton, 1991), resulting, for example, in great differences in the chrysomelid fauna between the swamp forest and other forest types (Fig. 14.5; Wagner, 1998, 1999, 2000). Overall, microclimatic conditions in the swamp forest were relatively constant throughout the year, and seasonal differences in arthropod faunal composition were accordingly few. This supports the observation that arthropod populations in environments with constant climatic conditions, such as those found in perhumid lowland rainforests, fluctuate less (Bigger, 1976; Wolda, 1978a; Hammond, 1990; Novotny & Basset, 1998).

The influence of seasonal change was greater in the primary and particularly in the selectively logged forests away from the influence of the stream. The understorey and ground vegetation was largely shaded by the well-defined upper canopy in the primary forest, even during the dry season, keeping the soil surface relatively moist. In the selectively logged forest, irradiation was more intensive on the ground, particularly during the dry season, and the decrease of air and soil moisture is greater on average than in mature forests (Brown & Lugo, 1990; Chapman & Chapman, 1990). However, the more intensive irradiation in the selectively logged forest understorey usually leads to higher leaf density and an increase in the production of young leaves (Brown & Lugo, 1990; Aide, 1993). This may explain the high number of arthropods, particularly on *Rinorea* trees.

Seasonal change in β-diversity

The faunal composition of curculionids and chrysomelids was highly seasonal, whilst the influence of tree species was very low. Faunal overlap of predatory Staphylinidae across collecting units was generally low and very different from that of phytophagous groups. This suggests a comparatively higher influence of local effects and habitat structure on the distribution of Staphylinidae, since a different staphylinid fauna was represented in each collecting unit.

Seasonal fluctuations in arthropod assemblages, and of beetle assemblages, are common in tropical biomes. As this study emphasizes, different seasonal patterns inside large forest plots can exist along a transect of only 4 km through different types of forest with different water supply and habitat structure. The seasonal changes were lower in the swamp forest where high levels of ground water persisted. However, tree species had a very low influence even on canopy-dwelling phytophagous beetles, which indicated a low number of specialists in terms of host–plant relationship (Wagner, 1998, 1999).

ACKNOWLEDGEMENTS

I am grateful to Andy Plumptre, Jeremy Lindsell, Chris Bakuneeta and Vernon Reynolds for the invitation and the opportunity to work and live in the Sonso Station of

Budongo Forest Project, to the Uganda National Council for Science and Technology and the Uganda Forest Department for allowing the field study in Budongo Forest, and to the National Geographic Society, Conservation International (Margot Marsh Fund), the Department for International Development and the Norwegian Agency for Development for supporting the Bundongo Forest Project. I thank Jakisa Deogracius, Ursula Göllner and Michael Schmitt for their help in collecting the arthropods; and Markus Cousin, Wolfram Freund, Ruth Hasenkamp, and Guido Velten for their valuable assistance with sorting and counting arthropod material from Budongo Forest. I thank, especially, Claudia Stuhllemmer who was intensively involved in collecting the data in Uganda, sorting material and analysing the data collected from *Cynometra* trees. Neil Springate and Yves Basset made valuable comments on the manuscript. This research was supported by Stifterverband für die Deutsche Wissenschaft and Deutsche Forschungsgemeinschaft (Grant no. Schm1137/2-2).

Appendix: Collected beetle taxa

Table 14A.1 lists the individuals and morphotypes of the collected beetle taxa in systematic order.

Table 14A.1. *Individuals and morphotypes of collected beetle taxa in systematic order. Each column represents the results of one collecting unit (eight conspecific trees)*

Taxa	Wet season												Dry season													
	Cynometra pf		sf		*Rinorea* pf		sf		swf		Sum		*Cynometra* pf		sf		*Rinorea* pf		sf		swf		Sum		Total	
	n	sp.	*n*	sp.	*n*	sp.	*n*	sp.	*n*	sp.	*n*	sp.	*n*	sp.	*n*	sp.	*n*	sp.	*n*	sp.	*n*	sp.	*n*	sp.	*n*	sp.
Larvae	28	–	34	–	54	–	25	–	285	–	426	–	28	–	45	–	14	–	24	–	60	–	171	–	597	–
Adephaga																										
Carabidae	11	8	20	9	14	8	11	5	25	10	81	20	92	6	212	5	87	7	126	11	106	9	622	12	704	23
Staphyliniformia																										
Histeridae	1	1	–	–	4	4	1	1	2	2	8	8	1	1	1	1	–	–	2	2	1	1	5	5	13	13
Ptiliidae	14	4	2	2	9	3	14	8	4	3	43	15	1	1	3	3	–	–	2	2	–	–	6	6	49	20
Scydmaenidae	4	3	3	3	5	2	13	4	11	2	36	10	–	–	1	1	–	–	3	3	4	4	8	7	42	16
Micropeplinae	3	1	1	1	–	–	–	–	–	–	4	1	–	–	2	1	–	–	–	–	–	–	2	1	6	2
Oxytelinae	68	3	138	3	68	2	124	2	23	2	421	3	1	1	4	2	1	1	–	–	1	1	7	4	429	4
Pselaphinae	7	5	8	3	48	5	45	5	52	8	190	17	6	2	–	–	12	3	17	3	34	4	69	4	259	18
Scaphidiinae	22	4	7	2	59	7	62	7	87	9	237	13	2	1	2	2	25	3	22	6	32	7	83	9	320	13
Staphylinidae (others)	173	28	190	51	160	43	774	55	336	53	1633	139	65	14	127	18	47	17	212	32	166	25	617	61	2250	171
Eucinetiformia																										
Clambidae	3	2	1	1	1	1	7	5	–	–	12	6	5	2	9	2	4	3	1	1	19	4	37	4	49	7
Eucinetidae	–	–	–	–	–	–	–	–	4	1	4	1	–	–	–	–	–	–	–	–	–	–	–	–	4	1
Scirtidae	–	–	1	1	1	1	5	3	4	1	11	3	2	1	2	1	2	1	8	4	6	5	20	6	31	9
Scarabaeiformia																										
Ceratocanthidae	9	2	33	4	60	4	126	5	75	5	303	5	–	–	2	1	26	3	146	4	49	4	223	5	526	5
Scarabaeidae	–	–	3	3	2	2	6	5	5	2	16	7	–	–	–	–	1	1	10	7	2	2	13	10	29	13
Elateriformia																										
Buprestidae	8	4	4	3	1	1	3	3	2	2	18	7	4	4	10	4	1	1	1	1	–	–	16	9	34	14
Cantharidae	5	2	3	3	–	–	–	–	6	2	14	5	3	1	–	–	–	–	2	1	–	–	5	1	19	5
Elateridae	49	15	15	7	49	16	41	11	176	17	330	29	21	9	153	11	45	13	126	14	33	9	378	24	708	36
Eucnemidae	3	1	–	–	–	–	–	–	–	–	3	1	–	–	1	1	12	3	1	1	–	–	14	5	17	6
Lampyridae	1	1	–	–	1	1	–	–	1	1	3	1	–	–	–	–	1	1	–	–	2	1	3	1	6	1
Limnichidae	–	–	–	–	–	–	–	–	1	1	1	1	–	–	–	–	–	–	–	–	–	–	–	–	1	1
Lycidae	16	6	25	4	15	8	16	8	2	2	74	9	–	–	–	–	2	1	13	3	2	2	17	4	91	9
Ptilodactylidae	10	1	2	2	1	1	2	1	10	2	25	3	1	1	–	–	–	–	5	2	5	2	11	3	36	5
Throscidae	1	1	–	–	2	1	1	1	–	–	4	2	2	2	–	–	–	–	–	–	1	1	3	2	7	3

Bostrichiformia																										
Anobiinae	27	9	83	11	10	5	17	8	71	8	208	16	30	7	31	10	72	13	48	10	82	7	263	22	471	26
Dermestidae	1	1	–	–	1	1	–	–	–	–	2	2	–	–	1	1	–	–	–	–	2	1	3	2	5	4
Cucujiformia/Cleroidea																										
Cleridae	7	5	13	7	4	4	9	7	9	6	42	12	1	1	8	5	3	3	14	6	1	1	27	10	69	15
Malachiinae	9	3	2	2	7	3	4	3	3	1	25	7	–	–	3	1	16	4	2	2	4	2	25	7	50	10
Melyrinae	–	–	–	–	1	1	3	2	–	–	4	2	–	–	–	–	–	–	3	1	–	–	3	1	7	3
Trogossitidae	2	2	15	5	1	1	3	2	4	1	25	8	1	1	4	3	6	2	10	3	1	1	22	6	47	10
Cucujiformia/Cucujoidea																										
Alexiidae	1	1	–	–	–	–	–	–	3	1	4	1	–	–	1	1	–	–	1	1	1	1	3	3	7	4
Aspidiphoridae	–	–	–	–	5	2	4	1	9	2	18	3	–	–	–	–	2	2	6	2	1	1	9	2	27	3
Biphyllidae	–	–	–	–	2	1	–	–	1	1	3	2	–	–	–	–	3	2	7	2	–	–	10	4	13	5
Brachypteridae	3	1	2	1	3	2	5	2	3	2	16	2	5	2	5	3	–	–	9	4	5	2	24	5	40	6
Cerylonidae	1	1	1	1	1	1	–	–	4	2	7	4	–	–	–	–	–	–	–	–	1	1	1	1	8	5
Coccinellidae	16	8	13	8	35	10	30	12	18	12	112	22	32	7	51	11	57	13	48	18	31	12	219	26	331	32
Corylophidae	450	13	162	11	262	12	240	11	132	9	1246	15	132	15	150	14	209	10	263	14	208	13	962	16	2208	17
Cryptophagidae	1	1	–	–	2	2	3	2	–	–	6	4	2	2	5	2	1	1	1	1	2	1	11	5	17	7
Discolomatidae	5	3	12	2	17	4	23	4	6	2	63	4	7	3	4	3	15	3	17	3	12	3	55	5	118	5
Endomychidae	2	2	1	1	–	–	3	2	1	1	7	3	–	–	–	–	1	1	2	1	4	1	7	1	14	3
Erotylidae	1	1	1	1	1	1	28	12	10	9	41	21	–	–	1	1	1	1	4	3	5	5	11	9	50	24
Languriidae	5	1	2	2	2	1	5	2	15	2	29	4	27	4	49	5	29	3	95	6	30	4	230	7	259	10
Latridiidae	283	4	687	3	509	3	1015	4	427	4	2921	5	37	3	54	4	36	4	82	5	20	2	229	5	3150	7
Laemophloeidae	–	–	–	–	2	2	1	1	8	1	11	2	1	1	4	4	1	1	9	2	3	2	18	8	29	9
Nitidulidae	1	1	1	1	1	1	–	–	–	–	3	3	2	1	1	1	2	2	8	4	7	1	20	7	23	9
Phalacridae	108	12	159	12	128	13	253	13	147	13	795	19	174	14	410	14	209	13	830	16	218	12	1841	16	2636	21
Propalticidae	1	1	2	2	1	1	1	1	–	–	5	4	–	–	–	–	–	–	1	1	–	–	1	1	6	5
Silvanidae	–	–	–	–	1	1	–	–	7	2	8	2	–	–	–	–	1	1	–	–	–	–	1	1	9	3
Cucujiformia/Tenebrionoidea																										
Aderidae	82	24	86	17	56	12	61	10	73	16	358	37	45	10	30	11	78	14	37	15	123	19	313	38	671	48
Anthicidae	9	2	22	3	32	2	30	3	47	4	140	4	7	1	3	2	46	1	22	2	25	2	103	3	243	5
Ciidae	–	–	–	–	1	1	1	1	–	–	2	2	1	1	–	–	4	2	3	2	5	2	13	8	15	8
Colydiidae	2	2	–	–	–	–	–	–	–	–	2	2	–	–	–	–	1	1	–	–	–	–	1	1	3	3
Melandryidae	3	3	2	1	4	2	–	–	4	2	13	4	3	2	–	–	–	–	1	1	–	–	4	2	17	5
Monommatidae	1	1	–	–	–	–	–	–	4	2	5	2	–	–	–	–	–	–	–	–	–	–	–	–	5	2
Mordellidae	3	3	15	8	5	5	19	2	3	2	45	19	7	4	2	2	8	6	6	5	10	6	33	12	78	23
Mycetophagidae	9	3	22	4	–	–	11	4	2	2	44	7	2	2	18	4	–	–	9	4	–	–	29	6	73	10
Mycteridae	4	1	–	–	–	–	–	–	–	–	4	1	–	–	–	–	–	–	–	–	–	–	–	–	4	1

Table 14A.1. *(cont.)*

Taxa	Wet season												Dry season												Total	
	Cynometra pf		sf		*Rinorea* pf		sf		swf		Sum		*Cynometra* pf		sf		*Rinorea* pf		sf		swf		Sum			
	n	sp.	*n*	sp.	*n*	sp.	*n*	sp.	*n*	sp.	*n*	sp.	*n*	sp.	*n*	sp.	*n*	sp.	*n*	sp.	*n*	sp.	*n*	sp.	*n*	sp.
Ripiphoridae	–	–	–	–	1	1	–	–	–	–	1	1	–	–	–	–	–	–	–	–	–	–	–	–	1	1
Salpingidae	27	2	17	2	5	2	–	–	1	1	50	3	2	1	8	2	13	3	17	3	6	3	46	4	96	5
Scraptiidae	1	1	6	3	2	1	4	2	4	4	17	7	2	2	2	2	–	–	5	3	–	–	9	5	26	10
Alleculinae	69	8	453	8	134	7	674	8	457	6	1787	11	9	2	22	4	9	3	32	6	15	2	87	9	1874	24
Lagriinae	44	7	71	6	68	11	68	10	32	11	283	16	9	7	6	4	5	4	43	13	4	4	67	20	350	25
Tenebrionidae (others)	96	20	71	22	68	21	28	24	46	20	309	62	34	8	26	10	52	6	26	13	33	8	171	26	480	69
Cucujiformia/Chrysomeloidea																										
Cerambycidae	3	3	11	8	15	9	7	4	15	13	51	26	1	1	5	5	7	6	9	8	2	2	24	19	75	40
Alticinae	108	11	193	9	189	10	252	12	178	19	920	32	378	18	592	20	747	20	853	26	667	25	3237	37	4157	56
Bruchinae	30	3	66	4	213	4	449	3	297	2	1055	6	41	4	19	4	4	3	12	3	13	5	89	7	1144	9
Chrysomelinae	34	1	–	–	63	1	–	–	13	1	110	1	11	1	2	1	39	2	9	2	99	3	160	3	270	3
Cryptocephalinae	8	1	11	1	4	2	–	–	–	–	23	2	–	–	1	1	–	–	–	–	–	–	1	1	24	3
Eumolpinae	369	22	107	16	95	18	144	23	92	18	834	41	44	14	63	15	38	17	118	17	69	26	332	42	1166	58
Galerucinae	233	19	133	12	62	10	36	8	108	19	572	33	39	10	53	10	18	10	75	9	48	12	233	22	805	45
Zeugophorinae	1	1	5	1	3	1	2	1	3	2	14	3	1	1	–	–	9	3	–	–	8	4	18	4	32	6
Other Chrysomelidae	1	1	1	1	1	1	1	1	6	5	10	7	–	–	5	2	2	1	15	3	4	2	26	5	36	11
Cucujiformia/Curculionoidea																										
Anthribidae	18	6	35	11	44	12	164	16	44	14	305	23	6	6	8	4	31	12	50	13	13	8	108	25	413	33
Apioninae	329	6	689	8	232	10	135	6	108	5	1493	11	313	9	329	11	798	13	430	12	240	10	2110	15	3603	17
Attelabidae	5	2	3	1	1	1	17	2	23	2	49	3	1	1	2	1	1	1	1	1	–	–	5	4	54	6
Brentinae	–	–	2	2	1	1	4	2	–	–	7	3	–	–	2	2	1	1	1	1	1	1	5	5	12	8
Scolytinae	7	3	5	4	7	2	2	2	4	4	25	14	4	4	20	11	5	2	–	–	4	4	33	20	58	33
Curculionidae (others)	267	28	198	38	495	47	282	39	421	42	1649	110	358	25	297	41	998	24	543	36	805	31	3001	83	4650	147
Total	3123	346	3870	362	3351	376	5314	401	3974	420	19 675	966	2003	241	2871	305	3858	292	4498	400	3355	333	16 583	779	36 256	1352

n, number of individuals; sp., number of morphotypes; pf, primary forest; sf, selectively logged forest; swf, swamp forest.

15

Seasonality and community composition of springtails in Mexican forests

José G. Palacios-Vargas and Gabriela Castaño-Meneses

ABSTRACT

Springtails or Collembola have wide distributions and occur at all latitudes and elevations, including at extreme environmental conditions. Some are associated with suspended soils, and others live in epiphytic plants, moss, lichens, bark crevices or on leaves and flowers. Their food resources are mainly vegetable matter and there are few predatory species. Springtails from the trees of Mexican forests have been studied, particularly those living in *Tillandsia* sp. (Bromeliaceae), but the canopy fauna has only recently been investigated with canopy fogging. These studies have showed that Collembola in the tropical canopy are the dominant arthropod group in abundance, in contrast to other geographical regions. The aims of the present study has been to compare the composition and communities of the springtails living in canopies of three different forest types in Mexico, and to analyse their seasonality. Specimens from central Mexico originated from epiphytic plants of the genus *Tillandsia*; other were obtained from fogging of a dry tropical forest on the Pacific coast and a tropical rainforest in Chiapas. The tropical dry forest was characterized by a high diversity of endemic and cosmopolitan species; most of them showing seasonal variation depending on humidity. There, 19 species of Collembola were collected; the most abundant and constant was *Salina banksi* (Paronellidae). The same species was very abundant in the tropical rainforest at Chajul. This species appears to be dominant in tropical Mexican forests, in contrast to the temperate forests in Hidalgo and Morelos states, where the most abundant species were *Sminthurinus quadrimaculatus* and *Pseudisotoma sensibilis*. There are important differences in composition between Collembola assemblages living in canopies and those of the soil and shrub layers. The structure of springtail assemblages varies significantly between seasons and forest types, depending on humidity and tree species.

INTRODUCTION

Since the nineteenth century, observations of arboricolous arthropods in tropical forests have been a major research focus in tropical biology (e.g. Bates, 1884). Technical considerations have almost always limited the study of these habitats, but the development of technologies for canopy access since the 1980s (Moffett & Lowman, 1995) has led to a rapid increase in analyses of the structure of arthropod assemblages (Erwin, 1995; Stork *et al.*, 1997a). These recent studies of natural forest canopies in temperate and tropical regions have showed that this habitat contains speciose, mostly unidentified and little-understood, associations of arthropods.

The fogging approach has been the most widely used research technique in recent years, presumably because of its simple design and practicality (Adis *et al.*, 1984; Watanabe & Ruaysoongnern, 1989; Stork, 1991; Guilbert *et al.*, 1995). This method was used in earlier studies conducted by Roberts (1973) to collect grasshoppers (Orthoptera) in Costa Rica. Since then, other researchers in different regions of the world have adopted and modified the approach.

Springtails or Collembola are usually associated with edaphic environments, where they are important for their sheer abundance and for the role that they play in nutrient cycling within the soil (Bardgett & Chan, 1999; Theenhaus *et al.*, 1999; Scheu & Falca, 2000). However, they can also be collected in great numbers from forest canopies (Palacios-Vargas *et al.*, 1998, 1999) and occupy various habitats, such as suspended soils (Delamare-Debouteville, 1948), epiphytic plants (Palacios-Vargas, 1981), moss, lichens, bark crevices (André, 1983), tree leaves and flowers. The role of

Collembola in the canopy can be as important as in the soil, particularly in sections of the canopy where large accumulations of organic matter occur. Distribution patterns of Collembola in the canopy vary spatially as well as temporally, and there is a strong vertical stratification stretching from the forest floor to the upper canopy (Rodgers & Kitching, 1998).

In this chapter, we present results of our studies and will compare the seasonal variations of the Collembola from the canopies of three different forests in Mexico: a tropical dry forest in Chamela, Jalisco; a tropical rainforest at Chajul, Chiapas; and a temperate *Quercus–Abies* forest in Central Mexico (Hidalgo and Morelos States).

METHODS

Study sites

Tropical dry forest, Chamela Biological Station

The Chamela region in Mexico is located on the coast of the State of Jalisco (19° 30′ N, 105° 03′ W), at 200 m above sea level (asl). It is limited in the north by the Sierra Madre del Sur, by the Pacific Ocean in the west, by the Transversal Neovolcanic system in the north-east and by Oaxaca State in the south (Bullock, 1988).

The region contains an ecological reserve of 3300 ha with a research centre run by the Universidad Nacional Autónoma de México. The dominant vegetation is tropical deciduous forest (Bullock, 1988). There are several small streams, and ephemeral rivers that flow during September (Cervantes, 1988). The slopes of the hillsides range mostly from 21° to 34° (Bullock, 1988). The soil includes sandy entisols of neutral pH and little organic matter (Solís, 1993b). The climate is Aw(x)i type: that is, warm subhumid with summer rains (Awx) but with low average rainfall and little thermal oscillation (less than 5 °C) occurring throughout the year (García, 1988). The average annual rainfall is 748 mm and 80% falls between July and October (Bullock, 1986). Hurricanes are frequent, resulting in erratic patterns of precipitation (García-Oliva *et al.*, 1991). The highest monthly average air temperatures (28.8–32.2 °C) occur from May to July, and the lowest monthly averages (15.9–22.6 °C) occur from December to February (Bullock, 1988).

The Station's flora consists of more than 780 species. The two families (among 107) of greatest diversity are the Leguminosae and Euphorbiaceae, followed by the Asteraceae, Convolvulaceae, Rubiaceae, Bromeliaceae, Malvaceae and Acanthaceae (Lott *et al.*, 1987). Trees reach a maximum height of 12 m and the main genera are *Lonchocarpus*, *Caesalpinia*, *Croton*, *Jatropha* and *Cordia* (Lott, 1985). These trees produce flowers and seeds when they lose their leaves, mainly during June and July (Bullock & Solís-Magallanes, 1990). There is a high diversity of epiphytes, mainly bromeliads of *Tillandsia* spp. (Lott *et al.*, 1987).

Tropical rainforest, Chajul tropical biology station

The station of Tropical Biology Chajul is 331 200 ha in area and is located in the southern end of the Montes Azules Biosphere Reservation (16° 4′–16° 57′ N, 90° 45′–91° 30′ W) in the north-eastern part of the State of Chiapas. The vegetation is mainly tropical rainforest (Gómez-Pompa & Dirzo, 1995). The climate is A(f)wi: warm and humid (Afw), more or less isothermal with an annual average temperature of over 22 °C, the average temperature being over 18 °C during the coolest month. Summers are rainy and affected by monsoon patterns, with annual precipitation reaching more than 2500 mm and up to 3000 mm in the most northerly part of the reserve.

The reserve is dominated by karst (limestone) relief patterns, with altitudes ranging from 300 to 1500 m. Lithosol soils prevail in the mountainous areas of the western part of the reserve. Because of their topographical characteristics, these soils are exposed to lixiviation, resulting in rather high soil acidity. More than 500 species of vascular plants have been recorded in the reserve (Rzedowski, 1988). Most of the reservation is covered by tropical rainforest, which ranges in altitude from 100 to 900 m asl. The highest trees reach 60 m and include the following species: *Terminalia amazonia* (J. F. Gmel.) Exell, *Lonchocarpus* sp., *Schizolobium parahybum* (Vell.) Blake, *Swietenia macrophylla* King and *Brosimum alicastrum* Swart.

Temperate Quercus–Abies *forest, 'El Chico' National Park and Chichinautzin lava flow*

The National Park El Chico is 2793 ha (Vargas Márquez, 1984) and is located on the north-eastern slope of the Valle de México (20° 13′ N, 98° 47′ W). The soil comprises mostly basaltic ashes, but andesitic and rhyolitic soils are also present. Organic matter is very high in the A horizon. Soil texture varies between sandy-clay and sand. Soils are for the most part quite deep and well drained (Aguilera, 1962). The climate of the area, as cited by García (1988) is a type Cb(m)(w)(i′)gw. This is temperate with long and fresh, humid summers and rainy winters. Annual average temperature ranges between 10 and 14 °C, with a minimum of −6 to −9 °C; annual rain averages between 600 and 1500 mm. The rainy season starts in May and continues through October to November. This period produces 80 to 90% of the annual precipitation, with the remainder falling during the winter and early spring (Melo & López, 1994).

The vegetation consists chiefly of forest dominated by *Abies religiosa* (H. B. K.) Schlecht. & Cham., standing between 20 and 30 m high. This forest is situated at elevations between 2700 and 3000 m, mainly on slopes where it is sheltered from strong winds and sun exposure. The *Quercus* forest occurs from 2300 to 3000 m in altitude and varies in appearance and composition. Dominant species include *Quercus laurina* Humb. & Blonpl., *Q. rugosa* Née, *Q. laeta* Liebm., *Q. mexicana* Humb. & Blonpl. and *Q. crassifolia* Humb. & Bonpl. Trees also reach heights of 20 to 30 m and support abundant epiphytic plants.

The Chichinautzin lava flow is located in Morelos State (19° 01′ N; 99° 09′ W) and occupies an area of approximately 300 ha. The altitude of the area varies between 2000 and 2650 m. In spite of its small area, two climate types are present in Chichinautzin: C(W_2)(W)b(i′)g at the northern end is one of the most humid of the temperate climates, with little variations in temperature (between 5 and 7 °C) through the year; and (A)C(W_2)(W)ig at the southern end, which may be considered intermediate between tropical and temperate. Annual precipitation varies between 1236 mm at the southern end and 1608 mm in the northern part, and the temperature range is from 16.1 to 12.1 °C.

In the northern part, between 2450 and 2650 m asl, vegetation is dominated by *Q. rugosa*, whereas, between 2150 and 2450 m asl, an association of *Hechtia podantha* Mez. and *Agave horrida* Lemaire ex Jacobi is present, in addition to one of *Quercus–Arctostaphylos polifolia*. In the southern part, from 2000 to 2150 m asl, *Bursera cuneata* (Schl.) Engl. is common as are several *Tillandsia* spp. (Espinosa, 1962).

Springtail collections and statistical analyses

Insecticide knockdowns were performed in the three forest types to study the fauna of Collembola in their canopies. In Chamela Biological Station, a total of seven such sampling events were carried out from August 1992 to May 1994, a period including both the rainy season (August and September 1992, July 1993) and the dry season (May and November 1993, and February and May 1994). In the Chajul Tropical Biological Station, samples were taken during July 1994 and July 1995. At both sites, a fogging machine (Dynafog™) was used, releasing natural pyrethrum insecticide (3% resmethrin in a kerosene solution) about 30 minutes before sunrise (04:00 to 06:00). For each fumigation, a projected area of 100 m^2 was delimited and 50 funnels (0.5 m diameter) were hung randomly in the shrub layer about 50 cm above the forest floor. Specimens that fell into the funnels were then collected by washing them with alcohol (80%), up to 5 hours after fogging. Specimens from the temperate *Abies–Quercus* forest were obtained from two fogging collections performed by Tovar (1999) in February and August 1997. He also used a natural pyrethrum but in aqueous solution and fogged three isolated trees of six *Quercus* species (*Q. castanea* Née, *Q. crassifolia*, *Q. crassipes* Humb. & Bonpl., *Q. greegii* Trel., *Q. laeta* and *Q. rugosa*), setting 10 square collecting surfaces (0.99 m^2 in surface) under each tree. All specimens collected were stored in 70% alcohol. Collembola were mounted in Hoyer's solution and identified to species.

For statistical analysis, data were transformed as their square root ($\sqrt{(x + 0.5)}$) to satisfy the assumption of normality. Comparisons between seasons were performed with Student *t*-tests. To compare the effects of vegetation (three levels: rainy forest, deciduous forest and epiphytes) on springtail abundance an analysis of variance (ANOVA) was used, and *post hoc* comparisons were made using Tukey's test for unequal sample sizes. Species diversity in the four localities was compared with *t*-tests (Zar, 1974). Presence–absence data for Collembola genera were used to calculate levels of similarity among the four localities by cluster analysis, using unweighted pair-group average (UPGMA) as the gathering rule. Analyses were performed with the Statistica™ Ver. 6 program (StatSoft, 1995).

RESULTS AND DISCUSSION

Assemblage composition of tropical dry forest at Chamela

At Chamela, 1 044 032 individuals (15 197 individuals m^{-2}) representing 19 species of Collembola were collected (Table 15.1). During the rainy season, Collembola represented 98% of the total canopy arthropods at Chamela, whereas in the dry season, they made up only 30% of the total (Fig. 15.1).

Salina banksi MacGillivray was the most abundant species, comprising 89.9% of all Collembola during

Table 15.1. *Average density of Collembola collected from the canopy at Chamela, Jalisco, Mexico and Student's* t*-test values for comparison between rainy and dry season*

Taxa	Average density (individuals m^{-2}) ±SD		
	Rainy season	Dry season	Student's t-test[a]
Hypogastruridae			
Ceratophysella gibbosa (Bagnall, 1940)	34 ± 56	3 ± 3	13.94*
Xenylla humicola (Fabricius, 1780)	76 ± 128	6 ± 9	13.12*
Brachystomellidae			
Brachystomella minimucronata Palacios-Vargas & Najt, 1981	26 ± 41	3 ± 5	13.82*
Neanuridae			
Neotropiella quinqueoculata (Denis, 1931)	31 ± 50	4 ± 5	11.07*
Pseudachorutes subcrassoides Mills, 1934	11 ± 15	2 ± 4	11.65*
Aethiopella sp.	32 ± 53	3 ± 5	13.10*
Entomobryidae			
Entomobrya californica Schött, 1891	31 ± 51	6 ± 9	5.54*
Lepidocyrtus finus Christiansen & Bellinger, 1980	62 ± 104	3 ± 4	12.59*
L. gr. *lanuginosus* (Gmelin, 1788)	115 ± 192	6 ± 9	12.78*
Seira bipunctata (Packard, 1873)	749 ± 1271	18 ± 30	14.17*
S. dubia Christiansen & Bellinger, 1980	771 ± 1312	22 ± 35	13.65*
S. purpurea (Schött, 1891)	1118 ± 1918	18 ± 29	13.89*
Paronellidae			
Salina banksi MacGillivray, 1894	31 420 ± 53 677	253 ± 487	14.90*
Katiannidae			
Sminthurinus ca. *radiculus* Maynard	32 ± 51	7 ± 3	11.34*
S. ca. *conchyliatus* Snider	20 ± 32	5 ± 2	10.34*
S. ca. *latimaculosus* Maynard	35 ± 57	5 ± 5	11.51*
Sminthurididae			
Sphaeridia pumilis (Krausbauer, 1898)	27 ± 46	6 ± 2	8.15*
Sminthuridae			
Sphyrotecha ca. *mucroserrata* Snider	35 ± 58	3 ± 3	11.82*
Bourletiellidae			
Deuterosminthurus maassius Palacios-Vargas & González, 1995	304 ± 507	15 ± 26	14.01*

SD, standard deviation.
[a]Degrees of freedom = 5.
*$p < 0.05$.

the rainy season with 925 438 specimens collected. However, its abundance decreased during the dry season, and only 983 individuals (65% of the total number of individuals) were collected at this time. Other species exhibited similar seasonal changes, decreasing considerably during the dry season (Fig. 15.2). Results of t-tests indicated significant differences between rainy and dry seasons for the abundance of total Collembola ($t = 2.17$, degrees of freedom (d.f.) = 131, $p = 0.03$), and for each species (Table 15.1). In contrast, the Shannon diversity index had higher values during the dry season than during the rainy season ($H' = 1.56$ and 0.52, respectively).

During the dry season, many arthropod groups migrate vertically from soil to canopy to encounter more suitable conditions of humidity and temperature. The

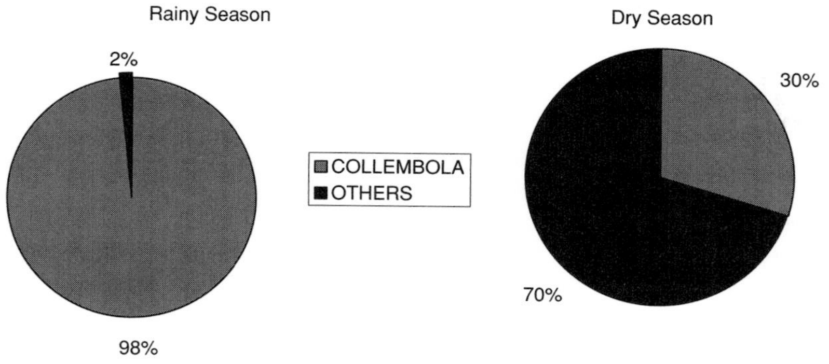

Fig. 15.1. Percentage of Collembola and other arthropods in the Chamela canopy during the rainy and dry seasons.

canopy at Chamela supports many epiphytic plants, such as *Tillandsia* spp. (Bromeliaceae), that provide a shelter from extreme climatic conditions. Eleven genera collected in the canopy at Chamela have been reported living in *Tillandsia* spp. elsewhere (Palacios-Vargas, 1981), and 12 species were collected both in the canopy and soil at Chamela (Palacios-Vargas & Gómez-Anaya, 1993). These vertical migrations and the lack of predators explain why species diversity of Collembola increases notably in the canopy during the dry season.

Assemblage composition of tropical rainforest at Chajul

In the canopy of the Chajul tropical rainforest, 26 species of Collembola were collected (Table 15.2). Of the

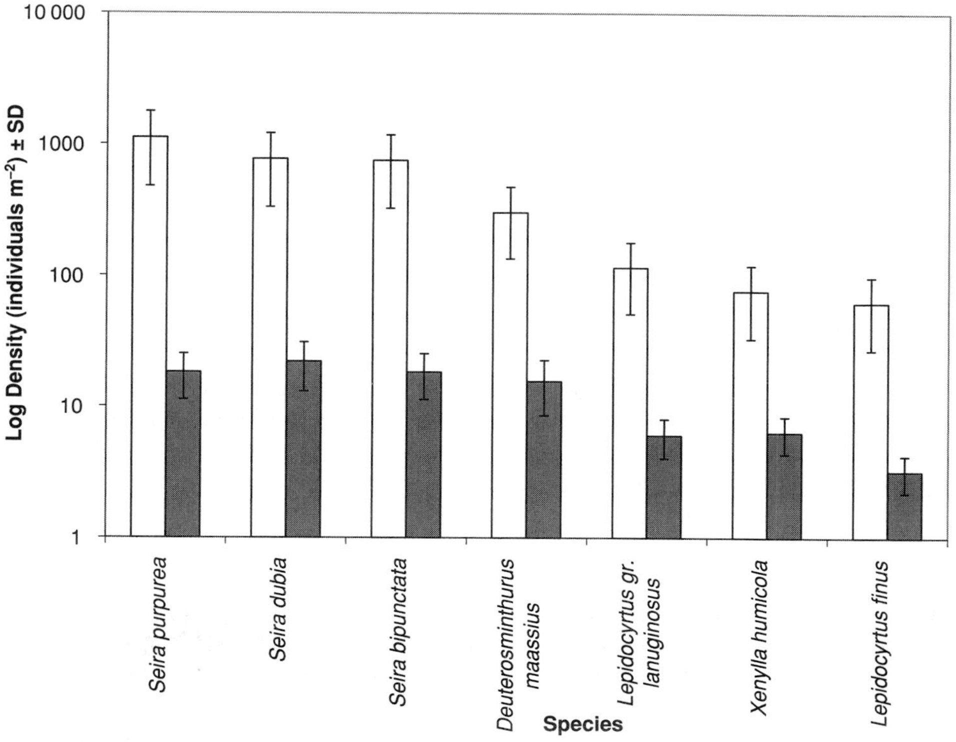

Fig. 15.2. Logarithm of seasonal density and standard deviation of the most abundant Collembola species at Chamela, during the wet (open bars) and dry (closed bars) seasons.

Table 15.2. *Collembola collected from the canopy at Chajul, Chiapas, Mexico*

Taxa	Species
Odontellidae	*Superodontella sp.*
Brachystomellidae	*Brachystomella surendrai* Goto, 1961
Neanuridae	*Arlesia sp.*; *Pseudachorutes indiana* Christiansen & Bellinger, 1980; *Neotropiella sp.*
Isotomidae	*Folsomides angularis* (Axelson, 1905); *Proisotoma sp.*
Entomobryidae	*Seira purpurea* (Schött, 1891); *Lepidocyrtus* ca. *finus*; *Willowsia sp.*; *Psudosinella sp.*; *Entomobrya sp.*; *Americabrya sp.*;
Sminthurididae	*Sphaeridia sp.*
Katiannidae	*Sminthurinus latimaculosus* Maynard, 1951; *Sminthurinus quadrimaculatus* (Ryder, 1878)
Sminthuridae	*Sphyrotheca mucroserrata* Snider, 1978; *Sminthurus sp.*;
Dicyrtomidae	*Calvatomina sp.*; *Dicyrtoma sp.*; *Ptenothrix sp.*;
Paronellidae	*Paronella sp.*; *Campylothorax sp.*; *Salina banksi* MacGillivray, 1894
Bourletiellidae	*Deuterosminthurus* ca. *tristani*; *Bourletiella sp.*

6644 arthropods (338 individuals m^{-2}) collected in the two fumigations of 1994 (3208) and 1995 (3436), 581 (30 individuals m^{-2}), 9% of the total number of arthropods, were springtails. This included 275 individuals (28 individuals m^{-2}) in 1994 and 306 individuals (31 individuals m^{-2}) in 1995 (Fig. 15.3). No significant differences were found between the two studied years ($t = 0.14$, d.f. $= 6$, $p = 0.09$) and the abundance of Collembola appears to be more or less constant across years in this forest.

The most abundant species in both samples at Chajul was *Deuterosminthurus* ca. *tristani*, followed by *S. banksi*. This situation is the opposite to that in the dry forest of Chamela, where *S. banksi* dominated and *Deuterosminthurus maassius* Palacios-Vargas & González ranked fifth in abundance during the rainy season. The Shannon diversity index recorded in July 1995 at Chajul was higher than that observed in Chamela ($H' = 2.03$, $J' = 0.69$, where J' is Pielou's evenness index), and species diversity differed significantly between the two localities ($t = 18.37$, $p < 0.05$).

Assemblage composition in temperate forests

In temperate forests, populations of canopy Collembola are significantly lower than in tropical zones. When fogging various oak species in three distinct locations in the Mexico Valley, Tovar (1999) observed that Collembola constituted 32.5% of all arthropods in epiphytic habitats during the rainy season, whereas in the dry season they made up only 8.5% of the total number of individuals. Acari were the dominant group in abundance in these locations (Fig. 15.4).

Samples obtained from *Tillandsia* spp. in the states of Morelos (Chichinautzin) and Hidalgo (El Chico) yielded a total of 36 species (Table 15.3). The most abundant species were *Pseudisotoma sensibilis* Tullgerg in Chichinautzin and *S. quadrimaculatus* (Ryder) in El Chico. However, *S. quadrimaculatus* was the most abundant species during the rainy season in both states, and it decreased significantly in abundance during the dry season. Nevertheless, in Morelos, the abundance of Collembola was higher during the dry season than during the wet season. Diversity increased during the rainy season, and abundance of *P. sensibilis* decreased (H'dry season $= 0.46$; H'rainy season $= 1.8$). During the rainy season, the Shannon diversity index was significantly higher in Chichinautzin than in El Chico (Table 15.4; $t = 64.85$, $p < 0.05$).

Differences between temperate and tropical forests

The assemblage structure of Collembola differs between temperate and tropical forests. In particular, the Paronellidae are frequent in tropical forests but not in the temperate forests of the Mexico Valley. Clusters of collembolan genera collected by fogging in Chamela and

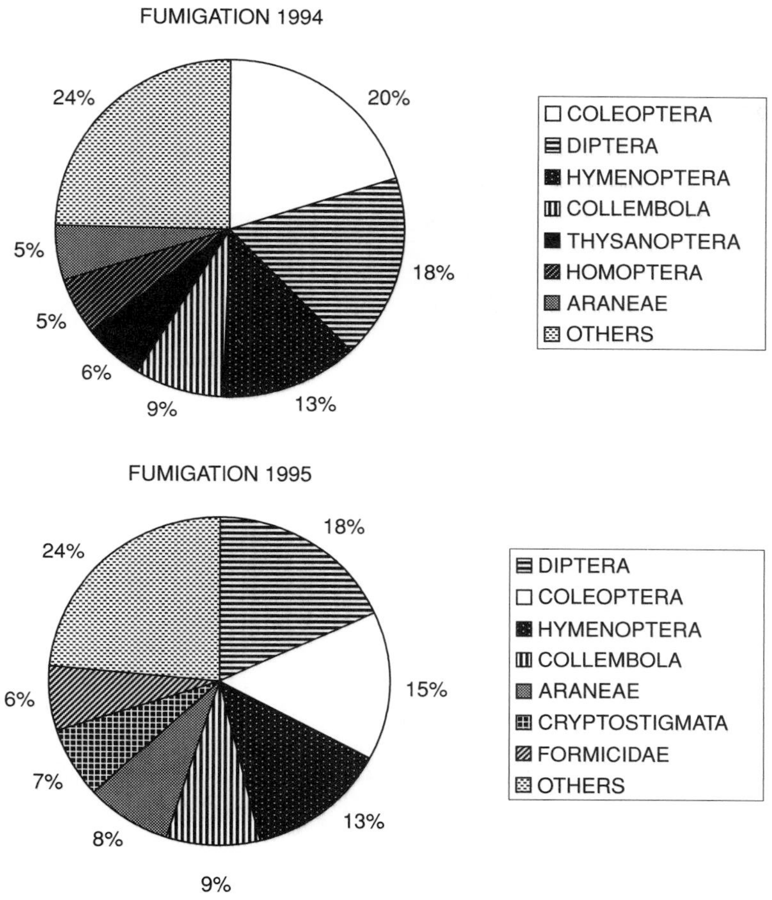

Fig. 15.3. Relative abundance of the most abundant arthropod groups in the canopy of the rain forest in Chajul, for samples obtained in 1994 and 1995.

Chajul showed greater similarity than did those collected in the epiphytic bromeliads of *Tillandsia* spp. collected on the Chichinautzin lava flow and in El Chico National Park (Fig. 15.5). This suggests a great habitat variety in the canopy and in the diversity of habitats offered by epiphytes occupied by specialized sets of arthropod. Table 15.4 details values of species diversity for all localities. Comparisons of the Shannon diversity index between Chamela and Chichinautzin show significant differences ($t = 48.48$, $p < 0.05$) as did that for Chamela with El Chico ($t = 142.06$, $p < 0.05$). The same comparison between Chajul with Chichinautzin and with El Chico also yielded significant differences ($t = 6.85$, $p < 0.05$ and $t = 10.79$, $p < 0.05$, respectively). A significant effect of vegetation type on springtail density was detected by ANOVA ($F = 4.31$, d.f. = 2,108, $p = 0.02$), and Tukey's test emphasized differences between epiphytes and rainy forest (Chajul, $p = 0.04$).

The sets of Collembola collected in the canopies of El Chico and Chichinautzin were more similar than those collected at Chamela and Chajul. The two last localities were faunistically very different, with only one species in common between the tropical dry forest and the tropical rainforest. When we compared similarities in species composition, assemblages of Collembola living inside the epiphytic *Tillandsia* at the two localities in Central Mexico were more similar than those living in the canopy and obtained by fogging. Nevertheless, there was a clear difference in the number of species and their abundance in *Tillandsia* in both localities. At the Chichinautzin lava flow, there were 18 species, whereas 23 species were present at El Chico. Among those present at the first locality, *P. sensibilis* was the most

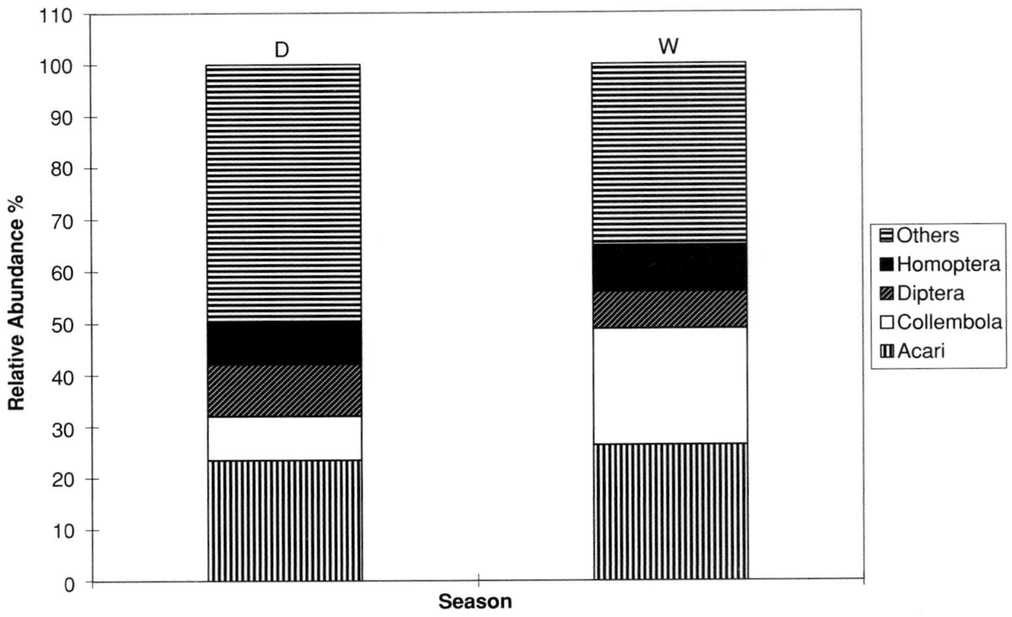

Fig. 15.4. Percentage of individuals collected in a *Quercus* spp. forest during the dry (D) and wet (W) seasons, detailed for the most common arthropod orders (modified from Tovar, 1999).

abundant. This species was also present during the rainy season but less abundantly. It is interesting to note that most of the species at Chichinautzin were more abundant during the dry season, with the exceptions of *Seira* sp., *S. quadrimaculatus* and *Sminthurus* sp. At El Chico during the rainy season, the most abundant species were *S. quadrimaculatus*, *Schoettella distincta* (Denis) and *Sphaeridia* gr. *brevilipa*. Species diversity was high during the dry season in Chichinautzin.

In contrast, at Chamela Biological Station, most species were better represented and more abundant during the rainy season. Here, the most abundant species were *S. banksi*, *D. maassius*, *Seira* spp., *Lepidocyrtus* sp., *Xenylla humicola* (Fabricius) and *Ceratophysella gibbosa* (Bagnall). Collembola were much more abundant during the rainy season, mainly in September, when they reached densities of about 3 million specimens per 100 m^2. Shannon–Weaver diversity, in contrast, was highest during the dry season.

Seasonality of Collembola

Seasonality is an important factor that determines the structure of collembolan assemblages in each forest. Despite different sampling methods being used in each locality, seasonal variation was detected and these differences were significant. Therefore, composition and abundance were significantly affected by seasonal factors.

Most species were more abundant during the rainy season in all the Mexican forest canopies studied and in the epiphytic *Tillandsia*. At Chamela, remarkably, *S. banksi* can reach populations of several millions in 1 ha. The temperate forest in middle Mexico (Hidalgo and Morelos States) includes a characteristic species during the rainy season – *S. quadrimaculatus* – whereas *P. sensibilis* is the most abundant species during winter (dry season).

The extraordinary abundance of *S. banksi* can probably be explained by the high level of humidity and the amount of organic debris in the canopy, which represents 75% of the canopy's vegetable matter (Martínez-Yrizar *et al.*, 1996). It may reflect the feeding habits of this species in the canopy – both detritivorous and mycetophagous – and the reduced numbers of predators in this environment.

Soil and litter are the primordial habitats of springtails. Nevertheless, a rich and very abundant fauna has been collected by insecticide knockdown in trees

Table 15.3. *Seasonal abundance of Collembola associated with* Tillandsia *spp. from Chichinautzin (Morelos State, abundance by plant), and El Chico National Park (Hidalgo State, abundance by 120 plants)*

Taxa	Chichinautzin lava flow			El Chico National Park rainy season
	Dry season	Rainy season	*t* value[a]	
Hypogastruridae				
Ceratophysella denticulata (Bagnall, 1941)	–	2	–	–
Ceratophysella succinea (Gisin, 1949)	–	–	–	20
Schoettella distincta (Denis, 1931)	–	–	–	3736
Xenylla grisea Axelson, 1900	–	–	–	879
X. proxima Denis, 1931	5	11	11.13*	–
Brachystomellidae				
Brachystomella barrerai Palacios-Vargas & Najt, 1981	4	1	5.33*	–
Neanuridae				
Friesea hoffmannorum Palacios-Vargas, 1982	44	34	5.92*	26
F. claviseta Axelson, 1900	–	–	–	1
F. sublimis Macnamara, 1921	52	13	2.29*	–
Pseudachorutes subcrassoides Mills, 1934	–	–	–	294
P. corticicolus (Schäffer, 1897)	–	–	–	24
P. boerneri Schött, 1902	45	16	5.53*	–
P. aureofasciatus (MacGillivray, 1893)	–	2	–	–
Onychiuridae				
Mesaphorura sp.	–	–	–	2
Isotomidae				
Folsomides angularis (Axelson, 1905)	–	5	–	24
Proisotoma ca. frisoni	–	–	–	63
Isotoma anglicana Lubbock, 1862	–	–	–	100
Pseudisotoma sensibilis (Tullgerg, 1876)	5099	316	5.89*	74
Cryptopygus sp.	–	–	–	3
Entomobryidae				
Entomobrya suzannae Schött, 1942	–	–	–	227
Americabrya arida (Christiansen & Bellinger, 1980)	11	68	3.34*	–
Seira sp.	53	93	11.78*	–
S. purpurea (Schött, 1891)	–	–	–	460
Pseudosinella ca. *fallax*	2	–	–	–
Pseudosinella sp.	–	–	–	128
Lepidocyrtus finensis Maynard, 1951	6	7	0.39	–
L. finus Christiansen & Bellinger, 1980	4	3	0.37	131
Willowsia sp.	–	–	–	–
Sminthurididae				
Sphaeridia sp.	55	18	1.99*	1760

(*cont.*)

Table 15.3. (*cont.*)

Taxa	Chichinautzin lava flow			El Chico National Park rainy season
	Dry season	Rainy season	*t* value[a]	
Katiannidae				
Sminthurinus quadrimaculatus (Ryder, 1878)	58	382	14.72*	14 349
Sminthuridae				
Sminthurus sp.	18	114	16.09*	4
Sphyrotheca confusa Snider, 1978	–	–	–	5
Dicyrtomidae				
Ptenothrix sp.	–	2	–	–
P. marmorata (Packard, 1873)	–	–	–	110
Dicyrtoma hageni (Folsom, 1896)	–	–	–	1

[a] *t* values compare the abundance of species between the dry and rainy seasons in Chichinautzin lava flow.
$^{*}p < 0.05$.

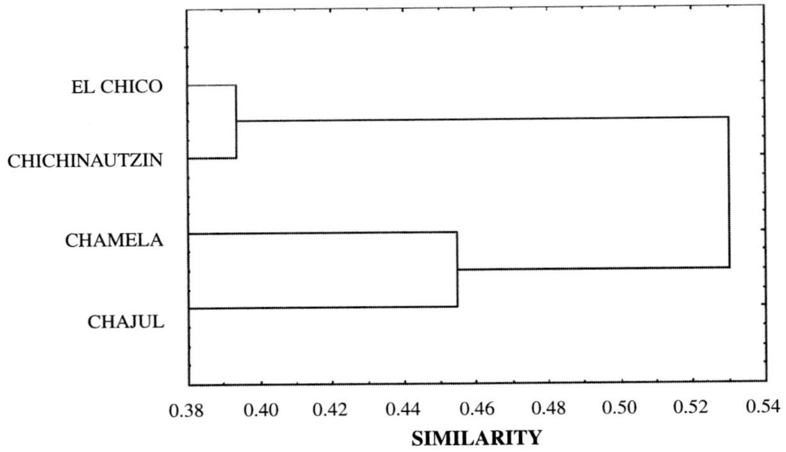

Fig. 15.5. Cluster tree (from unweighted pair-group average analysis) of the genera of Collembola collected in the four types of forest studied.

(Palacios-Vargas *et al.*, 1998, 1999) and from epiphytic plants (Palacios-Vargas, 1981; Palacios-Vargas *et al.*, 2000). Some authors (Watanabe, 1997) have suggested that most arboreal species are members of the soil fauna that have wandered up tree trunks. Despite the lack of specific information about the height on the trees that springtails can reach, we compared data on Collembola from Malaise traps, soil and litter in Chamela. Seven species were shared between Malaise traps and the canopy, but all of them were also collected in soil and litter. Soil and canopy shared three species, and litter shared only two with the canopy. There were eight species present in all vertical sections in the forest. The highest average density of Collembola from the Chamela canopy during the rainy season was represented by epiedaphic species such as *S. banksi* (31 420 ± 53 677 individuals m^{-2}) and *Seira purpurea* (Schött), which usually do not live in the soil or litter. *Seira dubia* Christiansen & Bellinger and *Seira bipunctata* (Packard) were also very abundant in the same place at the same time in the soil and litter.

CONCLUSIONS

By fogging the canopy, we have shown that there are important faunal differences in springtail assemblages that

Table 15.4. *Species richness, Shannon's diversity and Pielou's evenness indices for canopy Collembola at four localities, detailed for both the rainy and the dry seasons*

Diversity index	Chamela		Chajul rainy	El Chico rainy	Chichinatuzin	
	Rainy	Dry			Rainy	Dry
Species richness	19	19	26	23	17	14
Shannon's diversity index	0.52	1.56	2.03	1.27	1.8	0.46
Pielou's evenness index	0.18	0.53	0.69	0.41	0.63	0.17

depend on vegetation type. The most common species of Collembola in Mexican canopies are those of the genera *Salina*, *Seira*, *Sminthurinus*, *Deuterosminthurus*, *Pseudisotoma*, *Schoettella*, *Lepidocyrtus*, *Xenylla*, *Sphaeridia*, *Sphyrotheca* and *Ceratophysella*. Some of them are very mobile and feed on fungi growing between the bark and the trunk of the trees, or on moss and lichens such as *Schoettella*. Less common are species of *Neotropiella*, *Entomobrya*, *Brachystomella*, *Pseudachorutes*, *Americabrya*, *Friesea* and some dicyrtomids. Some taxa may be collected occasionally, such as *Mesaphorura* and *Crytopygus* spp., which are soil specialists.

Differences in the structure of collembolan assemblages from the canopies of tropical and temperate forests are clear. Tropical forests (dry or humid) are characterized by the presence of abundant specimens of *Salina*, which is completely absent from the temperate *Abies–Quercus* forest. The tropical dry forests have a rich diversity of Collembola well adapted to seasonal variations (Palacios-Vargas *et al.*, 1999). In each forest type, the main seasonal differences relate to the relative abundance of species that are, nevertheless, present during both the dry and the wet seasons. This affects significantly measurements of diversity (through changes in evenness values) across seasons.

Comparison between Chajul and Chamela proved very interesting, even though only two foggings were performed at Chajul (July 1994 and July 1995), and seven at Chamela. The number of species collected was very similar (26 and 19, respectively), but the composition and abundance were completely different. The only species that these tropical forest canopies shared was *S. banksi*. The following species were also well represented at Chajul (in decreasing order): *Deuterosminthurus ca. tristani*, *S. banksi* and *Willowsia* sp. At Chamela, the most abundant species in the canopy were *S. banksi*, *S. purpurea*, *S. dubia*, *S. bipunctata*, *D. maassius*, *Lepidocyrtus finus* , *X. humicola*, *Lepidocyrtus finensis* Maynard, *Ceratophysella gibbosa*, *Sphyroteca* ca. *mucroserrata* and *Sminthurinus* ca. *latimaculosos*.

In Chajul, species diversity was the highest of all localities, indicating a high species richness of springtails in tropical rainforests. Evenness was high, too, in contrast to Chamela, where it was low during the rainy season. The results obtained by hand collecting in *Tillandsia* spp. were very different from those collected by fogging, implying that the fauna living in *Tillandsia* is more diverse and more varied in feeding habits than that indicated by fogging.

There are many vegetation types in Mexico, and the structure of springtail assemblages in their canopies can vary substantially. At least for Central Mexico and in tropical dry forests, we now have good information on the genera present. The canopies of tropical rainforests in Mexico, in contrast, require much further investigation.

ACKNOWLEDGEMENTS

Dr Alfonso Pescador performed the foggings at Chamela, with the help of Alex Cadena, José A. Gómez, Alicia Rodríguez Palafox, Gerardo Ríos and Rogelio Villavicencio. At Chajul, the last two were in charge of the fogging with the assistance of Gloria Panecath and Arturo Juárez. Blanca E. Mejía, Carmen Maldonado, Leopoldo Cutz and Daniel Estrada sorted and prepared most of the specimens. Efrain Tovar shared his results and specimens from Mexico State.

16

Seasonal variation of canopy arthropods in Central Amazon

José Camilo Hurtado Guerrero, Cláudio Ruy Vasconcelos da Fonseca, Peter M. Hammond and Nigel E. Stork

ABSTRACT

The abundance and seasonal variation of the arthropod fauna associated with the canopy of nine tree species of the families Sapotaceae (*Ecclinusa guianensis, Micropholis guyanensis, Pouteria glomerata*) and Lecythidaceae (*Corythophora alta, Eschweilera atropetiolata, E. pseudodecolorans, E. rodriguesiana, E. romeu-cardosoi, E. wachenheimii*) have been evaluated using canopy fogging in the Adolpho Ducke Forest Reserve, Manaus, Amazonas, Brazil. A total of 40 trees were fogged twice (once in the dry season and once in the rainy season) from July 1995 to June 1996. This paper presents data for 10 trays per tree sample (787 samples for all trees). In total, 118 076 and 138 353 individuals were collected during the dry and rainy season, respectively. Arthropod density in each tree species ranged from 142.7 individuals m^{-2} in *E. atropetiolata* during the dry season to 877.4 individuals m^{-2} in *E. rodriguesiana* during the rainy season. Some species and genera of trees showed relatively large differences in arthropod composition and dominance between the two seasons. Overall, Formicidae was the most abundant group (dry season: 56.9%, rainy season: 47.3%) followed by Diptera, Psocoptera, Collembola and Hymenoptera. Percentages of Formicidae, Thysanoptera, Hymenoptera, Diptera larvae and Homoptera were higher during the dry season than during the rainy season, in contrast to the percentages of Diptera, Psocoptera, Collembola, Acarina and Araneida, which showed the opposite trend. Hypotheses are set forward to explain both the ordinal and seasonal differences observed in arthropod abundance.

INTRODUCTION

The forest canopy is one of the least studied biotopes on Earth and yet it is of vital importance as a habitat for large numbers of organisms, as one of the primary photosynthetic engines of the biosphere and because of its role in water and carbon fluxes. In addition, there is still considerable debate over whether the canopy of rainforests is the major source of global biodiversity (Erwin, 1982, 1991a; May, 1988, 1990, 1992; Stork, 1988, 1993, 1997, 1999; Hammond, 1992; Basset *et al.*, 1996a).

This chapter adds to previous studies of canopy arthropod and insect diversity in South and Central America (Wolda, 1979; Erwin & Scott, 1980; Erwin, 1983a,b, 1989; Adis *et al.*, 1984, 1997, 1998a; Adis & Schubart, 1984; Harada & Adis, 1997, 1998; Basset, 1999b, 2000; Basset *et al.*, 1999, 2001b) by examining the seasonal variation in abundance of different arthropod groups in the canopy of several species of trees of the families Lecythidaceae and Sapotaceae at a well-studied site, Reserva Ducke in Central Amazonia.

METHODS

Study area

The study was performed in the Adolpho Ducke Forest Reserve (Reserva Ducke) (2° 53′ S, 59 ° 59′ W), 26 km on the road Manaus-Itacoatiara (AM-010) to the northeast of Manaus. This reserve is 10 072 ha in area and has been managed by the Instituto Nacional de Pesquisas da Amazônia (INPA) since 1963. This reserve is representative of much of the primary forest of Central Amazonia and has been described in much detail (soil: Falesi & Rodrigues, 1969; flora: Alencar *et al.*, 1979; Alencar, 1986; Ribeiro *et al.*, 1994b, 1999b; fauna: Beck, 1968; Penny & Arias, 1982; Adis *et al.*, 1984, 1997, 1998a; Höfer *et al.*, 1994; Harada & Adis, 1997, 1998).

The study site was in a *floresta de platô*, a forest type found on the higher areas of the reserve and where the soils are clay or well-drained sand–clay (latosols). Trees here are between 25 and 35 m tall but a few

Table 16.1. *Density of arthropods collected on the tree species sampled during the rainy and dry seasons, and in total across both seasons*

Tree species	*n*	Dry		Rainy		Total	
		Mean	SD	Mean	SD	Mean	SD
Sapotaceae							
Ecclinusa guianensis Eyma	4	450.7	408.2	402.6	352.6	426.6	354.1
Micropholis guyanensis (A. DC.) Pierre	6	306.5	136.5	310.7	138.9	308.6	131.3
Pouteria glomerata Baehni	6	250.1	178.6	374.2	320.0	312.2	255.4
Lecythidaceae							
Corythophora alta R. Knoth	6	249.9	78.4	298.5	115.9	274.2	97.7
Eschweilera atropetiolata Mori	4	142.7	108.2	175.4	84.7	159.1	91.7
Eschweilera pseudodecolorans Mori	5	308.8	103.9	260.1	152.6	284.4	125.7
Eschweilera rodriguesiana Mori	2	331.5	65.7	877.4	323.5	604.4	368.4
Eschweilera romeu-cardosoi Mori	1	244.2	–	431.2	–	337.7	132.2
Eschweilera wachenheimii (Benoist) Sandwith	6	389.9	221.4	387.5	145.1	388.7	178.5
Total	40	300.1	188.4	350.6	234.6	325.3	213.0

SD, standard deviation.

emergents are taller than 45 m (e.g. *Dinizia excelsa* Ducke). This forest type has the largest biomass in the area (Ribeiro *et al.*, 1994b, 1999b). Monthly maximum (30–33° C) and minimum (21–22° C) mean temperatures vary little during the year but rainfall is highly seasonal, averaging 100–170 mm per month in the dry season (June to November) and 230–300 mm in the rainy season (December to May) (Ribeiro & Adis, 1984).

Arthropod sampling and processing

Forty trees were each sampled twice, once in the dry season and once in the rainy season, from July 1995 to June 1996. The identification of the trees and the number of individuals of each species sampled are detailed in Table 16.1.

Trees were sampled with a Swingfog™ fogging machine using a similar protocol to that of Stork (1987a,b, 1991) and Stork & Brendell (1990, 1993) except that each tree was fogged from a neighbour tree. The number of trees fogged each day depended on how many were located in a particular area and varied from two to three per day. The insecticide (pybuthrin) used contained 3.8% active ingredient dissolved in kerosene. This is a natural pyrethrum with low toxicity to vertebrates that breaks down rapidly in sunlight. Fogging began early in the morning, generally starting at 04:00 and finishing at 06:00. A minimum of 20 collecting trays each of 1 m^2 were suspended from a rope cradle under each tree. Two hours after fogging, the arthropods that had fallen into the trays were washed down into the central collecting pot using 70% alcohol from a garden sprayer. Samples were stored in 70% ethanol. Samples were sorted to order at Instituto Nacional de Pesquisas da Amazônia and at the Natural History Museum in London. Of 20 or more samples collected from each fogged tree, 10 samples were chosen randomly for ordinal counts. In total 787 samples (391 from the dry season and 396 from the rainy season) were sorted to ordinal level.

Statistical analyses

The abundance of each taxon was $\ln(x + 1)$ transformed to ensure normality and homogeneity of variance of the data. Paired-sample t-tests were used to compare between seasons and tree species. Analyses of variance (ANOVA) were used to compare the species of trees during the same season. We used Tukey's HSD (honestly significant difference) to test *a posteriori* significant differences. A two-way analysis of variance was also used to analyse seasonal differences in the number of different taxa comparing seasonal totals and tree

Table 16.2. *Mean density of arthropods for all taxa collected across all trees sampled (40 trees) during the dry and rainy season and during both seasons*

Taxa	Dry				Rainy				Dry and rainy			
	Mean	SD	%	*n*	Mean	SD	%	*n*	Mean	SD	%	*n*
Formicidae	170.8	133.6	56.9	67 317	165.7	142.2	47.3	65 497	168.3	137.1	51.7	132 814
Diptera	24.9	25.5	8.3	9 819	43.1	39.8	12.3	16 908	34.0	34.5	10.5	26 727
Psocoptera	14.8	12.8	4.9	5 766	36.1	33.6	10.3	14 212	25.5	27.4	7.8	19 978
Hymenoptera	19.5	15.6	6.5	7 651	18.8	15.3	5.4	7 405	19.1	15.4	5.9	15 056
Collembola	4.0	3.2	1.3	1 593	23.6	29.3	6.7	9 340	13.8	22.9	4.2	10 933
Thysanoptera	16.4	19.7	5.5	6 463	8.2	10.3	2.3	3 240	12.3	16.1	3.8	9 703
Homoptera	9.8	14.9	3.3	3 873	7.3	6.4	2.1	2 882	8.5	11.4	2.6	6 755
Diptera larvae	10.8	15.6	3.6	4 175	4.2	5.3	1.2	1 687	7.5	12.0	2.3	5 862
Araneida	5.0	3.1	1.7	1 957	8.7	5.8	2.5	3 416	6.8	5.0	2.1	5 373
Acarina	3.9	3.3	1.3	1 521	7.5	7.6	2.1	2 953	5.7	6.1	1.7	4 474
Orthoptera	3.7	2.5	1.2	1 443	5.2	2.7	1.5	2 050	4.4	2.7	1.4	3 493
Hemiptera	3.8	4.5	1.3	1 514	4.7	6.4	1.4	1 878	4.3	5.5	1.3	3 392
Coleoptera	3.7	2.3	1.2	1 445	4.9	2.8	1.4	1 978	4.0	2.8	1.2	3 178
Blattodea	3.3	2.4	1.1	1 303	4.4	2.6	1.3	1 737	3.9	2.5	1.2	3 040
Lepidoptera larvae	2.6	2.4	0.9	1 015	2.4	1.1	0.7	960	2.5	1.9	0.8	1 975
Archaeognatha	1.4	1.4	0.5	565	2.5	2.6	0.7	1 009	2.0	2.2	0.6	1 574
Coleoptera larvae	0.4	0.4	0.1	160	0.9	0.8	0.3	351	1.0	1.1	0.3	756
Lepidoptera	0.4	0.4	0.1	155	0.6	0.6	0.2	221	0.5	0.5	0.1	376
Isopoda	0.0	0.1	0.0	12	0.6	1.2	0.2	258	0.3	0.9	0.1	270
Mantodea	0.2	0.3	0.1	87	0.2	0.2	0.1	85	0.2	0.2	0.1	172
Isoptera	0.1	0.6	0.0	54	0.2	0.9	0.1	74	0.2	0.7	0.0	128
Dermaptera	0.1	0.1	0.0	22	0.2	0.2	0.0	59	0.1	0.2	0.0	81
Strepsiptera	0.1	0.2	0.0	34	0.1	0.1	0.0	25	0.1	0.1	0.0	59
Chelonethida	0.0	0.1	0.0	19	0.1	0.1	0.0	29	0.1	0.1	0.0	48
Neuroptera larvae	0.1	0.2	0.0	26	0.0	0.1	0.0	18	0.1	0.1	0.02	44
Trichoptera	0.1	0.3	0.0	21	0.0	0.1	0.0	12	0.0	0.2	0.0	33
Chilopoda	0.0	0.1	0.0	9	0.0	0.1	0.0	16	0.0	0.1	0.0	25
Plecoptera	0.1	0.4	0.0	24	0.0	0.0	0.0	1	0.0	0.3	0.0	25
Neuroptera	0.0	0.1	0.0	10	0.0	0.1	0.0	14	0.0	0.1	0.0	24
Phasmatodea	0.0	0.0	0.0	7	0.0	0.1	0.0	15	0.0	0.1	0.0	22
Embioptera	0.0	0.0	0.0	6	0.0	0.1	0.0	9	0.0	0.0	0.0	15
Diplopoda	0.0	0.0	0.0	3	0.0	0.0	0.0	5	0.0	0.0	0.0	8
Ephemeroptera	0.0	0.0	0.0	5	0.0	0.0	0.0	1	0.0	0.0	0.0	6
Scorpionida	0.0	0.0	0.0	1	0.0	0.0	0.0	4	0.0	0.0	0.0	5
Phalangida	0.0	0.0	0.0	1	0.0	0.0	0.0	2	0.0	0.0	0.0	3
Odonata	0.0	0.0	0.0	0.0	0.0	0.0	0.0	2	0.0	0.0	0.0	2
Total	–	–	–	118 076	–	–	–	138 353	–	–	–	256 429

SD, standard deviation.

species totals (Glantz, 1992; Zar, 1994). Bonferroni corrections were applied to simultaneous statistical tests.

RESULTS

A total of 256 429 arthropods were sorted from 787 samples. Insects dominated with 235 290 individuals. Formicidae were by far the single most abundant group followed by Diptera adults, Psocoptera, Hymenoptera, Collembola, Thysanoptera, Homoptera, Diptera larvae, Araneida and Acarina (Table 16.2).

Total arthropod abundance (mean number of arthropods per tray) varied considerably between tree species. For example, from 142.7 individuals m^{-2} in *E. atropetiolata* to 450.7 individuals m^{-2} in *E. guianensis* during the dry season, and from 175.4 individuals m^{-2} in *E. atropetiolata* to 877.4 individuals m^{-2} in *E. rodriguesiana* during the rainy season (Table 16.1). During the dry season, arthropod abundances for *E. wachenheimii* and *E. rodriguesiana* were significantly higher than those for *E. atropetiolata* (ANOVA, $p = 0.02$). However, during the rainy season total abundance was not statistically different between tree species ($p > 0.05$), suggesting that the abundance of arthropods in individual trees was more variable during the dry season than during the rainy season.

Arthropod abundances for all tree species during the dry and rainy seasons (Table 16.1), and for each tree species (Table 16.3) were not significantly different. However, even though there were apparently large differences in the average abundance of particular groups between season (Formicidae, for example, represented 56.9% of the catch in the dry season but 47.3% in the rainy season), these were significant for only a few groups, including Araneida, Collembola, Diptera larvae, Isopoda, Orthoptera, Psocoptera and Thysanoptera (Table 16.4). Some taxa, such as Araneida, Coleoptera larvae, Collembola, Isopoda, Orthoptera, Psocoptera and Thysanoptera, showed seasonal differences between tree species and seasons (Table 16.5), indicating that the densities of these groups were affected by seasonal changes.

Table 16.3. *Results of paired* t*-tests to compare arthropod densities between seasons and tree species; tree species with less than four individuals were not included in the analysis*

Variables	n	t	p
Seasons	40	0.755	0.455
Ecclinusa guianensis	4	0.155	0.887
Micropholis guyanensis	6	0.042	0.968
Pouteria glomerata	6	−0.368	0.728
Corythophora alta	6	−0.800	0.460
Eschweilera atropetiolata	4	−0.439	0.690
Eschweilera pseudodecolorans	5	0.566	0.602
Eschweilera wachenheimii	6	−0.195	0.853

Table 16.4. *Results of two-way analyses of variance (probabilities) to compare arthropod densities between seasons and tree species; only taxa with a minimum of 250 individuals across all samples were included*

Order	Season	Trees	Interaction
Acarina	0.081	0.121	0.830
Araneida	**0.004**	0.095	0.855
Archaeognatha	0.075	0.580	0.984
Blattodea	0.115	0.209	0.689
Coleoptera	0.106	0.193	0.955
Coleoptera larvae	0.176	0.260	0.980
Collembola	**0.001**	**0.046**	0.100
Diptera	0.083	0.688	0.974
Diptera larvae	**0.031**	0.486	0.647
Formicidae	0.603	0.117	0.970
Hemiptera	0.517	0.404	0.504
Homoptera	0.328	**0.038**	0.571
Hymenoptera	0.480	0.090	0.854
Isopoda	**0.001**	0.337	0.443
Lepidoptera	0.258	0.753	0.845
Lepidoptera larvae	0.633	0.328	0.334
Orthoptera	**0.016**	0.060	0.935
Psocoptera	**0.001**	**0.002**	0.382
Thysanoptera	**0.014**	0.130	0.659
All taxa	0.399	0.300	0.445

$p < 0.05$ indicated by bold.

DISCUSSION

The density of arthropods reported here is in the midrange of those reported by other researchers using similar canopy-fogging techniques in different areas of temperate and tropical forests (Southwood *et al.*, 1982a,b; Erwin, 1983b, 1989; Adis *et al.*, 1984, 1997,

Table 16.5. *Results of paired* t-*tests (Bonferroni adjusted probabilities) to compare arthropod densities collected in tree species during the rainy and dry seasons and across both seasons*

Order	C. a.	E. p.	E. w.	E. a.	P. g.	E. g.	M. g.	Seasons
Acarina	0.28	0.70	0.33	0.50	0.16	0.42	0.92	**0.034**
Araneida	0.19	0.96	0.27	0.27	**0.048**	0.40	0.24	**0.001**
Archaeognatha	0.06	0.79	0.45	0.41	0.84	0.85	0.10	**0.020**
Blattodea	0.92	0.90	0.14	0.50	0.17	0.69	0.51	**0.037**
Coleoptera larvae	**0.04**	0.64	0.07	0.42	0.52	0.68	0.11	**0.001**
Coleoptera	0.53	0.50	0.55	0.13	0.08	0.96	0.25	**0.024**
Collembola	**0.03**	**0.03**	0.09	0.19	**0.03**	0.05	0.19	**<0.001**
Diptera	0.31	0.71	0.42	0.50	0.21	0.82	0.37	**0.032**
Diptera larvae	0.11	0.33	0.92	0.89	0.85	0.33	0.15	**0.024**
Formicidae	0.68	0.33	0.36	0.74	0.52	0.59	0.86	0.546
Hemiptera	0.66	0.41	0.60	0.37	0.58	0.50	0.75	0.588
Homoptera	0.55	0.26	0.70	0.77	0.37	0.62	0.27	0.356
Hymenoptera	0.90	0.38	0.68	0.87	0.51	0.74	0.19	0.422
Isopoda	0.17	0.55	0.42	0.39	0.23	0.39	–	**<0.001**
Lepidoptera	0.40	0.65	0.30	0.60	0.69	0.27	0.27	0.204
Lepidoptera larvae	0.52	0.78	0.78	0.81	0.21	0.59	0.88	0.830
Orthoptera	0.41	0.33	**0.02**	0.32	0.20	0.32	0.96	**0.004**
Psocoptera	0.14	0.36	0.31	0.08	**0.03**	0.07	0.44	**<0.001**
Thysanoptera	0.08	0.47	**0.03**	0.47	0.75	0.63	**0.02**	**0.003**
All	–	–	–	–	–	–	–	0.778

C.a., *Corythophora alta*; E.p., *Eschweilera pseudodecolorans*; E.w., *Eschweilera wachenheimii*; E.a., *Eschweilera atropetiolata*; P.g., *Pouteria glomerata*; E.g., *Ecclinusa guianensis*; M.g., *Micropholis guyanensis*. Arthropod groups with less than 250 individuals were excluded from this analysis.
$p < 0.05$ indicated by bold.

1998a; Adis & Schubart, 1984; Stork, 1988, 1991; Stork & Brendell, 1990, 1993; Floren & Linsenmair, 1997a,b, 1998a) but are higher than those previously reported for forests in Central Amazonia. The possible reasons for these differences are manifold and include sampling effects (i.e. how effective canopy fogging was in different studies) and differences in the volume and density of foliage sampled.

In most studies that have sampled the arthropod canopy fauna of tropical rainforests, ants have usually been found to be the most abundant and dominant group. This study was no exception to this general rule. Ants represented, on average, 51.7% of the total catch. The high abundance of ants in the canopy has been the subject of intensive study in recent years since these insects have been assumed to be predators or scavengers. Tobin (1991, 1994) suggested that the most probable explanation was that many of these ants are herbivores. Davidson (1997, 1998) has confirmed that the high abundance of ants is largely the result of one or two very abundant species in each sample that feed on plant and homopteran exudates.

The second most abundant taxon in the samples was Diptera. Stork (1991) thought that the swarming in the canopy at dusk and dawn of some groups of Diptera normally associated with ground-level habitats (Cecidomyiidae, Chironomidae, Sciaridae and Ceratopogonidae), reported by Haddow & Corbert (1961a), might account for the large numbers of these insects in the samples. Also Stork (1991) and Didham (1997) suggested that Diptera might seek the canopy to avoid predators and climatically adverse conditions, and to use it as resting sites. In Amazonia, many Diptera breed in the soil in forests that are seasonally flooded

(Adis, 1997a,b). It is possible that the *terra firme* species of Diptera display the same behaviour. Diptera are trophically diverse and include groups that are predators, pollinators, fungivores and detritivores; few studies have examined their role in the canopy. In one such study in Amazonia, Adis *et al.*, (1997) found that there was an interaction between Cecidomyiidae and Formicidae (possibly *Crematogaster* spp.) and between Cecidomyiidae and Hymenoptera parasitoids.

So little is known of the biology of Diptera in the canopy that most Diptera families are often classified as 'tourists' in guild analyses (Moran & Southwood, 1982; Stork, 1987b). However, the relatively large numbers of Diptera larvae in the samples would suggest that many must be feeding in the canopy. It is clear that Diptera appear to be one of the least understood groups in the canopy and should be the focus of new studies.

Our analyses show that overall abundance of arthropods was not significantly different between seasons at the tree level but significant differences existed in seasonal abundances at the ordinal level for individual tree species and across all trees combined. One possible factor that might result in seasonal differences in abundance is the fact that the same trees were refogged in the rainy season having been fogged first about 6 months earlier in the dry season. However, a number of studies have shown that trees in closed rainforest are rapidly recolonized after canopy fogging and that 6 months is ample time for faunal recovery (Erwin, 1989; Stork, 1991; Stork & Hammond, 1997). The variety of biologies and population dynamics of different groups of insects is such that it is likely that seasonal differences in ordinal abundances may be related to many factors. Some groups of insects enter into a diapause in the dry season until the return of suitable environmental conditions, such as humidity, temperature and food resources (Solomon, 1980; Edwards & Wratten, 1981; Begon *et al.*, 1995). Taxa in our study that are more susceptible to desiccation were Diplopoda, Araneida, Isopoda, Chilopoda, Chelonethida, Acarina, Collembola and Psocoptera (Cloudsley-Thompson, 1980) and, in general, their abundances were higher in the rainy season. Taxa that may be more adapted to high temperatures and low humidity (Wootton, 1993), such as Thysanoptera, were more abundant in the dry season. It is notable that tree species with dense canopies have higher numbers of individuals than trees with less-dense canopies in both dry and rainy season (e.g. *E. guianensis; P. glomerata*). Insect species living in less-dense canopies would be more affected by drought than those inhabiting dense canopies, because the microclimate condition in the latter are less variable (Edwards & Wratten, 1981).

To conclude, this preliminary report indicates that a more in-depth analysis of different arthropod taxa is required to understand better arthropod seasonal variation and its biological determinants.

ACKNOWLEDGEMENTS

J. C. H. G. was supported by Instituto Nacional de Pesquisas da Amazônia (INPA), Conselho Nacional de Desenvolvimento Científico Technológico (CNPq) process numbers 141350/95-4, 300566/91-3 and 381547/00-0 and by the Darwin Initiative Fund, the Natural History Museum, London. J. C. H. G. thanks his field assistants Paulo Oliveira de Lima and Geraldo Pereira Lima Mesquita for their collaboration and help. Paulo Apostolo C. L. Assunção identified the tree species. Norma Cecilia Rodriguez Bustamante, Ana Felisa Hurtado Guerrero and Andrés Camilo Rodriguez Hurtado supported and helped J. C. H. G. during field work and the preparation of the thesis.

17

Arthropod seasonality in tree crowns with different epiphyte loads

Sabine Stuntz, Ulrich Simon and Gerhard Zotz

ABSTRACT

It has been proposed repeatedly that epiphytes influence the composition of arthropod assemblages in tropical tree crowns. To test this assumption, we conducted a 1 year survey of 25 *Annona glabra* trees within the Barro Colorado National Monument in Panama. Selected trees supported distinct epiphyte assemblages and were assigned to four different categories: three with different species of epiphyte, and an epiphyte-free control group. We collected arthropods continuously with three different types of trap. Tree phenology was monitored throughout the study year. In total, we collected 273 490 arthropods from 29 orders. There were no significant differences in arthropod abundances among the different tree/epiphyte categories, neither in total numbers nor for any particular taxon (with one exception: Diptera were slightly less abundant in trees with the large bromeliad *Vriesea sanguinolenta*). The relative proportions of the taxa were similar across categories. Arthropod abundance was independent of total epiphyte leaf area and biomass in individual trees, although the latter varied by almost two orders of magnitude. All taxa exhibited a strong seasonality, with highest abundances at the end of the dry season of 1998. Neither total abundance of arthropods nor herbivore abundance were synchronized with the phenology of the host trees or with epiphyte phenologies. The hypothesis that the presence of epiphytes leads to an increase in arthropod activity during the dry season by buffering environmental extremes could not be confirmed. In conclusion, there was no clear effect of epiphytes on the ordinal composition of the arthropod fauna at the level of entire tree crowns.

INTRODUCTION

Most of global biodiversity occurs in the canopy systems of tropical forests. Arthropods probably constitute the largest fraction of this species pool, but how large remains yet to be determined. This has prompted a vivid debate since the 1980s (Erwin, 1983a; May, 1986; Stork, 1988; Gaston & Williams, 1993; Mawdsley & Stork, 1997; Ødegaard, 2000c). Still, the primary focus is on making inventories of species (Leigh, 1999), while the mechanisms underlying the establishment and maintenance of arthropod diversity in tropical forest canopies remain rather unclear. In this respect, vascular epiphytes have been proposed to be of considerable significance (Benzing, 1990; Nadkarni, 1994; Rodgers & Kitching, 1998). Aside from increasing the structural heterogeneity of the canopy and mitigating climatic extremes, epiphytes could affect the occurrence and abundance of arboreal invertebrate species by supplying resources for herbivorous species. Ødegaard (2000c), for instance, estimated a total of 10 000 species of phytophagous beetles to be specialized on epiphytes.

The effect of epiphytes on arthropod diversity and abundance in tree crowns has not been thoroughly studied. In this chapter, we present the outcome of a 1 year survey of arthropods that was carried out in a tropical moist forest in Panama. We sought to answer two questions. Do vascular epiphytes affect arthropod assemblages in terms of relative abundance and faunal composition at the scale of entire tree crowns? How does arthropod abundance fluctuate seasonally, and are these fluctuations synchronized with the phenology of the host tree and/or its epiphytes?

Considering the complexity of most tree crowns, we chose a simple study system consisting of a small tree species and three species of epiphytes. The tree *Annona glabra* L. occurs abundantly along peninsular shorelines in the Barro Colorado National Monument (BCNM) and is often dominated by only one epiphyte species (Zotz *et al.*, 1999). Accordingly, tree crowns bearing distinct plant assemblages could be selected for a comparison of the respective arthropod faunas. We collected the arthropods with three different trap types to obtain an ample spectrum of the canopy fauna (Stuntz *et al.*, 1999). To account for seasonal variation, we sampled arthropods continuously throughout 1 year.

Our study area experiences a quite severe dry season from late December until May, which has profound consequences for plants and animals (e.g. Leigh & Smythe, 1978; Wolda, 1978b; Foster, 1982; Smythe, 1982; Leigh, 1999). Arthropods have been shown to synchronize their fluctuating abundances with the phenology of their host trees (e.g. Wolda, 1978b; Lowman, 1982b; Basset, 1991c; Aide, 1993). We monitored flowering, leaf flush and fruit fall in *A. glabra* in order to investigate whether the arthropod fauna adjusts to the seasonal rhythms of the host trees. The epiphytes we studied provide new leaves continuously over the year (*V. sanguinolenta* and *Tillandsia fasciculata*; G. Zotz & G. Schmidt, unpublished data) or during most of the rainy season (*Dimerandra emarginata* Zotz, 1998). Herbivory in epiphytes remains a poorly studied issue (Benzing, 1990), although it has been shown that one of our study epiphytes, *V. sanguinolenta*, may be strongly affected by leaf damage by lepidopteran larvae (Schmidt & Zotz, 2000). If phytophagous arthropods generally benefit from the continuous supply of young bromeliad leaves, their abundance might fluctuate less in epiphyte-laden trees than in trees devoid of them. In the same context, the abundance of herbivores could be increased, because epiphytes add to the green biomass in the canopy.

Here we will present the results at the ordinal level. We expected to detect a significant effect of epiphytes on the fauna at this high taxonomic level for a number of reasons. For instance, epiphytes indirectly increase the amount of organic matter (suspended soil) in the canopy by impounding leaf litter between their leaves or stems (Nadkarni, 1994; Rodgers & Kitching, 1998; Richardson, 1999; Benzing, 2000); they could, therefore, promote the occurrence of taxa that would otherwise be restricted to terrestrial habitats, such as Diplopoda, Chilopoda or Isopoda. The presence of ants could also be influenced positively by epiphytes: many ant species nest inside epiphytes (Schimper, 1888; Dejean *et al.*, 1992a; Blüthgen *et al.*, 2000a) or in the litter they suspend (Longino & Nadkarni, 1990). Epiphytes may even be the chief nesting sites for arboreal ants in tropical rainforest (Richards, 1996). Spiders often respond to the physical structure of environments (Duffey, 1966; Gunnarson, 1990; Halaj *et al.*, 1998). For example, Stevenson and Dindal (1982) and Uetz (1979) demonstrated that habitat complexity correlated with the diversity of hunting and web-building spiders. By increasing the structural heterogeneity of the canopy habitat (Benzing, 1990; Nadkarni, 1994), epiphytes might also influence spider species richness and abundance. Finally, epiphytes moderate climatic extremes in the canopy (Stuntz *et al.*, 2002a). Thus, the presence of epiphytes should act as a buffer for the activity of arthropods during the harsher dry season. This prompted the hypothesis that the arthropod assemblages of trees with epiphytes should be more diverse and show higher species abundances, in particular during the dry season.

METHODS

Study site

The study was conducted from April 1998 until April 1999 in the BCNM (9° 10′ N, 79° 51′ W) in the Republic of Panama. Study trees were located along the shores of Mainland peninsulas of Lake Gatún. The annual rainfall in this 'tropical moist forest' (Holdridge *et al.*, 1971) amounts to approximately 2600 mm. Detailed descriptions of climate, vegetation and ecology are reported by Croat (1978), Leigh *et al.* (1982) and Windsor (1990).

Study trees and epiphytes

The small tree *A. glabra* L. (Annonaceae), which grows along the lake shore, was well suited for our research goals. Roots and lower portions of the stems are inundated, impeding the access of terrestrial arthropods. The low canopy of *A. glabra* (height of the study trees: 4.9 ± 0.9 m (mean $\pm$ SD), $n = 25$) is easily accessible by boat but climatic conditions are similar to the upper forest canopy because of exposure to sun and wind along the shore (Zotz *et al.*, 1999). Moreover, Zotz *et al.* (1999) found *A. glabra* to be dominated frequently by a single epiphyte species, probably as a consequence of dispersal limitation among trees. This peculiarity allowed us to define distinct tree categories with uniform epiphyte assemblages. We defined four categories (Fig. 17.1). (i) Trees without epiphytes as control group. (ii) Trees with *D. emarginata* (G. Meyer) Hoehne (Orchidaceae), which grows in clusters of erect,

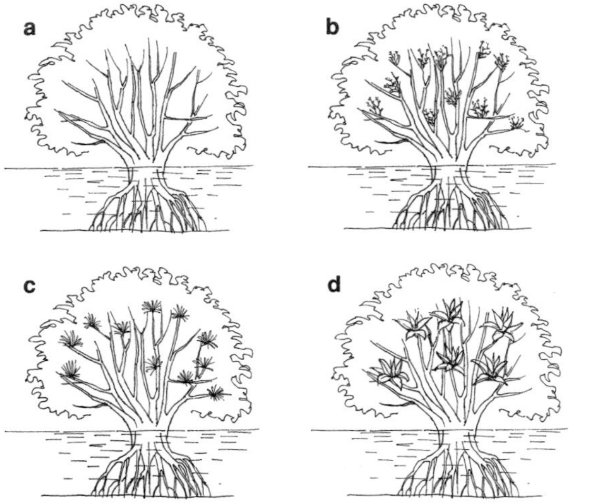

Fig. 17.1. Sketches of the four categories of study trees (*Annona glabra*); (**a**) epiphyte-free control tree; (**b**) tree with *Dimerandra emarginata* (Orchidaceae); (**c**) tree with *Tillandsia fasciculata* (Bromeliaceae); and (**d**) tree with *Vriesea sanguinolenta* (Bromeliaceae).

slender stems with linear distichous leaves (Fig. 17.2**a**). (iii) Trees with the large tank bromeliad *V. sanguinolenta* Cogn. & Marchal (Bromeliaceae), featuring broad leaves and tanks that can store rain water and considerable amounts of leaf litter (Fig. 17.2**c**). Organic matter decomposes among the basal portions of the leaves creating soil-like microhabitats. (iv) Trees dominated by *T. fasciculata* var. *fasciculata* Sw., a medium-sized bromeliad with numerous xeromorphic leaves and rather small tanks (Fig. 17.2**b**). It often occurs in dense clusters of several individuals.

These epiphyte species (thereafter addressed by their generic names) are adapted to the harsh conditions of the upper canopy and were locally abundant in the study area. Three of the four categories were replicated at seven sites distributed all over BCNM, wherever we could find trees of all categories in close vicinity. However, *Tillandsia*-carrying trees were found only at four of those sites. We sampled those trees only when arthropod abundance was expected to be high, and closed the traps during the second half of the rainy season, that is, from July to November 1998.

Sampling protocol

Different trapping techniques may yield very different animal assemblages and hence strongly influence the outcome of faunal studies (Basset *et al.*, 1997b; Stuntz *et al.*, 1999; Simon & Linsenmair, 2001). In order to obtain a reasonably broad spectrum of the arboreal fauna, we used three different types of trap: flight-interception traps (two per tree, approximate size 30 cm × 80 cm), branch traps as described by Koponen *et al.* (1997) (two per tree) and yellow colour traps (one per tree, diameter 15 cm). These are illustrated and described in Stuntz *et al.* (1999). We used 1% copper sulphate solution as killing and preservation liquid. The traps remained in the tree crowns for an entire year and were emptied every 2 weeks. We transferred the captured arthropods to 70% ethanol until further treatment in the laboratory. All animals were identified and counted at the ordinal level with the help of trained assistants.

Epiphyte biomass and tree phenology

To estimate the total biomass of epiphytes on a tree nondestructively, we measured the maximum leaf length of all bromeliads on a study tree or, respectively, the length of the youngest stem of each orchid stand. Because these parameters are strongly correlated with total plant biomass, the epiphyte load of each tree could be computed from known regressions (Schmidt & Zotz, 2001; G. Zotz, unpublished data).

Total leaf area of *Annona* was estimated from crown diameter and leaf area index (Zotz *et al.*, 1999). Leaf area estimates for the epiphytic vegetation were also obtained nondestructively from established correlations of plant

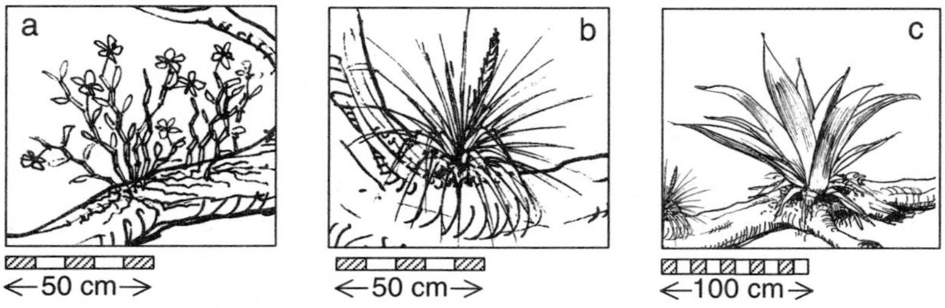

Fig. 17.2. Illustration of the study epiphytes, *Dimerandra* (**a**), *Tillandsia* (**b**) and *Vriesea* (**c**). Note the different scales. For comparison, see the specimen of *Tillandsia* in the lower left corner of frame (**c**).

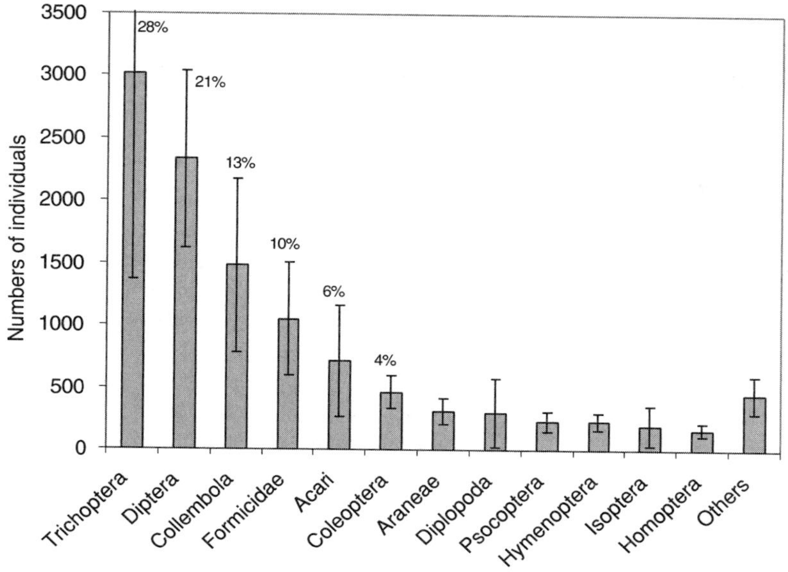

Fig. 17.3. Composition of the arthropod fauna collected in *Annona glabra* (means ± SD numbers of individuals collected per study tree ($n = 25$) during the 1 year trapping period). 'Others' includes 17 taxa that contributed less than 1% to the entire fauna.

size and leaf area (Zotz & Tyree, 1996; Zotz & Andrade, 1998; Schmidt & Zotz, 2001).

When surveying the traps, we recorded the phenological state of the host trees. We estimated the flushing of new leaves and the presence of flowers or fruit at a scale from zero to three (0, no new leaves/flowers/fruits; 1, very few; 2, present; 3, many). Because the performance of the 25 trees at a particular point in time was quite consistent, we believe that the estimates are reliable, despite the rather coarse scale. The dates for the beginning and end of the dry seasons 1998 and 1999 (in Fig. 17.4, below) were provided by the Panama Canal Commission.

Statistical analyses

Statistical analyses were performed using Statistica™ (StatSoft Inc., Oklahoma, USA). We compared the faunal assemblages of the four tree categories with Kruskal–Wallis analysis of variance (KW-ANOVA), Mann–Whitney U-tests and repeated-measures analyses of variance (RM-ANOVA), and used the Spearman rank coefficient to test for significant correlations among tree parameters and abundance of arthropods. Seasonal rhythms were analysed with circular statistics (Watson's U^2) (see Zar, 1999), using the program Rayleigh & Co. 3.1™ (Oxalis GmbH, Gütersloh, Germany).

RESULTS

Composition of the fauna

In total, we collected 273 490 arthropods belonging to 29 orders (Fig. 17.3). Micro-caddisflies (Hydroptilidae, Trichoptera) constituted the largest group of the captured arthropods with nearly a third (28%) of the total. This group was represented by a few species of the genus *Oxyethira*, which probably breed in Lake Gatún (O. Flint, personal communication). Two species, *O. circaverna* Kelley and *O. maya* Denning, were especially abundant. The second and third most abundant orders were flies (Diptera, 21%) and springtails (Collembola, 13%). Sixteen taxa were represented with less than 1% of the total fauna (in order of decreasing abundance): Lepidoptera, Thysanoptera, Hemiptera, Isopoda, Chilopoda, Blattodea, Orthoptera, Ephemeroptera, Trichoptera other than Hydroptilidae, Neuroptera, Odonata, Embioptera, Pseudoscorpiones, Dermaptera, Scorpiones, Strepsiptera and Mantodea. The last nine taxa constituted less than 0.1% of the catch.

Seasonality and phenology

The results of the phenological survey of the host trees are displayed in Fig. 17.4. Flowering and fruiting in *Annona* followed similar patterns. After a peak in April

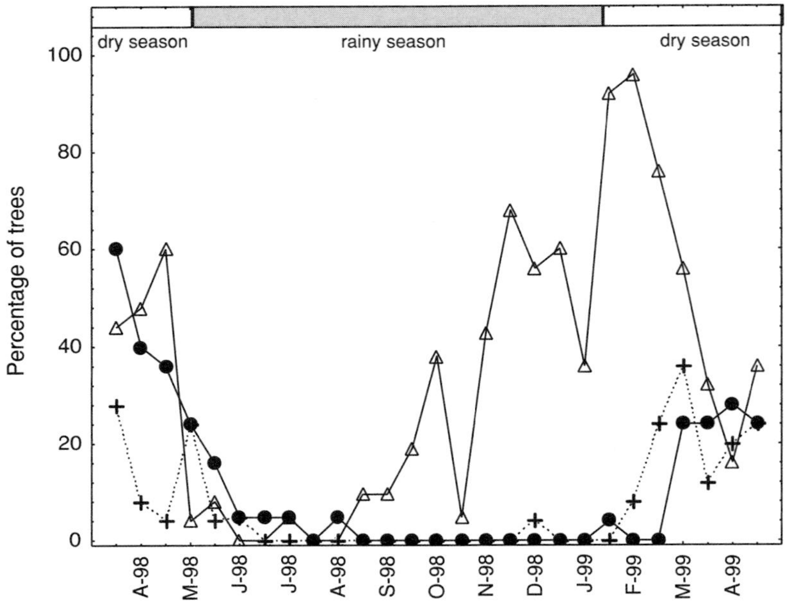

Fig. 17.4. Phenology of *Annona glabra* during the survey year (April 98 to April 99; month indicated by initial letter). Proportion of study trees in which new leaves (△), flowers (+) or fruit (●) were observed. In this chart we have included trees valued at a minimum score of 2, that is, having a fair amount of leaves/flowers/fruit visible.

and May of 1998, almost no flowers and fruit were observed until February 1999. Leaf flushes occurred several times during the study period, peaking in late January 1999, just at the onset of the dry season.

As expected, the arthropods exhibited strong annual fluctuations in their abundances (Fig. 17.5**a**). Trees with *Tillandsia* were not included in the following analysis, because they were not sampled continuously. On average, we caught 7232 individuals every 2 weeks (median, $n = 27$; range, 3450–21 733). There was a common peak shortly after the beginning of the trapping period, which coincided with the end of the dry season (Fig. 17.5**a**): in May and June of 1998, nearly a third of the catch of the entire year was collected (30.3%). This seasonal pattern was observed in all orders (data not shown), with few exceptions: in Homoptera, this initial peak was even sharper and lasted only 1 month (Fig. 17.5**b**). The same was true for termites (Isoptera): the vast majority were winged specimens flying almost exclusively after the first heavy rains following the dry season (Smythe, 1982). Conversely, Diplopoda, Chilopoda and Isopoda were abundant until August 1998 and dropped to almost zero by September for the rest of the sampling period.

These fluctuations in arthropod abundance were not synchronized with the leaf flushes of the host trees (Watson's U^2, $p > 0.05$; Fig. 17.5**a**). This was also true when considering only phytophagous taxa (Homoptera and Thysanoptera; Watson's U^2, $p > 0.05$; Fig. 17.5**b**).

Host tree traits

The median leaf area of the host trees was 30 m^2 (range 16–60) and did not differ among categories (KW-ANOVA, $p = 0.12$, $n = 25$; Table 17.1). The median leaf area of all epiphytes in a tree was 8 m^2 (range 0.2–28, $n = 18$, excluding control trees). Host-tree leaf area was weakly correlated with epiphyte leaf area (Spearman rank correlation, $p = 0.023$, $r^2 = 0.23$). However, this resulted from two trees with very high values of both tree leaf area and epiphyte leaf area. Removing these outliers led to loss of significance (Spearman rank, $p = 0.77$).

Both total biomass and leaf area of the epiphyte load of the trees varied greatly and ranged from 90 g dry weight (and 0.21 m^2 leaf area) in a tree with a sparse *Dimerandra* population to almost 4000 g dry weight in a tree abundantly laden with *Tillandsia*. The highest epiphyte leaf area was found in a tree with *Vriesea* (27.9 m^2; Table 17.1), corresponding to almost half of the host-tree foliage. Both epiphyte biomass and leaf area were significantly different across categories (KW-ANOVA, $p = 0.002$): trees with *Dimerandra* had lower epiphyte biomass and leaf area than trees with *Vriesea* or *Tillandsia*

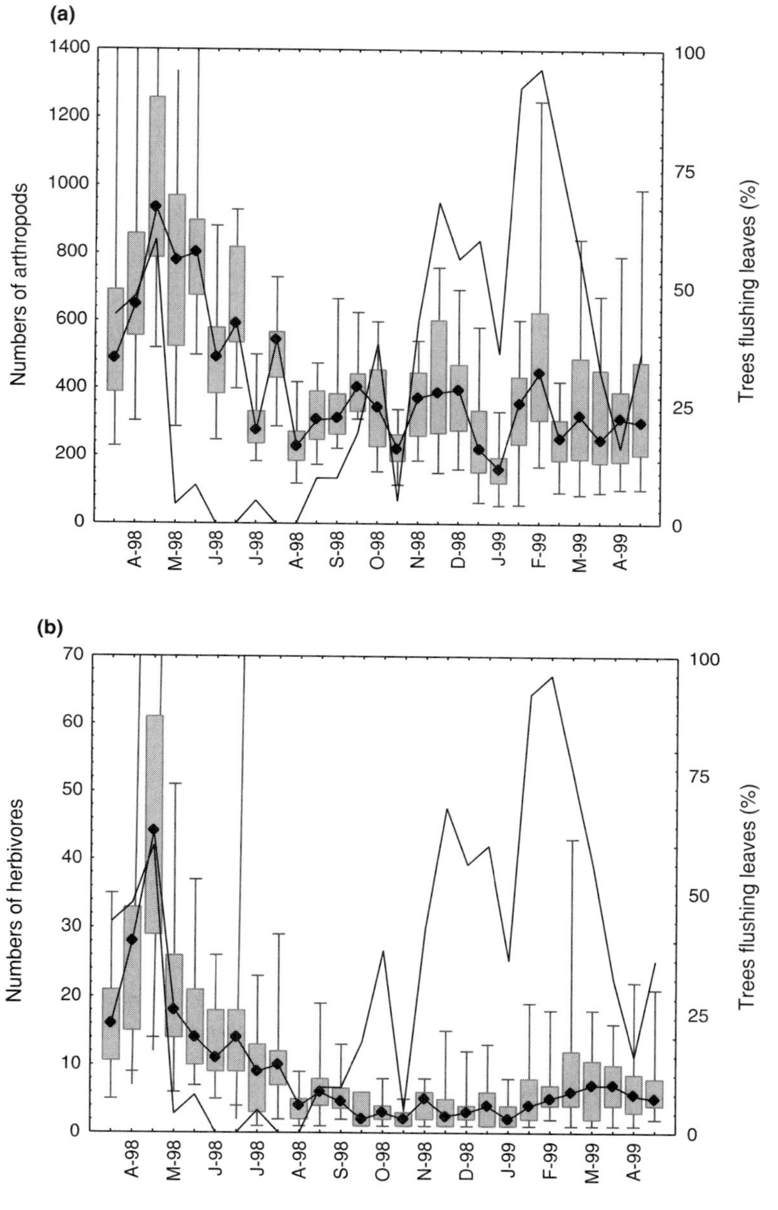

Fig. 17.5. Fauna collected in the study trees. (**a**) Box plot of total numbers of arthropods. (**b**) Box plot of the abundances of the herbivorous taxa Homoptera and Thysanoptera. The point is the median with the box indicating 25–75% confidence interval and the bar the minimum–maximum. The proportion of trees that were flushing new leaves is given as the other continual line. The month is indicated by the initial letter.

(U-test, $p < 0.01$). The latter two categories did not differ one from the other (U-test, $p = 0.12$; Table 17.1).

Differences among trees with different epiphyte loads

Epiphyte species identity

To account for seasonal fluctuations in animal abundance, we ran RM-ANOVA to consider the samples in their temporal sequence separately. There were no significant differences in numbers of individuals among the categories of trees: the analysis only confirmed a strong seasonality by yielding significant p values for the temporal factor ($p < 0.001$ for all taxa; Table 17.1). The epiphyte load of the trees had no significant influence on the abundance of any of the taxa, with one exception: among all orders representing at least 1% of the total fauna,

Table 17.1. *Host-tree traits (leaf area of host foliage; biomass and leaf area of its epiphyte load), numbers of arthropods and results of analyses detailed by tree category*[a]

Variable	Control trees	Trees with *Dimerandra*	Trees with *Vriesea*	Trees with *Tillandsia*	p value
Host tree leaf area (m^2)	28.6 (16.7–30.9)	25.3 (15.9–53.9)	33.3 (27.5–60.2)	32.1 (21.3–39.6)	KW-ANOVA, $p = 0.12$
Epiphyte leaf area (m^2)	0	0.6 (0.21–0.99)*	12.6 (6.34–27.9)**	10.9 (6.67–16.1)**	KW-ANOVA, $p < 0.001$
Epiphyte biomass (g dry weight)	0	318 (90–912)*	1670 (879–3853)**	3207 (2740–3828)**	KW-ANOVA, $p < 0.001$
Numbers of arthropods[b] per 2 weeks	370 (54–1621)	371 (92–2137)	303 (85–1317)	461 (113–1758)	RM-ANOVA, $p < 0.001$ (time), $p > 0.05$ (category)
Totals of arthropods[b] (32 weeks, n trees)	53 718	56 325	45 054	30 569	
Replicates (n)	7	7	7	4	

KW, Kruskal–Wallis; RM, repeated measures; ANOVA, analysis of variance.

[a]Values given as median with minimum–maximum in parentheses.

[b]Collected in 25 trees with 125 traps, emptied every 2 weeks over an 8 month period.

*and ** indicate significant differences among categories.

Diptera occurred in significantly lower numbers in trees with the large bromeliad *Vriesea* than in control trees and in trees with *Dimerandra* (RM-ANOVA, $p < 0.007$, post-hoc least significant difference test $p < 0.01$). The composition of the arthropod assemblages was similar across the four categories (ANOVA, $p > 0.1$).

Epiphyte quantity

To examine the variation in epiphyte load (Table 17.1), we investigated whether the quantity of epiphytes irrespective of species identity in an *Annona* crown had an influence on arthropod abundance, but found no correlation. Both the total arthropod yield of the study trees during the trapping period and the median values of the biweekly captures in each tree were independent of epiphyte biomass (Spearman rank correlation, $p > 0.1$; Fig. 17.6). This was also true for individual orders (all $p > 0.1$), again with Diptera as an exception. Their abundance was negatively correlated with epiphyte leaf area, albeit weakly (Spearman rank correlation, $p = 0.038$, $r^2 = 0.18$), but not with epiphyte biomass ($p > 0.1$). Arthropod abundance was also independent of host-tree leaf area (Spearman rank correlation, $p > 0.1$).

DISCUSSION

Faunal composition

The relative proportions of some taxa in our study differ from other samples of arthropod faunas of tropical tree canopies (e.g. Erwin, 1983a; Hijii, 1983; Stork, 1991; Kitching *et al.*, 1993; Höfer *et al.*, 1994; Guilbert *et al.*, 1995; Floren & Linsenmair, 1997b; Wagner, 1997; Adis *et al.*, 1998a). The most prominent peculiarity is the high abundance of micro-caddisflies (Trichoptera, Hydroptilidae), which contributed almost 30% to the arthropod assemblage in our study trees (Fig. 17.3). In most canopy-fogging studies, Trichoptera is either included in 'other arthropods' (Stork, 1991; Floren & Linsenmair, 1997b; Wagner, 1997) or not mentioned at all (Adis *et al.*, 1997). The majority of caddisfly larvae are aquatic, and most adults are weak fliers. Therefore, abundant Trichoptera species are a consequence of the locations of the focal trees along the shore of Lake Gatún. The exceedingly abundant species *O. circaverna*

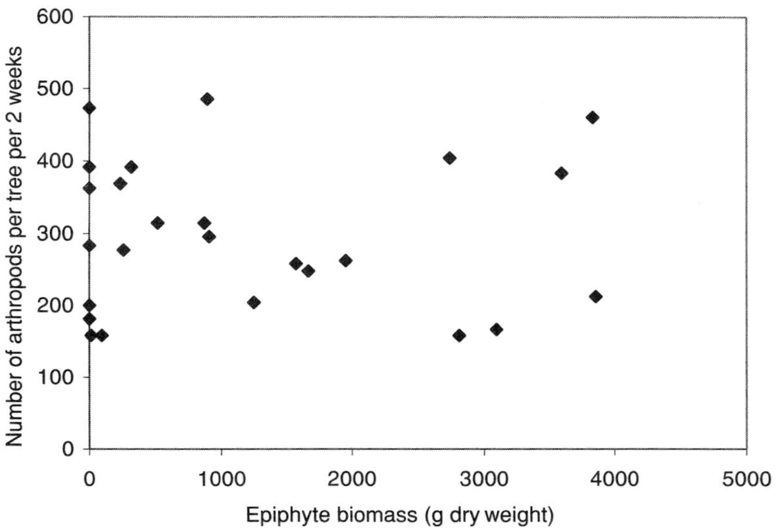

Fig. 17.6. Relationship of arthropod yield and epiphyte biomass in *Annona glabra* crowns. Each symbol represents the median value of numbers of individuals caught per study tree in 2 weeks from April 1998 until June 1998 and December 1998 until April 1999 (Spearman rank, $p > 0.72$).

and *O. maya* also dominated the caddisfly fractions in light-trap samples on Barro Colorado Island (O. Flint, personal communication). Lake Gatún apparently represents a very suitable habitat especially for these two species.

The proportion of ants in our samples (10%; Fig. 17.3) was low compared with the percentages reported by other authors (Stork, 1991: 18.2%; Tobin, 1995: 32%; Floren & Linsenmair, 1997b: 58%; Wagner, 1997: 36–49%; Adis *et al.*, 1998a: 45%). Methodological reasons partly explain this discrepancy: insecticidal fog used in these studies causes the majority of ants to abandon their nests and attempt to escape by dropping from the canopy (A. Floren, personal communication). Thus, ant colonies are sampled nearly quantitatively, while traps capture only a small fraction of each colony, that is, the workers that are active outside the nest. Considering the sizes of canopy ant colonies, these methodological differences can cause marked deviations in relative abundance of ants.

Despite the differences in methodology, the relative proportions and rankings of other taxa were consistent with those found by canopy fogging: the contribution of spiders and beetles, for example, were similar to those reported by Wagner (1997) and Höfer *et al.* (1994). Collembola was among the most dominant orders in tree crowns of other tropical forest canopies (Hijii, 1983; Guilbert *et al.*, 1995; Palacios-Vargas *et al.*, 1999; Ch. 15) but has sometimes been omitted from analyses (e.g. Basset, 1990). Diptera usually contributes a large fraction to the fauna (Stork, 1991; Kitching *et al.*, 1993; Wagner, 1997; Adis *et al.*, 1998a).

Diptera species are considered 'tourists' or 'transient species' that are not tightly associated with the trees in which they have been collected (e.g. Stork, 1987a). Interestingly, this was the only taxon in the present study with a significant correlation with the epiphyte load of a tree, albeit negatively. Diptera members decreased in numbers with increasing epiphyte leaf area and were less abundant in the category with the largest epiphyte, *Vriesea*. This might result from a certain reluctance of these flying insects to navigate through tree crowns densely laden with epiphytes, that could be obstacles in the flight path. However, this speculative argument does not apply for other orders with fast-flying members such as the Hymenoptera.

Seasonality

The movement of the Intertropical Convergence Zone causes a quite severe dry season in Panama. Leigh (1999) emphasized the importance of this cyclic alternation of seasons for the timing of tree phenologies in the BCNM remarking that it can 'cause feast and famine in successive years'. The study tree *A. glabra* exhibits a pronounced seasonality (Fig. 17.4) similar to most trees in the BCNM forest, approximately synchronizing with their major peaks of fruit fall, flowering and leaf flush (Leigh & Smythe, 1978; Foster, 1982; Leigh & Windsor, 1982). Contrasting to a study in Brazil, where several *Annona* spp. flowered during the rainy season (Gottsberger, 1989b), the trees studied here flowered only in the dry season (Fig. 17.4). Fluctuations of animal abundances, especially in phytophagous taxa, are sometimes correlated with host-tree phenology. For instance, abundances of herbivores frequently rise simultaneously with the production of new leaves, the preferred diet of most phytophagous insects (e.g. Wolda, 1978b; Lowman, 1982b; Basset, 1991c; Aide, 1993). In our study system, abundance of arthropods did not correlate with the flushing of new leaves in the tree crowns (Fig. 17.5**a**). This was also true for phytophagous taxa (Fig. 17.5**b**).

The bromeliads we studied produce new leaves continuously throughout the year (G. Zotz & G. Schmidt, unpublished data), but arthropod abundance changed significantly over time in all tree categories (Fig. 17.5**a**, Table 17.1). If herbivorous taxa were positively influenced by this predictable resource of young leaves in the bromeliad trees, then trees with *Vriesea* and *Tillandsia* should have smaller abundance fluctuations, or greater abundance of phytophagous taxa, or both. Neither was seen. Changes in arthropod abundance did not correlate with the phenology of *Dimerandra emarginata*, which flushes leaves approximately from April until August (Zotz, 1998): most taxa declined in abundance during this period.

Host-tree leaf area had no influence on the phytophagous taxa Homoptera and Thysanoptera, nor did the latter synchronize its seasonal rhythm in abundance with the phenology of *Annona*. This might indicate that the phytophagous arthropods studied here were rather weakly associated with the study trees. However, this notion must be regarded as hypothetical until data on species composition and ecology of the respective organisms are available. Recent studies have also reported low host specificity of phytophagous taxa in other tropical canopies (Stork, 1987a; Mawdsley & Stork, 1997). For example, Ødegaard (2000c), summarizing results of large arthropod inventories (Basset *et al.*, 1996a; Mawdsley & Stork, 1997; Ødegaard, 2000a) calculated a median of only four species of phytophagous beetles effectively specialized on a particular tree species.

It is also possible that our method underestimated phytophagous insects that are often sessile on the leaves on which they feed (e.g. Lepidopteran larvae). A prerequisite of being captured in traps in sufficient numbers is high activity of arboreal or aerial movement (or 'intensity of activity', see Adis, 1979). Thus, flight traps intercept increased proportions of highly mobile groups, in our case the 'tourist' taxa Diptera and Trichoptera. For less-mobile organisms, other methods, such as hand collecting or branch clipping, might be more useful for investigating small faunal differences (Köhler, 1997).

We recorded a pronounced peak of arthropod abundance at the beginning of the rainy season, and lower numbers during the dry season (Fig. 17.5**a**). This is consistent with earlier findings of Wolda (1978b) and Smythe (1982) for the forest on Barro Colorado Island. Leigh (1999) assumed that insects are generally most abundant in the early rainy season and much less so during the dry months. One could interpret this as an indication that arthropods are somewhat constrained by climatic conditions. However, the decrease in numbers during the second half of the rainy season (Fig. 17.5**a**) is not consistent with this argument. Moreover, if mitigation of microclimatic extremes increased arthropod abundance, we would expect to find more arthropods in trees with epiphytes because epiphytes substantially moderate the microclimate in tree crowns by reducing evaporation as well as air and surface temperatures in their immediate surroundings (Stuntz *et al.*, 2002a). However, arthropod abundances during the dry season in tree crowns with epiphytes were similar to those in trees devoid of them.

Do epiphytes influence arboreal arthropods?

At the ordinal level, the arthropod assemblages of entire tree crowns were not influenced by the presence of epiphytes. In fact, the four tree categories that we defined *a priori* (Fig. 17.1) had strikingly similar arthropod faunas in terms of relative and absolute abundance, although their epiphyte load differed significantly (Table 17.1). Furthermore, arthropod abundance correlated neither with epiphyte biomass (Fig. 17.6) nor with leaf area. Therefore, our initial hypothesis that epiphytes might act as a buffer for harsh climatic conditions and thus increase at least arthropod abundances during the dry season could not be confirmed (Table 17.1). At the level of individual epiphytes, however, epiphytes strongly influenced arthropod assemblages as a function of both plant species and biomass (Stuntz *et al.*, 2002b). The lack of a similar effect at the level of entire tree crowns could be an idiosyncrasy of the study system (a small tree living in inundated areas), in which aquatic taxa like the Hydroptilidae or other mainly transient taxa such as Diptera (Fig. 17.3) swamp out any signal from arthropods that are actually associated with the trees or epiphytes. We argue that this is not the case. First, many of the orders (e.g. Diplopoda, Isoptera) have no aquatic stages and still follow the same pattern when analysed individually. Second, in taxa with a finer level of identification (Coleoptera to family, Hymenoptera–Formicidae to species; Stuntz, 2001) there was also no detectable difference between the arthropod faunas on trees with and without epiphytes.

We conclude, therefore, that the effect of epiphytes on the canopy-dwelling arthropod fauna is obviously scale dependent: in spite of a pronounced effect at the level of individual epiphytes (Stuntz *et al.*, 2002b), entire tree canopies of *A. glabra* with and without epiphytes do not differ significantly in their faunal composition. It still remains to be shown, however, whether the small tree we studied can be taken as representative of tall forest trees. While there are obvious local idiosyncracies (e.g. a large number of aquatic insects), there are also striking similarities. For example, a majority of the common ant species found on *A. glabra* (Stuntz, 2001) were also common on emergent trees in adjacent forest (Yanoviak & Kaspari, 2000). This suggests that our study system may well be a useful model for the investigation of the effect of epiphytes on canopy arthropods in the tropics.

ACKNOWLEDGEMENTS

We thank Prof. K. Eduard Linsenmair, Universität Würzburg, for his advice in the initiation of this project. Assistants in the field were Edgardo Garrido, Pablo Ramos, Beatriz Wong, Sixto Martinez and Elizabeth Royte. José Rincón, Ricardo Concepción, Felix Matias and Michael Matzat helped in sorting arthropods. Oliver Flint from the National Museum of Natural History in Washington, DC identified the micro-caddisflies. Steve Paton from the Smithsonian Tropical Research Institute provided the climatic data. Ulrich Kern, TU München, contributed the drawings. This study was funded by the German Academic Exchange Service (DAAD), the European Science Foundation and the Deutsche Forschungsgemeinschaft (Graduiertenkolleg of the Department of Botany, Universität Würzburg).

PART IV

Resource use and host specificity in tropical canopies

Introduction

Since the mind-boggling calculations of Terry Erwin in the early 1980s about species diversity in the canopy, many authors have been interested in the host specificity of canopy arthropods as a parameter to estimate global diversity. In recent years, the research agenda has shifted more towards studying resource use and host specificity to understand complex food-webs in tropical rainforests and the maintenance of a high local diversity. The case studies presented in this section include two contributions originating from Australasia, one from Africa and four from the neotropics. There are also five other contributions in the synthesis section that discuss resource use and host specificity.

Changes in arthropod communities following anthropogenic disturbance are of increasing concern worldwide. For tropical rainforests, data are often available for ground-dwelling arthropod taxa or conspicuous – but not speciose – taxa such as butterflies. Floren & Linsenmair (Ch. 18) provide one of the few analyses of the effects of forest disturbance on canopy arthropods, targeting beetles in Sabah. They are particularly interested to study the mechanisms structuring these highly diverse assemblages, and their alteration with disturbance.

Savanna trees and their associated fauna are poorly known in the tropics. Mody *et al.* examine such a study system in the Ivory Coast (Ch. 19). Despite the apparent greater habitat homogeneity in savannas compared with rainforests, their research sensibly concentrates on the characteristics of individual trees as a set of very local factors structuring arthropod assemblages. They also compare these factors with other structuring factors acting at a larger scale, in a fashion similar to that used in Ch. 13.

Jaffé *et al.* (Ch. 20) are intrigued by the relationships between flowers and ants in Venezuela. From a phytocentric perspective, ants can be viewed either as potential exploiters of flower nectar or as mutualists that reduce herbivory and leaf damage. The authors ask whether flowers predominantly repel or attract ants. In doing so, they compare ingeniously these interactions in the canopies of a savanna and of a rainforest.

In Chapter 21, Ødegaard provides a comprehensive analysis of a local assemblage of phytophagous beetles in a dry forest in Panama. The author discusses beetle host specificity and compares in this regard the fauna associated with trees and lianas. One of the many merits of the study is the extensive taxonomic study of the material, which allowed estimates to be derived of the percentage of canopy species described among the material collected. Such data are still extremely rare in canopy studies.

In Chapter 22, Amédégnato considers arboreal grasshoppers in the Amazon basin and analyses their microhabitat distribution and food selection in the canopy. In particular, the extensive data also allow testing of the consistency of the observations at a regional scale, an outcome rarely achieved in canopy studies. In addition, microhabitat distribution is discussed in terms of oviposition, nymph and adult distribution, perhaps for the first time for any group of canopy arthropods.

With rare exceptions concentrating mostly on bees, the significance of flowering events in the canopy for local arthropod diversity has been rarely investigated. To close this section, Kirmse *et al.* (Ch. 23) compare such events for two species of trees in Venezuela. They target beetle visitors, analyse beetle similarities between the two hosts and discuss whether the species collected appear to be specialists.

18

How do beetle assemblages respond to anthropogenic disturbance?

Andreas Floren & K. Eduard Linsenmair

ABSTRACT

Our investigations of the mechanisms structuring highly diverse arthropod communities in tropical rainforests have shown that communities change drastically after anthropogenic disturbances. For example, assemblages of Formicidae in the mature forest could not be distinguished from random assemblages, whereas in the disturbed forest they showed a clear deterministic structure. In this chapter, we investigate whether similar changes in assemblages occur for Coleoptera. For this purpose, we sampled beetles from 19 *Aporusa lagenocarpa* trees (Euphorbiaceae) in a south-east Asian primary lowland forest using a selective fogging method. Beetle assemblages were compared with those in three disturbed forests that have been left to regenerate for 5, 15 and 40 years . From each forest type, 10 conspecific trees were fogged. In the primary forest, no abundant species occurred and an almost complete set of new species was collected within every fogging sample. Accordingly, it was impossible to distinguish Coleoptera assemblages from randomly composed assemblages. In contrast, beetle assemblages in the disturbed forests were significantly different: species diversity was lower than in the primary forest, the proportion of frequent species was higher and some species dominated in each forest type being collected regularly from most trees. These assemblage changes were similar to those observed for Formicidae and we suggest that beetle assemblages show the same transition in structuring mechanisms. However, a detailed theoretical analysis, such as for the ant assemblages, is much more difficult for the highly diverse Coleoptera and needs a much larger sample size.

INTRODUCTION

Coleoptera are one of the mega-diverse groups of arthropods in tropical rainforests and are particularly diverse in the canopy. How many species really exist is still not known, but there is no doubt that currently only a small proportion has been collected and described (Erwin, 1982; Hammond, 1994). In consequence, very little is known about distribution patterns of tropical Coleoptera, their biology or their importance in ecosystem functioning. In addition, most Coleoptera of primary forests are extremely rare and faunal overlap in samples is very low (references in Stork *et al.*, 1997a; Floren and Linsenmair, 1998a; Wagner, 1999; Linsenmair *et al.*, 2001). Accordingly, it is difficult to analyse tree-specific beetle communities and their dynamics. At present, little is known about how this high species diversity is maintained in tropical ecosystems and what mechanisms structure these communities (see e.g. Beaver, 1979; Basset, 1996; Novotny and Basset, 2000; Wagner, 2000). Here, we contribute to the understanding of beetle assemblage structure by investigating how arboreal Coleoptera respond to anthropogenic disturbance.

Until recently, our studies were mainly concerned with arboreal Formicidae. We demonstrated that ant assemblages in the primary forest could not be distinguished from randomly assembled sets, whereas they showed a clear deterministic pattern in disturbed forests (Götzke & Linsenmair, 1996; Floren & Linsenmair, 1997b, 1998b, 2000a; Floren *et al.*, 2001a). In this chapter, we investigate whether assemblages of Coleoptera show similar changes as a result of severe anthropogenic disturbances. In particular, we focus on the phytophagous Chrysomelidae and Curculionidae, because many of the species in these families are likely to be adapted to their host plants and their assemblage structure should be predictable to a large degree. This study provides information for basic research and nature conservation by analysing the changes of arboreal beetle assemblages in primary forests and forests disturbed by humans.

Table 18.1. *Forest types and tree species investigated (10 trees from each forest type were used for the comparison)*

Variable	SI	SII	SIII	Primary forest
Disturbance level	5 years of regeneration	15 years of regeneration	40 years of regeneration	None
Vegetation	Pioneer trees; ground covered by grass and bushes	Trees formed a thin forest; no grass; few bushes	Forest with thick growth of tree saplings; large amounts of dead wood	Multilayered primary lowland rain forest
Tree species fogged	*Melochia umbellata*	*Vitex pinnata*	*Vitex pinnata*	*Aporusa lagenocarpa*
Tree family	Sterculiaceae	Verbenaceae	Verbenaceae	Euphorbiaceae
Mean leaf cover of the fogged trees (%)	50 ± 10	60 ± 10	60 ± 14	60 ± 20
Mean area of collecting funnels (m^2)	23 ± 4	30 ± 8	34 ± 10	22 ± 3

SI, SII and SIII: plots clear-felled 5, 15 and 40 years ago, respectively.

METHODS

Coleopteran assemblages of understorey tree species were sampled by insecticidal fogging in primary and disturbed lowland rainforests of Kinabalu National Park, in Sabah, Malaysia. Foggings in the primary forest were carried out in a mixed dipterocarp forest at the substation Poring Hot Springs (6° 2.74′ N, 116° 42.2′ E), and the disturbed forest plots were at substation Sorinsim. A map of the area has been published elsewhere (Floren & Linsenmair, 2001). The disturbed forest plots were between 200 and 400 m wide and were close together. They finally merged into primary forest, thereby forming an anthropogenic gradient of disturbance. According to local people, these forest plots were clear-felled 5, 15 and 40 years ago (plots designated SI, SII and SIII, respectively); the large logs were removed and the dried undergrowth was burned. After 3 to 5 years of agricultural usage they were left for natural regeneration. From each forest type, 10 conspecific trees were fogged and beetle assemblages were compared across samples (Table 18.1). Distance between trees was 10 to 50 m within each forest type. Only trees that were similar in regard to habitat conditions were chosen; that is, we paid attention that no or only a few lianas covered the tree crowns and there were only occasional epiphytes in the trees (see Floren & Linsenmair, 1998b).

Canopy fogging is not only suitable for the measurement of diversity but also it allows the study of some of the dynamical processes that structure arthropod assemblages (e.g. Floren *et al.*, 2001a). Beetle assemblages were collected from understorey trees. In order to achieve tree-specific sampling in the multilayered primary forest, a large cotton roof was stretched out above the crown of the study tree the day before fogging. This prevented arthropods from higher canopy strata that were killed by the insecticide from falling into the collecting funnels, which were installed 1 m above ground level. Funnels always covered 80–90% of a respective crown projection area, equivalent to 25 to 50 m^2 of funnel area. This ensured that almost all arthropods that were living on the surface of leaves and bark were collected. Fogging was performed early in the morning for 10 minutes with natural pyrethrum as an insecticide and all arthropods caught within 2 hours following fogging were considered in the analyses.

The ten primary forest trees that were used for the comparison between forest types belonged to the species *Aporusa lagenocarpa* (Shaw), Euphorbiaceae. They are a subsample of a total of 79 foggings that were performed on 41 trees of 13 species (A. Floren, unpublished data). *A. lagenocarpa* was found in two clusters of up to 50 individuals that were separated by 3.5 km. From each cluster, five trees were fogged. Distance between trees varied between 20 m and a maximum of 300 m within a cluster (see Floren & Linsenmair, 2000a). Our aim was to study the same tree species in all forest types. This, however, was not possible because the primary forest trees were not found in the disturbed forests nor were the pioneer trees of SI growing in SII and SIII. It was necessary, therefore, to select two further tree species: from the 5-year-old forest this was the pioneer tree *Melochia umbellata* (Stapf), Sterculiaceae, which grew to 8 m height; and *Vitex pinnata* L., Verbenaceae, which was the only tree species that occurred in SII and SIII in sufficiently large numbers. As in the primary forest plot, *V. pinnata* reached an average height of 20 m. No tree species was flowering during the time of fogging. In all disturbed forests only a single canopy layer, not yet closed, had formed. Accordingly, it was not necessary to install the roof above the crown of the study trees to collect a tree-specific fogging sample (Floren & Linsenmair, 2001). We are fully aware that collecting arthropods from different tree species makes it difficult to distinguish tree taxonomic effects on assemblages from those induced by disturbance, and we discuss this problem in detail in the discussion.

Beetles were sorted to morphospecies and deposited in the collection of the first author. Most species were collected in low abundances. We define a species as 'more common' when it occurred with more than 10 individuals. For more details on study areas, forest types and collecting methods see Floren and Linsenmair (1997b, 2000a).

Forest types were compared using α- and β-diversity values (Magurran, 1988). Sample sizes were standardized by using rarefaction statistics (Hurlbert, 1971; Hayek & Buzas, 1997). For this purpose, beetles of all fogged trees per forest type were pooled and diversity was expressed as the number of expected species within an equal subsample size (this was equivalent to the 346 specimens of SII, which represented 91 species; compare with Table 18.2, below). Beta-diversity was calculated using Sørensen's quantitative index of similarity. Differences in means of β-diversity between forest types were tested with a Mantel test following a randomization test (Monte Carlo). The number of randomized runs was 1000. In all cases of multiple comparisons, the level of significance was adjusted using the Bonferroni correction (Lehmacher *et al.*, 1991).

RESULTS

Figure 18.1 illustrates changes along the disturbance gradient for the numerically dominant families of Coleoptera. The Corylophidae, living under bark or in rotting plant material and predominantly feeding on mould, dominated SI samples but were lacking in SII samples. In contrast, the Phalacridae were not found in SI but reached their maximum abundance in SIII. Adults and larvae live in and feed on fungi, but some species are primarily found on flowers, where they eat pollen. Scolytinae showed the greatest increase in numbers in SIII but did not occur in SI. This might reflect the increasing amount of dead wood in SII and SIII and could explain the correlated peak of Cleridae, of which many species prey on scolytines. In the disturbed forests, the abundant Anthicidae and Malachiidae, which are mostly found on foliage or flowers where they feed on insects, were numerically dominated by a few species. The analysed taxa of phytophaga – Chrysomelidae and

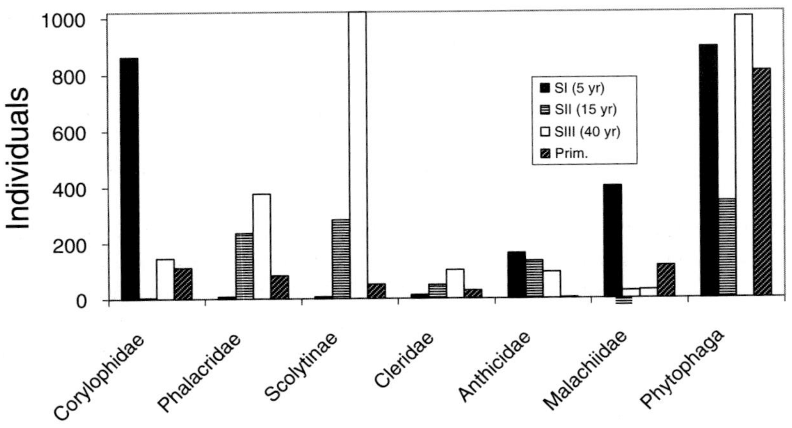

Fig. 18.1. Changes of Coleoptera at the family level in forests of different disturbance levels. SI, SII, SIII are 5-, 15- and 40-year-old regenerating forests respectively. Prim. is primary forest. Phytophaga includes Chrysomelidae and Curculionidae.

Table 18.2. *Comparison of arboreal communities of Chrysomelidae and Curculionidae between forest types, based on 10 trees fogged per forest type*

Variable	Forest type[a]			
	SI	SII	SIII	Primary
Relative proportion of Chrysomelidae and Curculionidae[b]	23.52	19.79	20.97	38.27
Number of species	84	91	186	285
Number of rarefied species	52	91	102	166
Standardized mean species number[c]	1.8 ± 0.5	0.9 ± 0.6	1.8 ± 0.7	3.8 ± 2.0
Abundance	895	346	1001	819
Standardized mean abundance[c]	7.5 ± 4.8	2.1 ± 2.3	5.1 ± 2.3	7.9 ± 6.6
Singletons	42 (50.0%)	53 (58.2%)	103 (55.4%)	167 (58.6%)
Abundance of species with $\geq$ 10 individuals	748 (83.6%)	199 (57.75%)	673 (67.23%)	335 (40.1%)
Abundance per species	10.66	3.80	5.38	2.87

[a]Regenerating forest of 5 (SI), 15 (SII) and 40 (SIII) years in age and primary undisturbed forest.
[b]Relative to total number of beetles collected per forest type.
[c]Standardized to a crown projection of 1 m^2 and leaf cover of 100%; mean $\pm$ standard deviation.

Curculionidae – dominated in numbers in all forests with the exception of SIII, where a similar number of scolytines (1020 specimens) were collected.

Diversity indices for Chrysomelidae and Curculionidae did not change uniformly during forest succession (Table 18.2). Their proportion relative to the entire beetle assemblage within a forest type peaked in the primary and in the 5-year-old regrowth forest, but significant differences were only found between SIII and the primary forest (Mann–Whitney U-test, $U = 14$, $p = 0.005$). The corrected level of significance was 0.0083. SI and SII were not significantly different from the primary forest after application of the Bonferroni correction (U-test, $U = 18$ and 10, $p = 0.015$ becoming $p = 0.02$). Chrysomelidae were always represented by at least twice as many specimens as Curculionidae. In the primary forest, only one species, an alticine (Chrysomelidae), was collected with more than 100 individuals. No other species occurred on all *A. lagenocarpa* trees (Floren & Linsenmair, 1998a). Species numbers increased during forest succession and rarefied values showed that this result was independent of the total number of beetles. However, after standardizing species numbers for a crown projection of 1 m^2 and a leaf cover of 100%, more species were found in SI than in SII. With regard to standardized number of species, forest types differed significantly from each other (all U-tests $p < 0.004$) with the exception of SII versus SIII (U-test, $U = 46.5$, $p = 0.796$). Beetle abundance was lowest in SII and comparisons of standardized abundance data showed again that only the SII data differed significantly from SI, SIII and the primary forest ($U = 10$, 11 and 13, $p < 0.004$). Since the variance in species numbers and abundances of beetles on individual trees within forest types was large, the proportions of singletons and more frequent species were calculated for each forest type. At least 50% of all specimens from each forest type was singletons. However, more species with higher levels of abundance were found in the disturbed forests than in the primary forest. On average, one species was represented by 10.7 individuals in SI, whilst the corresponding figure was 2.9 individuals in the primary forest, indicating that the number of individuals per species decreased during forest regeneration.

Changes in assemblages are reflected in the alpha of the log series and in Simpson's index (Fig. 18.2). The alpha of the log series, which considers primarily species numbers, showed significant differences only between SI and the primary forest (U-test, $U = 7$, $p < 0.001$). SII and SIII did not differ significantly from the primary forest after application of the Bonferroni correction ($p = 0.009$ becoming $p = 0.011$). This is because of increasing variance in species numbers and abundances in forest types during the course of forest

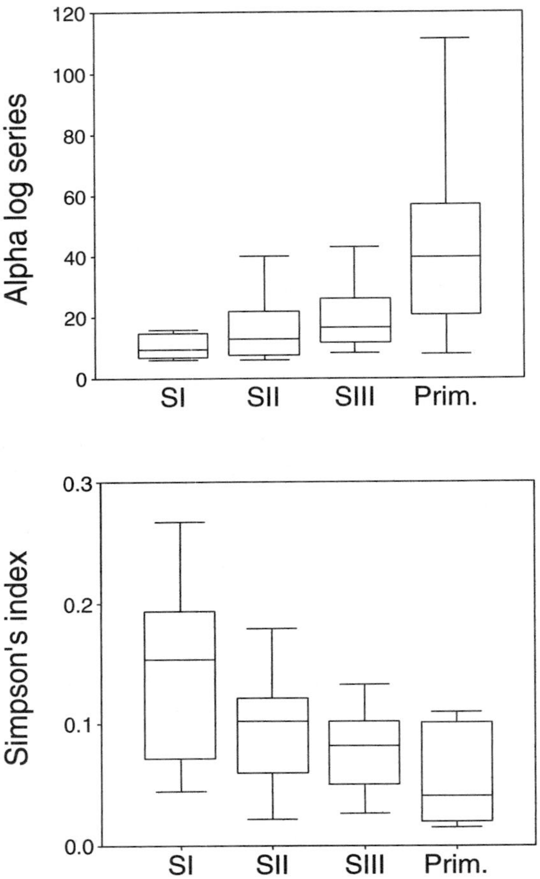

Fig. 18.2. Changes in α-diversity of assemblages of Chrysomelidae and Curculionidae between forests, indicated by the alpha of the log series and Simpson's index. Data are expressed as box plots showing the median and 50% of the values. SI, SII, SIII are 5-, 15- and 40-year-old regenerating forests, respectively; Prim. is primary forest.

succession. It also explains why Simpson's index, which is strongly biased by species abundances, did not differ significantly across forest types (Kruskal–Wallis test, $H = 8.070$, $p = 0.045$), although the medians showed a clear trend towards a decrease in frequent beetle species during forest succession.

Figure 18.3 shows how β-diversities of beetle assemblages varied across forest types. In the 5-year-old forest, the Sørensen index indicated high similarity in species composition of assemblages. However, during forest succession, assemblages became more and more dissimilar. Even after 40 years, species composition of beetle assemblages was clearly more similar than those of beetle assemblages of the primary forest. The Mantel test showed significant differences only between the 5-year-old forest and the 40-year-old forest ($r = 0.4497$, $p < 0.01$) and between the 5-year-old forest and the primary forest ($r = 0.3706$, $p = 0.01$). SIII did not differ significantly from the primary forest ($r = 0.30543$, $p = 0.05$) after applying the Bonferroni correction.

DISCUSSION

Coleoptera is often used as a target group to study ecosystem properties and how these change with human disturbance (for tropical ecosystems see Didham *et al.*, 1996, 1998; Lawton *et al.*, 1998). However, species diversity in tropical rainforests is so high that analyses are very difficult because usually neither regional species richness nor species turnover between sites is known. Therefore, changes in species composition of communities cannot be correlated with changes in habitat. This was also the case in our study, in which we analysed the influence of anthropogenic disturbance on arboreal assemblages of Coleoptera.

The disturbed forests selected for this investigation were all close together and growing under similar

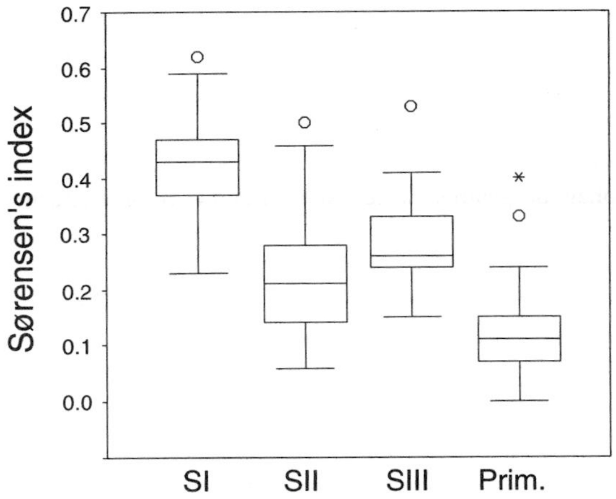

Fig. 18.3. Changes in β-diversity of assemblages of phytophaga between forests, indicated by Sørensen's index of similarity. A circle indicates extreme values between one and three times the box length, an asterisk, values larger than three times the box length. Abbreviations as in Fig. 18.2.

geomorphological conditions. According to the local people, these forests had been clear-cut for agricultural use 5, 15 and 40 years earlier. No large trees were left and the undergrowth was burned, allowing a few years of crop cultivation. Afterwards, the plots were abandoned and left to regenerate. In all of the disturbed forest types, only a single canopy layer had developed and the beetles could be collected easily in a tree-specific way by insecticidal fogging. By selecting only trees that were growing under similar habitat conditions (Floren & Linsenmair, 1998b), we tried to keep variability among study trees as low as possible. As in the primary forest, epiphytes were rarely observed in the tree crowns and only such trees were fogged on which no or only few lianas were growing. In the primary forest, the crowns of most study trees were isolated from higher canopy strata by a large cotton roof. Although this treatment certainly repelled some species, the composition of the arthropod and beetle communities could not be distinguished from those trees that were fogged without the roof installation (A. Floren, unpublished data; references referring to comparable studies in Stork *et al.*, 1997a; Linsenmair *et al.*, 2001). A more detailed discussion of the methodological problems and the effectiveness of fogging is given in Floren and Linsenmair (1997b, 1998a).

How anthropogenic disturbance changes beetle assemblages

In the primary forest, Chrysomelidae and Curculionidae, together with Staphylinidae, dominated all trees in terms of species numbers and species abundance, representing always more than 50% of all specimens (Floren & Linsenmair, 1998a). Although we sampled conspecific trees and restricted our analysis to phytophagous groups, which should show up tree-specific adaptations (Beaver, 1979; Schoeller, 1996), we could not find any species occurring regularly on conspecific host trees. High species diversity made it impossible to distinguish assemblages from randomly composed sets (Floren & Linsenmair, 1998a; Wagner, 2000). Currently we are analysing the beetle assemblages of all 79 foggings that were carried out in the primary forest. There is no indication that the above results will be modified by further analysis.

The difficulty of identifying a tree-specific fauna of herbivorous species is in sharp contrast with temperate forests, where a tree-specific beetle fauna can be identified across even a few conspecific trees (Southwood *et al.*, 1982a; McQuillan, 1993; Wagner, 1996b; Floren & Schmidl, 2000; A. Floren, unpublished data). Such differences lead to the still unanswered question of what are the mechanisms that structure highly diverse tropical communities (Beaver, 1979; Floren & Linsenmair, 1997, 1998a). The analysis of changes in communities that occur as a consequence of anthropogenic disturbance can give important indications (Floren & Linsenmair, 2001). For example, assemblages of arboreal Formicidae of a primary tropical lowland rainforest could not be distinguished from random assemblages although they showed a clear deterministic pattern in disturbed forest (Floren *et al.*, 2001a).

The diversity of ant species was greatly reduced in the disturbed forest compared with primary forest: some species occurred on all trees and became characteristic of a particular forest type. As a consequence, ant assemblages in the disturbed forests were to a large degree predictable. These results indicate strongly that nonequilibrium conditions mediate species coexistence in the primary forest (Wiens, 1984; Huston, 1994), whereas disturbed habitats favour deterministic dynamics (Floren & Linsenmair, 1999, 2000, 2001; Floren *et al.*, 2001a). Because of the extraordinarily high diversity of Coleoptera, a detailed analysis at the β-diversity level, as was made for the Formicidae, is much more difficult. However, changes in Coleoptera assemblages showed a very similar trend to those observed for ant assemblages, indicating a similar transition in the structure of the assemblages from primary to disturbed forest. In addition to Chrysomelidae and Curculionidae, other families of beetles occurred in greater abundance in different types of secondary forest. Total species diversity was distinctly lower than in the primary forest, individual species reached high numbers and some were collected from all trees (compare with data of Didham *et al.*, 1998; Lawton *et al.*, 1998; Wagner, 2000). As a consequence, the composition of the beetle assemblage was less variable than in the primary forest. Such results were never obtained in the primary forest, nor were they observed during the recolonization of fogged primary trees (Floren & Linsenmair 1998a; A. Floren & K. E. Linsenmair, unpublished data). In contrast to the situation in disturbed forests, we also detected some large aggregations of morphologically very similar beetles in the primary forest; these belonged to many different species, which we, therefore, termed 'beetle complexes' (Floren & Linsenmair, 1998a). Neither their existence nor their functional significance is currently understood.

During forest succession, beetle diversity and assemblage structure gradually became more similar to the conditions of the primary forest. This was reflected in the assemblage composition at the level of the family, the increase in species diversity and the decrease in species overlap among trees. Accordingly, beetle assemblages became more and more unpredictable. However, even after 40 years of forest regeneration, assemblages were clearly distinct from those of the primary forest. Since all forest types were close together, one can conclude that differences in beetle faunas did not result from lack of colonization. This suggests that anthropogenic disturbances have long-lasting effects on beetle diversity and assemblage structure. It is obvious that old secondary, nonisolated, forest remnants are of high importance for conservation, since they can preserve a much higher diversity than younger, more disturbed forest stands. This general statement refers to most arthropod taxa and is of crucial importance for eventual recolonization of untilled land.

Our data demonstrate that species rarity is not specific to the undisturbed primary forest. Even in the 5-year-old pioneer forest, more than 50% of all beetles were collected as singletons. This, however, might be because all forest types overlapped to some extent. Consequently, many of the rare species might enter from older forest types without being able to establish in the disturbed forest. Novotny and Basset (2000), who collected externally feeding leaf-chewing insects on 30 locally abundant species of trees and shrubs in primary and secondary lowland forests in New Guinea, found that the majority of species were rare on particular hosts while more common on other plants. They concluded that most of the rare insects collected from an individual tree are transient species. However, if many of the rare beetle species occur regularly in high numbers on other trees, then at least some species should have been found at higher frequency on the fogged trees. Frequent species, however, were never observed in the primary forest, even given the much larger sampling effort of 79 foggings that was performed in this area. A change in the distribution of species, however, is a typical side-effect of anthropogenic disturbance (e.g. Mason, 1996; Davis & Sutton, 1998; Dupuy & Chazdon, 1998; Vasconcelos *et al.*, 2000). With respect to the beetle assemblages, even disturbances that are older than 40 years may be recognized structurally, for example by the occurrence of more common species (Floren & Linsenmair, 1999, 2001, this study; Floren *et al.*, 2001a). Such changes in the distribution patterns of species may be a consequence of the lower plant diversity and the occurrence of common tree species in disturbed forests. We have suggested above that under such circumstances the mechanisms that structure arboreal arthropod communities may change. Such a change has not hitherto been taken into consideration.

Distinguishing generalists from specialists

According to the classic dogma only specialized species should be able to coexist (e.g. Pianka, 1966; Beaver, 1979). In particular, many phytophagous insects should be rather specialized because studies suggest that tropical plants are more poisonous than temperate plants (Coley & Barone, 1996). If this is true and if we assume that no insect species is able to detoxify the metabolites of all of the 4000 species of vascular plants occurring in Kinabalu Park (Beaman & Beaman, 1990), then, surely, herbivorous species must be specialized on particular food plants. Accordingly, it should be possible to collect them in greater abundance on their host plants, as is indeed the case in temperate regions (e.g. Southwood *et al.*, 1982a; Basset & Burckhardt, 1992; McQuillan, 1993; Wagner, 1996b; A. Floren, unpublished data). In Middle Europe, for example, more than 80% of all species of Chrysomelidae and Curculionidae are considered to be oligophagous (Schoeller, 1996). In addition, a high level of specialization of beetle species can be assumed for the group of 'xylobiontic' beetles (Floren and Schmidl, 2000): a term that encompasses all species that depend in both their trophic relations and their environmental requirements on wood, for example the degree of wood decomposition, the presence of fungi and so on (Geiser, 1994). However, in the rainforest, no such specialized communities can be identified, although, in total, we have performed 36 foggings on 19 *A. lagenocarpa* trees to date. *A. lagenocarpa* shows an aggregated distribution with up to 50 individuals per hectare. Two such clusters were found of which many trees were fogged, including direct neighbours (Floren & Linsenmair, 1998a, 2000; A. Floren & K. E. Linsenmair, unpublished data). Because it is assumed that specialized species would find it difficult to detect isolated host plants in a highly diverse forest when the plants are some kilometres apart, it seems reasonable to expect that at least some species would occur in higher numbers within these large aggregations of conspecific trees. The lack of evidence of specialized species across all our fogging samples contradicts this prediction. In fact, there is an increasing indication that the degree of specialization of beetles in rainforests might be much lower than previously thought. These conclusions have been reached by various authors (Basset, 1996; Fiedler, 1998; Floren & Linsenmair, 1998a; Basset & Novotny, 1999; Ødegaard, 2000a; Wagner, 2000; Ch. 21).

How to differentiate between tree-specific effects and forest succession?

We are fully aware that collecting beetle assemblages from three different tree species makes it difficult to separate the tree-specific effects on the assemblage composition from those caused by forest succession. First, we have to stress that the selection of different study trees was constrained by the field conditions; there were no single tree species growing in all forest types. However, we can provide several arguments that changes in beetle assemblages were indeed a consequence of disturbance and did not simply result from sampling different tree species. First, we found that beetle communities were structurally different across forest types. Second, they changed continuously in diversity and structure with increasing time of forest regeneration without any tendency toward tree-specific effects. Third, community patterns observed in disturbed forests were never detected in the primary forest. Finally, similar changes were observed in all other taxa currently under investigation, such as Formicidae (Floren et al., 2001a), Orthoptera (Floren *et al.*, 2001b); Ichneumonidae and Arachnida (A. Floren, unpublished data).

These results lead us to argue that the observed changes at the community level are the immediate consequences of anthropogenic disturbances and cannot be understood solely on the basis of host-plant effects. We hypothesize that the transition from random communities in the primary forest to deterministically structured ones in disturbed forests, as described in detail for the Formicidae (Floren & Linsenmair, 2000, 2001; Floren *et al.*, 2001a), is generally also true for species-rich arthropod groups such as Coleoptera.

ACKNOWLEDGEMENTS

We are greatly indebted to Datuk Ali Lamri, the director of Sabah Parks, who gave us permission to work in Kinabalu Park. We also thank the members of the park staff for their help, particularly Alim Biun and Andre Kessler, who were closely involved in field work. We are grateful to the following for their critical comments and/or discussion: Thomas Wagner, Yves Basset and an anonymous referee. This study was financially supported by the Deutsche Forschungsgemeinschaft (DFG), Li 150/13 1–4.

19

Organization of arthropod assemblages in individual African savanna trees

Karsten Mody, Henryk A. Bardorz and K. Eduard Linsenmair

ABSTRACT

This study describes and compares arthropod assemblages collected by insecticide knockdown in Comoé National Park, the Ivory Coast, and occurring on three savanna tree species, *Anogeissus leiocarpa* (Combretaceae), *Burkea africana* (Leguminosae), and *Crossopteryx febrifuga* (Rubiaceae). Differences between beetle assemblages indicate an effect of tree species on assemblage composition. Beetle density and species richness was highest on *Anogeissus* (maximum density: 113.7 beetles m^{-2}, maximum species number: 145 per tree) whereas *Burkea* and *Crossopteryx* assemblages were not distinguishable on this basis. Mean species similarity of beetles was higher for conspecific (22–28%) than for heterospecific trees (13–17%). Principal component analysis of the distribution of abundant beetle species clearly separated *Crossopteryx* assemblages and demonstrated highest variability for *Anogeissus* assemblages. Differences between tree species were not pronounced in terms of ant density and diversity, and mean species similarity of ant assemblages was comparable between conspecific (36–40%) and heterospecific host-tree species (30–39%). Density and diversity measures such as species richness, the log-series index, Berger–Parker dominance index, rarefaction and Shannon–Wiener index varied considerably among individuals of all three tree species for Coleoptera and Formicidae. Only the Berger-Parker index for ants showed little variation. A marked variation was also observed in the dominance of certain beetle families on single trees, and for the regularly aggregated distribution of beetle and ant species. These results were analysed with respect to (i) the hypothesis that conspecific trees provide similar habitats and are, therefore, inhabited by similar sets of arthropods, and (ii) the question whether stochastic colonization events and/or deterministic tree characteristics are more important in determining the assemblage structure of individual trees. Resampling the assemblages of the same trees (*Anogeissus*) at 2 months and at 1 year later indicated a high congruence of assemblage composition. At both high and low taxonomic levels (arthropod orders, beetle families, beetle species and *Camponotus* species) the assemblage structure of refogged trees correlated most strongly with those of the same tree in the previous year. Permutation of correlation matrices revealed that this finding was significant for arthropod orders, beetle families and beetle species. The results of this study, therefore, tend to reject the hypothesis that conspecific trees are very similar habitats: rather they stress the importance of individual tree characteristics for arthropod assemblage composition.

INTRODUCTION

Studies on assemblages of canopy arthropods have become a prominent part of taxonomic and ecological research since the 1980s (see contributions in Stork *et al.*, 1997a). Facilitated by new canopy-access techniques (Moffett & Lowman, 1995; Barker, 1997) and by effective arthropod sampling methods, including 'restricted and selective fogging' (Basset *et al.*, 1997b; Floren & Linsenmair, 1997b), these investigations have influenced fundamentally our perspectives upon global, regional and local patterns of biodiversity (Erwin, 1982; Stork, 1988; Basset, 1996; König & Linsenmair, 1996; Stork *et al.*, 1997a). However, they do not cover all terrestrial biomes evenly, and those of Africa are least explored (Erwin, 1995; but see, for example, Brown, 1961; Moran & Southwood, 1982; Samways *et al.*, 1982; Southwood *et al.*, 1982b; Jackson, 1984; Coe & Collins, 1986; Grant & Moran, 1986; West, 1986; Basset *et al.*, 1992, 2001a; Dejean *et al.*, 1994a, 2000e; Moran *et al.*, 1994; Krüger & McGavin, 1997, 1998a; Wagner, 1997, 1999; Mercier *et al.*, 1998; McGavin, 1999). For West Africa, quantitative investigations of arthropods in pristine habitats are generally scarce, and information obtained by canopy fogging is virtually nonexistent. This lack of information, together with the continuing but hardly recognized disappearance of natural lands in the forest–savanna mosaic of West Africa, prompted us in 1997 to start a canopy fogging study on the arthropod assemblages of savanna trees in the Comoé National Park, Republic of Ivory Coast.

Research on arboreal arthropods has covered a wide range of topics from investigations on the general composition of arthropod assemblages to studies aiming at understanding the observed assemblage patterns. Many factors influencing the composition of arboreal arthropod assemblages have been investigated, including intraindividual variation of plant characteristics, effects of season, disturbances, forest types and host-plant species. Considering this wealth of information, studies explicitly taking the individual tree as the study unit seem to be noticeably rare. This becomes especially evident in comparison with a related field of ecological research, that of animal–plant interactions. Such studies have shown that, for example, a great number of plant characteristics vary across individual conspecific plants (Marquis, 1992) and that these influence deterministically the distribution of herbivorous insects (Fritz, 1992). With regard to studies on arthropod assemblages on trees, the question arises whether particular findings are applicable to systems of (a few) herbivore species and their (often nonwoody) host plants or whether they apply to species-rich arthropod assemblages as a whole. On the one hand, it could be argued that an individual tree presents a much wider spectrum of characteristics than a small herb and that conspecific trees, therefore, provide very similar habitats for arthropods (Southwood & Kennedy, 1983). On the other hand, it is well known that even conspecific, syntopically growing trees are variable in arthropod-relevant parameters like phenology (Crawley & Akhteruzzaman, 1988; Mapongmetsem *et al.*, 1998; Stork *et al.*, 2001), architecture (Strong *et al.*, 1984), morphology (Basset, 1991e; Senn *et al.*, 1992; Schmidt, 1999) or chemistry (Edwards *et al.*, 1993; Barnola *et al.*, 1994; Butcher *et al.*, 1994; Hemming & Lindroth, 1995; Staudt *et al.*, 2001). Such differences could be taken to suggest that individual tree characteristics might be of more importance for the structure of arthropod assemblages than is reflected in the attention that has been paid to this topic hitherto.

In this study, we have investigated the structure of arthropod assemblages on three representative species of West African savanna trees, explicitly considering assemblage structure on tree individuals. Our specific objectives were (i) to compare assemblages on conspecific and heterospecific tree individuals for among-tree similarity; (ii) to test the hypothesis that conspecific trees are inhabited by similar arthropod assemblages as they provide similar habitats; and (iii) to identify the relative importance of deterministic tree characteristics versus stochastic events as factors structuring these assemblages.

METHODS

Study area and host trees

The study was conducted in Comoé National Park from June to August 1999 and from June to July 2000 during the rainy season. Comoé National Park is located in the north-east of the Republic of Ivory Coast, and covers 11 500 km^2 of Guinea- and Sudan-zone savanna. It supports a wide variety of vegetation types including gallery forest, island forest, savanna forest, shrub and tree savanna, grass savanna and areas almost free of vegetation. Porembski (1991), Poilecot *et al.* (1991) and Rödel (2000) provide more detailed descriptions of the Comoé National Park. The study area was situated within the Guinea-savanna in the southern part of the park (3° 49′ W, 8° 44′ N, *c.* 220 m).

The three tree species studied, *Anogeissus leiocarpa* Guillemin & Perrottet (Combretaceae), *Burkea africana* Hook (Leguminosae, Caesalpinieae *sensu* Mabberley, 1997) and *Crossopteryx febrifuga* Bentham (Rubiaceae), are widespread in West African savannas (Aubréville, 1950) and locally abundant in the study area (Hovestadt, 1997): we designate them by their generic names hereafter. In this contribution, data on the arthropod assemblages on six *Anogeissus*, six *Burkea* and seven *Crossopteryx* individuals are presented. The size of the trees was characterized by several parameters such as tree height (TH), crown width (CW; maximum crown diameter), crown depth (CD; maximum crown diameter perpendicular to CW), crown projection area (CPA; CPA = CW × (CD/4) × π) and tree volume (TV; roughly estimated as: TV = (CPA × TH)). In order to study the structure of the arthropod

Table 19.1. *Data on the 19 savanna trees studied including six* Anogeissus leiocarpa *(An.), six* Burkea africana *(Bu.) and seven* Crossopteryx febrifuga *(Cr.) individuals*

Tree code	TH (m)	CPA (m^2)	TV (m^3)	TVc (m^3)		Tray area (m^2)	
				1999	2000	1999	2000
*An.*1	10	67	670	220	310	22	31
*An.*2	15	113	1695	660	720	44	48
*An.*3[a]	12	10	120	120	–	16	–
*An.*4	15	99	1485	420	480	28	32
*An.*5	15	94	1410	690	690	46	46
*An.*6	12	133	1596	552	552	46	46
*Bu.*1	9.5	28	266	209	–	22	–
*Bu.*2	7	49	343	182	–	26	–
*Bu.*3	9	33	297	144	–	16	–
*Bu.*4	7	41	287	140	–	20	–
*Bu.*5	6.5	46	299	182	–	28	–
*Bu.*6	9	33	297	180	–	20	–
*Cr.*1	8	16	128	112	–	14	–
*Cr.*2	7	24	168	126	–	18	–
*Cr.*3	5	12	60	50	–	10	–
*Cr.*4[a]	9	18	162	180	–	20	–
*Cr.*5	7.5	16	120	90	–	12	–
*Cr.*6	8	22	176	128	–	16	–
*Cr.*7	8.5	16	136	136	–	16	–

TH, tree height; CPA, crown projection area; TV, tree volume; TVc, tree volume covered by collecting trays.

[a]Crown projection area is used instead of tray area for calculation of TVc.

assemblages, evaluate the sampling efficiency and observe changes in assemblage structure over time, the *Anogeissus* trees were fogged twice in 1999 (in June/July and in August) and once in 2000 (in June/July). One *Anogeissus* tree was excluded from sampling in 2000 since a colony of weaverbirds had established on this tree in the meantime. The *Burkea* and *Crossopteryx* trees were fogged once in 1999 (June/July). Trees selected for fogging seemed healthy, carried no epiphytes and were free of woody plant undergrowth. None of the trees bore ripe fruits or open flowers at the time of sampling. Details of the 19 trees are given in Table 19.1. To ensure that the arthropods collected were most likely associated with the target tree, only trees completely isolated from other canopies by several metres distance were used as study units. All studied trees were syntopically growing within a section of tree savanna 3 km^2 in area.

Sampling protocol

Canopy fogging of trees 8–15 m in height was carried out from the ground for about 5 minutes with natural pyrethrum (1% active ingredient), using a Swingfog™ SN-50. Natural pyrethrum is an insecticide that is very efficient for arthropods and that biodegrades within hours. In order to make the fog visible, a nontoxic highly refined white oil (Essobayol™ 82) was used as a carrier for the insecticide (Floren & Linsenmair, 1997b). Thus, fogging of the complete crown could be ensured. Sampling was performed 2 hours after sunset, the only time windless conditions and dry leaves could be expected on a regular basis. Falling arthropods were collected on 2 m^2 funnel-shaped trays made of smooth balloon-silk, fitted underneath with a collecting pot containing a detergent solution. The trays were arranged approximately 80 cm above the ground beneath each tree (the grass layer under the trees had been cut previously to a height of 10 cm). To ascertain that the crown projection area was representatively covered by trays, the number of used trays varied in relation to tree size (Table 19.1). A strong correlation between tray area and the physical parameters of tree size indicated that coupling tray area and tree size was effective (correlation between tray area and both crown projection area and tree volume: $r_s = 0.9$, $n = 19$, $p < 0.0001$). In order to incorporate tree size into the analyses of assemblage composition, number of individuals and species of arthropods sampled were related to number of trays used per tree (that is, 'samples per unit tray area') and tree volume covered by collecting trays (that is, 'sampled crown volume'). The latter was roughly estimated as: (tray area × TH). All arthropods falling during a drop time of 2 hours were sampled in the collecting pots, quickly washed with ethanol and transferred to storage bottles containing 70% ethanol. The beetle and ant assemblages were analysed for all three tree species. They were first sorted to families (beetles) or genera (ants) using the keys provided by Freude *et al.* (1965–1983), Delvare & Aberlenc (1989), Scholtz & Holm (1989) and Bolton (1994). They were assigned then to morphospecies (RTUs: recognizable taxonomic units), or determined to known species by specialists (see Acknowledgements). As *Crematogaster* spp., to date, have not been cross-checked for the samples of 1999 and 2000, only *Camponotus* spp. have been used for the interyear analysis. Additionally, all specimens collected on *Anogeissus* were sorted to higher taxonomic units, and the individuals were counted. Voucher specimens were deposited at the entomological collection of the Department of Animal Ecology and Tropical Biology, Würzburg, and will be transferred to public museums after completion of analyses. For description of assemblage composition, two general abundance definitions are used. These are (i) 'dominant', referring to beetle families contributing at least 5% to the total numbers on a single tree; and (ii) 'abundant', defined as species represented by at least 12 individuals found in 1 year on the same tree.

Data analysis

For the characterization of the arthropod assemblages, several diversity indices were calculated. Following recommendations given in Magurran (1988) and Southwood & Henderson (2000), α-diversity was described by a suite of measures such as species richness (S), the log-series index (α) (Fisher *et al.*, 1943), the Berger–Parker dominance index (d) (Berger & Parker, 1970) and rarefaction (Hurlbert, 1971). Since the Shannon–Wiener index (H', $\log_{10}$) is widely used, it was calculated to increase comparability with other studies, despite the well-known difficulties in its interpretion (May, 1975). Differences between several diversity measures of the three host-tree species were tested using one-way analyses of variance (ANOVA). Where necessary, data were transformed to meet the assumptions of homogeneity of variances and normal distribution (Sokal & Rohlf, 1995). In instances where an ANOVA was not applicable, the Kruskal–Wallis test was employed (Zar, 1999).

Beta-diversity was expressed by the Jaccard index (C_j) (Magurran, 1988). A comparison with the qualitative Sørensen index favoured by some other authors (e.g. Smith, 1986) showed no differences in the present study; therefore, *Cj* was used as it is the simplest, easiest to interpret and yet useful similarity coefficient. Faunal similarity was also investigated using correlation analyses (Spearman rank correlation, Sachs, 1997) and principal component analysis (Ludwig & Reynolds, 1988; Legendre & Legendre, 1998). These analyses were based on number of individuals within different taxonomic categories, such as orders, families and species (Basset *et al.*, 1996b). To detect relevant differences between groups of correlation coefficients, permutations of correlation matrices were conducted and all possible combinations were evaluated.

The distribution of individual beetle and ant species on the study trees was described by the index of dispersion (I_D) (Southwood & Henderson, 2000), with $I_D = s^2(\nu - 1)/m$, where m and s^2 are the sample mean and variance, respectively, and ν is the sample size. This index has been widely used (e.g. Morris *et al.*, 1992) and provides satisfactory results for small values of ν. I_D values significantly greater than the chi-square statistic for the 0.025 probability level with $(\nu - 1)$ degrees of freedom indicate an aggregated distribution of the taxa considered (Ludwig & Reynolds, 1988). The minimum number of individuals that are required to distinguish between a random and a nonrandom distribution varies with ν. For this reason, I_D values were not calculated when less than three individuals were encountered when evaluating a particular tree species ($\nu = 6$ for *Anogeissus* and *Burkea* and $\nu = 7$ for *Crossopteryx*) and when less than two individuals were encountered for a

Table 19.2. *Diversity measures for beetles and ants sampled from the 19 trees fogged in 1999*

Code for sampled trees	Total		Per tray area (m^2)		Per tree volume (m^3)		Fisher's α	Berger-Parker index	Rarefaction[a]	Shannon–Wiener index (log_{10} base)
	n	*s*	*n*	*s*	*n*	*s*				
Coleoptera (beetles)										
An.1	322	46	14.6	2.09	1.46	0.21	14.69	0.36	12.56	1.66
An.2	3221	145	73.2	3.30	4.88	0.22	31.23	0.53	10.18	2.16
An.3[b]	100	29	10.0	2.90	0.83	0.24	13.70	0.42	13.02	1.46
An.4	2445	83	87.3	2.96	5.82	0.20	16.62	0.77	6.40	1.92
An.5	1783	93	38.8	2.02	2.58	0.13	20.86	0.24	14.30	1.97
An.6	5228	142	113.7	3.09	9.47	0.26	26.95	0.26	12.44	2.15
Bu.1	69	27	3.1	1.23	0.33	0.13	16.32	0.17	15.44	1.43
Bu.2	104	27	4.0	1.04	0.57	0.15	11.84	0.18	13.89	1.43
Bu.3	222	32	13.9	2.0	1.54	0.22	10.26	0.66	7.87	1.51
Bu.4	360	39	18.0	1.95	2.57	0.28	11.12	0.55	9.45	1.59
Bu.5	250	37	8.9	1.32	1.37	0.20	12.00	0.18	12.39	1.57
Bu.6	215	33	10.8	1.65	1.19	0.18	10.88	0.46	11.36	1.52
Cr.1	109	28	7.8	2.0	0.97	0.25	12.19	0.19	13.35	1.45
Cr.2	133	44	7.4	2.44	1.06	0.35	22.97	0.18	16.43	1.64
Cr.3	27	12	2.7	1.20	0.54	0.24	8.28	0.19	12.00	1.08
Cr.4[b]	73	26	4.1	1.44	0.45	0.16	14.43	0.21	14.40	1.42
Cr.5	33	17	2.8	1.42	0.37	0.19	14.09	0.18	14.94	1.23
Cr.6	77	39	4.8	2.44	0.60	0.30	31.57	0.18	18.51	1.59
Cr.7	329	45	20.6	2.81	2.42	0.33	14.10	0.52	10.22	1.65
Formicidae (ants)										
An.1	302	11	13.7	0.50	1.37	0.05	2.33	0.49	6.49	1.11
An.2	3358	15	76.3	0.34	5.09	0.02	2.11	0.50	5.19	1.23
An.3[b]	1330	6	133.0	0.60	11.08	0.05	1.02	0.50	1.81	0.90
An.4	3334	12	119.1	0.43	7.94	0.03	1.70	0.50	3.57	1.15
An.5	266	9	5.8	0.20	0.39	0.01	1.96	0.49	6.97	1.04
An.6	1196	20	26.0	0.43	2.17	0.04	3.35	0.50	6.19	1.34
Bu.1	93	9	4.2	0.41	0.44	0.04	2.52	0.48	7.81	1.04
Bu.2	229	6	8.8	0.23	1.26	0.03	1.38	0.49	3.96	0.90
Bu.3	1403	8	87.7	0.50	9.74	0.06	1.31	0.50	2.86	1.00
Bu.4	218	6	10.9	0.30	1.56	0.04	1.39	0.49	4.24	0.90
Bu.5	324	8	11.6	0.29	1.78	0.04	1.68	0.49	3.61	1.00
Bu.6	177	11	8.9	0.55	0.98	0.06	2.63	0.49	7.05	1.11
Cr.1	401	7	28.6	0.50	3.58	0.06	1.42	0.50	5.17	0.95
Cr.2	415	9	23.1	0.50	3.29	0.07	1.79	0.50	6.59	1.04
Cr.3	54	7	5.4	0.70	1.08	0.14	2.29	0.47	7.00	0.95
Cr.4[b]	355	10	19.7	0.56	2.19	0.06	2.05	0.49	6.23	1.08
Cr.5	133	9	11.1	0.75	1.48	0.10	2.30	0.48	6.70	1.04
Cr.6	1430	7	89.4	0.44	11.17	0.05	1.16	0.50	1.81	0.95
Cr.7	555	14	34.7	0.88	4.08	0.10	2.65	0.49	4.47	1.20

An., *Anogeissus*; *Bu*., *Burkea*; *Cr*., *Crossopteryx*; *n*, number of individuals; *s*, number of morphospecies.

[a] Sample size for rarefaction (individuals) 27 for Coleoptera and 54 for Formicidae.

[b] Crown projection area is used instead of tray area for calculation of values considering area and volume.

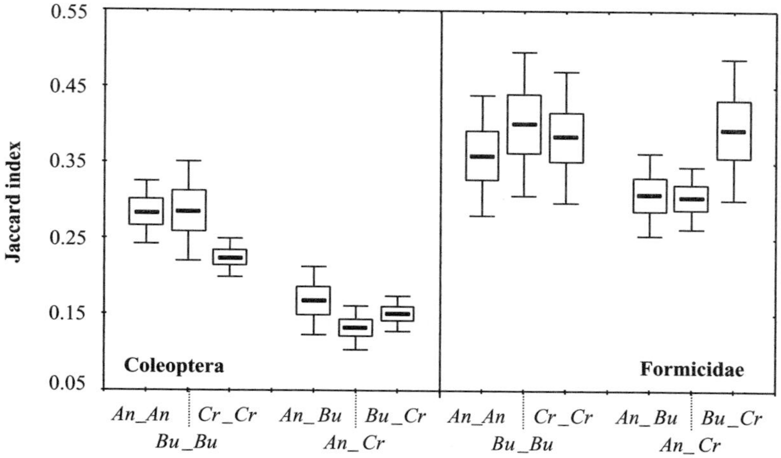

Fig. 19.1. Jaccard indices (mean, standard error and standard deviation) obtained for comparisons of beetle and ant assemblages of *Anogeissus* (*An*), *Burkea* (*Bu*) and *Crossopteryx* (*Cr*) trees.

consideration of all trees ($\nu = 19$). In addition to I_D, Green's Index (GI) was computed as $(s^2/m - 1)/(n - 1)$, where n is the number of individuals in the sample (Ludwig & Reynolds, 1988). The value of GI varies between 0 (for random) and 1 (for maximum clumping). It is considered to be independent of n and can be used to compare samples that vary in the total number of individuals, their sample means and the number of sampling units.

RESULTS

Beetle and ant assemblages on *Anogeissus*, *Burkea* and *Crossopteryx*

In total, 15 100 beetles and 15 573 ants were obtained from the sampling carried out in 1999 from trees fogged for the first time. The specimen number per tree varied clearly between conspecific and between heterospecific tree individuals (Table 19.2). In general, most beetles and ants were found on *Anogeissus*. Out of 322 beetle species sampled, *Anogeissus* hosted 78%, *Burkea* 29% and *Crossopteryx* 34% of species (one-way ANOVA, log-transformed, $F_{2,16} = 8.50$, $p < 0.01$). Relative beetle abundance was even higher on *Anogeissus*. Of all beetle individuals sampled, 87% came from *Anogeissus*, 8% from *Burkea* and 5% from *Crossopteryx* (one-way ANOVA, log-transformed, $F_{2,16} = 10.51$, $p < 0.01$). Differences between tree species for ants were not as distinct as for beetles. Out of 36 ant species sampled in total, 78% were collected on *Anogeissus*, 47% on *Burkea* and 58% on *Crossopteryx* (one-way ANOVA, $F_{2,16} = 2.62$, $p > 0.05$). Ant abundance was also not significantly different between tree species: 63% of all ants were obtained from *Anogeissus*, 16% from *Burkea* and 21% from *Crossopteryx* (one-way ANOVA, log-transformed, $F_{2,16} = 3.23$, $p > 0.05$). Species similarity of assemblages was quite low for beetles and much higher for ants (Fig. 19.1). Mean species similarity of beetle assemblages on conspecific trees ranged from 22 to 28%, and mean similarity on heterospecific trees from 13 to 17%. For ants, mean similarity was 36–40% for conspecific and 30–39% for heterospecific host-tree species.

When incorporating tree size into the analyses by use of 'samples per unit tray area' or 'sampled crown volume', significant interspecific differences were detected for beetle abundance, beetle species richness on a per tray area basis, and ant species richness. No significant interspecific differences were detected for beetle species richness on a crown volume basis and for ant abundance (Table 19.2). Beetle densities were highest on *Anogeissus* (one-way ANOVA, log-transformed, $F_{2,16} = 10.45$, $p < 0.01$ for beetles per tray area, and $F_{2,16} = 5.86$, $p < 0.05$ for beetles per crown volume). Beetle species richness was also highest on *Anogeissus* though only for tray area (one-way ANOVA, $F_{2,16} = 7.73$, $p < 0.01$ for beetles per tray area, and $F_{2,16} = 2.35$, $p > 0.05$ for beetles per crown volume). For other measures of beetle α-diversity the pattern was less clear. For indices influenced by evenness in species abundance (d and rarefaction), beetle α-diversity was highest on *Crossopteryx* but the differences were not significant (one-way ANOVA, $F_{2,16} = 1.94$, $p > 0.1$ for d, and

Table 19.3. *Abundance of beetle families, listed in decreasing order. Absolute number of beetles per family and their contribution to beetle communities of the single tree species are given*

Family	Total no. (%)	Median contribution to beetle communities (minimum–maximum)		
		Anogeissus sp.	*Burkea* sp.	*Crossopteryx* sp.
Anthicidae	4703 (31.1)	41.1 (0.3–79.0)	33.5 (0.0–66.7)	1.4 (0.0–57.8)
Chrysomelidae	4634 (30.7)	14.6 (8.3–64.1)	16.8 (4.1–30.4)	7.4 (2.4–36.7)
Nitidulidae	1020 (6.8)	4.8 (0.0–12.4)	0.0 (0.0–0.3)	0.0 (0.0–1.3)
Curculionidae	857 (5.7)	7.9 (0.9–16.5)	5.6 (1.6–24.0)	20.5 (6.1–29.6)
Corylophidae	625 (4.1)	3.4 (0.3–10.4)	3.7 (1.0–20.8)	6.8 (0.0–22.1)
Elateridae	577 (3.8)	4.1 (1.0–6.3)	1.0 (0.0–2.0)	1.8 (0.0–3.7)
Latridiidae	569 (3.8)	0.2 (0.0–24.0)	0.0 (0.0–4.3)	1.8 (0.0–20.5)
Coccinellidae	490 (3.2)	2.8 (1.6–5.5)	8.8 (0.0–11.7)	3.9 (1.5–9.1)
Scarabaeidae	203 (1.3)	1.3 (0.3–4.0)	2.7 (0.5–8.7)	1.5 (0.0–11.9)
Tenebrionidae	202 (1.3)	0.7 (0.4–6.2)	2.9 (0.0–4.2)	0.3 (0.0–6.8)
Staphylinidae	165 (1.1)	0.7 (0.2–4.0)	3.9 (1.1–6.7)	3.8 (2.4–11.7)
Alleculidae	102 (0.7)	0.5 (0.0–0.8)	0.6 (0.0–2.3)	1.8 (0.0–7.8)
Cybocephalidae	95 (0.6)	0.6 (0.1–3.1)	0.0 (0.0–1.4)	5.2 (0.0–18.5)
Cleridae	82 (0.5)	0.5 (0.1–1.0)	2.1 (0.6–7.2)	0.0 (0.0–1.4)
Lophocateridae	59 (0.4)	0.0 (0.0–0.2)	0.1 (0.0–1.4)	5.2 (0.0–18.0)
Silvanidae	48 (0.3)	0.1 (0.0–0.6)	0.0 (0.0–0.0)	1.5 (0.0–6.1)
Colydiidae	12 (0.1)	0.0 (0.0–0.1)	0.0 (0.0–0.0)	0.6 (0.0–9.1)

$F_{2,16} = 2.01$, $p > 0.1$ for rarefaction). Indices more strongly influenced by species richness, such as the α and H' indices, indicated highest beetle diversity for *Anogeissus*, which was only significant for H′ (one-way ANOVA, $F_{2,16} = 8.50$, $p < 0.01$ for H', and $F_{2,16} = 2.76$, $p > 0.05$ for α). *Anogeissus* hosted the highest ant densities per tray area or crown volume but there was no significant difference detectable (one-way ANOVA, log-transformed, $F_{2,16}= 1.58$, $p> 0.1$ for tray area and, not transformed, $F_{2,16}= 0.46$, $p> 0.5$ for crown volume). Ant species richness was highest for *Crossopteryx* for both tray area and crown volume data (one-way ANOVA, $F_{2,16} = 5.21$, $p < 0.05$ for tray area, and, log-transformed, $F_{2,16} = 11.01$, $p < 0.01$ for crown volume). Ant α-diversities were not different between the tree species (one-way ANOVA, $F_{2,16} = 1.11$, $p > 0.3$ for d, $F_{2,16} = 0.12$, $p > 0.8$ for rarefaction, $F_{2,16} = 2.38$, $p > 0.1$ for H', and $F_{2,16} = 0.25$, $p > 0.7$ for α).

All three tree species showed a high variability in most assemblage parameters between assemblages on conspecific trees (with the exception of d for ants; Table 19.2). Most pronounced differences could be found for beetle densities, with variation by orders of magnitude of 5.7 ($s^2 = 32.7$) on *Burkea* and 11.4 ($s^2 = 1745$) on *Anogeissus* for beetle density per tray area and 6.6 ($s^2 = 0.5$) on *Crossopteryx* and 11.4 ($s^2 = 10.5$) on *Anogeissus* for beetle density per crown volume. Data not corrected for tree size showed the highest variation for beetle density on *Anogeissus* (factor = 52, $s^2 = 3\,676\,064$) and the smallest variation for beetle density on *Burkea* (factor = 5.2, $s^2 = 11\,023$).

This marked variation was observed for assemblage parameters as well as for the dominance of certain beetle families on single trees, and for the distribution of particular beetle and ant species. The contribution of particular beetle families to the entire beetle assemblage was highly variable (Table 19.3, Fig. 19.2). However, this was not because of the dominance of single species of the families considered in this evaluation. Specimen number and species number were always highly correlated with the exception of Corylophidae (all beetle families indicated in Table 19.3: $r_s \geq 0.71$, $n = 19$, $p < 0.001$; for Corylophidae: $r_s = 0.4$, $n = 19$, $p = 0.07$). Although some beetle groups were very abundant and some differences between the tree species existed, it was not possible to rank the most abundant

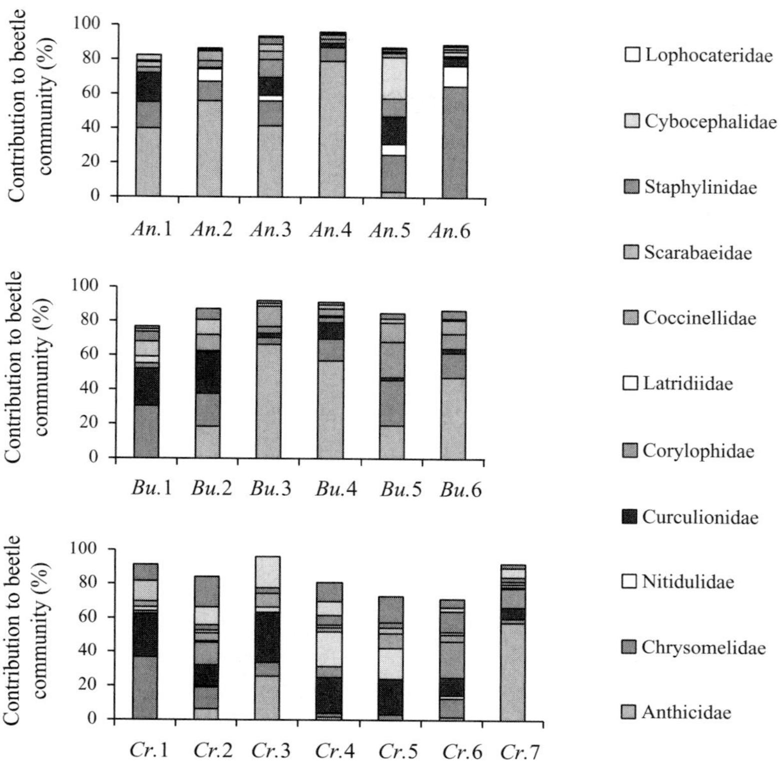

Fig. 19.2. Contribution of single beetle families to the entire beetle community of individual *Anogeissus* (*An.*), *Burkea* (*Bu.*), and *Crossopteryx* (*Cr.*) trees. Beetle families were included when they contributed at least 10% to the beetle assemblage of an individual tree.

beetle families universally. *Anogeissus* assemblages were regularly dominated by Anthicidae. However, this beetle family was almost absent on two trees, which were dominated by Latridiidae and Chrysomelidae, respectively (Fig. 19.2). In addition to Chrysomelidae, which were always an important group (contributing between 8–64% of specimens), beetles of five other families contributed at least 10% to the whole beetle assemblage of individual trees (Fig. 19.2). A similar pattern was observed for *Burkea* and *Crossopteryx* assemblages. *Burkea* assemblages were dominated by beetles of seven families, of which only Chrysomelidae showed a relatively stable contribution across all tree individuals. Beetle assemblages on *Crossopteryx* were dominated by beetles of 14 families. Out of these beetles, only Curculionidae were found regularly at higher densities. Anthicidae were again very abundant on some trees while completely lacking on others (Fig. 19.2).

Many beetle species showed an aggregated distribution (Fig. 19.3), that is, they were restricted to particular tree individuals while being absent from others, or they occurred in very different densities on individual trees. Taking into account all beetle species that could be used for calculation of I_D values, 74.1% of beetles were aggregated on individual *Anogeissus* trees, 35.1% were aggregated on individual *Burkea* trees and 60.5% were aggregated on individual *Crossopteryx* trees. When I_D values were calculated over all 19 trees, 72.3% of beetles were aggregated (Fig. 19.3**d**). Almost all ant species showed an aggregated distribution. When all ant species were considered, the proportion of aggregated ant species was lowest for assemblages on *Burkea* trees (66.6% for *Burkea*, 95% for *Anogeissus* and 100% for *Crossopteryx* versus 93.8% for the all-tree comparison; Fig. 19.3) as was the case for the beetles (see above). Computation of Green's index confirmed these findings. For both beetles and ants, Green's index varied significantly between tree species, with the lowest degree of clumping on *Burkea* trees: median Green's index for beetles: all trees, 0.30; *Anogeissus*, 0.35; *Burkea*, 0.12; *Crossopteryx*, 0.26 (Kruskal–Wallis test: $H = 30.56$, $p < 0.0001$) and for ants: all trees 0.52; *Anogeissus*, 0.53; *Burkea*, 0.31; *Crossopteryx*, 0.68 (one-way ANOVA, $F_{3,78} = 3.59$, $p < 0.05$).

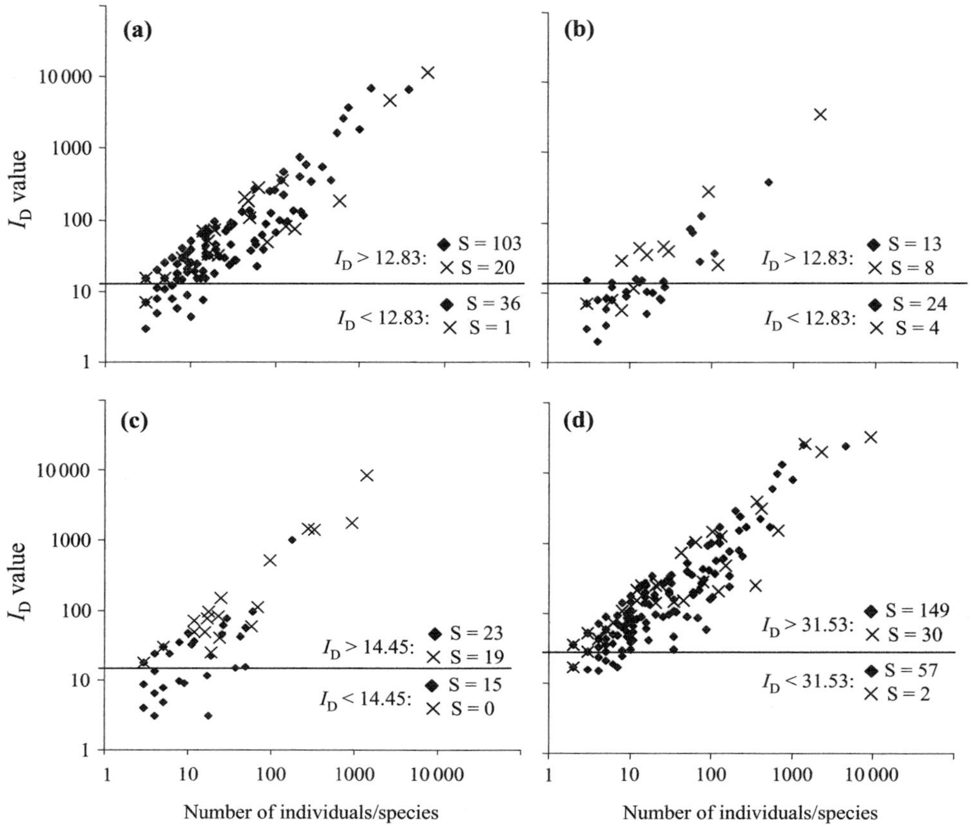

Fig. 19.3. Tendency of beetle (◆) and ant (X) species to occur aggregated on individual trees, indicated by I_D values greater than the chi-square statistic with sample size (ν) − 1 degrees of freedom. Distribution of I_D values is given for *Anogeissus* (**a**; $\nu = 6$), *Burkea* (**b**; $\nu = 6$), *Crossopteryx* (**c**; $\nu = 7$), and for all trees combined (**d**; $\nu = 19$). I_D values were not calculated when fewer than three individuals were encountered for **a–c** and when fewer than two individuals were encountered for **d**. S is the number of morphospecies showing an aggregated or a random distribution among trees (aggregated: above horizontal line separating I_D values greater or smaller than the critical chi-square statistic; random: below horizontal line).

Characteristics of resampled arthropod assemblages on *Anogeissus*

Beetle, and to a lesser extent ant, assemblages have turned out to be remarkably different among individual conspecific trees that were fogged in the same period. Resampling the assemblages of the same trees (*Anogeissus*) at 2 months and at 1 year later indicated a high congruence of assemblage composition with the results of the first fogging event. The data from the subsequent fogging in 1999 were quantified at the ordinal level. Comparing the first and the second fog, densities of many arthropod orders were rather similar for individual trees, or quantitative changes occurred while the relative position remained constant (Table 19.4). Densities and numbers per tray area or crown volume of Coleoptera, Formicidae, Hymenoptera (excluding Formicidae), Thysanoptera, and Orthoptera were significantly correlated for the first and the second fog. For Araneae and Lepidoptera larvae, correlation coefficients obtained for densities and numbers per tray area or crown volume were considerably different and only the latter were correlated. No significant correlation between the first and the second fogging event could be detected for Heteroptera, Diptera, Homoptera, Blattodea, Psocoptera and Thysanura. Though not significant, the negative correlation coefficients for Mantodea indicated a marked change of Mantodea densities on individual trees.

Table 19.4. *Correlation coefficients for comparisons of arthropod densities on individual trees between different fogging events at the ordinal level*

Taxa	Correlation coefficient for F1 versus F2 ($n = 6$)		Correlation coefficient for F1 versus F3 ($n = 5$)	
	Density per tray area	Density per crown volume	Density per tray area	Density per crown volume
Coleoptera	0.94**	0.94**	0.9*	0.9*
Formicidae	0.89*	0.94**	0.7	0.5
Araneae	0.77	0.94**	0.3	0.6
Heteroptera	0.54	0.54	0.7	0.7
Diptera	0.31	−0.35	0.1	−0.2
Hymenoptera[a]	0.94**	0.94**	0.9*	0.9*
Homoptera	0.64	0.54	0.87	0.9*
Blattodea	0.6	0.6	0.9*	0.9*
Lepidoptera (larvae)	0.77	0.89*	0.3	0.5
Neuroptera	0.66	0.77	0.2	0.2
Thysanoptera	0.93**	0.99***	0.6	0.9*
Orthoptera	0.94**	0.99***	0.8	0.56
Mantodea	−0.77	−0.77	−0.5	−0.5
Psocoptera	0.31	0.09	0.11	0.11
Thysanura	0.59	0.59	0.82	0.97**

F1, First fogging event June/July 1999; F2, second fogging event August 1999; F3, third fogging event June/July 2000.
[a]Excluding Formicidae.
$^{*}p < 0.05$; $^{**}p < 0.01$; $^{***}p < 0.001$.

Comparisons of samples obtained in 1999 and 2000 indicated that the composition of assemblages on individual trees was stable over time. At both high and low taxonomic levels (arthropod orders, beetle families, beetle species and *Camponotus* spp.) the assemblage structure of trees refogged in 2000 was most strongly correlated with the structure of the same tree in the previous year (Table 19.5). Permutation of correlation matrices revealed that this finding was significant for arthropod orders, beetle families and beetle species. Out of 120 possible combinations, mean correlation coefficients between assemblages of arthropod orders and beetle species on the same trees were higher than all other possible combinations ($p < 0.01$). For beetle families, only one other combination of correlation coefficients was higher than the correlation coefficients obtained for the same trees ($p < 0.01$). In contrast, the distribution of *Camponotus* spp. was less predictable on the basis of the assemblage structure in the previous year: 22 combinations calculated by permutation of the correlation matrix were higher than the correlation coefficients found for the comparison of the same trees ($p = 0.18$).

These results can be evaluated more precisely when considering the examined groups separately. At the ordinal level, a very high degree of constancy in specimen number of the respective orders was observed between years (median abundance 1999 versus 2000: $r_s = 0.9$, $n = 14$, $p < 0.0001$). Arthropod densities on individual trees were significantly correlated for Coleoptera, Hymenoptera (excluding Formicidae), Homoptera, Blattodea, Thysanoptera and Thysanura (Table 19.4). For Thysanoptera and Thysanura, correlation coefficients of densities on a square metre of tray and cubic metre of crown volume were considerably different and only r_s values on the former basis were correlated. Lowest r_s values between the 1999 and the 2000 foggings could be detected for Diptera, Psocoptera, Neuroptera, Lepidoptera larvae and Araneae. This

Table 19.5. *Correlation coefficients for comparisons between relative arthropod abundance of the same and different* Anogeissus *trees in 1999 and 2000 at different taxonomic levels*

Tree code	*An.*1_00	*An.*2_00	*An.*4_00	*An.*5_00	*An.*6_00	Median correlation coefficient (a.v.) other trees
Arthropod orders ($n = 14$)						
*An.*1_99	**0.79*****	0.77	0.81	0.69	0.57	0.73**
*An.*2_99	0.69	**0.72****	0.71	0.56	0.48	0.62*
*An.*4_99	0.85	0.94	**0.94*****	0.84	0.75	0.84***
*An.*5_99	0.55	0.83	0.73	**0.81*****	0.79	0.76**
*An.*6_99	0.74	0.88	0.90	0.82	**0.86*****	0.85***
Beetle families ($n = 9$)						
*An.*1_99	**0.79***	0.70	0.73	−0.13	−0.06	0.42[c]
*An.*2_99	0.38	**0.62**[c]	0.30	−0.14	−0.12	0.22[c]
*An.*4_99	0.76	0.93	**0.83****	−0.08	−0.13	0.45[c]
*An.*5_99	−0.18	−0.02	0.15	**0.80****	0.88	0.17[c]
*An.*6_99	−0.11	0.13	−0.07	0.83	**0.90*****	0.12[c]
Abundant beetle species[b] ($n = 29$)						
*An.*1_99	**0.53****	0.37	0.44	0.09	−0.30	0.34[c]
*An.*2_99	−0.02	**0.29**[c]	0.02	−0.09	−0.13	0.06[c]
*An.*4_99	−0.05	0.53	**0.39***	0.12	−0.10	0.11[c]
*An.*5_99	−0.09	0.36	0.32	**0.77*****	0.61	0.34[c]
*An.*6_99	−0.27	0.04	−0.10	0.36	**0.85*****	0.19[c]
***Camponotus* species ($n = 10$)**						
*An.*1_99	**0.48**[c]	0.67	0.74	0.65	0.75	0.70*
*An.*2_99	0.54	**0.68***	0.80	0.69	0.74	0.71*
*An.*4_99	0.80	0.72	**0.92*****	0.70	0.75	0.74*
*An.*5_99	0.66	0.77	0.76	**0.79****	0.69	0.73*
*An.*6_99	0.57	0.88	0.85	0.60	**0.94*****	0.73*

n, number compared; a.v., absolute value; *An.*1–*An.*6, *Anogeissus* trees 1–6; 99, sampled in 1999; –00, sampled in 2000.
[a]Values for comparison of the same tree are bold.
[b]At least 12 individuals found in 1 year.
[c]Not significant.
$^*p < 0.05$; $^{**}p < 0.01$; $^{***}p < 0.001$.

result corresponds with the findings of the within-year comparisons for Diptera, Psocoptera and Neuroptera. In contrast, densities of Araneae and Lepidoptera larvae were significantly correlated for tree volume-related values in 1999.

The species composition of ant assemblages was relatively similar for conspecific and heterospecific tree individuals (Fig. 19.1). Ant assemblages were dominated by *Crematogaster* and by *Camponotus* spp. in 1999 and in 2000. The composition of the *Camponotus* fauna was less distinct than that of the beetle fauna for individual trees. Although the composition of ant assemblages was significantly correlated over years for four out of five trees, the examination of median correlation coefficients supports the results of the permutation comparisons: similarities between the assemblages of the same trees were not higher than comparisons of different trees (Table 19.5).

The ranking of beetle families according to abundance on single trees was very stable for most of the dominant beetle families. The strongest correlation ($r_s \geq 0.9$, $p < 0.05$) was found for Anthicidae, Coccinellidae, Corylophidae and Tenebrionidae. No correlation existed for Elateridae ($r_s = 0.5$, $p = 0.4$), Scarabaeidae ($r_s = 0.3$, $p = 0.6$) and Carabidae ($r_s = 0.1$, $p = 0.9$), which were, however, not included in the previous analysis as they were never abundant. Comparisons at the level of 'abundant' beetle species revealed an even higher level of relationship between assemblages of the same tree during consecutive years. For four trees, the abundance of these common beetles, belonging to nine different beetle families, was most

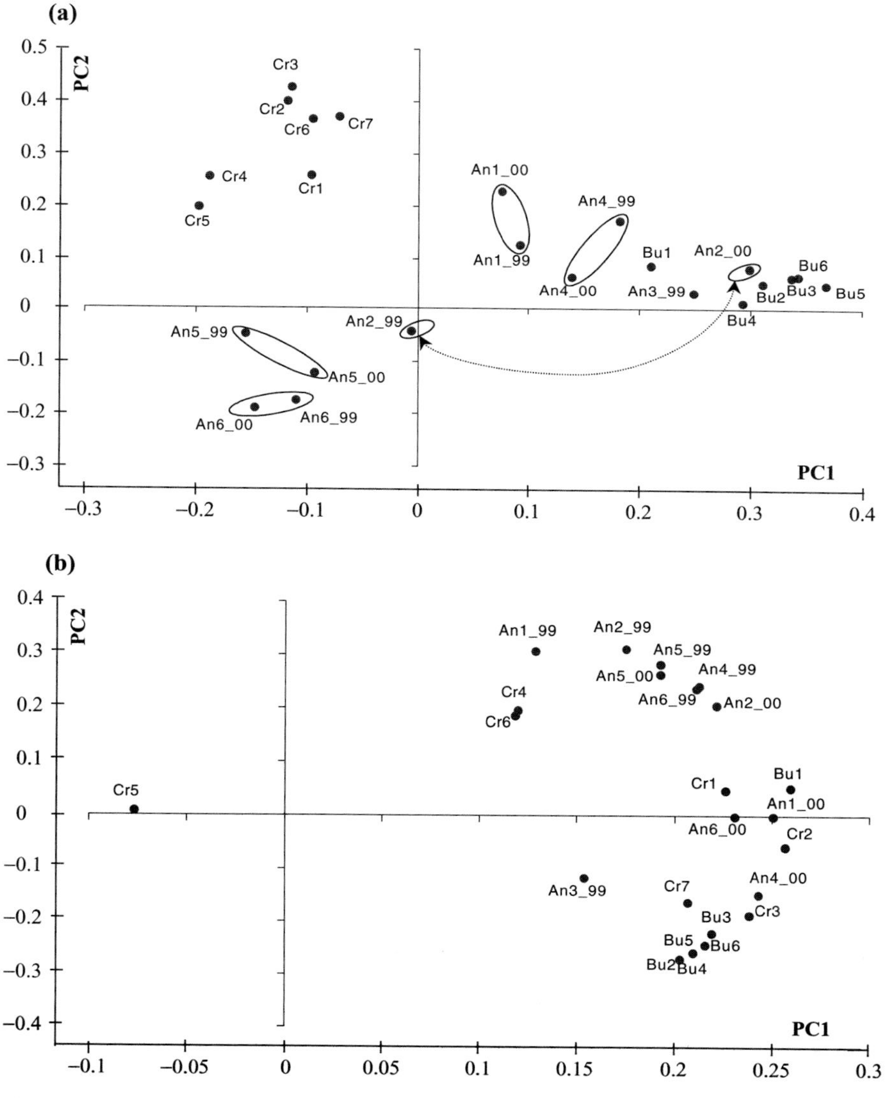

Fig. 19.4. Principal component (PC) analysis of relative abundance of 'abundant' beetles (a) and *Camponotus* species (b) following Table 19.5. In addition to the interyear comparison of *Anogeissus* (An), data of *Burkea* (Bu) and *Crossopteryx* (Cr) are included. The ordination position of individual *Anogeissus* trees in 1999 and 2000 is marked by ellipsoid circles for beetles. For *Camponotus* ants, communities of the same trees were not more similar than communities of different trees; they are, therefore, not marked.

strongly correlated with their abundance in the previous year. This resulted in beetle assemblages on the individual trees showing a high degree of constancy over time. Principal component analysis supported the results of the correlation analyses. Ordination position of beetle assemblages of resampled *Anogeissus* trees was very similar for four tree individuals (Fig. 19.4**a**). In addition to the consistent composition of beetle assemblages on individual trees, ordination results indicated a noticeable influence of tree species on assemblage composition. In particular assemblages on *Crossopteryx* were clearly separated from those on *Anogeissus* and *Burkea*. *Burkea* assemblages were not completely separated from those on *Anogeissus* but intraspecific groupings were, on average, much closer than interspecific groupings (Fig. 19.4**a**). The correlation analyses (Table 19.5) showed that *Camponotus* assemblages were not characteristic for individual trees (Fig. 19.4**b**). However, there appeared to exist an influence of the sampling date as intrayear samples were more closely grouped than interyear samples. In addition, *Camponotus* assemblages seemed to differ least among *Anogeissus* and *Burkea* trees while *Crossopteryx* assemblages were more variable.

DISCUSSION

The objective of this study was to describe and compare arthropod assemblages of savanna trees and to understand aspects of the organization of these assemblages. As several individuals of each of three different tree species were included in the analysis, the influence of traits specific to tree species as well as to tree individuals on assemblages of tree-associated arthropods can be discussed in more detail. Both groups of factors, interspecifically and intraspecifically variable tree characteristics, appeared to be of importance for the arthropod assemblages that we investigated. However, detailed consideration of both is beyond the scope of this contribution. Accordingly, we focus on the arthropod assemblages of individual trees, as understanding at this level of organization can form the basis for further far-reaching comparisons of, for example, the effect of tree species or region. By so doing, we stress the potential importance of small-scale variation as seen in the characteristics of individual trees. This is often indirectly reported (via the description of arthropod distribution, see below) but nonetheless is rarely accounted for directly (but see, for example, Crawley & Akhteruzzaman, 1988; Winchester, 1997a). A more detailed analysis of interspecific differences will be presented elsewhere, taking into account the influence of guild affiliation and single-species distributions (Stork, 1987a; Kitching & Zalucki, 1996; Basset & Novotny, 1999; Wagner, 2000). In order to understand the influence of individual tree characteristics, the results have been analysed with respect to the hypothesis that conspecific trees provide similar habitats and will, therefore, be inhabited by similar arthropod assemblages (Southwood & Kennedy, 1983). The relative importance of stochastic colonization events and/or deterministic tree characteristics in the determination of assemblage structure in individual trees is also discussed (Wiens, 1984; Linsenmair, 1990; Adis *et al.*, 1998a).

For all three tree species investigated, we have demonstrated that the composition of arthropod assemblages can differ substantially among syntopically growing individuals fogged within the same period of time. Such variation in arthropod density or diversity on individual conspecific trees seems to be a rather typical finding as it is in agreement with most studies that present arthropod assemblage data for individual trees (for example, Erwin & Scott, 1980; Stork, 1991; Allison *et al.*, 1993; Floren & Linsenmair, 1997b, 1998a; Krüger & McGavin, 1997; Mawdsley & Stork, 1997; Winchester, 1997a; Azarbayjani *et al.*, 1999; Basset & Novotny, 1999; Majer *et al.*, 2000; Wagner, 2000). These intraspecific differences engender variation in local arthropod density and diversity in canopy assemblages of temperate, subtropical and tropical trees. Only diversity measures obtained for beetle and ant assemblages of tropical rainforests were in general higher than the highest values of individual *Anogeissus* or *Crossopteryx* trees. This points, as do other studies, to interesting and widely discussed differences of arthropod assemblages of (pristine) tropical lowland rainforests on the one hand and tropical savannas, dry forests, montane forests and disturbed rainforests on the other. Combining the result of variable assemblage composition with data obtained in other comparable studies, the hypothesis that conspecific trees can be regarded as similar habitats should be rejected in its simple form. It may apply when the comparison of distantly related tree species is the objective. But even in this case, individual variation seems worth considering not just as a sampling artefact but also as a fundamental intraspecific property of tree species. Rejecting the hypothesis that assemblages on conspecific trees are *per se* similar, potential reasons for individual variation need to be evaluated. As this and other studies demonstrate, variation is not restricted to a few particular arthropod taxa or functional groups, although herbivores appear to be generally more influenced than predacious groups (Stork, 1987b). Differences were exhibited, for example, by beetles in general (Stork, 1991; Allison *et al.*, 1993; Floren & Linsenmair, 1997b), herbivorous beetles (Floren & Linsenmair, 1998a), herbivorous insects in general (Stork, 1987b; Basset, 1991e; Basset & Novotny, 1999), granivorous beetles (Miller, 1996), spiders (Stork, 1991; Floren & Linsenmair, 1997b; Azarbayjani *et al.*, 1999), dipterans (Floren & Linsenmair, 1997b; Azarbayjani *et al.*, 1999) and ants (Stork, 1991; Adis *et al.*, 1998a). However, no differences were found for fungal-feeders, predators, tourists, wood eaters, mesophyll feeders, Corylophidae and Lepidoptera by Basset (1992c). Supposing that the reported variability is not an artefact of small, unrepresentative samples, it suggests that different, intraspecific variable tree characteristics or stochastic colonization pathways (Wiens, 1984; Linsenmair, 1990; Adis *et al.*, 1998a) are influencing assemblage structure.

The distribution of insect herbivores has been related to several features of their host plants. For example, leaf nutrients, defence parameters and the proportion of young leaves have been shown to play an important role in determining food choice, abundance and distribution of herbivores (Strong *et al.*, 1984; Basset, 1991e). As these characteristics can vary among conspecific plants (Crawley & Akhteruzzaman, 1988; Macedo & Langenheim, 1989; Edwards *et al.*, 1993; Barnola *et al.*, 1994; Butcher *et al.*, 1994; Hemming & Lindroth, 1995; Mapongmetsem *et al.*, 1998; Staudt *et al.*, 2001; Stork *et al.*, 2001), the high potential of individual tree characteristics to influence the distribution of herbivorous insects is clear. Aside from plant traits, arthropod behaviour probably affects arthropod distribution; for example, when movement is more likely towards plants occupied by conspecifics (Turchin, 1987; Bach & Carr, 1990; Morris *et al.*, 1992; Loughrin *et al.*, 1995). The same can apply when considering fungus-feeders or saprophages, as there is much potential variation in fungus load and dead wood content across individual trees (Bull *et al.*, 1992; Clinton *et al.*, 1993; Lewis, 1997). Predatory arthropods may be tracking the (variable) abundance of their prey (Gagné & Martin, 1968). They may also be influenced by the availability of enemy-free space or suitable sites for web construction (Halaj *et al.*, 2000), both parameters being potentially variable among conspecific plants.

Comparisons of assemblages resampled from the same tree individuals were analysed to distinguish between individual tree characteristics and random colonization. The results of this study suggest that the distributions of Coleoptera, Hymenoptera (excluding Formicidae) and Thysanoptera were fundamentally influenced by tree characteristics, as the highest congruence was found consistently among assemblages from the same tree individual. Other groups such as (i) Formicidae, Araneae and Orthoptera and (ii) Homoptera, Blattodea and Thysanura were affected, in addition, by temporal effects. Members of the first group appeared to be particularly influenced by interyear parameters, probably linked to the recolonization of trees following the dry season. The distribution of members of the second group, however, seemed to be rather dependent on the season, showing no correlation within the same year but significant correlation between different years. For Heteroptera, Diptera, Neuroptera, Mantodea and Psocoptera, no relationship to individual trees was detectable, indicating that the distribution of these groups might be less tightly linked to individually variable tree characteristics than was the case for the other groups.

The interpretation that the distribution of many groups of arthropods is influenced by individual tree characteristics is in accordance with most studies that present resampling data, although these investigations mainly stress the random character of assemblage reorganization (Floren & Linsenmair, 1997b, 1998a; Adis *et al.*, 1998a; Azarbayjani *et al.*, 1999). The apparent contradictions in the interpretation of results of resampling studies may point to general differences in factors structuring assemblages between savannas and pristine rainforests (as is discussed for secondary and primary forests by Floren & Linsenmair, 2001). They may also be caused by methodological difficulties in assigning arthropod assemblages to individual trees in forests (see Erwin (1989) for 'horizontal highways'). However, these contradictions may be solved by considering different arthropod groups separately and by avoiding generalizations. Azarbayjani *et al.* (1999) studied small trees that could be sampled representatively. These authors found a high turnover between original and recolonizing assemblages of the same tree individuals. Therefore, they concluded that the suites of species on these trees were not a consequence of individually variable tree characteristics. Their interpretation appears meaningful in the framework of their data and it is, so far, in accordance with our data. Like Azarbayjani *et al.* (1999), we found no correlation–and, therefore, no detectable influence of individual tree characteristics – for the distribution of dipterans between different sampling events. The same is true for Araneae, the second taxon used in the analysis of Azarbayjani *et al.* (1999). Both studies indicate that Diptera and Araneae are generally weakly related to individual trees, being rather 'tourists' (Moran & Southwood, 1982; West, 1986; Stork, 1987b; Didham, 1997) or beneficiaries of structural or microclimatic properties provided by trees. However, having no close relationship to trees, these groups are not well suited for the derivation of general statements on the organization of arboreal arthropod assemblages. They should rather be considered examples of loosely connected assemblage components and should not be mixed up with other groups that are more closely linked to the tree. Floren & Linsenmair (1997b, 1998a) studied understorey trees of a pristine rainforest. They found a high congruence of re-established and original assemblages at the ordinal level but considerable differences at the level of ant and beetle species; they concluded that assemblages were mainly structured by stochastic factors and not by tree characteristics. Their findings for ants are in agreement with data of another refogging study (Adis *et al.*, 1998a) and also with our study (ant abundances were not correlated with individual trees between different years, and *Camponotus* assemblages showed the lowest congruence in permutation tests). Taken together, these investigations suggest that the composition of ant assemblages of trees may change rapidly in response to disturbances and stochastic recolonization processes, no matter whether the majority of ants nest in trees (in rainforests) or soil (in savannas). The general statement that beetle assemblages are randomly organized, however, seems not to hold true for savanna trees, where many families and species of beetles appeared to be closely linked to individual trees. The proof of individually characteristic assemblages requires a certain minimum number of individuals per species. Since densities of single arthropod species are generally very low on rainforest trees (Erwin & Scott, 1980; Morse *et al.*, 1988; Allison *et al.*, 1997; Floren & Linsenmair, 1997b; Wagner, 1997), it may be unreasonable to expect to detect characteristic assemblages with any reasonable sampling effort.

We conclude that individual tree characteristics are probably of greater importance in determining the structure and diversity of tree-associated arthropod assemblages than is reflected in current discussions on this topic. Whether or not the influence applies to all ecosystems and can also be demonstrated for forest trees within a closed canopy will need to be evaluated in further studies.

ACKNOWLEDGEMENTS

We thank Frank-Thorsten Krell, Thomas Wagner and Claus Wurst for their help in beetle identification and Michael Mazat for his help in ant sorting. Thomas Hovestadt and Martin Hinsch provided theoretical background and programming experience for performing permutation tests. Permission to perform the research was granted by the Ministère de l'Enseignement Supérieur et de la Recherche Scientifique, Republic of Côte d'Ivoire. An access permit to Comoé National Park was issued by the Ministère de la Construction et de l'Environnement. The manuscript benefited from comments by Yves Basset and two anonymous referees. The study was partly supported by scholarships of the DAAD (German Academic Exchange Service, No. 332 4 04 101), and by the DFG Graduiertenkolleg 200 to K. M., and by BIOTA (Biodiversity Monitoring Transect Analysis in Africa), German Federal Ministry of Education and Research (BMBF), subproject W06, 01LC0017.

20

Flower ecology in the neotropics: a flower–ant love–hate relationship

Klaus Jaffé, José Vicente Hernandez, William Goitía, Anaís Osio, Frances Osborn, Hugo Cerda, Alberto Arab, Johana Rincones, Roxana Gajardo, Leonardo Caraballo, Carmen Andara and Hender Lopez

ABSTRACT

A quantitative comparative study of ants visiting flowers in the canopies of savanna and forest in the upper Orinoco (Venezuela) showed that flowers in the savanna are more likely to display features for avoiding ant visits than do flowers in the forest canopy. In both savanna and forest canopy, some flowers have special features to attract ants, probably to reduce herbivore pressure, whereas others possess features that hinder the access of ants to the flowers. The number of plant specimens, and of species, displaying flowers of the former type is more abundant in forest canopy; whereas the latter predominate in the savanna.

INTRODUCTION

It has been commonly assumed that flowers somehow repel ants or make access to their reproductive parts difficult for them (Janzen, 1977; Guerrant & Fiedler, 1981; Herrera *et al.* 1984; Koptur & Truong 1998). However, the opposite often occurs, as several flowers have specialized features to attract ants. For example, the orchid *Spathoglottis plicata* Blume attracts ants using nectaries strategically placed on the base of flowers and on flower buds (Jaffé *et al.*, 1989), presumably to reduce herbivory on its flowers. Although there is an extensive literature on ant–plant relationships (see extensive reviews in Jolivet, 1996b, for example), not much is known about ant–flower relationships.

Plants produce flowers, in principle, for sexual reproduction. The adaptive value of flowers is to attract a specific species or a range of specific pollinators. In doing so, they offer nectar, pollen and scent, and they display odours, shapes and colours as reinforcing stimuli to attract pollinators.

Although ants are known to pollinate plants, especially orchids (Peakall, 1994), ants are poor pollinators, partly because they do not fly and are, therefore, very ineffective dispersers of pollen. Moreover, ants produce various antibiotics in their exocrine glands that inhibit pollen germination (Beattie *et al.*, 1984). Ants are known as generalist foragers, exploiting any resource available, especially nectars and sugary secretions from plants and animals. In this way, ants are potential exploiters of flower nectar, and plants have evolved mechanisms to prevent ants from distracting resources aimed at pollinators. Chemical repellents exist in flowers (Van der Pijl, 1954) but seem to be rare (Feinsinger & Swarm, 1978), suggesting that plants use other mechanisms to exclude ants from their flowers. The strength of these interactions may vary in different ecosystems. Specifically, ants living in the forest canopy may have evolved stronger adaptations to exploit plants and their flowers than ants living on the ground, which might have a wider range of resources at their disposal.

In order to get a closer understanding of the factors that determine the relationship between ants and flowers, we address in this paper the following questions. What are the main strategies flowers have to attract or repel ants? Do these strategies differ between plants flowering within the savanna and those of the forest canopy?

METHODS

We studied the flowers occurring in the upper layer of the savanna ('savanna canopy'), which consists of grasses, bushes and herbs, growing on sandy soils on a 5 ha plot next to the Humboldt Station at La Esmeralda, Estado Amazonas, Venezuela (3° 9′ N, 65° 36′ W). The savanna is of the type described for the Guyana Highlands by Huber (1986), with a characteristic low occurrence of species of Graminaceae. Most of the savanna is flooded once a year and burns, on average, every 2 years. Precipitation and temperature fluctuations at La Esmeralda are indistinguishable from those at Surumoni. The forest canopy study was performed at the Surumoni project site, located in the upper Orinoco region, Estado Amazonas, Venezuela, close to the blackwater river Surumoni, at an elevation of about 110 m (3° 10′ N, 65° 36′ W). The plot contains floodable and nonfloodable tropical lowland forest (*terra firme*) on sandy soils. Mean annual rainfall is about 2700 mm, with a pronounced seasonality, and the average temperature is 26.5 °C (further details in Nieder *et al.*, 2000; Szarzynski & Anhuf, 2001). A modified commercial crane with a swingable boom of 40 m, mobile on a 120 m rail, was set up in 1996. Researchers were able to reach the outer canopy layer in an area of 1.4 ha with an observation gondola (Blüthgen *et al.*, 2000b).

Plants were visited in the day and at night. The time when flowers opened was classified as a.m. (06:00–09:00) and p.m. (15:00–17:00) and this time interval was used to count the ants on the flowers of the plant. Times between 10.00 and 14.00 were extremely hot at the savanna site, with ant activity close to zero. Pollinator activity was assessed at these same times. The flowers were classified into three categories, according to the number of ants found on or near them (see results). Control sticks chosen in the savanna consisted of dry flower stems standing at the average height of the savanna canopy. In the forest, control sticks consisted of dry terminal branches of trees. The dates of visits made to both sites simultaneously were: 23–31 March 1997, 28 July to 18 August 1997, 17–28 August 1999, 15–21 July 1999 and 2–7 March 2000, totalling a minimum of 90 hours of observation for each habitat (savanna and forest canopy). The same pair of observers visited both sites at each visit but different visit dates had different pairs of observers.

RESULTS

The plants surveyed in the savanna were mostly grasses, herbs and small shrubs. In addition to the flowering plants reported in Table 20.1, the savanna contained small *Byrsonima* sp. trees (Malpighiaceae), which appear to have permanent associations with ants. These were not studied further. The plants found in the forest at Surumoni were trees and lianas. The species composition of both sites for both plants and ants was completely different (Tables 20.1 and 20.2). In both savanna and forest sites, we grouped the flowers (or inflorescences) into three distinct functional categories.

Category 1: flowers that repel ants. The frequency of visits by ants to these flowers is lower than of visits to a control wooden stick placed nearby. Alternatively, we might group flowers into this category if the number of ant visits to flowers were located in the lower quartile of that for all flowers sampled in that habitat. Both methods classified the same samples into category 1. These flowers have either physical or chemical structures that deter ants from visiting them. A clear example is *Chamaecrista desvauxii* (Collad.) Killip., which is pollinated by large bumble bees and thus produces large amounts of nectar. This plant protects its flowers from ants and other insects by secreting a sticky substance all along its stems, branches and leaves. Another example is *Polygala* spp. which display their flowers on long, very thin and waxy pseudostems, making it difficult for walking insects to access the flowers.

Category 2: flowers with no distinct feature for either attracting or repelling ants. Ants visit these flowers with the same frequency as they visit a control empty wooden stick placed nearby.

Category 3: flowers that attract more than twice the number of ants than that attracted to control sticks. Another way of grouping flowers into this category is if over twice the number of ants were attracted to the flower or inflorescence compared with the mean number attracted by all flower species sampled in that habitat. The attraction signal consisted mostly of 'extrafloral' nectaries on the flower stems, at the base of the inflorescence or flower, or at the external part of the calyx. Flowers of *Irlbachia* spp., for example, have nectaries on the base of the flowers and on the flower stems, and they secrete nectar when the flower is in bud or recently open. Some flowers occurring in the forest canopy that were classified into this category could have been classified as flowers of category 2 because we found no special feature for the attraction of ants on the flowers. However, we classified them as category 3 because large numbers of ants were found on the flowers, possibly because they were nesting nearby or were tending Homoptera feeding on the flower stems.

Table 20.1. *Pattern of ant visitation in savanna flowers of La Esmeralda. The species are ordered according to the number of ants visiting the flowers or inflorescences*

Family	Plant species	No. of ants per flower per minute	Ant species	Possible pollinator	Time flower opens[a]	No. of flowers	Category
Gentianaceae	*Irlbachia alata* = *Chelonanthus albus* (Spruce ex Progel) Badillo	3.40	*Camponotus* sp. *Pheidole* sp. *Crematogaster* sp. *Ectatomma ruidum* (Roger) *Cephalotes pusillus* Klug	?	a.m./p.m.	10	3
Gentianaceae	*Irlbachia* sp. 1 = *Chelonanthus* sp.1	2.29	*Camponotus* sp. *Paratrechina* sp. *Pseudomyrmex simplex* (F. Smith)	Bats	a.m.	14	3
Melastomataceae	*Tibouchina aspera* Aubl.	1.93	*Camponotus sp.*	Halictinae	a.m.	15	3
Bromelaceae	*Pitcairnia armata* Maury	0.42	*Ectatomma ruidum* (Roger)	Collibri (Aves) Apidae	a.m.	12	2
Caesalpinaceae	*Chamaecrista desvauxii* (Collad.) Killid	0.08	*Crematogaster* sp.	Apidae	a.m.	7	1
Polygalaceae	*Polygala* sp. 1	0.05	*Pheidole* sp.	?	a.m./p.m.	6	1
Melastomataceae	*Macairea* sp. 1	0.02	*Ectatomma tuberculatum* (Olivier)	?	a.m.	>20	1
Polygalaceae	*Polygala* sp. 2	0	–	Coleoptera	a.m./p.m.	>20	1
Polygalaceae	*Polygala* sp. 3	0	–	?	a.m.	>20	1
Melastomataceae	*Miconia* sp. 1 Ruiz et Pavon	0	–	Chrysomelidae	p.m.	3	1
Melastomataceae	*Pterogastra divaricata* (Bonpl.) Naud.	0	–	Curculionidae	a.m./p.m.	29	1
Cyperaceae	*Bulbostylis junciformis* (H. B. K.) C. B. Clarke	0	–	?	a.m./p.m.	5	1
Cyperaceae	*Rhynchospora globosa* (C. B. Clark) Kük	0	–	?	a.m./p.m.	5	1
Cyperaceae	*Rhynchospora barbata* (Vahl) Kunth	0	–	?	a.m./p.m.	5	1
Xyridaceae	*Xyris* sp. 1	0	–	?	p.m.	>20	1
Control	Wooden stick	0.1	*Camponotus* sp.	–	a.m./p.m.	20	2

[a] 06.00–09.00 is a.m.; 15.00–17.00 is p.m.

Table 20.2. *Pattern of ant visitation in canopy flowers at Surumoni. The species are ordered according to the number of ants visiting the flowers*

Family	Plant species	No. of ants per flower per minute	Ant species	Time flower opens[a]	No. of flowers	Category
Caesalpiniaceae	*Tachigali guianensis* (Benth.) Zarucchi et Herend.	30	*Azteca* sp.	–	1	3
Vochysiaceae	*Vochisia obscura* Warm.	15	*Azteca* sp. *Camponotus* sp. *Pseudomyrmex simplex* (F. Smith) *Cephalotes atratus* (L.)	a.m./p.m.	9	3
Caesalpiniaceae	*Chamaecrista negrensis* (H. S. Irwin) H. S. Irwin et Barneby	6	*Azteca* sp.	–	1	3
Caesalpiniaceae	*Sclerobium* sp. = *Tachigal* sp.	>5	*Camponotus* sp. *Cephalotes atratus* (L.) *Procryptocerus* sp. *Pseudomyrmex simplex* (F. Smith) *Pseudomyrmex* sp., *Dolichoderus bidens* (L.)	a.m./p.m.	8	3
Vochysiaceae	*Qualea trichanthera* Spruce ex Warm.	>5	*Pseudomyrmex flavidulus* (F. Smith) *Pheidole* sp. *Crematogaster* sp. *Camponotus* sp. *Solenopsis* sp. *Daceton armigerum* (Adis), *Cephalotes atratus* (L.) *Paraponera clavata* (Fabricius) *Dolichoderus* sp.	a.m./p.m.	6	3
Caesalpiniaceae	*Senna silvestris* (Vell.) H. S. Irwin et Barneby	5	*Crematogaster* spp. *Cephalotes spinosus* (Mayr)	a.m./p.m.	1	3
Caesalpiniaceae	*Dialium guianense* (Aubl.) Sandwith	3	*Cephalotes atratus* (L.) *Azteca* sp. *Pachycondyla villosa* (Fabricius) *Crematogaster* sp.	a.m./p.m.	10	3

Tiliaceae	*Mollia lepidota* (Baehni) Meijer	2	*Azteca* sp.	a.m/p.m.	1	2
Lauraceae	*Ocotea* cf. *amazonica* (Meisn.) Mez	2	*Azteca* sp.	–	1	2
			Paratrechina sp.			
			Cephalotes atratus (L.)			
			Dolichoderus bispinosa (Olivier)			
Dilleniaceae	*Pinzona coriacea* Mart. et Zucc.	1	*Azteca* sp.	a.m.	1	2
Caesalpiniaceae	*Peltogyne* cf. *parviflora* Spruce ex Benth.	1	*Daceton armigerum* (Adis)	a.m./p.m.	1	2
Vochysiaceae	*Ruizterania trichanthera* (Warm.) Marc.-Berti	1	*Crematogaster* sp.	a.m/p.m.	1	2
Annonaceae	*Xylopia venezuelana* R. E. Fr.	1	*Crematogaster* sp.	a.m./p.m.	1	2
			Cephalotes atratus (L.)			
Celastraceae	*Goupia glabra* Aubl.	1	*Azteca* sp.	a.m./p.m.	8	2
			Cephalotes atratus (L.)			
			Crematogaster sp.			
Lauraceae	*Aniba* sp.	0	–	a.m.	1	1
Clusiaceae	*Caraipa densifolia* Mart.	0	–	a.m./p.m.	1	1
Icacinaceae	*Dendrobangia boliviana* Rusby	0	–	a.m./p.m.	1	1
Apocynaceae	*Himatanthus articulatus* (Vahl) Woodson	0	–	a.m./p.m.	1	1
Control	Wooden stick	1	*Azteca* sp.	a.m./p.m.	20	2

[a] 06:00–09:00, a.m.; 15:00–17:00, p.m.

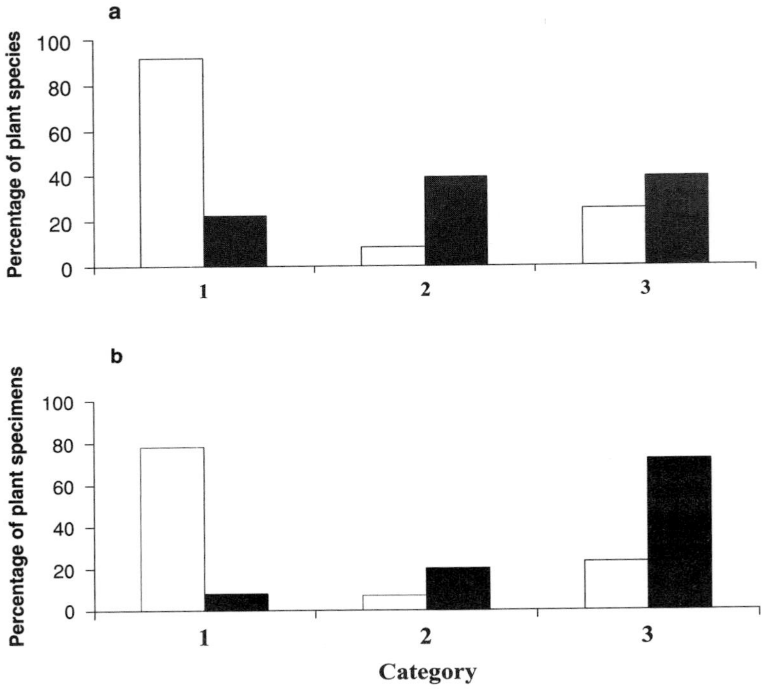

Fig. 20.1. Percentage of plant species observed (a) and percentage of individual plant specimens observed (b), assigned to categories 1, 2 or 3 with regard to ant visititation to flowers. Category 1, repels ants; category 2, indifferent to ants; category 3, attracts ants; open bars, savanna flowers; closed bars, canopy flowers.

The frequency of ants visiting the control structures was much higher in the forest than in the savanna. The flowers classified into the three categories are given in Table 20.1 for the savanna and in Table 20.2 for the flowers in the canopy. The frequency of occurrence of the different categories of flowers differs widely in the savanna compared with the canopy. In Figure 20.1**a**, we observe that most species in category 1 occurred in the savanna, whereas species in categories 2 and 3 were more common in the forest canopy. A similar pattern was observed when analysing the total abundance of the flowers per category (Fig. 20.1**b**).

The ant fauna in both sites was different (Tables 20.1 and 20.2) as was the proportion of plants distributed across the three categories. If we compare the number of plant species in categories 1 and 3 between both ecosystems with a chi-square test, we obtain $\chi^2 = 8.96$, $p = 0.0028$. The same comparison between the number of flower samples found in each category gives $\chi^2 = 77.0$, $p < 0.00001$. The most common ant species encountered in the forest was *Cephalotes atratus* (Linnaeus), whereas in the savanna *Camponotus* spp., *Ectatomma* spp. and *Crematogaster* spp. were the most common.

DISCUSSION

Coevolution, as defined by Ehrlich and Raven (1964) and elaborated by Janzen (1980b), is a process by which two organisms develop a close association over evolutionary time, by means of a series of reciprocal steps. Ehrlich and Raven developed this theory using the tight relationship between some butterflies and their host plants, and since then the theory of coevolution has been proposed as an explanation for the evolution of associations between many pairs of groups of organisms (Gilbert, 1983; Dettner & Liepert, 1994). Evidence to the contrary, that is, independent evolution of traits that allow for a close relationship between different types of organism, has also accumulated (Jermy, 1976; Janzen, 1980b; Miller, 1987; Fox 1988). Coevolution implies that each member of the association directly influences the evolution of the other organism. For example, in a plant–herbivore system, not only does the herbivore influence the evolution of the plant but the plant also influences the evolution of the herbivore. With this perspective, a broad and loose coevolutionary process between plant flowers suffering nectar robbing by ants and ants trying to exploit resources provided by flowers can be imagined. These two simultaneously occurring evolutionary processes may not necessarily qualify as coevolution but may be considered as an arms race between flowers and ants, where flowers sometimes switch to cooperation with ants when repelling them is impossible.

Our results suggest that flowers in the savanna are less indifferent to ants than those in the forest canopy. More plant species and more plants in the savanna repel ants than do flowers in the forest canopy studied. The most common mechanism flowers in the savanna use to make themselves inaccessible to ants is the production of long, thin and waxy flower pseudostems, making access to the flowers difficult for ants that normally walk on sand (Federle *et al.*, 2000). This hiding mechanism does not seem to be available to flowers in the forest canopy, as arboreal ants are harder to stop from climbing thin stems. Most ants in the canopy collect homopteran honeydew (Blüthgen *et al.*, 2000b) and thus are more likely to enter flowers in search of homopteran insects or plant nectar.

Other mechanisms that flowers have to avoid ants taking their nectar involve hiding the nectar in special organs, available only to birds or bats, or dispensing with nectar altogether, attracting the pollinator by offering only pollen or scent. The prevalence of ant avoidance mechanisms in the savanna compared with the forest canopy suggests that access to water limits nectar production on the savanna. However, arboreal ants seem to be more difficult to keep at bay than soil-dwelling ants. Consequently, flowers in the canopy are probably more tolerant to ants, dispensing with ant-specific repulsive mechanisms. One morphological feature that might provide a basis for this difference is that flowers in the forest canopy are more abundant, more globular in form and more accessible than those in the savanna, which are rather fewer and more tubular in shape. In Spain, Herrera *et al.* (1984) found that globular, accessible flowers were three times more likely to be visited by ants than tubular ones. They also showed that inflorescences with abundant flowers experienced more ant visits than those with one or a few flowers. Globular flowers having a bowl shape suffer relatively more desiccation than those that have a tubular shape (Kevan & Baker, 1983) and, accordingly, the nectar crystallizes much faster. These different environmental constraints may be correlated with the significantly higher levels of histidine found in nectars of herbaceous plants compared with trees and lianas (Baker and Baker, 1973).

Ants may benefit plants and their flowers by providing protection against herbivores. Some plants may even use ants as pollinators (Proctor *et al.*, 1996), although this is not very likely in the case of trees. Thus, plants may attract ('domesticate') certain ant species providing this provides protection to the flowers. It would seem that plants in the forest are more likely to have 'domesticated' ants than are their counterparts in the savanna, although further detailed ecological, morphological and physiological studies on the ant–flower interactions are needed in order to understand better these ecological interactions.

ACKNOWLEDGEMENTS

We thank the Concejo Nacional de Investigaciones Científicas y Tecnologicas (CONICIT) for financial help and the editors, two anonymous referees and D. Ajami for critical comments on earlier versions of the manuscript. This study was made possible through the collaboration of the community of La Esmeralda, the personnel of the Humboldt Station (Ministerio del Ambiente) and of the Surumoni Crane Project (Academy of Science of Austria).

21

Taxonomic composition and host specificity of phytophagous beetles in a dry forest in Panama

Frode Ødegaard

ABSTRACT

The species richness, taxonomic composition and host specificity of adult phytophagous beetles (Buprestidae, Chrysomeloidea and Curculionoidea) in the canopy of a tropical dry forest in Panama were studied. Beetle species associated with 26 liana species and 24 tree species were collected regularly over a 1-year period. The canopy was accessed with a tower crane. Host relationships were quantified by observations and probability-based methods. A total of 35 479 individuals belonging to 1165 species of phytophagous beetles were collected. Species regularly associated with the study plants were probably thoroughly collected. Transient species kept accumulating in the samples. The species list included 1020 and 411 species identified to generic and specific level, respectively. Approximately 40% of the captured species have been previously described. Chrysomelidae and Curculionidae were the dominant phytophagous families, in terms of both species richness and number of individuals. The species richness of Baridinae weevils is probably highly underestimated in tropical forests because they include many canopy specialists, and they are not attracted to light. In total, 2561 probable host records involving 697 beetle and 50 plant species were derived from the data collected in the study system. On average, 51.2 ± 32.0 (standard deviation (SD)) beetle species were associated with each plant species. The effective number of species associated with each plant species, accounting for the fauna shared between plant species, was 13.9 ± 10.0 (SD). A beta binomial abundance model was applied to estimate species richness and average host specificity of the beetle assemblage for a larger forest system. The forest, which may include 300–500 species of canopy plants (trees and lianas), was estimated to harbour 1600–2000 species of phytophagous beetles. Host specificity for phytophagous beetles in the canopy of this tropical dry forest was estimated to range from 7 to 10%. A relatively high proportion of described species combined with a beetle fauna less specialized than expected, contradict the prediction of 30 million arthropod species in the tropics.

INTRODUCTION

The canopy of tropical forests has been referred to as the last biotic frontier on Earth (Erwin, 1983a), and some scientists, considering the results of canopy fogging in tropical trees, have speculated that perhaps 30–100 million species of arthropods might exist in tropical forests (e.g. Erwin, 1982; May, 1988; Stork, 1993). These estimates are up to 100 times the actual number of named species on Earth.

A large proportion of tropical arthropods undoubtedly belongs to the order Coleoptera (Grove & Stork, 2000), and several recent studies have revealed the immense species richness in this group (e.g. Erwin & Scott, 1980; Stork, 1987b; Basset, 1991a; Basset *et al.*, 1996a; Kitching & Zalucki, 1996; Allison, *et al.*, 1997; Davies *et al.*, 1997; Wagner, 1997; Ødegaard, 2000a). Although there is an increasing number of datasets available, information on the taxonomic composition of these insect communities is poor (Hammond, 1994). Most commonly, specimens are sorted to higher taxonomic units or morphospecies without information on what proportion of the material belongs to previously described species. Accordingly, there are few data supporting the contention that species richness of tropical arthropods is several orders of magnitude larger than the number of named species. A major reason for this data shortage is poor knowledge of taxonomy of tropical beetles and the scarcity of taxonomists currently studying this material.

Even in well-studied tropical areas, the proportion of undescribed beetle species may be high. For example, in La Selva (Costa Rica), 13% of the Cerambycidae and 71% of the Zygopinae (Curculionidae) are undescribed (Hespenheide, 1993). The proportion of described to undescribed species in samples has been used to suggest that global species richness may be only a few times higher than the number of described species (Hodkinson & Casson, 1991). However, one has to be cautious of these estimates, because the proportion of described species in samples are highly variable, and it depends on factors like sampling effort, target taxon, geographic region and type of habitat, as well as the extrapolation techniques used. Nevertheless, an average sample of tropical arthropods needs to contain 99% undescribed species to support a figure of 100 million species. This consideration raises the fundamental question: where are all the unknown species? If they are in the canopy, then extensive samples of arthropods across a broad taxonomic spectrum should show 96–99% unnamed species in these habitats. Until such species lists become available, we cannot discount the occurrence of such high proportions of undescribed species in tropical forest canopies.

The first aim of this study has been to present a species list of phytophagous beetles associated with 50 plant species in the canopy of a tropical dry forest in Panama in order to assess how well the beetle fauna in this area is known.

The second aim has been to assess the magnitude of the local species richness in the forest. It is nearly impossible to obtain a complete sample of arthropod species at one site in tropical forests (Colwell & Coddington, 1994). As indicated by Janzen (1988a), it takes 5–10 years to characterize the insect faunas of several tree species in a tropical forest. Moreover, it has been shown that the addition of a new sampling method in a formerly satisfactorily sampled area results in the accumulation of new species (e.g. Basset *et al.*, 1997b). Consequently, the best characterization of species richness at one locality needs mass collection using several different sampling methods over many years. Accordingly, a satisfactory inventory is extremely expensive and time consuming (Janzen, 1996). Another approach that may give additional insight about the magnitude of the fauna that has *not* been observed may be derived from the application of statistical estimation techniques (Colwell & Coddington, 1994; Peterson & Slade, 1998).

Insect host specificity has been used to predict the magnitude of species richness at a global scale (Erwin, 1982; Thomas, 1990a; Ødegaard *et al.*, 2000). Host specificity is an even better predictor for *local* species richness (Mawdsley & Stork, 1997; Ødegaard, 2000c), which is the approach used in the present study. In this regard, however, it is important to define host specificity clearly. Host specificity is used as a general ecological term with application to attributes concerning trophic relationships, for instance between a host and its parasites, between a predator and its prey, or a herbivore and its host plant. Host specificity of a species of phytophagous insect can be defined as the taxonomic range of plant species used (Janzen, 1973b). A continuous spectrum exists between those insect species that feed only on one plant species, and others that feed on a wide range of plant species belonging to many different plant families. Feeding strategies are usually referred to as monophagy, oligophagy and polyphagy. However, there are no sharp boundaries between these categories. Host specificity is also used to describe the specialization of an insect species to different parts of a plant (Janzen, 1973b).

The average host specificity of the insect community associated with one or several host plant species is a third view of host specificity, and this is the focus of the present study. The average host specificity of an insect community can be calculated in terms of 'effective specialization' (May, 1990). Each insect species is weighted in inverse proportion to the number of host species on which they feed. The effective specialization of a set of beetles associated with one or several plant species can then be calculated if host associations within the community are known.

Host specificity is evolutionarily labile (Gould, 1979; Wasserman & Futuyma, 1981; Radtkey & Singer, 1995) and there is growing evidence for genetic differentiation in host use among populations (Bernays & Chapman, 1994). Host specificity, therefore, is a relative measure, dependent on both temporal and spatial scale (Fox & Morrow, 1981; Mawdsley & Stork, 1997). The evolutionary explanation for variation in host use may be that host plants are heterogeneous with respect to chemistry, morphology and abundance, both locally and along geographical gradients (Bernays & Chapman, 1994). One of the most interesting effects of this variation is the opportunity for sympatric speciation when local populations specialize upon new host plants (Maynard Smith, 1966; Futuyma & Mayer, 1980).

The magnitude of local host specificity is only known for particular ecological situations at restricted taxonomic and undefined spatial scales (Futuyma & Gould, 1979; Thomas, 1990a), because effective specialization is dependent on the number of available host plant species. Accordingly, the host specificity measured depends on the number of host plants sampled (Ødegaard *et al.*, 2000).

This study shows how the feeding specialization of adult beetles with respect to their host plants in a restricted community can be applied to estimate the magnitude of local species richness and the average host specificity of the beetle assemblage. Using a beta binomial model for trophic interactions among organisms, the expected number of beetle species can be predicted as a function of the number of plant species in the community (Diserud & Ødegaard, 2000).

METHODS

Study site and canopy access

The study site is located in the Parque Natural Metropolitano, province of Panama, Panama (8° 58′ N, 79° 33′ W), comprising 270 ha of tropical dry forest (Holdridge *et al.*, 1971) adjacent to Panama City. A larger complex of national parks is contiguous to the north. The total area of protected natural forests is 370 km^2 (Wright & Colley, 1994). The average annual rainfall is 1740 mm, of which 85% falls between May and November. Annual mean temperature is 28 °C (K. Kitajima, unpublished data). The flora at the site is characterized by a dominance of trees and lianas in the canopy; epiphytes are rather few and scattered. About 40 species of trees and 35 species of climbers (lianas, vines and hemiepiphytes) are within reach of a canopy access crane. In addition, the forest contains an unknown number of species of small trees, shrubs, herbs and epiphytes.

The study was carried out from a gondola connected to a tower crane. Access to all levels of the forest canopy is facilitated from the gondola. The crane is 42 m tall and has a horizontal boom of 51 m. The canopy at the site ranges from 20 to 38 m in height. Consequently, the crane facilitates the study of approximately 8000 m^2 of forest (Wright & Colley, 1994).

Beetles sampling

Coleoptera from the families Buprestidae, Cerambycidae, Chrysomelidae (including Bruchinae), Brentidae (including Apioninae) and Curculionidae (including Scolytinae and Platypodinae) were collected. These families constitute the major part of the phytophagous beetles present in these canopies.

All specimens were sorted to species. Genital dissections were performed systematically to reveal complex species groups and to understand intraspecific variation. As far as possible, described species were identified by comparison with museum reference collections, and by sending specimens to expert taxonomists. Some material was deposited at the synoptic insect collection at the Smithsonian Tropical Research Institute and the insect collection at the University of Panama. Some specimens were retained by the expert taxonomists, although most of the material was deposited in the author's collection.

The aim of the sampling protocol was to maximize the number and taxonomic range of plant species studied. The sampling protocol involved 37 individual trees and 40 individual lianas belonging to 24 and 26 species, respectively, representing 20 plant families (Ødegaard, 2000a). Each plant individual was visited approximately once a week, alternating visits by day and by night throughout 1 year. First, during each visit, 5 minutes was spent in recording insect behavioural observations such as feeding, oviposition and resting behaviours. Then, a nylon beating sheet (1 m^2) was used to collect cryptic beetles. In periods of leaf flushing and plant reproduction, the plants were sampled more frequently to detect seasonal species (Ødegaard, 2000a). Host observations were derived from feeding behaviour (host records) and probability-based methods following Flowers & Janzen (1997). These host occurrences were defined with three different levels of confidence: (i) 10 or more individuals recorded; (ii) 5–9 individuals recorded; (iii) 2–4 individuals recorded. The sum of host records and host occurrences was termed host observations, and these specific links between insect and host plant were termed host associations. Tropical samples of arthropods always include a large proportion of rare species (Peterson & Slade, 1998). In order to consider rare species, the weakest host occurrences were included in the analyses. Every host observation was evaluated carefully in accordance with the circumstances of each record (see Ødegaard, 2000a for details). In contrast to canopy fogging, this sampling procedure ensured that most insect species accidentally occurring on a plant would be omitted from the analyses. Such a beetle assemblage can never be completely sampled, but if the proportion of monophagous species is maintained constant after additional host observations within the crane perimeter, then additional sampling would not affect the magnitude of effective specialization significantly (Diserud & Ødegaard, 2000).

Species-accumulation curves

An accumulation curve is a plot of the cumulative number of species discovered with increasing sampling effort within a defined area. I used rarefaction (Colwell & Coddington, 1994), which calculates the randomization of n samples, all consisting of a fixed sample size of Y individuals each (Heck *et al.*, 1975).

The calculated accumulation curve can be extrapolated to its asymptote (e.g. Holdridge *et al.*, 1971; Soberón & Llorente, 1993) or it may be nonasymptotic (Palmer, 1990). The most useful information that may be obtained from extrapolated accumulation curves, independent of method, is likely to be a prediction of the expected increase in species richness with a given additional sampling effort rather than an estimate of total species richness for a defined area (Colwell & Coddington, 1994).

Estimation of effective specialization

Effective specialization of an insect community may be defined as the pooled number of plant-associated insect species in the sample divided by the number of host observations of those insects (Ødegaard *et al.*, 2000). Therefore, in a community, defined by T plant species, effective specialization (F_T) can be calculated as:

$$F_T = S_T/(\bar{S}_T T) \qquad (21.1)$$

where S_T is the total number of different insect species associated with T plant species, and $\bar{S}_T$ is the average number of insect species associated with each plant species (May, 1990; Thomas, 1990a; Mawdsley & Stork, 1997; Ødegaard *et al.*, 2000). Note that the magnitude of F_T depends on T. A dataset obtained from only one plant species gives an F_T of 1 (100% host specificity), because, then, $S_T = \bar{S}_T T$. In a dataset obtained from two plant species, F_T will decrease if some of the insects share the two host plants. Finally, an inverse relationship between F_T and T will appear when more plant species are included.

From our dataset, F_T can be calculated as a function of T. We let T range from 1 to the total number of plant species in our sample (50 plant species), and we let F_T be the average over all possible combinations of T plant species from our sample (Krylov, 1971). This nonparametric subsampling technique gives the magnitude of effective specialization for subsamples of $1, 2, \ldots, 50$ plant species in our dataset (Ødegaard *et al.*, 2000). To extrapolate, we apply a parametric model that allows us to predict effective specialization when we expand the community to include more plant species (Diserud & Ødegaard, 2000). This method enables us to estimate the host specificity for the community as a function of the number of potential host-plant species.

Estimation of local species richness

Many closely related procedures for the estimation of local species richness have been developed independently and have yet to be compared (Bunge & Fitzpatrick, 1993). In general, local species richness can be estimated as the asymptote of the sum of species-accumulation curves for species sampled by multiple methods. Estimation by parametric methods is also widely used; for example, by fitting different distribution models to the species-accumulation curve. Alternatively, there exists a series of nonparametric methods for estimation of species richness from samples; for example, with weighted abundance classes. According to Bunge & Fitzpatrick (1993), estimating the number of species is quite resistant to statistical solution, essentially because no matter how large the sample is, there may still be a large number of unobserved rare species. As an overall rule, it is recommended to try several different methods suitable for estimating local species richness and, then, determine how each of them fits the data.

A totally different approach is the host-specificity-based estimation method (Erwin, 1982). Erwin found approximately 1200 beetle species on one species of tropical tree. Further, he guessed that 20% of the 682 herbivorous species were host specific. Accordingly, 136 host-specific herbivorous species were associated with this tree species. Assuming an estimated 70 species of trees in 1 ha of tropical forest, the species richness of herbivorous beetles in 1 ha was estimated as $70 \times 136 = 9520$ species. This estimate was also used to estimate global arthropod species richness (Erwin, 1982). However, such estimates are better predictors of local than of regional species richness, because a global extrapolation includes serious problems associated with variation between areas and the scale extension (e.g. Stork, 1988; Thomas, 1990a; Basset *et al.*, 1996a; Mawdsley & Stork, 1997; Ødegaard, 2000c).

In the present study, the beta binomial model for trophic interactions among organisms (Diserud & Ødegaard, 2000) is applied for the estimation of local species richness. Here, the local insect species richness is predicted as a function of local plant species richness. The expected number of beetle species (i.e. predicted local species richness) is calculated by multiplying the number of tree species with the respective value of effective specialization and the average number of beetle species per tree species.

RESULTS

Species richness and composition

A total of 35 479 individuals belonging to 1165 species of phytophagous beetles was collected (see the Appendix to this chapter). More than half of the species belonged to the Curculionidae and 28% belonged to Chrysomelidae. Cerambycidae and Buprestidae constituted 7.9 and 6.7% of the species, respectively (Fig. 21.1**a**). When individuals were considered, Chrysomelidae were more abundant than Curculionidae. These two families together accounted for 91.7% of the total individuals (Fig. 21.1**b**). Among the Curculionidae, the material was heavily dominated by Cryptorhynchinae and Baridinae with 156 and 185 species, respectively. Zygopinae were also a large subfamily with 81 species. Among the Chrysomelidae, no subgroup dominated; Bruchinae, Cryptocephalini, Chlamisini, Eumolpini, Galerucini and Alticini were all represented with between 30 and 60 species in each case (Table 21.1).

In total, 1020 species (88%) were determined to generic level and 411 species (35%) were determined to specific level (Table 21.1). The confidence levels of the identifications were grouped into three categories:

1. Poorly determined reference collections were available for study; few or no specimens were examined by expert taxonomists
2. Well-determined reference collections were available for study; some material was examined by expert taxonomists
3. The material was examined by expert taxonomists.

At the two higher levels of confidence (2 and 3), 83% of the species were identified to named or unnamed species (Table 21.2). There is an obvious correlation between confidence level and the proportion of described genera and species (Table 21.2). It is also expected that the level of confidence will correlate well with the current level of knowledge of taxonomic groups. Thus, it is unlikely that a high proportion of species in low-confidence groups (1 and 2) could be described. A reasonable guess would be that some (10%) of the species in the groups 1 and 2 actually belong to named species ($n = 66$ species). Then, the total proportion of described species in this sample would be about 40%.

Accumulation of new species in the samples

A rarefaction curve (Fig. 21.2) showed that the total accumulation rate of new species started to level off after approximately 34 000 specimens were captured, but new species were still added at a rate of 0.14 species per 1000 individuals. In contrast, the rarefaction curve for insect species with known host associations accumulated new species at the lower rate of 0.05 species per 1000 individuals after approximately 33 000 specimens had been captured. This indicates that host-associated species were relatively well sampled within the crane perimeter.

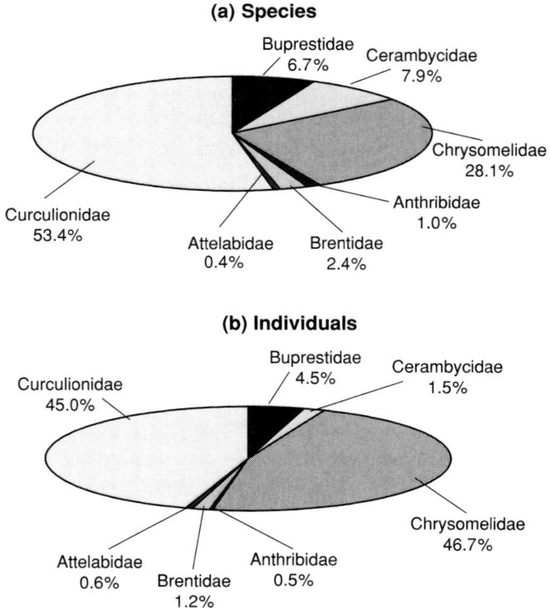

Fig. 21.1. Beetle distribution. **a** Proportion of phytophagous beetle species belonging to different families. **b** Proportion of phytophagous beetle individuals belonging to different families.

Table 21.1. *Taxonomic status of collected material, detailed by higher taxonomic categories*

Taxonomic group	Total No. species	Species identified to genera	Identified species	Confidence level[a]
Buprestidae	69	68	22	–
Buprestinae	4	4	2	3
Agrilinae	65	64	20	3
Cerambycidae	112	110	80	–
Disteniinae	3	3	1	3
Cerambycinae	20	20	16	3
Lamiinae	89	87	63	3
Megalopodidae	1	1	0	1
Orsodacnidae	1	1	1	2
Chrysomelidae	300	249	92	–
Bruchinae	44	44	38	3
Chrysomelinae	3	3	3	2
Criocerinae	2	2	1	1
Cryptocephalini	47	47	3	1
Clytrini	11	11	8	2
Chlamysini	37	37	0	1
Eumolpini	31	31	11	2
Megascelidini	2	2	0	1
Galerucini	30	20	9	1
Alticini	57	22	2	1
Cassidini	15	15	14	3
Hispini	14	8	3	2
Lamprosomatinae	7	7	0	1
Anthribidae	15	13	2	1
Attelabidae	5	5	3	2
Brentidae	46	46	21	–
Brentinae	9	9	5	2
Apioninae	37	37	16	3
Curculionidae	616	527	188	–
Brachyderinae	1	1	1	3
Otiorynchinae	4	4	3	3
Leptopiinae	2	2	2	3
Cylydrorhininae	1	1	0	2
Hyperinae	2	2	1	3
Hylobiinae	26	26	12	2
Erirrhininae	24	23	4	2
Otidocephalinae	33	33	4	2
Magdalinae	1	1	0	2
Anthonominae	33	32	18	3
Prionomerinae	9	9	2	2
Ceratopinae	2	2	1	2
Tychinae	5	5	3	3
Camarotinae	2	2	0	2
Cryptorhynchinae	156	153	68	3
Zygopinae	81	80	15	2
Tachygoninae	7	7	1	3
Ceuthorhynchinae	6	6	2	2
Baridinae	185	106	32	2
Rhynchophorinae	2	1	1	2
Cossoninae	4	2	1	2
Platypodinae	3	2	1	2
Scolytinae	27	27	18	3
Total	1165	1020	411	–

[a]For identification (categories 1–3); see text.

Table 21.2. *Identification status of collected material detailed by confidence levels for identifications*

Confidence level[a]	Total No. species (%)	Species identified to genera (%)	Identified species (%)
1	198 (17)	151 (76.3)	17 (8.6)
2	454 (39)	363 (80.0)	109 (24.0)
3	513 (44)	506 (98.6)	285 (55.5)
Total	1165	1020	411

[a]For identification (categories 1–3), see text.

Host specificity of the beetle community

A total of 2561 host observations of 697 species of beetles were obtained for the 50 plant species. There was, on average, 51.2 ± 32.0 (SD) beetle species associated with each plant species. Similarly, the effective number of beetle species associated with each plant species, which takes into consideration the fauna that is shared between plant species, was 13.9 ± 10.0 (SD). Accordingly, the effective specialization of this beetle community associated with 50 plant species was about 27% (13.9/51.2). A nonparametric calculation shows how the magnitude of effective specialization declines from 1 to 50 plant species in the dataset (Fig. 21.3). The different levels

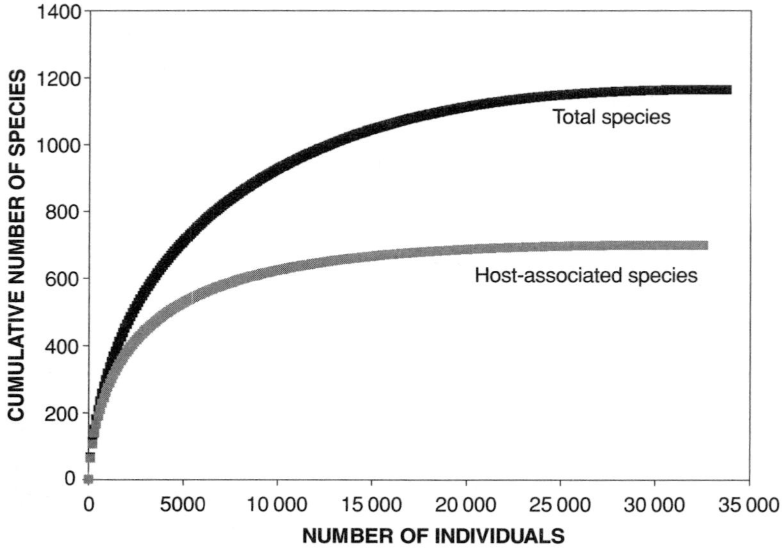

Fig. 21.2. Rarefaction curves of the cumulative number of phytophagous beetle species as a function of the number of individuals collected.

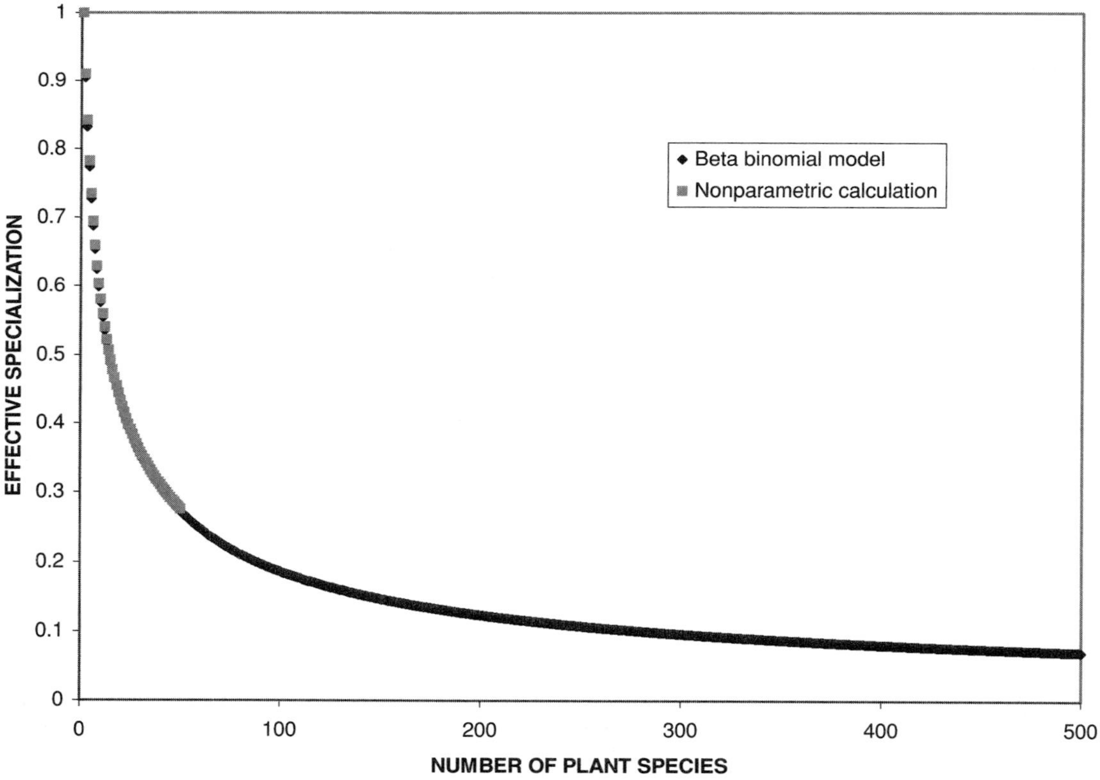

Fig. 21.3. Estimated effective specialization as a function of the number of plant species. The nonparametric curve represents the average effective specialization of a beetle assemblage on each of 1 to 50 plant species in the dataset. This curve was extrapolated by means of a beta binomial model where the parameters (k, β) were estimated to be −0.29 and 3.0, respectively.

of effective specialization are all theoretical, because the local assemblage of beetles in this forest exploits more than 50 host plants. This particular forest is expected to contain about 300–500 species of canopy plants (trees and lianas) (Ødegaard *et al.*, 2000). The modelled beta binomial curve appears as an extrapolation of the nonparametric curve (Fig. 21.3). Almost identical curves in the overlapping area indicate a good fit of the model. The estimated effective specialization in the forest ranges from 7 to 10%, dependent on the number of potential host plant species in the community (Ødegaard *et al.*, 2000). Therefore, the effective number of beetle species specialized per plant ranges from 3.6 to 5.1 species.

Local species richness

The expected number of phytophagous beetle species in the canopy of this forest is strongly correlated with the species richness of canopy plants (Ch. 25). Local species richness can be calculated from Fig. 21.3; multiplying the number of plant species by the respective value of effective specialization and the average number of beetle species per plant species (51.2 beetle species). Therefore, this forest consisting of 300–500 species of trees and lianas is estimated to contain 1600–2000 species of phytophagous beetles in its canopy (Fig. 21.4).

DISCUSSION

Taxonomic composition of phytophagous beetles

As in other studies, Chrysomelidae and Curculionidae dominated canopy samples in terms of both species and individuals (e.g. Erwin & Scott, 1980; Farrell & Erwin, 1988; Basset, 1991a; Basset *et al.*, 1996b; Davies *et al.*, 1997). Taxonomic composition at the family and subfamily levels was very similar to that recorded by Erwin & Scott (1980), who sampled beetles from one tree species in the same area, using canopy fogging.

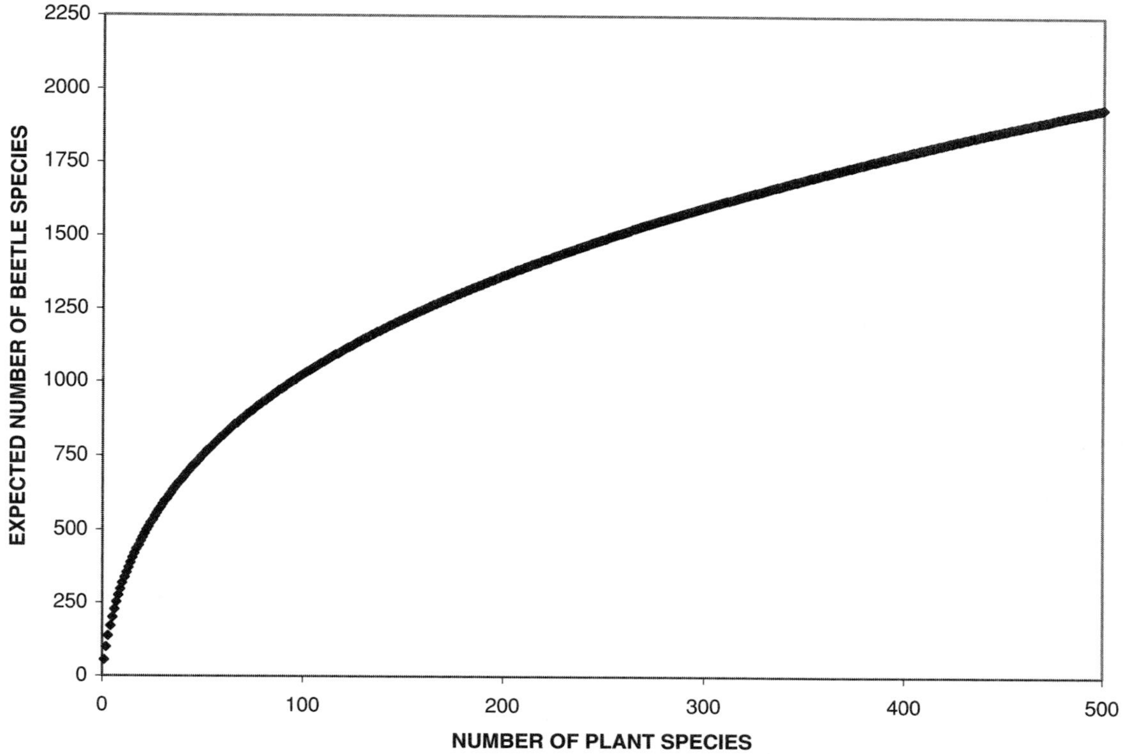

Fig. 21.4. Expected number of phytophagous beetle species in the canopy of a tropical dry forest in Panama as function of the number of tree and liana species.

The exceptions were the leaf-mining Buprestidae and seed-boring Bruchinae, which constituted 6% and 4%, respectively, of total abundance in this study, compared with 2% and 1%, respectively, in the study of Erwin and Scott. The reason for this discrepancy could be that these families include rather specialized species (Janzen, 1980a; Hespenheide, 1991). Therefore, samples obtained from many plant species will be more species rich for these taxa than comparable samples obtained from a single plant species. The species composition of beating/hand-collecting samples compared with fogging samples from the canopy is not very different, and both methods underestimate the concealed fauna (Basset *et al.*, 1997b; Stork & Hammond, 1997).

An advantage of hand collecting is the possibility of collecting species that stick to the substrate. In the present study, a large group of baridine weevils, feeding on canopy lianas, were able to attach to the tendrils of their host plants by tibial spines and ventral cavities (Ødegaard, 2000a; F. Ødegaard, unpublished data). These species would probably rarely appear in fogging samples. In comparison, Baridinae were more abundant than Cryptorhynchinae (Table 21.1). This is interesting because the latter group evidently represents the most species-rich subfamily of described weevils in the neotropics (O'Brien & Wibmer, 1982; Wibmer & O'Brien, 1986). For example, an extensive light-trapping regime at seven different sites elsewhere in Panama yielded a total of 385 species (9289 individuals) of Cryptorhynchinae and 19 species (137 individuals) of Baridinae (Wolda *et al.*, 1998). This suggests that the Baridinae are rarely attracted to light. Many species of Baridinae seem also to be restricted to flowers and lianas in the canopy. In addition, the subfamily is very poorly known taxonomically (R. S. Anderson, personal communication). For these reasons, baridine weevils may be highly underestimated in many tropical samples.

The dominance of certain taxonomic groups in the canopy compared with ground samples reflects habitat availability in the canopy. Ecological groups like twig-borers, leaf-miners and species associated with buds, flowers and seeds were comparatively species rich. Among the twig-borers, groups like Zygopinae, Cryptorhynchinae (Curculionidae), subgroups of Cerambycidae, and Agrilinae (Buprestidae) were important. The leaf-miners were dominated by Trachyini (Agrilinae, Buprestidae), Tachygoninae and Camarotinae (Curculionidae). Bud-eaters such as Anthonominae were common, while a high number of species from Baridinae, Erirrhininae, Bruchinae, Eumolpini, Galerucini and Alticini were regularly visiting flowers. All of these groups may be more common in the canopy as compared with ground samples.

The proportion of unnamed species

The approximately 40% proportion of named species in the present study is higher than that expected to support a prediction of 30 million arthropod species in the tropics. If 40% of the species were described, then total global species richness would probably be less than 3 million species. This figure is supported by a similar study of insects on *Ficus* spp. in Papua New Guinea, in which 76% of the Coleoptera were identified to generic level, and 36% at the specific level (S. E. Miller *et al.*, unpublished data). In contrast, if there were 30 million species of arthropods in the tropics, then only 3% of the species could be named. This discussion is, however, highly dependent on the availability of representative datasets. For instance, most tropical forests of Central America are probably better studied than most forests in the Amazon basin and Africa. Species lists from poorly known forests, as well as knowledge about the relative extent of poorly known versus well-known forests, are needed to improve these assessments. Although several arthropod species lists from tropical forests do exist, one has to be very careful when estimating the proportion of described species based on these data. The proportion of described species in a sample is highly dependent on particular taxa, and taxonomic expertise. Accordingly, the confidence level of identification of each taxon may be different (Table 21.1) and cannot be compared.

Host specificity

Few tropical studies have estimated the magnitude of host specificity for phytophagous beetles. Mawdsley & Stork (1997) calculated effective specialization indirectly based on samples from canopy fogging 10 Bornean trees of four species. By measuring turnover of beetle species within and between tree species, they could overcome sampling artefacts introduced by the fogging method. It was concluded that average effective specialization was less than 5% at the species level (Stork, 1997). A study of host specificity, based on feeding trials of insects associated with 10 tree species in Papua New Guinea, suggested that between 23 and 37 leaf-feeding beetles were monophagous in a system of 391 species, yielding an estimated host specificity of 3.3–5.2% (Basset *et al.*, 1996a). In comparison, the present study describes a slightly more specialized fauna, with 7–10% monophagous species (Fig. 21.3) but this figure is less than half of Erwin's value of 20% host specificity. A working figure for global host specificity in phytophagous insect communities may be presented as the median value of comparable data-based studies in different regions (Ødegaard, 2000c).

The quantification of host specificity is a complex topic involving several serious obstacles. Most important, effective specialization does not account for the phylogeny of the host, and the insect density per biomass of host plants (Ødegaard, 2000a). Accordingly, the measurement of effective specialization may be changed outside the crane perimeter, as they will include insect records on congeneric, confamilial or related host plants not encountered within the crane perimeter. Furthermore, the low number (one or two, on average) of individuals surveyed per plant species may introduce biases related to insect aggregation on individual plants, as insect herbivores are commonly aggregated in tropical forests (Novotny & Basset, 2000). Finally, host occurrences of the weakest level of confidence constituted more than half of the total number of host observations (Ødegaard, 2000a). This may introduce additional uncertainty. As pointed out by Diserud & Ødegaard (2000), it is important to be aware of the limitations when extrapolating the nonparametric curve describing the data on effective specialization of the beetle fauna (Fig. 21.3) across a variable subset of plant species. Such an extrapolation should not be done unless the study plant species are a random selection of plant species occurring in the overall area. If not, the extrapolation can only be performed for homogeneous vegetation types, since the parameters of the beta binomial model are expected to change if habitat barriers are encountered.

Local species richness

The discrepancy between the two species-accumulation curves (Fig. 21.2) indicates that host-associated species were relatively well sampled within the crane perimeter while the proportion of additional transient species steadily increases with sample size. The reason why species still accumulate in large samples is related to dispersal from hosts and habitats not included in the sampling universe (Novotny & Basset, 2000), in this case beyond the crane perimeter. For this reason, the recording of host observations of phytophagous beetles in an area may be cost-effective compared with compilation of pure species lists from the same area (e.g. from trap material), because the former method requires less sampling effort to attain the same level of completeness (as in Fig. 21.2).

Erwin (1982) estimated that there are 9520 species of herbivorous beetles in 1 ha of tropical forest in Panama based on canopy fogging of one tropical tree species. He collected 682 herbivorous beetle species and assumed that 20% of them were host specific to his study tree species, and that 1 ha of neotropical forest has on average 70 tree species. In the present study, 70 tree species yield about 900 beetle species (Fig. 21.4). This is still only about one fifth of Erwin's estimate (9520 species) if we assume 10% host specificity instead of 20%. The major discrepancies between this study and Erwin's assumption relate to the importance of transient species, the insect fauna shared among tree species and variation among tree species (see Basset *et al.*, 1996a). Effective specialization accounts for all these factors when applied to host observation data from the present study. Whereas Erwin assumed 136 monophagous species, the present study estimates only 3.6–5.1 species effectively specialized per plant species. The latter values of effective specialization are supported by similar studies elsewhere (Ødegaard *et al.*, 2000). In Papua New Guinea, 2.3–3.6 beetle species were effectively specialized to 10 tree species (Basset *et al.*, 1996a), and in Borneo 3.5–5.8 beetle species were effectively specialized to 10 trees (Mawdsley & Stork, 1997; Stork, 1997).

In the present study, a tenfold increase in plant species nearly tripled the estimated number of beetle species in the forest (697 host-associated species on 50 plants compared with the estimated 1937 host-associated species on 500 plants). Nearly 1200 species were collected within the study site, which may be a surprisingly high proportion of the estimated total species richness in the forest. However, if the estimated 1937 host-associated species in the forest ever happen to be collected, there would certainly be an additional collection of transient species from habitats outside the forest.

One of the most extensive studies of the beetle fauna of tropical forests comes from Dumoga-Bone National Park in Sulawesi, Indonesia (Hammond, 1990). In an area of 500 ha, more than 1 million beetles were sorted to about 4500 species (Hanski & Hammond, 1995). The forest is conservatively estimated to contain 6500 beetle species (Hammond, 1994). A comparative estimate of the total beetle fauna in the Panamanian forest based on the present data would be about 4500 (1937/0.428) species, if the herbivorous beetle families constitute 42.8% of all described beetle species (Lawrence, 1982). This comparison gives at least an indication of the order of magnitude of local species richness of beetles in tropical forests.

The size of the study area in the present study (0.8 ha) may be critically small for obtaining samples that are representative for a larger area of the forest. Because of the heterogeneity of tropical forests, a representative study site for tropical woody plants should be at least 50 ha (Condit, 1996). The same area would probably be needed when phytophagous insects are considered. Accordingly, the material in the present study may be heavily influenced by the distribution of plant individuals, species, genera and families within the small crane perimeter compared with that of the whole forest. The extrapolation, therefore, is likely to be biased towards fewer species. It is not known if this represents a serious problem as long as host-specificity measurements are the basic data for the extrapolation.

Most of the problems associated with estimation of local species richness relates to the completeness of local sampling (Janzen, 1988a; Colwell & Coddington, 1994). If one locality is sampled intensively for long enough, a large proportion of the regional species pool may be collected. In Denmark, 30% of all Danish beetle species was recorded in a forest of 170 ha (Jørum, 2000). In Finland, 23% of all Finnish beetle species was recorded from Oulanka National Park (200 km^2), and in England 38% of the British beetle fauna was recorded from a forest of approximately 200 km^2 (Hanski & Hammond, 1995). Although these examples originate from the temperate region, similar patterns may be expected in the tropics, at least if sampling could be comparatively complete in a sampling plot that is at least 50 ha. Accordingly, there may be a risk of overestimating regional species richness based on extensive material from one tropical site. However, the local representation of the regional fauna in the temperate region may be very different from that of tropical forests. The extent, to which geographical extrapolations are sound in tropical forests is probably smaller than in temperate regions because of frequent habitat shifts and high species turnover (Pianka, 1966; Stevens, 1990). An extrapolation up to 300–500 plant species, which is probably sound for the present study, represents a small area compared with, for instance, that of the whole country. This discussion, however, raises issues of species ranges and species turnover along geographical gradients that goes beyond the scope of the present study.

CONCLUSION

About 40% of the species sampled in the present study belonged to named species. This is a higher proportion than expected on the basis of the prediction of 30 million arthropod species in the tropics. A claim of fewer species in the tropics is also supported by host-use assessments. The basis for the '30 million prediction' is that 20% of the phytophagous beetles should be monophagous. Calculations based on effective specialization concluded that 7–10% of the phytophagous beetles in this forest were on average monophagous, a result indicating a less-specialized beetle fauna in tropical forests.

Based on host-plant observations for the beetles, it was estimated that this tropical dry forest in Panama contains between 1600 and 2000 species of phytophagous beetles. The estimate indicates that a relatively large proportion of the local or regional beetle fauna may be found at one site, a pattern that is typical of temperate regions. Sound extrapolations of arthropod species richness in tropical forests, however, are heavily restricted by sample size and area because of the heterogeneity of tropical vegetation. Improved knowledge of local species richness of arthropods in tropical forests should, therefore, come from studies of species turnover along geographical gradients.

ACKNOWLEDGEMENTS

This study was supported by the Norwegian Research Council and the Norwegian Institute for Nature Research. I am indebted to Wayne E. Clark, R. Wills Flowers, Edmund Giesbert, Henry A. Hespenheide, Bjarte Jordal, John M. Kingsolver, David G. Kissinger, Charles W. O'Brien, Paul E. Skelley, Henry P. Stockwell, Karl H. Thunes and Donald M. Windsor for their verification and identification of parts of the beetle material. Great thanks to Åslaug Viken for field assistance and help with handling data, and to Ola Diserud, Steinar Engen and Kaare Aagaard for our fruitful discussions on effective specialization.

I acknowledge Yves Basset, Raphael Didham, and S. Joseph Wright for valuable comments on the manuscript. I also thank the Smithsonian Tropical Research Institute for providing canopy-crane access and laboratory facilities during my stay in Panama.

Appendix: Species list of phytophagous beetles collected from 50 different plant species in the canopy of a tropical dry forest in Panama

The nomenclature follows Lawrence and Newton (1995) at the family level. Subgroup systematics follows O'Brien and Wibmer (1982) for the Curculionidae and Reid (1995) for the Chrysomelidae. Species are sorted alphabetically within each subgroup. The number of individuals (ind.) is detailed for each species.

BUPRESTIDAE

Buprestinae

Chrysobothris ?viridiimpressa LaPorte & Gory (2 ind.), *Euchlora gigantea* Linnaeus (1 ind.), *Halecia* sp. (1 sp., 4 ind.), *Halecia/Euplectalecia* sp. (1 sp., 1 ind.).

Agrilinae

Agrilaxia sp. (4 spp., 7 ind.), *Agrilus plagiatus* Waterhouse (17 ind.), *Agrilus* sp. (8 spp., 28 ind.), *Brachys kleinei* Obenberger (290 ind.), *Brachys laeta* Waterhouse (25 ind.), *Brachys* sp. (5 spp., 78 ind.), *Hylaeogena affinis* Obenberger (151 ind.), *Hylaeogena cincta* Waterhouse (39 ind.), *Hylaeogena festiva* Fisher (19 ind.), *Hylaeogena hydroporoides* Waterhouse (5 ind.), *Hylaeogena kirschii* Obenberger (125 ind.), *Hylaeogena lecerfi* Obenberger (201 ind.), *Hylaeogena pilosa* (Fisher) (126 ind.), *Hylaeogena rotundipennis* (Fisher) (27 ind.), *Hylaeogena thoracicus* (Waterhouse) (250 ind.), *Hylaeogena* sp. (1 sp., 1 ind.), *Leiopleura balloui* Fisher (3 ind.), *Leiopleura divisa* Waterhouse (18 ind.), *Leiopleura lateralis* Waterhouse (9 ind.), *Leiopleura* sp. (14 spp., 308 ind.), *Lius cuneiformis* Fisher (58 ind.), *Lius* sp.(4 spp., 63 ind.), *Pachyschelus bifasciatus* Waterhouse (1 ind.), *Pachyschelus collaris* Gory (4 ind.), *Pachyschelus cupricaudatus* Fisher (8 ind.), *Pachyschelus panamensis* Fisher (170 ind.), *Pachyschelus* sp. (8 spp., 94 ind.), unidentified genus (1 sp., 1 ind.).

CERAMBYCIDAE

Disteniinae

Distenia sallaei Bates (1 ind.), *Distenia* sp. (1 sp., 1 ind.), *Paracometes* sp. (1 sp., 15 ind.).

Cerambycinae

Anelaphus sp. (1 sp., 1 ind.), *Chlorida festiva* (Linnaeus) (2 ind.), *Coleoxestia* sp. (1 sp., 3 ind.), *Cosmisoma militaris* Giesbert & Chemsak (1 ind.), *Dihammaphora* sp. (1 sp., 10 ind.), *Eburodacrys laevicornis* Bates (5 ind.), *Euderces cleriformis* (Bates) (1 ind.), *Gnomidolon suturale* (White) (1 ind.), *Gorybia chontalensis* (Bates) (1 ind.), *Ibidion buqueti* Thomson (1 ind.), *Mecometopus ion* (Chevrolat) (2 ind.), *Mecometopus jansoni* Bates (2 ind.), *Neocompsa squalida* (Thomson) (3 ind.), *Obrium maculatum* (Oliver) (1 ind.), *Ommata minuens* Giesbert (1 ind.), *Psyrassa* sp. (1 sp., 1 ind.), *Rhopalophora versicolor* Chevrolat (1 ind.), *Stizocera poeyi* (Guérin–Méneville) (1 ind.), *Tetranodus ?xanthocollis* Chemsak (1 ind.), *Trachyderes succinctus* Linnaeus (5 ind.).

Lamiinae

Acanthoderes sp. (1 sp., 5 ind.), *Adesmus* sp. (1 sp., 1 ind.), *Adesmus ?ventralis* (Gahan) (1 ind.), *Adetus bacillarius* Bates (16 ind.), *Adetus curtulus* Bates (53 ind.), *Adetus analis* (Haldeman) (5 ind.), *Adetus costicollis* Bates (8 ind.), *Aerenea brunnea* Thomson (3 ind.), *Aerenea impetiginosa* Thomson (1 ind.), *Aerenea* sp. (1 sp., 3 ind.), *Alphus* sp. (1 sp., 1 ind.), *Anisopodus* sp. (4 spp., 24 ind.), *Ataxia fulvifrons* Bates (1 ind.), *Ataxia lineata* (Fabricius) (1 ind.), *Atrypanius cretiger* (White) Bates (3 ind.), *Atrypanius irrorellus* Bates (11 ind.), *Bactriola paupercula* Bates (25 ind.), *Bebelis ?furcula* (Bates) (1 ind.), *Bisaltes* sp. (1 sp., 5 ind.), *Cacostola* sp. (1 sp., 6 ind.), *Callia fulvocincta* Bates (2 ind.), *Callia* sp. (1 sp., 1 ind.), *Cacupira?* sp. (1 sp., 2 ind.), *Cephalodina capito* (Bates) (5 ind.), *Charoides* sp. (1 sp., 5 ind.), *Cherentes niveilateris* Thomson (1 ind.), *Cirrhicera sallei* Thomson (2 ind.), *Deliathis quadritaeniator* White (1 ind.), *Desmiphora canescens* Bates (9 ind.), *Desmiphora hirticollis* (Oliver) (31 ind.), *Desmiphora scapularis* Bates (4 ind.), *Desmiphora* sp. (1 sp., 1 ind.), *Dorcasta dasyera* (Erichson) (7 ind.), *Epectasis juncea* Newman (11 ind.), *Estola misella* Bates (5 ind.), *Estola* sp. (3 spp., 7 ind.), *Estoloides ?longicornis* Breuning (42 ind.), *Eumathes cuprascens* Bates (3 ind.), *Eupogonius infimus* Thomson (19 ind.), *Eupromera* sp. (1 sp., 12 ind.),

Hippopsis septemlineata Breuning (1 ind.), *Hydraschemopsis pugnatrix* Lane (3 ind.), *Lagocheirus araneiformis ypsilon* (Voet) (5 ind.), *Lagocheirus plantaris* Erichson (3 ind.), *Leptostylus decipiens* Bates (3 ind.), *Leptostylus gibbulosus* Bates (2 ind.), *Leptostylus leucopygus* Bates (3 ind.), *Leptostylus* sp. (2 spp., 6 ind.), *Lepturges callinus* (Bates) (1 ind.), *Lepturges inscriptus* (Bates) (5 ind.), *Lepturges sexvittatus* Bates (3 ind.), *Lochmaeodes tesselatus* Thomson (2 ind.), *Neoptychodes trilineatus* Linnaeus (16 ind.), *Nyssodrysina haldemani* (Le Conte) (5 ind.), *Nyssodrysina infima* (Bates) (6 ind.), *Nyssodrysina polyspila* (White) (2 ind.), *Nyssodrysina scutellata* (Bates) (3 ind.), *Nyssodrystes exilis* (Bates) (20 ind.), *Oedopeza ocellator* Fabricius (2 ind.), *Oncideres macra* Thomson (12 ind.), *Oreodera aerumnosa* Erichson (1 ind.), *Oreodera cretata* Bates (9 ind.), *Ozineus* sp. (1 sp., 1 ind.), *Pachypeza panamensis* Giesbert (1 ind.), *Panegyretes porsus* Galileo & Martins (1 ind.), *Paradesmiphora farinosa* (Bates) (3 ind.), *Phaea hogei* Bates (1 sp., 1 ind.), *Ptychodes niveisparsus* Bates (1 ind.), *Sarillus pygmaeus* Bates (3 ind.), *Stenolis ?circumscripta* (Bates) (5 ind.), *Stenolis inclusa* (Bates) (1 ind.), *Stereomerus ?lineatus* (Breuning) (1 ind.), *Sternycha paupera* Bates (1 ind.), *Taeniotes scalaris* Fabricius (13 ind.), *Trestonia pucherrima* Dillon & Dillon (24 ind.), *Urogleptes charillus* (Bates) (1 ind.), *Urgleptes mixtus* (Bates) (2 ind.), *Urgleptes ?laticollis* (Bates) (25 ind.), *Urgleptes* sp. (2 spp., 16 ind.), *Venustus zeteki* Dillon & Dillon (1 ind.), Unidentified genus (Acanthocini) (1 sp., 2 ind.), unidentified genus (Acanthoderini) (1 sp., 1 ind.).

MEGALOPODIDAE

Mastostethus sp. (1 sp., 1 ind.).

ORSODACNIDAE

Aulacoscelis melanocera Stål (2 ind.).

CHRYSOMELIDAE

Bruchinae

Acanthoscelides brevipes (Sharp) (1 ind.), *Acanthoscelides difficilis* (Sharp) (7 ind.), *Acanthoscelides hectori* Kingsolver (5 ind.), *Acanthoscelides lapsanae* (Motschulsky) (3 ind.), *Acanthoscelides puelliops* Johnson (1 ind.), *Acanthocelides puellus* Sharp (7 ind.), *Acanthoscelides taboga* Johnson (3 ind.), *Acanthoscelides* sp. (2 spp., 18 ind.), *Amblycerus alternans* (Pic) (42 ind.), *Amblycerus championi* (Pic) (6 ind.), *Amblycerus cistelinus* (Gyllenhal) (23 ind.), *Amblycerus dispar* (Sharp) (69 ind.), *Amblycerus marieae* Romero, Johnson & Kingsolver (5 ind.), *Amblycerus perfectus* (Sharp) (20 ind.), *Amblycerus rufulus* (Sharp) (152 ind.), *Amblycerus whiteheadi* Kingsolver (1 ind.), *Amblycerus* sp. (3 spp., 14 ind.), *Caryedes clitoriae* (Gyllenhal) (2 ind.), *Caryedes godmani* (Sharp) (3 ind.), *Caryedes grammicus* (Gyllenhal) (3 ind.), *Caryedes incensus* (Sharp) (1 ind.), *Caryedes juno* (Sharp) (1 ind.), *Caryedes longicollis* (Fåhraeus) (2 ind.), *Caryedes longifrons* (Sharp) (4 ind.), *Caryedes paradisensis* Kingsolver & Whitehead (7 ind.), *Megacerus bifloccosus* (Motschulsky) (1 ind.), *Megacerus cubicus* (Motschulsky) (23 ind.), *Megacerus ramicornis* (Erichson) (6 ind.), *Meibomeus apicicornis* (Pic) (1 ind.), *Meibomeus* sp. (1 sp., 2 ind.), *Merobruchus paquatae* Kingsolver (14 ind.), *Pachyomerus bactris* (Linnaeus) (1 ind.), *Pygiopachymerus lineolus* Chevrolat (8 ind.), *Pygiopachymerus theresae* Pic (1 ind.), *Sennius breveapicalis* (Pic) (8 ind.), *Sennius rufomaculatus* (Motschulsky) (1 ind.), *Stator generalis* Johnson & Kingsolver (7 ind.), *Stator limbatus* (Horn) (16 ind.), *Stator trisignatus* (Sharp) (1 ind.), *Zabrotes interstitialis* (Chevrolat) (1 ind.), *Zabrotes propinguus* (Sharp) (2 ind.).

Chrysomelinae

Calligrapha argus Stål (1 ind.), *Platyphora opima* (Stål) (6 ind.), *Platyphora ligata* (Stål) (3 ind.).

Criocerinae

Lema sp. (1 sp., 2 ind.), *Lema insularis* Jacoby (6 ind.).

Cryptocephalini

Cryptocephalus tesseratus Chevrolat (43 ind.), *Cryptocephalus* sp. (16 spp., 325 ind.), *Diachus auratus* (Fabricius) (171 ind.), *Diachus* sp. (1 sp., 243 ind.), *Griburius purpurascens* Suffrian (4 ind.), *Griburius* sp. (10 spp., 332 ind.), *Lexiphanes* sp. (10 spp., 505 ind.), *Pachybrachis* sp. (5 spp., 295 ind.), *Stegnocephala* sp. (2 spp., 176 ind.).

Clytrini

Anomoea ?nigropunctata Dejean (3 ind.), *Coleothropa* sp. (2 spp., 19 ind.), *Coscinoptera* sp. (1 sp., 1 ind.), *Euryscopa cingulata* Latreille (2 ind.), *Ischiopachys bicolor proteus* Lacordaire (11 ind.), *Megalostomis amazona* Jacoby (13 ind.), *Proctophana basalis*

Lacordaire (6 ind.), *Proctophana fulvicollis* Jacoby (1 ind.), *Temnodachys ?intermedia* (Dejean) (4 ind.), *Urodera quadrisignata* Lacordaire (4 ind.).

Chlamysini

Aulacochlamys sp. (16 spp., 218 ind.), *Chlamisus* sp. (21 spp., 261 ind.).

Eumolpini

Allocaspis ?submetallicus Jacoby (239 ind.), *Antitypona* sp. (3 spp., 4657 ind.), *Brachypnoea* sp. (3 spp., 141 ind.), *Caryonoda* sp. (1 sp., 638 ind.), *Cayetunya consanguinea* (Blake) (35 ind.), *Chalcophana mutabilis* Harold (1 ind.), *Chalcophana* sp. (2 spp., 11 ind.), *Colaspis confusa* Bowditch (4 ind.), *Colaspis femoralis* Oliver (3 ind.), *Colaspis inconstans* Lefevre (1 ind.), *Colaspis lebasi* Lefevre (45 ind.), *Habrophora ?maculipennis* Jacoby (307 ind.), *Habrophora* sp. (1 sp., 12 ind.), *Myochrous ?coenus* Blanke (1 ind.), *Percolaspis* sp. (3 spp., 399 ind.), *Phanaeta ruficollis* Lefevre (3 ind.), *Prionodera* sp. (1 sp., 13 ind.), *Rhabdopterus fulvipes* Jac. (296 ind.), *Rhabdopterus* sp. (2 spp., 97 ind.), *Spintherophyta* sp. (2 spp., 223 ind.), *Typophorus* sp. (2 spp., 5 ind.).

Megascelidini

Megascelis sp. (2 spp., 5 ind.).

Galerucini

Coelomera cajennensis Fabricius (1 ind.), *Coelomera godmani* Jacoby (24 ind.), *Diabrotica duplicata* Jacoby (15 ind.), *Diabrotica godmani* Jacoby (42 ind.), *Diabrotica haroldi* Baly (9 ind.), *Diabrotica nummularis* Harold (1 ind.), *Diabrotica viridula* Fabricius (2 ind.), *Diabrotica* sp. (9 spp., 19 ind.), *Isotes puella* (Baly) (15 ind.), *Masurius* sp. (1 sp., 42 ind.), *Neobrotica* sp. (1 sp., 1 ind.), *Trichobrotica sexplagiata* (Jacoby) (34 ind.), unidentified genera (10 spp., 1238 ind.).

Alticini

Epitrix sp. (1 sp., 10 ind.), *?Syphraea* sp. (1 sp., 1 ind.), *Alaginosa* sp. (1 sp., 20 ind.), *Capraita* sp. (1 sp., 8 ind.), *Glenidion/Sanguria* sp. (1 sp., 49 ind.), *Heikertingiella* sp. (4 spp., 433 ind.), *Homotyphus ?albomaculatus* Jacoby (3 ind.), *Homotyphus* sp. (2 spp., 94 ind.), *Hydmosyne* sp. (1 sp., 11 ind.), *Monomacra* sp. (1 sp., 3 ind.), *Stegnea* sp. (2 spp., 101 ind.), *Systena oberthüri* Baly (13 ind.), *Trichattica* sp. (1 sp., 110 ind.), *Walterianella* sp. (3 spp., 196 ind.), unidentified genera (35 spp., 2184 ind.).

Cassidini

Acromis sparsa Boheman (2 ind.), *Calyptocephala brevicornis* (Boheman) (7 ind.), *Charidotella högbergi* Boheman (1 ind.), *Charidotis abrupta* Boheman (15 ind.), *Charidotis incincta* Boheman (16 ind.), *Charidotis leprieuri* Boheman (3 ind.), *Charidotis pustulata* Champion (16 ind.), *Charidotis vitreata* (Perty) (1 ind.), *Charidotis* sp. (1 sp., 1 ind.), *Chersinellia heteropunctata* (Boheman) (230 ind.), *Cistundinella foveolata* Champion (18 ind.), *Coptocycla rufonotata* Champion (35 ind.), *Cyclocassis circulata* Boheman (7 ind.), *Ischnocodia annulus* (Fabricius) (5 ind.), *Parachirida subirrorata* (Boheman) (49 ind.).

Hispini

Cephalodonta sp. (1 sp., 6 ind.), *Octhispa elevata* (Baly) (8 ind.), *Octhispa* sp. (2 spp., 5 ind.), *Oxychalepus anchora*(Chapuis) (11 ind.), *Probaenia* sp. (1 sp., 3 ind.), *Sumitrosis terminata* (Baly) (2 ind.), *Uroplata* sp. (1 sp., 3 ind.), Unidentified genera (6 spp., 20 ind.).

Lamprosomatinae

Lamprosoma sp. (4 spp., 6 ind.), *Oomorphus* sp. (3 spp., 542 ind.).

ANTHRIBIDAE

Eugonus decorus Jordan (6 ind.), *Euparius* sp. (1 sp., 2 ind.), *Goniocloeus* sp. (2 spp., 8 ind.), *Gymnognathus* sp. (1 sp., 1 ind.), *Neanthribus pistor* Jordan (2 ind.), *Ormiscus* sp. (6 spp., 166 ind.), *Phaenithon* sp. (1 sp., 1 ind.), unidentified genera (2 spp., 5 ind.).

ATTELABIDAE

Eugnamptus ?laticeps Voss (1 ind.), *Eugnamptus* sp. (1 sp., 7 ind.), *Omolabus callosus* (Sharp) (30 ind.), *Pseudauletes* sp. (1 sp., 15 ind.), *Xestolabus corvinus* (Gyllenhal) (157 ind.).

BRENTIDAE

Arrhenodes ?flavolineatus Gyllenhal (1 ind.), *Brenthus armiger* Herbst (4 ind.), *Brenthus anchorago* Linell (13 ind.), *Brenthus* sp. (1 sp., 4 ind.), *Paratrachelius* sp.

(2 spp., 6 ind.), *Teramoceus* sp. (1 sp., 6 ind.), *Ulocerus sordidus* Sharp (123 ind.), *Ulosomus laticornis* Sharp (2 ind.), *Apion johnsoni* Kissinger (1 ind.), *Apion maceratum* (Sharp) (18 ind.), *Apion* sp. (2 spp., 10 ind.), *Apionion championi* (Sharp) (1 ind.), *Apionion deleon* Kissinger (25 ind.), *Apionion latipenne* (Sharp) (13 ind.), *Apionion samson* (Sharp) (2 ind.), *Apionion* sp. (1 sp., 5 ind.), *Bothryopteron* sp. (1 sp., 3 ind.), *Chrysapion chrysocomum* (Gerstaecker) (5 ind.), *Coelocephalapion adulcirostre* (Gerstaecker) (4 ind.), *Coelocephalapion luteirostre* (Gerstaecker) (100 ind.), *Coelocephalapion nodicorne* (Sharp) (2 ind.), *Coelocephalapion pedestre* (Sharp) (1 ind.), *Coelocephalapion spretissimum* (Sharp) (13 ind.), *Coelocephalapion* sp. (14 spp., 83 ind.), *Kissingeria seminudum* (Wagner) (10 ind.), *Neotropion lebasii* (Gyllenhal) (1 ind.), *Trichapion glyphicum* (Sharp) (5 ind.), *Trichapion hastifer* (Sharp) (37 ind.), *Trichapion* sp. (3 spp., 48 ind.).

CURCULIONIDAE

Brachyderinae

Polydacrys depressifrons Boheman (2 ind.).

Otiorynchinae

Compsus auricephalus Say (50 ind.), *Eustylus sexguttatus* Champion (75 ind.), *Exophthalmus sulcicrus* Champion (16 ind.), *Exophthalmus* sp. (1 sp., 18 ind.).

Leptopiinae

Hypoptus macularis (Jekel) (19 ind.), *Promecops unidentatus* Champion (28 ind.).

Hyperinae

Isorhinus sp. (1 sp., 1 ind.), *Phelypera distigma* Boheman (1 ind.).

Cylydrorhininae

Hormops sp. (1 sp., 1 ind.).

Hylobiinae

Arniticus sp. (2 spp., 14 ind.), *Heilipodus flavolineatus* (Champion) (1 ind.), *Heilipodus lutosus* (Pascoe) (33 ind.), *Heilipodus unifasciatus* (Champion) (12 ind.), *Heilipus draco* (Fabricius) (27 ind.), *Heilipus elegans* Guérin (17 ind.), *Heilipus* sp. (9 spp., 75 ind.), *Heilus bioculatus* (Boheman) (43 ind.), *Hilipinus latipennis* Champion (31 ind.), *Hilipinus ?punctatoscabrosus* (Boheman) (19 ind.), *Hilipinus sulcicrus* (Champion) (3 ind.), *Marshallius guttatus* (Boheman) (16 ind.), *Parabyzes angulosus* (Champion) (2 ind.), *Rhineilipus sulcifer* (Champion) (18 ind.), *Rhineilipus* sp. (1 sp., 1 ind.), *Sternechus* sp. (1 sp., 10 ind.), *Tylomus/Sternechus* sp. (1 sp., 1 ind.).

Erirrhininae

Andranthobius palmarum Champion (944 ind.), *Andranthobius* sp. (2 spp., 568 ind.), *Celetes* sp. (1 sp., 9 ind.), *Phyllotrox ?megalops* Champion (8 ind.), *Phyllotrox* sp. (6 spp., 837 ind.), *Tereris ?pilosa* Champion (2 ind.), *Tereris* sp. (8 spp., 38 ind.), *Udeus ?eugnomoides* Champion (464 ind.), *Udeus* sp. (2 spp., 1704 ind.), Unidentified genus (1 sp., 13 ind.).

Otidocephalinae

Erodiscus antilope Fabricius (8 ind.), *Erodiscus* sp. (4 spp., 32 ind.), *Myrmex grandis* Chevrolat (5 ind.), *Myrmex panamensis* Champion (94 ind.), *Myrmex* sp. (24 spp., 390 ind.), *Ptinopsis* sp. (3 ind.), *Toxophorus attenuatus* Fabricius (22 ind.).

Magdalinae

Laemosaccus sp. (1 sp., 1 ind.).

Anthonominae

Achia sp. (1 sp., 4 ind.), *Anthonomus abdominalis* Champion (63 ind.), *Anthonomus alboscutellatus* Champion (52 ind.), *Anthonomus caeruleisquamis* Champion (1 ind.), *Anthonomus cyanipennis* Champion (1 ind.), *Anthonomus denticrus* Clark (1 ind.), *Anthonomus dogma* Clark (111 ind.), *Anthonomus excelsus* Clark (14 ind.), *Anthonomus fischeri* Blackwelder (3 ind.), *Anthonomus marmoratus* Champion (2 ind.), *Anthonomus pruinosus* Champion (2 ind.), *Anthonomus puncticeps* Champion (12 ind.), *Anthonomus squamiger* Champion (1 ind.), *Anthonomus stockwelli* Clark (172 ind.), *Anthonomus subparallelus* Champion (9 ind.), *Anthonomus tantillus* Champion (1 ind.), *Anthonomus* sp. (7 spp., 190 ind.), *Loncophorus santarosae* Clark (2 ind.), *Melexerus* sp. (2 spp., 2 ind.), *Neomastix setulosus* Champion (55 ind.), *Neomastix* sp. (1 sp., 133 ind.), *Pseudathonomus muon* Clark (10 ind.), *Pseudathonomus* sp. (3 spp., 26 ind.), unidentified genus (1 sp., 1 ind.).

Ceratopinae

Ceratopus bisignatus Boheman (70 ind.), *Ceratopus* sp. (1 sp., 36 ind.).

Prionomerinae

Ectyrsus sp. (1 sp., 2 ind.), *Piazorhinus* sp. (4 spp., 112 ind.), *Odontopus aesopus* (Fabricius) (5 ind.), *Odontopus* sp. (2 spp., 5 ind.), *Themeropsis binodosa* Champion (4 ind.).

Tychinae

Plocetes faunus Clark (16 ind.), *Plocetes omniguttatus* Champion (71 ind.), *Plocetes pusillus* Champion (1 ind.), *Plocetes* sp. (2 spp., 2 ind.).

Camarotinae

Camarotus sp. (2 spp., 54 ind.).

Cryptorhynchinae

Acamptoides sp. (7 ind.), *Aeatus costulatus* Champion (7 ind.), *Aeatus ebeninus* Champion (6 ind.), *Aeatus* sp. (3 spp., 109 ind.), *Apteromecus* sp. (4 spp., 8 ind.), *Atriches* sp. (1 sp., 3 ind.), *Chalcodermus angulicollis* Fåhraeus (1 ind.), *Chalcodermus calidus* Fåhraeus (1 ind.), *Chalcodermus lineatus* Champion (52 ind.), *Chalcodermus* sp. (3 spp., 5 ind.), *Cleogonus armatus* Champion (18 ind.), *Cleogonus ruberta* Fåhraeus (10 ind.), *Coelosternus varisquamis* (Champion) (16 ind.), *Conotrachelus aristatus* Champion (1 ind.), *Conotrachelus brevisetis* Champion (8 ind.), *Conotrachelus cestrotus* Faust (2 ind.), *Conotrachelus cristatus* Fåhraeus (43 ind.), *Conotrachelus ?dentimaculatus* Champion (5 ind.), *Conotrachelus deplanatus* Champion (2 ind.), *Conotrachelus foveicollis* Champion (1 ind.), *Conotrachelus glabricollis* Champion (16 ind.), *Conotrachelus lateralis* Champion (8 ind.), *Conotrachelus nodifrons* Champion (1 ind.), *Conotrachelus planifrons* Champion (272 ind.), *Conotrachelus punctiventris* Champion (4 ind.), *Conotrachelus serpentinus* Boheman (23 ind.), *Conotrachelus sobrinus* Boheman (50 ind.), *Conotrachelus suturalis* Champion (2 ind.), *Conotrachelus turbatus* Faust (23 ind.), *Conotrachelus ?venustus* Champion (1 ind.), *Conotrachelus verticalis* Boheman (38 ind.), *Conotrachelus* sp. (19 spp., 572 ind.), *Cophes* sp. (1 sp., 33 ind.), *Cryptorhynchus alutaceus* Champion (1 ind.), *Cryptorhynchus bifenestratus* Champion (2 ind.), *Cryptorhynchus binoculatus* Champion (8 ind.), *Cryptorhynchus carinifer* Champion (1 ind.), *Cryptorhynchus concentricus* Champion (40 ind.), *Cryptorhynchus foveifrons* Champion (8 ind.), *Cryptorhynchus ?fraterculus* Champion (1 ind.), *Cryptorhynchus iniquus* Champion (31 ind.), *Cryptorhynchus ?latisquamosus* Champion (1 ind.), *Cryptorhynchus ?mistus* Champion (1 ind.), *Cryptorhynchus ?tirunnculus* Boheman (6 ind.), *Cryptorhynchus strigatus* Champion (56 ind.), *Cryptorhynchus* sp. (24 spp., 446 ind.), *Episcirrus ?propugnator* (Gyllenhal) (3 ind.), *Eubulomus ?multicostatus* Champion (1 ind.), *Eubulomus reflexirostris* Champion (21 ind.), *Eubulomus* sp. (2 spp., 211 ind.), *Eubulus ?biplagiatus* Champion (2 ind.), *Eubulus campestris* Champion (2 ind.), *Eubulus circumlitus* Champion (2 ind.), *Eubulus crispus* Champion (15 ind.), *Eubulus inaequalis* Champion (2 ind.), *Eubulus niveipectus* Hustache (4 ind.), *Eubulus punctifrons* Champion (80 ind.), *Eubulus* sp. (11 spp., 25 ind.), *Euscepes porcellus* Boheman (25 ind.), *Isus m-nigrum* Champion (3 ind.), *Macromeropsis binotata* Champion (1 ind.), *Metriophilus horridulus* Champion (173 ind.), *Metriophilus minimus* Champion (1 ind.), *Metriophilus ramulosus* Champion (24 ind.), *Metriophilus v-fulvum* Champion (2 ind.), *Metriophilus* sp. (2 spp., 9 ind.), *Microhyus* sp. (1 sp., 2 ind.), *Microxypterus suturalis* Champion (2 ind.), *Microxypterus* sp. (2 spp., 21 ind.), *Neoulosomus hispidus* Champion (2 ind.), *Rhynchus* sp. (1 sp., 45 ind.), *Rhyssomatus alternans* Champion (4 ind.), *Rhyssomatus nigerrimus* Fåhr (2 ind.), *Rhyssomatus* sp. (6 spp., 25 ind.), *Semnorhynchus fulvopictus* Champion (35 ind.), *Semnorhynchus planirostris* Champion (19 ind.), *Semnorhynchus tristis* Champion (15 ind.), *Semnorhynchus* sp. (2 spp., 20 ind.), *Staseas granulata* Champion (4 ind.), *Sternocoelus tardipes* Boheman (14 ind.), *Sternocoelus* sp. (1 sp., 6 ind.), *Thegilis baridioides* Champion (1 ind.), *Trachalus* sp. (1 sp., 14 ind.), *Tyrannion ochreopunctatum* Champion (3 ind.), *Zascelis ?affaber* Boheman (1 ind.), *Zascelis ?glabrata* Champion (27 ind.), unidentified genera (3 spp., 3 ind.).

Zygopinae

Copturominus sp. (4 spp., 23 ind.), *Copturus fulvosignatus* Champion (1 ind.), *Copturus verrucosus* Champion (42 ind.), *Copturus* sp. (3 spp., 95 ind.), *Eulechriops ornata* Champion (24 ind.), *Eulechriops* sp. (31 spp., 369 ind.), *Euzurus ornativentris* Champion (8 ind.), *Hoplocopturus nigripes* Champion (19 ind.), *Isotrachelus tibialis* (Champion) (223 ind.), *Isotrachelus* sp. (4 spp., 854 ind.), *Lechriops analis* Champion (86 ind.), *Lechriops aurita* Champion (51 ind.), *Lechriops canescens* Champion (2 ind.), *Lechriops parotica* (Pascoe) (69 ind.), *Lechriops vestita* (Boheman) (20 ind.), *Lechriops* sp. (16 spp., 410 ind.), *Philides* sp.

(1 sp., 71 ind.), *Pseudolechriops megacephala* Champion (33 ind.), *Pseudolechriops* sp. (1 sp., 1 ind.), *Pseudopinarus condyliatus* (Boheman) (2 ind.), *Psomus* sp. (5 spp., 206 ind.), *Neozurus aurivillanus* (Heller) (3 ind.), *Zygops tridentatus* Gyllenhal (6 ind.), unidentified genus (1 sp., 1 ind.).

Tachygoninae
Tachygonus semirufus Champion (5 ind.), *Tachygonus sp.* (6 spp., 18 ind.).

Ceuthorhynchinae
Hypocoeloides angulata Champion (16 ind.), *Hypocoeloides gibbicollis* Champion (18 ind.), *Hypocoeloides* sp. (4 spp., 16 ind.).

Baridinae
Acanthomadarus dirus Champion (14 ind.), *Ampeloglypter speculifer* Solari & Solari (10 ind.), *Ampeloglypter* sp. (4 spp., 25 ind.), *Baris aerea* (Boheman) (9 ind.), *Baris* sp. (2 spp., 31 ind.), *Cercobaris ?fortirostris* Champion (14 ind.), *Cercobaris* sp. (3 spp., 94 ind.), *Coelomerus* sp. (5 spp., 666 ind.), *Coelonertus nigrirostris* Solari & Solari (70 ind.), *Coelonertus* sp. (1 sp., 3 ind.), *Coluthus cribrarius* Champion (25 ind.), *Cylindrocerus comma* Boheman (9 ind.), *Cylindrocerus* sp. (1 sp., 3 ind.), *Cyrionyx* sp. (1 sp., 28 ind.), *Diorymerus serrulatus* Champion (16 ind.), *Diorymerus* sp. (15 spp., 279 ind.), *Eurhinus festivus* Fabricius (3 ind.), *Eurhinus viridicolor* Champion (1 ind.), *Geraeus biplagiatus* Champion (3 ind.), *Geraeus cemas* Champion (7 ind.), *Geraeus longiclava* Champion (39 ind.), *Geraeus tenuiclava* Champion (15 ind.), *Geraeus trilineatus* Champion (267 ind.), *Geraeus* sp. (16 spp., 186 ind.), *Glyptobaris ?rugosa* Boheman (2 ind.), *Glyptobaris* sp. (3 spp., 30 ind.), *Limnobaris* sp. (1 sp., 24 ind.), *Loboderes flavicornis* Champion (48 ind.), *Loboderes sulfureiventris* Champion (19 ind.), *Loboderes* sp. (4 spp., 57 ind.), *Madarellus rufomaculatus* Champion (1 ind.), *Madarellus* sp. (2 spp., 17 ind.), *Madarus ?clavipes* Champion (33 ind.), *Madarus distigma* Boheman (20 ind.), *Madarus eutoxoides* Champion (24 ind.), *Madarus excavatus* Champion (1 ind.), *Madarus* sp. (6 spp., 39 ind.), *Optatus fasciculosus* Champion (1 ind.), *Pachybaris ?stolida* Champion (6 ind.), *Parisoschoenus expositus* Champion (33 ind.), *Plocamus* sp. (1 sp., 3 ind.), *Pseudobaris binotata* Champion (2 ind.), *Pseudobaris sexguttata* Champion (1 ind.), *Pseudobaris* sp. (6 spp., 72 ind.), *Sibariopsis ?concinna* (LeConte) (8 ind.), *Solaria acutidens* Champion (2 ind.), *Sphenobaris quadridens* Champion (5 ind.), *Stegotes humeronotatus* Champion (1 ind.), *Stegotes* sp. (2 spp., 6 ind.), *Zygobaris* sp. (1 sp., 7 ind.), unidentified genera (79 spp., 906 ind.).

Rhynchophorinae
Sitophilus oryzae Linnaeus (1 ind.), Unidentified genus (1 sp., 4 ind.).

Cossoninae
Acamptus plurisetosus (Champion) (7 ind.), *Caulophilus* sp. (1 sp., 1 ind.), unidentified genera (2 spp., 2 ind.).

Platypodinae
Platypus parallelus Fabricius (1 ind.), *Platypus* sp. (1 sp., 1 ind.), unidentified genus (1 sp., 1 ind.).

Scolytinae
Araptus sp. (4 spp., 12 ind.), *Bothrosternus definitus* Wood (1 ind.), *Bothrosternus foveatus* (Blackman) (5 ind.), *Cnesinus blackmani* Schedl (2 ind.), *Cnesinus retifer* Wood (3 ind.), *Cnesinus* sp. (2 spp., 3 ind.), *Coccotrypes carpophagus* (Hornung) (1 ind.), *Corthylus pumilus* Wood (2 ind.), *Hypothemnus birmanus* Eichhoff (17 ind.), *Hypothemnus erectus* LeConte (1 ind.), *Hypothemnus eruditus* Westwood (6 ind.), *Hypothemnus serriatus* Eichhoff (27 ind.), *Pithyophthorus* sp. (1 sp., 1 ind.), *Pagiocerus frontalis* (Fabricius) (2 ind.), *Scolytodes atratus* (Blandford) (1 ind.), *Scolytodes ovalis* (Eggers) (11 ind.), *Scolytodes perditus* Wood (1 ind.), *Scolytodes punctifrons* Jordal (2 ind.), *Scolytodes* sp. (1 sp., 1 ind.), *Theoborus* sp. (1 sp., 7 ind.), *Xyloborus ferox* Blandford (1 ind.), *Xyloborus ferrugineus* (Fabricius) (38 ind.), *Xyloborus volvulus* (Fabricius) (15 ind.).

22

Microhabitat distribution of forest grasshoppers in the Amazon

Christiane Amédégnato

ABSTRACT

Organisms are strongly associated with their microhabitats, which can be simple or multidimensional. The microhabitat is more a determinant for the organism than is the macrohabitat as a whole. This is especially true for forest grasshoppers, which, during their life cycle, can live, eat and oviposit in either the same or in different microhabitats, which, in turn, can be more or less closely related or spatially distinct. The aggregate of these individual microhabitat preferences can lead to partitioning of both space and food resources. In the Amazonian forest, known for its structural diversity, the importance of microhabitats is perhaps greater than elsewhere. The main aim of this chapter is to characterize those structural components and parameters of the microhabitats important for grasshoppers, and to chart their distribution in the forest. To this is added a consideration of the multiple microhabitats used by some 390 taxa of grasshoppers. Together, the results cast light on the stratification and compartmentalization of microfaunas associated with particular resources and behaving as functional assemblages of species. Additionally, the distribution and relative frequency of taxa and of guilds has been compared at different sites across the whole of the Amazon basin. Points of difference are identified that could be the result of different histories of biodiversity: histories related or not to differences in environmental richness. The different types of microhabitat are compared in terms of their propensity to facilitate or inhibit the generation of diversity of the Acridomorpha.

INTRODUCTION

In recent years there has been a significant increase in the attention paid to relationships between plants and insects (Schoonhoven *et al.*, 1998) and, simultaneously, to life in forest canopies (Lowman & Moffett, 1993; Stork *et al.*, 1997a; Watt *et al.*, 1997a). The canopy biota, especially in tropical zones, is so rich and so diverse and at the same time so threatened that the need for its study is more pressing than ever. This is especially true of the herbivores, which make up half of all the insects, themselves constituting more than half the living world (Strong *et al.*, 1984). Although the acquisition of this knowledge is a formidable undertaking, it is of crucial importance for understanding the context in which phytophagous insects have evolved and for formulating hypotheses about the organization and evolution of herbivores within communities. Canopy grasshoppers (Orthoptera, Acridomorpha), which constitute a significant fraction of these insects, require special techniques as they are not captured in the general fogging studies that have provided most of our information on canopy arthropods (Adis *et al.*, 1997). We need to know how they are distributed through the forest as a whole, their vertical stratification and horizontal distribution, and their patterns of spatio-temporal utilization of the forest mosaic, at both local and larger scales.

Amédégnato (1997) examined the grasshopper assemblage of the mature forest and its canopy and showed how numerous sympatric species of the same genera shared food resources, especially the foliage of trees. Several guilds have been defined in terms of foraging location and food type: supracanopy (overstorey) species, species active at the canopy surface, canopy-foliage species (generalists and specialists), species dependent on epiphytes and lianas (generalists and specialists), species characteristic of trunks and branches, and specialists on palms (C. Amédégnato, unpublished data).

Since this earlier work (Amédégnato, 1997) three more north-west Amazonian forest canopies have been

Table 22.1. *Sites and samples used for the study*

Samples	Vegetation type	Number of trees or 500 m^2 plots[a]	Number of specimens	Arboreal species[b]	Low vegetation species[b]
Amazon (Colonia)	Succession	22 (0–40 years)	5053	82 (88) (144)	34 (42)
Amazon (Estiron)	Succession	14 (7–42 years)	1452	63 (88) (144)	34 (42)
Amazon (Estiron 1)	Forest (paleoterrace)	60	1324	72 (88) (144)	13 (22)
Amazon (Estiron 2)	Forest (plateau)	20	719	70 (88) (144)	19 (22)
Amazon (Estiron 3)	Forest (plateau)	25	992	72 (88) (144)	18 (22)
Yasuni E	Succession	7 (0–21 years)	1234	65	26 (44)
Yasuni E	Forest (plateau)	38	643	70	13 (22)
Yasuni W	Forest (plateau)	49	848	70	16 (24)
Napo	Forest (plateau)	20	412	57	15 (24)
Ucayali	Forest (plateau)	47	592	56	16 (?)
Pakitza	Forest (plateau)	50	397	50	? (20)
Manaus 1	Forest (plateau)	?	279	36 (41)	? (30)
Manaus 2	Forest (plateau)	45	250	35 (41)	5 (30)
French Guiana	Forest (?)	34	755	51	17 (32)

[a] Age of plots in parentheses.

[b] First number in parentheses is cumulative number for succession or forest only; second number in parentheses is cumulative number for both forest and succession.

sampled, along a transect that now crosses the whole of north-western Amazonia, from the Amazon to the upper Rio Napo in the foothills of the Andes. In addition, the regeneration communities that were only superficially examined in this western region now have been analysed in depth.

The recently acquired data, together with previous material (succession study: C. Amédégnato, unpublished data; other north-western Amazonian samples: Amédégnato, 1998) and careful re-examination of the previous samples (Amédégnato & Descamps, 1980b), make it possible to analyse the spatial distribution of the more important grasshopper genera at both local and trans-Amazonian scales.

To date, there has been no treatment of the microhabitats that, although not strictly part of the canopy, are still very much part of the forest environment, such as light gaps of different ages and the leaf litter, nor of the poorly understood links among the different compartments of the forest mosaic. Similarly, different forest communities have not been compared. In general, there are few data available on stratification and microhabitat occupation (Basset *et al.*, 1992; Hollier & Belshaw, 1993; DeVries e*t al.*, 1997) or on geographical comparisons. Where such data do exist, they refer usually only to vertebrates or plants (Emmons, 1984; Gentry & Emmons, 1987; Gentry, 1990). The aim of this chapter is to draw together the general information we now have about the way of life of forest grasshoppers in different sorts of canopy and other parts of the forest. Further, I consider individual species and the different microhabitats they occupy throughout their life cycle, from egg to adult.

METHODS

The taxonomic composition and ecological relations of forest grasshopper populations have been analysed at eight different sites, distributed over the Amazon basin (Fig. 22.1, Table 22.1). Various additional collections from sites widely distributed over the Amazon basin have not been used in this study, as the data they provide are inadequate for two reasons. First, some collections from several different places in the Colombian, Ecuadorian, Peruvian and Brazilian Amazon are incomplete. Second, relatively complete collections from localities in Brazil (Benjamin Constant, Tabatinga, Obidos, Santarem, Belem, Ouro Preto, Jutai) have not yet been analysed completely. These data are, nonetheless, useful in assessing the general biogeographical distribution of the forest grasshopper fauna. Among the sites studied in detail, six (Ucayali, French Guiana, Napo, East and West Yasuni, and Pakitza) are

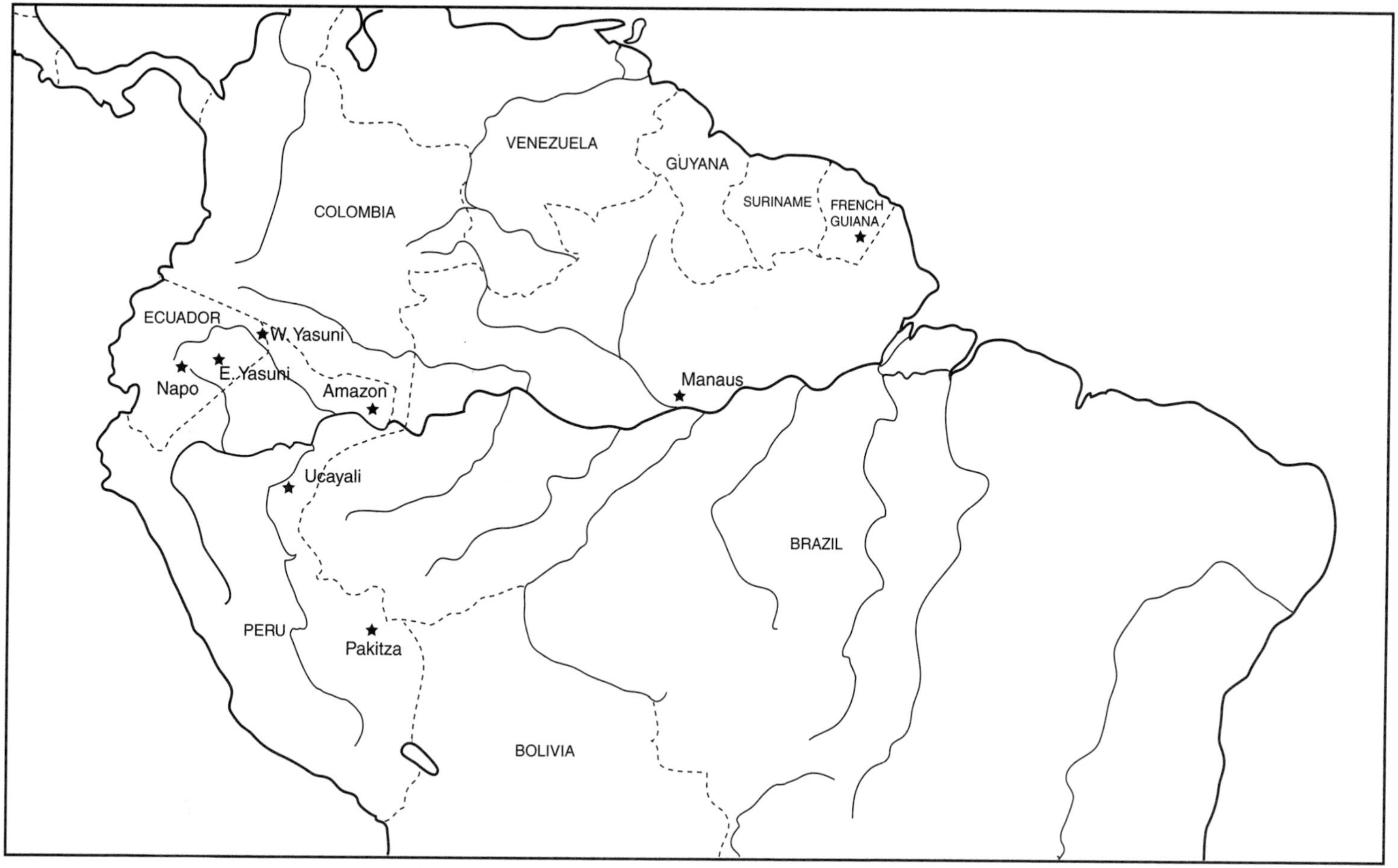

Fig. 22.1. Geographical origin of the samples used for the study.

represented by a single quantitative sample (supplemented by other collections). The site 'Amazon' includes several samples (see Table 22.1). All are situated along or close to the river Ampiyacu, which runs parallel to the Amazon at the foot of the first terrace: it limits the flood plain of the latter and acts rather as one of the banks of the Amazon than as an autonomous river basin. These samples were taken during different months and different years, two with fewer trees than the first sample: this was considered adequate to assess the location reliably. Three other forest samples designed for a different purpose are not used here. At the Manaus site, two samples were taken in different years, because the first sample was so poor. At both these sites, the results of replicated samples were consistent, as indicated by the Bravais–Pearson correlation coefficient calculated between the respective lists of the species abundances (Amazon, Estiron: $r = 0.859$; Manaus: $r = 0.869$).

Apart from the repeated samples beside the Amazon, all the remaining sites were sampled until the cumulative species curve levelled out. In a mature forest, a more or less similar number of trees were felled. They were cut at random within a circle of approximately 2 to 3 km in diameter. Alternatively, where regenerating vegetation was being investigated, the sample unit was two to five plots of 500 m^2 each per age. The latter procedure was carried out at the Rio Yubineto (near the E. Yasuni site), Colonia and Estiron (both near the Amazon). To assess understorey populations, insects were sought within the perimeter of the trees that were to be felled or, for secondary growth, after the trees or other vegetation had been cut. All insects emerging at the surface of the resulting mass of vegetation were collected, until no more appeared (three or four collectors during 2 or 3 hours or more, depending on weather conditions).

As the aim was a quantitative estimate, all specimens were collected in open areas or areas of regrowth after clearing the vegetation if necessary. In light gaps, and especially at their edges, this was not done, and for these communities data are only available on the presence or absence of different species and subjective impressions of their density.

At all the sampled sites, there are long-term research programmes in progress with extensive documentation; in the case of Pakitza this is gathered together as a book (Wilson & Sandoval, 1997). A comparison of many aspects of the different sites has also been published (Gentry, 1990).

Within these various habitats, the microhabitat occupied by particular grasshopper species had to be estimated indirectly, using the following information:

- the microhabitat occupied was determined by considering the stratum of vegetation in which the insect was found, its food plant preferences (see below) and the type of vegetation (i.e. the actual tree itself, epiphytes or lianas: these factors are tabulated in the Appendix under the headings life zone, vegetation structure and height level
- the food plant preferences were ascertained by microscopic comparison of the gut contents (after appropriate histological preparation) with epidermal samples from the major constituents of the collection of trees sampled (photographic catalogue of the epidermis): these are tabulated in the Appendix as food category and specialization
- the behavioural characteristics were assigned in the field using the standards generally applied in acridological studies, for example the type and site of oviposition, the type of escape behaviour shown, the degree of preference for light and degree of gregariousness.

The results led to the characterization of arboreal guilds, which are used to compare sites (surface-active taxa, tree folivores, liana specialists, liana and epiphyte generalists, bark-dwelling taxa, supracanopy taxa).

Detailed work of this sort was carried out only at the north-west Amazonian sites. At the central Amazonian and Guyanese sites, the gut contents were not preserved when the original collections were made. For these it was only possible to investigate the gut contents of dried specimens of the most abundant species. This approach was handicapped by the absence of local plant collections at these sites, so I have indicated only the main categories of food plant and not tried to identify them to plant species. The results were adequate to show that, with few exceptions, the same genus (or closely related genera) has the same behaviour everywhere. However, there is almost no information available about the populations of grasshoppers on regenerating vegetation in French Guiana.

For the main genera of Amazonian grasshoppers, a multifactorial analysis of the data obtained from all sites (see the Appendix) has been performed. The methodology was chosen to suit the data. Even after excluding genera and species that are apparently rare (or, at least, for which there are few data), there are still very large differences in abundance among the taxa that are characteristic of a given microhabitat. I have preferred, therefore, to characterize the taxa by their (almost exclusive) presence in their typical habitat, and to use reciprocal averaging, involving one matrix of all the data based on the χ^2 distance. This technique was preferred over canonical correspondence analysis, which uses two matrices, one for abundance and one for other attributes. The data matrix used included information on about 390, mainly arboricolous, species. The geographical distribution of the individual species is often so restricted that they do not participate usefully in such a global analysis: accordingly, these have been analysed at the level of the genus or species group. These species groups are either phylogenetic (*Poecilocloeus*) or functional (*Ophthalmolampis*, *Anablysis*) in nature, and their attributes are tabulated in the Appendix in the same way as those of species. The resulting matrix comprised about 130 individual taxa or sets of taxa. This allows one to visualize the space occupied by the grasshoppers in relation to the principal environmental factors.

The samples used in this study, covering the Amazon basin as a whole, have been compared in terms of the main forest guilds. It was originally intended to include the extreme margins of the Amazonian forest in this comparison, but it transpired that the faunas there were too different and really seem to represent another biota. At the extreme north and south of the Amazon basin, and close to the *cerrado* zone, we have insufficient data. Nonmetric dimensional scaling based on ranked distance (Sørensen measure) has been used for this purpose, as an ordination technique particularly suitable for data of this type (variation of relative abundance of guilds between the sites). The software packages used were PC-ORD™ and Statbox™.

RESULTS AND DISCUSSION

The main microhabitat complexes in relation to food-plant selection

The microhabitat concept and its important parameters for a forest grasshopper

The niche of a grasshopper, as of any other insect, cannot be equated with the main microhabitat used for feeding or mating, because this main area is often visited for only relatively short periods of time, with the rest of the time spent in other microhabitats that are essential refuges. The microhabitat of a species is, therefore, often complex and represents the sum of a variety of microhabitat elements. Apart from the hazards of dispersal (which are only important in low-density populations), it is primarily this range of habitat use that explains why a normally abundant species is sometimes not present on its normal host plant (e.g. the host plant has not reached its mature architecture, light conditions are bad or the normal refuge zone is not available). This does not apply, however, to specialists living colonially on their dispersed food plant in the light gaps of the Central American forests (Rowell, 1978; Rowell *et al.*, 1983) or in the montane forests of South America (C. Amédégnato, personal observation).

Different, complementary microhabitats may also be necessary for the different developmental stages of a species; however, as is usual for hemimetabolous insects, the larvae of most grasshoppers require more or less the same microhabitats and foodstuffs as the adults. In the case of some Romaleidae, which live in large gregarious bands in forest clearings, it has been shown that the food of the larvae influences the later choice of the adults after they disperse (Turk & Barrera, 1976). This is probably a general phenomenon, given that observation shows that this type of insect is somewhere between a polyphage and an oligophage. In the case of some arboricolous species, the mode of life of the larvae can be different from that of the adults: for example, the larvae of *Adelotettix*, a very mobile genus of the supracanopy, are cryptic and mimic bark or leaf litter. Members of this genus oviposit in the soil and the larvae climb to the canopy up the trunks, where we have trapped some of them. The genus *Saltonacris*, however, shows the same sort of differences in colouration between adult and larva, and there we have not noted any differences in life style between the larva and the adult.

In the forest ecosystem, the main zones likely to offer specific microhabitats for acridids are the trees themselves (leaf and woody systems); the associated epiphytes; lianas; the canopy surface; the space between the main canopy surface and the crowns of the emergent giant trees (overstorey); patches of smaller, regenerating trees; and light gaps, at various stages of succession. Contrary to what might be expected, the intermediate strata of trees, or at least their foliage, does not seem to provide an important habitat, as no significant population of grasshoppers appears to be associated with these secondary trees (Amédégnato, 1997). Further, there are marginal microhabitats interspersed within the forest, such as river banks, degenerate oxbow lakes, swamps or inselbergs: these are, in fact, very specific microhabitats, providing patches of large open space.

A multifactorial reciprocal averaging analysis including all the different places occupied by grasshoppers for whatever function (see the Appendix) shows the degree of occupation of the main microhabitats and their interrelationships. The resulting graph (Fig. 22.2) shows different aggregations of points, corresponding to the main biotic compartments of importance to forest grasshoppers. They are directly correlated with the main factors (stratification, life zone, etc.) that characterize the combination of microhabitats used for different purposes (e.g. as food, oviposition sites, as refuges, etc). The first axis is explained by five factors with high relative contributions (food category and height of the vegetation) whereas the second axis is explained by one important factor (oviposition type) and five other factors with a weaker impact (see the caption to Fig 22.2). The analysis suggests that the greater part of the fauna (to the left and around axis 2) lives in the canopy and is very diverse (as reflected in the large number of genera). Quantitatively speaking, however, the communities represented to the right and to the left on the graph are roughly equivalent in numbers of individuals. Reading from right to left, the following biota are represented

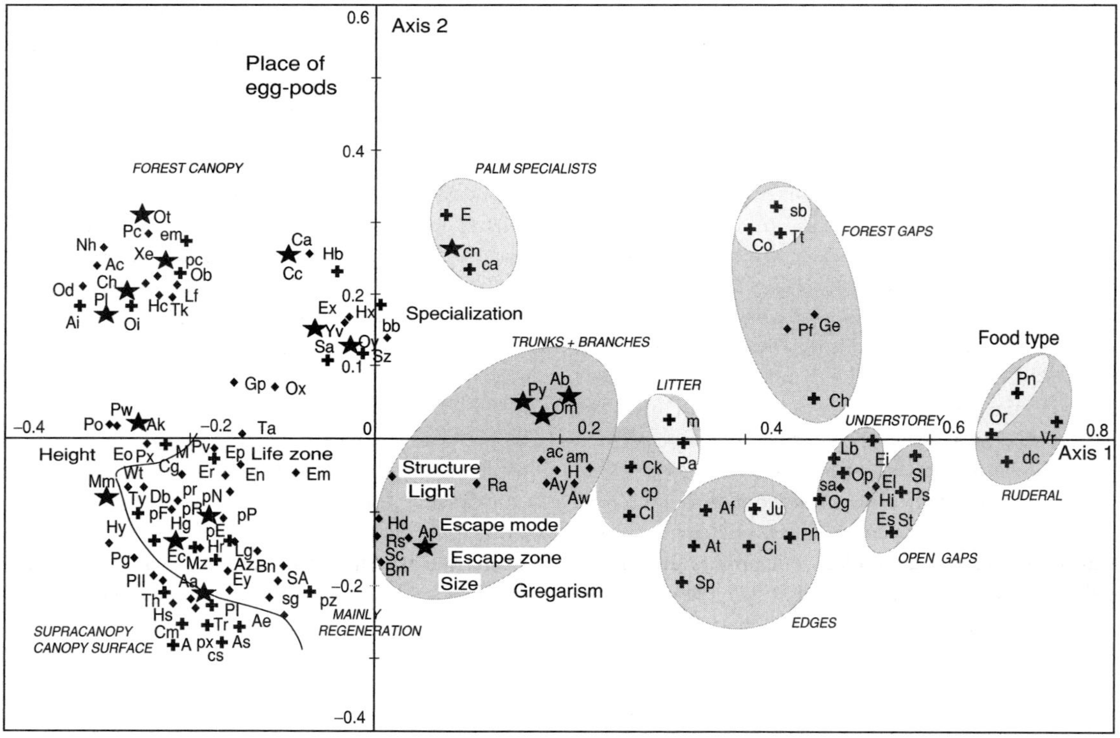

Fig. 22.2. Multifactorial analysis (reciprocal averaging) of spatial distribution and stratification of the main grasshopper genera within the Amazonian forest mosaic (see taxonomic list in the Appendix to Ch. 22 for the codes of taxa and attributes); small letters indicate habitat/microhabitat and behavioural attributes; capital letters indicate species associations (light-coloured areas within the right-hand aggregations: specialization on grasses); markers: ◆ West Amazonian genera; + Panamazonian genera; ★ Guyanese genera. Relative contribution to the axes: axis 1: food category (*F*) 0.923; height in the vegetation (*H*) 0.802; life zone (*Z*) 0.598; adult gregarism (*AG*) 0.415; nymphal gregarism (*NG*) 0.383; axis 2: egg-lying mode (*EG*) 0.793; size (*Sz*), 0.254; adult gregarism (*AG*), 0.235; escape zone (*EZ*), 0.236; food specialization (*S*), 0.185; nymphal gregarism (*NG*), 0.171.

on the graph: species of open sunny gaps, lower strata and edges; those of forest gaps and understorey (sun- and shade-loving variants); leaf litter species; grasshoppers of trunks and branches; and palm canopy genera. Insects of mature forest canopy are placed to the upper left and, finally, at the bottom left, there is the largest aggregation, consisting mainly of the fauna of the successional canopy (regeneration patches). In this last category, the external part of the aggregation comprises the supra- and surface-active canopy species, which share common characteristics with the successional fauna.

It is clear that the technique of combining the complementary microhabitats for each species is adequate to characterize the position of each within the environmental mosaic of the forest and to suggest possible relationships and links among the microhabitats. It also shows how some guilds requiring more than one microhabitat are affected and, potentially, how they interact with others. These species associations, linked to particular microhabitats, are themselves further subdivided according to the realized niches of each taxon, and they will now be reviewed in more detail.

Sunny open gaps with low (ruderal) vegetation

The typical ruderal assemblage inhabiting open sunlit gaps in the forest is composed of a small number of genera, all of very wide distribution in South America. All are of small size, lay a moderate number of eggs in bare soil (12 to 40) and have a rapid embryonic development (3 weeks from the egg to the last nymphal stage). Within this transient habitat, each genus occupies a particular microhabitat. *Dichroplus* is, at least partly, geophilous on bare ground, whereas *Vilerna* is found more at the edge of the gap where there is some leaf litter. *Peruvia* is also more or less geophilous but prefers dry places with small dry branches. *Orphulella* lives in grass (the several species of *Orphulella* have slightly different preferences, but two species always coexist in Amazonian light gaps, the commonest being *O. concinnula* Walker). As in the canopy, here also there are generalists and specialists – *Dichroplus* and *Vilerna* feed on small ruderal dicotyledons whereas *Orphulella* and *Peruvia* are grass specialists – but all are early pioneers of newly open gaps, as shown in different succession studies on the Rio Yubineto (Amédégnato & Descamps, 1980a) and near the Amazon (C. Amédégnato, unpublished data).

Sunny gaps with bushes

In open gaps with somewhat higher vegetation lives another category of pioneers, such as *Sitalces dorsualis* Giglio Tos and *Eusitalces vittatus* Bruner, polyphagous on low bushes. They are the successors of the early ruderal pioneers and become dominant in slightly later successional stages. Despite their small size, they colonize their habitats by flying (the two named species have each two morphs, winged and wingless).

True forest gaps and understorey

In this environment, light conditions vary from place to place during the course of the day. Accordingly, they harbour insects of diverse preferences. These species appear only later in the process of forest regeneration and are the characteristic fauna of tiny forest gaps. They fall into two distinct categories, the shade-loving species of gaps and the understorey, and the more sun-loving species which are the typical true forest-gap grasshoppers.

These populations, whether shade or sun loving, are frequent in the early part of succession (especially at around 7–10 years) but decrease sharply in later stages and in the forest understorey, where they are sparsely scattered, restricted to small patches of low vegetation. Most of these species are polyphagous. There is, however, a hygrophilous component, specializing on large monocotyledons such as *Heliconia, Canna* or sometimes small palms, most often found in the understorey or in gaps caused by small streams. These are all very specialized genera with endophytic egg pods and, in some taxa, are gregarious as nymphs (*Tetrataenia, Cornops* (see Turk, 1985)). The predominant genera range over the whole Amazon basin without diversification and belong to the subfamily Leptysminae. This group is found in open environments, associated in general with monocotyledons and, in one subdivision, specifically with grasses.

Edges of open gaps

These edges of gaps are inhabited by another kind of pioneer, a variant of the sun-loving ruderal assemblage described above. They are of larger size, live at a higher level in the vegetation and seem to have a somewhat slower development. All are very mobile and occur over virtually all of South America. Like the ruderal assemblage, they are mostly generalists, with one grass specialist, *Jagomphocerus* (an Amazonian vicariant of the well-known subtropical genus *Rhammatocerus*). The genus *Schistocerca* belongs to this group, represented by two species, one exclusively forbivorous, the other partially graminivorous.

Forest litter

A few genera are associated with forest litter. They do not, however, subsist on litter but feed on young plants. They are all rather large and consequently occur at low densities. They are heterogeneous both in aspect and in their way of life. Most eat dicotyledons, but two Phaeoparini (*Phaeoparia, Maculiparia*) are graminivorous or eat monocotyledons of the same structural type such as palm seedlings or Cyclanthaceae. *Rehnuciera* sp. (Clematodinini, Ommatolampinae) is marginally a member of this group. It specializes on mosses growing at the very base of tree trunks and, therefore, it lives at the level of the litter, going from the litter itself to its food source (in Fig. 22.2 it is an outlier, classified with the trunk and branch insects).

Trunks and branches

The genera of trunks and branches appear to be a subdivision of the ground fauna, with which they share many traits (e.g. oviposition habits, morphological and some behavioural adaptations). They occur on the entire woody part of the tree from the base to the canopy. Two closely related groups, the Vilernae and the Anablysae, fill this role in the Amazonian and Guianan forest. In western Amazonia the Anablysae inhabit mainly the upper levels (branches) whereas the Vilernae are dominant on the trunks and in the intermediate zone between the trunks and the understorey. In the eastern zone, however, the Anablysae occupy all levels. These genera depend on food sources related to their habitat. In the lower forest strata and on the trunks – wherever it is damp enough – the Vilernae are associated with mosses. At the lower levels they include some understorey dicotyledons as food, in proportions varying among genera and species. The Anablysae are still poorly understood. Usually their guts were found to be empty or to contain only highly comminuted remains of bark. Those living higher in the canopy feed on foliage (e.g. *Eurybiacris*) and some of these are even surface-active species (*Anablysis arboricola* Descamps in French Guiana). Within the genus *Anablysis*, then, there is a wide range of food-plant specialization, extending from foliage to bark and associated tiny plants, and from the lowest strata to the high crowns. If we take into account all aspects of the biology of the genera within this category (oviposition sites, mobility, food, etc.) they may be interpreted as exploiting trunks and branches as an extension of the soil.

Canopy regeneration patches

Most of the arboreal insects belonging to canopy regeneration patches are grouped in the bottom left of Figure 22.2. A summary of the attributes of this assemblage is given in Figure 22.3, which indicates the

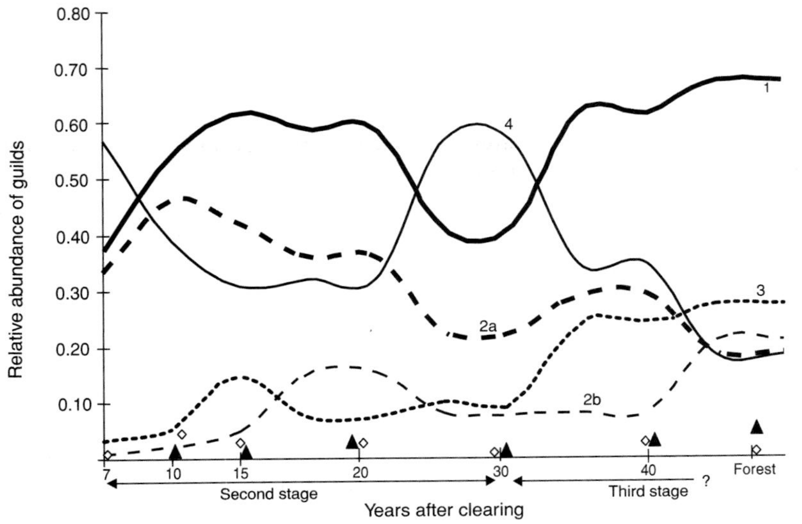

Fig. 22.3. Changes in the functional characteristics of the successional canopy fauna. 1, total of the intracanopy species (bold line); 2a, intracanopy foliage feeders with terrestrial oviposition; 2b, the same, with epiphytic oviposition; 3, feeders on intracanopy lianas and epiphytes; 4, canopy surface species (simple line); supracanopy species (◇); trunk and branch species (▲).

change in abundance of the various guilds over time during forest regeneration. These insects of successional forest are close to those of mature forest. Indeed, two groups of the latter (surface-active canopy species and supracanopy species) appear in Figure 22.2 grouped with the fauna of regenerating forest (forming the external margin of the cloud of points); this is because of numerous shared attributes, such as oviposition preferences, specialization on tree foliage as food, strong heliophily. The patches of successional forest have a vegetational structure that is much less complex than in the mature forest (e.g. fewer epiphytes, which appear at about 12 years into the succession (Gomez-Pompa, 1971) and which have a drier canopy microclimate) and does not offer the same diversity of microhabitats as does the mature forest. These patches, therefore, host different species. Most of the grasshoppers of regenerating forest are oligophagous on plant families typical of the successional flora (Flacourtiaceae, Melastomataceae), the foliage of which differs from that of mature forest (Whitmore, 1979, 1989). However, as succession is a dynamic process, species characteristic of successive stages follow each other, especially within a phyletic group (probably a function of light intensity, humidity or plant composition (C. Amédégnato, unpublished data)). Some specialists on a particular host plant, appear only at certain stages (*Cercoceracris*, *Taeniophora*, and some species of *Ophthalmolampis*). Even though sometimes common, these specialists are always a minority within the grasshopper assemblage.

The forest canopy

Several different guilds of grasshoppers occur within the true canopy and share the space, food and microhabitats associated with different types of foliage. In order to assure continuity with previous publications, I employ the term 'intracanopy', as opposed to 'canopy surface' to differentiate between two types of insect. Those of the latter occur principally at the external surface of the canopy and move considerable distances, often alone, on that surface. The intracanopy insects (intracanopy corresponding to 'canopy' of Nadkarni (1995)) sometimes occur singly but more often are grouped, in the case of tree folivores towards the ends of the branches, where the young leaves which they eat are found. Others live in the interior of their microhabitat (e.g. within large clumps of lianas or groups of small epiphytes).

Most of the supracanopy (overstorey) species (typically large Romaleinae) lay their eggs in the soil in gaps or along edges and have gregarious nymphs that may colonize trees typical of regeneration patches, such as *Inga* spp. Most of the surface-active species are commonest towards the end of the succesional process (see Fig. 22.3). These species are mostly polyphagous, though some are associated with Mimosaceae or *Vismia* (Guttiferae), especially among the Eumastacidae.

Within the intracanopy fauna, the high specificity of the microhabitats is illustrated by the way the clouds of points split, corresponding to tree-foliage eaters, liana-foliage eaters, specialist epiphyte eaters, all further linked to their host plant via endophytic oviposition, plus generalist epiphyte feeders with epiphytic oviposition.

Marginal microhabitats (not figured)

Some microhabitats, such as sandy or pebbly rivers, banks or the degenerating ox-bow lakes formed by the change in course of large rivers, are geographically included within the forest but do not really belong to it in terms of their fauna. They have a similar biota over the whole of the continent and constitute special segregated compartments, colonized by insects following the courses of the rivers (linear foraging).

The stony and sandy banks of the whole of Amazonia have been colonized by *Ommexecha*, a genus otherwise typical of the biota of the semideserts of the southern part of the continent. The ox-bow lakes and swamps are colonized by specific aquatic plants, including *Salvinia* and *Azolla* (ferns) and *Pistia* (Araceae). These have a specialized grasshopper fauna. *Paulinia acuminata* DeGeer and *Marellia remipes* (Uvarov) lay their egg pods on the submerged undersurfaces of the leaves (Carbonell & Arrillaga, 1959; Carbonell, 1964). Different species of *Cornops* (Leptysminae), with endophytic oviposition, also occur here, especially on the water hyacinth *Eichhornia crassipes* (Mart. & Zucc.) (Pontederiaceae) (Silveira-Guido & Perkins, 1974; Andres & Bennet, 1975). The escape mode of all these grasshoppers is diving, although all are very mobile; *Cornops* is always fully winged, whereas *Paulinia* and *Marellia* are winged or brachypterous, depending on unknown factors (Carbonell, 1964). Further, the grasses characteristic of these subaquatic environments are the domain of the subfamily Leptysminae, especially of the tribe Leptysmini, which are specialist grassfeeders, with endophytic oviposition in grass stalks (Hilliard, 1982; Nunes & Adis, 1992).

Interactions between the different microhabitats within the forest mosaic

Even if the microhabitats appear to be true entities, the importance of some of their interactions is evident in Figure 22.2. For instance, within the group of grasshoppers of trunks and branches there is no clear separation between those species inhabiting canopy branches and those in the understorey or even the leaf litter; all share some subgeophilous attributes, and some of them are very mobile. Again, the grouping together in Fig. 22.2 of the species of the supracanopy and of the canopy surface with those of the regenerating forest is because they all lay their eggs in the soil of gaps and forest edges. This emphasizes that there may be no relationship between the habitat in which the adults live and that in which they lay eggs, which may be very different. In such cases, the species can be totally dependent on a further microhabitat, in this instance edges and regenerating gaps.

In other cases, the analysis suggests a virtual absence of links with other microhabitats, notably for the grasshoppers of palms and those of open gaps, which appear to be very segregated. The genera that are confined to the crowns of trees, that move only within this zone and are trophically specialized for it (notably the epiphyte specialists), passing through their full reproductive cycle within it, also seem to be a case comparable with the palm specialists. This independence is, however, illusory, at least in the case of the palm specialists, for the palm trees depend on the canopy cover during their growth (closing gaps: Kahn & de Granville, 1992). Similarly, palm grasshoppers (group Copiocerae of the Copiocerini) need the surrounding tree canopy as a complementary microhabitat for refuge (a considerable proportion of them – about a third of the adults – was found in the surrounding trees, rather than in the palms themselves).

By comparison, the grasshoppers of such habitats as sandy river banks, swamps or of the open gaps resulting from clearing do not require any complementary forest microhabitats. They rely only on their dispersal abilities to find new habitats, or on the continuity of their present habitat.

This situation contrasts markedly with the situation in the forest, which presents a whole range of evolving microhabitats. Its inherent continuity may be visualized along the horizontal axis of Figure 22.2, where genera typical of some microhabitats mix with others characteristic of precisely defined regeneration stages.

Food selection, specialization and types of microhabitat

Within both simple and complex microhabitats, the compartmentalization of food specialization appears to follow the model forbs versus grasses (Fig. 22.2) in the gaps, or dicotyledenous trees versus monocotyledenous trees (palms) within the mature forest or advanced regenerating canopy. The forb-eating insects of the gaps are generalists, but the higher diversity and complexity of the tree foliage in the canopy leads to a range of specialization on various dicotyledonous tree families or genera (Amédégnato, 1997). Besides the large palm-eating contingent, there are also specialists on epiphytic monocotyledons; however, they appear to be rare (canopy Leptysminae).

Uniformity and diversity in the Amazon basin

Comparative studies from different sites appear to be scarce and to concern insects only rarely (Wagner, 2000). All the available information on grasshoppers could not be included in the present study, as some sites did not yield data adequate for the assessment of assemblage structure and its variations (see methods). The comparison of the complete samples used here, relatively well distributed through the Amazonian basin and reflecting the main traits of the local canopy fauna, should be of interest in this largely unexplored field. The incomplete data from the other important collections indicate at least a general similarity with the present samples.

It is clear that it is the homogeneity of the Amazonian fauna that permits us to compare the study areas: this would not be the case, for example, in the forests of Central America or in the montane regions. Nonetheless, there are differences among the Amazonian populations. These differences fall into two main categories: differences in abundance and density, and differences in species composition.

Faunistically, the sites mentioned belong to several, more or less distinct, biogeographic zones: northern subAndean (Napo), southern subAndean (Pakitza), western Amazonian, north of the Amazon (Amazon, Yasuni east and west), western Amazonian, south of the Amazon (Ucayali) and the Guyanan dispersal centre (Guyana & Manaus), with its subdivisions Guyanan and central Amazonian.

In relation to the western Amazonian region, which dominates the analysis, the other regions show characteristic differences in faunal composition. The northern subAndean zone includes, for example, some endemics and extends its influence into the western Amazonian dispersal centre (e.g. the genus *Galidacris*, Rhytidochrotinae, typical of the shrub layer of the montane forest). In the southern subAndean zone, the Ommatolampae are especially diverse in the same stratum of vegetation. The differences are smaller north and south of the Amazon (among the arboricolous species, the species groups within the genus *Poecilocloeus* are different, being more varied in the south (Amédégnato & Poulain, 1987)). The largest differences, both ecological and taxonomic, are found between western Amazonia and the Guyanan dispersal centre (Fig. 22.2). The Guyanan centre has many peculiarities (e.g. the dominance of the Anablysae on trunks and branches, and the endemic subfamily Eumastacopinae of the Eumastacidae).

Relative proportions of the different guilds

Comparison of the canopy samples studied (Fig. 22.4) indicates a high level of similarity in functional structure (effective guild organization). This must be seen in the perspective of the faunal similarity of the various sites, at the level of the generic group (listed in Table 22.2). Even for the most divergent faunas (Ucayali/Pakitza or Ucayali/south Peru versus French Guiana) a positive correlation (Bravais–Pearson correlation coefficient) was always obtained (for the quantitative data of Table 22.2, $r = 0.322$ and 0.384). This similarity in the basic faunal composition results on the one hand from the continuity of most of the microhabitats that make up the canopy environment over the whole Amazonian basin and, on the other, from the fact that the Amazon basin really is a biogeographic unit, in spite of the slight divergences mentioned above and of the pronounced endemic fauna of the Guianan dispersal centre (Amédégnato & Descamps, 1982).

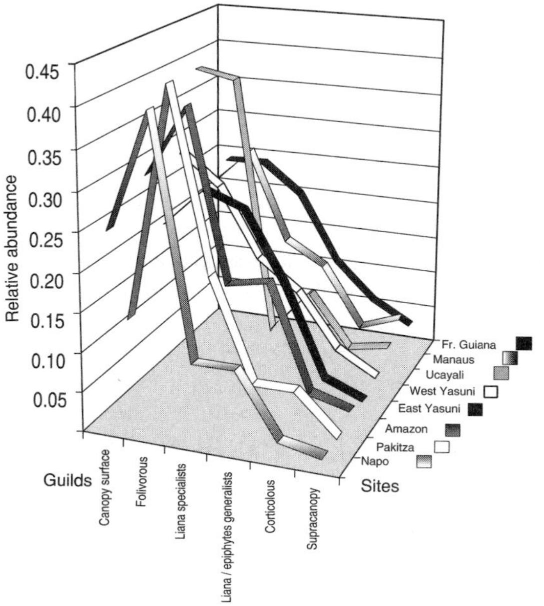

Fig. 22.4. Comparison of the forest canopy guilds or subguilds for the sites studied.

This overall taxonomic resemblance correlates with the similarity observed in functional structure. This is, however, not totally uniform. The most divergent samples were those from south Peru and Ucayali: in these localities, the canopy-surface species (a very sensitive component) had either too low or too high an index, relative to the mean.

However, the characteristics of the microhabitats are not exactly equivalent from one site to the next, particularly in regard to the proportions of the various trophic resources and its consequences for the different guilds. A comparison of the different sites for the canopy guilds (Fig. 22.5) shows that four sites are very poor in intracanopy foliage feeders. This may be because of local ecological factors at the sites, linked to long-term trends in ecosystem development. Two of the poorest sites for this guild (E. and W. Yasuni) are located in north-western Amazonia, thus having basically the same species composition as the other sites from this area (Napo and Amazon) but different ecological conditions. These are inland black water regions, rather than white water riverine forests, and are richer in inedible trees (mostly Moraceae, for Ecuador (Balslev & Renner, 1989)). The other two poor samples (Manaus and French Guiana) similarly belong to a common dispersal centre and are likewise very similar in species composition. Although the poverty of the ecosystem can be invoked to explain the situation in Manaus (where it is well recognized: see Fittkau & Klinge, 1973; Adis, 1988; Lovejoy & Bierregaard, 1990), the reasons for the apparent scarcity of the fauna in French Guiana is not so readily explained and requires further investigation.

Table 22.2. *Systematic composition of the samples studied at the group level (subfamily to genus group)*[a]

Group	Composition at site (%)							
	Amazon	E. Yasuni	W. Yasuni	Napo	Pakitza	Ucayali	Manaus	Fr. Guiana
Romaleidae: Bactrophorinae								
Ophthalmolampini								
Nautiae	5.56	20.48	21.68	8.29	18.86	2.87	12.9	8.19
Lagarolampae	8.63	5.87	6.22	7.56	2.33	9.46	5.24	0.72
Ophthalmolampae	16.59	10.00	11.43	21.22	12.92	18.92	18.95	20.83
Helicopacrae	0.05	0.16	0.17	3.9	12.66	0	0	0
Taeniophorini	2.07	0	0	0.49	0	0.17	1.81	3.59
Bactrophorini	1.02	0.79	0.84	2.68	0	0	1.61	0.14
Romaleidae: Romaleinae								
Tryblíophorus	4.59	2.54	2.69	2.68	8.27	2.36	0.81	8.33
Eurostacrae	2.16	2.54	2.69	1.46	0	0.84	2.42	0.86
Hisychiini	3.06	1.11	1.18	2.93	2.58	2.87	0	0
Tropidacrini	1.07	0.95	1.01	0.73	0.26	3.89	4.64	1.87
Acrididae: Ommatolampinae								
Ommatolampini								
Vilernae	0.68	0.79	0.84	1.22	3.1	0.51	1.21	0
Anablysae	1.24	2.54	2.69	1.46	0.78	2.2	3.23	16.52
Sciponacris/Agenacris	0.20	0.79	0.84	0	0	0.34	0	1.87
Syntomacrini								
Caloscirtae	8.96	13.49	14.45	9.76	12.92	7.77	4.84	10.34
Syntomacrae	1.90	3.17	3.19	4.39	7.22	0.17	0	0.86
Abracrini	0.05	0.48	0.50	0	0	0	0	0
Acrididae: Rhytidochrotinae	5.06	0.63	0.67	0.49	0	0.34	0	0
Acrididae: Leptysminae	0.07	0.48	0.50	0.24	0	0.34	0	0
Acrididae: Proctolabinae								
Coscineutini	3.72	0.48	0.50	0	0.26	2.36	0	0
Proctolabini								
Saltonacrae	3.33	0.32	0.50	1.46	0	0.68	1.01	3.3
Proctolabae	11.29	10.00	11.09	12.2	8.79	16.39	15.12	1.29
Eucephalacrae	0.34	3.02	3.19	1.95	1.81	0.34	3.63	7.47
Proscopiidae	8.21	5.24	5.38	0.98	3.62	20.95	1.01	2.44
Eumastacidae: Eumastacopinae	2.99	2.22	2.35	0	0.26	0	21.37	10.34
Eumastacidae: Pseudomastacinae	1.86	0.95	1.01	7.32	0	0.34	0	0
Eumastacidae: Eumastacinae	5.26	10.63	4.03	5.12	3.36	5.41	0	1.01
Other	0.04	0.33	0.36	1.47	0	0.48	0.2	0.03

[a]The value '0' corresponds to a true absence for Helicopacrae, Hisychiini, Rhytidochrotinae and Pseudomastacinae in the Guyanese centre; to a possible absence for Taeniophorini and Rhytidochrotinae in south Peru; and in other cases only reflects rarity and absence in the particular sample.

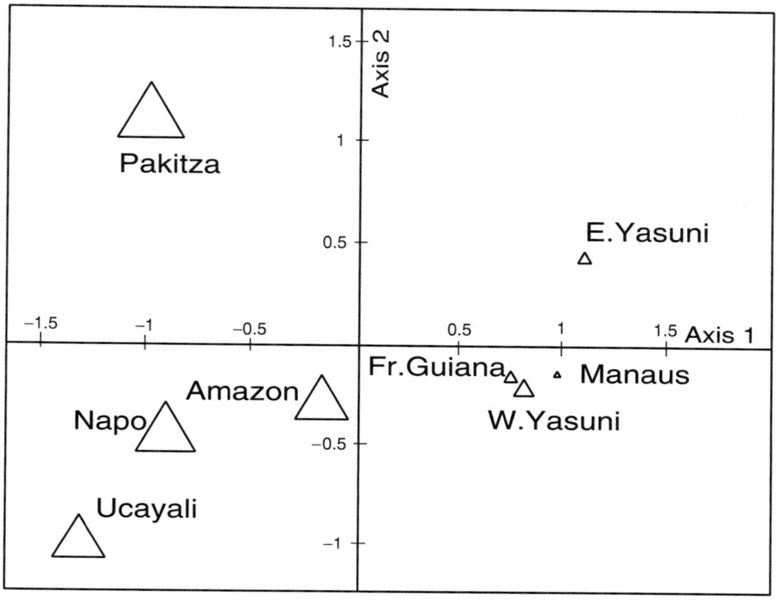

Fig. 22.5. Nonmetric multidimensional scaling of the guilds for the sites studied, showing the difference in relative importance of the tree foliage-eating grasshoppers (size of the triangles).

Generic diversity and occupation of microhabitats over the Amazonian basin

Specific diversity is too divergent at the extreme geographic points of the Amazonian forest to allow legitimate comparison. Generic diversity, which is more relevant to the present study, is an alternative way to approach the kind and intensity of exploitation of the microhabitats within the forest mosaic. Its patterns sometimes correspond to those of species diversity, but not always.

The different symbols used in Fig. 22.2 indicate the biogeographic range of the taxonomic units: West Amazonian, pan-Amazonian or Guyanese (extending from Manaus to French Guiana). From their distribution the following is clear:

- pan-Amazonian genera are to be found almost exclusively in the most open environments and among the fauna of low level strata, including leaf litter; very few are found in the arboreal habitat, with the exception of large mobile supracanopy insects
- the west Amazonian fauna is the largest and most diverse
- the Guyanese endemic fauna is smaller, even though recognizable vicariant groups or genera exist (trunk and branch species, canopy foliage species).

To what extent is generic diversity related to the kind of trophic resource used?

The fauna of open spaces, leaf litter, etc. inhabits relatively poor, very stereotyped microhabitats and is made up mostly of generalist forb feeders or graminivores of pan-Amazonian or even Continental distribution. Here there is a clear relationship between a low diversity of the environment and a low diversity of the acridid fauna. The same is true for the palm habitat, although that shows a very different ratio of species diversity compared with genus diversity.

In the canopy of the mature forest, globally, the same microhabitats in different parts of the Amazon seem to be populated (at least potentially) by similar communities, comprising the same genera or their vicariant variants. Diversity is high and seems to be subject to only two factors: differences in density because of ecological conditions, and differences in biogeographic composition, including the lower diversity of the eastern communities.

In the regenerating canopy, however, the same geographical parallelism does not seem to exist, as the Guyanan fauna is effectively absent. This could be a real phenomenon, as at least some specimens of pioneer canopy species have been captured there, or it may be artefactual as there is a total lack of quantitative data on regeneration and associated faunal changes in that area. Whether the fauna is really scarce, or whether simply our knowledge of it is inadequate, will be resolved only by further research.

CONCLUSIONS

We have seen that the observed differences in density and abundance within grasshopper assemblages are correlated to a large extent with differences in ecosystems. Apart from some local peculiarities, the microhabitats are occupied in a relatively homogeneous manner, the most variable parameter being the relative density of the different taxa.

This differing richness in taxa, particularly striking in western Amazonia, cannot be attributed solely to differences in the ecosystem, unlike the differences in density and abundance. It is probably more symptomatic of the importance of the historical component in the development of local fauna and thus of the present day state of the general population.

It is clear that in the course of evolution some microhabitats have been more influenced than others. Notably, the great radiation of the Ommatolampae in the southern subAndean zone seems to be closely linked to the appearance of the Andes; this event was simultaneously a direct cause of speciation and also a most important influence on the Pleistocene climatic oscillations in that region (Gentry, 1982).

The reasons for the relative poverty of the fauna of the Guyanan dispersal centre may be found to some extent in its partial isolation from the rest of the South American block during recent geological periods but also, indirectly, in the recent variation in the kind and extent of forest cover. The level at which one should characterize the majority of the divergences of the Guyanan fauna is unclear, but it is illustrated by the colonization of the branch and trunk microhabitat there by only one of the two groups that occupy it in Western Amazonia (even though both groups, Vilernae and Anablysae, are present in both regions). In the same way, the apparent marginal position of this centre has hindered the development of a major population of Proctolabinae, notably of the genus *Poecilocloeus*, which is elsewhere particularly prolific.

Future research should concentrate on both the comparative study of other regions (e.g. the periphery of the Amazon basin) and the analysis of the evolutionary processes occurring during the development of local ecosystems, including the historical component. This knowledge will aid greatly our current understanding of populations, especially in respect to their management and conservation.

ACKNOWLEDGEMENTS

The collection of the samples I refer to in this paper was made possible through the collaboration of numerous colleagues and institutions in various countries (Brazil: INPA; Peru: IIAP and NHM, University of San Marcos; Ecuador: PUCE) and also of local populations and indigenous federations (FUCUNAI). I am particularly indebted to the late Dr M. Descamps, who collected the samples from Manaus and French Guiana, and one from Pakitza, with Prof. C. S. Carbonell. I particularly thank S. Poulain, who collaborated with me in carrying out the other samples; I am also specially grateful to Jurg Gasché, anthropologist, who was of continuous help during field work in Peruvian northwestern Amazonia. I thank especially Dr H. Rowell, who took the trouble to translate the manuscript, and the anonymous reviewers for the improvements they suggested.

Appendix: List of taxa

Table 22A.1 lists the taxa considered in the study.

Table 22A.1. *List of taxa considered in this study*

Taxa	Code	Life zones and attributes											
		Z	L	EG	S	F	H	V	EZ	EM	NG	AG	Sz
Pyrgomorphidae													
Omura congrua Walker 1870	Og	3	1	1	1	5	2	2	3	1	2	2	1
Acrididae													
Cyrtacanthacridinae													
Schistocerca Stal 1893 (2 sp.)	Sp	4	2	1	1	5	3	2	6	2	2	2	2
Melanoplinae													
Dichroplus gr. *punctulatus* (Thunberg 1874) (2 sp.)	dc	1	2	1	1	5	1	2	1	2	2	2	1
Gomphocerinae													
Orphulella Giglio Tos 1894 (3 sp.)	Or	1	2	1	3	6	1	1	4	2	2	2	1
Peruvia nigromarginata Scudder 1875	Pn	1	2	1	3	6	1	2	1	2	2	2	1
Jagomphocerus amazonicus Carbonell 1995	Ju	3	2	1	3	6	3	2	6	2	2	2	2
Ommatolampinae													
Clematodinini													
Rehnuciera Carbonell 1969 (2 sp.)	Ra	9	1	1	2	3	1	2	2	1	2	2	1
Clematodina Gunther 1940 (2 sp.)	Ck	9	1	1	1	5	1	2	2	1	2	2	1
Abracrini													
Sitalces dorsualis (Giglio Tos 1898)	St	2	2	1	1	5	2	1	3	1	2	2	1
Eusitalces vittatus Bruner 1911	Es	2	2	1	1	5	2	1	3	1	2	2	1
Psiloscirtus Bruner 1911 (3 sp.)	Ps	2	2	1	1	5	2	1	3	1	2	2	1
Abracris flavolineata (De Geer 1773)	Af	4	2	1	1	5	3	2	3	2	2	2	1
Ommatolampini group Vilernae													
Vilerna Stal 1873 (2 sp.)	Vr	1	2	1	1	5	1	2	1	1	2	2	1
Locheuma brunneri Scudder 1875	Lb	4	1	1	1	5	1	1	2	1	2	2	1
Sciaphilacris alata Descamps 1976	sa	3	1	1	1	5	2	1	2	2	2	2	1
Aptoceras margaritatus Bruner 1908	am	9	1	1	2	5	3	2	3	1	2	2	1
Aptoceras Bruner 1908 (4 sp.)	ac	9	1	1	2	5	3	2	3	1	2	2	1
Hypsipages dives Gerstaecker 1889	Hd	9	1	1	2	3	4	2	5	1	2	2	1
Bryophilacris Descamps 1976 (2 sp.)	Bm	9	1	1	2	3	4	2	5	1	2	2	1
Acridocryptus pusillus Descamps 1976	Ap	9	1	1	2	3	3	2	5	1	2	2	1
Sclerophilacris Descamps 1976 (2 sp.)	Sc	9	1	1	2	3	4	2	5	1	2	2	1
Rhabdophilacris Descamps 1976 (3 sp.)	Rs	9	1	1	2	3	4	2	5	1	2	2	1
Ommatolampini group Anablysae													
Eurybiacris Descamps 1979 (3 sp.)	Ey	9	1	1	1	1	6	2	5	1	2	2	1
Agrotacris Descamps 1979 (4 sp.)	Aw	9	1	1	2	7	6	2	5	2	2	2	1
Anablysis teres Giglio Tos 1898	At	4	2	1	1	5	3	2	5	2	2	2	1
Anablysis arboricola Descamps 1978	Aa	9	2	1	1	1	6	2	5	2	2	2	1
Anablysis Giglio Tos 1898 (5 sp.)	Ay	9	1	1	2	7	6	2	5	2	2	2	1
Hysterotettix Descamps 1979 (3 sp.)	H	9	1	1	2	7	6	2	5	2	2	2	1
Ananotacris abditicolor Descamps 1978	Ab	9	1	2	2	7	6	2	5	2	2	2	1
Odontonotacris mimetica Descamps 1978	Om	9	1	2	2	7	6	2	5	2	2	2	1
Pseudhypsipages silvester Descamps 1977	Py	9	1	2	2	7	6	2	5	1	2	2	1

Table 22A.1. (*cont.*)

Taxa	Code	Life zones and attributes											
		Z	L	EG	S	F	H	V	EZ	EM	NG	AG	Sz
Ommatolampini group Ommatolampae													
Eulampiacris Carbonell & Descamps 1978 (2 sp.)	El	3	2	1	1	5	1	2	3	1	2	2	1
Hippariacris latona (Günther 1940)	Hi	3	2	1	1	5	1	2	3	1	2	2	1
Episomacris Carbonell & Descamps 1978 (3 sp.)	Ei	3	1	1	1	5	1	2	2	1	2	2	1
Ommatolampis perspicillata (Johannson 1763)	Op	3	1	1	1	5	2	1	3	1	2	2	1
Unassigned													
Agenacris Amédégnato & Descamps 1979 (2 sp.)	bb	9	1	3	3	4	6	2	3	1	2	2	1
Sciponacris Amédégnato & Descamps 1979 (5 sp.)	Sz	9	2	3	3	4	6	2	3	2	2	2	1
Syntomacrini group Syntomacrae													
Syntomacrella virgata (Gerstaecker 1889)	Sl	2	2	1	1	5	2	1	3	1	2	2	1
Syntomacris Walker 1870 (10 sp.)	SA	4	2	1	1	1	5	2	3	2	2	1	1
Syntomacrini group Caloscirtae													
Pseudanniceris Descamps 1977 (3 sp.)	pz	5	2	1	1	1	4	2	3	1	2	2	1
Stigacris rubropicta Descamps 1977	sg	6	2	1	1	1	5	2	3	1	2	1	1
Calohippus Descamps 1978 (4 sp.)	Ca	9	1	4	1	4	6	2	3	1	1	1	1
Caloscirtus cardinalis Gerstaecker 1873	Cc	9	1	4	1	4	6	2	3	1	1	1	1
Oyampiacris nemorensis Descamps 1977	Oy	9	1	4	1	4	6	2	3	2	2	2	1
Ociotettix Amédégnato & Descamps 1979 (2 sp.)	Ox	4	2	3	2	1	5	2	3	1	2	2	1
Leptysminae Tetrataeniini													
Tetrataenia surinama (Linné 1764)	Tt	2	2	3	3	6	2	2	4	2	1	1	1
Cornops frenatum (Marshall 1835)	Co	3	2	3	3	6	2	2	4	2	1	1	1
Chloropseustes Rehn 1918 (5 sp.)	Ch	2	2	2	1	5	2	2	3	1	2	2	1
Stenopola boliviana (Rehn 1913)	sb	3	1	3	3	6	2	2	3	1	2	1	1
Rhytidochrotinae													
Galidacris eckardtae (Günther 1940)	Ge	2	2	2	1	5	2	2	3	1	1	1	1
Galidacris purmae in litt	Gp	5	2	2	3	1	4	2	3	1	1	1	1
Paropaon laevifrons (Stal 1878)	Pf	2	2	2	1	5	2	2	3	1	1	1	1
Proctolabinae Coscineutini													
Coscineuta Stal 1873 (3 sp.)	cs	8	2	1	1	1	6	2	6	2	2	1	2
Proctolabinae Proctolabini group Proctolabae (32 sp.)													
Poecilocloeus pervagatus Descamps 1975 gr. fruticolus	pP	5	2	1	2	1	5	2	3	2	1	1	1
Poecilocloeus estironana Améd. & Poul. 1987, id.	pE	6	2	1	2	1	5	2	3	2	1	1	1

Table 22A.1. (*cont.*)

Taxa	Code	Life zones and attributes											
		Z	L	EG	S	F	H	V	EZ	EM	NG	AG	Sz
Poecilocloeus uncinatus Descamps 1980 gr. lacalatus	pN	7	1	1	2	1	5	2	3	2	1	1	1
Poecilocloeus ferus Descamps 1976 gr. ferus	pF	8	2	1	2	1	6	2	3	2	1	1	1
Poecilocloeus rubripes Desc. & Améd. 1970 gr. rubripes	pR	8	2	1	2	1	6	2	3	2	1	1	1
Adelotettix Bruner 1910 (6 sp.)	A	9	2	1	1	1	6	2	6	2	1	2	2
Cercoceracris Descamps 1976 (7 sp.)	Cg	6	2	1	3	1	5	2	3	2	1	1	1
Witotacris Descamps 1976 (3 sp.)	Wt	9	2	1	3	1	6	2	3	2	1	1	1
Halticacris Descamps 1976 (3 sp.)	Hr	9	2	1	1	1	6	2	3	2	1	1	1
Dendrophilacris Descamps 1976 (6 sp.)	Db	8	2	1	3	1	6	2	3	2	1	1	1
Proctolabinae Proctolabini group Saltonacrae													
Saltonacris Descamps 1976 (5 sp.)	Sa	9	2	3	2	4	6	2	3	2	1	1	1
Eucerotettix ludificator Descamps 1980	Ex	9	2	3	2	4	6	2	3	2	1	1	1
Harpotettix Descamps 1981 (2 sp.)	Hx	9	2	4	2	4	6	2	3	2	2	2	1
Ypsophilacris viduata (Descamps 1977)	Yv	9	2	3	2	4	6	2	3	2	1	1	1
Loretacris fascipes Amédégnato & Poulain 1987	Lf	9	2	4	2	2	6	2	3	2	1	1	1
Proctolabinae Proctolabini group Eucephalacrae													
Eucephalacris Descamps 1976 (6 sp.)	Ec	9	1	1	1	1	6	2	3	2	2	1	1
Pareucephalacris Descamps 1976 (3 sp.)	pr	9	1	1	1	1	6	2	3	2	2	1	1
Copiocerinae Copiocerini group Monachidiae													
Monachidium lunus Johannson 1763	Mm	9	2	1	2	1	6	2	3	2	1	1	2
Copiocerinae Copiocerini group Copiocerae													
Copiocerina formosa (Bruner 1920)	cn	9	2	4	3	6	5	2	3	2	1	1	2
Copiocera Burmeister 1838 (4 sp.)	ca	9	1	4	3	6	5	2	3	2	1	2	2
Eumecacris Descamps & Amédégnato 1972 (3 sp.)	E	9	1	4	3	6	5	2	3	2	1	1	1
Romaleidae: Bactrophorinae													
Ophthalmolampini group Lagarolampae													
Lagarolampis amazonica Descamps 1978	Lg	6	2	1	1	1	5	2	3	1	1	1	1
Tikaodacris elegantula Descamps 1978	Tk	6	2	4	1	1	5	2	3	1	1	1	1
Helolampis Descamps 1978 (9 sp.)	Hc	7	2	4	1	1	5	2	3	1	1	1	1
Habrolampis Descamps 1978 (2 sp.)	Hb	8	2	4	1	4	6	2	3	1	1	1	1
Othnacris surdaster Descamps 1978	Ot	9	1	4	3	2	6	1	3	1	1	1	1
Ophthalmolampini group Nautiae													
Euprepacris Descamps 1977 (3 sp.)	em	9	1	4	3	2	6	1	3	1	1	1	1
Pseudonautia Descamps 1978 (15 sp.)	Pc	9	1	4	3	2	6	1	3	1	1	1	1

Table 22A.1. (*cont.*)

Taxa	Code	Life zones and attributes											
		Z	L	EG	S	F	H	V	EZ	EM	NG	AG	Sz
Ophthalmolampini group													
Ophthalmolampae													
Apophylacris incondita Descamps 1983	Ai	9	2	4	1	1	6	2	3	1	1	1	1
Adrolampis Descamps 1977 (23 sp.)	Ac	9	2	4	3	1	6	2	3	1	1	1	1
Ophthalmolampis Saussure 1859 oligophagous sp. (12 sp.)	Ob	9	2	4	2	1	6	2	3	1	1	1	1
Ophthalmolampis, polyphagous sp. (10 sp.?)	Oi	9	2	4	1	1	6	2	3	1	1	1	1
Ophthalmolampis elaborata Descamps 1983	Od	8	2	4	3	1	6	2	3	1	1	1	1
Peruviacris Descamps 1978 (2 sp.)	pc	9	2	4	3	2	6	2	3	1	1	1	1
Poecilolampis saltatrix Descamps 1978	Pl	9	2	4	2	1	6	2	3	1	1	1	1
Chromolampis ornatipes (Bruner 1907)	Ch	9	2	4	2	1	6	2	3	1	1	1	1
Xenonautia concinna Descamps 1977	Xe	9	2	4	2	2	6	2	3	1	1	1	1
Nothonautia Descamps 1983 (2 sp.)	Nh	9	1	4	2	1	6	2	3	1	1	1	1
Taeniophorini													
Taeniophora Stal 1873 (2 sp.)	Ta	7	1	1	3	2	5	1	3	1	1	1	1
Hylephilacris Descamps 1978 (2 sp.)	Hg	9	2	1	2	1	6	2	5	1	1	1	1
Bactrophorini													
Hyleacris rubrogranulata Amédégnato & Descamps 1979	Hy	7	2	1	1	1	5	2	2	1	2	2	1
Bora nemoralis Amédégnato & Descamps 1979	Bn	7	2	1	1	1	5	2	2	1	2	2	1
Mezentia Stal 1878 (3 sp.)	Mz	9	1	1	1	1	6	2	2	1	2	2	2
Andeomezentia napoana Amédégnato & Poulain 1994	Az	9	1	1	1	1	6	2	2	1	2	2	2
Other Romaleidae													
Trybliophorus Serville 1831 (2 sp.)	Ty	9	2	1	2	1	6	2	2	2	1	1	1
Eurostacris Descamps 1978 (4 sp.)	Ep	9	1	1	3	2	6	1	3	1	1	1	1
Pseudeurostacris valida Descamps 1978	Pv	9	1	1	3	2	6	1	3	1	1	1	1
Pseudhisychius Descamps 1979 (3 sp.)	Pg	5	2	1	1	1	5	2	3	1	2	2	2
Hisychius Stal 1878 (3 sp.)	Hs	8	2	1	1	1	6	2	3	1	2	2	2
Prionolopha serrata (Linné 1758)	Ph	3	2	1	1	5	2	2	3	2	2	2	2
Colpolopha Stal 1873 (2 sp.)	Cl	9	1	1	1	5	1	2	2	2	2	2	2
aff. *Colpolopha* sp. (2 sp.)	cp	9	1	1	1	5	1	2	2	1	2	2	2
Chariacris Walker 1870 (3 sp.)	Cm	9	2	1	1	1	6	2	6	2	1	2	2
Prionacris Stal 1878 (5 sp.)	px	9	2	1	1	1	6	2	6	2	1	2	2
Aprionacris Descamps 1978 (2 sp.)	As	9	2	1	1	1	6	2	6	2	1	2	2
Titanacris Scudder 1869 (3 sp.)	Th	9	2	1	2	1	6	2	6	2	1	2	2
Tropidacris Scudder 1869 (2 sp.)	Tr	8	2	1	1	1	6	2	6	2	1	2	2
Phaeoparia Stal 1873 (3 sp.)	Pa	9	1	1	3	6	1	1	4	2	2	2	2

Table 22A.1. (*cont.*)

Taxa	Code	Life zones and attributes											
		Z	L	EG	S	F	H	V	EZ	EM	NG	AG	Sz
Maculiparia Jago 1980 (2 sp.)	m	9	1	1	3	6	1	2	2	1	2	2	2
Chromacris icterus (Pictet & Saussure 1867)	Ci	3	2	1	1	5	2	2	6	2	1	1	2
Proscopiidae													
Apiocelis sp. (4 sp.)	Ae	7	2	1	1	1	5	2	2	1	2	2	2
Proscopia group 1 (3 sp.)	PI	8	2	1	1	1	5	2	2	1	2	2	2
Proscopia group 2 (3 sp.)	PII	9	2	1	1	1	6	2	2	1	2	2	2
Eumastacidae													
Pseudomastax Bolivar 1914 (4 sp.)	Px	8	2	2	2	1	6	2	3	2	1	1	1
Eumastacops Rehn & Rehn 1942 (3 sp.)	Eo	8	2	2	2	1	6	2	3	2	1	1	1
Pareumastacops Descamps 1979 (2 sp.)	Pw	8	2	2	2	1	6	2	3	1	1	1	1
Pseudeumastacops Descamps 1974 (3 sp.)	Po	8	2	2	2	1	6	2	3	1	1	1	1
Maripa Descamps & Amédégnato 1970 (2 sp.)	M	8	2	2	2	1	5	2	3	1	1	2	1
Arawakella unca Rehn & Rehn 1942	Ak	8	2	2	2	1	5	2	3	1	1	1	1
Eumastax Burr 1899 (5 sp.)	Er	7	2	2	1	1	5	2	3	1	2	1	1
Eumastax vittata napoana Descamps 1982	En	6	2	2	1	1	4	2	3	1	2	1	1
Eumastax Burr 1899 (2 sp.)	Em	5	2	2	1	1	3	2	3	1	2	1	1
Other genera mentioned													
Rhammatocerus Saussure 1861 (Acrididae Gomphocerinae)													
Ommexecha Serville 1831 (Ommexechidae)													
Paulinia Blanchard 1843 (Pauliniidae)													
Marellia Uvarov 1929 (Pauliniidae)													

Taxa: wide-ranging species: full name; genera endemic or with numerous species: only generic name; highly diversified genera including different phyletic or functional species groups: groups represented by typical representative species.
Code: codes used in the multifactorial analysis (see text and Fig. 22.2).
Z, life zone: gap with partly bare soil (1), gap with vegetation (2,3), edge (4), regeneration (5 to 8), forest (9); L, light preference: sciaphilous (1), heliophilous (2); EG, egg-laying mode: soil (1), humus (2), endophyte (3), epiphyte (4); S, specialization: generalist (1), oligophagous (2), specialist (3); F, food category: tree foliage (1), liana foliage (2), mosses (3), other epiphyte (4), shrubs and forbs (5), grass, palm (6), diverse including bark (7); H, height: 0 to 1 m (1–3), >3 m (4), midcanopy (5), high canopy (6); V, vegetation structure: closed (1), branched (2); EZ, escape zone : soil (1), litter (2), bush or leafy branches (3), grass tuft (4), bare branches (5), space (6); EM, escape mode: jump (1), flight (2); NG, nymphal gregarism: present (1), absent (2); AG, adult gregarism: present (1), absent (2); Sz, size: <3 cm (1), >3 cm (2).

23

Flowering events and beetle diversity in Venezuela

Susan Kirmse, Joachim Adis and Wilfried Morawetz

ABSTRACT

Flowers in tropical rainforests are an important food resource for many arthropods. The flower-visiting beetle fauna has been investigated on two different mass-flowering tree species in a primary Amazonian rainforest in Southern Venezuela. Using a mobile crane system, beetles were collected with window traps throughout the flowering period in 1997 and, in addition, were observed and captured by hand. Ninety-nine morphospecies were collected with the traps on *Tachigali guianensis* (Caesalpiniaceae) and 114 on *Matayba guianensis* (Sapindaceae). On the latter tree species, an additional 65 beetle species were hand collected. Singletons represented more than 50% of the total number of species in trap collections from both trees. Study trees were more attractive for diurnal than nocturnal beetles, with respect to both numbers of species and specimens. The similarity of the beetle fauna between the tree species was low despite a comparable structure of flowers and the survey of subsequent flowering periods. Only 30 beetle species were common to both trees. Species most frequently collected on *M. guiane*nsis included Curculionidae, Cerambycidae, Chrysomelidae and Mordellidae. Flowers of *T. guianensis* were mainly visited by Curculionidae and Chrysomelidae. A comparison with similar data suggests a high degree of random organization in the species composition of flower-visiting Coleoptera assemblages. Only a few beetle species appear to be host specific. Such flower-visiting coleopteran assemblages are limited to the flowering period of study trees – attracted by food. In addition, many beetles visited the flowering trees either during day or night and had no constant presence in the tree crowns. Our results emphasize the importance of flowering events of individual trees as organizing phenomena for beetle migration, spatial distribution and the establishment of meeting concourses for members of the same species.

INTRODUCTION

Various methods for the investigation of forest canopies have been used since this habitat became increasingly a focus of scientific studies in tropical rainforests. Mobile crane systems are a convenient method to carry out observations and gather data without major disturbance of the ecosystem (Parker *et al.*, 1992; Morawetz, 1998). Integration of various research investigations within the same study site can improve our understanding of the diversity of organisms and processes in the canopy of tropical forests (Stork *et al.*, 1997b; Morawetz, 1998). In 1996, a multidisciplinary research project was implemented following the installation of an observation crane at Suromoni, Venezuela by the Austrian Academy of Science, the University of Leipzig and the Venezuelan government.

An important aspect of the 'Surumoni project' focusses on investigations of plant–animal interactions. Flowers in tropical forests represent an important source of food for many arthropods. Flowering trees are also important as meeting points for various conspecific flower visitors (Kugler, 1955; Dafni, 1992). Nearly all tropical angiosperms are pollinated by insects, birds or bats. The stimuli that allow pollinators to locate the flowers are colour, odour and shape of the flowers. Recently beetle-pollinated flowers usually have no or little nectar. They attract their visitors with large quantities of pollen, to be consumed for the most part by the insects (Howe & Westley, 1988). Typical beetle-pollinated flowers are mostly open during both day and night, are dull or white coloured, planiflorous or patelliform and radially symmetrical. They often smell fruity or aminoid (Howe & Westley, 1986). Only a few groups of angiosperms are pollinated predominantly or exclusively by beetles (Gottsberger, 1989a, 1990).

Pollinators include both generalists that pollinate dozens or hundreds of plants and specialists, which may be found only on a single species of plant (Real, 1983).

In the majority of cases, insects visit flowers while foraging for food and are not necessarily pollinators (Kugler, 1955; Dafni, 1992). Only mass-flowering plants will attract many flower visitors, because the insects usually concentrate their foraging on dense patches of flowers (Augspurger, 1980; Thompson, 1981). Mass-flowering is generally characterized by a short flowering period, during which an individual plant will produce flowers with both high amplitude and synchrony (Frankie, 1975; Baker *et al.*, 1983). The flower-visiting beetles of two mass-flowering tree species with comparable floral structure were collected and observed during the whole flowering period in 1997. The aim of this study has been to describe the composition and characteristics of flower-visiting beetle assemblages of such tree species.

METHODS

Study site

The study site of the Surumoni project lies about 15 km downstream from La Esmeralda, a small missionary village on the upper Orinoco River. The crane plot is located at 3° 10′ N, 65° 40′ W, 105 m above sea level on the bank of the blackwater river Surumoni, a tributary of the Orinoco River. The long-term average annual precipitation in La Esmeralda is about 2700 mm (1975–1995, Dirección de Hidrología y Meteorología, Caracas). However, an average of 3100 mm was measured within the study area from 1996 to 1998. The precipitation reaches a distinct maximum between April and July. There is, however, no clear dry season. The average annual temperature within the study area is about 26 °C, usually with slight variations between the coolest (25 °C) and the warmest month (26.5 °C). A daily range of 5–10 °C occurs frequently (Anhuf *et al.*, 1999).

The crane plot is situated on *terra firme* level and changes gradually into a blackwater inundation forest (*igapó*) at its outer margin. The vegetation is characteristic of neotropical moist lowland rainforest, with an intermediate level of tree species richness (J. Wesenberg *et al.*, unpublished data). Altogether there are over 800 trees, including many palms with a DBH (diameter at breast height) ≥10 cm within the crane plot. The upper canopy usually reaches a height of 25 to 27 m, with a few trees higher than 30 m. The canopy is not completely closed but is interrupted by several gaps.

The tower crane established at the study site is 42 m in height with an arm length of 40 m. This crane, running on 120 m of rails, allows access to an area of about 1.4 ha. Using an observation gondola, scientists can move among the trees in all directions. This system allows consistent observations and investigations of particular trees or sites within the canopy at any time. Furthermore, animals can be observed and studied *in situ*.

Study trees

Matayba guianensis *Aubl. (Sapindaceae)*

The distribution of *Matayba guianensis* includes Guyana, Brazil and Peru (von Martius, 1897). Six individuals with a DBH ≥ 10 cm were within the crane plot. A single specimen was studied, with a height of 26.5 m. Its irregular-shaped crown varied in length from 2.9 to 5.2 m and in width from 4.5 to 7.9 m.

The flowering period of the tree was about 21 days, from 18 September to 8 October 1997. Buds were observed until 23 September. The major part of the flowering was complete by the 5 October. The maximum number of open flowers (about 50% of an estimated one million flowers) was recorded between 25 and 28 September. This dioecious species generally flowers twice a year at the study site, with the main flowering period from August to November.

The disc-shaped flowers are relatively small (length 3–4 mm and diameter 3–3.5 mm) and are densely connected in complex paniculate inflorescences. They are yellowish-white in colour and pentamerous, with eight stamens overtopping the petals and sepals. The tree studied bore male flowers with reduced ovules. The flowers disperse a slight flowery scent and an extrastaminal discus offers nectar to potential pollinators.

Tachigali (Tachigalia) guianensis *(Benth.) Zarucchi & Herend. (Caesalpiniaceae)*

Tachigali guianensis can reach a height of up to 30 m and grows in nonflooded lowland rainforests, gallery forests and at lower elevations in montane rainforests. It occurs in Colombia, Guyana, Suriname, French Guiana, Peru, Brazil and Bolivia as well as in Venezuela and in the states of Amazonas and Bolivar (Berry *et al.*, 1998). There were four individuals with a DBH ≥ 10 cm within the crane plot. The specimen studied reached a height of 28 m. The dimensions of the crown were 16 m × 23 m.

The flowering period in 1997 lasted 18 days (8–27 November). The total number of flowers estimated was almost two million. Buds were mainly found until 12 November and most finished flowering by 24 November. The maximum number of open flowers (over one million) was recorded from 20 to 23 November. Within the study area, flowering periods have been observed once to twice a year, with the main flowering season recorded from October to November.

The creamy white flowers with radial symmetry are pentamerous and have 10 stamens. They are united to form multiflorous inflorescences that are compound racemes. They are cup-shaped or patelliform. The flowers are hermaphroditic and disperse a strong sweet scent. A single flower has a length of 6–7 mm and a diameter of 3.5–4.5 mm. Hereafter, the two tree species are denoted by their generic names.

Sampling and observation of the flower-visiting beetles

One window trap was set up in each tree crown studied, to collect beetles (Fig. 23.1). This type of trap is suitable for sampling flying insects (Basset *et al.*, 1997b), which collide with the transparent vertical panels and drop into the collecting funnel. Traps consisted of two discs of plexiglass, fitted crosswise into one another (height 25 cm; and width 30 cm). A funnel made of plastic foil with a detachable collecting container was fixed below the vertical panels. This container was filled with water mixed with detergent to diminish the surface tension. The traps were set up at the beginning of the period of anthesis at the first observed appearance of flower visitors. They were fixed on a thicker branch and remained in the crown until the flowers wilted. The traps were surveyed at intervals of 2 to 4 days. To clarify the dependence or otherwise of the beetles upon the flowering period, the beetle faunas of both trees were observed and collections made in several additional months of the year.

In addition, observations and hand collecting took place wherever possible on every second day and night, particularly on the flowers of *Matayba*. Furthermore, branch beating was carried out and beetles were collected in an entomological net (Fig. 23.1). Larger beetles were collected directly with the net or by hand. In this way, periods of activity as well as the presence or absence of many flower-visiting beetles could be determined.

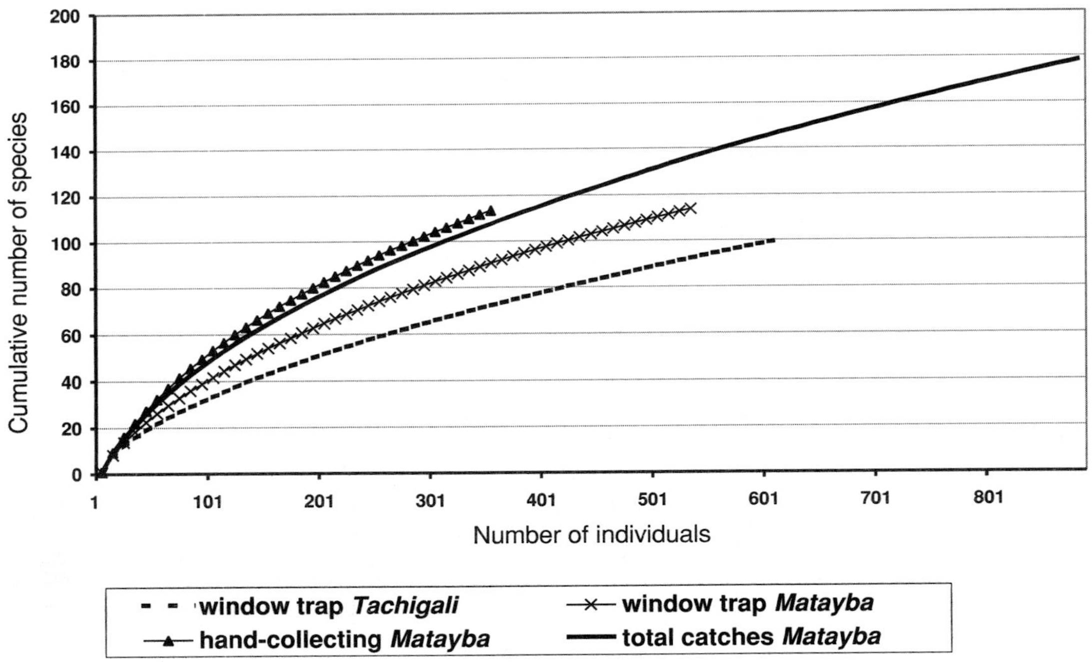

Fig. 23.1. Plots of the cumulative number of species against number of individuals for trap catches in *Tachigali guianensis* and *Matayba guianensis*, for hand-collecting on *Matayba*, and for total catches on *Matayba*.

The work at night was aided by the use of a halogen lamp.

Processing of samples and statistical analyses

Adult beetles were preserved in 70% ethanol. They were sorted and assigned to morphospecies on the basis of external characteristics. Several groups of beetles were subsequently identified by taxonomists (see Acknowledgements).

The beetles collected by trapping were divided on the basis of their diel activity patterns. They were classified as diurnal, nocturnal or indifferent. Diurnal (or nocturnal) beetles were found only during day (or night) in the crowns of the study trees. Indifferent beetles included beetles encountered at day and night within the tree crowns and beetles with unknown activity periods.

Both trap samples were compared using the complement of the Bray–Curtis measure (Bray & Curtis, 1957). Furthermore, the distribution of all individual beetle species collected in the traps was analysed using a chi-square test. The expected number of species was plotted against the number of individuals by rarefaction to assess the effectiveness of the trapping methods, including the hand collections on *Matayba*.

RESULTS

Flower-visitors on *Matayba*

Visits of beetles were limited to the 21 days of the flowering period. First flower visitors were observed from 18 September when the first buds opened. Most specimens were collected between 26 and 28 September (Fig. 23.2). This period corresponded to the maximum flowering of the study tree. When most of the flowers had wilted, no further beetles were observed or captured. Not one of these beetle species was found on the tree at months outside the flowering period. The most abundant families were collected regularly throughout the flowering period (Fig. 23.2).

The trap collected 533 individuals of 114 species. An additional 65 flower-visiting beetle species were collected by hand, bringing the total number of flower visitors on *Matayba* to 179 species representing 19 families (Table 23.1). Only one family (Tenebrionidae), represented by a single specimen, was not recorded in the trap.

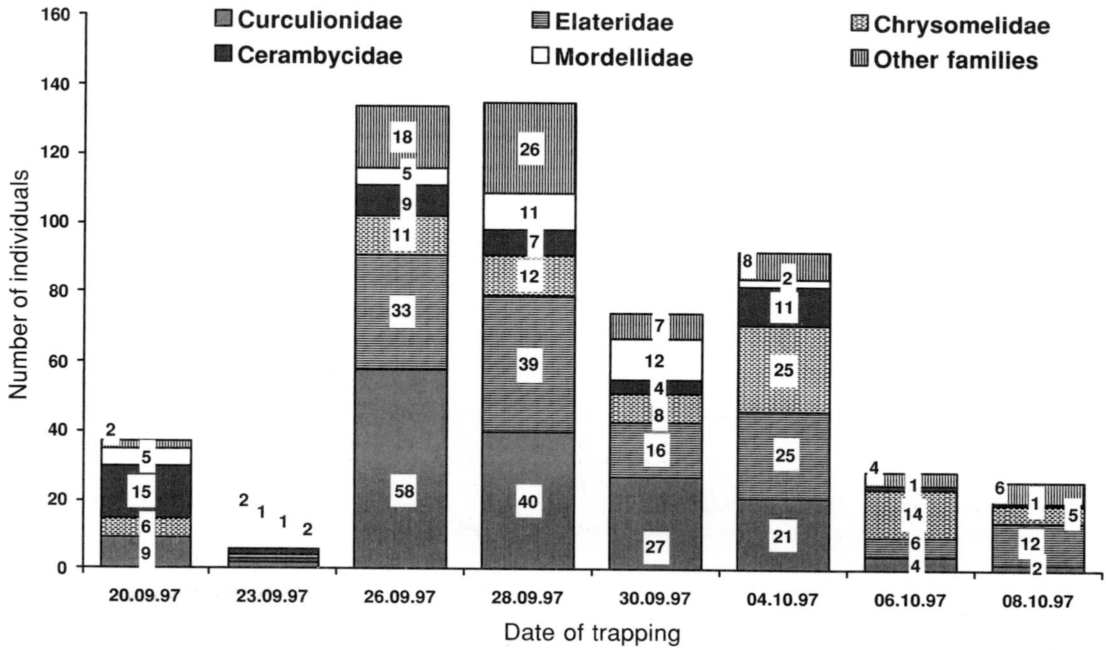

Fig. 23.2. Number of individuals, detailed by families, collected using the window trap on *Matayba guianensis* during the whole flowering period.

Table 23.1. *Individuals and species of beetles, detailed by family, collected by traps and hand collecting in the flowering crowns of the two study trees*

Taxon	*Tachigali* trap collection			*Matayba* trap collection			*Matayba*	
	Species	Singletons	Individuals	Species	Singletons	Individuals	Hand-collection	Total catches
Brentidae	0	0	0	1	1	1	4	5
Bruchidae	1	1	1	0	0	0	0	0
Buprestidae	0	0	0	1	0	2	2	3
Cantharidae	1	1	1	3	2	6	3	6
Carabidae	6	5	9	3	1	5	2	5
Cerambycidae	**1**	**0**	**4**	**19**	**12**	**49**	**21**	**40**
Cerambycinae	1	0	4	19	12	49	18	37
Lepturinae	0	0	0	0	0	0	3	3
Chrysomelidae	**27**	**9**	**372**	**17**	**8**	**82**	**6**	**23**
Alticinae	4	0	267	3	0	45	0	3
Cryptocephalinae	5	1	19	1	1	1	0	1
Eumolpinae	8	2	51	9	7	22	2	11
Galerucinae	9	5	34	4	0	14	4	8
Lamprosomatinae	1	1	1	0	0	0	0	0
Cleridae	1	1	1	3	1	11	0	3
Coccinellidae	4	3	6	1	1	1	0	1
Cucujidae	1	1	1	0	0	0	0	0
Curculionidae	34	23	105	34	16	163	16	50
Dermestidae	4	1	16	5	4	15	1	6
Elateridae	3	2	63	2	0	132	0	2
Elmidae	1	1	1	0	0	0	0	0
Endomychidae	0	0	0	1	1	1	0	1
Erytolidae	1	1	1	0	0	0	0	0
Eucnemidae	2	2	2	0	0	0	0	0
Histeridae	1	0	2	0	0	0	0	0
Leiodidae	0	0	0	1	0	2	0	1
Monommatidae	0	0	0	2	1	3	0	2
Mordellidae	1	0	3	12	6	36	3	15
Nitidulidae	3	1	10	5	2	20	1	6
Oedemeridae	0	0	0	1	1	1	1	2
Scarabaeidae	2	1	8	3	3	3	4	7
Scirtidae	2	2	2	0	0	0	0	0
Tenebrionidae	0	0	0	0	0	0	1	1
Trogossitidae	3	3	3	0	0	0	0	0
All Coleoptera	**99**	**58**	**611**	**114**	**60**	**533**	**65**	**179**

The most species-rich families were Curculionidae, Cerambycidae, Chrysomelidae and Mordellidae. A similar distribution of relative dominance across these families was maintained even after singletons were removed from consideration (Table 23.1). In terms of the number of individuals caught, Curculionidae was most abundant, followed by Elateridae, Chrysomelidae, Cerambycidae, Mordellidae and Nitidulidae. Within Chrysomelidae,

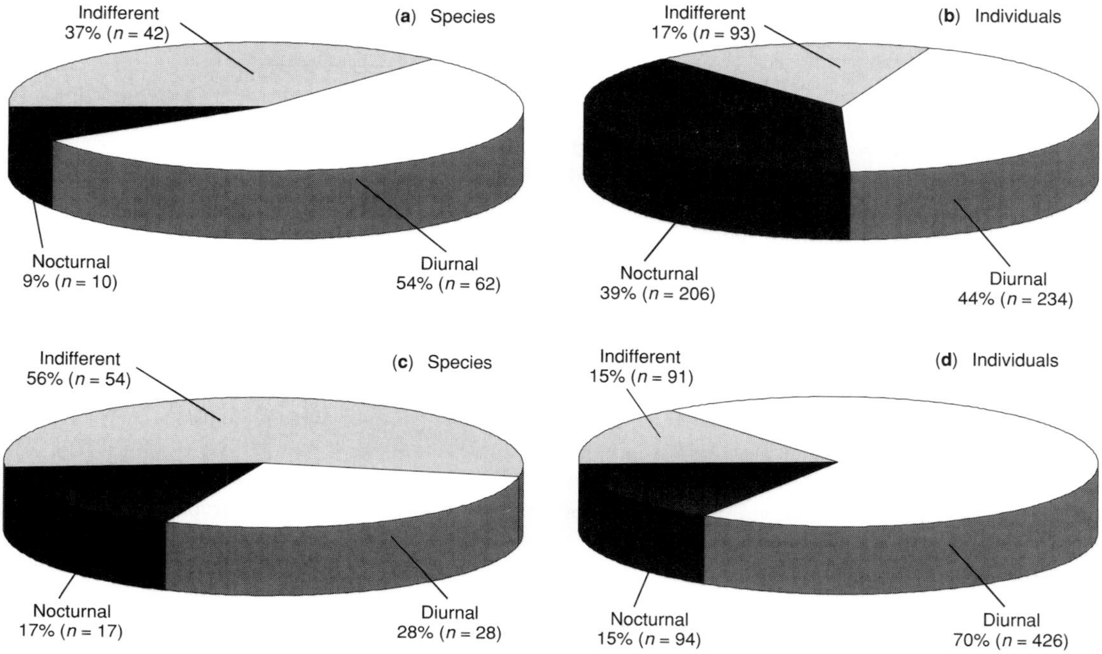

Fig. 23.3. Activity pattern of the flower-visiting beetle fauna sampled in window traps: (**a**) *Matayba guianensis* (number of species); (**b**) *M. guianensis* (number of individuals); (**c**) *Tachigali guianensis* (number of species) and (**d**) *T. guianensis* (number of individuals).

species of Alticinae, Eumolpinae and Galerucinae visited the flowers of *Matayba* frequently (Table 23.1).

Direct observations and hand collections of the flower-visiting beetles provided information on patterns of diel activity. Based on the number of beetles collected in the trap, *Matayba* attracted more diurnal species than nocturnal ones (Fig. 23.3**a**). The proportion of diurnal species was significantly higher than that of nocturnal species compared across the eight sampling periods (Mann–Whitney $U = 8.5$, $p < 0.05$). The relatively high proportion of nocturnal specimens was the result of high dominance of a single species of elaterid (Fig. 23.3**b**). There was no significant difference between the proportions of diurnal and nocturnal specimens (Mann–Whitney $U = 27.0$, $p = 0.599$).

Flower-visitors on *Tachigali*

The flowering period of *Tachigali* lasted 18 days, which also were the limits of the period of beetle visitation. The first visitors were observed at the beginning of anthesis on the 9 November. The highest number of beetles was collected between 22 and 24 November (Fig. 23.4), when the maximum number of flowers was open. From 24 November onwards there were still a few inflorescences with open flowers, but only two beetles were collected over the following 3 days. Only six of these flower-visiting beetle species were found on this tree in nonflowering periods. The family composition of visitors was relatively constant during the flowering period (Fig. 23.4).

Overall, 611 individuals of 99 beetle species belonging to 20 families were trapped during flowering (Table 23.1). Curculionidae and Chrysomelidae were especially diverse. Both of these families were also the most species rich, with the latter dominating when singletons were removed (Table 23.1). The most abundant family was Chrysomelidae, followed by Curculionidae and Elateridae. Within the Chrysomelidae, Alticinae, Eumolpinae and Galerucinae occurred most frequently (Table 23.1).

Most beetle species visited the flowers during daytime (Mann–Whitney $U = 0.0$, $p < 0.05$) (Fig. 23.3**c**). A distinct attractiveness of the flowers for diurnal species was shown on the basis of the number of individuals collected with the trap (Fig. 23.3**d**). More than four times the number of individuals were collected during the day than at night (Mann–Whitney $U = 5.0$, $p < 0.01$).

Comparison of tree species

The proportion of singletons (a species represented by one specimen) was relatively high in both traps and amounted to 58.6% ($n = 58$) on *Tachigali* and 52.6% ($n = 60$) on *Matayba* of the total number of species collected. On average, 6.17 and 4.68 individuals per species were collected on *Tachigali* and *Matayba*, respectively. In contrast to the earlier-flowering *Matayba*, the flowers of *Tachigali* appear to attract fewer beetle species. Diversity indices were often higher for catches on *Matayba* than on *Tachigali* (Table 23.2). Most species were distributed randomly among the samples (chi-square tests, $p > 0.05$). Only 24 out of 187 species showed an aggregated distribution in the samples (chi-square tests, $p < 0.05$).

A comparison of the trap samples between the two tree species showed an overlap of 22.7% (Bray–Curtis similarity). Only 27 species out of a total of 187 species were collected on both trees with the traps. Including the hand collections on *Matayba*, the number of shared flower visitors was 30 out of a total of 249 species. Consequently, more than two thirds of beetle appear to be restricted to one or other of the tree species. However, singletons should be excluded from these calculations, since nothing can be stated about their host specificity. Excluding singletons, 19 species were recorded exclusively on *Tachigali*, nearly 20% of all species collected. Similarly, 67 species (37%) were recorded exclusively on *Matayba*. Probably the proportion of exclusive species of this tree would have been lower if additional

Table 23.2. *Alpha-diversities of trap catches in flowering* Tachigali *and* Matayba *spp.*

Index	*Tachigali*	*Matayba*
Simpson's diversity (D)	0.09	0.08
Simpson's diversity ($1/D$)	10.86	12.63
Berger–Parker's dominance (d)	0.21	0.24
Berger–Parker's dominance ($1/d$)	4.81	4.20
Berger–Parker's dominance (d; %)	20.79	23.83
Shannon's H′ $\log_{10}$ base	1.38	1.53
Margaleff's M $\log_{10}$ base	66.76	68.21

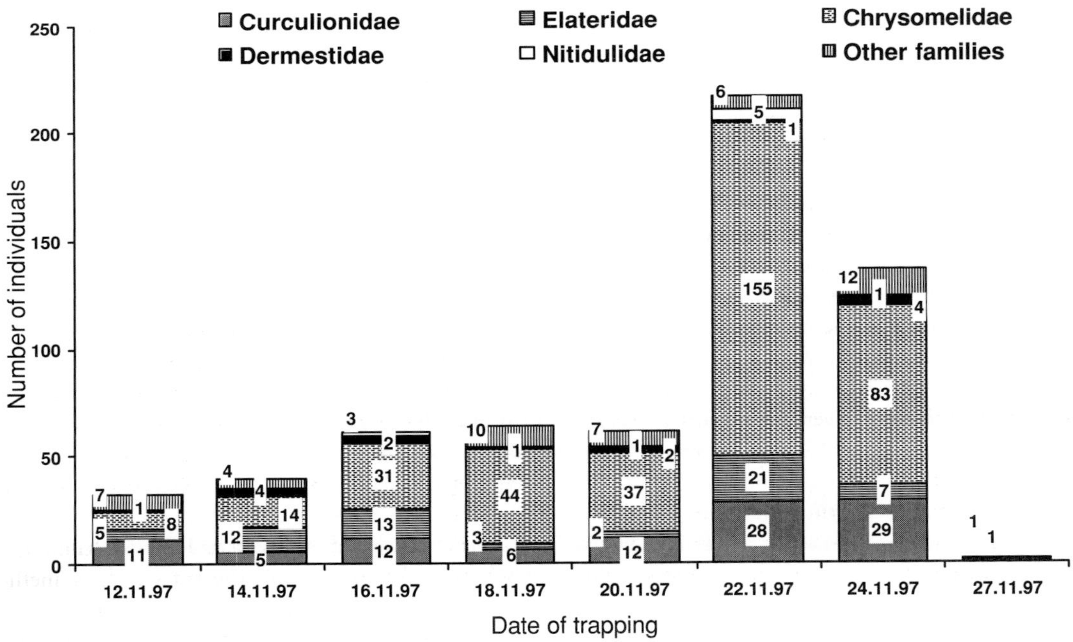

Fig. 23.4. Number of individuals, detailed by families, collected using the window trap on *Tachigali guianensis* during the whole flowering period.

extensive hand collections from *Tachigali* had been available for comparison.

Rarefaction curves were similar for both trap samples, as well as for hand collection on *Matayba* (Fig. 23.1). They did not approach an asymptote. Despite additional hand collecting on *Matayba*, the rarefaction curves did not level off.

DISCUSSION

Influence of tree phenology and composition of the flower-visiting beetle fauna

Although the study trees belong to different families, they resemble each other in flower colour and structure, the arrangement of the flowers in inflorescences, and have a similar flowering phenology. Both tree species may be categorized, morphologically, as entomophilous. A wide range of visitors frequent the flowering trees, which attract a high number of insects. Mass-flowering, whereby an individual plant produces a large number of flowers within a short period, is an appropriate strategy for canopy trees and lianas (Frankie *et al.*, 1974). This enables the trees to display distinct colour patterns to insects flying above the forest canopy (Frankie, 1975; Baker *et al.*, 1983). Nevertheless, odour is considered the primary stimulus for most beetle pollinations (Gottsberger, 1990; Gottsberger & Silberbauer-Gottsberger, 1991).

Beetles were not attracted to the study trees either before anthesis or after wilting of most flowers. The occurrence of nearly all beetles was strictly concurrent with flowering of the study trees. Very few if any beetle species visiting the flowers of study trees were also found on these trees out of the flowering period. The beetles were attracted to potential sources of food, which, possibly, they detected from a long distance. Additional observations within the study area indicated that most flower-visiting beetles were not attracted by the occurrence of single or of very few flowers. A decisive influence of the phenology of host trees on the occurrence and abundance of phytophagous beetles was also recorded by Basset (1991c) and Davies *et al.* (1997).

The composition of the beetle assemblages visiting the flowers of both species of study tree was similar to arboreal beetle communities of other tropical rainforests. In various studies, the beetle fauna is dominated by Chrysomelidae, Curculionidae, Anthribidae, Cerambycidae and Mordellidae (Erwin, 1983b; Floren & Linsenmair, 1994; Hammond *et al.*, 1997). A high similarity of the composition of the arboreal beetle fauna at the family level is characteristic of tropical rainforest trees (Hammond *et al.*, 1996; Floren & Linsenmair, 1997b). Accordingly, the abundant families were also collected repeatedly throughout the flowering period of both study trees.

Nevertheless, different groups of beetles did show preferences for one or other of the tree species. The flowers of *Matayba* attracted a more diverse spectrum of flower-visiting beetles than did those of *Tachigali*. Cerambycidae, Mordellidae and Cleridae frequented the flowers of *Matayba* more often than those of *Tachigali*. Chrysomelidae were common flower visitors on both tree species, but more so on *Tachigali*. Flowers of both tree species appeared to display similar attractiveness for species of Curculionidae, Dermestidae, Elateridae and Nitidulidae. Chrysomelidae, Curculionidae, Nitidulidae and Scarabaeidae have been recorded as important components of the pollinator fauna in Malaysian dipterocarp forests (Momose *et al.*, 1998; Sakai *et al.*, 1999a). Many reasons may influence the spectrum of flower visitors. Since nectar contains amino acids in sufficient concentrations to be nutritious for pollinators (Baker & Baker, 1983), the rich floral nectar of *Matayba* may contribute to its attractiveness across many groups of visitor. However, beetles are principally pollenophagous (Churchill & Christensen, 1970), although they also feed on all other flower parts (Momose *et al.*, 1998; Sakai *et al.*, 1999a).

Characteristics of flower-visiting beetle assemblage and beetle host specificity

Sampling by hand collecting and with window traps is largely selective (Basset, 1988). Window traps sample only a restricted area of the tree crown and catch a relatively small number of specimens. Actively flying species that move frequently from flower to flower may be sampled more frequently than species that crawl between flowers. However, this permanent sampling method is appropriate to obtain (semi-) quantitative samples at various sites and is not greatly influenced by random variables (Jessop & Hammond, 1993; Hammond, 1994). In a comparison of different sampling methods followed by hand collecting, Basset *et al.* (1997b) recorded the highest number of species using composite flight-interception traps. Accordingly, samples from these traps are suitable for comparing the visitor faunas of study trees and the combination of two collecting methods should have obtained a high percentage of the flower-visiting beetle species.

The rarefaction curves of the sampled beetle fauna did not reach an asymptote. The relatively steep incline of the curves suggests the existence of a larger pool of flower visitors than was recorded. However, additional hand collecting on *Matayba* did not diminish the incline of the curve or the proportion of singletons. The same feature of rarefaction curves has been reported by Allison *et al.* (1993) and by Floren and Linsenmair (1998a) after summarizing fogging samples from different trees of a single species. Rarefaction curves will reach an asymptote following increased sample size only in habitats (tree species) with a fixed number of species, just as the effectiveness of sampling methods cannot be assessed in habitats with a high percentage of fluctuating species. We conclude that a high proportion of the species visited the trees randomly, attracted by the rich offering of food. The high proportion of rare species also supports this interpretation. Most species occurred only in low abundance: the proportion of singletons represented more than half of the samples. This phenomenon accords with the results of other investigations and seems to be a general characteristic of arboreal beetle assemblages. The proportion of singletons, for instance, from studies in the Asian region amounted to 47.6% (Allison *et al.*, 1997) or 46.9% (Hammond *et al.*, 1997).

Diurnal flower visitors dominated on each tree species despite the rather different spectrum of visiting species and the fact that flowers of both trees remained open during both day and night. Most beetle species could be detected during either the day or the night on their chosen forage trees. Many beetle species obviously did not remain on the tree crowns. Only a few species occurred on the flowering trees both day and night. Because dusk in equatorial latitudes is only of short duration, both diurnal and nocturnal beetles may occur within this period. It is possible that most flower-visiting beetles visit their host tree only for feeding and do not stay permanently at their feeding place within the flowering period. Schowalter and Ganio (Ch. 28) also observed differences in the abundance of several arthropod taxa between day and night samples on two tree species in Puerto Rico. It is possible that this may minimize potential predation risk for many beetle species, perhaps reflecting a high predatory pressure from ants, which occur normally in high abundances in lowland rainforests (Floren & Linsenmair, 1997b; Wagner 1997, 1999; Adis *et al.*, 1998a; Harada & Adis, 1998; Novotny *et al.*, 1999a; Ch. 16).

Experiments on tree recolonization after fogging indicate high mobility and flexibility of beetle assemblages (Floren & Linsenmair, 1994; Allison *et al.*, 1997; Adis *et al.*, 1998a). Certainly, flower-feeding beetles have limited possibilities for specialization to single plant species as most tropical plants flower discontinuously, although within particular periods (Frankie *et al.*, 1974; Opler *et al.*, 1976). With a few exceptions, a particular plant species is pollinated by more than one species of visitor and each pollinator uses many plant species as its floral resource (Feinsinger, 1983). Beetles, in particular, are characterized as pollinators of a wide range of plant species that have variable reproductive strategies (Endress, 1994).

The flowering period of both tree species differed by just 1 month. Consequently, beetle species collected on *Matayba* had a high *a priori* probability of being collected also on *Tachigali* flowers. Based on the data presented here, this was not the case. Wolda *et al.* (1998) investigated Curculionidae in Panama and recorded the activity patterns of species that had only a short seasonal appearance up to species that occurred throughout the whole year. Many species showed similar seasonal patterns across years. Farrell and Erwin (1988) recorded the appearance of about half of the more frequently sampled leaf beetles in two of four seasons. This would suggest that at least some beetle species should have occurred continuously throughout the short period of our study. Farrell and Erwin (1988) also suggested the possibility of temporal changes in community composition.

In spite of all these factors (high mobility and flexibility, high probability of presence during both flowering periods, similar structure of the flowers and phenology of both tree species), the specificity of the visitor fauna appeared to be relatively high. Sample size certainly plays an important role in predicting host specificity (Novotny & Basset, 2000). Comparable information about host specificity from other studies is limited because flower visitors are a special case of phytophagous beetle assemblages. In addition, foraging by the beetle species collected is probably not restricted to flowers. Comparing various studies of arboreal beetle assemblages, the proportion of species that depends on a single tree species varies from 72% (Davies *et al.*, 1997) to an estimate for phytophagous beetles of 7–10% in a dry forest (Ch. 21). Amongst single families such as the Chrysomelidae, values of 19.7% have been suggested from Central Africa (Wagner, 1999). Host specificity is overall considered to be low (Novotny *et al.*, 1999b), although there are clear differences in the specificity of individual groups. Generally the host specificity of arboreal beetle assemblages here appears rather low, supporting the suggestion that many beetle species use a range of tree species.

CONCLUSIONS

A high number of beetles visited the flowers of both tree species. The similarity of the flower-visiting assemblages was relatively low, in spite of several factors that might have been expected to produce the reverse pattern. A comparison of the present study with similar data from other tree species from the same study site (S. Kirmse, unpublished data) suggests that the composition of flower-visiting beetle assemblages is influenced largely by stochastic events. Such random patterns were shown by Floren and Linsenmair (1997b, 1998a, b) for arboreal arthropod assemblages. These authors could neither find evidence of tree-specific communities of Coleoptera nor a stable composition at the species level. Most likely, the beetle communities studied within the Surumoni site consist of many species flying randomly in the area, being attracted by flower resources as and when they occur. The proportion of species obligatorily associated with the flowers of the study trees is probably rather low. Flower visitors may have a wider range of host plants than, for example, phyllophagous species. The beetle assemblage of both host trees appeared to depend on flowering phenology and was not characterized by a constant species occurrence on study trees. Mass-flowering events on tropical rainforest trees represent a conspicuous food resource, easy to locate by highly mobile arthropod species. Further, flowering events influence significantly the spatial distribution and dispersal of many insect species and represent important meeting points for conspecific insects.

ACKNOWLEDGEMENTS

The study was supported by a grant of the Stiftung der Deutschen Wirtschaft. Many thanks to J. Wesenberg, University of Leipzig, Germany for support and providing botanical data and descriptions and to M. Hartmann, Naturkundemuseum Erfurt, Germany for his help with identification of some beetle families. Beetles were determined by Dr C. L. Bellamy, Natural History Museum, Los Angeles, USA (Buprestidae), Dr U. R. Martins, Universidade de São Paulo, Brazil (Cerambycidae), Prof. V. Medvedev, Russian Academy of Sciences, Russia (Chrysomelidae), Prof. B. C. Ratcliffe, University of Nebraska, USA (Scarabaeidae), Dr S. A. Vanin, Universidade de São Paulo, Brazil (Brentidae & Curculionidae) and C. Wurst, Museum für Naturkunde Karlsruhe, Germany (Elateridae).

PART V

Synthesis: spatio-temporal dynamics and resource use in tropical canopies

Introduction

Canopy entomology is a young science and providing syntheses of the rather scarce data reflecting a multiplicity of forest types, arthropod groups and methodologies is rather challenging. We are motivated by the ever increasing interest of the scientific community in canopy arthropods and the certitude that such attempts will prompt others to refine our knowledge of these fascinating communities. This synthesis section includes contributions that either bear on study systems or organisms for which a good body of information already exists, or provide data and discussion that are particularly relevant to large-scale patterns of arthropod distribution or resource use in tropical canopies. The first three contributions discuss topics relevant to arthropod stratification. They are followed by two contributions more specifically addressing temporal variation in arthropod assemblages. Five other chapters consider either host specificity or resource use. The concluding chapter summarizes the trends observed in this volume and discusses promising research themes in future entomological studies of tropical canopies.

Prinzing and Woas review the information available about habitat use and stratification of microarthropods in tropical rainforests (Ch. 24). They consider three categories of factors (ecophysiological conditions, immigration from population pools and interspecific interactions) likely to influence the redistribution of species between habitats and, ultimately, their stratification.

In Ch. 25, Barrios takes a different approach and considers the herbivore fauna feeding on a similar resource in the understorey and upper canopy in Panama: conspecific saplings and mature trees of *Castilla elastica*, a pioneer tree. The author then reviews similar information available from other tropical locations and discusses vertical gradients of herbivore diversity in tropical rainforests.

Shaw and Walter review in Ch. 26 the often cryptic habitats, such as tree hollows or bird's nests, available to mites in the canopy. They are concerned with the isolation of such habitats and the implication for the dispersal of the arthropods that exploit them. They test Hamilton's hypothesis that inbreeding in these discrete, but complex habitats leads to a maintenance of primitive forms, for a number of mite taxa.

Basset *et al.* (Ch. 27) are concerned with arthropod diel activity in the canopy in Gabon and ask whether it is related to vertical migrations between forest strata, such as the understorey and the upper canopy. They analyse similarities of herbivore fauna collected with different collecting methods in the two strata during either day or night. They also briefly discuss whether a specific arthropod fauna is likely to occur in the upper canopy of closed wet tropical forests.

Schowalter and Ganio in Ch. 28, examine the interactions between forest disturbance and temporal patterns, such as arthropod seasonality and diel activity, in Puerto Rico and Panama. They analyse the data primarily by considering functional groups (guilds) of arthropods and herbivory. Like Didham and Springate in Ch. 4, they stress the importance of arthropod responses to temporal variation in environmental conditions.

The similarity of insect faunas of tropical trees and the factors shaping them are an increasing topic of interest in canopy entomology. In Ch. 29, Kitching *et al.* consider phytophagous beetles in Australia. Using top-down approaches, they contrast the faunas of a range of tree species of known intertaxonomic distance based on recent molecular phylogenies. In doing so, they review recent trends in work on host-plant specificity.

Perhaps one the best-studied arthropod groups (certainly at the family level) in the canopy is the ants. Dejean and Corbara provide a timely review of the organization of dominant ant mosaics in tropical environments (Ch. 30), in pristine forests, secondary forests or plantations. In particular, they discuss the factors leading to the distribution of dominant ant species over their supporting trees.

Another stimulating review is that of Ribeiro in Ch. 31, comparing the information available on herbivory and the abundance, diversity and host specificity of insect herbivores in tropical savannas and rainforests. In particular, monodominant tree populations and their associated herbivores can be profitably contrasted in the two forest types. The author also discusses the distribution patterns of endophytic and free-living insect herbivores and emphasizes the characteristics of the canopy in tropical savannas.

Roubik *et al.* (Ch. 32) discuss loose niches in plant pollinator systems within rainforest canopies, with examples drawn from Panama and involving the main flower visitors there, thrips and bees. The authors examine the consistency of flower visitation for many tree species at different sites over a 3 year period. In doing so, they stress the relevance of recent theories of biodiversity and biogeography to their data.

Janzen provides a comprehensive review in Ch. 33 of the pattern of food-plant use by caterpillars of Saturniidae in the canopy of a dry forest in Costa Rica. The author also discusses the geographical distribution of the species breeding in the study area and their immigration patterns. He contrasts the two concepts of ecological fitting and coevolution and discusses their relevance to his data and to elsewhere in the tropics.

In Ch. 34, Speight *et al.* argue convincingly that, despite most data on tropical canopy arthropods originating from pristine forests, many tropical forests have been, and will be in the future, subject to various degrees of management, including selective logging, clear-felling and conversion to exotic plantations. The authors review the likely impact of these management tactics on indigenous canopy insect species, with examples drawn from various taxa.

We hope that the concluding chapter, written by the editors, will help the reader to wade through the numerous and fascinating observations reported in this volume and stimulates further his/her interest in canopy arthropods and tropical canopies.

24

Habitat use and stratification of Collembola and oribatid mites

Andreas Prinzing and Steffen Woas

ABSTRACT

In this review, we first discuss habitat use by Collembola and oribatid mites ('microarthropods'). We show that these animals may differentiate among microhabitats (e.g. different cryptogam species) as much as between macrohabitats (e.g. different soil strata). Moreover, the habitat use of such microarthropods is not static. The animals redistribute among habitats constantly. Such redistributions are often caused by ecophysiological factors, such as climate, diet or toxic substances. Interestingly, this phenomenon does not occur in some species. Redistributions among habitats might also be caused by immigration from population pools or by interspecific interactions, although evidence is still scarce and controversial. Second, we discuss the patterns of faunistic stratification in forests, particularly in tropical rainforests. We show that the stratification of a forest is not linear and more resembles a sandwich. The fauna in the suspended soils at the top of the forest resembles the fauna in the soil on the forest floor. There is evidence that the faunistic difference between the canopy and the forest floor is larger in a (tropical) rainforest than in a temperate deciduous or coniferous forest. The 'sandwich' stratification can be explained by the degree of microhabitat differentiation in different strata and the ecophysiological living conditions, as well as by competition pressure from species of different life forms. The high degree of stratification in rainforests may be caused by lack of microhabitat differentiation and the temporary high saturation deficits at the bark surface of trees, the scarcity of humus on the forest floor and the abundance of ants in suspended soils. We point out that a higher degree of stratification in rainforests is one factor that is capable of increasing species richness in Collembola and oribatid mites.

INTRODUCTION

The investigation of habitat use is at the heart of animal ecology (Bell *et al.*, 1991). The way in which a species uses the habitats in a landscape characterizes both the species and the landscape. At which spatial scale does the species differentiate habitats? Is this differentiation of habitats static or dynamic? Which factors influence the redistribution of species among habitats: ecophysiological conditions, immigration from population pools or interspecific interactions? Finally, how does the habitat use of a set of species explain the broad faunistic patterns in a landscape, such as stratification? In the present chapter, we will discuss habitat use and stratification of Collembola and oribatid mites. These taxa are 'microarthropods', but of course the designation microarthropods comprises many taxa in addition to Collembola and Oribatei (Dunger, 1983). Collembola and Oribatei, however, are ecologically distinct from other microarthropods. Collembola and Oribatei graze on microphytes such as algae and lichens, or on detritus. Collembola and Oribatei do not inhabit the waterfilm and they are extremely species rich in most ecosystems of the world (Lawrence, 1953; Dunger, 1983; Hopkin, 1997; Walter & Proctor, 1999). Accordingly, it is worthwhile to investigate habitat use and stratification of Collembola and Oribatei.

We will first discuss the spatial scale at which microarthropods use habitats, and whether this habitat use is constant or not. Then, we will consider environmental factors that may influence habitat use. This review is largely based on studies from temperate regions. There are only very few studies of habitat use by microarthropods in tropical regions. Finally, we will discuss the stratification of tropical and temperate forests from the viewpoint of microarthropods and explain this stratification in terms of habitat use.

The topic of habitat use and stratification in microarthropods has been neglected in recent years. Although studies on habitat use and stratification have largely concentrated on flagship species for nature conservation (e.g. Meffe & Carrd, 1997; but see Basset *et al.*, 2001a), studies on microarthropods have largely concentrated on food webs, ecosystem functions and ecotoxicology (e.g. Verheof, 1996; Hopkin, 1997). Accordingly, it is hard to justify a study on habitat use of microarthropods just by writing that 'it has recently been shown. . . .' or referring to the topic as 'an exciting new area of research'. Nevertheless, we hope that our review may stimulate new research on this topic.

PATTERNS OF HABITAT USE

Macrohabitat use

Classically, entomologists have assigned microarthropod species to habitats that cover areas of several square metres and have a thickness of at least several centimetres: the soil layer, the litter layer, the layer of dead wood, the trunks of trees, etc. (Lawrence, 1953; Dunger, 1983; Eisenbeis & Wichard, 1987). We will call habitats of this dimension 'macrohabitats'. 'Microhabitats', in contrast, are structures within these macrohabitats and are discussed later on.

Patterns of macrohabitat use by microarthropods have been investigated using two different approaches. The first approach has included the description of assemblages (commonly referred to as 'communities') in different habitats. Such descriptions have usually been based on a sampling design that maximizes the range of environmental conditions sampled, often at the cost of a systematic replication. Therefore, the appropriate statistical tools for such investigations are exploratory techniques (Strenzke, 1952; Cassagneau, 1961; Travé, 1963; Gjelstrup, 1979; Ponge, 1993; but see Rodgers & Kitching, 1998 for a systematically replicated approach). Some authors have demonstrated that the composition of communities in different habitats correlates with environmental factors, such as pH (Loranger *et al.*, 2001). Other authors have focussed on the qualitative description of life-forms that dominate different habitats (Gisin, 1943; Klima, 1954). Physiologists then demonstrated the functional relationship between the life-forms and living conditions in different habitats (Verhoef & Witteven, 1980; Eisenbeis, 1989). These analyses at the community level, however, have not indicated whether or not an individual species significantly differentiates among macrohabitats. The second approach to the study of macrohabitat use has included tracking individual species. These studies were often based on a systematically replicated sampling design, across a limited range of environmental conditions (Christensen, 1980; von Allmen & Zettel, 1982; Vegter, 1983; Schenker, 1984; Wolters, 1985; Takeda, 1987; Wunderle, 1992b; Salmon & Ponge, 1999; Kampichler *et al.*, 2000). The results indicate that some species are clearly restricted to certain macrohabitats whereas others cover many macrohabitats (von Allmen & Zettel, 1982; Vegter, 1983; Schenker, 1984; Wunderle, 1992b).

Ideally, a study of macrohabitat use by microarthropods should include elements of both approaches: (i) description of the microhabitat composition of each macrohabitat; (ii) sampling of all major macrohabitats in a vegetation unit, for example within a forest; (iii) systematic replication; and (iv) repeated sampling under a range of different ambient conditions, such as day and night, different seasons or different altitudes. Only such an 'ideal' study would permit the assessment of whether or not most microarthropod species discriminate among macrohabitats.

Microhabitat use

Some authors have subdivided a given macrohabitat on a coarse level. Nicolai (1985) and Büchs (1988), for example, have investigated the north and south faces of tree trunks separately and reported differences in the abundance of certain Oribatei or Collembola species. However, the interpretation of such patterns is almost as difficult as is the interpretation of macrohabitat use. The north and south faces of a tree trunk, for instance, differ fundamentally in many ways, including the bark structure, the microclimate and the epiphyte cover (Stubbs, 1989).

In search of more interpretable patterns, some authors have tried to analyse the distribution of microarthropods on an even smaller scale – the scale of microhabitats. A microhabitat may be defined as a structure that has distinct food sources and distinct abiotic living conditions but is small enough that an animal can enter and leave it within only a few minutes. Microhabitats are patches at the scale of square centimetres or smaller. An individual conifer needle, for instance, is a microhabitat, or a lichen thallus or a leaf in the litter layer (e.g. Aoki, 1967; André, 1976; Anderson, 1978b; Prinzing, 1997; Prinzing & Wirtz, 1997; Edsberg & Hagvar, 1999). Investigation of such microhabitat use was usually possible only by drastically restricting the scope of the analysis: (i) to a single species (Bellido, 1979); (ii) to an explorative analysis of a species assemblage within a single patch of a microhabitat mosaic (for oribatids on a single rock: Bonnet *et al.*, 1975; for a single patch of soil surface: Hammer, 1972; and Berg, 1991); (iii) to a selection of the microhabitat types in a macrohabitat (for certain cryptogam species on tree trunks: André, 1975; for needles within a soil: Edsberg & Hagvar, 1999; for earthworm middens in a soil: Maraun *et al.*, 1999); or (iv) to certain classes of microhabitat (for corticolous cryptogams: André, 1983, 1985; for soil microstrata: Ponge, 2000). Despite these restrictions, the investigations indicated that the fauna of two microhabitat types may be as different as the fauna of two macrohabitats (Haarlov, 1960; André, 1983, 1985; but see Smrz & Kocourkova, 1999). The differences between microhabitats, however, are more interpretable. By detailed microscopic observations, it can be shown that the size of cavities and the availability of food sources differ between microhabitats, and that these differences correlate with the body size and the diet of the animals (Haarlov, 1960; Anderson, 1978a; Ponge, 1991, 2000; Edsberg & Hagvar, 1999).

The significance of microhabitats for microarthropods has also been investigated indirectly, without observing the animals in their microhabitats. The microhabitat structure of different macrohabitats has been quantified and then correlated with the microarthropod communities of these macrohabitats. Microhabitat structure was quantified, for instance, by measuring the average pore diameter, the fractal dimension or the Shannon–Weaver indices of microhabitat diversity or microclimate diversity. These measures of microhabitat structure have been correlated with the body size distribution, the species richness or the activity abundance of the microarthropod communities (Haarlov, 1960; Anderson, 1978b; Schenker & Block, 1986; Sharrocks *et al.*, 1991; Kampichler, 1999). However, these correlations are often rather loose. Such indirect investigations of microhabitat use were much less restricted in their scope than direct investigations where animals have been studied within their microhabitats (see above). Moreover, the indirect investigations have showed that some macrohabitats are much richer in microhabitats than others (Kampichler & Hauser, 1993).

Ideally, a study on microhabitat use by microarthropods should include elements of both the direct and the indirect approaches. The ideal study should include: (i) all the major microhabitats in a macrohabitat; (ii) an analysis at the level of individual species and microhabitat types; and (iii) a repeated sampling under a range of different ambient conditions, day and night, different seasons, or different macrohabitats. Moreover, the abundances of animals should be comparable across the types of microhabitat. Abundances per square centimetre basal surface of a microhabitat cannot be compared between a flat and a three-dimensional microhabitat, for instance, a crust-like cryptogam and a shrub-like cryptogam. In contrast, abundances per microhabitat volume, per search time or, ideally, per microhabitat surface area are much more comparable across microhabitat types. One of us tried to satisfy all these criteria, but the study was restricted to an extremely easily accessible macrohabitat, the trunks of trees (Prinzing, 1997). It turned out that all dominant microarthropod species differentiated significantly among microhabitats. For other less-accessible macrohabitats, it is still not clear whether or not the majority of microarthropod species differentiate among microhabitats. It is obvious, however, that the microhabitat scale should always be taken into account.

Redistribution among habitats

Despite our restricted knowledge on the patterns of habitat use, it is clear that these patterns are dynamic rather than static: microarthropods redistribute themselves. Redistributions may take place at a small scale, through the movement of individual animals, but it may also happen at a larger scale, as a result of shifts in birth and death rates. Small-scale redistributions among macrohabitats have been well studied; for example, the redistribution between the soil and the litter throughout the course of a day or a year (e.g. Riha, 1951; Vannier, 1970; Bellido, 1979; Funke, 1979; MacKay *et al.*, 1987; Takeda, 1987; Kampichler, 1990). By comparison, redistributions owing to more irregular climatic events such as rainfall have been rarely studied. Mostly, these studies have considered redistributions between the litter and the trunks of trees (Bowden *et al.*, 1976; Bauer, 1979). Large-scale redistributions among macrohabitats have only rarely been studied. For instance, some authors have examined the redistribution of microarthropods across forest strata with increasing altitude or changing topography (Pschorn-Walcher & Gunhold, 1957; Cassagnau, 1961; Travé, 1963; Aoki, 1971). The problem with these studies is that it is never clear to what degree the large-scale redistribution of the animals among macrohabitats reflects merely a redistribution of the constituent microhabitats. For instance, with increasing altitude, certain lichen species move from the tree trunks down to the ground (Wirth, 1980). In such a case, the macrohabitat may be the wrong scale to analyse the redistribution of microarthropods.

The redistribution of microarthropods among microhabitats has also rarely been studied. Most of these studies have been conducted at a small scale, across areas of a few square metres. Redistributions at such scales reflect primarily the movement of individual animals. For instance, cones of alders or the lower layers of a shrub-like lichen are only used during harsh seasons, such as winter (Hammer, 1972; Bellido, 1979). A crustose lichen species is used primarily on one face of a trunk, but not on the other (André, 1975). Changing exposure of trunk faces to weather conditions results in a redistribution of microarthropods across lichen species and bark crevices (Prinzing, 1997). Redistributions between microhabitats have hardly been studied at larger scales – across kilometres or more. As a notable exception, André (1983, 1985) found preliminary evidence that animals use different microhabitats in different landscapes. Further studies on the large-scale redistribution of microarthropods among microhabitats would be most useful. Such studies are likely to improve our understanding of the organization of species assemblages across large scales (see Wiens, 1989 for birds).

Overall, microarthropods redistribute between macrohabitats despite the large distances among different macrohabitats and the differences in their microhabitat structures. Microarthropods also redistribute among microhabitats, despite the complexity of most microhabitat mosaics. It would be interesting to know whether there are certain functional species types that redistribute primarily at the macrohabitat scale, and other types that redistribute primarily at the microhabitat scale.

ENVIRONMENTAL FACTORS THAT CAUSE REDISTRIBUTION AMONG HABITATS

We will now summarize the existing evidence on what may cause the redistribution of microarthropods among macro- and microhabitats. First, we will discuss ecophysiological factors, then the immigration from population pools and finally the effect of other species. Much evidence has accumulated since Anderson's (1977) pessimistic assessment of our understanding of habitat use of microarthropods but the evidence still remains anecdotal or speculative.

Ecophysiological factors

Changes of ecophysiological living conditions may induce changes of habitat use in microarthropods. Obviously, climate is relevant. The effects of desiccation, chill and inundation have repeatedly been demonstrated on both the macrohabitat scale (Riha, 1951; Vannier, 1970; Adis, 1984a; Hijii, 1987; Gauer, 1997a,b) and the microhabitat scale (Hammer, 1972; Bellido, 1979; Leinaas & Fjellberg, 1985; Bauer & Christian, 1993; Prinzing, 1997). Interestingly, certain ecophysiological factors have been investigated only for microarthropods on tree trunks, namely the availability of heat and access to wind dispersal (Nicolai, 1986; Farrow & Greenslade, 1992; Prinzing, 2001). Further, the response of microarthropods to short-term fluctuations in climate has only been studied on trunks (Bauer, 1979; Prinzing, 1997). In addition to climate, diet may be a relevant determinant of the redistribution of microarthropods among habitats. The suitability of diet often fluctuates with ambient climate. Many algae or fungi become more easily palatable or digestible when moistened, and this may explain at least partly the habitat shifts of microarthropods after precipitation (Hammer, 1972; Hassall *et al.*, 1986; Faber & Joosse, 1993; Prinzing & Wirtz, 1997; Prinzing, 1999). Moreover, fungi can grow very quickly during moist weather (Hassall *et al.*, 1986). The increased densities of fungi in moistened habitats might attract animals by olfactory cues across distances of 40 cm or more (Bengtsson *et al.*, 1994). Finally, toxic factors may be relevant. Application of pesticides onto the litter layer may induce a redistribution of species to lower levels within the soil (Beck *et al.*, 1988; Prinzing *et al.*, 2000).

There are a number of interesting counter-examples. Several species do not redistribute when the environment becomes dry, hot, inundated, cold or toxic (Poinsot-Balaguer, 1976; Adis, 1984a; Wallwork *et al.*, 1984; MacKay *et al.*, 1987; Vegter *et al.*, 1988; Kampichler, 1990; Franklin *et al.*, 1997; Gauer, 1997a,b; Prinzing *et al.*, 2000; Prinzing, 2001). Some of these 'static' species may tolerate or resist the implicit environmental stresses (Poinsot-Balaguer, 1976; MacKay *et al.*, 1987; Franklin *et al.*, 1997; Gauer, 1997b). Other static species may rely on their ability to reproduce rapidly and build up a new population after the environment has become favourable again (Prinzing *et al.*, 2000). In general, such static species may well be capable of redistributing, but they may be reluctant to do so because the costs outweigh the benefits (Rosenzweig, 1987). Overall, different microarthropod species respond very differently to ecophysiological factors. It would be interesting to test whether species that respond differently belong to different functional types, characterized by distinct sets of traits.

Immigration from population pools

The immigration of individuals from a population pool may determine whether or not a species uses a habitat. Changing immigration from population pools may determine the redistribution of the species among habitats. Immigration from population pools can be significant. For instance, fresh leaf litter or a defaunated herb layer are only colonized by microarthropods when large population pools exist in surrounding areas (Schaefer, 1974; Beck, 1993). However, there has been virtually no investigation on whether immigration from population pools influences the distribution of an individual species. One exception is a set of studies by Hertzberg and collaborators (1994) in tussock grasslands. They explained declines of population densities in increasingly fragmented habitat patches by a reduced immigration from population pools in other patches. However, later investigations demonstrated that some species did decline in fragmented environments even though they were excellent dispersers, whereas some poor dispersers did not decline (Hertzberg, 1997). Population declines in fragmented environments may be caused, in part, by the change of physiological living conditions as a result of decreasing habitat cover (Hertzberg *et al.*, 2000). Another exception is the study by Gonzales *et al.* (1998) on experimentally fragmented cryptogam crusts on rocks. Fragmentation, and the consequent isolation from population pools, caused a decline of species abundances. It is important to note, however, that both Hertzberg and Gonzales *et al.* compared different patches of a single habitat type, rather than different habitat types.

Travé (1963) suggested that immigration from population pools may determine the extent to which Oribatei use two different habitat types, rocks and tree trunks. He suggested that the number of species on rocks is larger than on trunks because rocks have a larger interface with the presumed population pools in the surrounding soil and litter. Rocks are accordingly less isolated from population pools than are trunks. However, investigations on the migration activity, walking speed and passive dispersal of Oribatei did not confirm this assertion. Oribatei are well capable of migrating between litter and trunk and back within days (Christensen, 1980; Wunderle, 1992b). In fact, the rapid fluctuations of Oribatei abundances on trunks suggest that these animals do temporarily leave and then recolonize tree trunks (Wallwork, 1976; Prinzing, 1996, 2001). In addition, the number of colonizers may not depend on the size of the 'population pool' in the adjacent soil (see Bowden *et al.*, 1976 for Collembola). Overall, the significance of immigration from population pools for the richness of microarthropod species on tree trunks remains controversial. But undoubtedly, immigration *is* relevant for the presence of particular species on trunks (Collembola; Bauer, 1979). Alvarez *et al.* (2000) reached a similar conclusion after studying arable fields, which represent a further physiologically extreme habitat. Immigration from adjacent hedgerows into the arable fields occurred for some Collembola species, but not for others.

Immigration of animals from population pools within a metapopulation has received much attention. This may be the major cause of the famous correlation between distribution and abundance of species (Gaston & Blackburn, 2000). For microarthropods, however, this correlation between distribution and abundance often does not exist (local scale: Haarlov, 1960; regional scale; Vegter *et al.*, 1988; Wauthy *et al.*, 1989). Immigration from population pools may thus be less important in microarthropods than in many other taxa.

Interaction with other species

The redistribution of a microarthropod species among habitats may be caused by the impact of other species. However, the evidence is scarce and controversial. Part of the evidence originates from experiments, but each of these experiments was rather artificial. Either the environmental conditions were strongly simplified or the population densities were extremely high. Anderson (1978a) observed two oribatid mite species in the laboratory and reported that, when kept together, the species shifted their vertical distribution away from each other. Interestingly, these species did not shift their microhabitat use. Faber and Joosse (1993) observed species of Collembola in a field experiment and found that pairs of species shifted their vertical distribution away from each other only rarely. Two other experiments showed more distinct interspecific responses among Collembola. Christensen (1980) kept Collembola in small vials in the laboratory and showed that the species can strongly repulse or attract one another by their pheromones. When Beck *et al.* (1988) poisoned the Collembola in the litter of a German forest, a species that was restricted to tree trunks under control conditions abruptly colonized the defaunated (i.e. competitor-free) litter.

Another part of the evidence on the impact of other species on habitat use comes from observational studies. These studies were not artificial but inevitably could only provide circumstantial evidence (Wiens, 1989). In most studies, species were grouped into guilds according to their body size or morphology. Species of the same guild were considered as potentially competitors. Negative correlations among the distributions of species from the same guild were then taken as circumstantial evidence for competition. Some authors indeed found such evidence (Cassagneau, 1961; Kaczmarek, 1975). But others authors found no (Takeda, 1987) or only mixed evidence (Wolters, 1985). Vegter (1987) took a different approach. He investigated all species of Collembola in a soil mass and inferred competition from community metrics such as species packing, average niche breadth and niche overlap. He found no evidence for competition. However, it should be noted that the utility of these metrics for detecting competition has been criticized (Wiens, 1989). Finally, Walter & Norton (1984) analysed the body sizes of congeneric oribatid mite species. The authors found that body sizes of coexisting species differed more strongly than body sizes of other species. This finding was taken as circumstantial evidence for competition.

A particularly promising line of research has focussed on ecosystem-engineer species (Jones, 1996). These species can modify the microhabitats of other species. Middens of earthworms for instance clearly affect the distribution of some microarthropod taxa (Loranger *et al.*, 1998; Maraun *et al.*, 1999). Interestingly, other taxa seem to be unaffected (Maraun *et al.*, 1999). Moreover, earthworms did modify the habitat preference of Collembola in an experimental setting (Salmon & Ponge, 1999).

The overall conclusion from this section is that, first, ecophysiological factors such as microclimate and diet clearly lead to a redistribution among habitats in many species of microarthropods. Despite their fragile nature and small stature, such animals are not passive victims of their abiotic environment. In contrast, some other species do not redistribute. Second, it is not clear whether isolation from population pools, or interspecific interactions, regularly cause redistributions among habitats in the field. These two factors deserve further investigation. Nevertheless, the available evidence is strong enough that these two factors must be considered in any discussion of faunistic stratification in forests.

STRATIFICATION OF FORESTS

The world is stratified. A seashore, or a mountain, consists of distinct habitat strata inhabited by distinct sets of species with distinct life histories and modes of interaction. Understanding the stratification of the biological world is a first step to understanding the life histories and interactions of species. Classical examples of stratification are linear. For instance, the higher one moves along a seashore, the more strongly the habitats and their inhabitants differ from those lower on the shore. Here, we will explore the stratification of forests, particularly tropical rainforests, from the viewpoint of microarthropods. We will show that the stratification of species is not linear in a forest. Rather, it resembles a sandwich. We will then discuss how such a 'sandwich stratification' may be understood from the habitat use of the species. We will also give evidence that the lower strata (the forest floor) and the higher strata (the trees) may be more distinct in a rainforest than in a temperate deciduous or coniferous forest. Accordingly, we will discuss how such a high degree of stratification of rainforests may be explained in terms of habitat use.

Stratification of forests from the viewpoint of microarthropods

The main strata recognized by entomologists in a forest are the soil (decomposed organic material, more or less mixed with inorganic material); the litter (freshly fallen leaves); the bark of trees; the foliage of trees; and the suspended soils in the crowns, for example in dead branches, in moss cushions or between the roots of epiphytic phanerogams (Delamare-Deboutteville, 1951). Some forests also have a herbaceous stratum, a shrub stratum or a moss stratum at the base of the trunks (Gjelstrup, 1979; Otto, 1994, Palacios-Vargas *et al.*, 1998).

Very few entomologists have studied microarthropods both in the lower and in the higher strata of a forest. The few exceptional studies always have showed that the fauna of the suspended soils in the tree crowns is more similar to the fauna in the forest-floor soil than to the fauna on the tree bark or in the litter layer. Such a pattern of similarity was demonstrated in a deciduous beech forest in Southern Germany (Wunderle, 1992b), in a secondary forest in the Ivory Coast (Delamare-Deboutteville, 1951) as well as in a Venezuelan cloud forest (Behan-Pelletier *et al.*, 1993). So, the top stratum of the forest resembles the lowest stratum. We call this a sandwich stratification.

We realize that our notion of sandwich stratification may turn out to be an oversimplification, since it neglects the foliage of trees. Until now, Collembola and Oribatei occurring in the foliage have hardly been investigated. Wunderle (1992b) studied the Oribatei of foliage and twigs. She found that the fauna resembled that of forest-floor soils, which confirms our notion of sandwich stratification. Wunderle, however, cautioned that her results may be strongly biased by the extraction method, a Berlese funnel. The investigation by Walter *et al.* (1994), as well as several more sporadic observations (Spain & Harrison, 1968; Walter & Proctor, 1999), indicate that the foliage is colonized by a highly distinct Oribatei fauna. This would partly contradict our thesis of sandwich stratification.

Possible causes for a sandwich stratification of forests

Why is the fauna of suspended soils more similar to that of the forest-floor soil than to that of the litter or the bark? The phenomenon can be explained by several of the above-mentioned determinants of habitat use.

1. **Options for redistribution among microhabitats.** The major option for redistribution in the forest-floor soil as well as in the suspended soil is to move between the surface and inner parts, a few centimetres below the surface (Haarlov, 1960; Takeda, 1987). A second option is to move between larger and smaller cavities (Haarlov, 1960). In contrast, at the bark as well as in the litter, the major option for redistribution is to move for some tens of centimetres across the complex mosaic-like surface (Hammer, 1972; Prinzing, 1997, 2001). This mosaic results from the conspicuous surface relief (which modifies the exposure to wind, sun and precipitation), and from the conspicuous microhabitats, such as lichens, bark crevices or different types of litter (Aoki, 1967; Nicolai, 1986).
2. **Ecophysiological conditions.** The cavities in the forest-floor soil as well as in the suspended soil are mostly small, dark and moist (Delamare-Deboutteville, 1951; Vannier, 1970). Such living conditions are used by Collembola species of a specific life-form: little pigmentation, small size, worm-like shape and absence of a furca (the 'springtail' of the Collembola; Dunger, 1983). Such species of Collembola can be found in both forest-floor soils and suspended soils (Delamare-Deboutteville, 1951). In contrast, on the bark as well as in the litter, small, dark and moist cavities are comparatively rare. This contrast in living conditions corresponds to the life-forms of the Collembola in these two strata: highly mobile, well-pigmented animals with a well-developed furca (above references).
3. **Other species.** As mentioned above, the life-forms of species in the litter or at the bark differ from those in the suspended soil or forest-floor soil. Different life-forms correspond to different life histories (Greenslade, 1983), and life histories affect the mode of interaction among species (Pianka, 1988). Fast-moving species may outcompete slowly moving species in the strongly fluctuating microcosm at the surface of trunks. In contrast, quickly reproducing species might outcompete slowly reproducing species in the much more constant soil microcosms (Southwood, 1988). So the fast-moving but slowly reproducing species (like many entomobryid Collembola: Greenslade, 1983; Joosse & Veltkamp, 1970) might be pushed out of the suspended soils and forest-floor soils by competitors within the soils. Correspondingly, the slow-moving but quickly reproducing species (e.g. many isotomid Collembola)

might be pushed into the suspended soils and forest-floor soils by competitors outside the soils.

The similarity between suspended soils and forest-floor soils cannot be explained by the immigration from population pools. Obviously, the sheer distance between the suspended soils and the forest-floor soils restricts the opportunities for migration between these strata. So, immigrations from population pools may be irrelevant in explaining sandwich stratification. A low degree of migration between strata would be consistent with the observation of evolutionary isolation of certain oribatid mite lineages in the suspended soil (Ch. 26). However, it is equally possible that such immigrations are so efficient across large distances that forest-floor soils and suspended soils can maintain similar faunas. The pathways of the upward migration would be below the bark (Delamare-Deboutteville, 1951) whereas the pathway of the downward migration would be simply by gravity (Christensen, 1980). A highly efficient large-distance migration would be consistent with the observation of Winchester & Behan-Pelletier (Ch. 10) that, within a stratum, even widely separated faunas can be very similar.

Stratification of tropical rainforests compared with temperate forests

The canopy layer of tropical rainforests has always been considered as a habitat of immense species richness (Erwin, 1995; Stork *et al.*, 1997a). Here, we will define the canopy *sensu lato* as starting above the forest floor (see Ch. 1). Knowledge of microarthropods in the canopy *sensu stricto*, the crown region or the canopy surface, is simply insufficient (but see Palacios-Vargas *et al.*, 1998). If the canopy of tropical rainforests is so special, one should expect that its microarthropod fauna would be highly distinct from the fauna at the forest floor, and more distinct than in temperate regions. We call this the 'distinctness hypothesis'. However, one might also expect the opposite. Because much of the canopy of a tropical rainforest is shaded, with apparently little fluctuation in abiotic living conditions, the tropical canopy might be a more suitable habitat for forest-floor microarthropods than the temperate canopy (Wallwork, 1976; Beck *et al.*, 1997). Occasionally, the microarthropods from the floor of a rainforest may even have to colonize the canopy because the floor becomes inundated by water (Adis, 1984a; Gauer, 1997a,b) or by divisions of army ants. So the canopy may be more suitable, or even indispensable, for forest-floor microarthropods in a tropical rainforest than in a temperate forest. If this is true, the fauna encountered in the canopy and on the forest floor should be more similar in the tropical rainforest than in temperate forests (Wallwork, 1976; Beck, 1963; Beck *et al.*, 1997). We call this the 'similarity hypothesis'.

There are only a few studies on microarthropods that can be used to test these two hypotheses as few authors have studied microarthropods on the forest floor as well as in the canopy to species level (Rodgers & Kitching, 1998). Even fewer authors applied comparable sampling techniques in both strata. Sampling techniques may affect the fauna sampled. In particular, Malaise traps, photo-eclectors and canopy fogging primarily sample the highly mobile species that move on surfaces and rarely retreat into crevices (Walter *et al.*, 1994, 1998). If the canopy and the forest floor are sampled with different techniques, differences in the results may be entirely a consequence of differences in the sampling techniques. But note that such a sampling approach is fully legitimate if the primary goal of a study is to obtain samples containing a large number of animals, including rare species (e.g. Palacios-Vargas & Gómez-Anaya, 1993).

Most of the studies that truly permit comparison of the forest floor with the canopy have focussed on oribatid mites. Three studies investigated the faunistic stratification in deciduous or coniferous temperate forests and reported faunistic similarities of 26, 38 and 71% (see Table 24.1). Four other studies investigated the stratification in tropical rainforests and reported faunistic similarities of 13, 17, 19 and 25% (see Table 24.1). So, the faunistic similarity between the floor and the canopy of tropical rainforests appears to be lower than in deciduous or coniferous forests from temperate regions. This appears to be a confirmation of the distinctness hypothesis.

There are two more studies from rainforests outside the tropics reporting faunistic similarities of 10% and, for parasitiform mites, 16% (Table 24.1). Overall, the large differences between canopy fauna and forest-floor fauna may be characteristic for broad-leafed rainforests in general.

The above comparisons of faunistic similarities between the floor stratum and the canopy stratum in different forest types are vulnerable to two statistical artefacts. The first is that lower similarities between

Table 24.1. *Overview of studies that compare the fauna of forest floors and forest canopies*[a]

Authors	Study area	Type of comparison	Faunistic similarity (%)[b]
Deciduous or coniferous temperate forests			
Wunderle, 1992b	German beech forest	Forest-floor soil versus suspended soil (dead wood)	38
		Forest-floor litter versus tree bark	38
Behan-Pelletier & Winchester, 1998	Canadian spruce forest	Forest floor versus canopy	71[c]
Woltemade, 1982	German oak forest	Forest-floor soil and litter versus tree bark	26
Tropical rainforests			
Beck, 1971	Amazonian *terra firme* forest	Forest floor versus suspended soil at a height of 10–25 m	19
Behan-Pelletier *et al.*, 1993	Venezuelan cloud forest	Forest-floor soil versus suspended soil	13
Wunderle, 1992a	Peruvian forests	Forest-floor soil and litter versus suspended soil and bark	25
Franklin *et al.*, 1997	Amazonian inundation forest	Forest floor versus canopy	17
Rainforests outside the tropics			
Walter *et al.*, 1994	Temperate Australian rainforest	Forest-floor litter versus tree bark	10
Walter *et al.*, 1998	Subtropical Australian rainforest	Forest-floor soil versus suspended soil	16

[a] All studies considered Oribatei, with the exception of Walter *et al.* (1998) who considered parasitiformes mites.
[b] Faunistic similarity between the floor and canopy is indicated with the Jaccard index (Krebs, 1999). Note the high faunistic similarities in deciduous or coniferous temperate forests.
[c] 34% if only moss cores were considered.

rainforest strata may be a consequence of insufficient sample sizes. If a species is only found once, (i.e. it is a 'singleton'; Novotny & Basset, 2000), it can only occur in a single stratum. Inclusion of such a species thus reduces the faunistic similarity between the strata (while exclusion results in a loss of information). Samples from rainforests are considered to be particularly rich in singletons (Novotny & Basset, 2000). However, for microarthropods this may not be true. The proportion of singletons in Wunderle's samples from suspended soils in a Peruvian rainforest was no larger than in a German deciduous forest (Wunderle, 1992a,b). Nevertheless, Wunderle (1992a) decided to exclude singletons from her estimate of faunistic similarity in Peru. Even then she found a faunistic similarity between forest floor and canopy of only 25%. Also the estimate of faunistic similarity by Walter *et al.* (1998) seems to be immune to the problem of insufficient sample size. The authors showed that the collector's curves approach a plateau for both the suspended soil and the forest-floor soil. Therefore, by increasing the sample size, only very few additional species would be found in either stratum. Overall, the low faunistic similarity across rainforest strata is not caused primarily by inadequate sample sizes. The second possible artefact is that low faunistic similarities between rainforest strata may be caused by high differences in the species richness of the strata. Differences in species richness generally reduce faunistic similarities (Wolda, 1981). But in our case, this problem is hardly relevant. On average, the species richness of the floor and the canopy in a rainforest differed by a factor of 2.2. The equivalent value for deciduous or coniferous forest in the temperate regions was 2.1 (see references above). Overall, the high degree of stratification of mite fauna in rainforests does not seem to be a statistical artefact. Nonetheless it should be kept in mind that the evidence is still scarce and tentative.

It is important to note that the above comparisons between strata were conducted at a local scale, within an individual forest or individual patches within a forest. At larger scales, the faunistic similarities between strata seem to be much higher, particularly in rainforests (Beck, 1963, 1971; Winter, 1963; Wunderle, 1992a). Apparently, the vertical distribution of species shifts from one locality to the other.

In addition to the above studies on mites, there is an important study on the stratification of Collembola (Rodgers & Kitching, 1998). The authors compared the fauna of the 'litter' in the canopy (i.e. suspended soils) and at ground level of a subtropical Australian rainforest and found a similarity of 52% (Jaccard index; Krebs, 1999). This is clearly higher than was reported for mites (see above). Possibly, Collembola can move more easily between the floor and the canopy than mites, given the ability of Collembola to walk quickly (Eisenbeis & Wichard, 1987). The ability of Collembola to leap may be less important for canopy-dwelling species (Bauer & Christian, 1987). It remains unclear, however, whether the 52% similarity between the strata of a rainforest is small or large compared with a temperate deciduous or coniferous forest. In these forests, the stratification of collembolan faunas has, to our knowledge, not been investigated so far.

Possible causes for a high degree of stratification in rainforests

If the canopy and the forest floor are more distinct in (tropical) rainforests than in deciduous and coniferous temperate forests, what might be the cause? We will discuss four possible causes, based on our review of habitat use above. Given the lack of knowledge on tropical canopy microarthropods, our discussion remains speculative.

1. **Options for redistribution among microhabitats may differ between the floor and the canopy of a rainforest.** The bark of many rainforest trees is very smooth (Vareschi, 1980). There are few crevices, or shrub-like cryptogams. Therefore, there is little shelter from radiation or from the numerous ants in a rainforest. Moreover, there is only little bark relief. Consequently, there is no mosaic of climatic exposures or of water-vapour fluxes comparable with that at the surface of temperate trees (Stoutjesdijk & Barkman, 1992; Prinzing, 1996, 2001). In contrast, the surface of the litter layer is much more heterogeneous. There is often an intense surface relief because of the large size of many rainforest leaves. Moreover, the horizontal distribution of fresh and decomposed leaves may be heterogeneous, reflecting the very heterogeneous litter input (Walter & Breckle, 1991). One tree crown may shed its leaves at a particular time while the neighbouring tree crown does not. Moreover, neighbouring tree crowns often belong to different tree species, producing different types of litter (Vareschi, 1980). In deciduous or coniferous forests of temperate regions, in contrast, the bark and the litter may be more similar in their degree of microhabitat differentiation. The bark, in particular, is more heterogeneous than in a rainforest because of the strong relief and the diversity of microhabitats (crevices, crustose lichens, fruticose lichens; Vareschi, 1980; Stubbs, 1989).
2. **Ecophysiological conditions may differ between the floor and the canopy of a rainforest.** The suspended soils and the forest-floor soils of a rainforest can be quite different in nature. The forest floor often contains little humus but many large roots (Vareschi, 1980; Walter & Breckle, 1991). The situation is reversed in the suspended soils. In a temperate forest, in contrast, the forest-floor soil and the suspended soil are much more alike (Otto, 1994). In both soils, the humus content is mostly high, and the root density is low. In addition, in a rainforest, the microclimate may differ strongly between strata. The popular notion that microclimates below the roof a rainforest are moist and constant (Hesse, 1924; Barbosa & Wagner, 1989; Beck *et al.*, 1997; Wunderle, 1992a) may be wrong from the viewpoint of a microarthropod. Even if the air humidity in a rainforest remains above, say, 85% (Walter & Breckle, 1991), this is already rather dry for many microarthropods. Microarthropods

'measure' air humidity in terms of saturation deficit (Stoujesdijk & Barmann, 1992). The saturation deficit at 85% relative humidity in a rainforest at 35°C exceeds the saturation deficit at 65% air humidity in a temperate forest at 20°C (Häckel, 1993). So from the viewpoint of a microarthropod, the air at the trunk surface within a rainforest may be drier than in a moderately moist temperate forest. In contrast, the air in the litter of the rainforest may be moist. Air in the litter layer is usually saturated with water vapour (Vannier, 1970). In a rainforest, the saturated air may be kept in the litter very efficiently because many leaves are large (Vareschi, 1980; Stoutjesdijk & Barkmann, 1992).

3. **Populations in the rainforest canopy may be isolated from population pools at the rainforest floor, and vice versa.** The lack of crevices in the bark of many rainforest trees may, on the one hand, prevent soil animals from migrating between the soils at the forest floor and the canopy (Delamare-Deboutteville, 1951). On the other hand, the sheer number of physical connections between the forest floor and the suspended soils is often large. There are many lianas and aerial roots that connect the two areas (Walter & Breckle, 1991) and many large animals that migrate between the forest floor and canopy which might act as vehicles for transporting microarthropods into the canopy (Hesse, 1924; Norton, 1980). Moreover, immigration is certainly not a general prerequisite for the colonization of either the floor or the canopy of a rainforest. The animals can reproduce in both strata (Beck, 1971).
4. **Interspecific interactions may differ between the floor and the canopy of a rainforest.** Ants use suspended soils in rainforests more than in temperate deciduous or coniferous forests (Delamare-Deboutteville, 1951; Beck, 1963). Ants are ecosystem engineers and modify the microhabitat structure of soils (Jones, 1996). Accordingly, ants may contribute to the faunistic differentiation between suspended soils and forest-floor soils in a rainforest.

CONCLUSIONS

Forests are stratified – just like the rest of the world. But contrary to much of the rest of the world, this stratification of forests is not linear. If we move up in a forest, the fauna does not become increasingly different from that at the floor. Rather, the fauna of the litter extends far onto the surface of the tree trunks, and the fauna of the soil at the forest floor has an outlet in the suspended soil of the crowns. In some tropical rainforests, it may even be more appropriate to consider the species-poor forest-floor soil as an outlet of the species-rich suspended soils (Behan-Pelletier *et al.*, 1993; but see Beck, 1971). There is preliminary evidence that the faunistic difference between the forest-floor stratum and the canopy stratum is larger in (tropical) rainforests than in temperate deciduous or coniferous forests. This phenomenon might be explained in terms of habitat use of the animals. Such a high degree of faunistic stratification in rainforests would contribute to the very large species richness of these forests.

ACKNOWLEDGEMENTS

We thank the editors for their invitation to contribute to this volume. We also thank Yves Basset and an anonymous referee for their valuable suggestions. Finally, we thank Adetola Badejo for correction of the English.

25

Insect herbivores feeding on conspecific seedlings and trees

Héctor Barrios

ABSTRACT

Assemblages of insect species feeding on conspecific saplings and trees in the tropics are poorly studied in comparison with similar assemblages in temperate forests. A comparative study of the relative importance of trees and saplings of *Castilla elastica* (Moraceae) as hosts for phytophagous insects was carried out in a dry forest in Panama. The arthropod fauna of 2000 understorey saplings and 12 canopy trees was surveyed within a 7 month period with particular reference to insect herbivores. Sample sizes in the canopy and understorey were equal; amounting, in both cases, to 364 m^2 of leaf area surveyed. Arthropod abundance was significantly higher in the canopy than in the understorey (9840 and 5234 individuals collected, respectively). A total of 120 morphospecies of insect herbivores were collected from both matures trees and saplings. For a similar leaf area sampled, insect herbivores were 19 times more abundant and 1.6 times more species rich on the foliage of mature trees than on that of saplings. Only one leaf-chewing species (Chrysomelidae) and 16 sap-sucking species (mostly Tingidae, Cicadellidae and Membracidae) overlapped between saplings and trees among the herbivore species. The present data are compared with a similar study performed on the insects feeding on conspecific saplings and trees of *Pourouma bicolor* (Cecropiaceae) in a wet forest in Panama. Patterns of insect distribution on saplings and mature trees were broadly similar in the two studies. I suggest that insect herbivores rarely colonize saplings from parent trees in rainforests, contrary to one key assumption of the Janzen–Connell model. Furthermore, the greater availability of food resources, such as young foliage, in the canopy than in the understorey may generate complex gradients of abundance and species richness of insect herbivores in tropical forests.

INTRODUCTION

Scientists have described and named some 1.4 million species of arthropods, of which more than half occur in tropical forests. Among them, insects are particularly abundant, accounting for 62% of total arthropod diversity (Wilson, 1988b). Although the estimation of species richness for tropical insects is the subject of intense debate (Basset *et al.*, 1996a), the ecology of these insects is virtually unknown (Janzen, 1988a; Basset, 1992b; Basset & Novotny, 1999). Given that over 26% of all insects and half of all beetle species are herbivorous (Strong *et al.*, 1984) and that leaf damage levels in tropical forests range between 6% and 15% (Hadwen *et al.*, 1998), the importance of arthropods in forest ecosystems deserves more attention than at present (Janzen, 1987b; Sutton & Collins, 1991).

In tropical forests, both primary production and biological diversity are distributed nonrandomly. Biotic conditions vary radically across the different strata of the forest. The upper canopy is directly exposed to intense solar radiation. Midday temperatures are 5–8 °C higher and relative humidities are 30% lower than on the cooler, shaded forest floor some 30 m below (Wright & Colley, 1994). Wind, fluctuation of relative humidity and water condensation at night are notably higher in the upper canopy than in the understorey (Blanc, 1990; Parker, 1995). Many mature tropical forest trees are taller than their conspecific seedlings by a factor of 100, compared with a factor of 30 for temperate tree species such as birch (Fowler, 1985). The foliage of mature tropical trees is likely to experience a different microclimate from that prevailing in the understorey. This may have important implications for the foraging strategies of insects, which, in tropical rainforests, can be affected profoundly by light regime and microclimate (e.g. Shelly, 1985; Roubik,

1993). Furthermore, leaf area density and the abundance of young leaves, flowers and seeds are also usually higher in the upper canopy than in lower strata (Parker, 1995; Hallé, 1998). Leaf buds in the upper canopy appear to be extremely well protected against desiccation and herbivory (Bell *et al.*, 1999). Accordingly, the variability of environmental and biological conditions from the ground to the upper canopy may affect the distribution of herbivorous insects (Schowalter *et al.*, 1986).

I hypothezise that, in tropical forests, mature trees will support more species of insect herbivores than conspecific seedlings or saplings. Many studies of tropical insects have reported a greater abundance, activity or diversity of insect herbivores in the upper canopy than in the understorey (e.g. Erwin, 1983a; Sutton, 1983; Basset *et al.*, 1992). These studies, however, compared whole forest strata, rather than the specific resources available to insect herbivores within strata. One way of studying vertical gradients of insect abundance and diversity in rainforests more rigorously is to compare the insect communities that feed on conspecific seedlings, saplings and mature trees. I compared the herbivore fauna foraging on saplings and mature trees of a light-demanding species, *Castilla elastica* Cerv. (Moraceae), in a tropical dry forest in Panama in an attempt to answer the question: do insect communities differ in composition, density, species richness and dominance between saplings and conspecific mature trees in the canopy?

The data are compared with a similar study of another light-demanding species, *Pourouma bicolor* Martius, growing in wet tropical forests in Panama. Studies of herbivory in tropical forests often focus on leaf damage, either in the understorey or in the canopy, often on nonconspecific tree species (Lowman, 1982a, 1984; Coley, 1983; Langeheim & Stubblebine, 1983; Lowman & Heatwole, 1992). Recently, Hadwen *et al.* 1998) compared leaf damage in mature trees and conspecific seedlings in Australia. One study on Barro Colorado Island, Panama (Barone, 2000) assigned leaf damage on conspecific trees and seedlings to particular herbivore morphospecies. Although these studies provide some information on the vertical stratification of folivory in tropical forests, they do not provide estimation of faunal turnover on conspecific hosts. This contribution provides an entomocentric perspective on plant–insect relationships, along a vertical gradient.

MATERIAL AND METHODS

Study site and canopy access

Censuses were conducted at the canopy crane site in Parque Natural Metropolitano, Panama Province, Panama (8°58' N, 79°33' W). This park comprises natural forest covering an area of 270 ha, close to Panama City. It is adjacent to Parque Nacional Soberania, forming together a total protected area of 370 km^2 (Wright & Colley, 1994). The vegetation is dry tropical forest (Holdridge *et al.*, 1971). The altitude of the site ranges between 20 and 138 m above sea level and, although generally flat, steep slopes occur locally. Average annual rainfall is 1740 mm, 85% of which falls between May and December. Average annual air temperature is 28 °C (K. Kitajima, unpublished data). The site has undergone substantial ecological modification, beginning in pre-Colombian times, but it has remained largely undisturbed since the 1920s. During that period, it has reverted from abandoned pasture to secondary forest (Illueca, 1985). Dominant canopy species on the study site are *Anacardium excelsium* (Bert. & Balb.) Skeels (Anacardiaceae), *Astronium graveolens* Jacq. (Anacardiaceae), *C. elastica* Cerv. (Moraceae), *Annona spraguei* Staff. (Annonaceae), *Enterolobium ciclocarpum* Griseb. (Fabaceae) and *Cordia alliodora* (R. & P.) Oken (Boraginaceae). The understorey is dominated by *Heliconia latispatha* Benth. (Heliconiaceae), *Piper reticulatum* L. (Piperaceae), *Psychotria* spp. (Rubiaceae), *Astronium graveolens* Jacq. (Anacardiaceae), *Cojoba rufescens* (Benth.) Britton & Rose (Fabaceae) and *Hirtella racemosa* Lam. (Chrysobalanaceae).

The study was carried out from a canopy crane employing a gondola suspended from the crane jib. A crane operator controlled the position of the gondola, from which insect collections were made. From the gondola, access to all levels of the forest canopy were possible. The crane is 42 m tall with an arm length of 51 m, allowing access to about 8000 m^2 of projected area. The height of the canopy under the crane ranges from 20 to 38 m, corresponding to a volume of forest of 200 000 m^3 (Wright & Colley, 1994). The system provides good access to mature trees and canopy lianas.

Study plant

Three species of *Castilla* (Moraceae) are known from Mexico to South America. *C. elastica* Cerv., 1794 grows in the understorey of secondary forest from Southern

Mexico to Colombia, and perhaps further south at low elevations (Hammell, 1986). *C. elastica* is one of the most abundant light-demanding species within the study area. This plant species was selected as part of a larger project, which compares the insect fauna of light-demanding and shade-tolerant plant species in Panama. *C. elastica* is a tree reaching up to 20–30 m in height at maturity. Leaves are oblong–obovate, 20–30 cm long, 10–14 cm broad, with both surfaces (but particularly the lower one) golden spreading-hirsute.

Sampling protocol

Mature trees and saplings of *C. elastica* were surveyed in the canopy and understorey at the study site. The sampling protocol ensured that leaf areas sampled in both strata were similar. I considered as 'mature', all *C. elastica* individuals taller than 15 m in height and accessible from the crane gondola (n = 12). Within the understorey, only seedlings and/or saplings below 1.5 m were sampled within an area adjacent to the crane (n = 2000). Insects were collected by beating in both strata (Lepointe, 1956). A total of 26 surveys were conducted between December 1998 and June 1999. Each survey consisted of 40 beating samples (see below) on seedlings/saplings and 40 beating samples on mature trees (total sampling effort 1040 beating samples for each stratum). All surveys occurred between 06:30 and 10:00.

Square beating sheets of 3970 cm^2 in area were used. These were of conical shape (slopes of 45°), ending in a circular aperture (7 cm of diameter), which was fitted with a removable plastic bag. Sheets were inserted below the foliage so that the layer of leaves above occupied more or less the entire area of the sheet (thus providing a standardized sample). The leaf area sampled was taken to be equivalent to the area of the beating tray. The insects were dislodged from the foliage with three or four strokes and gently brushed into the plastic bag, which was then closed and replaced by a new one ready for the next sample. A single beating in this fashion represented one sample, and 80 samples were collected every week (40 samples in the canopy, 40 in the understorey), representing one survey. During each survey three to four samples were obtained from each individual mature tree, in the upper canopy. These samples included young and mature leaves, but no flowers or fruits. Leaf age is an important factor explaining leaf damage in tropical forests and the majority of damage typically occurs on young leaves (Coley, 1980, 1983; Lowman, 1985). Accordingly, percentage of area occupied by young leaves was scored subjectively for each sample. On a scale from 0 to 10, for example, if young leaves occupied half of the beating tray, a score of 5 was assigned to the sample, whereas if they occupied all the surface of the tray, a score of 10 would be assigned.

Insect processing

The study targeted free-living insect herbivores (i.e. leaf-chewing and sap-sucking insects). Insects collected were placed in a cooler and taken immediately to the laboratory. Live leaf-chewing insects were placed individually in vials and fed with a young leaf of *C. elastica*. The feeding trials lasted a minimum of 3–4 days. Vials were examined every day for evidence of feeding. Specimens feeding were assigned to the 'feeding' category, others to the 'not feeding' category. Arthropods were counted and sorted to family level or higher taxonomic level. They were assigned to one of the arboreal guilds proposed by Moran and Southwood (1982) and Stork (1987b): leaf chewers, sap suckers, pollinators, fungal-feeders, insect predators, other predators, parasitoids, wood eaters, scavengers, ants, tourists and unknown. Eventually, all herbivorous (leaf-chewing, sap-sucking and wood-eating) insects were mounted and sorted to morphospecies, with the exception of free-living juveniles. When necessary, genitalia dissections were performed to separate sibling species. As far as possible, morphospecies were then identified by comparison with the insect collections of the University of Panama and Smithsonian Tropical Research Institute. Arthropod data were managed using the software Biota (Colwell, 1997a). All material was deposited at the Programa Centroamericano de Maestría en Entomología at the Universidad de Panamá.

Statistical analyses

Differences between mature trees and saplings in terms of the total number of individuals and/or species of major taxonomic groups of arthropods were tested using Mann–Whitney tests on data pooled from the 26 surveys. The same tests were used to compare the abundance of arthropod guilds (*sensu* Moran & Southwood, 1982) and of species richness for herbivorous guilds (leaf-chewing and sap-sucking insects) between mature trees and saplings. Seasonal changes in herbivore abundance between the canopy and the understorey were explored with a two-way analysis of variance (ANOVA) considering species with $n \geq 11$ individuals.

Vertical (canopy versus understorey) and temporal (over 7 months) gradients in the herbivore faunas were estimated using multiple indices and methods: Shannon and Simpson indices (Magurran, 1988; Colwell & Coddington, 1994), the number of individuals and similarity (Hurlbert, 1971; Magurran, 1988). Calculations were performed using the statistical program EstimateS (Colwell, 1997b). Differences in the structure of insect assemblages between the two strata (canopy and understorey) were quantified using the Morisita–Horn index and calculated among arthropod guilds and season (dry and wet). Jacquard's similarity coefficient (based on presence or absence data) was calculated for each survey to estimate the faunal overlap between the canopy and understorey (Hurlbert, 1971; Magurran, 1988).

Eventually, linear regression was used to examine the extent to which the occurrence of herbivorous insects coincided with the production of young leaves by *C. elastica*. Regressions were computed for each of the seven most abundant morphospecies ($n \geq 11$), using log abundance as the dependent variable and percentage of young leaves per sample as the independent variable. In a second set of analyses, the total number of insects, all species pooled, was considered separately for samples of the canopy and understorey. The log-transformed insect abundance was used as the dependent variable and percentage of young leaves as the independent variable.

RESULTS

Arthropod fauna

In total, 15 076 individuals were collected representing 248 morphospecies (131 and 117 morphospecies obtained from the canopy and the understorey, respectively). From these, 9840 individuals were collected from the mature trees and 5234 individuals from saplings. The total area of foliage sampled was similar for mature trees and saplings and estimated as 364 m^2. Young leaves represented 24.3% of the total leaf area surveyed. On average 7.2 ± 0.663 arthropods were collected per sample. Mean arthropod abundance per sample was significantly higher on mature trees than on saplings (9.8 ± 0.866 and 5.2 ± 0.170 individuals, respectively; Mann–Whitney $U = 5.9 \times 10^{-6}$, $p < 0.006$).

Table 25.1. *Total number of arthropod individuals collected on saplings and mature trees, detailed by guilds and the most common nonherbivorous taxa*

Taxa/guild	Number of individuals		Mann–Whitney *U*-test *p* value
	Saplings	Trees	
Leaf chewers	305	721	0.121
Sap suckers	226	5520	0.000*
Leaf-miners	1	19	0.814
Wood eaters	33	1	ID
Insect predators	118	515	0.000*
Staphylinidae	107	101	0.142
Coccinellidae	19	470	0.006*
Miridae	131	26	1.000
Ants (Formicidae)	1529	191	0.000*
Other predators	1072	1551	0.000*
Araneae	1007	1534	0.000*
Parasitoids	81	402	0.004*
Eulophidae	30	98	0.194
Fungal-feeders	79	251	0.000*
Endomychidae	15	151	0.018*
Scavengers	1510	491	0.000*
Entomobryidae	1061	80	0.000*
Isotomidae	156	7	0.222
Blattodea	60	237	0.129
'Tourists'	30	42	0.392
Unknown	174	75	ID

ID, insufficient data for the test.
*Probability significant.

Different patterns of distribution within strata were apparent for different arthropod taxa. Araneae, Coccinellidae and Endomychidae were more abundant in trees than in saplings whereas Entomobryidae and Formicidae showed the reverse pattern. Blattodea, Eulophidae, Isotomidae, Miridae and Staphylinidae did not differ significantly along the vertical gradient (Table 25.1). Herbivores represented 46.2% of all insects sampled, predators 33.0%, scavengers 13.3%; 7.5% of the material was represented by fungal–feeders, parasitoids and other guilds. Insect herbivores (leaf chewers and sap suckers), parasitoids and fungal-feeders were more abundant on trees than saplings

(Table 25.1): wood-eaters, ants and scavengers showed the reverse trend in abundance between strata.

Cumulative curves of species richness generated by the 26 surveys showed that the curves were shaped in a similar fashion for both strata. The total number of morphospecies collected was 131 for mature trees and 117 for saplings. I used EstimateS (Colwell, 1997b) to determine the asymptotic value of both curves when the same number of individuals had been collected. For 2280 individuals in each curve, 129 taxa had been collected in mature trees and 112 taxa in saplings.

Abundance and species richness of insect herbivores on saplings and matures trees

A total of 120 morphospecies of herbivorous insects was collected from mature trees and saplings. For an equal sampling effort, the species richness of herbivores was 1.6 times higher on trees than on saplings (species observed, see Table 25.3, below). Insect herbivores were 11 times more abundant on the foliage of trees than of saplings (respectively 6258 and 565 individuals; Mann–Whitney $U = 3.4 \times 10^{-5}$, $p < 0.001$). Of the four guilds of herbivores, sap suckers showed the highest difference between trees and saplings, being 24.4 times more abundant on the foliage of trees than on saplings (Table 25.1). Abundance of leaf-chewing insects feeding on *C. elastica* was 19 times higher on trees than on saplings, yet the Mann–Whitney test failed to detect a significant pattern (Table 25.2). Finally, nonfeeding leaf chewers were relatively similar in abundance across the two strata. At the family level, Curculionidae, Chrysomelidae, Nymphalidae, Noctuidae, Notodontidae, Cicadellidae, Thripidae, Membracidae and Flatidae were significantly more abundant on trees than on saplings, whereas Geometridae and Tingidae showed the reverse trend (Table 25.2).

I identified 17 morphospecies of leaf-chewing insects feeding on *C. elastica*, 13 on trees and five on saplings. Leaf-chewing insects feeding on canopy foliage included two species of Curculionidae, eight species of Chrysomelidae, one of Geometridae, one of Nymphalidae and one of Notodontidae. One of the Curculionidae, *Myrmex* sp., represented 90% of all leaf-chewing individuals feeding on *C. elastica* in the canopy. Leaf-chewing insects feeding on seedlings and/or saplings included five species of Chrysomelidae: one of them, *Rhobdopterus fulvipes* Jacoby, represented 60% of all leaf-chewing individuals feeding on *C. elastica* in the understorey. Sap-sucking insects were represented by 103 morphospecies, 70 on trees and 48 on saplings (Table 25.2). Sap-sucking insects collected from the canopy included 24 species of Cicadellidae, 15 species of Membracidae and five species of Flatidae. In the understorey, sap-sucking insects included eight species of Tingidae, seven Cicadellidae and six Membracidae. Seventeen herbivore species, one leaf-chewer and 16 sap-suckers, were found in both trees and saplings: three species of Tingidae, two species of Membracidae, two species of Cicadellidae and one species each of Chrysomelidae, Cercopidae, Dictyopharidae, Flatidae, Issidae, Pentatomidae, Psyllidae, Rhopalidae and Thripidae. The herbivore assemblage on mature trees was dominated by three sap suckers (one of Thripidae, THRI-01, and two of Cicadellidae, CICA-27 and CICA-01), two leaf chewers, (*Myrmex sp.*, Curculionidae and larvae of one species of Notodontidae, NOTO-01). The herbivore community on saplings was dominated by one of the above xylem feeders (Cicadellidae CICA-27) and the chrysomelid *Rhobdopterus fulvipes*.

I used different measures of diversity to characterize the assemblages of insect herbivores in the canopy and the understorey. Overall, more species were collected on mature trees than on seedlings and this higher species richness was associated with higher dominance, as shown by Simpson's index (Table 25.3). Further examination of species richness and the values of Simpson's index suggest that this general pattern was the result of changes in the assemblage of sap suckers rather than that of leaf chewers (Table 25.3). As Shannon's index is affected by both species richness and evenness, the difference in the former tended to decrease the differences between this index for trees and saplings, and this for all herbivore guilds. For both the overall herbivore assemblage and the sap-sucking insects, the Morisita–Horn index indicated a high similarity in species composition between the canopy and understorey (Table 25.3). A different pattern emerges for the leaf-chewing insects, for which trees and saplings showed low levels of similarity according to the Morisita–Horn index.

Seasonality and herbivorous communities

Changes in abundance of seven herbivore species represented by at least 11 individuals were analysed through

Table 25.2. *Number of individuals and total number of species collected on saplings and mature trees detailed by higher taxa of insect herbivores*

Taxa/guild	Number of individuals		Mann–Whitney *U*-test *p* value	Number of species		Mann–Whitney *U*-test *p* value
	Saplings	Trees		Saplings	Trees	
Chewers	305	718	0.121	69	62	0.065
Leaf chewers: feeding	22	413	1.000	5	13	0.000
Leaf chewers: not feeding	283	302	0.301	68	54	0.641
Chrysomelidae	164	247	1.000	29	30	0.000
Curculionidae	91	416	1.000	34	22	0.000
Geometridae	6	5	1.000	1	2	0.102
Nymphalidae	0	1	ID	0	1	ID
Noctuidae	0	1	ID	0	1	ID
Notodontidae	0	13	ID	0	2	ID
Others	41	38	–	5	9	–
Sap suckers	226	5520	0.000	48	70	0.000
Cicadellidae	84	2022	0.000	7	24	0.341
Thripidae	41	3321	0.000	1	1	1.000
Membracidae	6	27	1.000	6	15	0.412
Flatidae	5	49	0.747	1	5	0.004
Tingidae	14	9	1.000	8	4	0.452
Others	76	44	–	27	21	–

ID, insufficient data to complete the test.

Table 25.3. *Estimators of species richness for different herbivore guilds on saplings and trees of* Castilla elastica

Guild	No. of species observed			Shannon's index (SD)		Simpson's index (SD)		Morisita–Horn index
	Total	Saplings	Trees	Saplings	Trees	Saplings	Trees	
All herbivores	120	53	83	1.4 (0.01)	2.1 (0.8)	2.5 (0.01)	5.8 (3.3)	0.71
Leaf chewers (feeding)	17	5	13	0.7 (0.01)	0.9 (0.3)	1.4 (0.01)	1.9 (0.7)	0.0
Sap suckers	100	48	70	1.2 (0.01)	1.9 (0.9)	2.2 (0.01)	4.9 (2.8)	0.73

time for both the canopy and the understorey. These species include three of Cicadellidae (CICA-01, CICA-24, CICA-27), one of Chrysomelidae (CHRY-07), one of Curculionidae (*Myrmex* sp.), one of Notodontidae (NOTO-01) and one of Thripidae (THRI-01). Analyses of variance indicated that the abundance of different species (i.e. number of individual per species) was significantly different in the canopy and the understorey ($F = 146.6$, $p < 0.0001$). The abundance of the seven morphospecies of herbivorous insects was between 25 and 30 times higher in the trees than on saplings. Although the analysis failed to detect a temporal trend in insect abundance ($F = 1.57$, $p = 0.60$), this variable varied significantly with species ($F = 32.67$, $p < 0.001$). Throughout the season, the thripid (THRI-01) was the most abundant species with an average of 129 individuals per sample and the chrysomelid (CHRY-07) was the least abundant.

Table 25.4. *Estimators of species richness for different herbivore guilds during the wet and dry season on saplings and trees of* Castilla elastica

Guild	No. of species observed			Shannon's index (SD)		Simpson's index (SD)		Morisita–Horn index
	Total	Dry	Wet	Dry	Wet	Dry	Wet	
All herbivores	120	72	81	1.4 (0.01)	1.3 (0.01)	2.5 (0.1)	2.2 (0.1)	0.78
Leaf chewers (feeding)	17	8	15	0.7 (0.1)	0.7 (0.1)	1.4 (0.1)	1.3 (0.1)	1.0
Sap suckers	103	64	67	1.2 (0.1)	1.1 (0.1)	2.2 (0.1)	1.9 (0.3)	0.77

Given the absence of a temporal trend for insect abundance, I examined the relationship between the mean number of individuals per sample and the production of new leaves. First, data from the seven most abundant species of herbivores were pooled to examine the relationship between insect abundance and production of new leaves either for trees or for saplings. The linear regressions were not significant, with r^2 values of 0.038 and 0.025, respectively. However, the abundance of two species, a cicadellid (CICA-24) and *Myrmex* sp., depended significantly on the production of new leaves. In the case of the cicadellid, 60% of the variation in abundance can be explained by the percentage of new leaves; whereas that percentage was 51% for *Myrmex sp.* None of the other species showed any significant linear relationship. Graphical examination of the data suggests that abundance might be related to leaf production for some of the other species, albeit in a nonlinear way. Another species of cicadellid (CICA-27) was apparently the only species in which abundance was completely independent of leaf production. Three morphospecies, two cicadellids (CICA-01, CICA-24) and a notodontid (NOTO-01), showed a clear peak in abundance when the percentage of new leaves was, respectively, 30%, 30% and 10%. Both *Myrmex sp.* and a thripid (THRI-01) were found to increase in abundance steadily as the percentage of new leaves increased.

Temporal trends at the community level were also characterized by species richness, as reflected in both Shannon's and Simpson's indices. Species number was, overall, slightly higher in the wet then the dry season with similar patterns of dominance and evenness as shown by both Shannon's and Simpson's indices (Table 25.4). The wet/dry season difference in the herbivore assemblage is probably caused by an increase in diversity of the leaf chewers feeding on *C. elastica*, which are represented by nearly twice as many species in the wet than the dry season. Despite such large difference in species richness for leaf chewers, dominance patterns appeared to respond little to seasonality (Table 25.4). The Morisita–Horn index of similarity between the dry and wet seasons was the highest for the sap-sucking insects, with similar values for all herbivores and leaf chewers (Table 25.4).

DISCUSSION

The main objective of this study was to compare tropical insect communities along a vertical gradient while controlling for the identity of the plant host. In order to do so, I studied in detail the fauna collected on *C. elastica* over a 7 month period overlapping the dry and wet season in Panama. This experiment was part of a larger project examining also *P. bicolor* (Basset, 2001a), *Luehea seemannii* Triana and Planch, and *Manilkara zapota* (L.) van Royen in an attempt to uncover general patterns of stratification in tropical forests. Our interest in studying insect diversity along vertical gradients stems from basic species-diversity theory. It is believed that the number of insect species associated with one host plant depends on the geographic range and local abundance of the plant species (Southwood, 1960; Lawton & Schröder 1977; Fowler & Lawton, 1982); on the intrinsic characteristics of the plant, such as size, structural complexity (e.g. Lawton, 1983); and on biochemical properties (e.g. Bernays & Chapman, 1994). Accordingly, given strong environmental gradients from the forest floor to the canopy of tropical forests (Blanc, 1990; Wright & Coley, 1994; Parker, 1995) and differences in leaf area abundance and quality (Parker, 1995; Hallé, 1998; Bell *et al.*, 1999), it is likely that vertical stratification will be observed for herbivorous insects (Schowalter *et al.*, 1986). Although other studies have reported faunal differences between the understorey and the canopy (Cachan, 1964; Amédégnato, 1997; Hammond *et al.*, 1997; Rodgers & Kitching, 1998), they did not dissociate the effects of stratification from that of changes in plant species composition at different heights in the canopy. By comparing insect diversity on conspecific saplings and mature trees of *C. elastica*, the present study provides a more direct test of vertical stratification in a dry tropical forest.

My data appear to support vertical stratification since insect diversity was higher on the more complex mature trees than on saplings (Lawton, 1983). Arthropods in general were more abundant on the foliage of trees than on that of saplings, and many arthropod guilds or taxa showed a higher species richness on trees than on saplings. Notable exceptions were taxa well represented in the soil habitat, such as the scavenging fauna. Ants, for example, were more abundant on saplings than on trees. For a similar leaf area sampled, I found 1.6 times more species of insect herbivores on the foliage of mature trees than on that of saplings. Our companion study on *P. bicolor*, a light-demanding species of Cecropiaceae growing in wet lowland forests of Panama, indicated that insect herbivores were 1.5 times as species rich on the foliage of trees than on that of saplings (Basset, 2001a). The most extreme difference between the fauna of saplings and that of trees was found for leaf-chewing species with a proven ability to feed on the foliage of *C. elastica*. This guild was 2.6 times as species rich on trees than on saplings, whereas sap suckers were only 1.5 times as species rich. Furthermore, a single species of leaf-chewing insect was present in both trees and seedlings of *C. elastica* whereas 15 species of sap suckers were found in the canopy and the understorey. A similar pattern was observed on *P. bicolor*, where leaf-chewing feeding insects were 3.5 times as species rich on trees than on saplings (Basset, 2001a). This contrasts with the results of Le Corff and Marquis (1999), who recorded significantly more leaf-chewing species feeding on the saplings than in the canopy of *Quercus* spp. and did not encounter any leaf-chewing species restricted to the canopy. For both *C. elastica* and *P. bicolor*, the vertical gradient I studied spanned 35–40 m whereas that reported by Le Corff and Marquis (1999) was 15–20 m. It has also been reported that differences in illumination between the canopy and the understorey are more marked in tropical than temperate forests (Blanc, 1990). It is likely, therefore, that the insects feeding on *Quercus* were less stratified than those in Panama. For both *C. elastica* and *P. bicolor*, few differences between saplings and trees existed when nonfeeding species of leaf-chewing insects were considered. This emphasizes the importance of testing the ability of arthropods to feed on the foliage of the study plants before drawing general conclusions (Basset, 1999a). For example, the Janzen–Connell model suggests that specialist insects disperse from parent tree to offspring plants, that is, that insects are more likely to attack seedlings within the vicinity of the parent tree in a density-dependent fashion (Janzen, 1970; Connell, 1971; Leigh, 1994). The present results, those on *P. bicolor* (Basset, 2001a) and of Basset *et al.* (1999) in Guyana cast a serious doubt on this proposition.

Although I found similar vertical gradients in species number for *C. elastica* and *P. bicolor*, stratification of abundance was very different across the two host species. In *C. elastica* insect herbivores were 23.9 times more abundant on mature trees than on saplings, but in *P. bicolor* this difference in abundance was only 2.5 times (Basset, 2001a). In *C. elastica*, the large difference in abundance between sapling and mature trees was explained by Thripidae, with 3321 individuals representing a single species compared with only 41 individuals collected on saplings. The numbers of Thripidae collected on *C. elastica* are themselves larger then the total number of insect herbivores collected on *P. bicolor* (3084 insect herbivores). In the temperate zone, it has been proposed that the length of the growing season of plant species also influences the number of herbivorous insect species exploiting it through altered exposure time (Lawton, 1978). The situation is somewhat analogous to dry tropical forests, since the dry season often corresponds to a halving in leaf production. In such forests, the amount of young leaves is positively correlated with leaf damage (Coley, 1983), while abundance of herbivorous insect species also correlates with the presence of young leaves because these leaves have soft and easily digestible material, as well as a high nitrogen content (Basset, 1991b). The present study shows that, although *C. elastica* produces new leaves for long periods throughout the year, the young leaves produced at any time represent, on average, less than 5% of the total leaf area of the trees or saplings. There was a peak in production of young leaves from 26 March to 18 June 1999, resulting in 95–100% new leaves for trees and saplings. My results indicated that different insect species reacted differently to leaf production in *C. elastica*. Leaf production at a particular time strongly correlated with the distribution of certain species, such as a cicadellid (CICA-24) and *Myrmex* sp. Such a relationship, however, did not explain the high abundance of Thripidae, probably because they originated from the flowers of *C. elastica* and spread to the foliage of mature trees. Flowers as a resource provided by the different host stages may explain a substantial part of the variance in insect abundance/diversity between the canopy and understorey.

In tropical rainforests, clear differences in insect assemblages feeding on conspecific mature trees, saplings and seedlings are expected, because the amount and quality of resource, as well as the number of habitats provided by these various host stages, are likely to be drastically different, unlike in temperate forest. However, other insect variables may differ between conspecific mature trees and seedlings, such as the density of herbivores, the proportion of specialist or generalist species, as well as the dominance of particular species within the insect assemblage. The abundance of insect predators and parasitoids was higher on trees than on saplings; ants showing the reverse trend. Consequently, it is difficult to comment on the relative importance of enemy-free space on saplings and trees for insect herbivores. Some of these differences could have important implications for tree regeneration in tropical forests, both from an ecological and an evolutionary perspective.

Many studies on tropical herbivory (Lowman, 1982a, 1984; Langeheim & Stubblebine, 1983; Coley, 1983, Lowman & Heatwole, 1992; Hadwen *et al.*, 1998, Barone, 2000) suggest that canopy trees suffer high levels of leaf damage compared with that in their conspecific seedlings and/or saplings. My entomocentric results suggest that herbivory data based on leaf area might not always provide a good indication of vertical stratification in insect diversity. The vertical patterns of herbivory, based on leaf damage, may be an artefact of which species is responsible for the damage in a particular stratum. A single species of caterpillar can cause a great loss of leaf area, unlike a species of Curculionidae such as *Myrmex sp.*, which is responsible for very little damage. Therefore, it is obvious that the majority of leaf area loss is related to the species of caterpillar and not that of the Curculionidae. One may conclude, in turn, that most damage occurs in the stratum preferred by that particular species of caterpillar and does not depend on the type of stratum *per se*.

My data show clearly that different herbivore assemblages exploit different host stages, generating a vertical stratification of organisms on *C. elastica* and *P. bicolor* in Panama (Basset, 2001a). Several studies have reported convincing faunal differences between the understorey and the canopy of wet closed tropical rainforests (Cachan, 1964; Amédégnato, 1997; Hammond *et al.*, 1997; Rodgers & Kitching, 1998; Walter *et al.*, 1998; Basset *et al.*, 1999). The present work, however, appears to be one of the first to demonstrate such stratification on a discrete resource (*C. elastica*), rather than the totality of the forest, suggesting a mechanism by which biodiversity may be greatly enhanced in tropical closed rainforests.

ACKNOWLEDGEMENTS

This study was supported by the University of Panama (Masters Program in Entomology) and the Smithsonian Canopy Crane Project. I acknowledge the efforts of the team leader Joe Wright without whom this project would not have been possible. Enrique Medianero provided valuable help in the field. Special thanks to Yves Basset who shared with me his *Pourouma* data and his great knowledge of canopy and insect ecology. Catherine Potvin helped by analysing some data and commenting on the manuscript. Helpful comments on the manuscript were also made by Gloria Dangerfield.

26

Hallowed hideaways: basal mites in tree hollows and allied habitats

Matthew D. Shaw and David E. Walter

ABSTRACT

More than 20 years ago, the late W. D. Hamilton (1978) pointed out a pattern of dead trees and allied habitats hosting a disproportionately high frequency of 'primitive' insects and apterygotes. While examining this pattern, he also suggested its cause: rapid evolution owing to inbreeding in these discrete, but complex habitats. His argument can be applied to any habitat that tends to host isolated and inbred lineages. Here we highlight the high number of early derivative mites associated with the decaying wood complex including canopy tree hollows, emphasizing mite groups that might be suitable for a thorough test of Hamilton's hypothesis. As this hypothesis suggests, the 'primitive' forms found in tree hollows often show 'unexpected' derived characters such as transitions between fluid feeding and particulate feeding or predation to scavenging, insect symbiosis or fungivory. Mite diversity has also accrued in tree hollows that are used as nests by vertebrates. Nests, in general, have prompted many evolutionary transitions from free-living life styles to ectocommensal or parasitic habits. Nests in tree hollows, distinguished by their longevity and their colonization by obligate cavity-nesting vertebrates, may have allowed speciation within and even genesis of certain parasitic mite lineages.

INTRODUCTION

Large units of decaying wood, such as stumps, logs and tree hollows, are often treasure troves of bizarre and 'primitive' mites. This is especially true of the suborder Mesostigmata, where basal lineages such as Sejina, Microgyniina, Uropodina and Cercomegistina are strongly associated with logs and tree hollows (Trägårdh, 1942, 1950; Camin, 1953b, 1955; Gilyarov & Bregetova, 1977; Krantz, 1978; Walter & Krantz, 1999) or with arthropods or vertebrates that live under bark, in logs, in fungal sporocarps growing on logs or in hollows in dead and living trees (e.g. Camin, 1955; Kinn, 1971). A similar pattern is present in the suborder Astigmata, where many early derivative forms are associated with wood-inhabiting beetles.

The trend for decaying wood and allied habitats to harbour coppice-like offshoots from the bases of acarine evolutionary trees may also occur within the basal Heterostigmata (i.e. Tarsocheylidae, Heterocheylidae), which are primarily inhabitants of logs (Lindquist, 1986). Other early derived prostigmatans and some members of the early derived paraphyletic grouping 'Endeostigmata' (Walter, 2001) are common in, although not restricted to, woody litter and tree hollows. In contrast, Opilioacariformes, usually considered the most 'primitive' of mites are not associated with decaying wood. Similarly, with few exceptions, early derivative oribatid mites (Oribatida) show no strong phylogenetic pattern with respect to decaying wood, despite being abundant in soil and other decomposer habitats.

METHODS

First, we review and interpret Hamilton's (1978) theoretical model for the evolution of arthropods in a decaying wood habitat. We slightly extend the scope of his ideas by including tree hollows as an important subset of the woody litter complex. We also point out the phenomenon of facultative phoresy in mites, and its potential importance for maintaining breeding structures that favour localized inbred demes and hence rapid evolution. These ideas all relate to the process of evolution. We then search for evidence of a consequent pattern: early derived mites associated with decaying wood. We look twice through the Acari for such phylogenetic pattern.

First we review the Acari with respect to decomposer assemblages, some of which occur in woody litter. Second, we pay special attention to mites associated with mammals and birds, some of which inhabit tree hollows.

RESULTS AND DISCUSSION

Hamilton's hypothesis

Why should early derived mites be associated with decaying wood? Hamilton (1978) not only recognized this pattern in insects but then supplied a compelling explanation for it based on breeding structure.

Hamilton's reasoning started with the observation that many striking occurrences of male haploidy, flight polymorphism, social behaviour and sexual dimorphism in insects are associated with decaying wood. These, he claims, are clear testament to a relatively high level of inbreeding in these habitats. He suggested that decaying wood habitats tend to force such a breeding structure because of the need of its inhabitants to live under bark and hence live enclosed and isolated. The isolation of demes is further suggested by the patchy distribution of decaying wood. Because of the longevity of these habitat patches, arthropods inhabiting this decaying wood complex typically spend many generations in one patch, further increasing relatedness among patch members. Here we extend his ideas by including forest canopy tree hollows – habitats that are highly patchy in space but persistent across time.

If one is prepared to accept Hamilton's assertion of inbreeding as a common quality of arthropods living in decaying wood, at least those that tend to live under bark, then Hamilton's (1978) next claim was for the special and rapid evolutionary potential of such inbred populations. Although, on average, inbred and isolated demes face a higher probability of extinction, where there are many replicated demes, overall, there should be a greater rate of evolution. A précis of this argument was provided by Trivers (2000): 'More or less closed spaces created by rotting wood imposed a system of small, inbred subpopulations in insects inhabiting it, leading to a great diversity of homozygous forms, often with arbitrary novel characters such as a second complete metamorphosis in many male scale insects'.

So, rapid evolution in rotting wood was predicted. Rapid evolution gives a higher probability of cladogenesis and one long-term phylogenetic consequence of this process should be a greater frequency of early derivative lineages associated with the habitat. Another consequence from rapid evolution might be dense bush-like speciation, but this is not observed nor is it expected in the homogeneous habitat of decaying wood. Rapid evolution could be evinced as accelerated specialization of other sorts such as in the formation of symbiotic relationships (see Thompson, 1994) or changes in feeding behaviour, and many examples like these can be found in mite lineages associated with rotting wood. Initiation of these deep phyletic divergences in decaying wood is expected to produce evolutionary trees with coppice-like branches from their bases (i.e. long branch lengths). Note that the central prediction of Hamilton's hypothesis is rapid evolution. Certain topologies of evolutionary trees, although easier to recognize, are secondary predictions from this.

Hamilton recognized the need to reconcile a seeming paradox here too – an abundance of seemingly primitive forms alongside a prediction of rapid evolution. He argued that the paradox is more apparent than real. The plesiomorphies of 'primitive' taxa are often taken as evidence for conservatism or even evolutionary stagnation, but Hamilton urges us to reconsider our prejudices. By refocussing our attention on the often-overlooked morphological, physiological and behavioural autapomorphies of such 'primitive' taxa we may start to uncover evidence of rapid and innovative evolution. In the Insecta, Hamilton's ideas have continued to attract some interest (Taylor, 1978, 1981; Lattin, 1999; Ronquist, 1999), but the idea has not been examined by acarologists.

Hamilton's hypothesis applied to mites

To examine Hamilton's hypothesis, a two-pronged approach is necessary. First, we need to look for evidence of a pattern of early derived mites associated with decaying wood. Although acarine phylogenetics are at an early stage, some hypotheses of mite evolution based on phylogenetic principles are available (e.g. Norton *et al.*, 1993). Second, we need to find evidence of inbreeding in decaying wood. Direct evidence (e.g. from allozyme or other biochemical markers) is lacking. However, there is much indirect evidence from life-history data.

We devote some space to discussing life histories because how and when mites disperse is probably critical in influencing the extent of gene flow among populations. In particular, we are intrigued by the phenomenon of facultative phoresy, common in decaying wood mites, which can allow many generations to remain in the one patch before dispersal. In facultative phoresy, the life stage that undertakes dispersal, typically the adult female in haplodiploids and the deutonymph in diplodiploids (Norton *et al.*, 1993), has two morphs: one sedentary and the other adapted for hitch-hiking on larger host animals. It contrasts with the more derived condition of obligate phoresy found in some Uropodina and Astigmata, where the deutonymph has become monomorphic. Facultative phoresy in mites is a clear analogy to wing polymorphism in insects. Hamilton (1996) emphasized examples of wing polymorphism in insects as a marker for inbreeding.

While we believe certain life-history traits (e.g. facultative phoresy) may be very important in promoting inbred populations of mites, we are aware that inferences concerning population genetics must ultimately be tested with genetic data. Hence the current contribution is more a suggestion as to which mite lineages may be useful for testing Hamilton's hypothesis, than a test itself. The latter will require a combination of phylogenetic, genetic and life-history studies.

Hamilton signalled the probable applicability of his hypothesis to mites, a suggestion we consider below. We selectively highlight many examples of early derivative mites associated with the decomposer communities of decaying wood including tree hollows (Table 26.1, below).

We then revisit the orders Mesostigmata and Astigmata to examine those especially associated with vertebrates and tree-hole nests. Taxonomic categories follow Evans (1992), except in placing Astigmata with the Oribatida.

Uropodina

Early derivative uropodine groups retain many plesiomorphies, such as a great number of setae, pores and body plates. However, they also boast distinctive autapomorphies (unique derived characteristics) such as signet ring mating in the polyaspidoid *Caminella* sp., where extensive secretions from the female adhere to her body and are moulded by her mate to channel his sperm (Compton & Krantz, 1978). Along with the Microgyniina and Sejina (see below) they attach to hosts using a stalked anal pedicel in (phoretic) deutonymphs. Derived Uropodina have radiated onto arthropods, producing remarkable specializations such as precisely mimicking the feet of army ants (Trachyuropodidae) (Gotwald, 1996). Similarly, Diarthrophallina, closely allied to the Uropodina, live under the elytra of passalid beetles (Johnston, 1982).

Uropodina have a close relationship with rotting wood and tree hollows and also with other nutrient-rich decomposer systems such as nests (especially ant nests and nests in tree hollows), compost, tropical forest litter and beach wrack. They can produce high population densities in such patches. The Thinozerconoidea are considered the most early derivative uropodines (Athias-Binche, 1982), and although they are best known from beach wrack, others are associated with logs. The Polyaspididoidea are also early derived (Fig. 26.1**d**) (Evans & Till, 1979) and are closely associated with tree hollows (Camin, 1953b). The Metagynellidae (Uropodoidea) are found in tree hollows (Camin, 1953a) and rotting logs (D. Walter, unpublished data) and is distinguished from other uropodoid families by the uniquely derived posterior position of its genital openings.

Many Uropodina have two forms of deutonymphs: a specialized dispersing phoretic form (phoretomorph) and a nondispersing form that is similar morphologically and behaviourally to other life stages (Evans, 1992). The retention of dimorphic deutonymphs (i.e. facultative phoresy) is common in species from nonstatic habitats such as manure, compost, rotting tree stumps and tree holes. According to Siepel (1994), species using facultative phoresy are generally species of long-lasting, discretely distributed biotopes. Switching between these two types of deutonymph may occur in response to changes in environmental conditions, such as a decline in habitat quality. This potentially allows many generations to be completed within a patch, followed by eventual migration to another patch. These factors lead us to predict inbreeding as a common consequence of facultative phoresy.

Athias-Binche *et al.* (1993) demonstrated that monomorphism in uropodid deutonymphs could lead to a high degree of carrier specificity and hence, presumably, reproductive isolation and even inbreeding. However, we do not think it is necessary to assume that the reverse is true, that is, that facultative phoretics are outbreeders. Aside from the potential to spend many generations within the one patch, the use of many carriers does not preclude the development of some ecological specificity and hence even carrier specificity

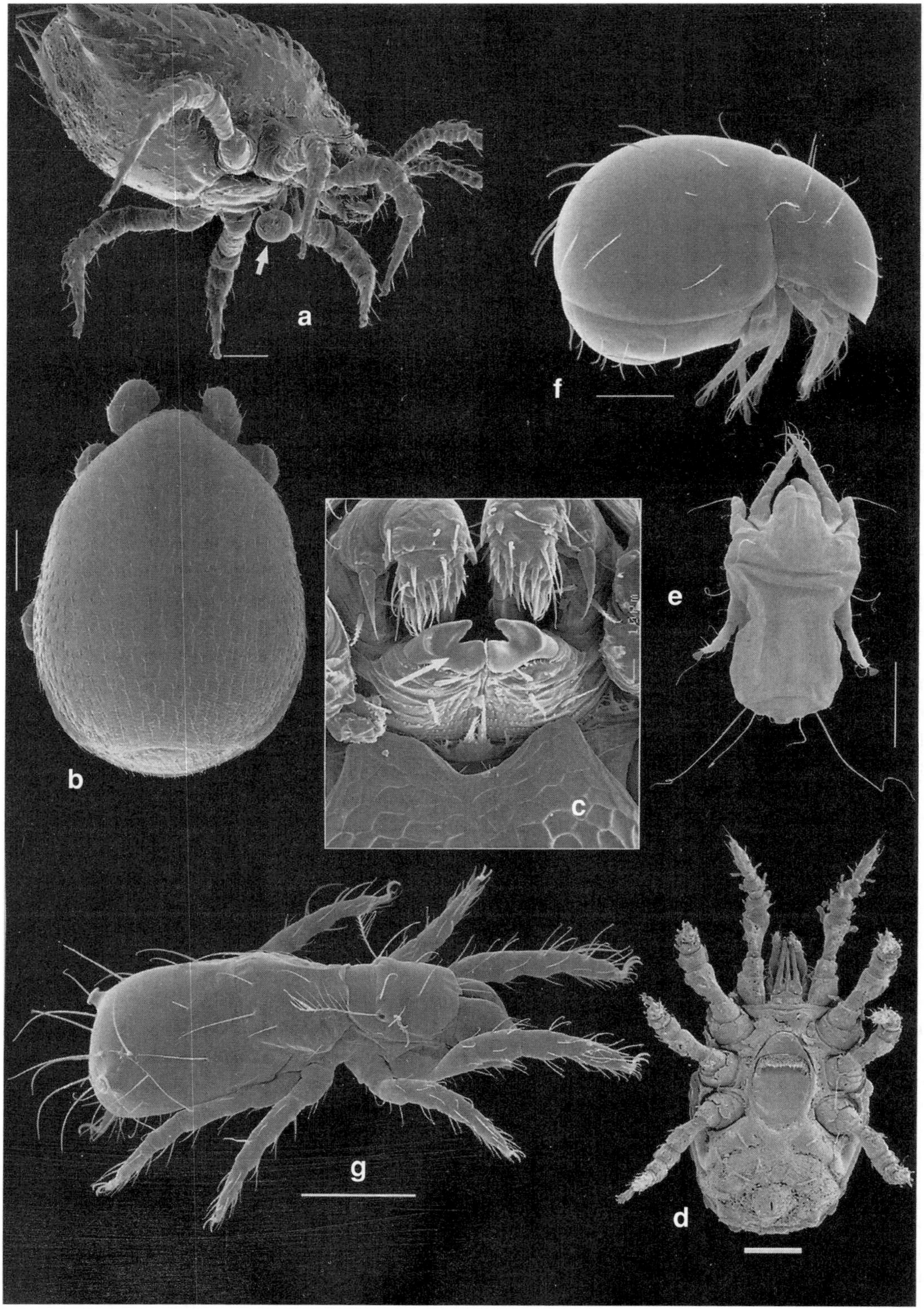
a
b
c
d
e
f
g

for certain habitat successional stages (Athias-Binche, 1993).

Levels of homozygosity are affected by many factors. Deutonymphal loads per carrier and other factors contributing to founder effects are important (Norton *et al.* 1993). Also the number of generations spent in a patch before dispersal would be critical in increasing levels of homozygosity within a patch. Although facultative phoretics may use many carriers, this is a concomitant of remaining in a patch for long periods with phoresy being initiated by environmental cues, rather than carrier cues.

Sejina

Sejines (Fig. 26.1**a**–**c**) are considered to be among the earliest derived mesostigmatans (Johnston, 1982; Evans, 1992). Sejids are found in litter and decaying wood, tree holes and nests, including those of seabirds. Perhaps the most primitive looking of these, species of *Asternolaelaps* (Fig. 26.1**b**,**c**), have been extracted from debris in a British tree hole, from Italy in humus and moss and from Sweden in the nest of a scoter duck (Evans & Till, 1979). In Australia, this genus is known from a cave inhabited by bats, from the tree-hollow den of a brush-tailed possum (*Trichosurus vulpecula*) (M. Shaw, unpublished data) and from several non-nest tree hollows (D. Walter, unpublished data). Unlike other Sejina, *Asternolaelaps* is a particulate-feeding predator and fungivore (Walter & Proctor, 1998). Within the otherwise fluid-feeding Mesostigmata (Walter & Proctor, 1998), particulate feeding is a striking and original adaptation that has occurred only a handful of times (see p. 298).

Microgyniina

Another group allied to Sejina–Uropodina is also strongly associated with decaying wood and tree hollows. The suborder Microgyniina superficially resembles the Sejina and is associated with tree hollows and stumps and also rotting wood (Evans & Till, 1979). In Australia, *Nothogynus* spp. occur in logs and in tree hollows (Walter & Krantz, 1999).

Trigynaspida

Many other taxa within Mesostigmata are early derivative and lovers of decaying wood, such as groups within the cohort Cercomegistina, the large cohort Antennophorina and, more arguably, the diplopod-associated Heterozerconina. Within these cohorts, plesiomorphies such as tocospermous reproductive systems (where sperm is transferred from the male genital pore directly to the female; Fig. 26.1**a**) and many genital plates underpin autapomorphies such as giant males in *Megisthanus* and jumping in *Saltiseius* (Walter, 2000). Tree hollows and logs also support early derivative triplogyniids (Seeman & Walter, 1997). Deutonymphs of the early derivative *Philodana johnstoni* Kethley (Parantennuloidea, Philodanidae) are associated with litter in tree hollows and the adults are found on tenebrionid beetles (Kethley, 1977).

Mesostigmatid associates of bark beetles and passalid beetles

We now focus on two especially claustral and tellingly fecund habitats: the galleries of bark beetles and passalid beetles. Here, cloistered life histories and infrequent dispersal of host beetles co-occur alongside a wealth of early derivative mite associates. Scolytine bark beetles live under bark in colonies. According to a recent phylogenetic analysis, their underbark life style evolved first, later resulting in haplodiploidy. Either haplodiploidy or the advent of fungal feeding, or both, sparked a subsequent speciose radiation (Normark *et al.*, 1999; Jordal *et al.*, 2000). Scolytinae hosts an impressive array of

Fig. 26.1. (**a**) Ventrolateral view of a recently mated female Sejina (*Epicroseius* sp.) demonstrating the primitive mating system (tocospermy) for Mesostigmata (Mesostigmata: Sejina: Sejidae). Arrow points to the spermatophore attached to a notch in the female's genital shield. (**b**) Dorsal view of *Asternolaelaps* sp. from a treehole in Michigan (Mesostigmata: Sejina: Icththyostomatogasteridae). (**c**) Close-up of the modified corniculi (arrow) of an *Asternolaelaps* (Mesostigmata: Sejina: Icththyostomatogasteridae) from a tree hole in Queensland. These structures are used to snip the edges of food items (fungi, arthropod bodies) into pieces small enough for ingestion. (**d**) Ventral view of an early derivative Uropodina, *Polyaspis* sp. (Mesostigmata: Uropodina: Polyaspididae), from a tree hole in Queensland. (**e**) Dorsal view of a *Dermatophagoides* sp., a member of a lineage that has radiated extensively in association with nest-building vertebrates (Oribatida-Astigmata: Psoroptidia: Pyroglyphidae). (**f**) Lateral view of *Mesoplophora* sp., an oribatid mite associated with logs and passalid beetles (Oribatida: Enarthronota: Mesoplophoridae). (**g**) *Parhypochthonius aphidinus* Berlese, a 'primitive' oribatid mite characteristically associated with tree hollows (Oribatida: Parhyposomata: Parhypochthoniidae). Scale bars = 0.1 mm.

mesostigmatans such as Cercomegistina, Antennophorina (e.g. Celaenopsoidea), Digamasellidae and *Eugamasus* (Parasitidae) (Trägårdh, 1950; McGraw & Farrier, 1969; Kinn, 1971).

Remarkably, the appearance and radiation of scolytine beetles, and hence presumably their associated mites, probably occurred no earlier than the Miocene (Jordal *et al.*, 2000). So although these beetles belong to ancient lineages, celaenopsoids and cercomegistines on bark beetles may be relatively modern associations. This perspective should challenge notions of these mites as mere stagnant relics. Hamilton (1978) referred to the contrast between the rapid evolution he predicted and the seeming 'primitivity' of decaying wood lineages as a paradox. To illustrate, he pointed out the striking autapomorphies that many early derivative insects have developed. Similarly, we point out the elaborate symbioses (including mutualisms (e.g. Kinn, 1980)) and specialized body forms developed by cercomegistines (e.g. Kinn, 1971) and digamasellids (e.g. *Longoseius*) found under bark (Lindquist, 1975). The capacity of beetle-associated mites to develop inbred demes is highlighted by laboratory studies showing that the digamasellid *Dendrolaelaps neodisetus* (Hurlbutt) could complete four to six generations for every generation of *Dendroctonus* bark beetles (Kinn, 1984).

Outside of bark beetle galleries, the Digamasellidae are also associated with soils and litter worldwide, including coastal or littoral situations and seabird guano or nests. This also applies to its putative sister group Rhodacaridae and to the two most early derived digamasellid genera *Dendroseius* and *Digamasellus* (Lindquist, 1975). *Dendrolaelaps* is eurytopic, being found in salt marshes, pasture soils, leaf litter, humus, compost, manure and ant nests; in addition, 100 *Dendrolaelaps* spp. are associated with bark beetles and their galleries. Other still more-derived digamasellids such as *Longoseius* also inhabit bark-beetle galleries (Lindquist, 1975). A phylogeny of the Rhodacaroidea, especially resolution of relationships within *Dendrolaelaps*, could provide a useful test of Hamilton's hypothesis.

Passalid beetles, early-derived scarabaeoids, live in decaying forest logs. They are known to host at least 24 families of mites, more than any other known family of arthropods (Hunter, 1993; Seeman, 2001). The high preponderance of early derivative forms associated with passalids is also striking. The probably parasitic Diarthrophallina have already been mentioned. Other examples come from within the early derivative Antennophorina, where mites of the families Megisthanidae, Fedrizziidae and Klinckowstroemiidae are essentially restricted to passalid beetles. Only adults of these latter mites are found on passalids; immature mites live in the galleries of their host. The confined, cryptic habits of passalids could be another example of Hamilton's population structure, and the subsocial breeding system of passalids themselves implies inbreeding. We suggest that it is the population structure forced onto passalid associates that has led to their innovative evolution. Hence we conclude that mesostigmatan mites associated with passalids and scolytine beetles provide strong support for Hamilton's hypothesis.

Prostigmata

A strong indication of a Hamiltonian pattern occurs with early derivative Heterostigmatans such as Tarsocheylidae (Fig. 26.2: rotting wood, holes in deciduous trees and decaying leaf litter) and Heterocheylidae (on passalid beetles) (Lindquist, 1976, 1986; Johnston, 1982). These two families are basal to the large and ecologically diverse Tarsonemina, which parasitize insects, feed on plants or inhabit mammal nests or decaying wood. Within the Tarsonemina, Hamilton (1978) mentioned the facultative phoresy exhibited by pygmephorid mites in support of his hypothesis, although pygmephorids are in fact not early derivative relative to the taxa we are discussing here. The Heterostigmata show the right pattern but are limited in usefulness as a replicate in a test of Hamilton's evolutionary model because the initial entry into dead wood has not been tracked. If a sister group to the Heterostigmata could be diagnosed and its ancestral habitat characterized, then a better test could be formulated.

The pattern of early derived taxa appearing in decaying wood is not strong elsewhere in the suborder Prostigmata. However, we will argue that the exceptions prove a rule with regard to phoresy. Many Prostigmata occasionally found associated with decaying wood, such as representatives of Anystina (Caeculidae), Eupodina (Eupodidae) or Raphignathina (Stigmaeidae, Rhapignathidae), are not considered especially early derivative. The telling contrast seems to be that none of these prostigmatans form phoretic relationships with other animals. Perhaps the possibility of reliably exploiting highly patchy and localized habitats is lacking

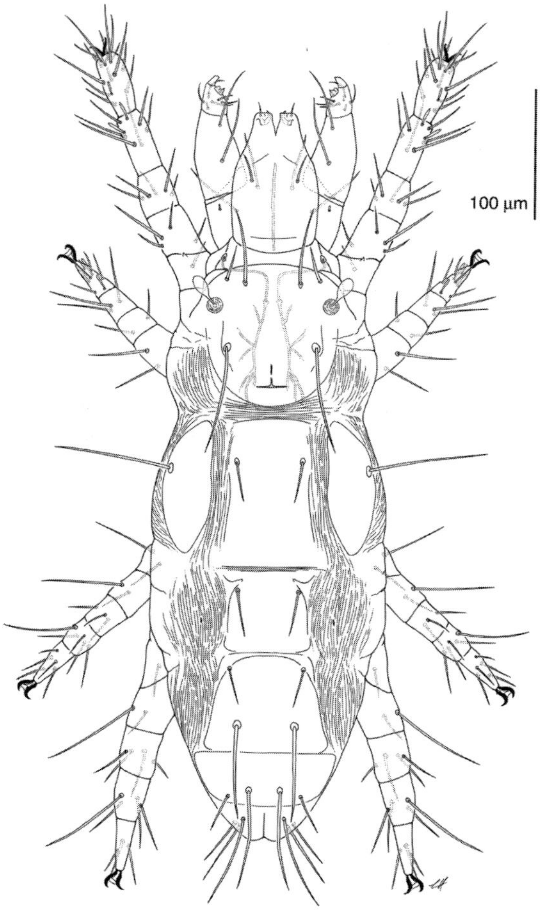

Fig. 26.2. Dorsal view of a tarsocheylid mite from a log in an Australian rainforest showing early derivative heterostigmatan morphologies, e.g. numerous shields and unconsolidated capitulum (Prostigmata: Heterostigmata: Tarsocheylidae).

because these predators and fungal-feeders must undertake their own dispersal.

However, the prostigmatan cohort Parasitengona do hitch-hike on larger animals, albeit parasitically on hosts. In addition, some occur in decaying wood (e.g. Smarididae). So why can we not see a Hamiltonian pattern here? Their dispersal may be of the wrong kind as it occurs in every generation, negating the possibility of developing isolated and inbred populations.

Oribatida

Parhypochthonius aphidinus Berlese (Fig. 26.1**g**) is an early derived oribatid (Paleosomata) that lives primarily in treeholes (R. A. Norton, personal communication; D. E. Walter, unpublished data). Some lohmanniids burrow in decaying wood and another group of Enarthonota, the genus *Mesoplophora* spp. (Fig. 26.1**f**), is closely associated with decaying wood and regularly clasp onto the hairs of passalid beetles (Norton, 1980). Enarthronotan oribatids are early derivative, with a fossil history extending back to the Devonian. Admittedly, however, lohmanniids and *Mesoplophora* are rather later derivative examples of this early branch (Norton *et al.*, 1988, 1993; Norton, 2001). Aside from the exceptions just mentioned, the majority of early derivative oribatids are not found in wood but in soils, reflecting the dominance of nonastigmatan oribatid mites in soil ecosystems worldwide.

Nonastigmatid oribatid mites do not provide strong support for Hamilton's hypothesis but they do show an interesting, perhaps even relevant, pattern with respect to phoresy. Although phoresy is an especially rare occurrence in nonastigmatid oribatids, nearly all available examples are from decaying wood habitats. In fact, of the several genera of oribatids reported from insects by Norton (1980), all of them were from insects associated with woody litter for all or part of their life cycle. Certain oribatids (e.g. *Scheloribates*) are transported on small mammals too (Miko & Stanko, 1991), although these mammals tend to use ground burrows rather than tree holes.

Even if a definite pattern of early derived oribatids in decaying wood does exist, the inbreeding that we are seeking to ascribe would not be the sort we have previously invoked, since many early derivative oribatids (e.g. Enarthonota and Paleosomata) are suspected or actual parthenogens (Norton & Palmer, 1991). Therefore, putative ancestors of many modern oribatids, dead wood dwellers or not, were very possibly all-female populations. Hamilton's hypothesis rests upon the population structure forced on the inhabitants of dead wood, but for asexuals such population structure is genetically irrelevant. So we are not sure whether oribatids can be said to conform to a Hamiltonian pattern and even if they do we are not sure why they would. However, we can say that rotting wood has provided a back-drop for some rare developments of oribatid phoresy; especially notable are *Mesoplophora* (just mentioned), and also the earliest derived Astigmata, which are primitively phoretic (see next). Considering ancestors of Astigmata, semiaquatic habitats have been suggested as another possible ancestral habitat (Norton, 1998), which recalls Hamilton's (1978) discussion of the overlap between rotting wood and aquatic habitats.

Astigmatan oribatids

The astigmatan oribatids, hereafter simply called the Astigmata, inherited particulate feeding from earlier oribatid ancestors. The first Astigmata are assumed to be fungal-feeders, as are many modern representatives. They are also thought to have evolved in a decaying wood habitat because the earliest derivative lineages retain associations with this habitat (OConnor, 1982). Following the phyletic sequence provided in Norton *et al.* (1993), the most early derived superfamily, Schizoglyphoidea, has a single representative phoretic on a tenebrionid beetle from New Guinea (OConnor, 1982; Houck & OConnor, 1991). The next most derived superfamily is the Histiostomatoidea. This large filter-feeding group has many associations with tree hollows and wood-dwelling insects. Next in this sequence is the Canestrinoidea, also strongly associated with several decaying wood insect groups including bark beetles and passalid beetles.

Thus early derivative astigmatids seem to fit Hamilton's hypothesis rather well. However, further comment is required for the Histiostomatoidea. Here there is a good fit with the Hamiltonian phylogenetic pattern but perhaps not the hypothesized process. Histiostomatoids have an unusual combination of highly *r*-selected life histories, haplodiploidy and obligate production of phoretic deutonymphs. The high level of outbreeding which this life style implies argues against operation of the hypothesis under discussion. We could brush aside obligate phoresy in histiostomatoids by suggesting that it was preceded by inbreeding and haplodiploidy, and indeed haplodiploidy is thought to evolve under conditions of inbreeding. However, using other criteria, an opposite suggestion has been made, namely that obligate phoresy preceded the advent of haplodiploidy in this taxon and that the former had become phylogenetically fixed (Norton *et al.* 1993).

Particulate-feeding convergence in mites from decomposer communities

To complete our review of the Acari for examples of mites associated with dead wood, we highlight a common functional convergence, namely particulate feeding. Particulate feeding is unusual in the Arachnida, most members of which are fluid-feeding predators (Walter & Proctor, 1998). However, particulate-feeding mites include the most early derived of all mites, the Opilioacarida, and also basal Sarcoptiformes such as the Oribatida and many of the Endeostigmata. Particulate feeding such as in the Terpnacarida (Endeostigmata) may reflect the ancestral condition for the Acari. Two thelytokous taxa from the endeostigmatan infraorder Terpnacarida are common in tree holes, namely *Terpnacarus* and a species of a derived genus, *Grandjeanicus*, that has been found high inside a decaying standing tree in Australia (Walter, 2001). The Mesostigmata was originally fluid feeding, as demonstrated by the two fluid-feeding lineages basal to it, the Holothyrida and Ixodida (Walter & Proctor, 1998). Hence the appearance of two particulate-feeding mesostigmatans, the sejine *Asternolaelaps* (which is strongly associated with tree holes) and the ascid genus *Proctolaelaps* (associated with flowers and a wide variety of decomposer habitats including nests and bark-beetle galleries), invites examination. Perhaps they developed particulate feeding in decaying wood habitats?

One explanation for the development of particulate feeding would be as an adaptation to fungal-feeding. Fungi are the principal agents of decay in wood. Among other convergences, Hamilton (1978) noted several instances of fungal-feeding in decaying wood insects. Such convergence is easily appreciated in *Asternolaelaps*, where particulate feeding is aided by the possession of massive chelicerae and large, notched corniculi (Fig. 26.1c). These latter structures parallel the rutellum found in many endeostigmatan mites. More needs to be explained here than just fungal-feeding though. Many fungal-feeders are also facultative predators (e.g. *Asternolaelaps, Grandjeanicus, Proctolaelaps*) and similarly opportunistic feeding is also found in fluid feeders. For example, *Androlaelaps casalis* (Berlese) can be fungivorous (OConnor, 1984) as well as predatory and occasionally parasitic. And many Ascidae are both predatory and fungivorous (Walter & Lindquist, 1989). So there is some nonexclusive overlap between the phenomena of particulate feeding, fungivory and non-specialized feeding. Seeking a common thread of explanation in all of this we see particulate feeding as a useful adaptation for making complete use of available fungal biomass by ingesting particles of fungal hyphae and whole spores. In contrast, fluid feeders gain energy rich food but are restricted to immediately liquefiable cell contents and spore coatings. Presumably fluid feeders exhaust their digestible fungal supplies sooner whereas particulate feeding may allow longer intervals before starvation or forced emigration. Facultative feeding could be an additional, optional strategy that allows a mite to remain on the one patch, switching between substrates as circumstances require.

Tree hollows and vertebrate-associated mites

Vertebrates may use tree hollows as nests or roosts. In such hollows, an overlay of vertebrate-associated mites appears in addition to other mites. It is believed that nest-dwelling mites were the progenitors of most parasitic mite lineages (Radovsky, 1985). If Hamilton's hypothesis has general utility, then a similar acceleration of evolution should occur in tree-hollow nests relative to other nest types.

There are at least two genuinely early derivative mites associated with vertebrates. *Asternolaelaps*, already mentioned, are phoretic on mammals and otherwise live in tree holes or caves. Radovsky (1969) suggested that the bizzare bat-attaching spelaeorhynchids were related to *Asternolaelaps*. This intriguing hypothesis implies an *Asternolaelaps*-like ancestor of spelaeorhynchids attaching to bats phoretically in tree hollows or perhaps caves. This seemingly short hypothetical journey would be lengthened by a necessary reversal: *Spelaeorhynchus* is probably fluid feeding judging by its small anal opening, and as previously noted, *Asternolaelaps* is particulate feeding.

A second clearly parasitic and early derivative mite lineage is the tick (Ixodida). Application of Hamilton's hypothesis would suggest the most basal tick lineages should be associated with tree-hollow nests. This is eminently possible but not proven. Knowledge of the position of argasid ticks relative to other parasitiformes and the position of tree-hole dwelling argasids relative to other argasids is necessary. We note that argasids have especially large sperm cells (Alberti, 1991), a peculiar development that is paralleled in some insects of decaying wood (Taylor, 1981).

Vertebrate-associated mites are mostly not early derivative in the broad context of the Acari, apart from the two exceptions just noted. But tree-hollow-nest associates may often be early derived forms within their own circumscribed lineages.

Vertebrate-associated Mesostigmata

The hugely successful and diverse Laelapidae allegedly arose from a hypoaspidid taxon exemplified by *Androlaelaps* (=*Haemolaelaps* (Halliday, 1998)) (Radovsky, 1985). The Laelapidae include many vertebrate parasites. The generalist hemiparasite *Androlaelaps casalis* (Berlese) is found in nests, especially bird nests, and may be predominantly arboreal (e.g. McGraw & Farrier, 1969; Mrciak & Sixl, 1979; Pung *et al.*, 2000). However, it is not yet possible to point to an ancestral form or habitat in this assemblage, as *Androlaelaps* spp. of generalized morphology occur in nests and on hosts of all sorts, even including invertebrates (e.g. Karg, 1991). There are several vertebrate parasitic groupings associated with tree hollows, however, such as the *A. mesopicos* group found with tree-hole nesters such as woodpeckers and sun squirrels (Till, 1963; Domrow, 1981) and the *A. ulysses* group known exclusively from Australian possums (Domrow, 1964, 1980). According to Tipton (1960), *Androlaelaps* spp. are mostly associated with squirrels (sciuromorph rodents with many tree-hole nesting species); however, there are large numbers of *Androlaelaps* spp. found on other rodents also (Strandtmann & Wharton, 1958). Finding out where these putatively tree-hole-associated taxa are positioned in a phylogeny of *Androlaelaps* would constitute an excellent test of Hamilton's hypothesis.

The Hirstionyssinae (Laelapidae) are parasitic on mammals. *Trichosurolaelaps* on Australopapuan possums and bandicoots has been implied to be the most early derived genus (Radovsky, 1985). Possums are mostly (and certainly originally) tree-hole nesting. A phylogeny of the Hirstionyssinae would allow another replicate in a test of Hamilton's hypothesis.

The Macronyssinae also seem to be derived from within the Laelapidae (Radovsky, 1985). The earliest derived members of this lineage are associated with pteropid bats, so no role for tree hollows is postulated here. The Macronyssinae has transferred to birds and it is worth noting that the most 'primitive' member of the widespread bird-infesting macronyssine genus *Pellonyssus* is found on a tree-hollow-nesting African woodpecker (Radovsky, 1994).

Rhinonyssines are permanently parasitic intranasal parasites of birds. Hence their radiations should be independent of the nest environment. However, the origins of rhinonyssines clearly lie within the nest-inhabiting

Macronyssinae (Radovsky, 1994). Within the Rhinonyssinae, the most early derivative genus is *Tinaminyssus* (Pence, 1979). *Tinaminyssus* spp. are associated with hole nesters such as parrots and barbets and also a variety of open-nesting nonpasserines. Curiously, *Tinaminyssus* spp. plesiomorphically retaining a tritosternum include three species associated with parrots and one species with a kingfisher (Domrow, 1988). Again a starring role for a tree-hollow nest may be implied but is unproven.

Another interesting family with many tree-hole associations is the Dermanyssidae. Dermanyssid mites are exemplars of parasites that remain in nests for long periods with only short feeding visits to the host. Radovsky (1985) claimed, on the basis of host relationships, that the probably early derived dermanyssid genus *Liponyssoides* evolved with mammals. However *Liponyssoides* is also known from birds such as swiftlets (Crouther, 1983), a burrowing seabird (Fain & Galloway, 1993), and two species of tree-hollow nesting Australopapuan treecreepers (Domrow, 1979). Also, an undescribed dermanyssid from the hollow-nesting New Zealand saddleback (Callaeidae) (M. Shaw & D. Walter, unpublished data) further clouds current taxonomic and ecological distinctions made between *Liponyssoides* and *Dermanyssus*. The female of this species has many plesiomorphies such as retained J3 and J2 setae, a holodorsal shield and a sternal shield bearing three setae. Unlike typical *Liponyssoides* spp. it has a U-shaped sternal shield. We consider the ancestral host and habitat of dermanyssids to be of considerable interest for testing Hamilton's hypothesis.

Vertebrate-associated Astigmata

The earliest derivative taxon within the mammal and bird-associated Astigmata are the Glycyphagoidea, which has radiated extensively with mammals. Intriguingly, the earliest derivative glycyphagoid is the monotypic Euglycyphagidae, known only from the (tree-hollow) nest of a screech owl *Otus asio* (Fain & Philips, 1977; OConnor, 1982).

Turbinoptids live in the nares of birds and are patchily distributed among open nesters such as frogmouths, babblers, quails, gulls and waders. Some of these are *Passerrhinoptes*, the most derived genus. However it is within the two orders of cavity nesters, the Coraciformes (rollers, kingfishers, etc.) and Piciformes (woodpeckers and their allies), where approximately half of all known species occur (e.g. *Congocoptes* spp.; Hyland, 1979). If the ancestral turbinoptid was found to be free living, then application of Hamilton's model suggests a tree-hollow-nest origin.

The family Pyroglyphidae contains most house dust mites (Fig. 26.1e). The earliest derived pyroglyphids are the Paralgopsinae, quill parasites of New World parrots (Gaud & Atyeo, 1996). This subfamily is permanently parasitic so no role for the nest environment need be postulated. However, if this lineage originated with nidicolous habits, as with many parasitic mites, a tree-hole nest is again strongly implicated.

If it can be demonstrated that tree-hole nests of birds or mammals have, on average, sparked a greater than expected cladogenesis of parasitic mite lineages, it would be necessary to examine a competing (and often favoured) hypothesis that early derived mites are simply associated with early derived hosts. Unfortunately, there is some overlap here with predictions from Hamilton's hypothesis because key early derived vertebrates tend to be tree-hole dwelling. Especially with birds, a great many early derived forest forms (e.g. parrots, owls, toucans, woodpeckers, barbets, trogons, hornbills, kingfishers, rollers) are tree-hollow nesters (Feduccia, 1999). A similar problem could be found with squirrels, which may also be early derived relative to other rodents (Huchon *et al.*, 2000). The need for detailed phylogenies of hosts and mite associates is clear.

Theoretically, the behaviour of tree-hollow-nesting birds may be conducive to relatively higher levels of mite inbreeding. Tree-hollow-nesting birds tend to have long developmental times between hatching and fledgling – for example, 9 weeks in many cockatoos. Several generations of parasitic mites could be completed in that time span. Additionally, tree-hollow nests are often re-used year after year. In contrast to birds, mammals may show less fidelity to particular tree holes (Cockburn & Lazenby-Cohen, 1992; Lindenmayer *et al.*, 1996). Even though occupation of a given hollow may be fairly regular, at least in some instances, there seems a greater chance that mites will be transferred between hollows by mammals moving among tree hollows, and hence mite inbreeding would be disrupted.

We mention a final example to highlight the uniqueness of canopy habitats. During canopy fogging at Lamington Plateau in southeast Queensland (Walter *et al.*, 1998), a new and truly bizarre dermanyssoid mite species was collected. The female has tick-like chelicerae

Table 26.1. *Phylogenetic patterns in the Acari relevant to Hamilton's hypothesis*

Early derived taxa associated with tree hollows or woody litter	Early derived relative to[a]	Comments and how supports Hamilton (1978)	Reference
Mesostigmata			
Polyaspidoidea	Other Uropodina excluding Thinozerconoidea	Many but not all polyaspidoids are restricted to tree holes; phylogenetic relationships within Uropodina are poorly known but polyaspidoids are recognized as 'Lower Uropodina'	Evans & Till, 1979
Asternolaelaps	Other Sejina	Best known from tree hollows but also caves; occasionally found on or with mammals	Walter & Proctor, 1998
Uropodina-Sejina	Other Monogynaspida	Uropodines and sejines are often found in tree hollows but also many other decomposer habitats	Norton *et al.*, 1993
Diarthrophallidae	?	Only known from under the wing-covers of passalid beetles; now included within the Uropodina	Norton *et al.*, 1993
Microgyniina	Other Monogynapsida	Often in tree holes	Walter & Krantz, 1999
Trigynaspida (cercomegistines, antennophorines): diverse grouping includes many associates of decaying wood or larger arthropods found in logs, etc.	Other Mesostigmata?	Great diversity in body form and strong association with dead wood; with greater phylogenetic resolution many possibilities for testing Hamilton's hypothesis	Norton *et al.*, 1993
Argasidae found in tree-hole nests but also other types of nest and roost	Other Ixodida?	The phylogenetic position of argasids relative to other ticks has not been settled	Black *et al.*, 1997
Trichosurolaelaps best known from tree-hole nesting possums but is also reported from bandicoots	Other hirstionyssines? occur on a wide variety of mammals.	Radovsky (1985) implied that *Trichosurolaelaps* was early derived as it was the closest genus morphologically to the laelapines, their putative ancestors	Radovsky, 1969, 1985
Pellonyssus biscutatus (Hirst, 1921)	Other *Pellonyssus* spp.	Nest parasites of birds	Radovsky, 1994
Tinaminyssus aprosmicti (Domrow, 1964), *Tinaminyssus kakatuae* (Domrow, 1964), *T. tanysipterae* (Wilson, 1966) *T. trichoglossi* (Domrow, 1964)	Other *Tinaminyssus* spp.? and Rhinonyssinae	Intranasal parasites of birds; the *Tinaminyssus* spp. mentioned are all from tree-hole nesters; their phylogenetic position suggested here is based solely, and hence unreliably, upon their plesiomorphic retention of a tritosternum	Domrow, 1988
Liponyssoides	*Dermanyssus?*	These may either support or refute Hamilton's hypothesis depending on resolution within and between genera	Radovsky, 1985

Table 26.1. (*cont.*)

Early derived taxa associated with tree hollows or woody litter	Early derived relative to[a]	Comments and how supports Hamilton (1978)	Reference
Prostigmata			
Tarsocheylidae (rotting wood, holes in deciduous trees and decaying leaf litter, etc.) and Heterocheylidae (on passalid beetles)	Other Heterostigmata (Tarsonemina)	–	Lindquist, 1986
Non-astigmatan Oribatida			
Parhypochthonius aphidinus	Non-Paleosomata oribatids	This species is mostly but not entirely found in tree holes	Norton *et al.*, 1993
Certain Enarthonota viz. *Mesoplophora*, which are phoretic on passalid beetles and some lohmanniids, which burrow into wood	Brachypyline oribatids	–	Norton *et al.*, 1993
Astigmata			
Schizoglyphoidea monotypic and phoretic on a tenebrionid beetle	Other Astigmata	Early derived astigmatans in general are phoretic and associated with rotting wood	Norton *et al.*, 1993
Histiostomatoidea	Other Astigmata with exceptions above	Many histiostomatoids have radiated extensively outside of rotting wood but this is probably their ancestral habitat (see next)	Norton *et al.*, 1993
Scoliianoetinae phoretic on passalid beetles and *Bonomoia* under bark	Other Histiostomatoidea	–	OConnor, 1982
Canestrinoidea	Other Astigmata with exceptions above	Canestrinoids are strongly associated with wood-inhabiting beetles; however, the earliest derived is known from African *Tefflus* carabid beetles	Norton *et al.*, 1993
Euglycyphagidae only known from a tree-hole owl nest	Glycyphagoidea diverse group found on many mammals especially as fur mites	Euglycyphagidae are phoretic on nest-inhabiting *Trox* beetles	OConnor, 1982
Congocoptes spp. intranasal parasites of woodpeckers, kingfishers and other cavity nesters	Other turbinoptids? Intranasal parasites of birds including passerines	Known extant taxa and likely relatives are all permanent symbionts so no role for nest environment can necessarily be supported	Hyland, 1979
Paralgopsinae quill parasites of New World parrots	Other Pyroglyphidae: mostly nidicoles plus some skin parasites	All known Paralgopsines are permanent parasites so no role for nest environment can necessarily be supported	Gaud & Atyeo, 1996

[a] A question mark following the relative phylogenetic position indicates that hypotheses of relationships are lacking but, once developed, will be relevant for testing Hamilton's hypothesis.

and the male has highly attenuated mouthparts. In these respects, this species closely resembles the monotypic New World Alphalaelaptinae. We suspect it is a nest parasite and as it exhibits the type of evolutionary extravagance predicted by Hamilton's hypothesis, we predict that its habitat will be a tree hollow.

CONCLUSIONS

Are the groups of mites associated with decaying wood so numerous and so generally basal that Hamilton's model can be supported? We believe there is a strong pattern showing that decaying wood supports a disproportionately high number of early derived lineages and that this requires some explanation. If Hamilton's model is valid, then the origin of many lineages should be attributable to the decaying wood habitat. Especially, the association of many basal mesostigmatans, astigmatans and heterostigmatans with decaying wood is very striking and suggestive. However, although we can appreciate intuitively Hamilton's model here, there are problems in actually testing it. For instance, if all the basal lineages of a group are associated with decaying wood there are no obvious sister groups to compare them against. For example Sejina–Uropodina and its descendants probably started in, and have often remained in, decaying wood.

Because of this problem in testing Hamilton's model, we require instances where decaying wood was clearly entered from other habitats so that we can estimate changes in evolutionary tempo associated with this shift. The microevolutionary (population genetic) processes that Hamilton's model rests upon would best be tested by tracking the changes in evolutionary tempo in a series of discrete lineages that have entered decaying wood. (e.g. Digamasellidae). Any thorough test of these ideas will rely on properly diagnosed sister groups being compared; with one group having entered decaying wood and the other a different habitat. The prediction is that entry into dead wood will promote rapid evolution, perhaps discernible by estimating divergence in a molecular phylogeny. Many currently recognized mite groupings are paraphyletic so refinements in phylogeny are a crucial prerequisite before proceeding.

We predict that mites with life histories that mirror the long patch persistence of dead wood habitats are most likely to conform to Hamilton's model (see summary in Table 26.1). In particular, facultative phoresy found in mites in decaying wood, such as in Uropodina, Sejina and various Astigmata, allows many generations to remain on one patch and, presumably, increase inbreeding in that patch, followed by eventual dispersal.

ACKNOWLEDGEMENTS

Thanks to Catherine Harvey who drew the tarsocheylid figure and to Yves Basset, Jenny Beard, Fred Beaulieu and Brenton Peters who commented on an earlier draft of this manuscript. We also thank two anonymous reviewers who made valuable suggestions.

27

Arthropod diel activity and stratification

Yves Basset, Henri-Pierre Aberlenc, Héctor Barrios & Gianfranco Curletti

ABSTRACT

Many studies of canopy arthropods in rainforests rely on methods that do not permit comparison of day and night activities of the target fauna: accordingly, data on arthropod diel activity are scarce. Nevertheless, such information as is available suggests significant dissimilarities during the course of the day and a greater activity of insect herbivores during the day than at night. We studied faunal exchanges, particularly of insect herbivores, between day and night, and between the upper canopy and understorey of a lowland rainforest in Gabon. In total 14 161 arthropods were collected by beating, flight interception and sticky traps from six canopy sites during mid-January to mid-March 1999. Diel activity explained about 6–9% of the total variance in arthropod distribution, depending on the sampling method. In general, arthropod activity was higher during the day than at night. In particular, Thysanoptera, Psylloidea, Membracidae, Curculionidae, Scelionidae and Apidae were notably more active during the day in the upper canopy. Similarity values between day and night faunas from particular strata were often higher than between faunas of the understorey and the upper canopy at a particular time of day. This suggests that there are fairly distinct suites of herbivores foraging in the understorey and in the upper canopy. Faunal turnover between day and night was higher in the upper canopy than in the understorey, suggesting that changes in the microclimatic conditions between day and night in the upper canopy may be more severe than in the understorey, and that only a well-adapted fauna may be able to cope with these changes. In addition, certain groups, such as the Chrysomelidae, may migrate during the night from the understorey to the upper canopy, although there was no evidence of such mass insect movement overall. Our data suggest that insect herbivores of the upper canopy may be resident and well adapted to environmental conditions there.

INTRODUCTION

Although most entomologists would agree that different insect faunas can often be collected during the day or at night in the same location, there are surprisingly few data available on the diel activity of arthropods in tropical rainforests (other than biting flies), and, particularly, on that of canopy arthropods. Many entomological studies rely on methods such as pyrethrum knockdown and light traps, which are unsuitable for such day/night comparisons. Few protocols have been designed specifically to study the diel activity of canopy arthropods. Hammond (1990) used Malaise traps in Sulawesi and reported the extremely low levels of nocturnal flight activity in all major insect groups. Basset and Springate (1992) set up flight-interception traps within a species of canopy tree growing in a subtropical rainforest in Australia and showed that the activity of associated insect herbivores was significantly higher during the day than at night. Springate and Basset (1996) reached a parallel conclusion using similar traps operated in different tree species in Papua New Guinea. Davis *et al.* (1997) studied the activity of dung beetles in the canopy of a forest in Borneo using baited traps. They showed that most species were only active during the day, with clear differentiation in activity between two species groups, one peaking at dawn–dusk, the other at midday. Compton *et al.* (2000) used sticky traps in Borneo to show that the overall activity of Chalcidoidea was higher during the day than at night, in contrast with the Agaoninae, which preferred to fly over the canopy at night. Although arthropod activity often appears to be higher during the day than at night, the validity of this statement varies across particular insect species or guilds (e.g. Springate & Basset, 1996). Of particular interest, some species of insect herbivores concentrate their activity during the day, others at night (e.g. Johnson & Mueller, 1990; Duan *et al.*, 1996).

The distinct microclimates in the understorey and canopy of tropical forests (e.g. Blanc, 1990; Parker, 1995; Barker, 1996) may well influence the diel activity of arthropods in these strata. In particular, the volume immediately below the canopy surface, the upper canopy, experiences much larger fluctuations in air temperature, wind, relative humidity and water condensation than the understorey does (e.g. Blanc, 1990; Parker, 1995). However, these seemingly adverse conditions may be offset by the greater availability of foliage, flower and fruit resources for insect herbivores (e.g. Basset *et al.,* 1992; Basset, 2001a).

Insect herbivores foraging and feeding in the upper canopy of closed tropical forests may encounter serious hygrothermal stress during the day, and threatening water condensation at night. Three strategies may overcome these obstacles:

1. The development of a specialized, distinct fauna well adapted to the extreme microclimatic conditions of the upper canopy – this could include day-active herbivores resistant to desiccation or night-active herbivores untroubled by water condensation, or even a blending of several adaptations
2. Interchanges of fauna between the upper canopy and lower layers, for example, in which individuals resting in lower layers during the day may move to the upper canopy to feed at night, perhaps taking advantage of air movements (e.g. Haddow & Corbet, 1961a; Sutton, 1989; Compton *et al.*, 2000)
3. A combination of both strategies 1 and 2.

Hereafter, the first two hypotheses are designated by their respective numbers. If hypothesis 1 is correct, little overlap of herbivore faunas will be expected between the upper canopy and the understorey, both during the day and at night (Fig. 27.1**a**). In this case, the highest faunal similarities are likely to be between the day and night faunas of the upper canopy on the one hand, and the day and night faunas of the understorey on the other (Fig. 27.1**a**). If hypothesis 2 is correct, a high similarity (Fig. 27.1**b**) will be expected between the fauna in the understorey during the day and that in the upper canopy at night. Further, a relatively high herbivore turnover will be expected between day and night in the upper canopy (Fig. 27.1**b**). In addition, the abundance and activity of herbivores in the upper canopy at night will be similar to those in the upper canopy during the day. If hypothesis 3 is correct, the results will be intermediate and difficult to interpret.

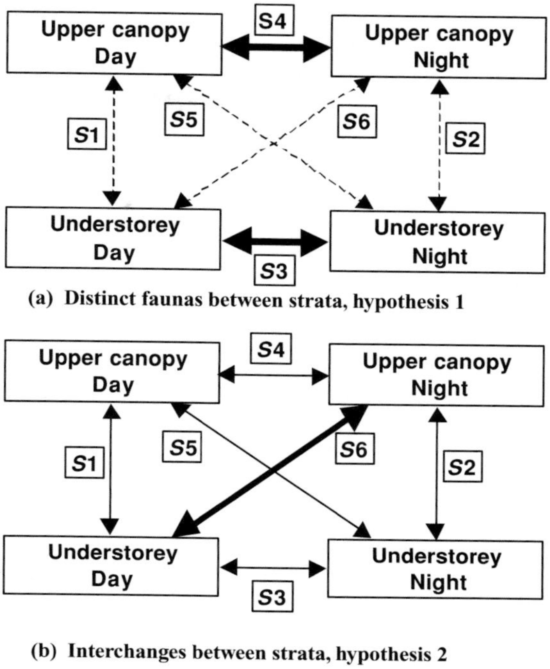

Fig. 27.1. Hypothetical faunal relationships between the understorey and upper canopy during the day and at night within a closed tropical rainforest: (**a**) distinct faunas between strata, hypothesis 1; and (**b**) interchange between strata, hypothesis 2 (see Introduction). *S1* to *S6* refer to coefficients of similarity among the faunas of the four situations. A bold line denotes a particularly high similarity, a broken line a particularly low similarity.

In this contribution, we present data on the diel activity of arthropods in the understorey and upper canopy of a lowland rainforest in Gabon. In particular, we focus on the insect herbivores, estimate their turnover between strata and time of day, and discuss the results against hypotheses 1 and 2.

METHODS

Study sites and canopy access

Arthropod samples were obtained from a lowland tropical rainforest in the Forêt des Abeilles, near the station of La Makandé, Gabon (0° 40′ 39″ S, 11° 54′ 35″ E, 200–700 m above sea level). The height of the upper canopy varies between 35 and 45 m. The main features of the forest are described in Doucet (1996), Fréty and Dewynter (1998) and Hallé (2000). The canopy was accessed from mid-January to mid-March 1999 using the canopy raft, the sledge and the treetop bubble (see Ch. 2). The canopy raft is a 580 m^2 hexagonal platform consisting of air-inflated beams from which is suspended a platform of Aramide™ (PVC) netting. An air-inflated dirigible raises the raft and sets it upon the canopy. The raft is positioned at specific sites upon the canopy and moved every fortnight by the dirigible. Access to the raft is provided by single-rope techniques (Ebersolt, 1990; Hallé & Blanc, 1990). The sledge is a triangular platform of about 16 m^2 that is suspended below the dirigible and which 'glides' over the canopy at low speed (Ebersolt, 1990; Lowman *et al.*, 1993a). The treetop bubble is a 180 m^3 helium balloon, 6 m in diameter, that runs along a fixed line set up in the upper canopy by the dirigible (Cleyet-Marrel, 2000).

Five sites (coded A to E), separated by at least 100 m and at most by 4 km, were sampled for arthropods. The sites occurred within similar forest types and at similar altitudes. For collection purposes, a site included the portion of foliage directly accessible in the upper canopy from either the raft (sites A, B and D) or the bubble (sites C and E), and the projected area of the raft or transect of the bubble (*c.* 100 m) in the understorey below. In addition, samples were obtained from the sledge at various locations in the upper canopy early in the morning and equivalent samples were obtained at various locations in the understorey for direct comparison (sixth 'site', coded L). Canopy cover was similar for all sites but differed in floristic composition. Their main characteristics and that of arthropod collections are detailed in Basset *et al.* (2001a).

Sampling methods

The sampling methods assessed the following at all sites: (i) the density of arthropods per unit area of foliage obtained by beating; (ii) the relative activity of arthropods along a transect of three flight-interception traps, situated at ground level, in the canopy and in the upper canopy; and (iii) the relative activity of arthropods collected with sticky traps. These methods are complementary and provide a better assessment of the overall arthropod fauna present than use of any single method (e.g. Basset *et al.*, 1997b). The sampling methods employed were intended to collect macroarthropods, particularly insect herbivores.

Arthropods were collected on square beating sheets 0.397 m^2 in area, conical in shape (with internal slopes of 45°), ending in a circular aperture (7 cm in diameter) that was fitted with a removable plastic bag. Sheets were inserted below the foliage so that the layer of leaves above occupied approximately the entire area of the sheet. Arthropods were dislodged from the foliage with three sharp strokes of the beating stick and gently brushed inside the plastic bag, which was then closed and replaced by a new one. At each site, 20 samples were obtained per stratum (upper canopy or understorey), either during the day (between 13:00 and 16:00) or the night (between 21:00 and 24:00). Upper canopy samples were taken from the periphery of the raft, or with the sledge. Understorey samples were collected at a height below 2 m and originated either from immediately below the projected area of the raft, or from sampling at random in the understorey for comparison with the samples obtained with the sledge ($n = 78$ samples). The size of the samples obtained by beating averaged 2492 ± 267 cm^2 of leaf area.

Nonattractive flight-interception traps, combining features of Malaise and window traps, were used at the raft and bubble sites. The main body of the trap consisted of two intersecting rectangular cross-panels of black netting (mesh width 0.5 mm, double-sided collecting surface of $1.2\,m \times 1.4\,m \times 4 = 6.7\,m^2$) with a roof of the same black netting, connected to a vertical duct and collecting jar. A clear plastic funnel was attached below the main body of the trap (upper diameter of 1.12 m) connected to a large collecting jar. A plastic grid with a 2 cm mesh covered the plastic funnel, preventing larger debris, but not arthropods, from falling into the suspended jar. Collecting fluids were 70% alcohol in the upper jar and water saturated with salt in the lower jar. A similar trap model is described elsewhere (Springate & Basset, 1996).

At each site, one vertical transect of three flight-interception traps was operated for at least 3 days. The traps were set on a rope, with a pulley system that allowed convenient surveying and resetting of the traps in the same position. On the transect, the third trap was set immediately below the canopy raft or within the upper canopy at bubble sites (upper canopy trap), the second one 6 m below (canopy trap) and the first at 2 m above ground (understorey trap). Day and night catches were segregated by surveying the three traps at sunset (18:00) and sunrise (06:00), respectively. A sample represented the pooled catches of the upper and lower collecting jars of each trap for 12 hours.

At each site, 21 sticky traps (Temmen GmbH, Hattersheim, Germany) were established in the upper canopy and in the understorey. Each trap was yellow, with glue (Tanglefoot™) on both faces, each 29 × 12.5 cm in area (total collecting area per trap 725 cm^2). In the upper canopy, traps were set up in the foliage along the periphery of the canopy raft (maximum distance available 84 m) or along the transect of the bubble (*c.* 100 m). In the understorey, traps were set up along a transect line of 80 m situated below the raft or below the transect of the bubble, at a height of 1.5 m. At each raft site and for each stratum, traps were operated for 3 hours in the afternoon (13:00–16:00) then replaced by fresh and inactive traps (with their protection sheet in place) at the same location; these were later operated at night for 3 hours (21:00–24:00). A similar protocol was used at the bubble sites (C and E) but, for logistical reasons, traps had to be surveyed at 07:00 and 17:00, both in the understorey and upper canopy. As a result, traps at sites C and E ran for 10 hours during the day and for 14 hours at night. A sticky trap sample represented the catches of a trap standardized to catch for 3 hours (see below).

Processing of arthropod material and statistical methods

Arthropods were counted and sorted to family or higher taxonomic level. Adults of insect herbivores (*sensu lato*: leaf-chewing, sap-sucking and wood-eating insects) from beating and flight-interception trap samples were mounted and sorted to morphospecies (hereafter termed species). The poor quality of the material collected with sticky traps did not justify this approach and, in this case, specimens were only sorted to the level of the family. Scolytinae were considered as distinct from Curculionidae for the analyses. Collections of insect herbivores were deposited at the Laboratoire Entotrop (Faunistique-Taxonomie) of the Centre de Coopération Internationale en Recherche Agronomique pour le Dévelopment (CIRAD-Amis), Montpellier, France.

Four sample sets relevant to hypotheses 1 and 2 presented in the Introduction were considered: from the understorey during the day; the upper canopy during the day; the understorey at night; and the upper canopy at night. For each sampling method, data on the 10 most common higher arthropod taxa are presented. For beating and flight-interception trap data, measurements of faunal similarity for herbivores across each pair of sample sets (Fig. 27.1) were estimated using the Morisita–Horn index (Magurran, 1988) and the Shared Species Estimator, V (Chen *et al.*, 1995). In the latter, the estimator V augments the observed number of shared species by a correction term based on the relative abundance of shared, rare species (Colwell, 1997b). Further, the rarefied number of species present in a sample of n individuals was estimated with Coleman's curve (e.g. Colwell & Coddington, 1994). The Morisita–Horn index, Shared Species Estimator and Coleman statistics (based on 50 randomizations) were calculated using the program Estimates (Colwell, 1997b). The evenness of herbivore communities was estimated with the index E, proposed by Bulla (1994).

For all sampling methods, we tested differences in the abundance of the higher taxa in the upper canopy between day and night using Mann–Whitney tests (beating data) and Wilcoxon signed ranks tests (flight-interception and sticky trap data). Following Stewart-Oaten (1995), we judge Bonferroni's correction for multiple tests uninformative and do not use it. An additional index of 'abundance change', AC, was calculated in order to estimate the magnitude of change in taxa abundance in the upper canopy between day and night, as follows:

$$AC = (X_d - X_n)/(X_d + X_n) \qquad (27.1)$$

where X_d and X_n are the means of numbers of individuals collected per sample of a particular taxa with a particular method in the upper canopy during day and night, respectively. AC may vary from -1 to $+1$.

To account for the longer exposure of sticky traps at sites C and E, arthropod catches at these sites were corrected by a factor of 0.3 for day catches ($10 \times 0.3 =$ 3 hours) and a factor of 0.214 for night catches ($14 \times 0.214 =$ 3 hours). Analyses were performed using these corrected data. Unless otherwise stated, means are presented with the associated standard errors throughout the text. In the interests of clarity standard errors of means in the Figures are not indicated.

RESULTS

Overview

In total, 14 161 arthropods were collected across the three sampling methods (beating: 2469 individuals; flight-interception traps: 6450 individuals; sticky traps: 5242 individuals). The material collected in the upper canopy included 5148 individuals collected during the day, and 1724 individuals collected at night. Insect herbivores collected in the upper canopy included 1791 individuals representing 187 species. The variable 'time of day' explained 6 to 9% in the total variance in arthropod distribution, depending on the sampling method (strata effects: 40–70% of variance; site effects: 20–40% of variance; Basset *et al.*, 2001a). The salient features and limitations of the sampling programme at La Makandé are detailed and discussed elsewhere (Basset *et al.*, 2001a).

Beating samples

In total, 195 beating samples were obtained from the upper canopy, representing 48.6 m^2 of leaf area sampled. They included 145 and 50 samples obtained during day and night, respectively. Significantly more arthropods were collected per sample during the day than at night (mean per sample was 7.02 ± 0.590 and 2.94 ± 0.279 individuals, respectively; Mann–Whitney $U = 4957.0$, $p < 0.001$; see Fig. 27.2 for the most common taxa). In particular, Formicidae and Apidae were more abundant during the day than at night (Mann–Whitney tests, both with $p < 0.001$; Fig. 27.2; Formicidae not displayed).

The abundance of insect herbivores, however, was not significantly higher during the day than at night (Table 27.1). There was no indication of a large vertical movement of insect herbivores from the understorey to the upper canopy at night ($AC = 0.22$). The Chrysomelidae were more abundant during the night than the day, Curculionidae and Psylloidea showed the reverse trend, and other taxa showed no particular affinity for time of day (Fig. 27.2). Collectively, sap-sucking insects were more abundant during the day than at night in the upper canopy, whereas leaf-chewing insects were not (Mann–Whitney tests with $p < 0.01$ and $p = 0.436$, respectively).

Although many more species of insect herbivores were collected in the upper canopy during the day ($n = 102$) than at night ($n = 25$), this simply reflected the higher sampling effort during the day, particularly with the sledge. Rarefaction estimates calculated with 50 individuals indicated similar species richness in the upper canopy during the day and at night ($n = 27 \pm 4$ (SD) and 25 ± 1 (SD), respectively). The proportion of singleton species collected in the upper canopy during both day and night was high and not significantly different (63% and 76%, respectively; Fisher's exact test, $p = 0.346$). The evenness of herbivore communities in the upper canopy tended to be lower during day than

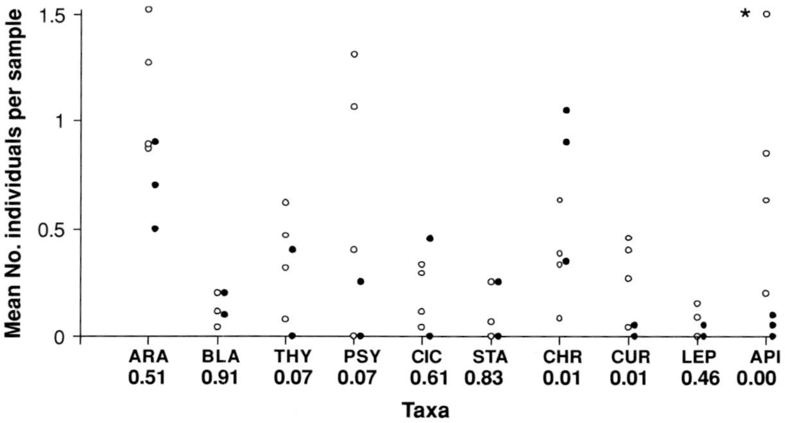

Fig. 27.2. Mean number of individuals collected per beating sample during the day (open circles, four sites) and at night (closed circles, three sites) in the upper canopy for the most common higher arthropod taxa. Numbers below each taxa code refer to the p value of Mann–Whitney tests. ARA, Araneae; BLA, Blattodea; THY, Thysanoptera; PSY, Psylloidea; CIC, Cicadellidae; STA, Staphylinidae; CHR, Chrysomelidae; CUR, Curculionidae; LEP, Lepidoptera; API, Apidae; *out of scale, with 2.92 individuals per sample.

Table 27.1. *Faunal similarities among the understorey and upper canopy samples, during the day and at night (refer to Fig. 27.1 for similarity codes), for the most common herbivore taxa in samples obtained by beating*

Taxa	Total No. species collected	Morisita–Horn index (Shared Species Estimator)[a]						Change of abundance index[b]	Mann–Whitney test p value[b]	Interpretation[c]
		S1	*S2*	*S3*	*S4*	*S5*	*S6*			
All herbivores	154	0.146 (10)	0.069 (4)	0.750 (12)	0.375 (19)	0.044 (7)	0.307 (6)	0.22	0.231	1, 2 less evident
Psylloidea	4	0.805 (–)	1 (–)	1 (–)	0.805 (–)	0.805 (–)	1 (–)	0.79	0.071	?
Membracidae	14	0 (–)	0 (–)	0 (–)	0.292 (–)	0 (–)	0 (–)	0.24	0.607	?
Cicadellidae	20	0 (–)	0 (–)	0 (–)	0.739 (–)	0 (–)	0 (–)	0.10	0.608	Probably 1
Elateridae	7	0 (–)	0 (–)	0.921 (–)	0 (–)	0 (–)	0 (–)	0.17	0.972	Probably 1
Chrysomelidae	40	0.423 (8)	0.274 (1)	0.397 (2)	0.648 (9)	0.183 (3)	0.649 (6)	−0.09	0.010	2 and 1
Curculionidae	35	0 (–)	0 (–)	0.925 (–)	0 (–)	0.004 (–)	0 (–)	0.89	0.006	Probably 1

[a]Insufficient data for calculation indicated by –.

[b]For differences between day and night in the upper canopy (see Methods).

[c]Whether results are consistent with hypotheses 1 or 2 (see text) or cannot be evaluated (?).

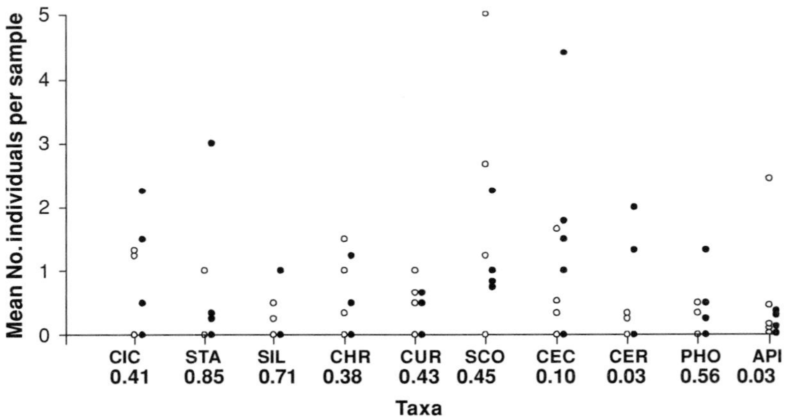

Fig. 27.3. Mean number of individuals collected per flight-interception trap sample during the day (open circles, five sites) and at night (closed circles, five sites) in the upper canopy for the most common higher arthropod taxa. Numbers below each taxa code refer to the p value of Wilcoxon signed-rank tests. CIC, Cicadellidae; STA, Staphylinidae; SIL, Silvanidae; CHR, Chrysomelidae; CUR, Curculionidae; SCO, Scolytinae; CEC, Cecidomyiidae; CER, Ceratopogonidae; PHO, Phoridae; API, Apidae. For the sake of clarity, data for Scolytinae, Cecidomyiidae and Apidae are scaled down by factors 2, 5 and 100, respectively.

during night, but not significantly so ($E = 0.628$ (95% confidence limit (CL) 0.667–0.588) and 0.733 (95% CL 0.822–0.644), respectively).

Most coefficients of similarity calculated for herbivore taxa using the data generated by beating were low (Table 27.1). In many cases, low sample size precluded a clear interpretation of the data. Overall, the similarities of the assemblages of insect herbivores across day and night in either stratum (coefficients *S3* and *S4*) were higher than the similarities between strata at a particular time of day (*S1* and *S2*), favouring hypothesis 1. The coefficient *S6*, however, was also relatively high for all herbivores (Morisita–Horn index = 0.307), indicating that movements of fauna do occur between the understorey during the day and the upper canopy at night. This was evident for the Chrysomelidae (*S6*: Morisita–Horn index = 0.649), which further showed significantly higher abundance in the upper canopy at night than during the day (Table 27.1, Fig. 27.2).

Flight-interception trap samples

During the 16 trapping days at the five sites, the trap set up in the upper canopy provided 28 samples, 14 obtained during the day and 14 at night. Overall, arthropod activity in the upper canopy during the day was high, but not significantly higher than at night (mean number of individuals collected per sample was 105.6 ± 38.6 and 53.08 ± 9.47, respectively; Wilcoxon test, $Z = -0.941$, $p = 0.347$; see Fig. 27.3 for the most common taxa). The Ceratopogonidae, however, were more active at night than during the day and the Apidae showed the reverse trend (Wilcoxon tests, both with $p < 0.05$; Fig. 27.3).

Overall, insect herbivores were not significantly more active during the day than at night in the upper canopy and this was also true of the particular herbivore taxa that were well sampled with flight-interception traps (Table 27.2, Fig. 27.3). There was no indication of a large vertical migration of insect herbivores from the understorey to the upper canopy at night ($AC = 0.15$). The number of herbivore species collected in the upper canopy during the day and at night was similar ($n = 47$ and 44, respectively) and so were the rarefactions calculated to 50 individuals (day: $n = 182 \pm 151$ (SD); night: 120 ± 67 (SD)). The proportion of singleton species collected in the upper canopy during the day and at night was high and not significantly different (83% and 75%, respectively; Fisher's exact test, $p = 0.441$). The evenness of herbivore communities in the upper canopy tended to be lower during the day than at night, but not significantly so ($E = 0.703$ (95% CL 0.769–0.638) and 0.783 (95% CL 0.853–0.714), respectively).

Scolytinae, the most abundant herbivore group in flight-interception traps, greatly influenced the similarity values across the assemblages of insect herbivores (Table 27.2). Several species of Scolytinae appeared indiscriminantly with regard to either stratum or time

Table 27.2. *Faunal similarities between the understorey and upper canopy, during the day and night (refer to Fig. 27.1 for similarity codes), for the most common herbivore taxa collected using flight-interception traps*

Taxa	Total No. species collected	Morisita–Horn index (Shared Species Estimator)[a]						Change of abundance index[b]	Mann–Whitney test p value[b]	Interpretation[c]
		S1	*S2*	*S3*	*S4*	*S5*	*S6*			
All herbivores	158	0.758 (0)	0.601 (15)	0.748 (20)	0.752 (13)	0.699 (119)	0.586 (0)	0.15	0.528	Neither 1 nor 2
Psylloidea	7	0 (–)	0.571 (–)	0 (–)	0.333 (–)	0.400 (–)	0 (–)	0.33	0.480	?
Cicadellidae	33	0.284 (–)	0.241 (1)	0.746 (1)	0.550 (1)	0.442 (5)	0 (–)	−0.15	0.414	1
Derbidae	11	0 (–)	0 (–)	0.396 (–)	0 (–)	0 (–)	0 (–)	−0.30	0.564	Probably 1
Elateridae	9	0 (–)	0 (–)	0 (–)	0 (–)	0 (–)	0 (–)	−0.30	0.705	?
Chrysomelidae	15	0.063 (–)	0.133 (–)	0.800 (1)	0.520 (2)	0 (–)	0.158 (–)	0.21	0.380	1
Curculionidae	20	0 (–)	0 (–)	0 (–)	0 (–)	0 (–)	0 (–)	0.43	0.429	?
Scolytinae	23	0.884 (–)	0.842 (4)	0.979 (–)	0.900 (8)	0.881 (–)	0.813 (–)	0.32	0.447	Neither 1 nor 2

[a]Insufficient data for calculation indicated by –.

[b]For differences between day and night in the upper canopy (see Methods).

[c]Whether results are consistent with hypotheses 1 or 2 (see text) or cannot be evaluated (?).

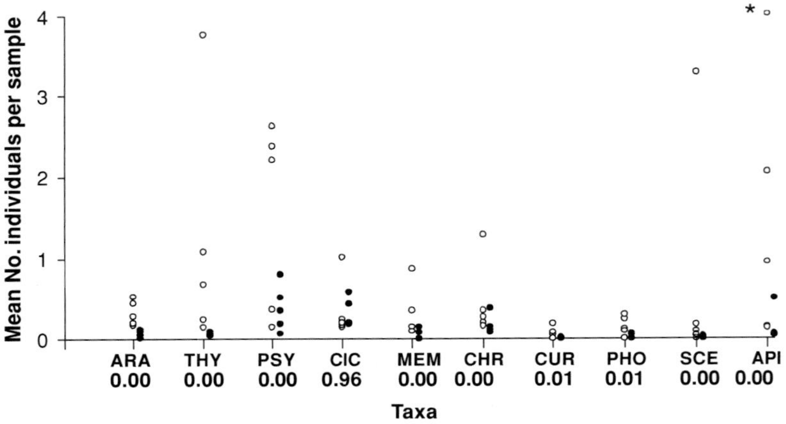

Fig. 27.4. Mean number of individuals collected per sticky trap sample during the day (open circles, five sites) and at night (closed circles, five sites) in the upper canopy for the most common higher arthropod taxa. Numbers below each taxa code refer to the p value of Wilcoxon signed-rank tests. ARA, Araneae; THY, Thysanoptera; PSY, Psylloidea; CIC, Cicadellidae; MEM, Membracidae; CHR, Chrysomelidae; CUR, Curculionidae; PHO, Phoridae; SCE, Scelionidae; API, Apidae; *out of scale, with 18.5 individuals per sample.

of day. The percentage of individuals, for example, of one undetermined species of *Xyleborus* collected in the understorey during day and night, and in the upper canopy during day and night, were 17, 35, 33 and 15%, respectively. In this case, neither hypothesis 1 nor 2 pertains. More individuals of Cicadellidae, Derbidae and Elateridae were collected in the upper canopy at night than during the day, but differences were not statistically significant (Table 27.2). Further, similarities calculated for these groups were not consistent with hypothesis 2. Similarities for Cicadellidae, Chrysomelidae and, possibly, Derbidae were more consistent with hypothesis 1.

Sticky-trap samples

A total of 192 sticky traps were recovered from the upper canopy at five sites, with 100 traps operating during the day and 92 at night. On average and correcting for longer exposure at sites C and E, 10.8 ± 0.78 arthropods were caught per trap during 3 hours of exposure in the upper canopy (day and night data pooled). This corresponded to catch rates of about 3.6 arthropods per trap per hour, or of 2.5 arthropods per 500 cm^2 per hour. Overall, sticky traps set in the upper canopy collected significantly more arthropods during the day than at night (mean 17.7 ± 1.10 individuals and 3.4 ± 0.29, respectively; Wilcoxon rank test, $Z = -8.09$, $p < 0.001$). The pattern was similar for insect herbivores (day: 4.5 ± 0.51 individuals, night: 1.12 ± 0.12 individuals, $Z = -6.92$, $p < 0.001$). The overall levels of activity of many taxa were higher during the day than at night, including those of the Araneae, Phoridae, Brachycera, Scelionidae, Apidae and, among herbivorous taxa, Thysanoptera, Psylloidea, Membracidae, Chrysomelidae and Curculionidae (Fig. 27.4). Catches of herbivores did not increase notably in the upper canopy at night ($AC = 0.60$), suggesting that no strong influx of herbivores from lower strata occurred at that time.

DISCUSSION

Each of the sampling methods used in this study has limitations and these are discussed in Basset *et al.* (2001a). In particular, the reflectance of yellow sticky traps and their efficiency may be higher during the day than at night. In addition, their efficiency may be higher in the canopy than it is in the understorey, although the reverse is also plausible. Other factors, such as wind speed and air temperature, may further complicate the interpretation of levels of arthropod occurrence as measured by flight-interception and sticky traps. Overall, beating data indicated real differences in the occurrence of sedentary taxa from site to site. Data from the flight-interception traps reflect the flight activity of larger, heavier arthropods, whereas sticky-trap data reflect the flight activity of smaller arthropods, perhaps increasing the magnitude of the differences observed here, although to what extent is not clear.

Our sampling methods were also less efficient for larger herbivores, such as certain Orthoptera and Phasmoptera. It is well known that the activity of these taxa tend to segregate between day and night (e.g. Lockwood *et al.*, 1996). It is probable that more efficient collection of these taxa would have resulted in more contrast in overall arthropod activity between day and night.

Beating data identified significant differences in the abundance of certain taxa between day and night. These differences were often higher when estimated with sticky traps – which are commonly used in economic entomology to monitor the diel activity of particular herbivore pests (e.g. Johnson & Mueller, 1990; Weintraub & Horowitz, 1996). In contrast, flight-interception traps showed few differences in arthropod activity between day and night. This result may reflect the different sampling regime represented by the three methods. Whereas collecting using beating or sticky traps occurred for 3 hours in the middle of the afternoon or of the night, flight-interception traps ran continuously from 06:00 to 18:00 and from 18:00 to 06:00, respectively. A simple segregation of flight-interception trap material into day and night catches may have resulted in fewer differences between day and night, particularly if a high proportion of catches occurred either at dusk or dawn. Data from the flight-interception traps were also influenced by abundant catches of Scolytinae, of which several species showed no clear preferences for either strata or time of day. Some of these beetle species may well forage preferentially at dawn or dusk.

We conclude, therefore, that arthropod activity is indeed higher during the day than at night in the upper canopy. This parallels the outcomes of other studies elsewhere in the tropics, as indicated in the Introduction to this chapter. In particular, taxa such as the Thysanoptera, Psylloidea, Membracidae, Curculionidae, Scelionidae and Apidae were notably more active during the day than at night in the upper canopy. Many Brachycera (Diptera) that were not sorted to family level appeared also to be particularly active during the day in the upper canopy.

Referring back to hypotheses 1 and 2, it is inevitable that a low sample size (with regard to particular herbivore species) will result in low similarities of herbivore faunas, as in the present study. However, it is instructive to compare relative similarity values obtained with comparable sampling effort. Similarity values between day and night faunas in particular strata (*S3* and *S4*) were often higher than between faunas of the understorey and the upper canopy at a particular time of the day (*S1* and *S2*). This suggests that there are rather distinct herbivore faunas foraging in each stratum, an observation consistent with hypothesis 1.

Assemblages of insect herbivores in the upper canopy during the day were species rich but unevenly distributed, with just a few species dominating the communities there. In addition, similarities in *S3* (understorey, similarity between day and night) were often higher than *S4* (upper canopy, similarity between day and night). In particular, the data obtained by beating suggested that faunal turnover between day and night was higher in the upper canopy compared with that in the understorey. This suggests that the magnitude of changes in the microclimatic conditions between day and night in the upper canopy may be more severe than in the understorey, and that only a well-adapted fauna may be able to cope with these changes. It is well known that many insect taxa of tropical rainforests show behavioural and physiological adaptations that result in thermal guilds, such as 'light-seeking' or 'shade-seeking' insects (e.g. Shelly, 1985; Hood & Tschinkel, 1990; Roubik, 1993). Further, it is known that certain insect pests show a positive correlation between activity and air temperature, others a negative one (e.g. Johnson & Mueller, 1990; Weintraub & Horowitz, 1996). All of these behavioural differences may be important in shaping the different assemblages of insect herbivores in different forest strata at different times of the day. In contrast, arthropod diel activity in the understorey may be less clearly differentiated on a day/night basis (Davis, 1999a; Basset, 2000).

The present data also indicate that certain groups, notably the Chrysomelidae, may move at night from the understorey to the upper canopy (evidenced by high values of *S6*), which is also consistent with hypothesis 2. This observation may reflect the activities of just a few species, since none of the sampling methods showed evidence of mass migration from the understorey to the upper canopy at night. As stressed by Compton *et al.* (2000), however, insect flight in the overstorey (i.e. above the canopy) for dispersal may be significant, especially at night. Future work should determine the ecological characteristics of *resident* species in the upper canopy, of migrating species from the understorey to the overstorey, as well as of species foraging indiscriminately in these different layers.

There are two obvious implications of the present results. First, if faunal turnover is high in the upper canopy between day and night, comprehensive entomological surveys should ensure that arthropods are sampled during the full daily cycle. A popular sampling method such as pyrethrum knockdown is usually performed early in the morning when airflow is reduced (e.g. Adis *et al.*, 1998b). It would be advisable to assess the efficiency of this method for the collection of taxa active predominantly in either the day or the night. Second, our data suggest that many insect herbivores of the upper canopy may be resident and well adapted to environmental conditions there. Since faunal stratification in tropical rainforests may be at an optimum, leading to a diverse fauna in the upper canopy only in closed and undisturbed lowland rainforests, the implications for the conservation of tropical rainforest arthropods are also significant.

ACKNOWLEDGEMENTS

We thank the support teams of Océan Vert, Pro Natura International and Les Accro-Branchés: Francis Hallé, Gilles Ebersolt, Dany Cleyet-Marrel, François Collignon, Laurent Pyot, Olivier Pascal, Roland Fourcaud, Carole Megevand, Thierry Aubert, Lionel Picart and Jean-Yves Serein. Jean-Michel Bérenger, Jean-Pierre Vesco, Philippe Causse, Andréa Haug, Anne-Sophie Hennion, Loïc Lesobre, Florent Marquès and Robert O'Meara helped with insect collecting. Enrique Medianero and Anayansi Valderrama helped to sort the insect samples. Neil Springate commented on the manuscript. Financial help was provided by a Tupper fellowship at the Smithsonian Tropical Research Institute in Panama (Y. B.), Bayer-Agro (H.-P. A.), Secretaria Nacional de Ciencia y Tecnología (H. B.), and Dr Franco Tassi of the Parco Nazionale d'Abruzzo and 'Progetto Biodiversità' (G. C.).

28

Diel, seasonal and disturbance-induced variation in invertebrate assemblages

Timothy D. Schowalter and Lisa M. Ganio

ABSTRACT

Canopy invertebrates are a diverse group in forest ecosystems. They represent important food resources for other animals, and they can alter canopy structure and, therefore, canopy atmosphere and canopy–forest floor interactions. Accordingly, factors that influence spatial and temporal patterns in the structure of invertebrate assemblages and herbivory are important in the understanding of ecosystem structure and function. We sampled canopy invertebrate assemblages and leaf area missing (LAM) on representative early and late successional tree species, during day and night, in wet and dry seasons, and in plots representing light and severe hurricane disturbance at the Luquillo Experimental Forest Long Term Ecological Research (LTER) Programme site in Puerto Rico, and during wet and dry seasons at the Fort Sherman Canopy Crane site in Panama. We expected to find significant variation in canopy invertebrate abundances and assemblage structure among tree species and between diel, seasonal and annual sampling periods. Many species and functional groups showed the expected differences in abundance among tree species, reflecting feeding preferences. Contrary to our hypothesis, no species showed significant diel or seasonal patterns of abundance, although LAM showed significant seasonal and annual variation. Significant interaction among tree species and years was observed for four of five Coccoidea analysed (*Ceroplastes rubens*, *Coccus acutissimus*, *Protopulvinaria pyriformis* and *Vinsonia stellata*), for combined Coccoidea, a mirid (*Itacoris*) species, and for three of four functional groups (sap suckers, predators and detritivores) analysed, primarily reflecting disturbances by hurricane or drought. The fifth species of coccoid (a pseudococcid species) was significantly affected by tree species. A species of Tropiduchidae (*Ladella stali*), combined Lepidoptera and the folivore functional group approached significant responses. In particular, the sap-sucker functional group and individual species of Coccoidea and *Itacoris* were most abundant following hurricanes (1991 and 1999). However, some species of Homoptera showed opposite responses to changes in environmental conditions. The detritivore functional group (primarily Psocoptera and Blattidae) was most abundant during the drought period (1994–1995). Predators were more abundant following hurricanes or during drought, reflecting elevated prey availability. Data on the responses of assemblages of canopy invertebrates to temporal variation in environmental conditions are necessary to predict effects of environmental changes on canopy communities and consequent effects on ecosystem structure and function.

INTRODUCTION

Canopy invertebrates represent a major portion of the biodiversity in forest ecosystems and play important ecological roles (e.g. Stork *et al.*, 1997a). Herbivores in forest canopies frequently cause loss of large amounts of foliage, either directly through consumption or indirectly through foliage deformation and premature abscission. Consequently, herbivores affect canopy–atmosphere interactions (including photosynthesis, evapotranspiration and the interception of light, precipitation and atmospheric nutrients) and canopy–forest floor interactions (including water and nutrient fluxes to the forest floor, and soil/litter decomposition and respiration), as well as the growth and survival of host plants (Schowalter & Lowman, 1999; Schowalter, 2000). Accordingly, factors that influence spatial and temporal patterns in invertebrate assemblage structure and herbivory (including abiotic conditions, host-plant condition and predation) are important in our understanding of ecosystem structure and function. Relatively few studies have addressed the effects of factors influencing canopy communities (e.g. Majer & Recher, 1988; Springate & Basset, 1996; Schowalter & Ganio, 1998, 1999).

Tropical forests are receiving increased attention because of their occurrence over a large portion of the land surface of the Earth (Terborgh, 1985), the accelerating rate of their conversion to commercial plantations or pastures (Vitousek *et al.*, 1997) and recognition of their contributions to global carbon flux and potential impacts on global climate (Brown & Lugo, 1982; Salati, 1987). Clearly, the effects of canopy communities on tropical forest structure and function may have important global consequences.

A number of studies have examined tropical canopy invertebrates (e.g. Janzen, 1973a; Wolda, 1978b; Majer & Recher, 1988; Springate & Basset, 1996; Stork *et al.*, 1997a; Schowalter & Ganio, 1999; Ch. 27), but few have addressed factors that affect temporal patterns in invertebrate abundances or assemblage structure. For example, few studies have compared diel patterns of abundance in tropical forests using comparable techniques (Janzen, 1973a; Springate & Basset, 1996; Compton *et al.*, 2000; Ch. 27). A number of studies have demonstrated seasonal variation in invertebrate occurrence or abundance in tropical, as well as temperate, forests (Janzen, 1973a; Wolda, 1978b; Recher *et al.*, 1996a; Schowalter & Ganio, 1998; Chs. 15 and 16). These temporal patterns result from seasonal patterns of precipitation and foliage production (Coley & Aide, 1991; Lowman, 1985, 1992; Murali & Sukumar, 1993). Fewer studies have provided long-term data on the effects of environmental changes or disturbances on annual variation in canopy invertebrate abundances or assemblage structure (e.g. Schowalter & Ganio, 1999).

This paper describes diel, seasonal and interannual dynamics of canopy assemblages and herbivory in neotropical rainforests in Puerto Rico and Panama. Our hypothesis has been that tropical invertebrate abundances and assemblage structure will show diel patterns, seasonal patterns (with peaks in the wet season) and annual patterns (related to hurricane and drought disturbances). Data from such studies can be used to predict the effects of environmental changes on canopy communities and consequential effects on ecosystem structure and function.

METHODS

Site descriptions

This study was conducted at El Verde Field Station in the Luquillo Experimental Forest (18° 19′ N, 65° 49′ W) in Puerto Rico and at the Fort Sherman Canopy Crane site (9° 18′ N, 80° 02′ W) near Colón, Panama. These two sites have similar climates and share several tree genera, including *Manilkara* and *Cecropia*, permitting comparison of temporal changes in arthropod assemblages on these trees.

The Luquillo Experimental Forest is administered by the USDA Forest Service Caribbean National Forest and Southern Forest Experiment Station and by the University of Puerto Rico. El Verde Field Station of the US Forest Service is located 10 km south of Rio Grande, Puerto Rico at 500–600 m elevation. Mean monthly temperatures range from 21 °C in January to 25 °C in September (Brown *et al.*, 1983). Annual precipitation averages 3700 mm and varies seasonally, with 200–250 mm per month from January to April (dry season) and 350 mm per month the remainder of the year (wet season) (McDowell & Estrada-Pinto, 1988).

Vegetation at Luquillo Forest is subtropical evergreen wet forest. Dominant tree species include *Dacryodes excelsa* Vahl. (Burseraceae), *Manilkara bidentata* (A. DC.) A. Chev. (Sapotaceae) and *Sloanea berteriana* Choisy (Elaeocarpaceae). *Prestoea montana* (Graham) Nicolson (Arecaceae) is a major subcanopy species. Canopy height averages 20 m, and small light gaps occur infrequently in an otherwise closed canopy in undisturbed forests. *Cecropia schreberiana* Miq. (Cecropiaceae) and *Casearia arborea* (Rich.) Urb. (Flacourtiaceae) are important early successional tree species that grow rapidly in large gaps.

Hurricanes that cause significant defoliation recur frequently, but hurricanes with sufficient force to damage or topple trees recur on average about every 60 years. Hurricane Hugo (September 1989) left severely disturbed patches (30–60 m diameter), with nearly complete treefall (gaps) interspersed with less-disturbed patches where all trees remained standing but lost most of their foliage and smaller branches (relatively intact stands). Rapid refoliation and shoot replacement in intact stands and seedling and sprout regrowth in gaps began during the wet season of 1990 (Frangi & Lugo, 1991). Thickets of *Cecropia* saplings and other early successional plants, especially *Heliconia* spp., developed rapidly in treefall gaps. Similar canopy opening and recovery were observed following Hurricane Georges (September 1998). Three hurricanes during 1996 caused some defoliation but little treefall. Drought represents another disturbance that can affect tropical, as well as temperate, ecosystems. Dry conditions prevailed during 1991–1996, with annual precipitation averaging only 72% of the long-term mean. Drought was particularly pronounced in 1994, during which precipitation was only 41% of the long-term annual average. Precipitation in 1995 was 3500 mm.

The Fort Sherman canopy crane site is located 12 km south-west of Colón, Panama, 5 km west of the Canal, in a low valley at about 100 m elevation. Average annual temperature is constant at 26–27 °C. Annual precipitation averages 3300 mm, with 40–120 mm per month from January to April (dry season) and 280–500 mm per month the remainder of the year (wet season). The wet season of 1998 was affected by a pronounced drought. The vegetation is classified as tropical lowland wet forest.

The Smithsonian Tropical Research Institute constructed a canopy crane at this site in 1997 to study primary forest (about 300 years old). The crane tower is 56 m in height and the jib is 54 m long. The 1 ha plot centred on the crane includes 83 tree species and 487 individuals >10 cm DBH (diameter at breast height), all accessible from the crane gondola. Dominant species include *Brosimum utile* (H. B. K.) Pitt. (Moraceae), *Guatteria dumetorum* R. E. Fries (Annonaceae), *Licania hypoleuca* L. H. Benth. (Chrysobalanaceae), *Manilkara zapota* (L.) Royen, *Tapirira guianensis* Aubl. (Anacardiaceae), and *Virola sebifera* Aubl. and *Virola nobilis* A. C. Sm. (Myristicaceae). Canopy height averages 25–30 m. *Cecropia* sp. was represented by two canopy subdominant individuals in gaps.

Sampling methods

Replicate trees at both sites were sampled using the branch bagging method (Majer & Recher, 1988; Blanton, 1990; Schowalter & Ganio, 1998, 1999). This technique emphasizes arthropods associated with a particular plant at the time of sampling and provides data on density and effects on canopy structure, but typically it underrepresents highly mobile taxa, which are able to escape capture.

Two tree species, *C. schreberiana* and *S. berteriana*, were sampled day and night seasonally during 1999 at El Verde (four replicates for each nocturnal sample date; six replicates for each diurnal sample date). Two additional tree species (six replicates per sample date) were sampled seasonally during the day in treefall gaps and intact forest during 1991–1999 at El Verde. Four tree species (two to five replicates per sample date) were sampled seasonally during July 1998 and March 1999 at Fort Sherman. The canopy at El Verde was sampled from the ground using a long-handled insect net fitted with a plastic bag that could be closed with a drawstring (Schowalter & Ganio, 1999). Samples could be collected up to a height of 10–12 m. Data reported here for *C. schreberiana* in Puerto Rico are for seedling through tree development in canopy gaps; data for other tree species in Puerto Rico are for trees in intact stands. At Fort Sherman, branch bagging was done by hand from the gondola. Samples were collected at the upper and lower crown levels for each sampled tree.

At both sites, a 10–30 g sample was quickly enclosed in the plastic bag; the bag was sealed and returned to the laboratory, where invertebrates were sorted and counted by taxon. This branch-bagging technique has two advantages over other sampling techniques for ecosystem studies: arthropod abundances can be expressed as intensity (or density with adequate data on plant biomass) and the sampled foliage can be examined for effects of herbivory. Plant material in the samples was dried at 50 °C, weighed and leaf area missing (LAM) as a result of herbivory was measured. Arthropod abundances were standardized as number per kilogram of plant material.

Statistical analyses

We examined responses of individual taxa and some pooled taxa, with sufficient numbers and frequency of representation among samples, to tree species and environmental changes through time using two-way analysis of variance (ANOVA) (Steel & Torrie, 1980). In addition, we pooled species into four functional groups (folivores, sap suckers, predators and detritivores). For assessment of annual variation, abundances were pooled by plot and season. The log-transformation was used as necessary to meet the assumptions of the statistical analyses. In order to account for the correlation between observations made on the same plots but in different years,

Table 28.1. *Significance levels for responses of abundant arthropods to tree species–year interaction in Puerto Rico*[a]

Taxon or group	Degrees of freedom		F	p value
	Numerator	Denominator		
Folivores	18	20	1.9	0.088
Lepidoptera	18	20	0.9	0.56
All sap suckers	18	20	3.2	0.007
Ceroplastes rubens	18	20	6.6	0.0001
Coccus acutissimus	18	20	2.5	0.027
Protopulvinaria pyriformis	18	20	2.3	0.037
Vinsonia stellifera	18	20	53	0.0001
Pseudococcidae species	18	20	0.9	0.56
All Coccoidea	18	20	6.7	0.0001
Ladella species	18	20	1.8	0.097
Itacoris species	18	20	5.6	0.0001
Predators	18	20	2.7	0.016
Detritivores	18	20	5.2	0.0003

[a]Tree species were *Dacryodes excelsa, Manilkara bidentata* and *Sloanea berteriana* in intact forest and *Cecropia schreberiana* in treefall gaps; years were 1991–1992, 1994–1995, 1997–1999.

we defined a correlation matrix to model data relationships across years. Results were considered significant at $p < 0.05$. SAS software (SAS, 1982) was used for ANOVA. We also carried out multivariate analysis (detrended correspondence analysis) on annual trends in species and functional group composition (McCune & Mefford, 1999).

RESULTS

A total of 155 taxa were analyzed over this study period. These comprised primarily morphospecies but some families and orders for insects, and families for spiders and mites. The vast majority of species occurred too infrequently for statistical analysis and were pooled and are represented as families or orders in the Appendix to this chapter. Some common and mobile species, such as *Cyrtoxipha gundlachi* Saussure (Gryllacrididae) were widely distributed and showed no significant responses to tree species or time period. Only seven species (all sap suckers: *Ceroplastes rubens* Maskell, *Coccus acutissimus* (Green), *Protopulvinaria pyriformis* (Cockerell), *Vinsonia stellifera* (Westwood) and a Pseudococcidae sp. (all Coccoidea), *Ladella stali* Fennah (Tropiduchidae) and *Itacoris* sp. (Miridae)) had sufficient numbers and frequency of representation among samples for statistical analysis.

Diurnal variation in arthropod communities

Abundances of major taxa in diurnal and nocturnal samples are shown in the Appendix (Table 28A.1). No taxa showed statistically significant differences in abundance between samples collected during the day or night. A number of nocturnally active species, including species from Lampyridae and Gryllidae, were consistently collected from foliage during the day. *P. pyriformis* is sessile, making apparent differences in its abundance between diurnal and nocturnal samples an indicator of sample variation. *Phyllophaga* sp. (Coleoptera: Scarabaeidae) were seen only at night, chewing on *Cecropia* leaves, and appeared to be responsible for much of the relatively high foliage loss for this (and perhaps some other) tree species, but this beetle occurred only once in our samples.

Table 28.2. *Functional group abundances on* Cecropia *(regenerating in canopy gaps),* Manilkara *and* Sloanea *(in intact forest) spp. in Puerto Rico over a 9 year period following Hurricane Hugo (1989) and in Panama during 1998–1999; a major drought occurred in 1994 and Hurricane Georges occurred after the 1998 season in Puerto Rico*

Functional group	Abundances (No. kg^{-1} dry foliage)[a] in Puerto Rico							Abundances (No. kg^{-1} dry foliage)[b] in Panama 1999
	1991	1992	1994	1995	1997	1998	1999	
***Cecropia* sp.**								
Folivores	37 (21)	70 (20)	66 (16)	130 (85)	80 (29)	22 (19)	9 (9)	14
Sap suckers	190 (53)	64 (16)	93 (17)	140 (51)	0	0	200 (91)	24
Predators	180 (36)	260 (71)	150 (44)	150 (36)	63 (21)	150 (44)	220 (87)	228
Detritivores	3 (3)	19 (10)	171 (24)	240 (91)	43 (17)	150 (53)	290 (61)	10
***Manilkara* sp.**								
Folivores	49 (13)	38 (12)	23 (9)	96 (26)	40 (23)	70 (29)	29 (15)	18
Sap suckers	530 (55)	180 (55)	56 (13)	110 (44)	0	0	150 (23)	21
Predators	130 (30)	59 (16)	71 (16)	220 (83)	55 (20)	200 (100)	53 (15)	74
Detritivores	24 (12)	12 (4)	50 (15)	66 (13)	32 (21)	42 (18)	35 (14)	167
***Sloanea* sp.**								
Folivores	77 (14)	72 (24)	78 (21)	85 (24)	42 (20)	76 (25)	95 (40)	–
Sap suckers	58 (9)	46 (8)	22 (10)	38 (18)	0	0	27 (8)	–
Predators	140 (30)	90 (13)	150 (44)	42 (20)	60 (17)	52 (18)	160 (55)	–
Detritivores	41 (14)	38 (15)	170 (39)	100 (25)	150 (120)	99 (32)	94 (24)	–

[a]Averaged over season and then over replicate plots; given as mean (SE).
[b]Averaged over season; given as mean.

Seasonal variation in arthropod communities

Seasonal abundance of major arthropod taxa on four canopy dominant tree species in Puerto Rico and Panama, including two tree genera common to both sites, *Cecropia* and *Manilkara*, is summarized in the Appendix (Table 28A.2). Although herbivory varied seasonally (see below), no taxa showed statistically significant differences in abundance between wet and dry seasons at either site.

Annual variation in arthropod communities

Abundances of major taxa on *Cecropia, Manilkara* and *Sloanea* spp. in Puerto Rico from 1991 to 1999 are shown in the Appendix (Tables 28A.3–28A.5, respectively). Five of the seven species (four Coccoidea spp. and an *Itacoris* sp.), combined Coccoidea, and three of four functional groups (sap suckers, predators and detritivores) showed significant responses to the tree species – year interaction (Tables 28.1 and 28.2). In addition, *Ladella* and folivores approached significant responses at $p = 0.097$ and 0.088, respectively. *Ceroplastes, Coccus, Protopulvinaria*, and *Vinsonia* spp. generally were most abundant during posthurricane years (1991 and 1999), although some species on some hosts were also abundant during the drought years (1994–1995). *Protopulvinaria* sp. on *D. excelsa* was most abundant during 1991 and 1999 and least abundant during 1995, whereas *Protopulvinaria* sp. on *C. schreberiana* showed the opposite trend, suggesting undetected differences in *Protopulvinaria* spp. among host trees, or differential responses to host conditions. The trend toward higher abundances during posthurricane years was masked to some extent at the level of combined Coccoidea, reflecting these different trends among constituent species. *Itacoris* sp. was most abundant during posthurricane years, as were pooled sap suckers. By contrast, detritivores (primarily Psocoptera and Blattidae) were most abundant during the drought years and least abundant during posthurricane years; predators were more abundant during posthurricane and drought years.

Detrended correspondence analysis also suggested responses to hurricane or drought disturbances (Fig. 28.1). Posthurricane years (1991 and 1999) tended to cluster toward the left and nondisturbance years (1997–1998) toward the right, with drought years (1994–1995) in between. This was most apparent for *Manilkara* and *Sloanea* spp. The greater distinction between 1991 and 1999 for *Cecropia schreberiana* perhaps reflects the severe damage to this species during Hurricane Georges (1999) but not during Hurricane Hugo (1991).

Abundances of arthropod functional groups on *Cecropia* and *Manilkara* spp. in Panama are included in Table 28.2 for comparison. The abundances of folivores and predators are similar between the two sites, but abundances of sap suckers were much higher in Puerto Rico, probably reflecting hurricane disturbance.

Herbivory

Seasonal and annual patterns of herbivory have been observed in tropical, as well as temperate, forests (Lowman, 1985, 1992; Coley & Aide, 1991; Ribeiro *et al.*, 1994a). Changes in leaf area associated with herbivory potentially alter canopy–atmosphere and canopy–forest floor interactions (Schowalter, 2000).

Patterns of LAM on *Cecropia*, *Manilkara* and *Sloanea* spp. in Puerto Rico are shown in Fig. 28.2. Seasonal patterns in LAM differed between tree species, with LAM generally peaking during the dry season for *Cecropia* but during the wet season for *Sloanea*. Seasonal dynamics apparently were influenced by annual variation. LAM generally peaked during the season following drought on all three tree species, concurrent with peak folivore abundances (Table 28.2), but, thereafter, crashed abruptly. Although LAM generally was low (<6% foliage area), it approached or exceeded 10% during some seasons for all three tree species, suggesting associated pulses of light, water and nutrient fluxes to the forest floor and reductions in photosynthesis and transpiration.

LAM in Panama increased from wet (4%) to dry (8%) season for *Cecropia* but was constant at 3% for *Manilkara*. An unquantified irruption of several folivores occurred at the Fort Sherman site during August 1998, associated with extended drought conditions during the wet season (S. J. Wright, personal communication).

DISCUSSION

Most taxa showed distinct patterns of representation and abundance among tree species, as expected from known biochemical interactions between herbivores and plants, and between predators and prey. These patterns of abundance can be scaled up to the ecosystem level, given data on foliage biomass for each tree species, to provide arthropod density data (number per unit area).

We expected to find significant diel and seasonal variation in abundances. However, most taxa were represented in both diurnal and nocturnal samples, except for some rare and highly mobile taxa. For example, *Phyllophaga* sp. frequently was observed feeding on *C. schreberiana* leaves, only at night, but was not sufficiently abundant to be represented, given our relatively small size and number of samples. Our failure to detect significant diel and seasonal variation in invertebrate abundances probably reflects small sample size and low abundance and the consequent low frequency of occurrence among our samples.

Annual variation in invertebrate abundances was statistically significant and pronounced, reflecting the influence of hurricane and drought disturbances, as well as the value of higher abundances and frequency of occurrence among pooled samples. We limited the number of species or groups analysed in order to minimize the risk of type 1 errors, which can result from analysis of a large number of groups. We expected that no more than one significant effect out of the 13 species or groups analysed would occur by chance (at $p = 0.05$), the same rate as if data were reported in separate papers for individual species. Combining these responses in a single paper maximizes the convenience of comparison among species and illuminates the consistency of significant responses to disturbance among the species or groups analysed.

Folivore, sap sucker, and detritivore functional groups showed complementary responses to disturbances. Folivore and detritivore abundances were reduced by canopy opening but increased by drought, whereas sap sucker abundance was increased by canopy opening, as reported previously for temperate and tropical forests (Schowalter *et al.*, 1981; Waring & Cobb, 1992; Schowalter, 1995; Schowalter & Ganio, 1999).

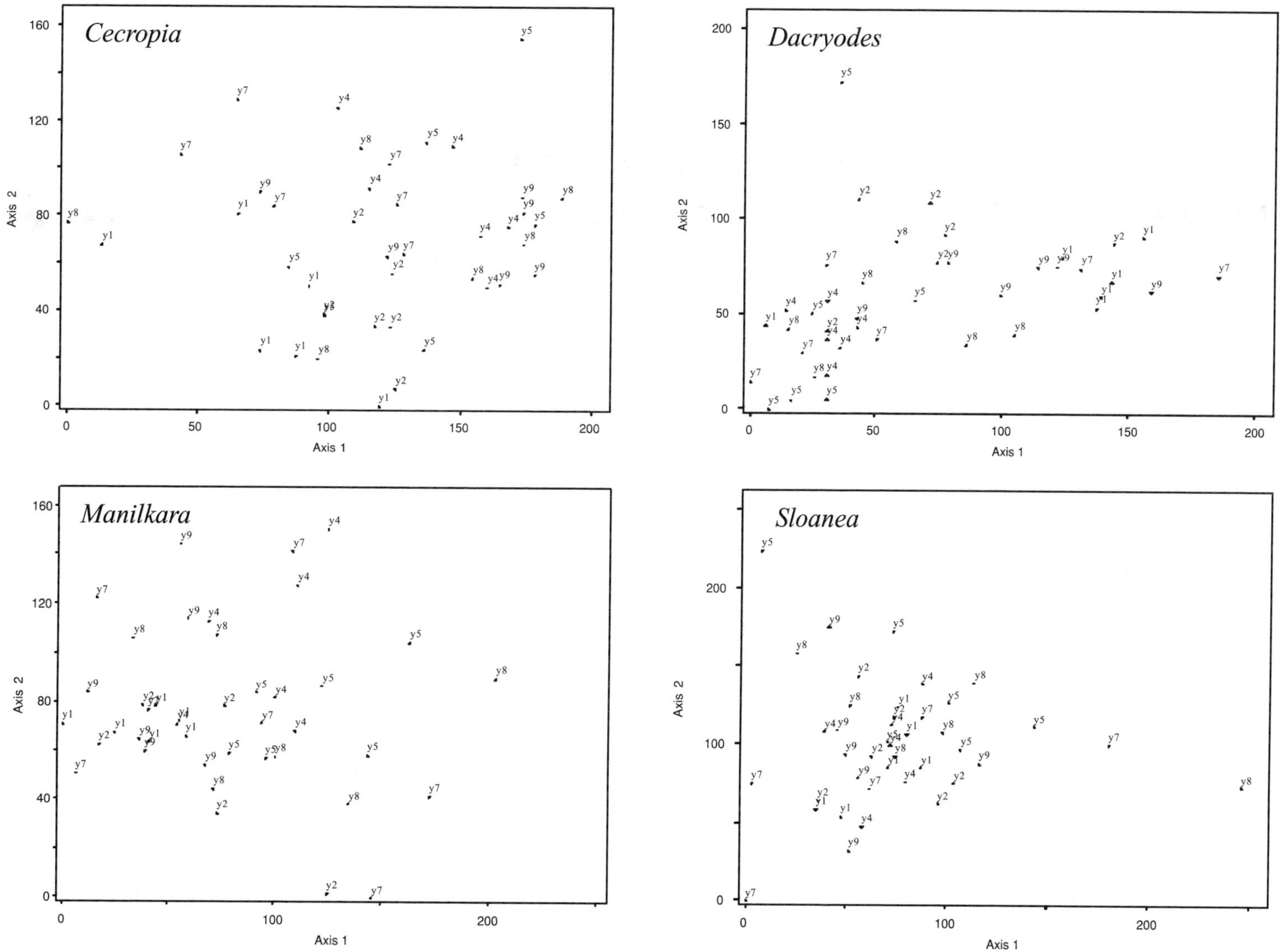

Fig. 28.1. Detrended correspondence analysis ordination of functional group (folivore, sap sucker, predator and detritivore abundances) organization during 1991 to 1999 for four tree species (*Cecropia* from seedling to tree in gaps resulting from Hurricane Hugo, *Dacryodes*, *Manilkara* and *Sloanea* in intact forest) at the Luquillo Experimental Forest in Puerto Rico. Data are the last digit for each year, for example y1 = 1991; $n = 6$. Hurricane disturbances occurred in September 1989 and 1999 and drought disturbance in 1994.

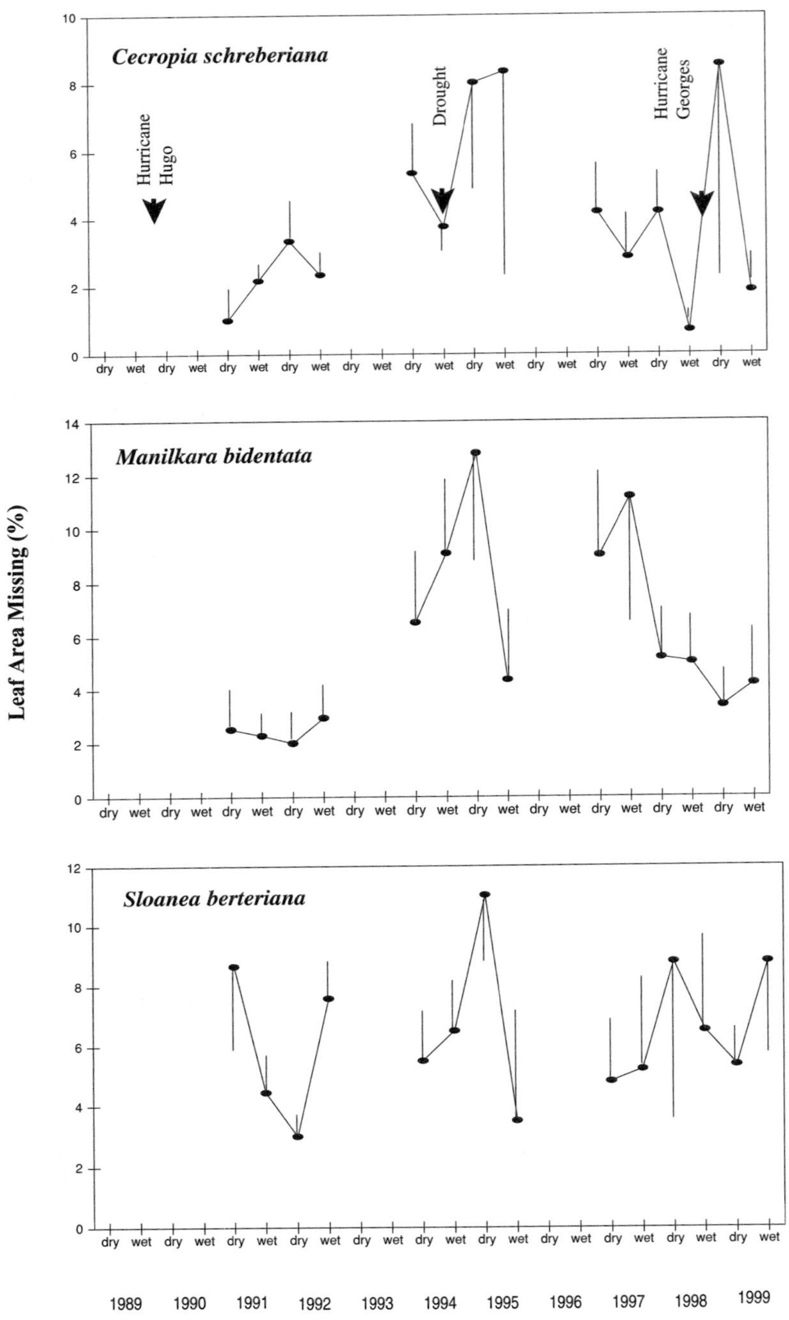

Fig. 28.2. Temporal patterns of herbivory on three tree species (*Cecropia* from seedling to tree in gaps resulting from Hurricane Hugo, *Manilkara* and *Sloanea* in intact forest) at the Luquillo Experimental Forest in Puerto Rico. Hurricane disturbances occurred in September 1989 and 1999 and drought disturbance in 1994. Vertical lines represent one standard error of the mean; $n = 6$.

These results were significant for sap suckers and detritivores, but not for folivores. Elevated predator abundances following hurricanes or drought probably reflect elevated abundances of these prey groups. We note that differences in results reported here and in Schowalter & Ganio (1999) reflect the longer time period of this study. The apparent posthurricane recovery of predator abundance reported by Schowalter & Ganio (1999) was offset by subsequently reduced abundance until after Hurricane Georges. Individual species within these functional groups also showed complementary responses, with some species increasing and others declining or showing no change in abundance during the period of this study. As a result, data for pooled taxa or functional groups may mask responses by individual species, depending on the consistency of response to environmental gradients among constituent species. Taken together, our data for the abundances of species and functional groups suggest that hurricane and drought disturbances are important aspects of temporal variability in environmental conditions that contribute to species population dynamics and to community structure.

Herbivores had variable but important effects on canopy structure, depending on tree species. Some tree species experienced negligible rates of leaf area reduction, but other species showed substantial leaf area reduction, especially during drought conditions. Variation in nutrient concentrations among tree species and foliage classes (e.g. Schowalter & Ganio, 1999) influences both herbivory and the spatial and temporal patterns of nutrient turnover from foliage to the forest floor. Furthermore, spatial patterns of leaf area reduction influence canopy porosity. Increased canopy opening increases the penetration of light, water, and airflow through the canopy. These processes, in turn, increase availability, uptake and turnover of water, carbon and other nutrients in the canopy via perched soils, epiphytes and canopy roots, as well as on the forest floor (e.g. Coxson and Nadkarni, 1995).

In conclusion, our data demonstrate that the responses of arthropod species to temporal variation in environmental conditions, especially disturbance, significantly influence assemblage structure (e.g. relative abundances of functional groups) and rates of foliage loss. The effects of changing arthropod assemblage structure and foliage loss on ecosystem processes are being investigated in current studies.

ACKNOWLEDGEMENTS

Coccoidea species were identified by G. L. Miller and Orthoptera by D. A. Nickle, Systematic Entomology Laboratory, USDA. Yanli Zhang performed detrended correspondence analysis. Research was performed under National Science Foundation grants BSR-8811902 and DEB-9411973 to the Terrestrial Ecology Division, University of Puerto Rico and International Institute of Tropical Forestry as part of the Long-Term Ecological Research Programme in the Luquillo Experimental Forest. Additional support was provided by the Forest Service (USDA) and the University of Puerto Rico; Oak Ridge Associated Universities (Travel Grant S-3441); the Canopy Biology Program, Smithsonian Tropical Research Institute and United Nations Environment Program; the Oregon Agricultural Experiment Station and Forest Research Laboratory, Oregon State University.

Appendix: Abundance data

The abundances of dominant invertebrates in hurricane recovery plots (Table 28A.1) and in the wet and dry season (Table 28A.2) are listed for sites in 1999. Tables 28A.3, 28A.4 and 28A.5, respectively, list athropod abundances on *Cecropia schreberiana*, *Manilkara bidentata* and *Sloanea berteriana* during a 9 year period following Hurricane Hugo in 1989. A major drought occurred in 1994 and Hurricane Georges occurred after the 1998 season and broke many *Cecropia* stems.

Table 28A.1. *Abundances of dominant invertebrates collected during the day and night from* Cecropia schreberiana *and* Sloanea berteriana *in hurricane recovery plots at El Verde Field Station in Puerto Rico during 1999*

Taxon	Abundances[a] on *Cecropia* sp.		Abundances[a] on *Sloanea* sp.	
	Day	Night	Day	Night
Folivores				
Cyrtoxipha gundlachi (Gryllacrididae)	0	14 (14)	70 (23)	17 (10)
Chrysomelidae	0	20 (20)	0	0
Arctiidae	0	49 (38)	18 (9)	7 (7)
Sap suckers				
Protopulvinaria pyriformis	0	10 (10)	25 (21)	0
Pseudococcidae spp.	66 (66)	88 (44)	0	7 (7)
Derbidae	0	26 (15)	8 (5)	0
Ladella sp. (Tropiduchidae)	0	0	7 (7)	7 (7)
Flatidae	0	0	15 (15)	13 (13)
Itacoris sp. (Miridae)	35 (21)	120 (51)	0	0
Predators				
Wasmania sp. (Formicidae)	3 (3)	91 (42)	37 (20)	22 (13)
Coccinellidae	23 (23)	47 (47)	0	0
Hymenoptera	10 (10)	39 (39)	11 (11)	7 (7)
Theridiidae (Araneae)	0	0	19 (12)	0
Salticidae (Araneae)	13 (13)	26 (15)	0	29 (19)
Thomisidae (Araneae)	37 (13)	0	5 (5)	40 (25)
Araneidae (Araneae)	83 (60)	25 (25)	0	24 (8)
Detritivores				
Blattidae	0	36 (22)	18 (14)	42 (19)
Collembola	110 (45)	140 (110)	14 (9)	51 (32)
Psocoptera	150 (50)	10 (10)	12 (8)	27 (19)
Oribatida (Acari)	25 (25)	0	21 (21)	24 (24)
Miscellaneous				
Chironomidae	0	0	0	24 (15)
Culicidae	0	15 (15)	0	0
Mollusca	0	20 (20)	0	0

[a]No. per kg dry foliage averaged over season for each tree and then averaged over replicate trees; given as mean (SE).

Table 28A.2. *Abundances of dominant invertebrates by season on* Brosimum utile, Cecropia *spp.*, Dacryodes excelsa, Manilkara *spp.*, Sloanea berteriana, *and* Virola *spp. at El Verde Field Station in Puerto Rico and the Fort Sherman Canopy Crane site in Panama during June–July 1998 (wet season) and March–April 1999 (dry season)*

Taxon	Puerto Rico abundances[a] on								Panama abundances[a] on							
	Cecropia		*Manilkara*		*Dacryodes*		*Sloanea*		*Cecropia*		*Manilkara*		*Brosimum*		*Virola*	
	Wet	Dry	Wet	Dry	Wet	Dry	Wet	Dry	Wet	Dry	Wet	Dry	Wet	Dry	Wet	Dry
Folivores																
Gryllacrididae spp.	0	0	74 (42)	21 (16)	28 (18)	22 (17)	58 (30)	97 (39)	0	0	0	0	17 (17)	0	0	0
Arctiidae	13 (13)	0	6 (6)	8 (8)	0	8 (8)	0	0	0	0	0	0	0	0	0	0
Tortricidae	0	0	0	0	0	0	0	0	0	0	0	5 (5)	0	0	0	0
Sap suckers																
Ceroplastes rubens	31 (31)	0	140 (110)	92 (62)	4 (4)	9 (9)	0	0	0	0	0	0	0	0	0	0
Coccus acutissimus	0	0	0	10 (10)	62 (41)	45 (29)	0	0	0	0	10 (10)	0	110 (110)	4 (4)	350 (310)	0
Coccus species 2	0	0	0	0	0	0	0	0	0	0	0	0	100 (36)	10 (10)	0	0
Protopulvinaria pyriformis	7 (7)	0	0	17 (17)	16 (10)	36 (13)	0	0	0	0	0	0	0	0	0	0
Pseudococcidae sp.	7 (7)	130 (130)	0	0	0	0	0	0	0	0	0	0	0	0	0	0
Vinsonia stellata	0	0	24 (17)	23 (23)	0	0	0	0	0	0	20 (20)	0	0	0	27 (27)	0
Psyllidae	0	0	0	0	0	0	0	0	0	0	0	0	0	0	65 (49)	100 (100)
Itacoris sp.	46 (33)	69 (42)	0	0	0	0	0	0	0	0	0	0	0	0	0	0
Predators																
Azteca ant	0	0	0	0	0	0	0	0	260 (89)	0	0	0	0	0	0	0
Wasmania ant	20 (20)	6 (6)	36 (36)	0	10 (6)	0	0	0	0	0	0	0	0	0	0	0
Formica ant	0	0	0	0	0	0	0	0	0	30 (30)	0	51 (51)	140 (87)	4 (4)	0	0
Parasitic Hymenoptera	7 (7)	0	16 (16)	0	22 (11)	5 (5)	0	9 (9)	0	0	9 (9)	0	0	8 (8)	11 (11)	0
Coccinellidae	50 (50)	0	0	17 (11)	5 (5)	8 (8)	0	20 (13)	0	0	0	0	0	0	0	0
Salticidae	0	26 (26)	0	0	5 (5)	41 (16)	8 (8)	33 (19)	0	0	6 (6)	0	5 (5)	0	11 (11)	0
Thomisidae	0	73 (28)	0	0	0	8 (8)	0	0	0	0	0	0	0	0	0	8 (8)
Araneidae	22 (22)	10 (10)	6 (6)	27 (27)	6 (6)	23 (15)	0	28 (14)	56 (56)	0	0	0	5 (5)	0	0	0
Detritivores																
Blattidae	0	0	23 (15)	0	0	0	7 (7)	21 (13)	0	0	0	0	26 (26)	4 (4)	0	0
Collembola	25 (25)	200 (80)	0	9 (9)	0	0	0	34 (34)	0	0	0	0	9 (9)	0	56 (56)	0
Psocoptera	0	290 (100)	0	8 (8)	0	5 (5)	67 (67)	47 (23)	19 (19)	0	27 (16)	0	9 (9)	0	0	0
Oribatida	46 (46)	0	0	9 (9)	0	0	8 (8)	19 (19)	0	0	270 (190)	0	0	0	0	0

[a] No. per kg dry foliage averaged over replicate trees.

Table 28A.3. *Arthropod abundances on* Cecropia schreberiana *during development from seedlings to trees over a 9 year period following Hurricane Hugo (1989) in Puerto Rico. A major drought occurred in 1994 and Hurricane Georges occurred after the 1998 season and broke many* Cecropia *stems*

Taxon	Abundance[a]						
	1991	1992	1994	1995	1997	1998	1999
Folivores							
Cyrtoxipha gundlachi	1 (1)	34 (17)	26 (19)	25 (22)	46 (22)	3 (3)	0
Chrysomelidae	14 (11)	9 (6)	0	0	0	0	0
Arctiidae	24 (24)	18 (10)	28 (14)	17 (14)	0	19 (16)	0
Geometridae	18 (18)	0	0	14 (14)	12 (9)	0	0
Sap suckers							
Ceroplastes rubens	1 (1)	0	1 (1)	0	0	16 (16)	0
Protopulvinaria pyriformis	0	1 (1)	5 (5)	21 (21)	2 (2)	3 (3)	0
Pseudococcidae sp.	0	23 (20)	0	17 (15)	45 (32)	3 (3)	66 (66)
Ladella sp.	3 (3)	0	8 (7)	10 (5)	14 (12)	13 (9)	0
Sibovea coffeacola	10 (10)	2 (2)	0	0	0	19 (19)	5 (5)
Itacoris sp.	145 (50)	23 (9)	33 (15)	70 (23)	19 (9)	26 (16)	34 (21)
Predators							
Wasmania sp.	57 (20)	177 (82)	14 (12)	25 (17)	31 (21)	21 (19)	3 (3)
Parasitic Hymenoptera	3 (3)	0	9 (7)	5 (5)	0	3 (3)	10 (10)
Coccinellidae	3 (3)	5 (3)	16 (8)	4 (4)	7 (5)	30 (25)	23 (23)
Thomisidae	1 (1)	0	11 (11)	0	0	5 (5)	37 (14)
Salticidae	0	0	16 (13)	19 (13)	0	28 (19)	13 (13)
Oxyopidae	5 (5)	6 (3)	16 (8)	9 (6)	5 (5)	0	0
Theridiidae	2 (2)	8 (6)	9 (6)	14 (12)	0	0	0
Araneidae	69 (32)	49 (18)	27 (15)	37 (19)	0	14 (11)	83 (60)
Detritivores							
Blattidae	0	0	17 (9)	29 (29)	0	6 (4)	0
Psocoptera	3 (3)	9 (6)	87 (23)	62 (30)	17 (12)	77 (29)	145 (50)
Collembola	0	9 (6)	37 (21)	128 (83)	7 (7)	37 (24)	112 (45)
Corylophidae	0	0	0	5 (5)	0	0	0
Euglenidae	0	0	8 (8)	0	0	3 (3)	3 (3)
Melandryidae	0	0	5 (5)	2 (2)	0	0	0
Oribatida	0	0	8 (5)	0	18 (18)	23 (23)	25 (25)
Miscellaneous							
Chironomidae	19 (19)	0	1 (1)	0	7 (7)	0	0
Mollusca	0	15 (15)	7 (5)	0	0	0	0

[a] No. per kg dry foliage averaged over season for each tree and then averaged over replicate trees; given as mean (SE).

Table 28A.4. *Arthropod abundances on* Manilkara bidentata *over a 9 year period following Hurricane Hugo (1989) in Puerto Rico*

Taxon	Abundance[a]						
	1991	1992	1994	1995	1997	1998	1999
Folivores							
Cyrtoxipha gundlachi	21 (7)	27 (8)	9 (5)	50 (25)	30 (17)	59 (26)	21 (8)
Chrysomelidae	12 (8)	0	2 (2)	0	0	0	0
Arctiidae	5 (3)	0	0	0	0	3 (3)	4 (4)
Geometridae	0	0	0	6 (6)	0	0	0
Sap suckers							
Ceroplastes rubens	270 (33)	110 (50)	15 (4)	5 (3)	28 (10)	75 (55)	72 (28)
Coccus acutissimus	14 (6)	3 (3)	2 (2)	3 (3)	0	0	11 (5)
Protopulvinaria pyriformis	0	0	0	0	0	0	14 (9)
Vinsonia stellata	200 (42)	51 (33)	21 (11)	76 (50)	17 (9)	57 (23)	38 (17)
Derbidae	0	0	5 (5)	0	0	0	0
Ladella sp.	4 (3)	0	2 (2)	5 (5)	18 (13)	0	0
Flatidae	13 (13)	0	0	0	0	0	0
Cixiidae sp. 1	8 (6)	10 (5)	5 (5)	0	4 (4)	0	0
Predators							
Wasmania sp.	38 (32)	15 (9)	2 (2)	10 (7)	4 (4)	18 (18)	0
Parasitic Hymenoptera	8 (3)	6 (4)	6 (4)	31 (20)	3 (3)	17 (8)	0
Coccinellidae	1 (1)	2 (2)	0	5 (5)	4 (4)	0	9 (5)
Thomisidae	6 (6)	3 (2)	0	0	0	12 (9)	0
Salticidae	7 (4)	15 (15)	9 (7)	12 (8)	0	0	6 (4)
Oxyopidae	5 (2)	1 (1)	0	5 (5)	4 (4)	5 (5)	4 (4)
Theridiidae	15 (9)	21 (21)	23 (17)	27 (12)	20 (16)	9 (6)	7 (7)
Araneidae	0	5 (4)	12 (5)	26 (9)	0	3 (3)	13 (9)
Detritivores							
Blattidae	5 (3)	3 (3)	6 (4)	5 (5)	0	12 (8)	8 (8)
Psocoptera	4 (3)	6 (4)	34 (11)	31 (14)	16 (16)	20 (10)	4 (4)
Collembola	3 (2)	0	5 (2)	10 (6)	0	5 (5)	5 (5)
Euglenidae	0	0	1 (1)	5 (5)	0	9 (9)	0
Melandryidae	0	0	0	11 (11)	0	7 (7)	0
Oribatida	9 (5)	2 (2)	0	4 (4)	13 (13)	0	5 (5)
Miscellaneous							
Culicidae	1 (1)	0	2 (2)	0	0	0	0
Mollusca	1 (1)	0	0	2 (2)	3 (3)	7 (7)	0

[a] No. per kg dry foliage averaged over season for each tree and then averaged over replicate trees; given as mean (SE).

Table 28A.5. *Arthropod abundances on* Sloanea berteriana *over a 9 year period following Hurricane Hugo (1989) in Puerto Rico*

Taxon	Abundances[a]						
	1991	1992	1994	1995	1997	1998	1999
Folivores							
Cyrtoxipha gundlachi	36 (4)	51 (21)	32 (13)	60 (21)	28 (22)	51 (20)	62 (22)
Chrysomelidae	2 (2)	2 (2)	20 (12)	0	0	4 (4)	0
Arctiidae	7 (3)	0	7 (4)	6 (6)	0	0	10 (7)
Geometridae	0	7 (7)	0	0	0	0	0
Sap suckers							
Ceroplastes rubens	0	2 (2)	0	0	0	0	0
Coccus acutissimus	2 (2)	3 (3)	5 (5)	0	3 (3)	0	5 (5)
Protopulvinaria pyriformis	7 (5)	0	0	0	0	0	0
Derbidae	2 (2)	8 (6)	0	0	0	0	4 (4)
Ladella sp.	6 (3)	6 (4)	10 (7)	18 (12)	4 (4)	0	2 (2)
Flatidae	3 (2)	1 (1)	0	9 (9)	0	11 (11)	0
Cixiidae sp. 1	7 (6)	8 (7)	5 (3)	7 (7)	0	0	4 (4)
Predators							
Wasmania sp.	21 (18)	13 (4)	6 (6)	0	2 (2)	4 (4)	0
Parasitic Hymenoptera	5 (3)	6 (4)	5 (4)	3 (3)	9 (6)	4 (4)	13 (9)
Coccinellidae	0	3 (3)	0	0	0	0	15 (10)
Thomisidae	24 (15)	0	7 (6)	3 (3)	5 (5)	0	24 (13)
Salticidae	12 (3)	28 (12)	19 (7)	0	4 (4)	8 (8)	21 (12)
Oxyopidae	10 (6)	14 (5)	8 (5)	0	6 (6)	0	0
Theridiidae	16 (9)	8 (4)	12 (3)	0	0	11 (11)	4 (4)
Araneidae	4 (3)	8 (4)	12 (9)	10 (5)	2 (2)	0	19 (7)
Detritivores							
Blattidae	21 (11)	19 (13)	18 (8)	7 (5)	4 (4)	3 (3)	28 (14)
Psocoptera	5 (3)	5 (3)	82 (40)	43 (15)	45 (23)	79 (31)	24 (12)
Collembola	2 (2)	2 (2)	39 (15)	30 (13)	100 (100)	9 (9)	20 (17)
Corylophidae	1 (1)	0	4 (3)	10 (7)	4 (4)	0	0
Euglenidae	0	0	0	4 (4)	0	4 (4)	0
Melandryidae	0	0	3 (3)	7 (7)	0	0	0
Oribatida	11 (5)	11 (7)	15 (8)	6 (6)	0	4 (4)	19 (19)
Miscellaneous							
Chironomidae	7 (6)	4 (3)	0	3 (3)	0	12 (12)	0
Culicidae	0	2 (2)	6 (4)	8 (8)	0	0	0

[a]No. per kg dry foliage averaged over season for each tree and then averaged over replicate trees; given as mean (SE).

29

Tree relatedness and the similarity of insect assemblages: pushing the limits?

Roger L. Kitching, Karen L. Hurley and Lukman Thalib

ABSTRACT

The similarity of insect fauna of tropical trees should reflect not only the coevolved specializations of phytophagous insects and the associated higher trophic levels but also the degree of relatedness of the trees themselves. We have explored these ideas using top-down approaches to the determination of crown faunas in a design that contrasts the faunas of a range of tree species of known intertaxonomic distance based on recent molecular phylogenies. We have focussed on assemblages of Coleoptera and on the phytophagous subsets of these. Target trees have included species of *Araucaria* (Araucariaceae), *Eupomatia* (Eupomatiaceae), *Wilkiea, Doryphora* (Monimiaceae), *Brachychiton, Argyrodendron* (Malvaceae), *Cryptocarya, Neolitsea* (Lauraceae), *Acronynchia, Acradenia* (Rutaceae) and *Orites* (Proteaceae). Understorey trees have been used in order to obtain samples from 'pure' crowns rather than those 'contaminated' by vines, epiphytes or adjacent trees. Our results show a clear decline in similarity as the inter-tree taxonomic distance increases, whether measured on the basis of species presence or absence or by incorporating measures of relative abundance. Variances of these measures around the slope of decline are large, as chance effects play greater and greater roles as insect samples are taxonomically subdivided progressively (and, accordingly, get smaller and smaller). We discuss the results and the shortcomings of the approach in general. We review recent trends in work on host-plant specificity. We identify alternatives but note the increasing demands both in terms of sample size and cost associated with more precise and useful measures.

INTRODUCTION

Attempts to estimate the size and composition of an insect fauna within a forest canopy will almost always involve extrapolation from the knowledge of the fauna of a few species of trees within that forest to estimate the overall fauna (Erwin & Scott, 1980; Erwin, 1982; Stork, 1988; May, 1990; Hammond, 1992; Basset *et al.*, 1996a; Ødegaard, 2000c). This general statement is predicated upon the fact that there are many fewer species of plant than animal and knowledge of the diversity and taxonomy of the angiosperms, at least, is very much more complete than that of the arthropods – which make up the vast preponderance of any crown fauna. This proposition has special weight when we consider highly diverse and poorly known ecosystems such as the rainforests of the world (May, 1990).

The canonical approach in this regard involves the intensive collection of the insects associated with the crowns of one or a few tree species, the determination of a proportion of that fauna which may be considered endemic to that tree species and the multiplication of that endemic number by the number of tree taxa in the forest, region or even the entire Earth.

There are many pitfalls associated with extrapolations of this kind.

Perhaps the most obvious is that of assuming that any endemic insect species has only one host tree (Basset, 1992a; Mawdsley & Stork, 1997; Ødegaard, 2000a; Ch. 5). This is clearly not the case, but trying to reduce all the varying degrees of oligophagy and polyphagy to a single multiplier is very challenging. Approaches through the use and application of complex statistical distributions of the degree of host association are promising (Diserud & Ødegaard, 2000) but demanding of biological information that is, at best, hard to come by.

A second problem in using a tree-based extrapolation is that the many species of herbs, scandents, epiphytes and saprophagous plants, all of which will carry more or less specific arthropod faunas, are not included within the final estimate (Benzing, 1983; Gentry & Dodson, 1987; Ødegaard, 2000a). We generally have no idea of how great an error this creates, but even allowing for the differentials in size and structural complexity across plant types, it must be large. We note also that guild studies of canopy Coleoptera show large proportions of fungivorous species, underlining the importance of the fungus-related guilds (Stork, 1987a; Kitching *et al.*, 1994; Hammond *et al.*, 1996).

Lastly we question the treatment of tree taxa (or indeed of any other plants) as independent entities. The trees of a tropical forest ecosystem represent a set of interrelated species where we should, *a priori*, expect more similar faunas on trees of the same genus, within the same family and so on (see, *inter alia*, Stork, 1987a; Basset *et al.*, 1997a).

We set out to test the hypothesis underlying this third criticism: that arthropod assemblages on the crowns of particular tree species will show levels of similarity with those associated with other tree species that reflect the taxonomic relatedness of the tree species being compared. We report the results of these tests here.

In making these carefully structured observations, we also came to an understanding of the likely limitations of approaches to the assessment of crown faunas and host specificity that use remote 'knockdown' approaches to assemblage sampling. These limitations and their implications are also discussed here.

METHODS

Our initial design and set of observations in 1998 involved the sampling of at least eight individuals of nine species of tree (Table 29.1), each chosen to represent three sets of three, each set comprising two congeners and one other within each of three angiosperm families. The endemic angiosperm *Eupomatia laurina* (Eupomatiaceae) and the gymnosperm *Araucaria cunninghamii* (Araucariaceae) were added to the three sets of trees sampled to act as 'out-groups' for the subsequent analyses. An index of relatedness of these trees was calculated using the phylogenies in Soltis *et al.* (2000) with additional detail on the Gymnospermae from Hill (1998). In order to obtain this index, we counted the number of branching points occurring between the stems of the families involved, as depicted in the phylogenies. We added two details in addition to this. We added a fixed score to the angiosperm/gymnosperm differences so that any angiosperm/gymnosperm pair was regarded as at least as distant from each other as the most disparate angiosperm/angiosperm pair. This was in recognition of the depth and age of separation of these two higher taxa. The family level phylogeny we used is summarized in Fig. 29.1.

Second, in the absence of detailed, within-family phylogenies, we regarded two species within a single genus as having a relatedness factor of 1, and species in separate genera within a single family as having a value of 3. This huge simplification, regarding the within-family phylogenies as merely a shallow two-branching tree, will, in fact, have little impact on the index values used as the number of between-family steps always greatly exceeds the within-family steps.

When this first analysis of the arthropod faunas proved insensitive (see below), we carried out a second set of comparisons in which we sampled a further set of trees in order to obtain a second separate dataset with larger numbers of pairwise species-to-species comparisons at a range of taxonomic ranks (Table 29.1). In this second set of samples, six trees of each selected species were sampled (except for *Acronychia pubescens*, for which only five individuals could be located). Wherever possible we chose individuals from the tree species that had been sampled in the first analysis. Additional or alternative tree species were selected in order to obtain a sufficient number of congeneric and confamilial pairs.

In each instance, we sampled understorey trees without attached epiphytes or vines. If surrounding plants threatened to overlap the sampled area, they were separated from the specimen using guide ropes. We were also very careful to minimize disturbance to the specimen plant before spraying. This method enabled us to obtain insect samples that we could associate, unequivocally, with particular tree crowns. All individuals sampled were growing under similar habitat conditions. To standardize the data, we estimated the crown volume sampled in each case by measuring the principal dimensions of the crown and calculating the crown volumes by approximating the particular crowns to appropriate geometrical solids in each case (cylinders, hemispheres, ellipses, cones, etc.).

Crowns were sampled using standard pyrethrum knockdown techniques (Kitching *et al.*, 2000). We sprayed each crown for 30 seconds using a Stihl™ backpack mister using a mixture of natural pyrethrins (Pyrethrins 2EL™) with piperonyl butoxide. The

Table 29.1. *The tree species studied*

Subdivision	Order	Family	Genus	Species	Acronym	Number of individuals sampled 1998	Number of individuals sampled 2000
Gymnospermae	Coniferae	Araucariaceae	*Araucaria*	*cunninghamii* Aiton ex D. Don	Ac	10	0
Angiospermae	Magnoliales	Eupomatiaceae	*Eupomatia*	*laurina* R. Br.	El	10	0
	Laurales	Lauraceae	*Neolitsea*	*australiensis* Kosterm	Na	10	6
			Neolitsea	*dealbata* (R. Br.) Merr.	Nd	10	6
			Cryptocarya	*obovata* R. Br.	Co	10	0
		Monimiaceae	*Doryphora*	*sassafras* Endl.	Ds	8	6
			Wilkiea	*austroqueenslandica* Domin	Wa	8	6
			Wilkiea	*huegeliana* (Tul.) DC.	Wh	10	6
	Malvales	Malvaceae	*Argyrodendron*	*actinophyllum* (Bailey)	Aa	10	0
			Argyrodendron	*trifoliolatum* F. Muell.	At	10	0
			Brachychiton	*acerifolius* F. Muell.	Ba	10	0
	Proteales	Proteaceae	*Orites*	*excelsa* R. Br.	Oe	0	6
	Sapindales	Rutaceae	*Acradenia*	*euodiiformis* (F. Muell.)	Ae	0	6
			Acronychia	*pubescens* Bailey (C. White)	Ap	0	5
			Acronychia	*suberosa* C. White	As	0	6

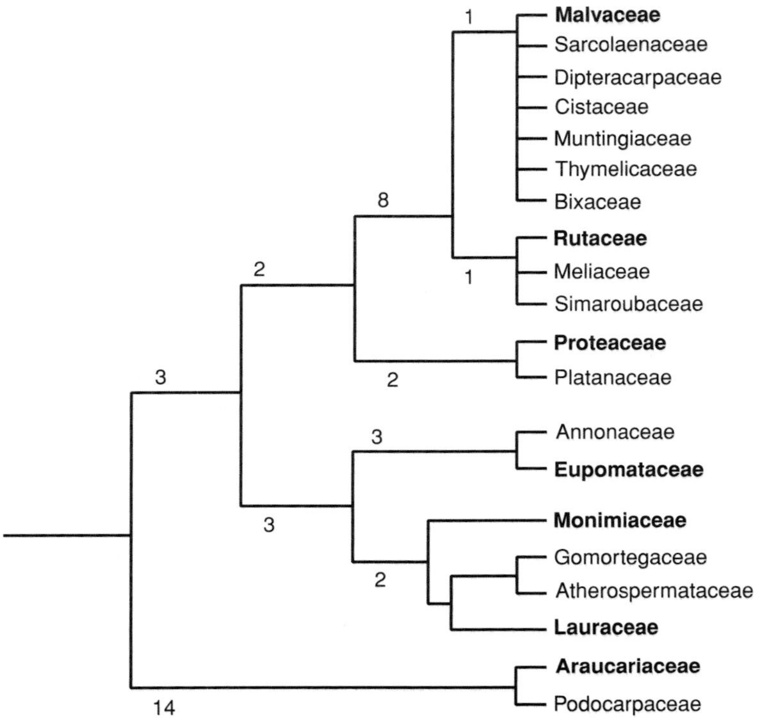

Fig. 29.1. The phylogenetic relationships among the plant families sampled in this study based on the molecular analyses of Soltis *et al.* (2000). The figures adjacent to particular branches in the tree indicate the number of branch junctions recognized by Soltis *et al.* but omitted from the figure.

sprayer was adjusted to maximize projection whilst maintaining a fine mist, and this enabled the entire crown of each tree with a medium sized crown to be sprayed. Samples were collected after a 1 hour period in four to five 0.5 m^2 collecting funnels suspended around the base of the tree before spraying. These collecting funnels covered around 80 to 90% of the area directly under the canopy of each tree. Each tree was shaken before the funnels were gently brushed down and the sample removed.

The trees sampled were all located at Green Mountains, Lamington National Park (28° 13′ S, 153° 07′ E, 900 m above sea level). All trees were located within a subtropical complex notophyll vine forest (Webb *et al.*, 1984). The flora of the region is described by McDonald and Thomas (1990) and a detailed assessment of the vegetation of a 1 ha area of the forest adjacent to the sampling sites is presented by Laidlaw *et al.* (2000). The region has a marked wet season, with peak average rainfall of approximately 500 mm in February, and a winter minimum of approximately 100 mm in August. Mean maximum temperatures range from 16 °C in July to 25 °C in January, with mean minima of 8 °C in July to 16 °C in January.

The first set of samples was taken in December 1998 and the second set in February 2000. The insect samples were sorted to major taxa (generally to order) and the Coleoptera were subsequently dry mounted and sorted to family and morphospecies. Results were analysed graphically. Multivariate analysis of the results from the first set of observations was carried out using multivariate scaling techniques in order to detect any groupings of the tree faunas in *n*-space. The similarities between the assemblages of Coleoptera from all pairs of trees were calculated using both Sørensen and Morisita–Horn indices (Magurran, 1988). Sørensen indices, based only on the presence or absence of species, were calculated as:

$$SI_{ij} = 2c/a + b \quad (29.1)$$

Where c is the number of species shared by samples i and j, a is the total number of species in sample i and b is the total number of species in sample j.

The Morisita–Horn index, which takes into account the relative abundances of species, is calculated (following Magurran, 1988) as:

$$MH_{ij} = \frac{2\sum(an_i \times bn_i)}{(da + db)(aN \times bN)} \qquad (29.2)$$

where aN is the abundance of individuals of all species in tree species a; bN is the abundance of individuals of all species in tree species b; an_i is the abundance of individuals of species i in tree species a; bn_i is the abundance of individuals of species i in tree species b; and

$$db = \frac{\sum bn_i^2}{bN^2} \qquad (29.3)$$

Abundance levels in each case were adjusted to reflect the approximate volume of foliage sampled in each case. The values obtained for both these indices of similarity were calculated on all beetle morphospecies recognized and, separately, on the herbivorous species only.

In an attempt to obtain larger samples and to increase the number of tree/tree comparisons, the second set of samples of an intersecting set of tree species was taken in February 2000 (Table 29.1). Again, understorey trees were targeted but larger, denser canopies were selected deliberately to try to increase the size of the catches. Coleoptera were extracted from these samples, mounted and sorted to families and morphospecies as before. Overall similarity between the taxonomic structure of the Coleoptera obtained was estimated using correlation analyses, and then Sørensen and Morisita–Horn indices were calculated as before.

RESULTS

A total of 6198 arthropods from 106 trees were sampled in 1998 and 3986 from 53 trees in 2000. Of these there were 681 adult beetles in the 1998 samples and 594 in the 2000 samples. The distribution of the beetle species by family across tree species is summarized in Table 29.2. The general taxonomic structure of the two separate sets of samples was similar, with a Pearson rank correlation value of 0.72 comparing numbers of individuals in each of the coleopteran families encountered. Four of the five most abundant families in each of the two sets of samples were common (Staphylinidae, Ptiliidae, Curculionidae, Corylophidae). Thirty-three families of beetles were encountered in total, 27 in 1998 and 25 in 2000. Nineteen of these families were common to both sets of samples and the remaining 14 were represented by six or less individuals in each case. The distribution of numbers of individuals across families is summarized in Fig. 29.2.

The key relationship between the inter-tree similarity and the inter-tree taxonomic distance is shown in Fig. 29.3. Figure 29.3**a**,**b** shows the results for the first set of observations calculated using the Sørensen and Morisita–Horn indices. Figure 29.3**c**,**d** shows the comparable plots from the second set of observations. Figure 29.3**a**,**b** shows significant declines in similarity between faunas as the taxonomic distance between the trees increases (Sørensen index $r = -0.519$, $p < 0.001$; Morisita–Horn index $r = -0.345$, $0.001 < p < 0.01$). Probability values for the correlation coefficients, r, relating the similarities of the beetle assemblage to the taxonomic distance between the trees was calculated as a Mantel test in which the actual values of each index were randomized 1000 times and the r values recalculated against the tree relatedness values each time. We note (and return in the Discussion to) the wide variance of values about the trend line. The plots for the 2000 values of both the Sørensen and Morisita–Horn indices showed no significant decline with increasing inter-tree taxonomic distance.

When the analyses were repeated using only unequivocally herbivorous families, the negative trend in the 1998 data was still apparent but the relationship was weaker. Not surprisingly, no trend was apparent in this subset of data for the smaller 2000 samples. Figure 29.4 shows the relative proportions of unequivocally herbivorous species, compared with the entire sample of Coleoptera, on each species of tree for the set of observations from 1998 and 2000 (Fig. 29.4**a** and Fig. 29.4**b**, respectively). Although no statistical credence can be given to particular differences, we note the substantial tree species-to-species differences in the ratios of both numbers of individuals and of species. In the 1998 dataset, the proportion of species that is herbivorous ranges from 80% in *Doryphora sassafras* down to less than 20% in *Wilkiea austroqueenslandica* and *Brachychiton acerifolius*. In terms of numbers of individual beetles, about 50% were herbivorous on *A. cunninghamii*, yet as few as 14% on *Argyrodendron actinophyllum*. The 2000 data showed a similar variety. The numbers of herbivorous species varied from 35% in *Acradenia euodiiformis* to as little as 10% in *Neolitsea*

Table 29.2. *The number of beetle species, organized by families, encountered in each species of tree sampled*[a]

	Ac 98	Eu 00	Na		Nd		Co 98	Ds		Wa		Wh		Aa 98	At 98	Ba 98	Oe 00	Ae 00	Ap 00	As 00	Total Coleoptera	
			98	00	98	00		98	00	98	00	98	00								98	00
Aderidae	2	–	1	–	1	–	1	–	–	–	–	–	–	–	–	–	–	–	–	–	5	0
Anobiidae	–	–	–	–	–	–	–	–	–	–	–	–	–	–	–	–	–	–	1	–	0	1
Anthicidae	1	–	–	–	–	–	–	–	–	–	–	–	–	–	–	–	–	–	–	–	1	0
Anthribidae	–	–	–	–	1	–	–	–	–	–	–	–	–	–	–	–	–	–	–	–	1	0
Apionidae	–	–	–	–	–	–	–	1	–	–	1	–	2	–	–	–	–	–	1	–	1	4
Buprestidae	–	–	–	1	–	–	–	1	–	–	–	–	–	–	–	–	–	–	–	–	1	1
Cantharidae	1	2	1	3	1	3	1	–	1	2	5	1	4	1	2	1	2	2	2	3	11	27
Carabidae	–	–	–	2	–	1	–	–	–	–	1	–	1	–	1	–	1	–	–	1	1	7
Cerambycidae	–	–	–	–	–	–	1	–	–	–	1	1	1	–	1	–	–	–	–	3	3	5
Chrysomelidae	6	5	3	–	4	2	3	4	–	2	2	3	3	3	–	–	3	6	3	1	28	25
Cleridae	–	–	–	–	–	–	–	–	–	–	–	–	–	–	–	–	–	–	–	1	0	1
Coccinellidae	1	1	1	2	–	3	–	3	2	–	3	1	4	–	–	–	6	5	–	8	6	34
Corylophidae	3	2	2	7	1	1	3	2	1	–	9	2	5	2	2	1	3	–	2	5	18	35
Curculionidae	9	3	2	3	1	4	1	1	1	1	16	4	7	2	8	5	8	8	5	10	34	65
Elateridae	–	1	–	–	–	–	–	–	–	–	–	–	–	1	–	–	1	–	–	1	1	3
Histeridae	–	–	–	–	–	–	–	–	–	1	–	–	–	–	–	–	–	–	–	–	1	0
Laemophloeidae	–	–	–	–	–	1	–	–	–	–	1	–	1	–	–	–	–	–	–	–	0	3
Languriidae	–	–	–	–	–	–	–	–	–	–	–	–	–	–	–	–	–	–	1	–	0	1
Latridiidae	2	1	–	1	–	2	–	–	1	–	4	–	1	–	3	–	1	1	1	2	5	15
Leiodidae	–	–	–	–	–	–	1	–	–	–	–	–	–	–	–	–	–	–	–	–	1	0
Limnichidae	–	–	–	–	–	–	–	–	1	–	–	–	–	–	–	–	1	–	–	1	0	3
Mordellidae	1	–	–	–	–	–	–	1	–	–	–	–	–	–	–	1	–	1	–	–	3	1
Nitidulidae	–	–	–	–	–	–	–	–	–	–	–	–	–	1	–	–	–	–	–	–	1	0
Phalacridae	1	–	–	–	1	2	–	–	–	–	4	–	1	–	1	1	–	6	1	4	4	18
Platypodinae	–	–	–	–	–	–	–	–	–	–	1	–	–	–	–	–	–	–	–	–	0	1
Pselaphidae	–	1	1	7	2	3	1	–	–	1	6	2	1	–	2	1	–	1	2	2	10	23
Ptiliidae	1	4	4	7	4	2	3	2	–	4	12	2	–	3	3	4	–	5	6	10	30	46
Scarabaeidae	–	1	–	–	–	–	1	–	–	–	–	1	–	–	1	–	–	–	–	2	3	3
Scolytinae	3	1	–	–	–	2	–	–	–	1	–	1	1	–	–	1	–	–	1	–	6	5
Scydmaenidae	1	–	–	–	–	1	1	–	–	–	3	–	1	–	1	–	–	–	2	–	3	7
Silvanidae	–	–	–	–	–	–	–	–	–	1	–	1	–	1	–	–	–	–	–	–	3	0
Staphylinidae	8	8	4	11	11	10	8	1	2	4	10	6	9	6	13	10	4	20	11	18	71	103
Zopheridae	–	–	–	–	–	–	–	–	–	–	–	–	–	–	2	–	–	–	–	–	2	0
Totals	40	30	19	44	27	37	25	16	9	17	79	25	42	20	40	25	30	55	39	72	–	–

[a] A total of 33 Coleoptera families were encountered in the two sets; the first set was collected in 1998 (98) and the second in 2000 (00). The tree species are indicated by two letter codes: Ac, *Araucaria cunninghamii*; Eu, *Eupomatia laurina*; Na, *Neolitsea australiensis*; Nd, *Neolitsea dealbata*; Co, *Cryptocarya obovata*; Ds, *Doryphora sassafras*; Wa, *Wilkiea austroqueenslandica*; Wh, *Wilkiea huegeliana*; Aa, *Argyrodendron actinophyllum*; At, *Argyrodendron trifoliolatum*; Ba, *Brachychiton acerifolius*; Oe, *Orites excelsa*; Ae, *Acradenia euodiiformis*; Ap, *Acronychia pubescens*; As, *Acronychia suberosa*.

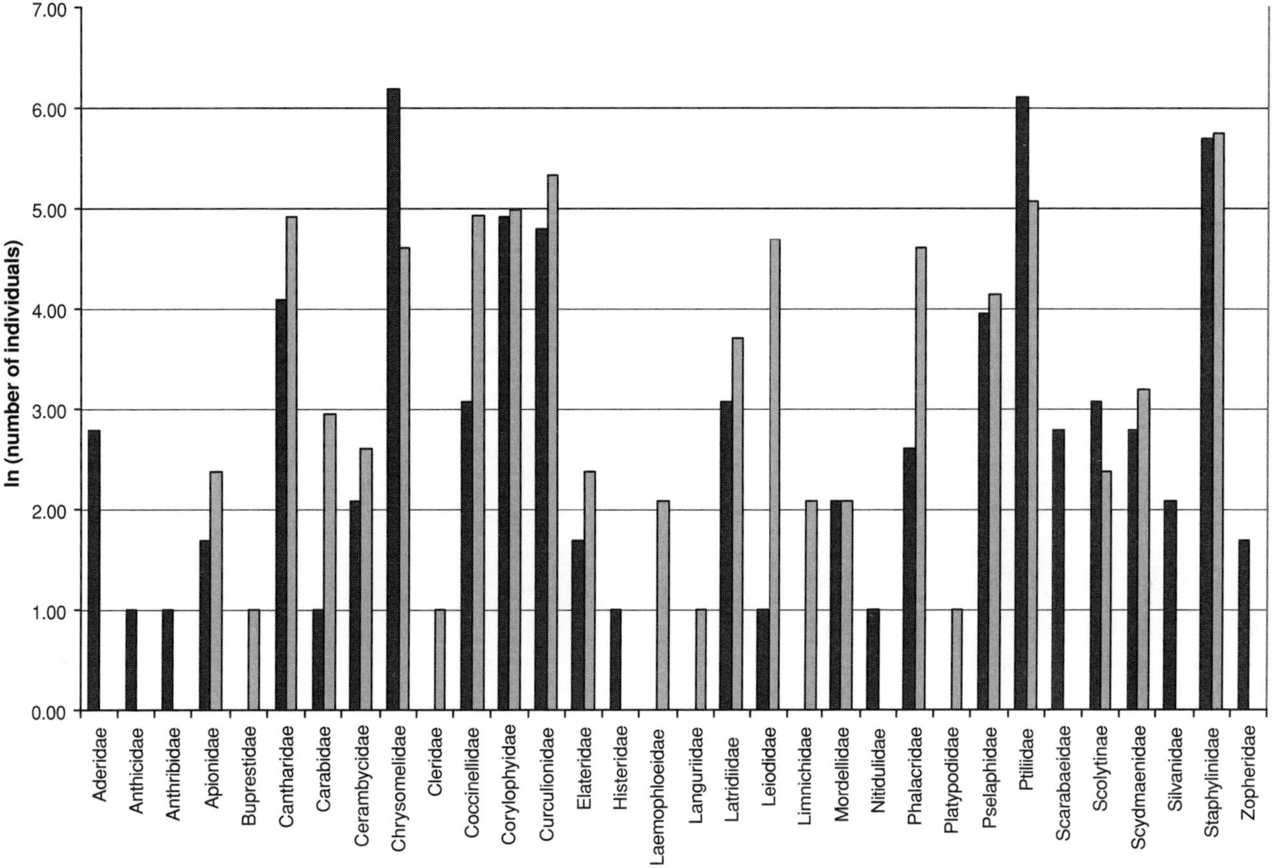

Fig. 29.2. The numbers of individuals of each family of Coleoptera encountered in our studies. The black bars are the 1998 data, the grey bars the 2000 data.

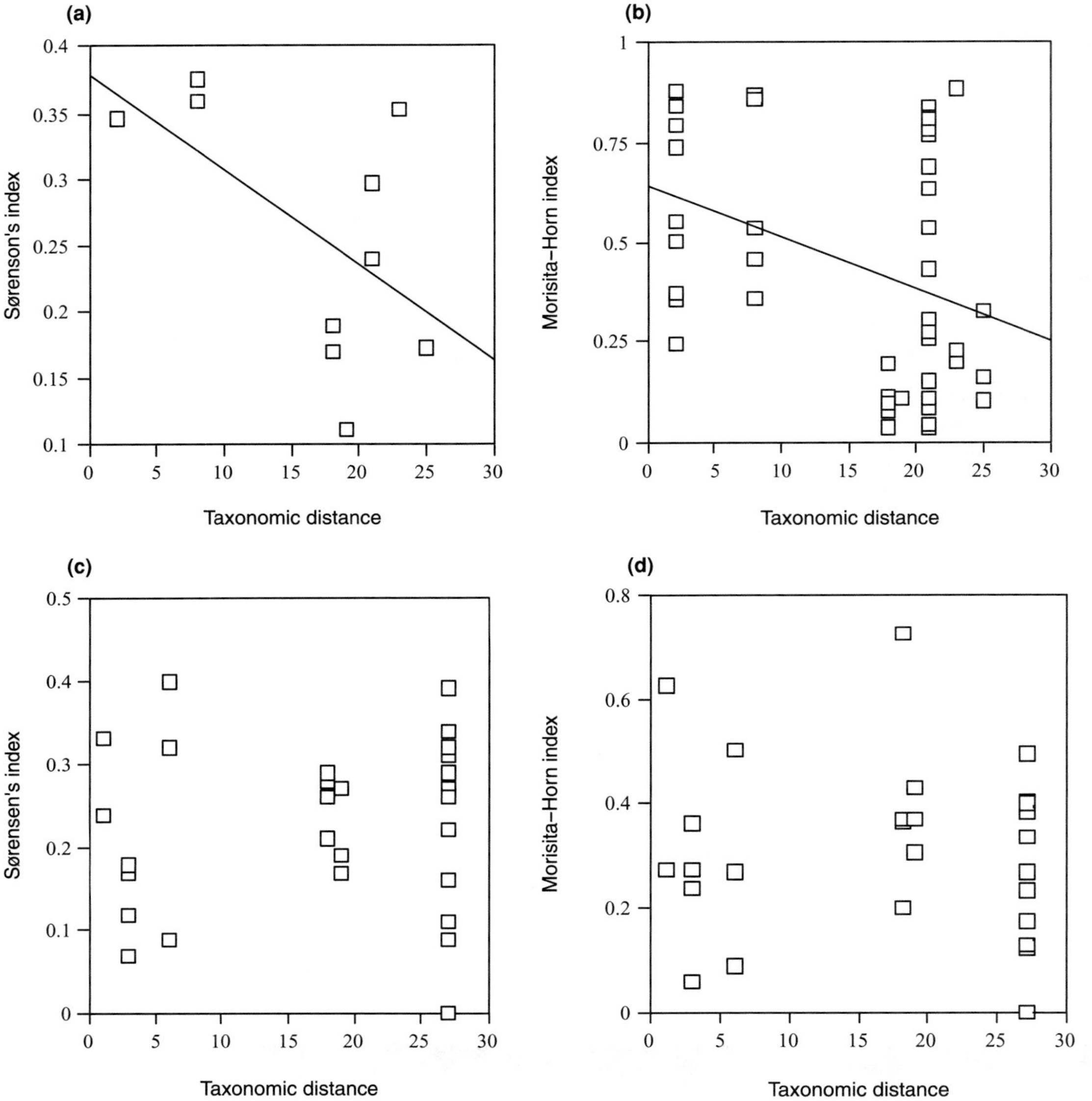

Fig. 29.3. Plots of the values of the Sørensen and Morisita–Horn indices for coleopteran assemblages from all pairs of tree species in the study against the taxonomic distances between the selected trees. (**a**,**b**) 1998 data; (**c**,**d**) 2000 data.

australiensis. In terms of numbers of individuals collected in 2000, the proportion of herbivores ranged from 46% in *W. austroqueenslandica* to 31% in *N. australiensis*.

DISCUSSION

In summary, our results have supported those others who have claimed lower rather than higher values of specificity for beetle assemblages on rainforest trees. Along with Wagner (1999, 2000), Kitching & Zalucki (1996), May (1990) and Stork (1987a, 1988), we find little support for the high levels of host specificity implied in some earlier works. For a variety of reasons, reviewed below, it is simply not realistic to expect a single-figure conversion factor that allows the prediction of numbers of species of host-specific insects within any patch of

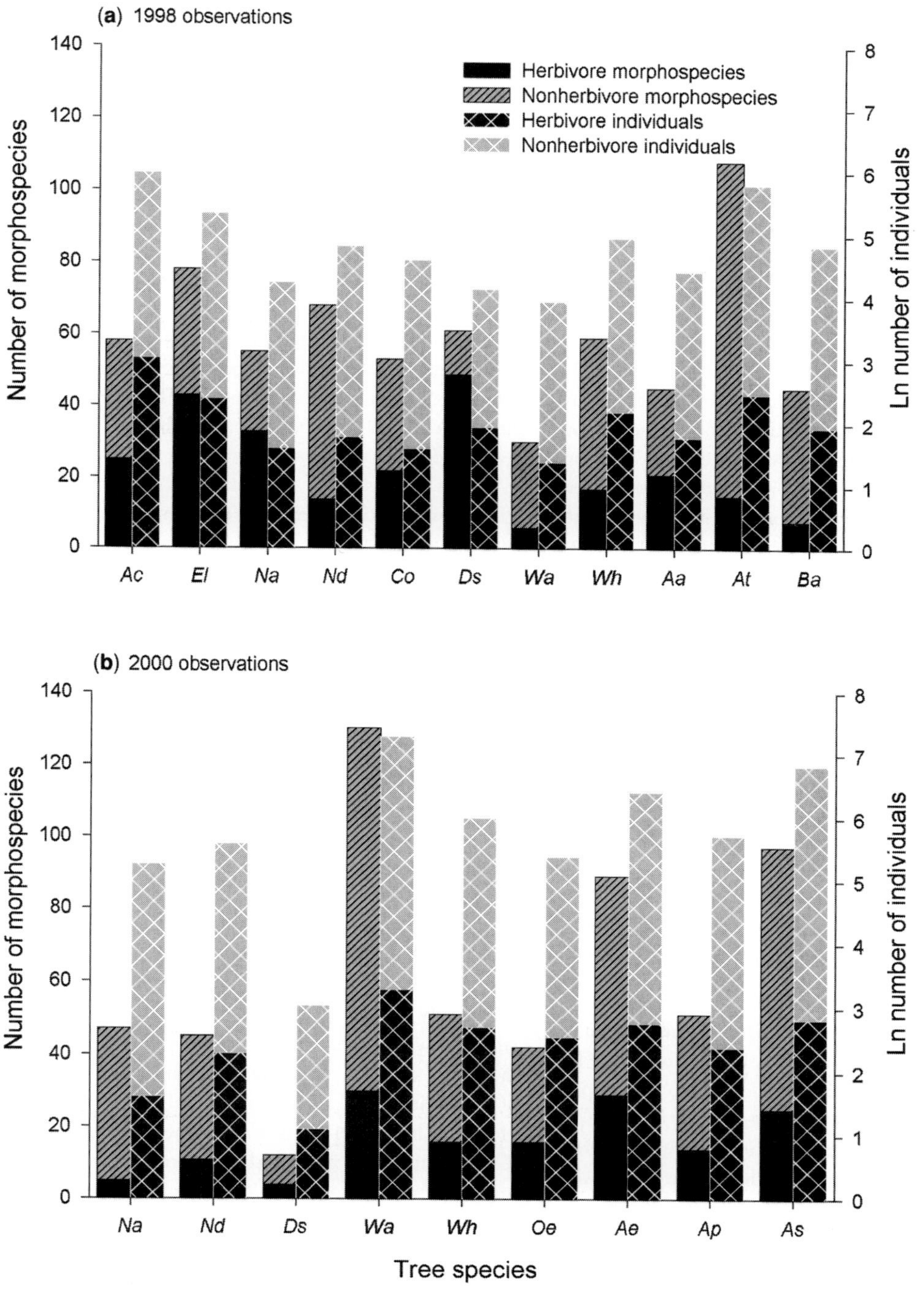

Fig. 29.4. The proportion of species (left hand columns) and numbers of individuals (right hand columns) that were designated as herbivorous in (**a**) the 1998 dataset and (**b**) the 2000 dataset. Tree abbreviations are the first letter of the generic and species name; refer to Table 29.1 for the full name.

forest. Further, the sort of approach we (and others) have taken is methodologically limited by an almost unavoidable set of truisms. In order to obtain insects from known 'pure' canopies one must use highly targeted sampling methods – of small crowns or segments of crowns. Inevitably this results in smaller rather than larger samples. Yet the more similar any two assemblages are, the larger the sample, and the smaller the variance between samples needs to be in order to quantify these, presumably, subtle differences. We visit this conundrum in more detail in the later part of this discussion.

So, there are two major issues raised by the results that we have presented here. The first is the set of biological factors that might produce the trend we have identified. In other words, what might be the reasons underlying the decreasing similarity between beetle assemblages with increasing taxonomic distance between their host trees. Equally important, however, are the methodological issues which clearly show that there are major impediments to using the sort of 'top-down' approach we have used in the estimation of host specificity. These point the way to productive future work.

The determination of assemblage similarity

The presence of a particular species of beetle in a crown sample from a particular species of tree will reflect any one of a number of factors or, indeed, any combination of these. The 'biodiversity templet' of Kitching *et al.* (1997) provides a synthesis of the many evolutionary, ecological and biogeographical processes that may act as determinants of biodiversity. In the present context, we must focus on the ecological interaction between the tree and the insect. The species encountered on the tree may be:

- a herbivore intent on eating parts of the living plant
- a nonherbivore that, nevertheless, is using the tree as a 'concourse' (*sensu* Elton, 1966) to obtain prey, hosts, shelter or some other resource
- a 'tourist' and its presence in the crown implies no direct ecological interaction with the tree (Southwood *et al.*, 1982b).

Those who have used top-down approaches for the estimation of host specificity have generally obtained an estimate for the numbers of herbivores associated with a tree crown and have added an additional multiplier to encompass the nonherbivores that are sampled simultaneously (Erwin, 1982; Stork, 1988; May, 1990; Thomas, 1990a). These relatively naive approaches have stimulated extended work on host specificity (Basset, 1992a, Gaston, 1992; Marquis & Braker, 1993; Kitching & Zalucki, 1996; Basset *et al.*, 1997a; Mawdsley & Stork, 1997; Ødegaard, 2000a) and should not be dismissed out of hand: they were of their time and the fact that the biodiversity debate has moved on does not negate their earlier importance. Host-plant specificity has now been shown to have several dimensions, which must be incorporated into any assessment or extrapolation from one crown to another. Most important of these are questions of polyphagy, of insect taxon-to-taxon differences, and of the relatedness of tree species within the resource range of species. The first two of these have been discussed in Ch. 5. The final factor has been of particular relevance in the present study.

Phylogeny of the plant resource

Many authors have suggested or implied indirectly that host 'specificity' is more readily apparent in herbivorous insects at the level of the plant genus or even family (Ward & Spalding, 1993; Basset *et al.*, 1997a; Mawdsley & Stork, 1997) rather than being a species-level phenomenon. In addition, authors have attempted to explain similarity between insect (particularly beetle) assemblages in tropical forests on the basis of the taxonomic relatedness of the pairs of trees being compared (e.g. Stork, 1987a; Leps *et al.*, 2001). Further, it is generally supposed that there has been a coradiation of phytophagous insects with that of the higher plants, and that this is most clearly evidenced within the major phytophagous families of the Coleoptera (Farrell, 1998) and Lepidoptera (Strong *et al.*, 1984). From such general considerations, we expect that the more phylogenetically distant any two plants are, the fewer herbivores they are likely to share. The practical difficulty in testing such a hypothesis is that simple comparison of tree taxons is flawed by the differing concepts of taxonomic level even within the vascular plants. The availability of more and more definitive molecular phylogenies allows us, as in the present analysis, to circumvent this taxonomic arbitrariness by using actual quantitative estimates of genetic distances among the plants being studied. In the present case, this is done based on a published phylogeny. As molecular typing becomes more and more accessible, we expect, in the near future, to be able to calculate the actual genetic distance between the species of trees selected for study.

The limitations of top-down approaches

In addition to the many ecological and evolutionary complications of determining host specificity in rainforest situations, there remain almost insurmountable methodological barriers to the top-down approach.

Fundamental to such approaches is the need for large samples from each tree species targeted. This need is underscored when, as has been repeatedly shown to be the case, the differences between the assemblages associated with particular tree canopies are subtle. Without large samples, real differences are simply swamped by noise associated with both sampling error and the preponderance of 'tourist' species. Further, the smaller the target taxon the larger the initial sample needs to be. Differences from tree to tree in total numbers of Coleoptera, say, will be easier to detect than differences associated with particular families or genera of Coleoptera. Three sets of difficulties present themselves: those associated with choice of sampling method, the choice of sampling time, and the need to obtain a sample that really is from the target plant species.

Sampling methods

Mass sampling of tree canopies can be done in a variety of ways depending upon available time, equipment and ease of access. These are well reviewed by Basset *et al.* (1997b). These authors present data comparing the phytophagous beetles caught using four commonly used methods (beating/hand picking, branch clipping, flight-interception trapping, canopy fogging). They show that each method has a different target set. The hands-on methods (beating/hand picking and branch clipping) targeted a greater proportion of beetles, which, subsequently, were shown to be actually feeding on the sampled tree (about 47% of the phytophagous beetles). Of the more remote approaches (interception traps and fogging), fogging produced a much higher number of proven feeders (29%) than did the interception traps (19%). Last they showed that the average size of the beetles sampled by fogging was smaller than those targeted by any other method (4.62 ± 0.268 mm; the next nearest method, branch clipping, sampled beetles of average size 5.10 ± 0.883 mm). We conclude that if many trees are to be sampled, if human resources are limited, and if comparisons are required at a point in time (see below) then a 'remote' method is inevitable. Once this decision is made, then fogging (or misting as in our case) is the technique of choice. If a more complete sample is required and the resources to collect and process such samples are available, then we join Basset *et al.* (1997b) in recommending that a diversity of methods should be used.

Fogging procedures have been reviewed at length by Erwin (1989), Adis *et al.* (1997), Paarmann & Kerck (1997) and Stork & Hammond (1997).

Time of sampling

We carried out our sampling in the middle of the wet season (midsummer in subtropical Queensland). The spring flush of growth had generally ended by this time and leaves, in general, were hardening off. However, as in any study involving trees of differing species, not all species would be in the same phenological stage. Further, seeking between six and ten replicates of each of 11 species (as in the 1998 study), all of which needed to be both able to be isolated from their neighbours (see below) and accessible for sampling, restricts the choice of trees considerably. Accordingly, it was not always possible to obtain trees with fresh dense foliage in each case. The scarcity of individuals (even of these relatively common trees) would also inhibit a cumulative approach where the selected tree species are sampled over a period of time, especially if this were to be done without resampling the same crowns – a desirable condition when pyrethrum knockdown methods are being used.

We know that the canopy fauna of trees is affected substantially by the phenological stage of the trees. Basset (1992c, 1994) showed the impact of leaf flush and mast-seeding on the canopy fauna in *Argyrodendron actinophyllum*, and we have many unpublished data demonstrating the great enrichment of the canopy fauna during episodes of flowering.

All of these considerations, together with inherent differences in tree architecture and leaf anatomy, add considerably to the difficulties of designing truly comparative, cross-species sampling programmes.

Sample 'purity'

A sample derived from fogging an overstorey tree will include, of course, material derived from the target assemblage of species that are at least temporarily associated with the canopy of the tree species in question. In rainforest situations, it will also include material from the assemblages of species associated with the lianas, vines, epiphytic angiosperms, ferns, bryophytes and lichens. In addition, there will be material knocked down from the segments of adjacent canopies, usually of different species, which interdigitate with the target canopy. Accordingly, ascribing the organisms encountered in a particular sample to the target tree is difficult.

We circumvented this problem by selecting relatively small understorey examples of the trees we were sampling. Others have placed fabric roofs above the crown of the study trees (Floren & Linsenmair, 1997b) or carried out large numbers of 'restricted area fogging' samples, which rely on bagging individual branches of the target tree (Basset & Kitching, 1991). In each case, the cost of attempting to focus the sampling on the targeted tree is likely to be smaller samples or massively greater sampling effort.

The way forward

There remains no doubt that the degree of specificity of phytophagous insect assemblages on rainforest trees remains a topic of considerable importance. The role of this specificity in allowing sensible extrapolation for the estimation of species richness at a broader scale is, in fact, of secondary importance. The importance of phytophagous insects as part of the forest's functional dynamics is of far greater significance. Accordingly, estimating the numbers of species associated with particular species of tree and their degree of fidelity to that tree remains an important question. Top-down approaches of the kind described here are a first step in defining these canopy assemblages but, for reasons outlined above, are unlikely to extend our understanding much further. So what is the way forward?

We know from the few faunas for which we have near complete documentation – those of western Europe and eastern North America – that there is no substitute for extended natural historical observation in which meticulous records of occurrence and feeding activity on host plants are accumulated over many years. In tropical forests, we, and others, have taken steps to short-cut this expensive and time-consuming observational work. Such work, however, now remains the only way forward. As several contributions in this volume attest, it is possible through intensive, highly designed and meticulously documented observation to build up a detailed picture of both the insect species associated with particular trees and the degree of host specialization exhibited by those insects beyond the focal tree species. The work of Basset and Novotny in New Guinea, and of Ødegaard in Panama are models of their kind. Placed together with detailed information on tree assemblages, such information can be used to make informed estimates of the total size of local faunas (V. Novotny *et al.*, unpublished data). Of course, the assembly of such detailed and extended datasets on herbivory requires longer time commitments and more people power than the more or less straightforward sampling of tree crowns by insecticide knockdown. The latter remains an important tool for rapid assessment but the former will become the approach of choice if we are to understand the underlying processes that connect insect assemblages to their plant hosts in rainforests.

ACKNOWLEDGEMENTS

This study was funded by the Cooperative Research Centre for Tropical Rainforest Ecology and Management. It benefited in the design phase from comments by Dr M. G. Hopkins and by statistical advice and assistance from Dr Heather Proctor and Mr Cameron Hurst. Assistance in the field was ably provided by Terry Reis, David Froggatt, Melinda Laidlaw, Guy Vickerman and Peter O'Reilly Sr. The O'Reilly family kindly gave us permission to work on their land. Work in the Lamington National Park was carried out under a permit issued by the Queensland National Parks and Wildlife Service.

30

A review of mosaics of dominant ants in rainforests and plantations

Alain Dejean and Bruno Corbara

ABSTRACT

The rainforest canopy is considered to be a habitat where a high proportion of the Earth's biodiversity resides. Nevertheless, ant species richness is relatively low in the canopy, although with their high abundance ants represent a major part of the animal biomass. The most abundant ant species, called 'dominants', are characterized by large colonies with territories that are distributed in a mosaic pattern. Dominant species tolerate within their territory 'nondominant' species, which have less populous colonies. An intermediary status, 'subdominant', also exists, which corresponds to species generally recorded as nondominants but which, under certain conditions, are able to develop larger colonies and to defend territories in much the same way that dominants do. By studying African forests of different ages as well as tree crop plantations, we have been able to contribute to the knowledge of ant mosaics and the protective role of these ants *vis-à-vis* their supporting trees. The distribution of dominant ant species on trees is not merely the result of chance. Choice tests permitted us to demonstrate that *Oecophylla longinoda* workers are attracted by citrus or mango trees rather than cocoa trees, whereas the contrary is true for *Tetramorium aculeatum*. Nevertheless, environmental influences can modify the choice of founding queens or workers (the latter play a role during colony budding) through imprinting. Although most dominant ant species are good predators that protect their supporting trees, they mostly feed on sugary substances such as extrafloral nectar and honeydew produced by hemipterans. In the latter situation, the negative effect of sap sucking is offset by the positive effect the ants have on the plant by eliminating defoliators. However, although some ant species tend hemipterans that do not affect their supporting tree, others may be associated with hemipterans that transmit diseases to these trees.

INTRODUCTION

Studies on canopy-dwelling arthropod species in tropical rainforests show that ants are a major component, often representing 90% of the individuals and 50% of the biomass (for 'canopy', see definitions in Moffett, 2000). Tropical arboreal ants are characterized by a very high abundance, but a moderate diversity (Adis *et al.*, 1984; Stork, 1991; Tobin, 1994, 1995; Davidson & Patrell-Kim, 1996; Brühl *et al.*, 1998). This remarkable ecological success reflects the adaptation of arboreal ant species to this particular, tridimensional environment (as opposed to the bidimensional environment of ground-dwelling ants). Among the main constraints of this environment are the limited number of nest sites and food resources, climatic factors and the fact that rainforest canopies are rather dry compared with the ground (Adis *et al.*, 1984; Wilson, 1987; Stork, 1991; Davidson & Patrell-Kim, 1996; Davidson, 1997; Floren & Linsenmair, 1997b).

The fogging technique has enabled researchers to gather information on the presence of ants in tree crowns in terms of abundance and species richness (Adis *et al.*, 1998a; Brühl *et al.*, 1998; C. H. Schulze & T. Wagner, unpublished data). As a result, this technique is particularly useful for comparative studies on the impact of extended clearing on arboreal arthropod communities (Watt *et al.*, 1997b, 2002; Lawton *et al.*, 1998; Floren & Linsenmair, 1999, 2001; Floren *et al.*, 2001a). The ants sampled may include true canopy dwellers and also ground-nesting species that can develop colonies in hanging soil and epiphytes. This is frequent in the neotropics where epiphytic bromeliads and orchids as well as araceous hemiepiphytes of the genus *Phylloden-dron* abound (Dejean *et al.*, 1995; Blüthgen *et al.*, 2000b). In Africa, epiphytic ferns and araceous hemiepiphytes of the genus *Rektophyllum* play the same role, even if they are less frequent (Ngnegueu & Dejean, 1992; Djiéto-Lordon *et al.*, 2002). Other ants, only represented by a few or even one worker, may correspond to small cryptic species (such as those that nest in the crevices of rough bark).

In contrast, other sampling techniques that depend on the use of exploration devices, such as canopy cranes, the canopy raft or the canopy sledge, are inadequate for estimating the total ant species richness (see details of the techniques in Ch. 2). In southern Cameroon, for example, both the raft and the sledge led us to note only 28 ant species on 167 trees of an old forest (Dejean *et al.*, 2000e), whereas fogging permitted Watt *et al.* (2002) to record 97 ant species in plantations of *Terminalia ivorensis* A. Chev (Combretaceae), a rapidly growing timber tree. Among the latter, the 22 most common species together made up 99% of the total number of ant individuals sampled.

Because they facilitate direct observation, canopy cranes, the canopy raft and the canopy sledge are best suited to behavioural and ecological research at the colony level (many species have polydomous nests built with carton that are easy to observe or to gather). Both cranes and the canopy raft permit repetitive long-term studies, whereas the canopy sledge permits the collection of data over a relatively large number of trees – up to 200 trees in a few days in Gabon (Blüthgen *et al.*, 2000b; Dejean *et al.*, 2000b). In this chapter we shall emphasize those sampling techniques that allow the study of arboreal ants at the colony level. Specifically, we examine the mosaic-like distribution of some of these canopy ants.

ANT MOSAIC: HISTORY AND DEFINITIONS

Wilson (1958) was the first to describe a patchy distribution of the arboreal ant fauna in tropical rainforests. The notion of ant mosaic appeared later in a series of studies conducted in African cocoa tree plantations (Room, 1971; Majer, 1972, 1976a,b; Leston, 1973; Taylor, 1977; Jackson, 1984). Cocoa plantations provide an easily attainable canopy that allows the rapid and efficient identification of the ant species occupying each individual tree. The concept of ant mosaic was later generalized to include American, Papuan and Australian forests, as well as tree crop plantations (Room, 1975; Leston, 1978; Majer, 1990, 1993; Paulson & Akre, 1991; Adams 1994; Andersen & Reichel, 1994; Majer *et al.*, 1994; Medeiros *et al.*, 1996; Armbrecht *et al.*, 2001).

These studies have shown that the canopies of tropical forests and of tree crop plantations are occupied by dominant ant species. These are characterized by extremely populous colonies (several hundred thousand to several million individuals), the ability to build large or polydomous nests (carton builders, carpenter ants and weaver ants) and a highly developed intra- as well as interspecific territoriality. As a consequence of their territorial behaviour, dominant ants are distributed in a three-dimensional mosaic pattern in the forest canopies (Leston, 1973). These dominant arboreal ants are also called territorial dominant or simply territorial species (Vepsäläinen, 1982; Davidson, 1998), the definition remaining the same (i.e. ant species that defend space *per se*, absolute spatial territories, usually against both intraspecific and some interspecific enemies). In comparison, in ground-dwelling species, one can distinguish (i) numerical dominance, or the predominance of a species in number, biomass and/or frequency of occurrence in the ant community; (ii) behavioural dominance, or dominance in interspecific competition owing to superior fighting and/or recruitment abilities; and (iii) ecological dominance, which applies to invasive species or the combination of both numerical and behavioural dominance (Davidson, 1998).

Hölldobler & Lumdsen (1980) were the first to demonstrate that colonies of the African weaver ant *Oecophylla longinoda* Latreille obtain considerable advantages by excluding competitors from their territory rather than only from their nests and food resources in the immediate surroundings. As a result, they introduced the concept of absolute territory. In such a situation, the territory is marked with persistent pheromones or landmarks that can last for more than 1 year (Beugnon & Dejean, 1992). Moreover, such a territory is defended, especially at the periphery, both day and night against conspecific aliens and other dominant ant species. Small colonies of nondominant species are nevertheless tolerated. The mutual exclusion of alien members from adjacent colonies during aggressive interactions leads to the formation of unoccupied zones that constitute a 'no ant's land' (Hölldobler & Wilson, 1978, 1990a; Hölldobler, 1979).

Defending absolute territory, although energetically costly, does not pose a problem as the species involved exploit sugar-rich extrafloral nectar (EFN) and honeydew produced by trophobionts. These are mostly hemipterans of the former suborder Homoptera (i.e. aphids, coccids, pseudococcids, membracids, tettigometrids, etc.), although heteropterans are sometimes attended (see Dejean *et al.*, 2000c; Gibernau & Dejean, 2001). As a result, energy is not a limiting factor in the defence of territory. Moreover, the EFN when present and the trophobionts able to develop on the different tree species can shape the ant mosaic (Davidson, 1997; Blüthgen *et al.*, 2000b; Dejean *et al.*, 2000e; Hossaert-McKey *et al.*, 2001). The need for sugary substances is illustrated by the absence of the ant mosaic in the lower canopy trees of a pristine forest in Borneo where ant-attended hemipterans were largely absent (Floren & Linsenmair, 2000a), whereas ant mosaics were observed in the upper-level trees of different Asian forests, including in Borneo (N. Stork, personal communication).

RELATIONSHIPS BETWEEN DOMINANT AND OTHER ARBOREAL ANT SPECIES

Dominant arboreal ants tolerate the presence of nondominant species on their territory. Nondominant ants have less populous colonies (up to a few thousand individuals) and generally depend on pre-existing botanical structures for nesting (hollow branches, rough bark, epiphytes). An intermediary status, known as 'subdominant', has also been described. It corresponds to species that generally act as nondominants but which are able, under certain conditions, to defend territories in the same way as do dominants (Majer, 1972, 1993; Leston, 1973; Hölldobler & Wilson, 1977, 1978; Hölldobler, 1979, 1983; Majer *et al.*, 1994).

We noted in Cameroon that, when large *O. longinoda* colonies develop sexual broods, the need for protein is so great that certain workers hunt on secondary territories at ground level (Dejean 1990b, 1991). For example, numerous workers of a large colony sectioned a 2.5 m long snake (about 3 kg) killed by farmers in 1 week, while others ambushed flies that tried to lay eggs. The third day, in 10 minutes we counted 255 workers retrieving a piece of flesh, 122 for a maggot and 64 for a fly. Also, workers can attack and plunder the brood of the *Camponotus* spp. and *Polyrhachis* spp. colonies that they normally tolerate within their territory (Mercier *et al.*, 1998; A. Dejean, personal observation). In such a situation, the nondominant species serves as a kind of food reserve for the dominant one.

Dominant species can also rob prey from nondominant ones (the same being true among nondominant species). This is the case for *Polyrhachis laboriosa* F. Smith and *Polyrhachis weissi* Santschi sharing trees with *O. longinoda*. To avoid prey robbing, some *P. laboriosa* workers acting as guards release allomones around their captured prey, while nestmates cut up these prey on the spot. Intimidation displays from the guards, plus allomones deposited around the perimeter where the prey is cut up, allow direct confrontations between workers of these species to be avoided when they share the same trees (Mercier *et al.*, 1998).

Majer (1976a,b) noted that certain nondominant ant species, which he called subdominant species, develop colonies that can occupy an entire tree crown, behave as dominants and exclude from their trees other ant species, including known dominants. *P. laboriosa* is a typical subdominant species, which is able to occupy entire cocoa trees or pioneer trees along forest edges (Majer, 1976a,b; Dejean & Gibernau, 2000). This species has developed a limited aggressiveness through ritualized behaviour, both at the intraspecific level and when confronted with *Camponotus brutus* Forel. In the latter case, both species can share the same territory, with *P. laboriosa* being active during the day and *C. brutus* at night, thus permitting the permanent occupancy of the territory (Mercier & Dejean, 1996; Mercier *et al.*, 1997). Two species sharing the same territory that behave as dominant was called 'codominance' by Majer (1976b) and frequently results in a complementary rhythm of activity of the two species, one species being diurnal, the other nocturnal (see also Mercier & Dejean, 1996; Dejean & Olmsted, 1997; Mercier *et al.* 1998; Hossaert-McKey *et al.* 2001).

The occurrence of codominant species in the canopy of the French Guianian forest is relatively frequent compared with that in African forests (Dejean *et al.*, 1999). Among the codominant species we note the frequent presence of 'parabiotic' species such as *Camponotus femoratus* Fabricius and *Crematogaster limata parabiotica* Smith, which share the same trails and the same nests (ant gardens) but shelter in different cavities of the nests (see Wilson, 1987; Davidson, 1988; Dejean *et al.*, 2000a).

THE PREDATORY BEHAVIOUR OF ARBOREAL ANTS

In ants, it is generally considered that arboreal life was acquired secondarily (Hölldobler & Wilson, 1990a). We mentioned above that arboreal ants exploit energetic sugary sources such as EFN and honeydew from hemipterans, so 'energy' is not a limiting factor. However, nitrogen can be a limiting factor; consequently most dominant arboreal species have workers with a thin cuticle and nonproteinaceous venom (Davidson, 1997; Orivel & Dejean, 1999). A large proportion of the nitrogen needed is acquired by harvesting attended trophobionts in excess of their honeydew needs (Way, 1963; Carroll & Janzen, 1973), and through types of predation that are adapted to arboreal life, for prey are generally more likely to escape when they are on a tree. The predatory behaviour of the weaver ant *O. longinoda* has been the most studied and so serves as a reference. This dominant ant hunts diurnally in groups. Prey detected by sight from a relatively long distance are seized by an appendage and immobilized by a first worker, which then attracts nestmates with the aid of a pheromone. Recruited nestmates, in turn, seize an appendage of the prey and pull backward, resulting in the spread-eagling of the prey. This behaviour, which is used even for relatively small prey (only very small prey are captured by a single worker), also permits the capture of large insects (Dejean, 1990a,b; Hölldobler & Wilson, 1990a). Entire prey are retrieved cooperatively, including heavy prey such as small birds (Wojtusiak *et al.*, 1995). This form of prey capture and retrieval requires that the workers adhere to the substrate by means of very powerful adhesive pads and claws, a characteristic that seems general in arboreal species (Wojtusiak *et al.*, 1995; Federle *et al.*, 2000; Djiéto-Lordon *et al.*, 2001c; Orivel *et al.*, 2001a; Richard *et al.*, 2001).

Other dominant ants exhibit a similar behaviour based on the spread-eagling of the prey. Detection may occur from a short distance or even by contact; venom is generally used in order to subdue the prey before cutting it up on the spot and transporting it in small pieces. This is the case for *Atopomyrmex mocquerisi* André (Ngnegueu & Dejean 1994), *Crematogaster* sp. (Richard *et al.*, 2001), *Myrmicaria opaciventris* Emery (Kenne *et al.*, 2000), and *Tetramorium aculeatum* (Mayr) (Djiéto-Lordon *et al.*, 2001c). As a result, among dominant African arboreal ants, the existence of predatory behaviour only needs to be verified in *Technomyrmex* spp., noted as dominant by Room (1975) and Watt *et al.*, (2002).

Workers of the subdominant species *P. laboriosa* hunt singly and are able to subdue large prey. When a worker cannot retrieve a prey item, it recruits nestmates at long range. Recruited workers cut up the prey on the spot and generally transport the pieces alone (Dejean *et al.*, 1994b). Solitary hunting, noted in nondominant tropical species, seems widespread (Dejean & Corbara, 1990; Orivel *et al.*, 2000; Djiéto-Lordon *et al.*, 2001a). It probably depends on very powerful venoms, as noted for *Pachycondyla goeldii* (Forel) (Orivel & Dejean, 2000; Orivel *et al.* 2001b). In addition, in certain species, after a hunting worker captures a large prey, the entire colony can move to the prey, which is consumed on the spot (Djiéto-Lordon *et al.*, 2001b).

FACTORS INFLUENCING DOMINANT ANT DISTRIBUTION

We note first that the ant mosaic is not a permanent structure of forest and plantation canopies. For instance, ant mosaics were not found in the lower canopy trees of a mature forest in Borneo (Floren & Linsenmair, 2000a). In forests and plantations of northern Australia, the tree crowns between blocks occupied by dominant ants were devoid of ants, occupied by nondominant arboreal ants or visited by ground-nesting species (Majer & Camer-Pesci, 1991). Researchers may overlook the presence of certain dominant species, independently of the technique used. This is the case for instance for *Paraponera clavata* Fabricius, a nocturnal neotropical ponerine ant that nests in the ground at the base of extrafloral nectary-bearing trees. During the day, the trees can appear devoid of ants or be occupied by *Crematogaster* spp., but at night numerous giant workers of this species forage in the foliage (Young & Hermann, 1980; Breed & Harrison, 1989; Hölldobler & Wilson, 1990b; A. Dejean, personal observation).

It is known that ant aggressiveness is responsible for the mosaic, but comparisons between different, cultivated tree species also show different associations. In fact, both the presence of extrafloral nectaries and honeydew-producing insects shape the species distribution of dominant ants (Blüthgen *et al.*, 2000b; Dejean *et al.*, 2000e; Hossaert-McKey *et al.* 2001). In tree-crop plantations, the proportions of the different dominant ant species vary with the cultivated tree species. For example *O. longinoda*, like the Asian species *Oecophylla smaragdina* Fabricius, is relatively frequent on citrus and mango trees, less frequent on cocoa tree plantations and rare on palm trees (Majer, 1976a,b; Jackson, 1984; Dejean *et al.*, 1997; Mercier, 1997; Way & Bolton, 1997).

In Cameroon, ethological studies conducted on *O. longinoda* and *T. aculeatum*, two species that compete for nesting sites in pioneer formations and in plantations, allowed us to demonstrate that the selection of host plants by winged females (dissemination of colonies) and workers (colony budding, i.e. new colonies founded by workers and queens originating from a previous colony) can take either of two paths. The two species differ in terms of 'innate' attraction to nesting site plants. *O. longinoda* individuals selected leaves of citrus or mango trees rather than the leaves of cocoa and guava trees. The reverse was true for *T. aculeatum* individuals. Nevertheless, by using winged females and workers originating from one of the plants to be tested, we showed in both species the existence of a familiarization process that can supersede 'innate' attraction. Individuals bred in contact with one of the tested plants during larval life (pre-imaginal learning) or the first days of adult life (early learning), or both, have a tendency to choose leaves of the plant species with which they were in contact. After several days of adult life, this conditioning is not possible. There is, therefore, a sensitive period after which the influence of the environment ceases, suggesting a true imprinting process (Djiéto-Lordon & Dejean, 1999).

DYNAMICS OF THE ASSOCIATIONS

In the very old rainforest of Akok, Campo Forest Reserve, Cameroon (167 trees sampled), we recorded only three dominant ant species. One of them, *Crematogaster depressa* (Latreille) occupied 87.4% of the trees (N. Stork, personal communication, also noted the over-representation of one ant species in an Asian forest). We also found *Crematogaster* sp.1 (1.8% of the trees) and *O. longinoda* (6%). In the Makandé forest, Gabon, which is a secondary forest owing to the presence of the Okoumé tree (*Aucoumea klaineana* L. Pierre; Burseraceae), *T. aculeatum* was the most frequent (on 27.5% of 200 trees) among 18 other ant species. We obtained very similar results in a 50-year-old secondary forest situated at Matomb, Cameroon. Along the forest edges, *O. longinoda* and *T. aculeatum* were the most frequent species, but ground-nesting, arboreal-foraging species competed with them (Dejean *et al.*, 1994a, 2000b,e; Dejean & Gibernau, 2000).

In 1990, M. Kenne (unpublished data), who studied citrus and guava plantations in Cameroon, noted that insecticide treatments favoured ground-nesting, arboreal-foraging ant species. These ants occupied 94.2% of the citrus trees ($n = 593$) during periods of insecticide treatment. Workers of *M. opaciventris* were noted on all trees (see also Kenne & Dejean, 1999), while *Camponotus acvapimensis* Mayr and *Paratrechina longicornis* (Latreille) were noted on 51.4% and 39.3% of the trees, respectively. These three species tolerate each other and were often noted on the same trees. True arboreal species occupied only 5.8% of the trees. A concomitant survey conducted in the guava tree plantation, where insecticide treatment had been stopped 2 years earlier, indicated that arboreal-dwelling ant colonies occupied 65.3% of the trees ($n = 414$), from which they had excluded ground-nesting, arboreal-foraging ant species.

Mercier (1997) studied a mango tree plantation over 8 years (1989–1996) after insecticide treatment ceased. It was noted that the mosaic established itself progressively. Two years after insecticide treatment was stopped, 73% of the trees were already occupied by arboreal ant species, and after 6 years this level reached 100% ($n = 284$). Among these species, *T. aculeatum* was the first dominant ant to install itself in a large proportion of the trees (28.5% were occupied by this species in 1990). This percentage then decreased and stabilized to 19% of the trees in 1995. The number of trees occupied by *O. longinoda* first increased from 9.5% of the trees occupied in 1990 to 35.2% in 1994; it then decreased and remained stable at 15.8%. *Atopomyrmex mocquerisi* (dominant status) followed a pattern similar to that of *O. longinoda*, with which it can share trees, and was noted on 11.3% of the trees in 1990, 24.3% in 1993 and only 13% in 1996. During the same time, *Crematogaster* spp. occupied an increasing number of trees, from only one species noted on 16.9% of the trees in 1990 to 75.5% of the trees occupied by four species at the end of the experiment in 1996. On the contrary, *Tetramorium africanum* (Mayr) (dominant status), a species that inflicts an extremely painful sting, occupied 2.5 to 3.2% of the trees during the first 4 years, after which it was progressively eliminated by *Crematogaster* spp. The colonies of these *Crematogaster* spp. install themselves on the territories of other dominant ants. Further studies are needed in order to know whether the queens and the workers of incipient colonies are tolerated by the previous dominant ant species.

Subdominant species were also recorded as able to occupy an entire tree. This was the case for *P. laboriosa*: occupancy increased from 5.3% to 29.6% of the trees between 1990 and 1994, and then decreased to 11.6% in 1996. In contrast, *Tetraponera anthracina* (Santschi) was noted on an increasing number of trees (always occupied by a dominant species) throughout the entire study. During the study period, the overall number of nondominant species increased.

This set of observations indicates that the cessation of insecticide treatment was followed in succession by three groups of ant species: (i) Ground-nesting, arboreal-foraging species were initially favoured by the treatment; (ii) these were replaced by *T. aculeatum*, *T. africanum*, *O. longinoda* and *P. laboriosa*; (iii) group 2 species were replaced in turn by carton-building *Crematogaster* spp. (see also Bigger, 1993).

The same ant species are noted in relative proportions in various plant formations. Pioneer plant formations favour ground-nesting, arboreal-foraging species, particularly *M. opaciventris* (Kenne & Dejean, 1999). Forest edges and secondary forests favour *T. aculeatum*, *O. longinoda* and *P. laboriosa* (Dejean *et al.*, 1994a; Dejean & Gibernau, 2000), but certain carton-building *Crematogaster* such as *C. clariventris* Mayr are also present. Finally, old forests are mostly occupied by carton-building *Crematogaster* spp. (Dejean *et al.*, 2000e).

PLANT PROTECTION BY DOMINANT ANT SPECIES: THE USE OF ANTS AS BIOLOGICAL CONTROL AGENTS

The use of dominant ant species as biological control agents is traditional in southern Asia, where *O. smaragdina* is tolerated or even introduced into citrus tree orchards (Huang & Yang, 1987; Way & Khoo, 1992; van Mele & Cuc, 2000). This ant species has also been noted as an efficacious pest control agent for Australian cashew trees (Peng *et al.*, 1997). In Africa, the efficacy of dominant arboreal ants against defoliators has been noted in plantations of coconut, cocoa and palm trees (Way, 1953; Majer, 1976a,b, 1993; Dejean *et al.*, 1991, 1997; Bigger, 1993). Under natural conditions, they have been noted as good protectors of their supporting trees against gregarious locusts (Dejean, 2000; Dejean *et al.*, 2000d).

From studies on cocoa tree plantations, Majer (1976a,b; 1993) introduced the notion of 'manipulation of the ant mosaic'. Indeed, the plants support honeydew-producing hemipterans attended by dominant ants and, in return, are protected against other herbivores. In such a context, dominant ants may be an effective biological control agent against the pests of cultivated plants, on the condition that the attended hemipterans do not damage the plant. This satisfactory situation is demonstrated by the use of *O. smaragdina* in Asian citrus tree orchards.

In Africa, numerous studies on ant mosaics have been conducted on cocoa tree plantations to determine the impact of the different ant species on pest control (Majer, 1976a,b, 1993; Taylor, 1977; Dejean *et al.*, 1991; Bigger, 1993). *T. aculeatum*, *O. longinoda* and several species of carton-building *Crematogaster* spp. constitute the base of the mosaic in the canopy of the cocoa tree plantations. *Crematogaster* spp., which tolerate mirids (Heteroptera) and tend pseudococcids that transmit diseases to cocoa trees, should be excluded, while *O. longinoda*, which tends stictococcids that do not cause problems to this plant, needs to be favoured (Majer, 1976a,b). The contrary is true for oil palm trees, where *Crematogaster gabonensis* Emery is able to limit attacks by a leaf-mining chrysomelid beetle, while *T. aculeatum* is ineffective and *O. longinoda* is uncommon (Dejean *et al.*, 1997).

In citrus tree orchards of subtropical and temperate regions, ground-nesting, arboreal-foraging ants attend large numbers of aphids, pseudococcids and coccids that affect fruit production and the health of supporting trees. As these ants hunt all kinds of insect, including the predators and the parasitoids of their trophobionts, they have to be excluded (Samways, 1990; James *et al.* 1999).

In conclusion, we now know that (i) most dominant ant species protect their supporting trees, although they generally attend sap-sucking Hemiptera; (ii) in plantations, some species of Hemiptera attended by dominant ants can affect certain species of tree and not, or only slightly, other species; (iii) a succession of ant species follows the cessation of insecticide treatment, at least in central African regions; (iv) dominant arboreal ants can be attracted by some plant species rather than others and (v) we can condition ants to be attracted to a targeted plant through imprinting. As result, we now have the theoretical basis for permitting the ant mosaic to be manipulated as predicted by Majer (1976a,b). Studies are necessary in neotropical and Asian regions in order to improve the identification of dominant ant species and to verify if they really defend absolute territories (see Davidson, 1997) and to determine if nest site selection is based on the tree species.

ACKNOWLEDGEMENTS

We wish to thank Barry Bolton (Natural History Museum, London) whose identifications of our African ant samples has always been very rapid and for his numerous comments concerning ant taxonomy, permitting us to know African ants in detail and to conduct biological, behavioural and ecological studies. We are also grateful to Yves Basset and an anonymous referee for their helpful comments on the former version of the manuscript, and to Andrea Dejean for improving the English.

31

Insect herbivores in the canopies of savannas and rainforests

Sérvio P. Ribeiro

ABSTRACT

Insect herbivory, species richness and distribution in the canopy of tropical savanna is compared with that of tropical rainforests by compiling published information from various savanna and forest habitats in the neotropics, Africa and Australasia. An overview of herbivory patterns is presented, contrasting species-rich plant assemblages with natural monodominant tree populations. Data suggest that host trees suffer greater herbivory in savannas than in rainforests and in dry forests compared with wet forests. However, monodominant populations of pioneer *Tabebuia aurea* (Bignoniaceae), in the Brazilian wetlands of *Pantanal Matogrossense*, show much lower herbivory than observed on this species in the *cerrado* (Brazilian savanna). The environmental stochasticity in the wetland seems to decrease negative biotic pressures, such as herbivory. Conversely, monodominant tree populations within rainforests in South America and Africa show higher herbivory rates than adjacent species-rich forests. Tropical monodominant forests seem to be associated with the early successional process, but the monodominant populations of *T. aurea* are persistent in time. Regarding the insect fauna, rare species are common in both vegetation types, but the distribution of species richness and density of specialists seem to differ. Cumulative species richness of free-feeding herbivores per host species does not differ between savannas and rainforests but species richness per tree sampled tends to be greater in rainforest. However, the published data are difficult to compare quantitatively. Nevertheless, the number of species of specialist beetles on the crowns of a rainforest tree was twice that of specialist lepidopteran or all free-feeding species on tree species in savanna. The canopies of tropical savannas support relatively denser populations of specialist insects than do canopies of rainforests. Gall-forming insects are much more common in neotropical savanna habitats, particularly in the *cerrado*, which has been explained by the hypothesis of the harsh environment, HHE (Fernandes & Price, 1988). This hypothesis predicts a greater number of gall species will occur in xeric habitats, such as scleromorphic savannas, than in mesic habitats, such as rainforests or gallery forests, and an inverse pattern should be observed for free-feeding herbivores. However, I argue that data analysis to support the HHE was based on comparisons between the canopy of savannas versus the understorey of gallery forests and did not consider properly plant species composition and food quality across these two vegetation strata. Therefore, the HHE is challenged, and the importance of an evolutionary response of gall species to sclerophylly is emphasized. In the conclusion, the state-of-art research on insect species distribution and abundance in the canopy is discussed. The consequences of canopy openness, discontinuity, spatial organization of foliage, dehydration and solar radiation for the population dynamics of insect herbivores are considered, contrasting semideciduous forests and savanna habitats with closed rainforests. The concept of canopy roughness is discussed and applied to savanna vegetation.

INTRODUCTION

The distribution of insect species in the canopy of tropical savannas is a poorly understood aspect of tropical ecology. For tropical forests, research since the early 1980s has shown striking differences between habitats and continents in terms of insect species composition and diversity (e.g., Erwin, 1982; Stork, 1991; Lowman, 1995; Basset & Samuelson, 1996; Krüger & McGavin, 1997), as well as levels of herbivory among host species

(Lowman, 1995; Coley & Barone, 1996; Lowman & Wittman, 1996). However, patterns of herbivory may not reflect fully important aspects of insect species distribution and their population densities.

In this chapter, the insect fauna in the canopy of tropical savannas is compared with that of tropical rainforests. After a short overview of patterns of herbivory, insect species richness and distribution in savannas and rainforests are reviewed and compared, with emphasis on free-feeding herbivore guilds. Despite difficulties in comparing the variety of data in the literature, similarities and contrasts between these two vegetation types are apparent.

For neotropical savannas, gall-forming species play an important role in the insect assemblage and deserve a separate analysis. Price *et al.* (1998) have proposed a mechanistic hypothesis, based on differential mortality rates of galls (higher in mesic than xeric habitats) to explain the global pattern of species diversity of gall-formers. The hypothesis supports higher numbers of gall-forming species in scleromorphic, Mediterranean-type vegetation than in mesic vegetation. Mendonça (2001) has challenged their argument recently, based on an evolutionary approach in which high rates of host shift caused by synchronized resprouting could have resulted in great radiation of galling species. I review patterns of gall species diversity that have been described since Fernandes and Price (1988) and present a reinterpretation of previous hypotheses to explain insect gall-forming and free-feeding species diversity in the savannas. The concept of 'specialist herbivore' is used for insects identified by studies that have carried out feeding tests to identify herbivores which use restricted food sources, or for guilds that are classically known as specialists, that is, gall-forming insects.

METHODS

In the first part of this chapter patterns of herbivory are presented and discussed, although the lack of details in some original articles or previous reviews does not allow the development of rigorous, quantitative comparisons. Most of the articles about insect species richness and distribution on host trees do provide information that can be compared statistically. Estimates of mean and variance in the number of insect species per host species and per sampled trees can be derived from: number of chewing species per clipping samples in Basset (1996); number of species per number of trees, averaged by sampling period, in Ribeiro and Pimenta (1991), Diniz and Morais (1997) and Price *et al.* (1995); or straight from the appendices and tables in Ribeiro *et al.* (1999a), S. P. Ribeiro and V. K. Brown (unpublished data) and Ødegaard (2000c). The taxa discussed varied among published studies: specialist beetles only (Basset, 1996; Ødegaard, 2000c), specialist lepidopteran species (Price *et al.*, 1995; Diniz & Morais, 1997), the specialist herbivore fauna (Ribeiro & Pimenta, 1991) or all free-feeding herbivores (Krüger & McGavin, 1997; Ribeiro *et al.*, 1999a; S. P. Ribeiro & V. K. Brown, unpublished data). In their articles about gall-forming species, Fernandes & Price (1988, 1992) did not provide enough information to allow comparisons with other literature, and hence gall species will be considered separately. Although conclusions obtained from such comparisons are constrained by the lack of a comparable control response, a mixed model analysis of variance (ANOVA) was constructed from the calculated variance amongst articles (log-transformed) to test the hypothesis that savanna and rainforest vegetation support different numbers of species of herbivore, both per host species and per individual tree crown. Savanna and rainforest were compared as fixed categories, and articles were assigned as random experimental blocks when testing differences in cumulative number of insect species. This analysis aimed at emphasizing patterns derived from the literature but is not based on an exhaustive review, although most of recent literature for savannas is covered (but see Grant & Moran, 1986; Davies, 1999). Many more articles describing insects on trees in tropical rainforests are available but are not explored in details (see Marquis, 1991; Moran *et al.*, 1994; Novotny *et al.*, 1999b).

In the second part of the chapter, I revisit Fernandes & Price's data and hypotheses in their 1988, 1991 and 1992 papers, where they attempt to explain the higher gall-forming species diversity in xeric savanna than in mesic gallery forest habitats in the Brazilian savanna – *cerrado* vegetation. Recent data are then used to discuss new hypotheses related to differential plant colonization and survivorship. Finally, the latest research on insect species distribution and abundance in the canopy is discussed in terms of the consequences of canopy openness, discontinuity, spatial organization of foliage, dehydration and solar radiation for the population dynamics of insect herbivores.

RESULTS AND DISCUSSION

Comparison of herbivory and insect herbivore diversity between savanna and rainforest

Herbivory in tropical vegetation

Previous work suggests that savannas exhibit higher levels of herbivory than do forests, both overall and when compared with different types of forest and different strata within a forest. Ribeiro & Brown (1999) compared herbivory levels for two species of *Tabebuia* (Bignoniaceae) in Brazilian *cerrado* vegetation with data from tropical forests (e.g. Coley, 1983; Coley & Aide, 1991; Basset & Höft, 1994; Filip *et al.*, 1995; Lowman, 1995; Coley & Barone, 1996). Despite the inability to carry out any statistical analyses on these data, herbivory levels were found to be higher in the *cerrado* than in dry forests (18% and 14%, respectively). Dry forests are usually supposed to experience higher levels of herbivory than do wet forests (Filip *et al.*, 1995; Lowman, 1995; Coley & Barone, 1996). Moreover, herbivory on tree crowns in the savanna canopy tended to be greater than those in the understorey of wet forests, which is usually also greater than in the high canopy of the forest (Coley & Barone, 1996).

In contrast with the high levels of herbivory on *Tabebuia* spp. observed in the *cerrado*, one of the studied species, *Tabebuia aurea* (Manso) Bentham, occurs also as large monodominant populations (i.e., monospecific tree stands) in the wetlands of *Pantanal Matogrossense*, Brazil. In these wetter environments, herbivory levels were as low as 10% of leaf area loss per year (Ribeiro & Brown, 1999). In the wetlands, habitat stochasticity may affect, negatively, pathogens and herbivores and, in consequence, be associated with the long-term maintenance of these large monodominant populations (Ribeiro, 1999; Ribeiro & Brown, 1999). Conversely, monodominant populations within tropical rainforests are susceptible to insect outbreaks and to high levels of defoliation. Monodominant rainforest species normally experience rapid successional processes (Nascimento & Proctor, 1994; Gross *et al.*, 2000). Low seedling and sapling mortality has been recorded for both the monodominant *T. aurea* on the savannas (Ribeiro, 1999; S. P. Ribeiro & V. K. Brown, unpublished data) and the monodominant *Peltogyne gracilipes* Ducke (Fabaceae) in rainforest (Nascimento & Proctor, 1997), regardless of their distinct susceptibility to folivory and successional dynamics. Seedling survivorship is considered an important mechanism for the persistence of both populations in nature.

Although contrasting herbivory patterns were found across savannas and rainforests, further quantitative comparisons are needed to generate consistent conclusions and for the construction of hypotheses to explain the observed contrasts.

Contrasting insect herbivore diversity and distribution between tropical savanna and rainforest

Ecologically sound surveys have shown that some aspects of insect species richness and taxa composition are similar in savannas and tropical rainforests (Table 31.1). For instance, both savannas and rainforests have high Coleoptera diversity, and a high frequency of certain families such as the Cicadellidae, Chrysomelidae, Miridae and Formicidae (Stork, 1991; Basset, 1992c; Carneiro *et al.*, 1995; Krüger & McGavin, 1998a; Novotny & Basset, 2000). There may be phylogenetic rather than ecological reasons for these similarities, since these are all common families in many ecological communities, even outside the tropics (Strong *et al.*, 1984; Farrell & Mitter, 1993). However, there are some differences in the frequencies of certain taxa, which also may have a phylogenetic basis. For instance, the Chrysomelidae in the neotropical region is mainly represented by species of Alticinae, which is the most important subfamily in the Brazilian *cerrado* (Carneiro *et al.*, 1995; Ribeiro *et al.*, 1998; Ribeiro, 1999), and in rainforests of Central America (Ødegaard, 2000c). Conversely, the Eumolpinae and Galerucinae seem to be the most speciose and abundant subfamilies in the tropical rainforests of Papua New Guinea (Basset & Samuelson, 1996).

For both savanna and rainforest, several insect species are rare, that is, they are sampled once (singletons) or only a few times, and many individual trees of a host population do not support specialized insects in their crowns (Basset, 1992c, 1996; Price *et al.*, 1995; Diniz & Morais, 1997). In the Brazilian *cerrado*, for several species of *Erythroxylum* (Erythroxylaceae), the percentages of trees without any specialist lepidopterans varied from 8 to 18% (Price *et al.*, 1995). More dramatically, in a companion study of Lepidoptera on species of

Table 31.1. *Habitat, host species, and insect data*[a]

Habitat	Host plant	Insect taxa or target group	Tree size range (m)	Sampling method	Sample size (number of trees)	Total insect species per host species (average)	Cumulative insect species per host tree (± variance)[b]	Reference
Cerrado (Brazilian savanna)	*Byrsonima*, *Qualea*, *Erythroxylum* spp.	Lepidoptera (adult only)	0.5–2.5	Direct sampling and rearing	1107	137 (57)	0.02 ± 0.008	Diniz & Morais, 1997
Cerrado	*Erythroxylum* spp.	Lepidoptera	1–2	Direct sampling and rearing	502	31 (12)	0.12 ± 0.07	Price *et al.*, 1995[c]
Cerrado	*Tabebuia ochracea* (Cham) Standley	Free-feeding herbivores collected in more than 10% of sampling days	1–6	Weekly direct observation on marked individuals over 25 months	3	16	5.3 ± 1.04[d]	Ribeiro & Pimenta, 1991
Cerrado	*Tabebuia* spp.	All herbivores	3–20	Beating plus mist fogging, clipping and direct observations over two wet seasons	47	53 (27)	1.9 ± 1.2	S. P. Ribeiro & V. K. Brown, unpublished data
Cerrado and transition to high-altitude grasslands	*Miconia*, *Tibouchina* spp.	Free-feeding herbivores	1–3	10 minutes of observation on marked plants	100	25 (4.8)	1.9 ± 1.4	Ribeiro *et al.*, 1999a
Cerrado (xeric and mesic habitats)	Various	Gall-formers	0.5–2	Direct observation	1300	–	6[e]	Fernandes & Price, 1988
African savanna	*Acacia* spp.	Free-feeding herbivores[f]	?–10	Mist fogging over 1 year sampling	31	272 (48)	15.5 ± 11.9	Krüger & McGavin, 1997
Dry tropical forest	24 tree species and 26 liana species	Phytophagous beetles (specialist green feeders only)	20–38	Beating and hand collecting from a canopy crane	77	320 (6.4)	4.1 ± 2.5	Ødegaard, 2000c
Rainforest	10 tree species	Specialist chewing	2.5–40	Hand collecting, beating, and branch clipping	157	185 (18.5)	6.1 ± 2.3[g]	Basset, 1996

[a] Means, if not provided, were estimated from available data in original articles.

[b] Variances were estimated from replicate sample efforts per host species and/or individuals, as available in the original articles.

[c] Data taken only from samples of 1992: average species per plant is the mean value presented in the original article obtained by summing up the four *Erythroxylum* spp. together.

[d] Value is the number of species occurring per host plant per sampling period to correct for the effect of recounting the same species over time.

[e] Fernandes & Price (1988) did not present data that could allow an estimate of variance.

[f] The sum of species from the Orders Orthoptera, Phasmodea, Hemiptera, Coleoptera and Lepidoptera, as the authors did not define values for feeding guilds directly.

[g] Value represents the cumulative number of species of chewers per host sampled.

Erythroxylum and *Byrsonima* (Malpighiaceae), Diniz and Morais (1997) recorded as high as 80% of host trees with no caterpillars at all (out of 16 000 sampled trees), even though the overall richness of insect species was high for all host-tree species. In this study, 179 species of Lepidoptera were collected on 80 host species from the *cerrado*. From 91 identified lepidopteran genera, 11 were oligophagous. Seven out of these eleven genera were recorded as having more than one species per host-plant species. Accordingly, the trend towards large numbers of rare species, as is usually described for rainforests (Stork, 1991; Basset *et al.*, 1996a; Novotny & Basset, 2000), may also occur in savannas, at least for assemblages of Lepidoptera (Price *et al.*, 1995; Diniz & Morais, 1997).

Recently, many studies have focussed on host-species specificity and insect–plant interactions within the canopy (Stork, 1991; Basset, 1992c, 1999a; Krüger & McGavin, 1997, 1998a; Adis *et al.*, 1998a; Basset & Novotny, 1999; Novotny & Basset, 2000). In rainforests, as in savannas, many species of insect herbivores occur at low densities per tree crown and occupy only a few trees in a host population (Novotny & Leps, 1997). As a consequence, insect specificity is very hard to define. Basset (1999a) showed that transient species are a very important component of the fauna collected by beating *Castanopsis acuminatissima* (Bl.) A.D.C. (Fagaceae) trees, in a tropical rainforest of Papua New Guinea. Actually, low proportions of specialist insect herbivores and a significant number of transient species have been found consistently in tropical rainforest assemblages (Stork, 1987b, 1991; Basset, 1992c, 1999a; Chey *et al.*, 1997; Basset & Novotny, 1999; Novotny & Basset, 2000; Ødegaard, 2000c).

Species richness and the abundance of specialist herbivores are apparently higher in savannas than in rainforests. Ribeiro & Pimenta (1991) consider that 15% of sampled herbivores on *Tabebuia ochracea* (Chamisso) Standley were specialists (16 out of 104 species). Their estimate was much higher than the 3.0–4.5% found by Basset (1992c) for *Argyrodendron actinophyllum* F. Muell. (Sterculiaceae) in an Australian subtropical rainforest; the 2.3–3.6% recalculated from Erwin's classical paper by Basset *et al.* (1996a), using data from 10 unrelated tree species in Papua New Guinea; or the 2.8–5.0% estimated by Ødegaard *et al.* (2000), from 24 trees and 26 lianas in Panama. In addition, S. P. Ribeiro and V. K. Brown (unpublished data) showed that 41% of the herbivorous beetle species observed feeding on the trees of *T. ochracea* studied by Ribeiro and Pimenta (1991) were still observed feeding on these same tree crowns 10 years later, suggesting an even higher proportion of specialists in the system. Ribeiro and Pimenta (1991) estimated the number of specialists based on a collection made over 3 years with weekly samples plus direct counting on a set of three trees of *T. ochracea* and surrounding vegetation (see also Ribeiro *et al.*, 1994a). Data on leaf-chewing species observed on *Acacia* trees in the Tanzanian savanna suggest the existence of a great number of specialists, although the authors did not discuss the problem of specificity (Krüger & McGavin, 1997, 1998a; McGavin, 1999).

In general, analysis does not suggest that a tree species in the rainforest could accumulate more insect species than another tree in the savanna. Nevertheless, it is likely that more species per tree crown occur in the rainforest than in the savanna. Although the overall richness of free-feeding herbivores per host species did not differ between vegetation types (ANOVA, $F_{1,6} = 0.65$, $p > 0.05$), the richness of specialist beetles per sampled tree in the rainforest was twice as high as for specialist lepidopteran or for all free-feeding species on a tree in the savanna vegetation, even though the large variance among articles generated only a marginally significant result when comparing insect species per sampled tree (ANOVA, $F_{1,6} = 4.4$, $p < 0.08$). Populations of specialist insects in rainforests may respond to host trees that occur at lower densities than are encountered in the savanna, thus forcing the insects to stay wherever a host tree exists or, alternatively, to occupy any individual tree found. This hypothesis predicts a greater species richness per tree in the rainforest than in the savanna. Conversely, tree species in both African and Brazilian savanna occur in high densities, patchily distributed within particular soil types (Sarmiento, 1983; Richards, 1996). In spite of abundant resources, effective insect population size may not be large enough to generate broad tree occupancy because of the operation of other regulatory factors. Such factors may include low nutritional value of resources, associated mechanical defences and differential predation. Accordingly, specialist insect herbivores may be also patchily distributed, becoming abundant on only a few trees (Fernandes & Price, 1992; Ribeiro *et al.*, 1994a, 1999a; Espírito-Santo & Fernandes, 1998).

In the Brazilian *cerrado*, most leaf-chewing specialists (particularly chrysomelid beetles) and sap-sucking and gall-forming herbivores on *T. ochracea* and *T. aurea* were abundant when they occurred on a single tree (Ribeiro & Pimenta, 1991; Ribeiro *et al.*, 1994a; Ribeiro, 1999). However, absence of particular insect taxa was often recorded for a percentage of trees on any sampling occasion. Ribeiro (1999), in one sampling season in 1996 (from 46 individual trees), found 82% and 70%, respectively, of *T. aurea* and *T. ochracea* trees without species of Chrysomelidae. Comparable results were obtained by Diniz and Morais (1995), Morais *et al.* (1995) and Price *et al.* (1995) for lepidopteran species in the *cerrado*.

There are, clearly, great numbers of endophagous species of herbivores on the savannas, particularly in the neotropical region. In the Brazilian *cerrado*, leaf-mining and gall-forming insects occur as large populations on leaves of several host plants (Fernandes & Price, 1988, 1992; Lara & Fernandes, 1996; Fernandes *et al.*, 1997; Price *et al.*, 1998; Ribeiro, 1999). The previous analysis showed a trend towards fewer species of insect herbivore per tree crown in the savannas than in the tropical rainforest, but that result referred to free-feeding herbivores only. It did not include guilds, such as gall-forming insects, which are extremely diverse in the *cerrado*. The most ubiquitous pattern in the guild distribution of insect herbivores in the *cerrado*, and other sclerophyllous vegetation with open canopies, is the high diversity of gall-forming species. The evolutionary and ecological reasons for the success of this guild need to be better understood in order to clarify what may be the key differences between the canopies of savannas and rainforests.

The *cerrado* is a biome with a high diversity of tree, shrub and herbaceous species (Goodland & Ferri, 1979; Sarmiento, 1983), which make up large and complex enclaves of xeric and mesic arboreal vegetation. Most of the mesic vegetation is gallery forests distributed along rivers and streams (Eiten, 1978). This juxtapposition allows exploration of the mechanisms, such as humidity and canopy openness, which may generate the greater number of gall species and abundances in the xeric *cerrado* proper compared with the mesic gallery forests.

The studies of Fernandes & Price (1988, 1992) concerning variation in the diversity of gall species along a hygrothermal gradient was a first attempt to compare wet forests and dry *cerrado*. These studies gave rise to the hypothesis to explain the global distribution of gall species, which claimed that environmental harshness and sclerophylly (rather than water stress) are the likely determinants of high diversity (Price *et al.*, 1998: the hypothesis of the harsh environment (HHE)). Although adaptation to sclerophylly may be the mechanism behind the increasing diversity of gall species in the *cerrado*, Price *et al.* (1988) suggested that higher mortality rates in wet habitats (by fungi and parasitoids) may lead to lower numbers of galls in these habitats and so contribute to the observed patterns of occurrence. They maintained that there are more gall species in xeric than mesic habitats, even when sclerophyllous vegetation dominates the latter. More recent data on the causes of gall mortality are evaluated below against this hypothesis.

There have been two attempts to propose alternative hypotheses to explain the xeric/mesic contrast in species richness of gall-formers. One considered soil types and specific taxa of host plants as key factors driving the pattern (Blanche & Westoby, 1995). The other more complete reassessment proposed that resource synchronization by seasonal resprouting after fire was a likely causes of higher rates of speciation (by host shifts) in Mediterranean vegetation types, which includes *cerrado* and other savannas (Mendonça, 2001). In the next section, I reassess the hypotheses of Fernandes and Price and reject their original conclusions, based on recent gall mortality data and bottom-up effects on gall insect oviposition choices.

The hypothesis of harsh environment (Fernandes & Price, 1998): a re-appraisal

Insect herbivores in canopies of dry savanna and wet gallery forest in the cerrado: *the sclerophylly effect*

The *cerrado* is a persistent biome, occurring in a geologically old region (Sarmiento, 1983). There is strong evidence that this vegetation has covered the Central Brazilian Plateau since the Tertiary period (Freitas, 1951; Rizzini, 1997). It is a mosaic plant assemblage, varying from grassland to dense arboreal savannas in response to a gradient of fertility (Eiten, 1972; Goodland & Pollard, 1973). Like other savannas, the vegetation of the *cerrado* is frequently affected by fire. *Cerrado* is a type of sclerophyllous vegetation, *sensu* Loveless (1962), as it is the result of nutrient poor soils rather than of drought (Goodland, 1971; Goodland & Pollard, 1973; Eiten, 1978; Goodland & Ferri, 1979).

The studies of Fernandes and Price (1988, 1991, 1992) and Ribeiro *et al.* (1998, 1999a) were conducted in the National Park of Serra do Cipó, in *campo cerrado* (woody savanna) or *cerrado sensu stricto* (savanna woodland), depending on altitude. Serra do Cipó is a mountain region within the *cerrado* domain, where most of the mesic habitats are thin gallery forests surrounded by xeric savannas and high-altitude grasslands. Its vegetation is highly sclerophyllous, with many species endemic to small areas isolated by rocky soils, marshes or sandy grasslands (Giullieti & Pirani, 1988).

The effect of leaf sclerophylly on insect feeding guilds

Sclerophyllous plants have a high-carbon and low-nitrogen content, which results in structures such as thick cell walls and tough leaves (Juniper & Jeffree, 1983; Mole *et al.*, 1988; Salatino, 1993; Turner, 1994). These scleromorphic traits may have a direct impact on insect oviposition, feeding and survivorship. However, the likely origin and causes of the evolution of sclerophylly is thought to be physical factors rather than defence against insects (Haslam, 1988; Salatino, 1993). Sclerophyllous plants are adapted to nutrient-poor and water-deficient soils, and low levels of phosphate appear to be one of the most relevant factors associated with the phenomenon (Loveless, 1962; Fernandes & Price, 1991). Loveless (1962) considered sclerophylly to be a response to nutrient stress rather than to water stress. In the Brazilian *cerrado*, sclerophylly is also related to high levels of aluminium in the soil, which causes the so-called aluminotoxic scleromorphism (Goodland & Ferri, 1979).

High concentrations of polyphenols occur in scleromorphic plants (Salatino, 1993). Phenolic compounds, such as tannin and lignin, result from alternative pathways in protein formation, particularly when there is inadequate availability of phosphate and nitrogen (Harborne, 1980; Haslam, 1988; Perevolotsky, 1994; Turner, 1994). Tannins are known to be antifeedants (Feeny, 1976; Harborne, 1980; Bernays *et al.*, 1989). Accordingly, highly sclerophyllous plants are expected to be protected against insect herbivores. However, despite a general low species richness and abundance of insect herbivores (Ribeiro *et al.*, 1998), no direct effect of tannins on patterns of insect species distribution has been demonstrated in the *cerrado* (Madeira *et al.*, 1998; Ribeiro *et al.*, 1999a).

The effect of sclerophylly on guilds of insect herbivores is an evolutionary and, therefore, historical phenomenon. Nonetheless, an evolutionary response to sclerophylly in the *cerrado* may be reflected in the great diversity of gall-forming species (Fernandes & Price, 1991; Lara & Fernandes, 1996; Fernandes *et al.*, 1997; Price *et al.*, 1998). Larvae of gall insects manipulate the plant tissue around them genetically, causing changes in the allocation of nutrients and tannins. Nutrients are concentrated in the meristematic tissue in contact with the larvae, and the tannins are transported to external tissues, creating a chemical barrier around the gall (Fernandes, 1998). This insect–plant interaction supposedly explains their success in stressed, harsh and sclerophyllous environments. Moreover, tough scleromorphic leaves tend to be long lived, leading to low gall mortality through leaf shedding (Fernandes & Price, 1991).

Gall species distribution on plant crowns in the cerrado*: is there really a gradient effect?*

A classical approach to ecological evolutionary questions is the search for patterns in species distribution along physical gradients. Gall-forming insects, it is suggested, are diverse in harsh habitats. In other words, their species richness increases with increase in plant water stress and atmospheric dehydration along altitudinal and hygrothermal gradients (that is, from mesic to xeric habitats). This pattern has been shown, recently, to be global, associated with Mediterranean-type (that is, scleromorphic) vegetation in temperate warm climates with well-defined wet/rainy seasons (Price *et al.*, 1998).

An interesting exception may be the African savannas. There are no data on gall diversity for subSaharan savanna, in contrast with the southern African Cape floristic region. A broad search in WebSPIRS for the last 20 years generated no references on gall species on native savanna trees and I know of no articles about gall insects or galls on African savanna trees nor of references to such insects in recent literature on free-living insects in canopies in the African savanna (Krüger & McGavin, 1997, 1998a,b, 2000, 2001; McGavin, 1999; Gross *et al.*, 2000; Ch. 19). It is not clear if the lack of data reflects an actual lack of galls. If galls are indeed absent, then the ecological or evolutionary reasons need investigating. Richards (1996) has pointed out that some African savannas are similar in their responses to soil conditions to the Brazilian *cerrado*. Geographically distinct forces acting on African savannas need to be considered, such as the strong effect of vertebrate grazing and herbivory (Keesing, 2000; Oba *et al.*, 2000).

Idiosyncratic trends: are the actual causes of insect distribution properly considered?

Studies by Fernandes & Price (1988, 1991, 1992) in the Serra do Cipó National Park region have been extremely important in our awareness of the global pattern of distribution of gall species. Several idiosyncrasies in relation to this global pattern are apparent. For instance, although there is a clear direct relationship between scleromorphy and gall species richness at a regional and local scale, for various host–insect systems studied there was no direct effect of tannins on gall mortality or levels of free-feeding herbivory (Fernandes *et al.*, 1996; Espírito-Santo & Fernandes, 1998; Madeira *et al.*, 1998; Ribeiro & Brown, 1999; Ribeiro *et al.*, 1999a).

In their studies of six species of Melastomataceae in Serra do Cipó, Ribeiro *et al.* (1999a) have shown that polyphenol concentration is as high in mesic as in xeric habitats and across altitudinal extremes. In addition, for five species of plants, they found that tannin concentrations were as high or higher than those found in species considered sclerophyllous, or for mean values of sclerophyllous vegetation around the world. In other words, high tannin concentration was correlated with sclerophylly, but it did not vary among habitats or along physical gradients in the *cerrado*. Despite the general trend towards high concentration, polyphenol contents varied significantly between species (Ribeiro *et al.*, 1999a). Similarly, Madeira *et al.* (1998) found that tannin concentration and sclerophylly did not vary among altitudes for flowers, fruits or leaves of *Chamaecrista linearifolia* (G. Don) H.S. (Fabaceae). Although the data point to similar levels of sclerophylly in mesic gallery forests and the open *cerrado*, the number of species of galling insects in the former habitat was lower. In addition, although the number of gall species fell with increasing altitude in the xeric habitats, it remained constant in the mesic gallery forests, which suggests that mesic conditions are more important in restricting gall colonization and survival than sclerophylly itself (Fernandes & Price, 1988, 1991).

The effect of sclerophylly on populations of gall species was studied by relating tannin concentration to insect distribution. *Neopelma baccharidis* Buckhardt, is a species of gall-forming homopteran occurring on *Baccharis dracunculifolia* D.C. (Asteraceae). This species was shown to respond to host-plant sex, but no relationship was found between population density and tannin concentration (Espírito-Santo & Fernandes, 1998; but see Faria & Fernandes, 2001). On the basis of gall physiology and the ability of gall formers to re-allocate tannins, this compound was expected to be one of the most important traits affecting positively the occurrence of gall-forming insects. However, the high α- and β-diversity of gall species (up to 17 species) on *B. dracunculifolia*, which is a host plant with a relatively low concentration of tannins, is clearly contrary to this expectation (Fernandes *et al.*, 1996). The relative importance of this plant in regional patterns of gall distribution is remarkable and shows that factors other than tannin concentration need to be considered in looking for general explanations of herbivore patterns in the *cerrado* (Espírito-Santo & Fernandes, 1998).

Returning to the more specific question of what does cause differential mortality of insect larvae in galls in xeric and mesic habitats, it was found that differential growth rates among populations of gall-forming species may change their survivorship rates. Large galls, for instance, become more vulnerable to predation and fungal contamination (de Souza *et al.*, 1998). Likewise, the toughness of gall walls may increase survivorship during the first month by decreasing infestation by parasitoids (Fernandes *et al.*, 1999). In addition, Fernandes (1998) and Fernandes & Negreiros (2001) have shown that plant hypersensitivity reactions are the main cause of mortality in several gall-forming species.

These hypersensitivity reactions seems to form a general pattern, likely to have evolved several times in many host plants, and are not habitat specific (Fernandes, 1998). Fernandes & Negreiros (2001) showed that hypersensitivity was the main cause of gall death in seven out of eight tree host species in the *cerrado* that were taxonomic and geographically independent. This general pattern undermines the hypothesis of Fernandes and Price (Fernandes & Price, 1992; Price *et al.*, 1998), which appeared to be supported by early data analyses. Although subsequent studies have provided interesting data about ecological mortality factors, none is clearly related to the hygrothermal gradient (Espírito-Santo & Fernandes, 1998; Fernandes, 1998; Fernandes & Negreiros, 2001) and there remains no evidence of a mechanism leading to increased vulnerabilty of gall larvae in mesic forests (such as the gallery forests adjacent to the *cerrado*).

Is the gallery forest poor in other herbivore guilds?

Based on the assumptions of HHE, a contrasting pattern of species distribution may be expected for free-feeding herbivores. These species should be affected severely by drought, dehydration and nutrient-poor conditions of plants in harsh xeric habitats and should be more diverse in the more climatically favourable mesic habitats. Caution is needed here, as the idea of a truly favourable habitat may be illusory for the thin, soil-poor wet forests within the *cerrado*, and the xeric/mesic contrasts may relate to the understorey to canopy gradients. If this is the case, sampling only in the canopy of both habitats would not show any pattern.

Ribeiro *et al.* (1998) studied xeric and mesic habitats using a sweep net to sample isolated crowns (of trees and shrubs) in the *cerrado* on the upper canopy and fringing foliage of gallery forests, summing up 6600 sweeps from 43 sample sites. Insects were sampled on leaves that had experienced similar exposure to solar radiation and dehydration, for both the open xeric savanna and the closed, wetter gallery forest. As expected, and in contrast to the original data in support of HHE, no significant differences were found in free-feeding insect richness and abundance between xeric and mesic habitats along 800 m of altitudinal gradient. Ribeiro *et al.* (1998, 1999a) confirmed that physical conditions could be as harsh in the mesic as in the xeric habitats, at least in terms of plant nutrient and polyphenol concentrations. Ribeiro *et al.* (1999a) proposed that the occurrence of sclerophylly was consistent with overall low species richness of free-feeding insects and the high tannin concentrations in host plants along altitudinal and hygrothermal gradients. Accordingly, for free-feeding insects, it makes sense to consider that a mesic habitat within the *cerrado* is an environment as scleromorphic as a nearby xeric savanna.

Fernandes & Price (1988) demonstrated a higher diversity and abundance of free-feeding herbivores in the mesic understorey than in the xeric *cerrado* canopy. All available data (Fernandes & Price, 1988; Ribeiro *et al.*, 1998) suggest a higher diversity and abundance of free-feeding insects, in general, in the understorey than in the canopy of gallery forests (although such a comparison was not made by these authors). This pattern is the opposite to that found by Basset *et al.* (1992) in rainforests in Cameroon. High proportions of young foliage and enhanced activity owing to temperature were positive factors related to high abundance of insect fauna in the canopy of the rainforests of Cameroon. Such factors do not appear sufficient to raise the numbers of insect herbivores in the gallery forests in Serra do Cipó. This may reflect the occurrence of sclerophylly in these latter canopies. Marques *et al.* (2000) has shown leaf plasticity for two species of *Miconia* (Melastomataceae) in the *cerrado* in response to light levels. Leaves of plants fully exposed to the sun were denser, thicker and with higher trichome density than leaves of the same plants in the shade. All these traits are deterrents to insect herbivory and may moderate the growth of insect populations in the canopies of gallery forests. A thorough examination of the data suggests a large negative effect of sclerophylly, which results in high and low species richness of gall-forming and free-feeding insects, respectively. In this scenario, low numbers of gall-forming species originally found in the mesic habitats may be a spurious pattern induced by an inappropriate comparison between understorey and canopy. To a certain extent, the contrast in plant architectures between xeric and mesic habitats presented by Fernandes and Price (1988, 1992) ignored an important aspect: in the xeric habitats, the foliage that was sampled belonged to low tree and shrub species but represented the upper canopy of such a discontinuous arboreal habitat; in the gallery forests, the samples taken at the same height and from the same plant taxa represented the understorey.

Do the patterns change from the understorey to the canopy?

Price *et al.* (1998) have shown for two different tropical regions that the number of gall species on canopies would not be expected to differ from that in the understorey (or if it did, it would be smaller in the former). Such a pattern reflects the stronger effect of nutritional conditions over microclimate in defining adequate habitats for the accumulation of gall species, that is, scleromorphic vegetation. However, the understorey of the thin gallery forest in Serra do Cipó may somehow represent a particular habitat type. First, many plant species found in the mesic understorey are also found in the xeric habitat (Ribeiro *et al.*, 1999a). Therefore, the pattern of herbivore distribution in the understorey may not be mainly (or only) defined by vertical colonization of specialists from the canopy trees onto their supposed seedlings and saplings (Barone, 2000). The gallery forests within the *cerrado* are composed of populations of riparian species, plus marginal individuals from plant populations from the xeric *cerrado*. For the latter, the low numbers of gall-forming insects may reflect a response to bottom-up effects such as slower growth rates, lower leaf nutrient quality and stress owing to light deficiency. These may result in a lower probability of successful encounter by adult insects with suitable oviposition sites. The contrast with the same species of plants and their associated insects in the xeric habitat, to which they are adapted better, may well account for the observed patterns (Price, 1991; Herms & Mattson, 1992; Karban & Baldwin, 1997).

Little sampling of the canopy of these forests has been carried out since Fernandes and Price (1988, 1991) suggested that plant architecture is not an important determinant for gall species richness. However, the present status of our knowledge of gall species distribution in the *cerrado* suggests that further detailed analysis of the canopy should be made, searching, for instance, for variation within and among tree crowns of different host taxa (see Basset, 1991b). The hypothesis here proposed is that scleromorphic conditions found in the canopy of the mesic forests provide a similar physical background for gall establishment to that found on the canopies in the open savanna. In contrast to the insects that colonize plants in the understorey, those in the crowns would likely suffer as much nutritional stress and solar radiation as those growing in open savannas. This conclusion agrees with the statement of Bell *et al.* (1999) that branch construction at the canopy surface of tropical rainforest is more akin to chaparral shrub vegetation than to understorey vegetation.

In addition, the patterns of ant diversity and activity among the crowns of trees in the xeric and mesic habitats must be taken into account as part of a broader understanding of herbivore species distribution, and this may be a more important, and unexplored, top-down effect in the *cerrado*. For instance, the number of ant species detected increased significantly in mesic compared with xeric habitats and was related to higher plant species number and more complex vegetation architecture (Fernandes *et al.*, 1997).

Free-feeding insects and gall larvae predation in the cerrado*: an unexplored study field*

Arboreal ant species are important predators and key species in arboreal assemblages. The ant assemblage in the *cerrado* is extremely diverse and not as well studied as it is in the tropical rainforest. Del-Claro & Oliveira (1999) described 21 ant species attending the membracid *Guayaquila xiphias* Fabr. on the *cerrado* tree *Didymopanax vinosum* March. (Araliaceae). Ribeiro (1999) collected 17 species of ants on 20 individuals of *T. ochracea* from the *cerrado* of Minas Gerais, and another 13 different species on 18 individuals of the sympatric congener *T. aurea*. Most of these species belonged to the genus *Pseudomyrmex* (Myrmicinae). A higher number of arboreal ant species was found in mesic than in xeric habitats in Serra do Cipó (Fernandes *et al.*, 1997), but their impact on insect herbivores and their variation between canopy and understorey has been little studied. In an African savanna, for instance, Krüger & McGavin (1998a) observed that insect diversity responded significantly to ant biomass in a sample of 31 *Acacia* trees of six different species. The effects of predation and 'negative patrolling' by ants on oviposition and survivorship of species of gall insects could be an important factor reducing the diversity of this guild. Clearly, this factor could exacerbate the negative effect of plant hypersensitivity and then explain the pattern of low gall abundance in the mesic understorey, which cannot be explained by fungi contamination (Espírito-Santo & Fernandes, 1998).

CONCLUSIONS: DISCONTINUITY OR A COMPLEX STRATIFICATION IN THE *CERRADO* CANOPY?

I have reviewed here patterns of insect herbivory, diversity and distribution among savannas and dry and wet tropical forests. Focussing on the Brazilian *cerrado* and African savanna, peculiarities of the insect fauna in this type of vegetation with an open canopy have been described. Higher herbivory levels, fewer insect species per tree crown and larger populations of specialists, particularly of endophagous insects, are the main differences between tropical savanna and rainforest. In addition, because of the importance of gall-forming insects in savanna and other harsh types of habitat, I have reviewed patterns in the distribution of gall-forming insects and the interactions with their host plants, parasites and predators. As it is one of the most diverse biomes in the world for gall species, the *cerrado* vegetation has been my focus. Following this review, I revisited Fernandes & Price's hypotheses of harshness and hygrothermal gradients. The original explanation for low gall species diversity in mesic habitats, based on greater levels of mortality, has been rejected. Bottom-up effects and lower rates of oviposition in the understorey compared with the canopy are proposed as important factors in determining the low numbers of gall species in the understorey of mesic forests within the *cerrado*.

One plausible explanation for the pattern of higher specificity and population densities of (mainly endophagous) herbivore species in savanna habitats compared with rainforests may be the combination of less-palatable but more available resources over evolutionary time. The radiation of gall species may be related to greater adaptability to sclerophyllous vegetation, as described in the 'hypothesis of sclerophylly' (Ribeiro *et al.*, 1998, 1999a). However, to what exactly they adapted in the evolutionary process is uncertain. Evolutionary response to synchronizing resprouting induced by fire may well be the main mechanism behind such radiation (Mendonça, 2001), but this explanation does not completely exclude the sclerophylly hypothesis. Recently, Hawthorne and Via (2001) have studied the genetic architecture of *Acyrthosiphon pisum pisum* (Harris) (Aphididae) and have demonstrated how closely linked (pleiotropic) quantitative traits related to resource use and mate choice may result in speciation. Although the hypothesis of Mendonça (2001) needs to be tested, genetic linkage could be the basic explanation. The fact that many plant populations have patchy distributions in the *cerrado* could contribute to host species apparency and predictability. In addition, the discontinuous canopy of savanna vegetation makes a tree crown considerably more apparent. As a consequence, fidelity to the tree species' crown could increase the probability of a conservative evolutionary response. In this sense, the mechanism behind speciation could be based on rates of host shift on a large spatial and temporal scale, as proposed by Mendonça (2001), followed by local and ecological maintenance through *a posteriori* host selection and fidelity. The debate continues and further empirical studies are urgently needed to clarify the mechanisms behind the pattern.

Experience with tall equatorial rainforest might lead one to conclude that savanna, or even low-stature semidecidous tropical rainforest, does not have a proper canopy surface (outer canopy *sensu* Moffett, 2000), and consequently there would be no such a thing as an 'upper' canopy (see Ch. 1 for terminology) as there would be no clear layers of leaves resulting from closely distributed tree crowns. In open arboreal vegetation, the discontinuity allows an even climate from the canopy into the understorey and to very close to the ground, which breaks down the idea of a canopy structure (Y. Basset, personal communication). However, experience of sampling and researching in this habitat leads to a different view of the canopy and of its effect on insect populations. When a very sparse tree distribution generates a set of totally isolated crowns, insect specificity may be favoured, together with a strong interaction with the epigeal fauna, particularly patrolling ants such as *Camponotus* spp. (Formicinae). This notwithstanding, it is important to remember that not all types of savanna are so clearly discontinuous, at least not in the Brazilian *cerrado*. In most arboreal *cerrado* vegetation, there is some proximity among tree crowns, and, occasionally, a milder climate in the understorey can be detected. Of course, no obvious understorey-to-upper canopy microclimatic gradient may occur. Factors such as soil nutrients, solar radiation and dehydration will be constraining to insect populations in any stratum, more or less equally. In this sense, neither a clear canopy microclimatic structure (Basset, 1992c) nor hot-spot regions for herbivory (Lowman, 1995) should be expected, except *within* each tree crown. The broader definition of canopy proposed by Moffett (2000) clearly encompasses this type of more open, less-differentiated canopy.

Canopy derives from a Greek word for bed sheet. Such a symbolism works well with the upper, and even better with the outer, canopy (*sensu* Moffett, 2000). Nevertheless, in many arboreal savannas, the outer canopy may be envisioned as a corrugated surface, going up and down across the tree crowns. Strong dehydration, usually found in the outer canopy, goes up and down in the *cerrado* following the shape of each tree crown. Instead of leaf layers in a well-defined understorey–canopy gradient with distinct microclimates in each stratum, the savanna is better described in terms of its great complexity and heterogeneity of microhabitats based on highly stratified tree crowns, various levels of crown proximity and canopy continuity, plus the lack of a real understorey. The corrugated aspect of the canopy has been described as 'canopy roughness' (Richards, 1996: p. 46). Considering that canopy discontinuity does not address aspects of vertical structure but only horizontal interruption in the foliage, the concept of canopy roughness is useful to explain such habitats as savanna canopies. As pointed out by Birnbaum (2001), the topography of a canopy surface is defined by a complex system of tree crowns and groups of tree crowns and gaps. Both the topography (Birnbaum, 2001) and the tree architecture (Sterck *et al.*, 2001) of a canopy surface can be described in detail. However, if patterns in the undulation of a given canopy are of interest, the development of broader concepts that group canopy types with similar traits is of great importance for our science.

The ecological significance of the shape and variation of the outer canopy is a poorly explored subject. For instance, how vulnerable are the insect assemblages in a fragmented forest with natural canopy roughness compared with forests with closed canopies? Could edge effects be less detrimental to insect populations adapted to a heterogeneous canopy surface (Foggo *et al.*, 2001; Théry, 2001)? Or could selective logging cause less damage in a forest with a high degree of canopy roughness, as the degree of light incidence and patterns of rainfall may vary little from natural to the logged stage (Chappell *et al.*, 2001)? Even in a natural forest with highcanopy roughness, the lack of a strong vertical gradient in light and temperature could decrease population sizes of those insect species sensitive to dehydration (particularly generalist free-feeding species). However, high levels of roughness of a canopy surface will result in vertically stratified tree crowns, and thus there will be a greater proportion of foliage biomass exposed to solar radiation and dehydration, per unit of plant biomass. For those insects adapted to dehydration and low nutritional contents of leaves, such as gall-forming species, the area of suitable habitat should be larger and more apparent than that provided by crowns in a dense rainforest canopy. Along with host-plant suitability for different insect species, it seems quite likely that canopy structure and phenological dynamics may influence species radiation in certain feeding guilds and, therefore, may structure insect assemblages.

ACKNOWLEDGEMENTS

I thank Milton Mendonça Jr, Cláudia Malafaia, Greg Masters and two anonymous referees for their comments on the manuscript and Yves Basset for his support to the proposal of this review and fruitful discussions. G. Wilson Fernandes and Maurício Faria contributed with inspiring ideas for the revisitation of the Fernandes & Price hypothesis. My studies were funded by FAPEMIG, CNPq and Universidade Federal de Ouro Preto.

32

Canopy flowers and certainty: loose niches revisited

David W. Roubik, Shoko Sakai and Francesco Gattesco

ABSTRACT

It is the business of evolutionary ecology to assess variation in mutualism and antagonism. We studied ecological certainty using forest canopy observations of insects visiting flowers and using collections of thrips in flowers fallen on the ground or near ground level, at four lowland sites in central Panamanian forests. The nonthrips studies were made using canopy cranes operated by the Smithsonian Tropical Research Institute in a primary and a secondary forest. Loose niches in plant-pollinator systems call for radical shifts in abundance and sporadic participation by interacting organisms. Thrips and bees were the principal flower visitors and potential pollinators studied, here discussed for 30 plant species monitored for two or three sequential flowering seasons. Thrips, in general, shifted greatly in abundance from year to year for half the plant species but were (as a group) consistently associated with host plants. For example, *Frankliniella parvula* (Thripidae) on *Gustavia superba* (Lecythidaceae) remained the dominant species, when thrips in the flowers were abundant both years. For bees and other visitors, the observed species remained the dominant flower visitors for two or three seasons in half the study plants. In contrast, many plants possessed loose pollination niches and the African honey bee had a prominent role in their visitation. This bee appeared at 25% of over 100 species, implying dispersal–assembly of the association (not close coevolution) and transitory competition in such systems (rather than hard competition/niche assembly). Specialized flowers pollinated by large bees, however, had the least variable pollinator species and the largest variance among individual species, that is, the highest individual bee species dominance. Nonetheless, a log-normal-like distribution of species abundance, both for specialized and generalized flowering plants, follows predictions of the multinomial neutral theory of biodiversity presented by Hubbell (2001). We conclude that loose niches are a quantitative phenomenon not a qualitative 'coming and going' of individual species, that such systems are common and that they often involve generalist plant–pollinator relationships.

INTRODUCTION

Canopy insects most often fall outside the consideration of evolutionary ecology because of the associated sampling problems: data are limited by few, scattered observations, sometimes taken in the wrong place or at the wrong time. Nearly all tropical trees, lianas and shrubs have flowers visited, if not pollinated, by insects, birds, bats or other vertebrates. However, many tree species do not flower every year (e.g. Sakai *et al.*, 1999b) and most are locally rare; consequently, their associates are thought not to be narrow specialists (Roubik, 1989; Bawa, 1990; Hubbell, 2001). Similar holes in knowledge about herbivores and their hosts have led to inquiries grappling with interannual variation in antagonistic relationships (Basset, 1992a, 1999b; Colwell, 1995; Barone, 1998; Herrera *et al.*, 1998; Minckley *et al.*, 1999; Curran & Webb, 2000). To the extent that flower visitors are almost never purely mutualists but exploit floral rewards and floral parts themselves as herbivores, such studies overlap in content and theory. In some forests, observations at relatively exposed or lower canopy flowers are deemed adequate for general overviews of floral visitation at trees and lianas by pollinator guilds (e.g. Frankie *et al.*, 1983; Bawa, 1990). Spatio-temporal change in canopy activity of pollinators is, however, noteworthy (Roubik, 1993). If the major events in plant reproduction occur exclusively in the upper canopy, then comparative data across years remain critically scarce. This exacerbates the fact that we know too little about reproductive ecology in tropical forest trees, pollination biology at the metapopulation or local community levels, or the behaviour of flower visitors.

In order to discover the rules governing plant-pollination systems, an attempt to formulate a testable hypothesis from existing information led to the suggestion that many pollinators and their flowers maintain 'loose niches' (Roubik, 1992). Sustained competition has not led to strict partitioning of different floral species among a guild of ecologically similar visitors, or vice versa. Genera, taxonomic tribes or families may have predictable and sustained interactions, upheld by seasonality, while species may 'come and go'. In addition, core species (i.e. those responsible for most of the interaction) were postulated to change from year to year. This dynamic concept resembles the 'ecological drift' explicit in a general theorem on relative species abundance and taxonomic richness in biological communities (Hubbell, 2001). As we will show, some tenets of island biogeography and Hubbell's formulation may also apply to tree canopy utilization by flower visitors.

Here we explore the consistency of association among canopy flowers and insects, using our observations from 2 or 3 successive years and seasons in four lowland moist and wet forests in central Panama. Our purpose is to examine the certainty with which flowers are associated with visitors, but we also highlight the exotic honey bee, *Apis mellifera scutellata*. Having arrived in central Panama in 1984, this unusual bee, which makes large but migratory colonies, has in recent decades become established throughout the neotropics, with barely explored biological consequences (Roubik, 1991, 2000; Roubik & Wolda, 2001). As an exotic invader of intact forests, it gives valuable perspective on the organization of communities and the study of interactions among species with a much longer history in the habitat.

METHODS

The four study sites were large forested nature preserves in central Panama, all at elevations under 200 m. We used two canopy cranes (construction cranes permanently installed within the forest) that occupy two sites. One is in primary forest approximately 10 km from a still-forested Caribbean coast, and one is on the outskirts of Panama city, within 2 km of the Pacific Ocean but connected to extensive secondary forest merging eventually with lowland Caribbean forest some 50 km north. We recorded flower visitors generally in the top of the canopy, or thrips present in flowers of lower tree branches or on canopy-level flowers recently fallen to the ground (F. Gattesco & D. W. Roubik, unpublished data). In canopy crane work, dominant visitors were judged not just by number but also by the tempo of flower visitation. Those that visited more flowers and sustained a greater visitation tempo increased in their visible dominance. The canopy cranes each had a gondola suspended from a jib 50 m long at a maximum height that reached the top of all trees at 30–40 m, or at any lower level within their circumference, encompassing a total area of some 0.8 ha. We recorded visitor number during one or more observation periods of 30 minutes to 1 hour, between 07:00 and 12:00 at the Parque Metropolitano (PM) and Fort Sherman (CS) canopy observation cranes, from 1999 to 2001. We usually repeated observations at particular trees or species several times in one morning. As field observers, we continued monitoring flowers until we felt confident of having observed the participants in interactions within canopies available to us. Thrip surveys were made during 1995–1996 along several extensive trails in the closed canopy forest on Barro Colorado Island (BCI) and along a 12 km section of Pipeline Road (PR) within Parque Nacional Soberania, in lowland forest situated in the centre of the isthmus of Panama and near the Panama Canal. We walked along trails and collected flowers and thrips during the morning on BCI (08:00–12:00) and slightly later in the day along PR. Much of the forest near the Parque Soberania access road (PR) consists of second growth, much of it under 70 years of age, as does forest at PM canopy crane. However, extensive primary and older (> 100 years) secondary forests comprise both the BCI and CS sites and also surround and are continuous with both the Parque Metropolitano and Parque Soberania areas in which we worked.

Our thrips and general visitor surveys were not made at the same time, although some included the same tree species for both dynamic airborne flower visitors and tiny thrips active within the flowers or inflorescences. Voucher specimens for both studies are at the Smithsonian Tropical Research Institute (STRI) and in the collections of the authors. When possible, primarily with bees, visitors were identified to species as observed on flowers, thus avoiding disturbance by sweeping with an insect net. Thrips were collected from 20 sample flowers placed in a zip-lock plastic bag by floating them in a detergent–water medium and then removing them with a fine brush or forceps for preservation (F. Gattesco & D. W. Roubik, unpublished data). For the thrip observations, we include only a sample of taxonomic information obtained by collecting and preserving nearly 30 000 adults and larvae. Most of the information here concerns thrip 'infestation' (see Toy, 1991) in successive years, with additional details given on thrip taxonomy in association with one species that provided many samples, *Gustavia superba* (H.B.K.) (Lecythidaceae), and the distribution of the common associated species among the other 265 plants included in our survey. Taxonomic identifications of bees were made with the reference collections of D. W. R. at STRI in Panama, whereas most other insects are not identified here beyond family or genus. Analysis of flying visitors was necessarily focussed on the bees, which in canopy plants were the major visitors and often the only potential pollinators of angiosperms (Bawa, 1990; see also Rincon *et al.*, 1999).

RESULTS

Our results can be presented as abundance classes for visitors, as shown in Tables 32.1, 32.2 and 32.3, and

Table 32.1. *Woody plant species and their flower visitors in canopy studies using the crane access systems*

Family	Tree	No. trees observed	Primary visitors and seasons[a]
Anacardiaceae	*Tapirira guianensis* Aubl.	2	*Apis* 12/18/99, 5/6/00, 5/13/00, 4/28/01
Anacardiaceae	*Anacardium excelsum* (Bert. & Balb.) Skeels	6	*Tetragonisca angustula* (Lat.), Syrphidae 2/10/00, 2/24/00, 3/22/00, 7/4/01, 27/4/01, 12/5/01, 13/6/01
Anacardiaceae	*Spondias mombin* L.	1	*Tetragonsica angustula, Scaptotrigona luteipennis* (Friese), *Trigona muzoensis* Schwarz 4/23/00, 5/4/00, 5/12/01
Apocynaceae	*Forsteronia* sp.	1	*Apis* 5/13/00, *Trigona cilipes* (Fab.) 2/17/01
Asteraceae	*Mikania leiostachya* Benth.	1	*Trigona muzoensis, Tetragonsica angustula* 2/24/00, *Apis* 1/28/01
Bignoniaceae	*Arrabidea chica* (H. & B.) Verlot	1	*Euglossa crassipunctata* Moure, *Paratetrapedia calcarata* Cresson 5/27/00, 7/26/00, *Trigona muzoensis, Centris analis* (Fab.) 2/24/01, 5/12/01
Boraginaceae	*Cordia alliodora* (Ruiz & Pav.) Oken	2	*Halictus hesperus, Apis*, Diptera 3/12/00, 4/29/00, 3/10/01, 4/27/01
Clusiaceae	*Clusia flavida* (Benth.) Pipoly	2	*Eufriesea pulchra* (Smith), *E. ornata* (Mocsary) 8/28/99, 9/18/99, *Eufriesea pulchra, E. purpurata* (Mocsary) 5/29/01, 6/14/01
Meliaceae	*Guarea kunthiana* A. Juss.	3	*Apis* 10/9/99, 2/17/01, *Melipona micheneri* Schwarz, *Partamona peckolti* (Friese) 4/28/01
Rubiaceae	*Pittoniotis trichantha* Griseb.	2	*Apis, Tetragonsica angustula* 4/23/00, 4/29/00, 4/27/01
Sapotaceae	*Manilkara bidentata* (DC) Chev.	1	*Partamona peckolti, P. musarum* (Ckll.), *Trigona cilipes, Nannotrigona perilampoides* (Cresson) 12/18/99, 6/14/01 *Apis*, 6/14/01
Tiliaceae	*Luehea seemannii* Tr. & Pl.	3	*Apis* 2/10/00, 2/16/00 *Scaptotrigona luteipennis* 2/10/01
Tiliaceae	*Apeiba membranacea* Spruce ex. Benth.	1	*Epicharis metatarsalis* Friese, *Centris flavilabris* Mocsary 7/9/99 *Xylocopa fimbriata* Fab., Rutelinae 6/25/00 *Epicharis metatarsalis*, Rutelinae 5/29/01
Vochysiaceae	*Vochysia ferruginea* Mart.	4	*Epicharis metatarsalis, Xylocopa* spp. 6/14/99, 7/9/99, 6/25/00, *Apis, Xylocopa fimbriata* 4/28/01, 5/10/01, 5/29/01, 6/14/01

[a]Dates given as month/day/year.

Table 32.2. *Recorded visitors that were not dominant species at flowers of plants listed in Table 32.1*

Family and tree	No. trees observed	Secondary visitors
Anacardiaceae		
Anacardium excelsum	6	*Trigona corvina* Ckll., *T. fulviventris* Guerein Melenville, *T. muzoensis, T. fuscipennis* Friese, *Cephalotrigona capitata* (Smith), *Trigonisca schulthessi* (Friese), *Paratrigona lophocoryphe* Moure, *Halictus hesperus, Scaptotrigona luteipennis, Frieseomelitta nigra* (Cresson), *Exomalopsis, Apis, Heliconius,* Hesperiidae, Riodinidae, Muscidae, Halictidae
Spondias mombin	1	Diptera, Coleoptera, Homoptera, Lepidoptera, wasps, *Trigona nigerrima* Cresson
Tapirira guianensis	2	*Hoplostelis, Oxytrigona mellicolor* (Packard), *Melipona compressipes triplaridis* (Ckll.), *Tetragonisca angustula, Partamona musarum, Plebeia frontalis* (Friese), *Trigona cilipes, Cephalotrigona capitata, Partamona peckolti, Rhinetula dentricus* Friese, *Nannotrigona mellaria* (Smith), Tachinidae
Asteraceae		
Mikania leiostachya	1	*Tetragonisca angustula, Trigona muzoensis*
Bignoniaceae		
Arrabidaea chica	1	*Euglossa townsendi* Ckll., *Paratetrapedia calcarata, Tetragona dorsalis* (Friese), *Paratetrapedia lineata* (Smith), *Apis, Ceratina, Megachile, Halictidae, Scaptotrigona luteipennis*
Boraginaceae		
Cordia alliodora	2	*Paratetrapedia,* Lepidoptera, *Paratrigona opaca* (Ckll.), *Exomalopsis*, Coleoptera
Clusiaceae		
Clusia flavida	2	*Frieseomelitta nigra, Trigona fulviventris, T. ferricauda* Ckll., *T. cilipes, Aparatrigona isopterophila* (Schwarz), *Scaura longula* (Lep.), *Paratetrapedia lineata*
Meliaceae		
Guarea kunthiana	3	*Melipona panamica, Partamona musarum, Geotrigona subgrisea, Melipona compressipes, Tetragona dorsalis, Trigona nigerrima, Trigona fulviventris, T. cilipes, Nannotrigona mellaria, Ceratina*, Hesperiidae, beetles, wasps
Rubiaceae		
Pittoniotis trichantha	2	*Scaptotrigona pectoralis*, Syrphidae
Sapotaceae		
Manilkara bidentata	1	*Geotrigona kraussi, Trigonisca schulthessi, Neocorynura*
Tiliaceae		
Luehea seemannii	3	*Tetragonisca angustula, Apis*
Apeiba membranacea	1	Cetoniinae, *Paratetrapedia lugubris* (Cresson), *Eulaema meriana* (Olivier), *Eufriesea schmidtiana* (Friese)
Vochysiaceae		
Vochysia ferruginea	4	Wasps, Papilionidae, Uraniidae, Hesperiidae, Riodinidae, *Tetragonisca angustula, Melipona panamica* (Ckll.), Trochilidae, *Eufriesea pulchra, Eufriesea purpurata, Trigona fulviventris, Epicharis albofasciata* Friese, *Euglossa championi* Cheesmann, *Euglossa cognata* Moure, *Euglossa sp., Ctenioschelus, Centris flavilabris* Mocsary, *Xylocopa* spp., Scoliidae, Sphingidae

Table 32.3. *Thrip flower hosts observed for multiple years (1995–1996) and seasons*

Families at sites	Taxon	Season observed[a]
Barro Colorado Island		
Apocynaceae	*Lacmella panamensis* (Woods.) Markg.	2 flowering events, 1 thrip year
Apocynaceae	*Tabernaemontana arborea* Rose	2 flowering events, 1 thrip year
Apocynaceae	*Odontadenia macrantha* (R. & S.) Markg.	2 flowering events, 1 thrip year
Bignoniaceae	*Arrabidaea candicans* (L. C. Rich.) DC.	3 flowering events, many thrips
Bignoniaceae	*Arrabidaea patellifera* (Schiecht.) Sandw.	2 flowering events, many thrips
Bignoniaceae	*Jacaranda copaia* (Aubl.) D. Don.	2 flowering events, many thrips
Bignoniaceae	*Tabebuia guayacan* (Seem.) Hemsl.	2 flowering events, many thrips
Bignoniaceae	*Tabebuia rosea* (Bertol.) DC.	2 flowering events, many thrips
Bignoniaceae	*Phryganocydia corymbosa* K. Schum.	2 flowering events, 1 thrip year
Bignoniaceae	*Cydista aequinoctialis* (L.) Miers.	2 flowering events, many thrips
Bignoniaceae	*Ceratophytum tetragonalobum* (Jacq.) Sprague & Sandw.	2 flowering events, 1 thrip year
Bignoniaceae	*Paragonia pyramidata* (L. C. Rich.) Bur.	2 flowering events, 1 thrip year
Bombacaceae	*Pseudobombax septenatum* (Jacq.) Dugand	2 flowering events, many thrips
Convolvulaceae	*Maripa panamensis* Hemsl.	2 flowering events, 1 thrip year
Fabaceae	*Clitoria javitensis* H.B.K.	2 flowering events, 1 thrip year
Lecythidaceae	*Gustavia superba* (H.B.K.) Berg.	2 flowering events, many thrips
Tiliaceae	*Trichospermum galeottii* (Turcz.) Kosterm.	2 flowering events, many thrips
Pipeline Road in Parque Soberania		
Apocynaceae	*Odontodaenia macrantha* (R. & S.) Markg.	2 flowering events, 1 thrip year
Asteraceae	*Heterocondylus vitabae* (DC) K. & R.	2 flowering events, many thrips
Bignoniaceae	*Jacaranda copaia*	2 flowering events, many thrips
Bignoniaceae	*Phryganocydia corymbosa*	2 flowering events, 1 thrip year
Bignoniaceae	*Paragonia pyramidiata*	2 flowering events, 1 thrip year
Bignoniaceae	*Xylophragma seemannianum* (O. Kuntze) Sandw.	2 flowering events, 1 thrip year
Bombacaceae	*Pseudobombax septenatum*	2 flowering events, some thrips
Caricaceae	*Carica cauliflora* Jacq.	2 flowering events, many thrips

Table 32.3. *(cont.)*

Families at sites	Taxon	Season observed[a]
Cochlospermaceae	*Cochlospermum vitifolium* (Willd.) Spreng.	2 flowering events, many thrips
Convolvulaceae	*Maripa panamensis*	2 flowering events, 1 thrip year
Fabaceae	*Clitoria javitensis*	2 flowering events, 1 thrip year
Fabaceae	*Dioclea guianensis* Benth.	2 flowering events, many thrips
Fabaceae	*Mucuna mutisiana* (H.B.K.) DC	2 flowering events, many thrips
Lecythidaceae	*Gustavia superba*	2 flowering events, many thrips
Sapindaceae	*Serjania cornigera* Turcz.	2 flowering events, many thrips
Solanaceae	*Solanum hayesii* Fern.	2 flowering events, many thrips
Sterculiaceae	*Bytternia aculeata* Jacq.	2 flowering events, 1 thrip year
Tiliaceae	*Trichospermum galeottii*	2 flowering events, many thrips

[a]Many thrips signifies several adults and larvae usually on most flowers; 1 thrip year, adults or larvae were relatively rare during one season but were always present.
From: F. Gattesco & D. Roubik, unpublished data.

in quantitative form for selected data on species with the largest samples (Fig. 32.1). The thrips at two different sites were 'abundant' on a given flower species for two consecutive flowering seasons in 16 of 30 cases (6 of 18 on BCI, 12 of 22 on PR). The other results were 'scarce' both years ($n = 7$), scarce 1 year and abundant in another ($n = 6$), and scarce with no adults sampled

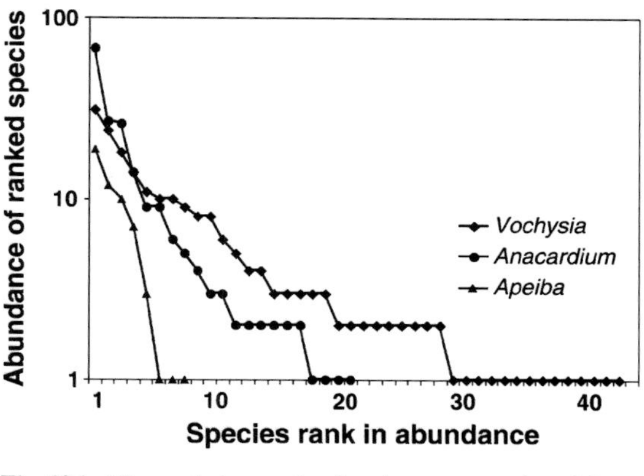

Fig. 32.1. Three relative species abundance curves plotted for flower visitors at three tree species observed with canopy cranes at two sites in lowland Panamanian forests. *Vochysia ferruginea* and *Anacardium excelsum* were studied at multiple individuals, while *Apeiba membranacea* was studied at a single tree (S. Sakai & D. W. Roubik, unpublished data).

but larvae recovered from collected flowers ($n = 1$). In addition, the single plant species for which we analysed the thrip species present, *G. superba* (Lecythidaceae), showed that *Frankliniella parvula* Hood (Thripidae) was a major thrip species quantitatively and qualitatively in 1 year (68% of all thrips) on BCI. Although it was the most abundant single species during both years, with two other *Frankliniella*, *F.* cf. *invasor* Sakimura and *F. gardeniae* Moulton also abundant, it constituted the majority of thrips only one year in two. During the second year on BCI, 29% of all thrips were *F. parvula*. Averaging 1995 and 1996, 43% of all collected thrips on this tree were *F. parvula*. In 1995, when the flowers were collected from several individuals on PR, 64% of thrips were *F. parvula*; in 1996, a single sample yielded only thrips of this species. Further, *F. parvula* seemed relatively specialized because, although associated with at least nine other plant species (Anacardiaceae, Apocynaceae, Bignoniaceae, Convolvulaceae, Lecythidaceae, Lythraceae, Solanaceae), it constituted only 6% of all thrips ($n = 153$) identified in these flowers. It was the major single thrip species on *G. superba* in both years and, perhaps significantly for the theoretical treatment of dominance, it was dominant in different ways among years.

The main visitors to the canopy flowers were the 'big four' tropical bee tribes in the neotropics:

Meliponini, Xylocopini, Centridini and Euglossini. The invader Apini (*A. mellifera* Lep.) already had a dominant role that eclipsed many native bees. Its workers were among the most common bees in 8 of 14 species surveyed for more than 1 year, and sometimes were more common than all other species combined (Tables 32.1 and 32.2). *Apis* spp. were minor visitors to three additional species, thus appeared at 79% of plants studied in multiple seasons. *A. mellifera* visited 25 of the 104 plant species monitored during the canopy crane survey in 1999–2001 (Table 32.4, S. Sakai & D. W. Roubik, unpublished data).

A number of pollinator species were rare and observed only once, even among the most contrasting hosts (species rich with visitors versus relatively species poor) for which ample data were available (Fig. 32.1, Tables 32.1 and 32.2). A steeper curve in the figure shows more variance in species abundance, as the total number of visitor species decreases. In this case, trees with more visitor species displayed smaller differences in visitor relative abundance. Our sample of 14 species observed with the canopy cranes (Table 32.1) showed mixed results in that some were associated with the same primary visitors in all years studied, and some were not. There were equal numbers in the strict tally. Seven species remained the same from year to year, whereas seven changed. However, the exotic African honey bee was a dominant visitor of five species that showed substantial yearly variation, where it occasionally replaced the primary native pollinators. The strict tally showed that *A. mellifera* was a major visitor of five and a minor visitor of one of the seven species that showed considerable variability, while it was a major visitor of two species that were primarily visited and pollinated by native bees but showed negligible annual variation.

Table 32.4. *Plants visited by the African honey bee (*Apis mellifera scutellata*) total survey dates*

Families at site[a]	Taxon	Season observed[b]
Fort Sherman[c]		
Anacardiaceae	*Tapirira*	5/13/00, 5/29/01
Apocynaceae	*Forsteronia*	5/13/00
Araliaceae	*Dendropanax*	7/16/00
Boraginaceae	*Cordia*	4/8/00
Burseraceae	*Protium*	11/7/00
Fabaceae	*Inga* 1	10/9/99
Fabaceae	*Inga* 2	11/11/99
Fabaceae	*Inga* 3	1/15/00
Meliaceae	*Guarea*	10/9/99
Sapotaceae	*Manilkara*	12/18/99, 6/14/01
Tiliaceae	*Trichospermum*	11/27/99
Verbenaceae	*Petrea*	4/22/00
Parque Metropolitano[d]		
Anacardiaceae	*Spondias*	5/4/00
Araliaceae	*Schefflera*	10/4/99
Asclepiadaceae	*Fishcheria*	4/6/00
Asteraceae	*Mikania*	2/24/00, 1/28/01
Bignoniaceae	*Arrabidaea*	7/26/00, 5/10/01
Bombacaceae	*Pseudobombax*	1/4/00
Boraginaceae	*Cordia*	3/12/00, 4/29/00
Fabaceae	*Enterolobium*	3/12/00
Phytolaccaceae	*Trichostigma*	2/16/00, 3/22/00
Piperaceae	*Piper*	1/6/01
Rubiaceae	*Pittoniotis*	4/23/00, 4/29/00, 4/28/01
Sapindaceae	*Serjania*	3/15/00
Tiliaceae	*Luehea*	2/10/00, 2/16/00, 2/10/01

[a]In all the bees visited 25 species in 23 genera and 19 families.
[b]Dates given as month/day/year.
[c]Caribbean lowland forest; host plants formed 12 of 60 species observed.
[d]Pacific lowland forest; host plants formed 13 of 45 species observed.

DISCUSSION

Data on thrips within flowers ($n = 16$ plant species), and insects or vertebrates visiting flowers ($n = 14$) showed similar trends. Abundant mutualists/herbivores in one season were abundant species in the next on approximately half the plant species, but the continuity of association was largely preserved from year to year. Many rare species were included in visitor guilds, making a log-normal-like distribution of species abundance (Fig. 32.1), as predicted for metapopulation models with large source areas and biota (summary by Condit *et al.*, 1996; Hubbell, 2001). For the thrips, dominance could be absolute or relative, the latter occurring when the most common species did not constitute a majority for its guild. This is similar to generalized flower-visiting guilds or plant-pollinator systems, where loose niches are expected. Generalized bees were the most frequent visitors of flowering plants, a conclusion also reached by somewhat different methods in a single-year ground-level survey that included several hundred flowering species in a forest of south-eastern Brazil (Wilms *et al.*, 1996).

The African honey bee either caused displacement or rapidly filled a 'gap' in local visitors to 71% of the most variable pollination systems. In fact, during three different years *A. mellifera* was the single primary visitor of *Tapirira guianensis* and thus it is likely that it had displaced visitors that now appear as 'secondary visitors', among them megachilids, halictids and Meliponini (Table 32.2). In the present study, *A. mellifera* appeared at 79% of plants studied in multiple seasons. On a community level, however, the invasive African honey bee was recorded only on flowers of 25 of the 104 woody plants surveyed in the present canopy crane work, agreeing closely with the general estimate of its presence on flowering plants calculated by Roubik (1989, 1996a). This confirms that, both in old forest and in secondary growth, flowering species are easily invaded by exotic visitors. The implication is that exclusivity and tight relationships among such plants and flower visitors are indeed rare. Despite ecological replacement by the honey bees and often substantial pollination impact, no native bee species has shown a decline as a result of continuous competition with invading African honey bees in forest settings (Roubik, 1996b,c; Roubik & Wolda, 2001). Because the plant-pollinator community is evidently not constructed by niche-assembly rules, it may then presumably be governed by dispersal-assembly rules (Hubbell, 2001). Here, as one might expect, loose niches are common (Roubik, 1992), and competition is ephemeral rather than 'hard' (losses at resource patches immediately translated into fitness losses).

Log-normal-like dominance/abundance curves may result from biotic saturation and zero-sum ecological drift (Hubbell, 2001). Although the Hubbell theta was not fitted to the data points, the curves shown in Fig. 32.1 have a remarkable resemblance to the broadest patterns among ecological communities discussed by Hubbell (his Fig. 1.1). The similarity probably results from sampling of metacommunities that contain many potential visitor species to a given flowering tree, depending on nesting substrates available, nectar and pollen sources or other factors. Dispersal to the particular habitat in question leads to a temporal coexistence and the relative abundance patterns in which many individual species are rare (corresponding to a metapopulation model) rather than to islands largely cut off from other metapopulations that can easily colonize them. The main difference, in our view, is that the relative abundance of pollinators can shift greatly from year to year; consequently, the relative versus absolute dominance in the visitor guild is apparent at short intervals, whereas relative dominance by, for example, tree species in a 50 ha plot changes much more slowly with immigration or local extinction. Many flower visitors may be locally or temporally rare. When loose niches are the rule, species importance in the system undergoes frequent and large changes. For example, secondary and primary visitors may often exchange roles.

Although the qualitative data did not support the loose niches hypothesis, quantitative observations seemed to fit this model. Did specialized systems have relatively fewer rare species? The comparison of *Apeiba membranacea* Spruce ex. Benth, and *Vochysia ferruginea* Mart, in Fig. 32.1 make us believe they did not. A log-normal-like species dominance seemed to hold both for generalized and specialized flowers. Shifting abundance patterns of flower visitors may be attributable in part to the generalized floral structure of some flowering plants, which, therefore, attract a greater number of species (Fig. 32.1). Flowers of *A. membranacea* are large, fleshy and relatively closed and they are exploited primarily by large bees that can manipulate them successfully. A few species were the dominant visitors of this and another local species in all observations (Table 32.1 and D. W. Roubik, unpublished data). In contrast, the flowers of *V. ferruginea* were visited primarily for nectar by a wide variety of both relatively short- and longer-tongued bees, many butterflies, a few moths and several hummingbirds (see also Oliveira & Gibbs, 1994).

G. superba, which had many thrips both in 1995 and 1996, had one principal species, *F. parvula*, which, because of the small size of *G. superba* pollen, may be one of its pollinators (F. Gattesco & D. W. Roubik, unpublished data). However, in 1 year on BCI, a mix of other thrip species comprised the bulk of flower inhabitants. Like the flower visitors (Tables 32.1 and 32.2), abundance shifted, but unlike these visitors, species numbers shifted without an apparent shift in dominance by the most abundant species. Considering that most thrips on the flowers were unlikely to be pollinators (F. Gattesco & D. W. Roubik, unpublished data), our limited sampling of individual species interactions only indicate that the floral thrip guild was as variable across years as was the dominance of particular flower visitors and pollinators.

Bee colonies that recruit can take over flowers and finely adjust their forager numbers to current resource availability on a canopy tree. In contrast, the thrips that land on flowers and possibly reproduce there are probably more limited to chance encounter with hosts, and their association with a host was generally more predictable than that of bees. Their dispersal events may occur relatively infrequently, or they may be more specialized than the other insect flower visitors. Further taxonomic data are needed to clarify the degree of specialization or generalization by thrips. The presence of literally millions of Meliponini and Apini within the reach of any tree canopy in tropical forest (Roubik, 2002) make it likely that flowering trees will often be dominated by them, and that the degree of dominance will rapidly adjust to resource levels. When resource density is relatively low, different species will then dominate, including the solitary bee species (when fewer highly eusocial bees are present). In our multiyear study, only one tree genus, *Apeiba*, had no highly social bee visitors (see also Fig. 32.1), whereas highly social bees dominated 71% of species. We conclude that loose niches are a quantitative phenomenon involving rare species but not a qualitative 'coming and going' of individual species for plant-pollinator systems, at least not on an annual basis, and that most loose niches involve generalist plant-pollinator relationships.

ACKNOWLEDGEMENTS

We thank the canopy crane crew at STRI, in particular Vibeke, Oswaldo and Mirna for logistic help and botanical taxonomic assistance. F. Gattesco thanks the University of Milan for a graduate fellowship, and S. Sakai acknowledges the support of a Japanese Institute for Science postdoctoral fellowship.

33

How polyphagous are Costa Rican dry forest saturniid caterpillars?

Daniel H. Janzen

ABSTRACT

The caterpillars of the 31 species that breed in the dry forest of the Area de Conservación Guanacaste (ACG) in northwestern Costa Rica feed on the foliage of 77% of the 66 woody dicotyledenous families, 51% of the 240 woody dicotyledenous genera and 47% of the 370 woody dicotyledenous species available to them in the ACG dry forest. Their degree of polyphagy ranges from that of *Automeris zugana* caterpillars feeding on 84 species in 66 genera in 26 families to that of 10 caterpillar species having only a single food plant species each (e.g. *Copaxa moinieri* on *Ocotea veraguensis*, *Schausiella santarosensis* on *Hymenaea courbaril*, *Syssphinx mexicana* on *Acacia collinsii*, *Copiopteryx semiramis* on *Manilkara chicle*, *Caio championi* on *Bombacopsis quinatum*). The extreme polyphages, which are almost entirely Hemileucinae, display a two-stage strategy of oviposition on 5–10 species of food plants followed by penultimate and last instar caterpillar movement onto many additional species of food plants. The pattern of food-plant use in the ACG dry forest displayed by these ecologically and geographically widespread large caterpillars is hypothesized to be derived from ecological fitting of immigrant species as they arrive, rather than through contemporary and sustained coevolution among the caterpillars themselves and their food plants. A caterpillar–food plant interaction of this nature is probably characteristic of any diverse tropical forest canopy and many large taxa of caterpillars.

INTRODUCTION

Ongoing biodiversity inventory of the species of macrocaterpillar, their food plants, and their parasitoids of the Area de Conservación Guanacaste (ACG) in northwestern Costa Rica began in 1978. The overarching goal is to find all of the 7000+ species of macrocaterpillar, associate them with their adults, document one or more of their food plants and find all parasitoids possible (as well as facilitate the exploration of caterpillar gut microbiodiversity by other researchers).

The results of this inventory are continuously updated in a website database (http://janzen.sas.upenn.edu) and also in taxonomic infrastructure papers (e.g. Lemaire, 1980, 1988, 2002; Janzen, 1982; Clarke, 1983; Hodges, 1985; Sharkey & Janzen, 1985; Gauld & Ward, 1987; Gauld, 1988, 1991, 1997, 2000; Smith, 1992; Solis, 1992, 1993a; Gauld & Janzen, 1994; Woodley & Janzen, 1995; Burns, 1996; Dangerfield *et al.*, 1996; Corrales & Epstein, 1997; Miller *et al.*, 1997; Zitani *et al.*, 1997, 1998; Gentili & Solis, 1998; Pogue & Janzen 1998; Burns & Janzen, 1999, 2001). The goal is to complete the inventory in several more decades through the application of substantially more person-years than in the past. However, in addition to the species-specific data of who eats what, and how to identify the ACG caterpillars and parasitoids, this growing pool of tri-trophic documentation displays many emerging supraspecific patterns (e.g. Janzen, 1984a,b, 1986, 1987a–c, 1988a,b, 1993b, 2003; Bernays & Janzen, 1988; Flowers & Janzen, 1997; Janzen & Gauld, 1997; Janzen *et al.*, 1998; Burns & Janzen, 2001; Schauff & Janzen, 2001; Sittenfeld *et al.*, 2002).

These patterns may be extracted directly from the website database by any interested party, and considerable self-discipline has been required to keep the focus on the inventory and not be diverted into the exploration of the ecology of these patterns. However, it may be useful to draw attention to some of them and this essay is such an effort.

What is the emerging pattern displayed by caterpillar–food plant relationships in the ACG? The initial ACG inventory emphasis was on the dry forest of the Pacific side of the ACG. It expanded to the adjacent cloud forests and rainforests in 1998. To date, only the dry forest portion of the inventory is thorough enough to make this pattern worth display and discussion. Second, the inventory started with the largest (and, therefore, most easily located) caterpillars. Intensive searching continues, even in the dry forest, for the smaller ones. Consequently, only Saturniidae, Sphingidae and Notodontidae are sufficiently thoroughly inventoried to be worthy of an initial analysis. To keep the focus on the complexities of describing food-plant specificity, I restrict this essay to the Saturniidae. The question then becomes 'What is the emerging pattern of caterpillar–food plant relationships displayed by saturniid caterpillars in the ACG dry forest?' Since Saturniidae do not feed as adults, a parallel question cannot be asked of the adults.

A similar analysis will eventually be applied to all other families of macrolepidoptera in the ACG and all its ecosystems, but this will require both many more years of inventory and the ordering of both terminology and concepts of caterpillar–food plant relationships.

This is not a review paper of saturniid caterpillar biologies or the terminology of caterpillar–food plant relationships. I offer my apologies for not tying this essay to the publications of other authors who have grappled with the same class of questions. I have had to choose between energy invested in caterpillar inventory in the forest and the within-library effort of cross-author and cross-species comparisons, and have opted for the former.

METHODS

The study site: dry forest in the Area de Conservación Guanacaste

The 150 000 ha ACG (Janzen, 2000, 2002; Allen, 2001, and see http://www.acguanacaste.ac.cr) begins 6–19 miles out in the Pacific Ocean. Its 60 000 ha of dry forest extend inland eastward across 15–40 km of lowland coastal plain, mesas and hills to the western lower slopes of Volcán Orosí, Volcán Cacao and Volcán Rincón de la Vieja (Cordillera Guanacaste). Here, these extremely seasonal forests (Janzen, 1993b, 2003) merge with the mid-elevation wetter forests leading into upper elevation cloud forest (1000–2000 m) and thence down into the ACG northern and eastern boundaries in Caribbean lowland rainforest at 400–600 m elevation. The ACG is bounded on the north and south by agricultural landscapes. The 85 km long conserved transect contains the three widespread neotropical vegetation types – dry forest, cloud forest and rainforest – and many intergrades (this covers eight Holdridge life zones (Holdridge, 1978) and a plethora of transitions). The underlying physical substrate ranges from serpentine barrens that have been above the sea for 85 million years to marine limestone and volcanic deposits that range in age from tens of millions of years to only a few days.

The ACG dry forest ecosystem contained between the evergreen forests on the volcanic slopes and the coastal mangroves has 5 to 7 continuously rain-free months in the December–May period, a total rainfall of about 800–2800 mm (contemporary 15 year average of about 1500 mm), and a short 0–6 week dry season (July–August) in the middle of the rainy season. Nocturnal temperatures in this dry forest generally range between 18 and 22 °C, with lowest values on cloudless dry season nights. The diurnal temperatures range between 26 and 36 °C, with hottest days during the relatively wind-free last 2 months of the dry season (late March to mid-May) and coldest days during the second half of the rainy season (October–November) during heavy rains of hurricane origin. The most abrupt temperature change is the abrupt cooling that comes with the first rains at the end of the long dry season (Janzen, 1984a, 1993b, and see detailed weather records at http://www.acguanacaste.ac.cr).

The contemporary dry forest ecosystem of the ACG is nearly contiguous north to the foothills of the Sierras east of Mazatlan, Mexico (e.g. Janzen, 2003) and south at least to the Panama Canal, with rainforested breaks on Pacific coastal Guatemala (San José area) and in the Golfo Dulce region of southern Pacific Costa Rica. There is also a desert break at the Rio Balsas basin in western Mexico. There are extensive intrusions of this dry forest ecosystem into the coastal lowlands of the Gulf of Mexico (Tampico to Yucatan and eastern Guatemala).

Some physical characteristics of the ACG that are key for caterpillar ecology are:

- its dry forest lies within a few tens of kilometres of both (cooler) high-elevation cloud forests and (wetter) low-elevation rainforests, and its margins intergrade with these other ecosystems
- the dry season is severe enough to create deciduous and semideciduous forest types coupled with

very strong seasonal phenology in plant growth and reproduction

- the internal rain-fed aquatic systems are severely dried out by the long (and hot) rain-free season, but the adjacent volcanoes create ever-flowing rivers that cross the dry forest as linear seasonal oases on their way to both the Atlantic and Pacific
- the diverse ages and chemistries of the ACG rock and soil substrates, combined with major meteorological variation in space and time, sustain a complex array of plant species complexes, phenologies and life-forms as food and shelter for Lepidoptera
- despite four centuries of European-style agriculture and ranching, and millenia of indigenous use, the relatively low-quality soils and long sociopolitical distances from Costa Rican and Nicaraguan social centres have protected the general area from having been thoroughly cleaned of its natural vegetation and, therefore, of much of its caterpillar fauna: what is described here is the plant and caterpillar biota that would have greeted the invading Spaniards.

The ACG old-growth forest remnants range from several thousand hectares of dwarf forest on serpentine on the western tip of the Santa Elena peninsula to patches of a few tens of hectares on volcanic soils, with the remainder being a very complex mosaic of secondary successional stages that did not begin the long trek to old-growth status until free-ranging anthropogenic fires were eliminated after the mid-1980s.

Biodiversity of the ACG dry forest

The ACG dry forest contains at least 800 species of angiosperm in 115 families (Janzen & Liesner, 1980 and unpublished collections and corrections since that date). However, there is uncertainty as to where the dry forest stops and the wetter forests to the east begin, since these two major ecosystems intergrade in many ways. To set aside this source of variation, I here consider only the approximately 750 species in the 110 families that occur unambiguously and naturally within the approximately 60 000 ha of dry forest of Sector Santa Rosa, Sector Santa Elena, Sector Murcielago, Sector Horizontes and Sector Pocosol (all non-native species are excluded from the analyses in this essay). At least 80% of these species may be found in any area of 20 km^2 of ACG dry forest. The great majority currently have quite broad distributions within the ACG dry forest (as well as extending in dry forest far to the north and south of the ACG). Once the entire ACG has been restored to 'old growth' status (in five centuries?), some of these species will have narrower local distributions, but others will have broader distributions.

Since this analysis deals only with the Saturniidae, the 'available' food plant list for the ACG dry forest is further restricted to the 66 families that contain woody perennial dicotyledons. This is because ACG saturniid caterpillars almost always feed on the foliage of woody dicotyledons (the exception being five records of *Automeris zugana* and one record of *Automeris io* feeding on a grass). This non-use of herbs and herbaceous shrubs/vines is in strong contrast to the other large caterpillars in this habitat. It presumably stems at least in part from saturniid caterpillars being specialists on plants that are rich in digestion inhibitors (e.g. polyphenols, terpenoids, protease inhibitors). Sphingids, by comparison, very commonly feed on plants rich in toxic small molecules (e.g. cyanogenic glycosides, alkaloids, uncommon amino acids: Janzen, 1984a; Bernays & Janzen, 1988). These 66 families contain approximately 370 woody species in 240 genera. In short, this is an examination of the distribution of the ACG dry forest saturniid caterpillars over 370 plant species in 240 genera in 66 families. Given the thorough plant inventory to date (http://www.acguanacaste.ac.cr), coming decades of search will increase these numbers of species and genera by no more than 5%.

The inventory of the Saturniidae of the ACG dry forest was first conducted with intensive light trapping (15 W black and inflorescent lights hung in exposed and forest interior sites) from 1978 to 1991. Because all 31 breeding species (Table 33.1) were found as adults in the first 5 years of this inventory (e.g. Janzen, 1982, 1986, 1993b), there is no reason to think that any breeding species were overlooked. Occasional waif males (*Hylesia hamata*, *Rothschildia triloba*, *Automeris hamata*) from the wetter eastern end of the ACG have been encountered in ACG dry forest, but they do not breed in the ACG dry forest and are not considered further here. Species that breed in the dry forest to wet forest interface (e.g. *Automeris pallidior*, *Arsenura batesii*), but not in the main body of ACG dry forest, are not considered in this analysis. Twenty-eight per cent of Costa Rica's 112 species of Saturniidae (http://janzen.sas.upenn.edu) breed in the ACG dry forest.

Caterpillars of all but 3 of these 31 species were also discovered during the first 5 years of caterpillar inventory of the ACG dry forest (1978–1982). This search process was by day and night, by direct inspection of trees, shrubs, vines and herbs of all heights (1–35 m), and by searching above fresh frass found on the ground below plants. The three exceptions have the following history. Adults of the very rare *Automeris zurobara* have only been found at lights, but by trial and error it was discovered that its caterpillars develop very well only (to date) on *Inga vera* (Fabaceae), the sole species of *Inga* in the ACG dry forest. It is assumed that this is its only food plant there. *Adeloneivaia jason* caterpillars have been found feeding on six species of *Inga* and one species of *Pithecellobium* (Fabaceae) in the wetter portions of the ACG, but only adults have been found in the ACG dry forest – specifically, in the canyon of the Rio Pozo Salada. It is assumed there is a breeding population on *I. vera* in this canyon. Food sources for caterpillars of *Othorene verana* remained unknown after those of all other species had been found. First instar caterpillars were then offered 200 species of woody plant and were found to eat only oak, *Quercus oleoides* (Fagaceae). Subsequent directed search found the caterpillars of *O. verana* to be common in the crowns of oak trees, and they have been found on no other plant in the ACG dry forest.

Caterpillar search process and negative records

Between 1978 and the end of 1999, the ACG caterpillar inventory process found and reared more than 10 500 wild-caught saturniid caterpillars in the ACG dry forest (the records for 2000 are still being polished, are not yet available on the website, are not considered here and would not change the information in Table 33.1). This number is approximate because in some cases a single record represents a clutch of siblings rather than a single caterpillar. Saturniid caterpillars have been found during both the directed and the haphazard search of all ACG species of food plant for all species of caterpillar. All wild saturniid caterpillars encountered are captured, recorded and reared until they die of disease, produce one or more parasites or produce an adult (see http://janzen.sas.upenn.edu for details).

This search process for all species of caterpillar on all species of plant at all times of year, night and day, allows both the determination of the normal wild food plants of each species in the area and a reasonable estimate of what plant species are not being used by that species. For example, *Copiopteryx semiramis* not only feeds only on *Manilkara chicle* (Sapotaceae) in this forest but also does not feed on many hundreds of species of other woody dicotyledon. Though the other two Sapotaceae (*Chrysophyllum brenesii* and *Pouteria campechiana*) are obvious suspects as possible additional food plants for *C. semiramis*, intensive search of the crowns of both species has yet to yield these caterpillars.

It is my opinion that further search of the ACG dry forest food plants will not change the pattern displayed in Table 33.1 other than to increase the number of species fed on by the saturniid species with the greatest number of food-plant species. For example, *Syssphinx mexicana* will never be found on any of the five species of *Acacia* besides the one it eats, whereas *Eacles imperialis* will be found on a few more species of plant.

Table 33.1 contains detailed records in the hope that it will stimulate others to summarize their food-plant records for the caterpillars of other tropical areas, so as to begin to enrich the literature with the basis for among-site and among-taxa comparisons.

Taxonomy

All the species of ACG Saturniidae have been identified initially from adult specimens studied by C. Lemaire, and the names follow those used in his revisions of neotropical Saturniidae (Lemaire, 1978, 1980, 1988, 2002) and Janzen (1982). Each name in Table 33.1 unambiguously represents a single population within the ACG dry forest (leaving undiscussed the question of whether single species having different-appearing morphs on the wet and dry ends of the ACG are really one species or two). Photographs of adults and larvae are available at http://janzen.sas.upenn.edu, and voucher specimens for all species are deposited in the national inventory collections at INBio, Santo Domingo de Heredia, Costa Rica. All plant names follow Janzen and Liesner (1980) and INBio's national inventory at http://www.inbio.ac.cr. Because the taxa have been well documented, author names are not included in this paper.

RESULTS

The species of ACG dry forest saturniid caterpillars are not equally polyphagous (Table 33.1). The four most polyphagous – *A. zugana*, *Hylesia lineata*, *Periphoba*

Table 33.1. *Distribution of saturniid moth caterpillars across the Area de Conservación Guanacaste dry forest woody dicots (370 species, 240 genera, 66 families)*

Saturniidae moth species[a]	Food species[b]			Food genera[b]			Food families[b]		
	Eaten	Available	%	Eaten	Available	%	Eaten	Available	%
Automeris zugana (H)	84	129	65	66	128	52	26	66	39
Hylesia lineata (H) dry	73	122	60	58	142	41	25	66	38
Periphoba arcaei (H)	49	86	57	43	145	28	31	66	47
Automeris tridens (H) dry	48	95	51	44	123	36	18	66	27
Eacles imperialis (C)	23	37	62	20	103	19	12	66	18
Automeris io (H) dry	16	39	41	11	57	19	3	66	5
Rothschildia lebeau (S)	11	17	65	10	43	23	7	66	11
Citheronia lobesis (C) dry	8	11	73	8	33	24	8	66	12
Molippa nibasa (H) dry	7	22	32	7	37	19	1	66	2
Syssphinx molina (C)	7	20	35	6	37	16	1	66	2
Automeris metzli (H)	6	20	30	6	52	12	4	66	6
Hylesia dalina (H) dry	6	15	40	6	67	9	6	66	9
Syssphinx quadrilineata (C)	6	27	22	4	37	11	1	66	2
Dirphia avia (H)	5	5	100	5	42	12	3	66	5
Adeloneivaia isara (C) dry	2	2	100	1	37	3	1	66	2
Arsenura armida (A)	2	2	100	2	10	20	2	66	3
Ptiloscola dargei (C)	2	16	13	2	37	5	1	66	2
Rothschildia erycina (S)	2	3	65	2	15	1	1	66	2
Syssphinx colla (C) dry	2	9	22	2	37	5	1	66	2
Adeloneivaia jason (C)	1	1	100	1	37	3	1	66	2
Automeris zurobara (H) dry	1	1	100	1	37	3	1	66	2
Caio championi (A)	1	1	100	1	4	25	1	66	2
Citheronia bellavista (C)	1	2	50	1	2	50	1	66	2
Copaxa moinieri (S) dry	1	1	100	1	1	100	1	66	2
Copiopteryx semiramis (A)	1	1	100	1	5	20	1	66	2
Dysdaemonia boreas (A)	1	2	50	1	4	25	1	66	2
Othorene purpurascens (C)	1	1	100	1	5	20	1	66	2
Othorene verana (C)	1	1	100	1	1	100	1	66	2
Schausiella santarosensis (C) dry	1	1	67	1	37	3	1	66	2
Syssphinx mexicana (C) dry	1	6	17	1	37	3	1	66	2
Titaea tamerlan (A)	1	1	100	1	4	25	1	66	2

[a]Species names followed by 'dry' are restricted to tropical dry forest (as based on 20 years of Costa Rican saturniid inventory) while the others range widely in neotropical rainforest as well; H, Hemileucinae; C, Ceratocampinae; S, Saturniinae; A, Arsenurinae.

[b]Numbers and percentages of species, genera and families 'available' to a moth species are determined by counting how many species of woody dicot are in the genera fed on, how many genera are in the families fed on, and how many of the 66 families containing woody dicots are fed on.

arcaei, *Automeris tridens* (13% of the 31 species) – are Hemileucinae. This subfamily is famous for being highly urticating. *H. lineata* passes the day hidden in silk-leaf shelters (Janzen, 1984b), and the other three have bright warning colours (see Janzen, 1984a for discussion of the high correlation between polyphagy, slow growth, strong personal defences and warning colouration among Saturniidae). The list of plant species eaten by these four species will probably increase by 10–20% as the years pass and more samples accumulate.

The more intermediately polyphagous species – *Citheronia lobesis* through *Dirphia avia* (32% of the 31 species) – are members of Ceratocampinae, Hemileucinae and Saturniinae. They have less noticeably slow growth rates, are less well defended and are more cryptically coloured. The list of species eaten by these species might increase by 1–3% as the years pass and more samples accumulate.

The least polyphagous species (55% of the 31) – *Adeloneivaia isara* to *Titaea tamerlani* – have only one or two species of food plant and contain only one Hemileucinae, all the Arsenurinae and a mix of Ceratocampinae and Saturniinae. If they feed on two species then the plants are closely related. This group of caterpillars contains the most cryptic species, the most rapidly growing species and the most poorly defended species (except for *A. armida*, see Costa *et al.*, 2002). I cannot, however, call them monophagous since some do have more than one food plant; since their 'monophagy' is undoubtedly to some degree set by ecological availability of suitable species; and since many will have other food plant species over their entire geographic range.

As a general rule, the species that feed on many species of woody ACG dicotyledon also eat many genera and families of woody ACG dicotyledon. At the opposite extreme, 61% of the species feed on only one of the 66 available families, and about 40% feed on only one species of plant.

The percentages eaten of the woody genera in the 'available' families (Table 33.1) is extremely complex. However, a high percentage value generally occurs when the caterpillar is quite polyphagous (e.g. the species in the upper part of Table 33.1) or feeds on a genus or family that contains only one or a few taxa in the ACG dry forest (e.g. *Copaxa moinieri*, *O. verana*, all species of Arsenurinae).

What proportion of the geographically 'available' food plants are not being eaten? On a caterpillar species-by-species basis, Table 33.1 presents the percentage of the available plants (defined as occurring within a taxon that contains at least one member that is fed on) that are eaten. The remainder are those not eaten. It is clear that other than the top four polyphages, the saturniid caterpillars of the ACG dry forest are not using the great bulk of the plant species geographically available to them.

Altogether, 23% of the 66 ACG dry forest woody dicotyledonous families are not fed on by any species in the Saturniidae (this is not derivable from Table 33.1 but rather from the database at http://janzen.sas.upenn.edu). Similarly, 49% of the 240 woody dicotyledonous genera and 53% of the 370 woody dicotyledonous species in these 66 families are not fed on by any species in the Saturniidae.

DISCUSSION

Different kinds of polyphagy

Finding and rearing these caterpillars has generated a large array of natural history observations that, taken as a whole, indicate that saturniid caterpillars use two of the three distinctive behavioural patterns associated with degrees of polyphagy in ACG caterpillars (see below). Second, there is polyphagy as expressed where caterpillars are found feeding in nature, and a more restricted polyphagy as expressed by where their mothers oviposited. I do not discuss 'host specificity', since all ACG caterpillars are host specific – they simply vary in the number of species that are in the list of species for which they are host specific.

Species to some degree polyphagous, individuals monophagous

All ACG Ceratocampinae, Saturniinae and Arsenurinae are found only on the species of food plant (and normally the individual food plant) on which the female oviposited. In other words, for these three subfamilies, the degree of polyphagy expressed in Table 33.1 is essentially a perfect match with adult female 'polyphagy'. The caterpillars of these three subfamilies resist releasing their hold on 'their' food plant with great tenacity. They will eat a food-plant's leaf crop down to the point of total defoliation before voluntarily leaving it. They cannot live many days without feeding and often have poor ability to switch from one known food plant to another food plant when forced to do so in captivity. They are quite cryptic on their food plant (except for the aposematic *A. armida*, Costa *et al.*, 2002) and are not likely to locate another food plant if forced to search on their own through the general vegetation.

If one of these species has more than one food-plant species, the newly hatched first instar caterpillars grow normally on any one of these plants (though there are obvious differences in growth rates in some cases, not discussed here).

If first instar caterpillars of the species in these three subfamilies are forced onto an array of species of 'nonfood plants', the probability that one species will support development rises as the number of species in the first column of Table 33.1 rises. This has been examined in more detail with *Rothschildia lebeau* than any other species (Olson, 1994b). This species can develop on at least four other species of ACG woody dicotyledons in addition to those tabulated in Table 33.1 (though the caterpillars have never been found on these species in nature). However, there are no indications that such artificial feeding would result in more than doubling the number of ACG dry forest woody dicotyledonous species that could be eaten by any of the species in these three subfamilies. *E. imperialis* has the greatest potential to achieve such a doubling. The bottom line is that these three subfamilies are feeding on only 39% of the 66 woody families of dicotyledons in the ACG, and only 16% of the 370 species in those families.

Species quite polyphagous, individuals polyphagous

The Hemileucinae of the ACG dry forest lay their eggs on many fewer species than their caterpillars are found to feed upon. That is to say, the Hemileucinae oviposition records, and feeding records for first and second instars, are only a small subset of the feeding records tabulated in Table 33.1. This subset becomes proportionately larger the more species that are fed on by the species. For example, *D. avia* has only five food plants in Table 33.1 and its eggs are laid on all five species. The eggs are laid on the tree trunk 1–3 m above the ground, and the last and penultimate instar caterpillars feed each night in that same tree's crown and pass the day resting on the tree trunk.

At the other extreme, females of *A. zugana* and *H. lineata* usually oviposit on four to five species of ACG plant. The caterpillars, if unmolested, feed gregariously on the oviposition plant through third instar, and then the fourth and fifth instars 'wander' widely and solitarily, feeding for variable periods on the many species of plant tabulated in Table 33.1 (and probably a few more). These caterpillars, however, if restricted in captivity to their different food plants display noticeably different growth rates on them (e.g. Sittenfeld *et al.*, 2002). If the caterpillars of these polyphages are molested on their natal food plant, they commonly let go of the substrate and fall to the ground. They then wander in search of food plants and only occasionally find a member of the natal species. If they are moved in this manner to a new food plant, they appear to grow about as well as they did on their natal food plant, though there are obvious differences in rate. If they are confined in captivity to the 'new' food plant on which they were found (the majority of the species tabulated in Table 33.1), they complete their development to pupation and adult normally.

The length of the food plant lists for *H. lineata* compared with *A. zugana* and *P. arcaei* is partly a function of the inventory process. This is because there was a two-generation population explosion of *H. lineata* at the end of the 1978 growing season and in the first half of the 1979 growing season, terminated by a viral epidemic at the end of the 1979 growing season (Janzen, 1984b). The millions of caterpillars and newly oviposited egg masses that occurred allowed a very extensive test of both oviposition sites and caterpillar feeding tastes derived solely from wild-caught caterpillars. Most of the different species and family food plant records tabulated for *H. lineata* in Table 33.1 were first recorded in 1979. Over the subsequent 22 years, almost all of these records have been repeated, but with only 1–10 records each. More than 80% of the postoutbreak wild-caught records have been from plants within the four ovipositional hosts most commonly used by *H. lineata*: *Casearia corymbosa* (Flacourtiaceae), *Lonchocarpus minimiflorus* (Fabaceae), *Annona purpurea* (Annonaceae) and *Calycophyllum candidissimum* (Rubiaceae). No outbreak for *A. zugana* or *P. arcaei* has occurred (yet); I suspect that, if one does, the number of species in its host list will increase another 10% at least.

It would be incorrect to conclude that even very polyphagous species in the upper quarter of Table 33.1 can feed on any woody ACG plant. There are about 200 species of common woody plant in the ACG dry forest on which no saturniid caterpillar has ever been found feeding, and I suspect that most of them never will be food plants (without further evolution or the input of species new to this forest).

The extreme polyphagy of Hemileucinae differs from that displayed by two common species of ACG dry forest Arctiidae, *Ecpantheria icasia* and *Ecpantheria suffusa*. These two species have hemileucine-sized caterpillars (the familiar 'woolly bears' of northern climates) that wander widely from the time of hatching, commonly feeding for a few hours on one plant and then moving on. No attack by a carnivore is required for them to leave a plant on which (in captivity) they can complete their development. Collection records of caterpillars of these two species (http://janzen.sas.upenn.edu) display polyphagy for species, genera and families equal to that of the Hemileucinae in the upper quarter of Table 33.1. Strikingly, however, there is almost no overlap between the two food plant lists for these two species. The Hemileucinae polyphagy in Table 33.1 should also not be confused with that displayed by the only very polyphagous hesperiid butterfly in the ACG, *Astraptes fulgerator*. This exceptional skipper oviposits on, and its caterpillars develop on, 17 species in seven quite unrelated families (http://janzen.sas.upenn.edu).

Numbers of caterpillar records per food plant species

In constructing Table 33.1, the food-plant records were not weighted by their frequency of occurrence within a given species. For example, *R. lebeau* has 11 food-plant species in the ACG dry forest, but well over 90% of the caterpillar population lives on just three of them: *C. corymbosa* (Flacourtiaceae), *Spondias mombin* (Anacardiaceae), and *Exostema mexicanum* (Rubiaceae).

Placement in the canopy

How many of the woody dicotyledonous plants of the ACG dry forest are 'canopy' plants (Moffett, 2000), in keeping with the general theme of this book? Since the ACG is a horrendous mix of patches of regenerating forest of all sizes ranging from 0 to at least 400 years of age, essentially all of the woody plant species discussed here are members of a canopy somewhere between 1 and 35 m above the ground. Even when the entire ACG dry forest is old growth, a half millennium from now, this will be the case. The great bulk of ACG saturniid caterpillar species and individuals pass their feeding stages in the sunnier to lightly shaded parts of this canopy, though a few (e.g. the first four species in Table 33.1) may be nearly as common in the shadier parts of the canopy (i.e. below the canopy) as in its more insolated parts.

Since caterpillars are feeding on leaves, and leaf nutrient content is at least partly positively correlated with insolation, if only nutrients were a consideration, most caterpillars (including saturniids) might be expected to be found in the more insolated parts of their host plants (irrespective of the heights of those plants). However, a host plant is more than its chemistry (Janzen, 1985a). At least half of the species of ACG dry forest saturniid caterpillar are found regularly at heights of only 1–3 m above the ground, a portion of airspace that is poorly frequented by birds and primates and, therefore, probably relatively free of at least one major set of carnivores. Those caterpillars that frequent the fully insolated crowns of large trees (5–30 m tall) mostly belong to species that are moderately to highly cryptic (and poorly protected as individuals).

Anticipated contrasts with other families and other ecosystems

Space is not available here to contrast the ACG dry forest with its adjacent cloud forest and rainforest with respect to data such as those in Table 33.1. This will not be appropriate until more years of accumulated food-plant records are available from these wetter ecosystems. However, a multispecies caterpillar–food plant interaction of this nature is characteristic of any species-rich tropical forest canopy. Both the ACG cloud forest and the rainforest clearly have extreme polyphages, intermediate level species and essentially monophagous species among the Saturniidae. The differences among the three ACG major ecosystems will be found to lie in the relative proportions of these degrees of polyphagy. The same may be said of the other species-rich families of ACG macrocaterpillars (Sphingidae, Notodontidae, Noctuidae, Geometridae, Pyraloidea, Hesperiidae, Nymphalidae, Riodinidae, Lycaenidae, etc.). I suspect also that when the ACG is contrasted with other parts of the world, the relative proportions of degrees of polyphagy will be found to vary both among adjacent ecosystems and among major faunas and floras.

How did the pattern in Table 33.1 come about?

One extreme would be to view the results presented here as the current status of an on-site long-time multispecies evolutionary interaction between a very large array of plants and their saturniid caterpillars (and their competitors and carnivores). In this view, the patterns in Table 33.1 reflect the evolutionary outcome of interspecific interactions through the millennia, where the interacting species are what we see today and the current structure is maintained by an ongoing evolutionary interaction. Taken overall, nothing in the biology of the ACG supports this viewpoint.

The other extreme is to view the pattern in Table 33.1 as the contemporary, serendipitous outcome of the independent arrivals of 31 species of saturniid over millennia, each species bearing its current traits upon arrival. Equally, each has arrived at its demographic and interacting status by virtue of what it carries with it genetically (and what other species are already there with which it may interact). It is my opinion that the pattern in Table 33.1, and the analogous patterns to be displayed by any other family of caterpillars in the ACG and, indeed, those of all the caterpillars taken as a whole, is the result almost entirely of the latter rather than the former extreme (Janzen, 1985b, 1986).

In essence, this is to say that the pattern in Table 33.1 was generated by 'ecological fitting' (see Janzen, 1985b for the development of this term) of 31 species of incoming saturniid caterpillar onto an array of plants and among an array of other caterpillars and carnivores, all of which are/were there through processes and traits not significantly influenced by saturniid caterpillars. Even the parasitoids of these caterpillars probably came with them or arrived later by the same kind of immigration.

I find it far more productive and realistic to view the situation in this manner, with a few possible deviations, than to view the structure reflected in Table 33.1 as the product of pair-wise coevolution or community-level evolutionary adjustments. Furthermore, if there are interspecific interactions among the saturniid caterpillars that lead to traits as gross as what species of plant are eaten, or even their presence/absence in the ACG dry forest, these will be mediated through shared carnivores rather than through evolutionary interactions over or with food plants. In short, the traits we see today evolved somewhere, probably under some form of ecological insularity of a population, but then the owner of the trait spread out over some much larger 'ecological continent', including the ACG dry forest.

The ACG was cleaned of rainforest in a volcanic blast approximately 1.5 million years ago (Chiesa *et al.*, 1998). With the exception of *Schausiella santarosensis* (see below), all of the ACG saturniid caterpillars have very large multicountry and multiecosystem distributions (Lemaire, 1978, 1980, 1988, 2002; and subsequent Costa Rican-level inventory). Undoubtedly, they simply immigrated back into the new rainforest as it appeared on the devastated site. As the ACG dry forests developed in the rain shadow on the Pacific side of the newly emerging volcanic cordillera 20 000–50 000 years ago (Chiesa *et al.*, 1998), they were undoubtedly put together by dry forest plants immigrating on-site from the north and south, where they occur still. The saturniid caterpillar fauna ate what was available, based on ovipositional and gustatory abilities. Some rainforest species would have dropped out because of the disappearance of their rainforest food plants (or the increasingly intense dry season, or the shifting abilities of carnivores, or, or, or . . .). In a few cases, a widespread rainforest species may have spun off a dry forest isolate, but the only two candidates for this scenario are the endemics *S. santarosensis* (probably derived from the widespread rainforest *Schausiella denhezorum*, which ranges from Costa Rica to Colombia (Lemaire, 1988)) and *Copaxa moinieri* (probably derived from the widespread rainforest *Copaxa rufinans*, which ranges from Mexico to Ecuador (Lemaire, 1978)).

The three sorts of polyphagy displayed in Table 33.1 – extreme polyphagy (which is expandable with more caterpillar wild-caught records), intermediate low polyphagy and essentially monophagy – can also be examined in the context of ecological fitting.

S. santarosensis and *C. moinieri*, two of the 12 ACG saturniids restricted to dry forest, illustrate the very different ways of having only one food plant. *C. moinieri* is a dry forest local Costa Rican species (it squeaks into the same habitat in northern Pacific Panama (Lemaire, 1978)). Its widespread congener, *C. rufinans*, has to date been reared from 18 species in the Lauraceae in five genera in ACG wet forest (http://janzen.sas.upenn.edu). It is unambiguously a specialist on the Lauraceae (as are other species of *Copaxa*, e.g. Wolfe, 1993). Why does *C. moinieri* have only one species of food plant in Table 33.1? Because the ACG dry forest contains only one native species of Lauraceae, *O. veraguensis*. Strikingly, as this dry forest *Ocotea* sp. slides into the margin of the wetter forests of the eastern ACG, *C. moinieri* can be found side by side with *C. rufinans*, both feeding on the same food plant. Here, where there are other Lauraceae available (and used by *C. rufinans*), *C. moinieri* apparently does not use them. In the ACG dry forest, however, *C. moinieri* females oviposit on introduced *Persea americana* (avocado) and the caterpillars grow faster and larger than on *O. veraguensis*.

The monophagy of *S. santarosensis* cannot be explained by the absence of related plants. It has at least 67 other woody dicotyledonous species in 37 genera of Fabaceae (including many caesalpinoid legumes) available as potential food plants. This moth is an ACG dry forest endemic very likely derivative from the more widespread rainforest *S. denhezorum* (found in the ACG north-eastern rainforest). The food plant of *S. santarosensis* is *Hymenaea courbaril* (Fabaceae). This large tree, the source of Oligocene neotropical amber, has an enormous range from lowland Mexico to central South America. The only other known caterpillar food plant of the genus *Schausiella* is *Hymenaea altissima* in Brazil (Lemaire, 1988). However, *S. denhezorum* does not use this food plant since there are no *Hymenaea* spp. in the ACG rainforest (there, it probably feeds on some other large caesalpinaceous tree).

I suspect that in both of the above cases of monophagy, some event isolated a population of each moth species in the Pacific lowland dry forest, and it 'evolved onto' whatever plant species minimally offered oviposition cues to the females and sustained the digestive abilities of the caterpillars. The presence of few or many other species of plant was about as relevant as the presence of a field of rocks. The addition of more species of plant to the ACG would threaten their monophagous status only if those species had very similar traits to those eaten today, and even then, they would not necessarily be used. Put another way, were a third species of *Lysiloma* to invade the ACG dry forest, *A. isara* might, but not necessarily would, increase the number of species on which it feeds from two to three, but there is no indication that the other 66 species of legume matter at all to it.

A. armida offers an example of yet a third kind of minimal polyphagy. In the ACG dry forest, it has just two food plants in different (but related) families: Sterculiaceae and Bombacaceae. Since it has been found eating other Bombacaceae and Tiliaceae in other parts of Costa Rica, and since two other species of ACG rainforest *Arsenura* feed on Bombacaceae, Malvaceae and Sterculiaceae, it is easy to infer that *Arsenura* in general is a Malvales specialist. However, just on the margin of the ACG dry forest, where it intersects with mid-elevation rainforest on the west slope of Volcán Rincán de la Vieja, *A. armida* feeds on *Rollinia membranacea* (Annonaceae); it ignores the five species in three genera of Annonaceae in the dry forest proper.

The species of intermediate polyphagy in Table 33.1 (*A. io* with 16 food plants to *D. avia* with four food plants) characteristically feed on a small group of noncongeners and even nonconfamilials. For example, *D. avia* eats *H. courbaril* (Fabaceae), *Q. oleoides* (Fagaceae), *Cedrela odorata*, *Swietenia macrophylla*, and *Guarea excelsa* (Meliaceae). All of these caterpillar species with intermediate polyphagy have very broad latitudinal ranges (*A. io* ranges from Canada to north-western Costa Rica and presumably feeds on an enormous list of species over that range). Each of them presumably arrived in the ACG with its ovipositional preferences and caterpillar digestive abilities that we encounter today (as well as had the ability to withstand the carnivores associated with each host plant used (Janzen (1985a)). However, only a very few of the ACG species are in fact 'usable' to their caterpillars. Were more plant species to move into the ACG over time, presumably some of these would become food plants for these species. However, having said all that, I should note that both *A. io* and *Hylesia dalina* are low-density moths, with only 100+ and 200+, respectively, rearing records. It is my suspicion that by the time 500+ of each have been found, these two species may 'graduate' to the category of extremely polyphagous.

With respect to ovipositional behaviour, the extremely polyphagous species in Table 33.1 are not really very different from the intermediately polyphagous species. Five to ten different species of plant are oviposited on by each species (with minor overlap). However, the caterpillars of these extreme polyphages move onto a variety of other food plants in their penultimate and last instars and will even do it earlier if forced to by carnivores or by simply falling off the food plant. It is reasonable to view the length of the species list fed on by the extreme polyphages as a function of how many plant taxa are present in the forest.

This generalization, however, should not be taken to imply that they are able to eat all species. Informal feeding trials with *A. zugana*, *A. tridens* and *H. lineata* indicate that at least half of the woody plant species in the ACG dry forest are not acceptable as food plants to their caterpillars. Often, what is rejected by one species is readily eaten by another (e.g. *A. zugana* can eat *Cordia alliodora* leaves but barely survives on it, while *A. tridens* consumes it as a favoured food and grows very well on it). In addition, congeneric plants are not equally acceptable or used in nature by the different species of extreme polyphages.

In effect, the extreme polyphages have invented a two-stage feeding strategy. The winged adult specializes in locating a small number of species of plant that release some set of odours that function as ovipositional cues. Once caterpillar development has reached an intermediate state, the caterpillar itself launches into the habitat in search of presumably 'better' food plants (some of which may be the same as those on which the mother oviposited).

One consequence is that as the food plants are tallied for such a species over a large geographic range, the tally may include many dozens of ovipositional food plants, but they are likely to fall in a fairly narrow subset of higher taxa. However, a second consequence is that over such a wide range the penultimate and last instar caterpillars may accumulate records of many hundreds of species of food plant, depending on the plant diversity in the ecosystems and latitudes occupied. *E. imperialis*, *R. lebeau* (called *Rothschildia forbesi* in the Texas portion of its range) and *A. io*, as well as the top four in Table 33.1, will fall into this pattern. At the opposite extreme, it is likely that *O. verana* is a *Quercus* (oak) specialist throughout its range from dry forest to cloud forest to rainforest (Colombia to northern Mexico: Lemaire, 1988), all of which falls within the distribution of the genus *Quercus*. In Costa Rica, it ranges from the oaks of the ACG dry hot Pacific lowlands to the oaks living in the swirling mist of Atlantic cloud forest.

ACKNOWLEDGEMENTS

The ACG caterpillar inventory has been supported by NSF grants BSR 90-24770, DEB 93-06296, DEB-94-00829, DEB-97-05072, and DEB-0072730, by taxonomists of the Smithsonian Institution and the Systematic Entomology Laboratory of the USDA, and by financial, administrative and logistic support from Costa Rica's INBio, the government of Costa Rica, the Area de Conservación Guanacaste and CONICIT of Costa Rica. Many individuals have supported the development of all stages of the project in a multitude of ways. I wish to thank particularly W. Hallwachs, C. Lemaire and J. Corrales for identifications. I would like to thank the following for caterpillar hunting and husbandry, R. Moraga, G. Sihezar, G. Pereira, L. Rios, M. Pereira, O. Espinosa, E. Cantillano, M. Pereira, R. Franco, H. Ramirez, F. Chavarria, M. M. Chavarria, E. Olson, C. Moraga, P. Rios, C. Cano, D. Garcia, F. Quesada, E. Araya, E. Guadamuz, R. Espinosa, R. Blanco, A. Guadamuz, D. Perez, R. Blanco, F. Chavarria, C. Camargo, H. Kidono, A. Masis, W. Haber and W. Hallwachs.

34

Influences of forest management on insects

Martin R. Speight, Jurie Intachat, Chey Vun Khen and Arthur Y. C. Chung

ABSTRACT

Most work on the occurrence and ecology of insects in rainforest canopies has been concentrated on natural, primary forests. However, a great many rainforests around the world have been, and will be in the future, subject to a variety of management tactics that subtly or dramatically alter both the abiotic and biotic habitats that their canopies provide. These management tactics include selective logging, which results in secondary growth and the appearance of vegetation types associated with early successional stages, clear-felling and complete clearance, leading to a total destruction of natural habitats, and the conversion of natural forests to plantations, either of indigenous tree species or, more commonly, of exotics. The exotics include commercial timber species such as pine, eucalyptus and acacia, as well as cash-crops such as oil palm. Work described here, mainly from Peninsular Malaysia and Sabah (East Malaysia) has investigated the insect assemblages that occur in these managed forests and comments on the likely impact of forest management on naturally occurring insect species. Moth and beetle assemblages were sampled using light traps, insecticidal knockdown and flight-interception traps. While the actual species occurring in each habitat type does change, the overall abundance, richness and species diversity change rather little. As management becomes more intense, and resulting ecosystems become less and less related to natural ones, alterations in species richness, abundance and, most importantly, guild structure are detectable. Even so, some exotic plantations, such as eucalypts, still maintain a relatively rich insect fauna, while those that do not allow the continuation of natural understorey, such as oil palm, show the most extreme changes.

INTRODUCTION

Most work on the occurrence and ecology of insects in rainforests has been concentrated on natural, primary forests. However, a variety of management tactics can alter both the abiotic and biotic habitats that their canopies provide. Selective logging results in secondary growth and the appearance of vegetation types associated with early successional stages; clear-felling and complete clearance leads to a total destruction of natural habitats. Natural forests can be converted to plantations of indigenous tree species or, more commonly, exotics such as the commercial timber species pine, eucalyptus, teak and acacia or cash-crops such as oil palm. Exotic tree species have been used in Kibale Forest, Uganda to help with the rehabilitation of indigenous forest (Chapman & Chapman, 1996).

All these manipulations are likely to have influences on the habitats that tropical forest provide for animals and plants. Indeed, the forest itself changes after management activities such as logging. Clearly, some plant species disappear to be replaced by others (usually ones associated with earlier successional stages), and the physical and botanical structure of the ecosystem may be altered severely. In Indonesia, for example, the ranges of girth and heights of forest trees were found to be lower in logged forests, resulting in more even age classes of dominant vegetation (Hill *et al.*, 1995). In fact, large trees are usually rare in regenerating forest. In central Amazonia, logging operations were estimated to remove 50% of the basal area of commercially valuable tree species, equivalent to around eight trees per hectare (Vasconcelos *et al.*, 2000). Furthermore, these situations take a relatively long time to revert to more natural conditions. In Pasoh, Peninsular Malaysia, for instance, it takes 35 years or so after logging before tree basal areas within the forest are returning to normal (Manokaran, 1996).

Not all components of the plant association are so affected by management. The richness of species of woody climbers, again in Malaysia, was not significantly different between logged and unlogged forest, though large individual climbers were rare in logged forest (Gardette, 1996). So, botanical diversity may not vary tremendously, but the size range of individuals may be reduced significantly. Interestingly, other members of the tropical forest community are less affected by management. For example, the species diversity of ectomycorrhizal fungi may show little difference between logged and unlogged areas, perhaps because remnants of natural forest such as small dipterocarp trees and saplings are able to persist after logging the big trees, (Watling *et al.*, 1996).

Forest disturbance, resulting in the formation of gaps and fragments, can be a natural event. Hurricanes, for example, can seriously affect forest composition and the animals that live within them. For example, Schowalter & Ganio (1999) found that gaps created by a hurricane in Puerto Rico yielded higher numbers of sap-feeding insects and molluscs compared with intact forest, whereas defoliating insects and detritivores were more abundant in undisturbed habitats. However, forest management also results in somewhat similar disturbance events, one of which is exemplified by fragmentation, where large blocks of forest are split into much smaller patches, often isolated one from another by roads, fields or plantations of exotic plant species. If forest management causes habitat fragmentation and hence isolation, then animal groups vary in their responses to this. Various examples of this effect are available. Species richness of small mammals and frogs, for example, increased after fragmentation in Amazonian Brazil, although species richness of birds and ants decreased (Gascon *et al.*, 1999). Mammals such as primates, tree shrews and squirrels mostly seem to be able to survive at least one commercial thinning, though specialist species do not like fragmentation (Laidlaw, 1996). However, some mammals do better in logged than in natural forests (e.g. squirrels and tree shrews), and undoubtedly, large herbivorous mammals such as elephants and rhinos are likely to do better in secondary forest than in primary, by virtue of the provision of much more luxuriant vegetation within easy reach. In Uganda, it has been concluded that low-intensity selective logging benefits primate conservation, though high-intensity logging has the reverse effect (Chapman *et al.*, 2000). One extreme result of forest management is the conversion of rainforest to exotic plantations, wherein native flora and fauna may not be at home. The nature of the plantation also may be important. In Malaysia, *Acacia mangium* Willd. plantations are almost completely devoid of primates, whereas oil palm, which provides large quantities of food for monkeys, is a very good home for at least one species (M. Speight, personal observation).

Many studies have attempted to use insects as measurements of habitat change in managed forests compared with natural ones. Insects are often used as indicator groups, the distribution of which predicts the overall importance of the biodiversity of candidate areas (Howard *et al.*, 1998). In other words, if certain groups of insects are studied, then the variations in their numbers and species richness between habitat types can be related directly to changes in the ecosystems in which they live. Thus, indicator groups of insects can be used as 'early warning' measures of habitat change, especially if their populations can be measured readily (Brown, 1997). However, it is important to study the impacts of habitat management on insects for their own sake, to investigate changes in biodiversity and to help with their conservation (Watt *et al.*, 1997b; Basset *et al.*, 1998). Certainly, forest management, as mentioned for mammals above, can have predictable affects on insect assemblages. Differences in habitat structure among forest and savanna ecosystems in Venezuela had the greatest influence on ant taxonomic composition (Netuzhilin *et al.*, 1999). In the neotropics, for instance, butterfly diversity often increases with disturbance, but some sensitive species and genes are eliminated at very low levels of interference (Brown, 1997). This loss of ultraspecialist species is to be expected, but even severe tropical forest manipulation may not be as devastating as dogma might lead one to expect. If, given time and some assistance, heavily perturbed tropical forest sites are allowed to regenerate, then insect assemblages may be relatively little affected. For example, ant assemblages in tropical monsoon regions of Brazil converted to bauxite mining and then allowed to recover mainly using mixed plantings of species of native forest trees, were characterized by proportionally more generalist species and fewer specialists. However, ant species richness in eucalyptus/acacia plantations was as great as in the areas rehabilitated with native vegetation. Around 25% of ant species were only found in the rehabilitated areas, hence 'new' diversity is promoted (Majer, 1996).

The following sections describe some of the collaborative research carried out during the 1990s, employing links among various institutions, namely (in no particular order), the Forest Research Institute of Malaysia (FRIM), Kepong, Kuala Lumpur, Peninsular

Malaysia, the Forest Research Centre (FRC), Sepilok, Sandakan, Sabah, the Natural History Museum (NHM) London, UK, and the Department of Zoology, University of Oxford, UK. Three separate doctoral studies are combined to illustrate the effects of tropical forest management on various types of insect and other arthropod living in these habitats, with particular reference to the forest canopy.

METHODS

A particular problem in the present context involves the actual definition of forest canopy (see Ch. 1). Many workers, the present authors included, consider that forest canopies begin at, or even below, ground level and continue all the way to the uppermost reaches of the treetops. Insects and their relatives are of course nonrandomly distributed through this extremely broad habitat called 'canopy', and any sampling system can only represent certain subdivisions of this. So, for instance, invertebrate abundance and higher taxon richness were greatest at ground level (compared with rainforest canopy) in flight-interception traps in North Queensland, whereas other workers find the insect diversity in light traps is greatest in the canopies (Hill & Cermak, 1997). The studies discussed in this chapter employ various techniques for sampling insect populations, and it must be accepted that none can be completely canopy specific. Instead, they all provide information on the general effects of managing tropical forest communities at several levels within the forest canopy.

Three major sampling techniques were employed in these Malaysian studies: light trapping, flight-interception trapping and canopy knockdown employing a petrol-driven mist-blower. Rothamstead-type light traps were used in Peninsular Malaysia, fitted with 250 W clouded lamp bulbs as opposed to the more usual 200 W tungsten bulbs, the former being more readily available in Malaysia. Each trap was run for 5 hours per night, starting at dusk (Intachat, 1995). In Sabah, a 250 W mercury-lithium (HgLi) bulb was mounted on top of and in front of a vertical white sheet. Macromoths attracted to the light were hand collected by a team of collectors working full time for 4 hours per night beginning at 18:30 on each occasion. All other insects attracted to the light were allowed to redisperse back into the forest (Chey *et al.*, 1997).

Mist-blowing is a simple and cheap method for sampling many different types of arthropod from relatively low forest canopies; in the opinions of the authors, it has several advantages over the more widespread practice of canopy fogging. The equipment is easier to use than a thermal fogger, produces less downwind drift and is more easily targeted into localized canopies. Its biggest drawback is the lack of penetration into high closed tree canopies, and hence sampling has to be restricted to 10 m or so above ground (Chung *et al.*, 2001). The mist-blower used in Sabah was a Japanese-made Maruyama™ MD300, with a ULV nozzle fed directly from a small reservoir attached to the emission pipe. A non-residual knockdown pyrethroid, Pybuthrin™ 2/16 (Wellcome Foundation/Roussel Uclaf) was sprayed into the canopy of sample trees for 1 minute. Pilot studies using ultraviolet tracers showed that the maximum height reached by mist-blown chemical from the ground was approximately 10 m; consequently only lower canopy or smaller trees were sampled. Arthropods falling from the canopy were collected on 1 m^2 white sheets or funnels laid under the trees (Chey *et al.*, 1998; Chung, 1999; Chung *et al.*, 2001).

For beetle sampling in Sabah, flight-interception trapping was used in addition to mist-blowing, pitfall trapping and Winkler sampling (results from the latter two techniques are not presented here). The flight-interception traps consisted of a 1 m × 2.5 m fine-mesh black nylon net supported vertically on poles. Insects hitting the net fell into aluminium trays beneath, containing preservative (chloral hydrate solution). Each sample was taken over a 24 hour flight period (Chung, 1999).

Statistical methodologies for analysing data relied where possible on techniques of parametric analyses of variance (ANOVA), normalizing the data by transformations when required. The diversity index employed both for insect and vegetational samples was the alpha statistic (α) derived from the log-series distribution ($S = \alpha \log_e(1 + N/\alpha)$, where S is the total number of species and N the total number of individuals in a sample (Fisher *et al.*, 1943). This index has regularly been employed in investigations of tropical forest insects and has been found to be relatively insensitive to sample size, or overly rare or abundant species (Intachat & Holloway, 2000). As well as insects, this index has been employed for investigating other species including

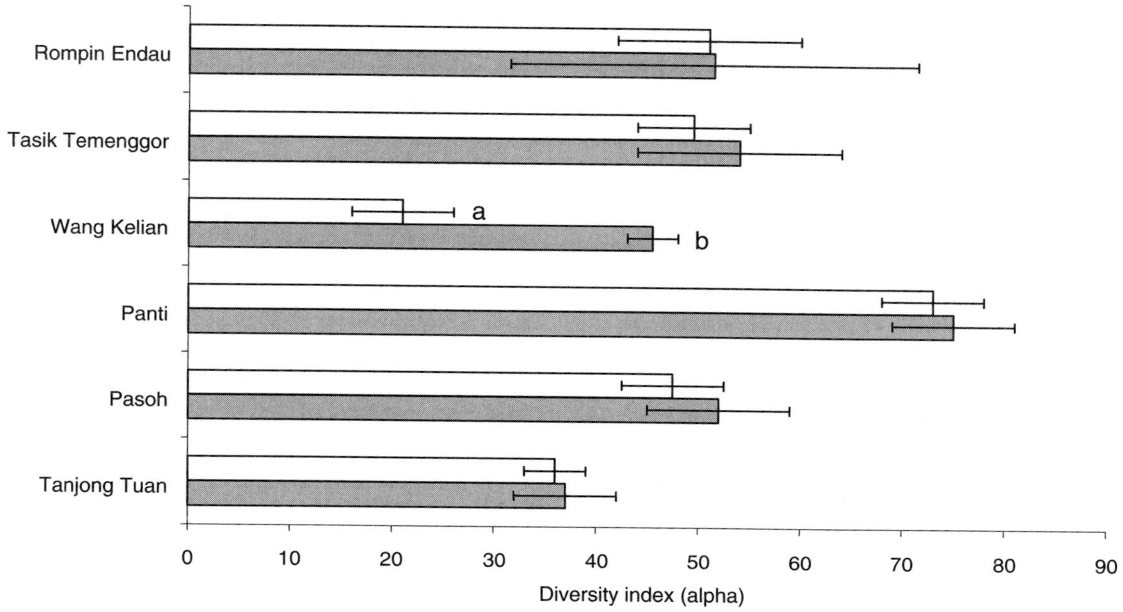

Fig. 34.1. Species α-diversity indices ($\pm$ SE) for paired primary (open columns) and secondary (hatched columns; selectively logged) sites in Peninsular Malaysia. Means with different letters are significant at $p < 0.05$. (From Intachat, 1995.)

birds (van Heezik & Seddon, 1999) and spiders (Russell-Smith & Stork, 1995).

Finally, the similarities between forest types and the arthropods associated with them was investigated using detrended correspondence analysis (DCA) and canonical correspondence analysis (CCA). These two techniques are usually used in combination to see whether the measured environmental variables (in this case, the types of forest) can account for the variations in the sampled species data (Chung *et al.*, 2000 provide details of the analyses).

RESULTS

The sections below are separated into different types of forest management tactic for convenience.

Selectively logged forests

In various parts of Peninsular Malaysia, six sites were identified where primary tropical forest had been selectively logged in part, leaving some primary forest situated adjacent to equivalent logged forest. Light traps were set up in pairs of such forests, and the numbers of species of moth, and their abundances, determined over the same trapping period in each case. Three replicate traps were run simultaneously in each pair of forests, and the data analysed as paired comparisons in a randomized block design using ANOVA. As shown in Figure 34.1, the species α-diversity index calculated from these paired samples did not differ significantly from primary to logged forest in five of six cases. In fact, species diversity tended to be slightly higher in logged patches of forest compared with their primary equivalent. In the one case where differences were detectable statistically (Wang Kelian, a forest in the extreme north-west of Peninsular Malaysia dominated by *Shorea* spp. (Dipterocarpaceae) belonging to the White Meranti group), the secondary forest exhibited approximately double the diversity of the primary (Intachat, 1995; Intachat *et al.*, 1999). Clearly, logging has little or no effect on the α-diversity index in most cases: if any effects can be detected, they reflect an increase in the species diversity of logged areas. However, though the diversity remains little changed, the actual species composition varied considerably between pairs of logged and unlogged sites. Table 34.1 summarizes shared and unique species from all six localities. In primary forest, *Idaea* sp. was the most abundant genus whereas, in secondary forest, *Ourapteryx fulvinervis* Warren was the commonest species. Species thought to be associated

Table 34.1. *Total number of species and abundance collected with light trapping in primary and secondary forests*

Variable	No. species	Abundance
Total primary	136	350
Total secondary	104	196
Unique to primary	180	2424
Unique to secondary	316	1757
Shared by primary and secondary	284	1213

with dipterocarps such as *Ornithospila* spp. were unique to primary forests.

Disturbance by logging, thereby creating secondary regenerating rainforest, must of course alter the environment in tropical forest canopies, in particular by opening up the canopy and allowing plants associated with early successional stages to proliferate. Insects such as Lepidoptera that specialize on grasses, creepers and early climbers are, consequently, likely to become much more abundant.

Clear felled and replanted natural forests

Tropical forest management is often more intense and hence potentially serious for insects than a simple and obviously rarely harmful selective logging programme. Clear-felling essentially removes all canopy habitat and, depending on the future manipulation of the site (leaving it to recover over time, or, alternatively, planting it with new trees), may well have a much greater influence on the animals and plants living there. In order to test the effects of more intense and varied types of management, three sites near Kuala Lumpur, Peninsular Malaysia were sampled, again using light traps. Table 34.2 summarizes the ecological characteristics of the sites. Site 1, the clear-fell, had little real forest canopy even 20 years after the manipulation. Plant species diversity, measured by α-diversity, remained relatively low and was indicative of early successional stage tropical forest. Site 2, a selectively logged area, was similar in ecology to those forests sampled in the previous example. As would now be expected, plant diversity was high, with succession advanced. Finally, site 3, situated at Kepong, consisted of dominant tree species representative of natural primary dipterocarp forest, though in fact the forest had been planted in the 1920s. The forest was a late successional stage, and as Table 34.2 shows, the plant species diversity was reduced relative to the secondary forest site (site 2) (Intachat *et al.*, 1997). Numbers of species, abundance and species diversity of moths totalled over a 14 month light-trapping period are presented in Fig. 34.2, which illustrates that though the clear-felled site tended to have fewer moth species, and an overall lower α-diversity value than the other two, the selectively logged forest was rather similar to the native plantation. It might be expected that after 60 years or more, a dipterocarp plantation should approach the ecosystem of an untouched primary dipterocarp-dominated rainforest, and it is significant to note that insects found within it are not at all dissimilar to those in a much newer selectively logged forest. Clear-felling, even when allowed to regenerate naturally, maintains moth assemblages clearly different from even semi-natural rainforest even after nearly a quarter of a century.

Table 34.2. *Summary site descriptions of three managed forests in Peninsular Malaysia*

Site No.	Vegetation characteristics	Plant species α-diversity	Successional stage
1	Clear felled in 1970s, subsequently abandoned. Mainly grasses and shrubs, little forest canopy	4.0 ± 0.3	Early
2	Selectively logged in the 1970s	23.5 ± 3.2	Advanced
3	Mixture of dipterocarp species in a plantation, established in 1927	10.7 ± 1.6	Late

From: Intachat *et al.*, 1997.

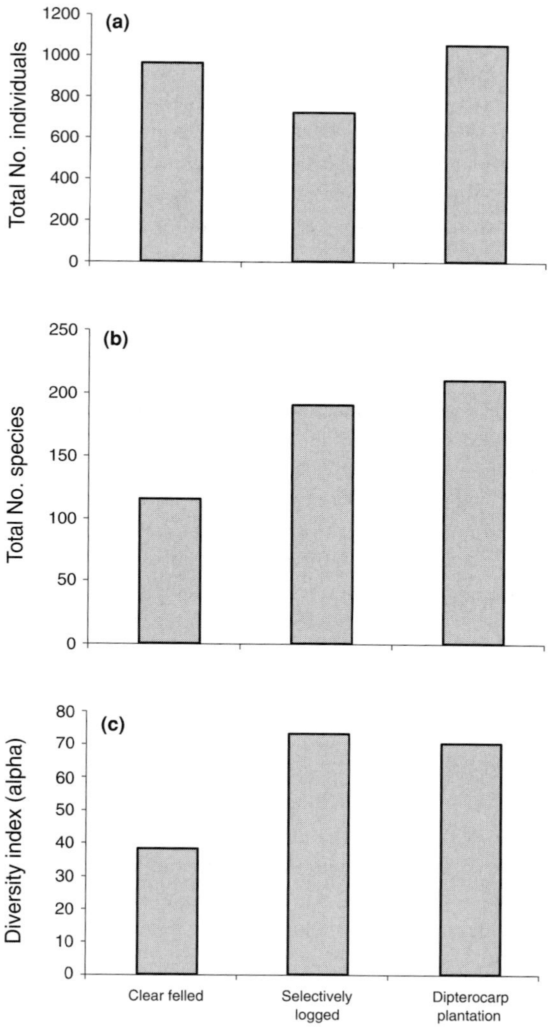

Fig. 34.2. Macromoths caught in light traps in three types of managed forest in Peninsular Malaysia over a 14 month period. (**a**) Total numbers of individuals; (**b**) total numbers of species; and (**c**) overall α-diversity index. (From Intachat *et al.*, 1997.)

Commercial forest plantations

Very often, rainforest is clear-felled as in the above examples but, instead of being left to regenerate or, indeed, being given a helping hand by the planting of indigenous tree species, the sites are planted with fast-growing commercial species, which are in the main exotic. These include teak, acacia, eucalyptus and pine amongst others, none of which occur naturally in most of south-east Asia. So widespread is this type of planting that it is frequently impossible to find any remaining primary forest in the vicinity to use as a baseline for comparison. As a result, the several studies described below have had to use local secondary, selectively logged sites for comparison instead. Note, however, that, as discussed above, secondary and primary forest are often not significantly different one from another in terms of the macromoth populations which they support.

In Sabah, east Malaysia, two distinct types of insect sampling have been used to compare populations in different types of commercial plantation: light traps and mist-blowing. As discussed earlier, light traps placed on or near the ground are not sampling canopy insects alone by any means, but since many types of managed forest do not have a canopy as such, this type of sampling regime is a reasonable system for comparison of natural forests in their entirety with heavily modified ones. As before, adult Lepidoptera are the only group of insects to be studied quantitatively using light traps, but in order to investigate all the major groups of arthropods in tree canopies, Chey Vun Khen and colleagues employed, in addition, mist-blowing in the lower canopies of various plantation forests and compared their samples with those from indigenous, albeit secondary forest.

Figure 34.3**a** summarizes the results of large-scale moth trapping in six types of forest in Sabah. All sites were situated at Brumas in the south-west of the state, within a few kilometres of each other. Results from a primary forest site at Danum Valley near Lahad Datu, Sabah are also depicted to provide an overall comparison, though this area is some 100 km distant. As can be seen, species α-diversity values for adult moths caught on sheet light traps in the seven forest sites show some significant differences. Firstly, the mean value for the primary site is the lowest of all. Holloway (1998) has in fact measured diversity values for Lepidoptera in the same primary forest at Danum as Chey *et al.* (1997) and found values approaching 200. Even if the present values from primary forest are an underestimate, it is clear that moth diversity is not particularly depressed even in monocultures of exotic tree species. The figure also shows that lepidopteran diversity remains high in exotic plantations, compared with natural secondary forest. Eucalyptus stands recorded α-diversity values not significantly different from those of the secondary forest, indicating that in circumstances where tropical forest is replaced by commercial species, biodiversity in general terms may be substantially maintained.

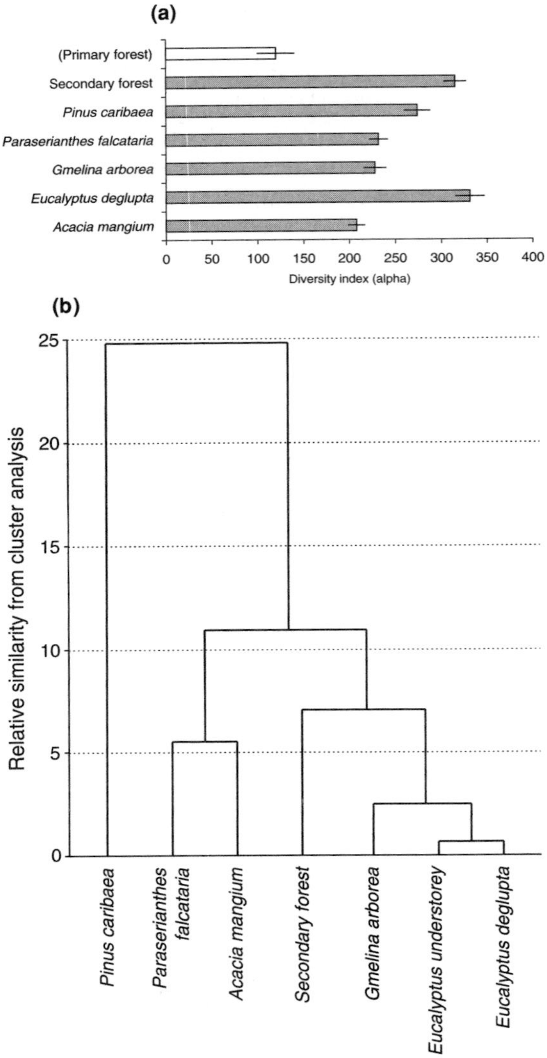

Fig. 34.3. Light trap catches in managed forests. (**a**) Species α-diversity indices ($\pm$95% confidence intervals) for moths caught in light traps in six different managed and one primary forests in Sabah; (**b**) hierarchical cluster analysis (nearest neighbour/single linkage) on the numbers of individual arthropods in each major family from seven managed forest types. (From Chey *et al.*, 1997.)

However, species diversity measurements do not provide the whole truth. It is possible to maintain such figures at high levels by replacing insect species typical of primary rainforest with others more typical of secondary or early successional stage habitats, and the numbers of individual species shared by these different forest types gives an insight into the impact of forest management. Table 34.3 shows that only a proportion of moth species caught by light trapping is shared by each habitat. Of the 1048 species caught in the secondary forest in Sabah, only 62% were also found in *Eucalyptus* for example (Chey, 1994). The missing species are likely to be the host-plant specialists that are unable to survive in the exotic plantations. Those that do occur in both natural and exotic habitats are able to feed on secondary plant regeneration within the plantations (see also below).

Lepidoptera are thought of as useful indicator species, being relatively host specific as larvae. However, much more information can be obtained using more general sampling systems, especially those that do not rely (as light traps do) on insect activity but are able to sample a variety of arthropod groups from forest canopies. Chey *et al.* (1998) used mist-blowing in the same six forest types at Brumas, Sabah as in the previous example, where all arthropods were collected from low canopies in replicated samples. One extra habitat was also sampled, that of the natural understorey that proliferated under the thinly planted *Eucalyptus deglupta* Blume. Table 34.4 presents some of the results for the major arthropod groups collected. Note that for the analysis portrayed in the table, *Pinus caribaea* Morelet is omitted because of low sample frequency. Numbers of individuals of Coleoptera, Hymenoptera (especially Formicidae), Hemiptera and Orthoptera were significantly higher in the secondary forest, and Coleoptera and Hemiptera were also more species rich in this habitat. Interestingly, lepidopteran abundance and richness was significantly higher in the *Paraserianthes falcataria* (L.) Nielsen plots than anywhere else. This leguminous tree species is known to support large populations of herbivores in its canopy by virtue of its relatively nutritionally rich and poorly defended foliage (Speight & Wylie, 2001).

Once again, the similarities among forest types in terms of the arthropod assemblages they support is of interest. Figure 34.3b shows a cluster analysis performed on the knockdown data. *P. caribaea* is clearly the habitat with least similarity to any of the others. *Paraserianthes* and *Acacia* spp., both legumes but with very different leaf characteristics, are closely linked, whereas secondary forest is associated with the group containing *Gmelina arborea* and, in particular, both the *Eucalyptus* sp. canopy itself and its associated native understorey. In the last, it would appear that as long

Table 34.3. *Numbers of species of macromoths caught in light traps shared between various forest types in Sabah*

	Acacia mangium	*Eucalyptus deglupta*	*Gmelina arborea*	*Paraserianthes falacataria*	*Pinus caribaea*	Secondary forest	Total no. per site
Acacia mangium	–	487	444	483	452	527	726
Eucalyptus deglupta	487	–	480	545	507	653	872
Gmelina arborea	444	480	–	459	452	503	675
Paraserianthes falcataria	483	545	459	–	474	586	782
Pinus caribaea	452	507	452	474	–	515	778
Secondary forest	527	653	503	586	515	–	1048

From: Chey, 1994.

Table 34.4. *Comparisons of mean number of individuals and mean number of species in eight major groups of arthropods collected in knockdown samples at Brumas, south-west Sabah*

Taxa	Mean No. collected					
	Acacia mangium	*Eucalyptus deglupta*	*Eucalyptus* understorey	*Gmelina arborea*	*Paraserianthes falcataria*	Secondary forest
Individuals						
Lepidoptera	2.8	2.3	4.3	3.5	19.0*	7.0
Coleoptera	4.0	2.0	9.3	4.0	3.3	14.0*
Formicidae	82.5	24.5	67.3	8.0	52.5	242.3*
Hymenoptera	3.3	2.5	3.5	2.0	1.0	31.3*
Diptera	5.5	1.8	6.0	2.5	2.0	9.5
Hemiptera	4.8	5.8	4.8	6.0	4.5	19.3*
Orthoptera	3.3	1.3	3.3	2.5	2.3	9.5*
Araneae	6.0	6.5	4.8	4.8	2.0	7.5
Species						
Lepidoptera	2.5	2.3	4.3	2.8	11.8*	5.5
Coleoptera	3.8	1.8	5.5	4.0	2.8	10.0*
Formicidae	8.0	6.5	7.8	5.3	4.3	8.8
Hymenoptera	3.0	1.0	2.8	2.0	1.0	5.3
Diptera	4.3	1.5	3.5	2.3	4.5	5.8
Hemiptera	4.0	5.0	3.8	4.8	3.8	14.0*
Orthoptera	2.3	1.3	2.5	2.3	2.3	6.5*
Araneae	4.5	5.8	4.0	4.8	2.0	5.5

*Significant differences between means between sites for each taxon ($p < 0.05$) for comparisons carried out on $\log_{10}$ transformed abundance data, *P. caribaea* excluded.
From: V. K. Chey & M. R. Speight, unpublished data.

as regeneration of native plant species is possible under the canopy of exotic plantations, then a substantial representation of natural arthropod fauna may be retained.

In a different part of Sabah, at Sepilok, work has also been carried out comparing in detail the insect assemblages in various types of managed tropical forest. In this case, Coleoptera were studied intensively in order to investigate various feeding guilds (carnivores, herbivores and detritivores) within a single insect order. In a relatively simple investigation, the beetles from the canopies of selectively logged forest with good regeneration were compared with those from the canopies of *A. mangium*, the most widely planted exotic tree species in many parts of Asia. A 20 m aluminium tower was used to gain access to the canopies of both forest types, and mist-blowing was employed again to collect insects at various heights up the tower. The sampling trays in this case were positioned in the canopies, directly below the misted region, so that specimens falling from the trees were caught immediately, obviating the problem of many individuals being trapped by intervening foliage or, as in the case of small individuals, being blown away in the wind. As can be seen in Figure 34.4**a**, though the magnitude of collections was small, natural forest canopy yielded significantly more individuals and species of beetle compared with the *A. mangium* plantation (Chung *et al.*, 2001). In addition, the distribution of coleopteran families differed between forest types. In the natural forest, certain curculionids and chrysomelids were most abundant, whereas in *A. mangium*, tenebrionids and elaterids were more common (Chung, 1999).

Chung then carried out much more extensive studies on four types of forest: primary rainforest, secondary (selectively logged) rainforest in the same area, plantations of *A. mangium*, and plantations of the oil palm (*Elaeis guineensis* Jacq.). The final habitat type provided an extreme example of forest manipulation, where natural forest is completely removed and replaced with an exotic cash crop, which is intensively managed. Several different sampling techniques were employed, each one designed to investigate coleopteran assemblages from different parts of the forest ecosystem. Most relevant to canopy assemblages were flight-intercept traps and mist-blowing. Three flight-intercept traps and 10 mist-blown trees per site were utilized. Catches in flight-intercept traps did not differ significantly either in species richness or individual abundance between primary and secondary forest (Fig. 34.4**b**). Mist-blowing showed somewhat lower numbers of beetles in the primary habitat compared with secondary. *Acacia* sp., and especially oil palm, sites showed very reduced captures in the flight-interception traps, though the results from mist-blowing were less conclusive. Beetle abundance was not significantly less in oil palm than in the primary forest, though richness was reduced considerably (Chung *et al.*, 2000).

It was thought for the purposes of this study that the actual changes in species composition between forest site were less important than the functional changes reflected by the coleopteran guilds. Accordingly, more complex analyses of coleopteran assemblages were carried out (Chung *et al.*, 2000). In Figure 34.5**b** and Tables 34.5 and 34.6, the results of canonical correspondence analysis (CCA) on the mist-blowing data are shown. The centroid of significant nominal environmental variables, indicated in the figure by a black diamond, is the position in ordination space of the variable that splits samples into categories. Table 34.5 shows that the primary forest and oil palm variables are significant ($p < 0.05$); hence it is clear that the composition of the beetle assemblage from primary forest is distinctly different from that in oil palm plantations, whereas the logged forest and the *Acacia* plantations are very similar. Exactly equivalent results were found using flight-interception traps (Fig. 34.5**a**). Clearly, there is a considerable shift in the composition of beetle assemblages as primary forest is modified, whether simply by selectively logging or by the seemingly much more extreme replacement with an exotic tree species. As discussed in general above, species richness and abundance do not provide much, if any, information about the functioning of ecosystems and the communities within them. In contrast, the relative proportion of trophic groups or guilds is of great significance in terms of community structure. Figure 34.6 shows such an analysis for the four forest types in Chung *et al.* (2000), using Hammond's (1990) guild assignments for coleopteran families. All major habitats in the forest were included, from litter to canopy. From these results, it is clear that primary rainforests overall are typified by a relatively large proportion of predatory beetles, with smaller but roughly equal numbers of fungivores and saprophages. Herbivores make up not much more than 10% or so of species, and less than 10% of individuals. Logged forest appears to differ only slightly in guild structure. Individual

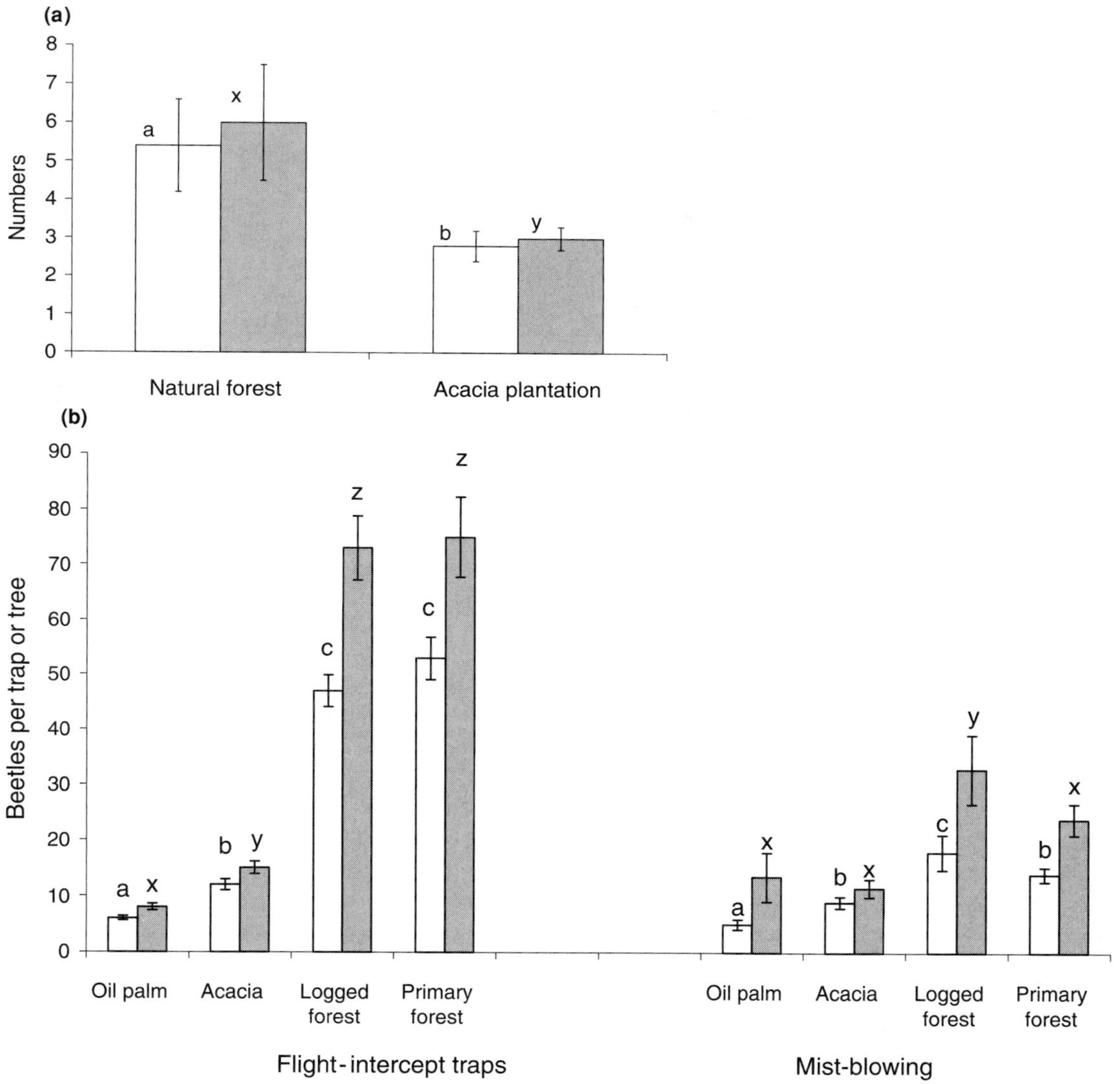

Fig. 34.4. Mean numbers of beetle species (open columns) and mean number of individual beetles (black columns) ±SE in different tree habitats. (**a**) Beetles caught by mist-blowing in tree canopies (different letters denote significant differences in paired samples at $p < 0.05$ level). (From Chung *et al.*, 1999.) (**b**) Beetles sampled from four habitat types in Sabah (different letters denote significant differences ($p < 0.05$). (From Chung *et al.*, 2000.)

species may disappear, to be replaced by others, but community function, as assessed by guild structure, probably remains the same. *Acacia* sp. shows some reduction in predators, but the major changes are evident in the oil palm assemblages, where the percentage of herbivores increases dramatically at the expense of predators. The implications of this shift in trophic relationships may be serious for population stability and regulation.

DISCUSSION

The results of this work in Malaysia suggest that rainforest management, manipulation and conversion may

Table 34.5. *Environmental variables that account for significant variation in beetle species composition between sites (assessed by mist blowing) by their marginal and conditional effects on beetle species, as obtained by the canonical correspondence analysis forward selection procedure*

j variable	Marginal effects (forward: step 1)			Conditional effects (forward:continued)		
	Step	λ_1	*p* value[a]	Step	λ_a	*p* value[a]
PF	1	0.87	0.009	1	0.87	0.009
OP	2	0.85	0.037	2	0.83	0.031
CanopMB	3	0.75	0.070	3	0.65	0.141
LF	4	0.73	0.084			
A	5	0.68	0.128			
Period	6	0.51	0.713	4	0.49	0.434

λ_1, eigenvalue (fit) with variable *j* only; λ_a, increase in eigenvalue (additional fit); PF, primary forest, OP, oil palm plantation, CanopMB, canopy cover, LF, logged forest, A, acacia plantation; Period, sampling period.

[a] Significance level of effect, as obtained with a Monte Carlo permutation test under the null model with 999 random permutations (i.e. minimum *p* value is 0.001).

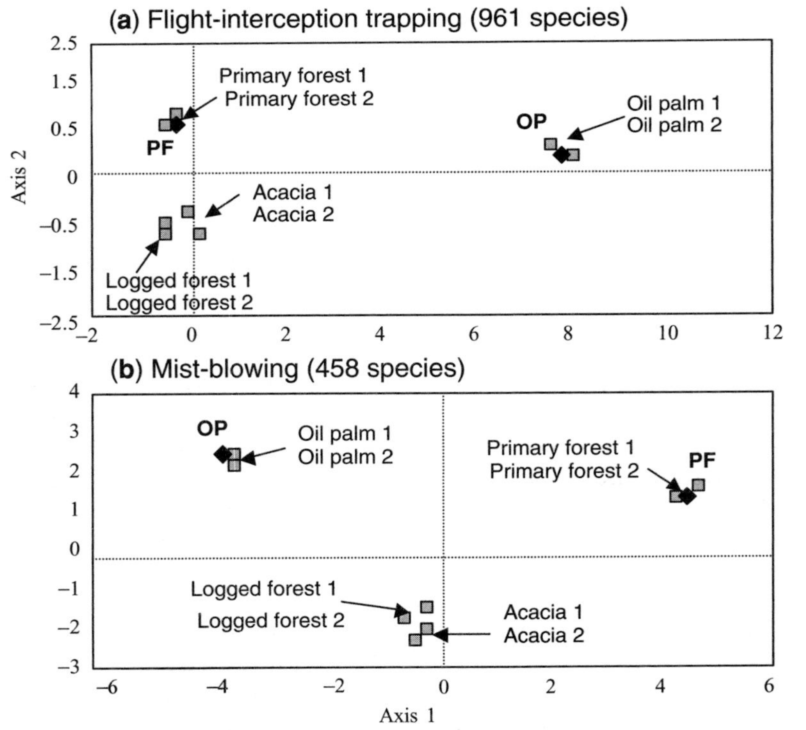

Fig. 34.5. The distribution of beetle samples collected in four habitat types in Sabah. Canonical correspondence analysis (CCA) ordination diagrams on axis 1 and 2 with beetles samples (grey boxes) and the centroid of significant nominal environmental variable (black diamonds) (see text and Tables 34.5 and 34.6). (From Chung *et al.*, 2000.)

Table 34.6. *Comparison of eigenvalues by correspondence analysis of beetle species composition as assessed by mist-blowing*

Correspondence analysis	Axis 1		Axis 2	
	Eigenvalue	Correlation coefficient[a]	Eigenvalue	Correlation coefficient[a]
Detrended	0.888		0.583	
Canonical	0.886	1.000	0.816	0.998

[a] Species–environment intraset correlation coefficient of CCA.

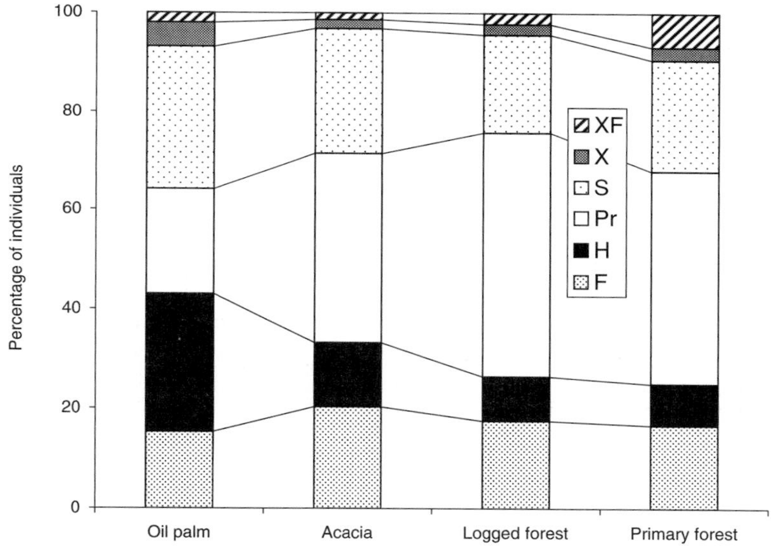

Fig. 34.6. Proportions of beetle trophic groups (guilds) caught using a combination of methods in four forest types in Sabah. XF, xylo-mycetophages; X, xylophages; S, saprophages; Pr, predators; H, herbivores; F, fungivores. (From Chung, 1999; Chung *et al.*, 2000.)

have only limited impact on insect biodiversity. Certainly, as new types of forest tend towards extreme monocultures of exotic tree crops under intensive management (such as oil palms), then insect populations and guild structures do change, but less extreme management such as selective logging or even replacement with timber tree species may be less of a concern in terms of biodiversity loss. In several of the cases studied here, species richness, abundance and α-diversity may remain unaltered in managed forest, or even increased somewhat. Individual insect species, however, may disappear under management, to be replaced by others better adapted to the modified habitat conditions.

These conclusions agree with much of the published literature, but not with all of it. For example, in a study of Indonesian butterflies, species richness, abundance and evenness were all significantly higher in unlogged monsoon lowland forest than in sites that had been selectively logged 5 years previously (Hill *et al.*, 1995). A word of caution is appropriate here; butterflies may not be good indicators of forest change since they are thought to be rather marginal in deep forest habitats by virtue of their heliophilic lifestyles (Basset *et al.*, 1998). By comparison, large moth families such as the Geometridae seem to be much more reliable in the context of assessing impacts on closed tropical forest. Thus, in another Indonesian example (Sulawesi), Holloway (1998) used light traps to survey assemblages of Lepidoptera, including the geometrids, and found that conversion of natural forest systems to modern cropping systems (such as softwood plantations) resulted in a major reduction in moth diversity and its quality in terms of faunistic composition, endemism and so on. However, Holloway commented that diversity can recover to near its original level as cultivated (or replanted) sites mature.

In the main, most results suggest less of a general problem and, as mentioned above, conclusions from our studies in Malaysia which suggest that diversity may not decrease markedly after forest management are supported by published work from various tropical countries. The richness, evenness and abundance of ant species in Amazonia did not vary significantly between natural and logged forests, and most of the species found in natural forest also occurred in logged plots. Only their abundance dropped for 10 years or so after logging (Vasconcelos *et al.*, 2000). These authors believe that the persistence of ant assemblages typical of disturbed forests is likely to depend on the amount of structural damage following logging. If this is true for other countries, then it might be expected that planting of exotic species in Malaysia as reported here will have more impact on native insects than 'mere' selective logging. Ants in the canopy at Pasoh Malaysia (sampled by fogging) yielded similar species richness samples in both unlogged and logged forest. However, some genera were recovered from natural forest, but not from regenerating forest, while other genera were only found in the logged forest (Bolton, 1996).

Rather than managed tropical forests causing a reduction in insect species richness and abundance, logged forest, and especially plantations, may positively encourage certain insect groups. In Kibalae Forest, Uganda, Nummelin & Fuersch (1992) found that logged forest sites and, indeed, exotic pine plantations, had higher species numbers of Coccinellidae than virgin forest sites. Davis & Sutton (1998) studied the assemblages of arboreal dung beetles in Sabah, in an area close to that described in the present studies. These authors found that logged and especially plantation forest showed a marked increase in dung beetles living in canopies, which they related to a proliferation of suitable microclimatic conditions in the modified habitats. However, more recent work suggests that only edge-specialist dung beetles are found in plantations (Davis *et al.*, 2000). In general, it seems clear that the same mechanisms may well apply to the various arthropod orders studied here. Forest plantations of exotic tree species such as *Eucalyptus* provide for rapid growth and high productivity. In many countries, this production relaxes the pressure somewhat on the logging of native forests (Majer & Recher, 1999), though high-value native hardwoods are still not easily grown in plantations (Evans, 1996). Countries where indigenous plant species proliferate that are related to the new exotics are likely to provide a rich fauna to colonize the new plantations; this mechanism may contribute to burgeoning diversity in the new habitats but may represent, also, potential for pest problems (Speight & Wylie, 2001). For instance, lepidopteran abundance was higher, and natural enemy abundance lower, in eucalypt plantations in Brazil compared with the interior of adjacent native forests. Associating native vegetation with eucalypt plantations might reduce outbreaks of lepidopteran pests by increasing the numbers of hymenopterous parasitoids (Braganca *et al.*, 1998). Pest insects such as gall-forming psyllids may be more common or active in large forest openings or in plantations compared with unlogged forests (Nichol *et al.*, 1999). The potential for pest problems in exotic plantations such as *Eucalyptus* or *Acacia* spp. needs to be considered further. As Figure 34.6 suggests, potential pests (the herbivores) increase in abundance from primary forest to plantations. This phenomenon has also been reported in the literature. Risks from colonizing native herbivores in eucalypt plantations in Brazil are thought to be considerable (Majer & Recher, 1999), while back in Malaysia, lepidopteran larvae reached high levels of abundance in logged forests relative to primary forest (Floren & Linsenmair, 1999).

The removal of commercial tree species by selective logging undoubtedly results in significant decreases in canopy cover (Vasconcelos *et al.*, 2000), which results in increases in understorey vegetation. This new growth, however, will provide food and habitats for many animal groups and will clearly redress some of the balance of diversity and abundance thought to be lost as a result of the management activity. The understorey of eucalypt plantations are indeed a case in point. It has to be borne in mind that insects may not forage on both seedlings and canopies of mature trees (Basset *et al.*, 1999), but spatial heterogeneity is a fundamental provider of mixed habitats for flora and fauna, and work in Borneo on forest butterflies indicated that logged forest was equally suitable as natural forest for these insects (Hill, 1999). Some groups of insect forage daily through tropical forest canopies but have a nest or colony elsewhere in the forest, often on the ground. The giant ant of

Borneo, *Camponotus gigas* Latreille, for example, has subterranean nests from which workers venture forth at dusk to forage in the canopy (Pfeiffer & Linsenmair, 2000). Changes in canopy structure and botanical composition may not influence such species unduly, as long as their nests are left comparatively undisturbed by initial forest operations and food items are still abundant in the nocturnal tree canopy. There is also an added complication in the relationship between natural forest understorey and plantation canopies. The close association between understorey and *Eucalyptus* sp. in the present studies has been reported earlier, and it is clear from other work that processes such as herbivory in understorey sometimes closely match those in the tree canopies above (Vasconcelos, 1999). If the arthropod fauna of natural understorey can freely migrate into plantation canopies then, on the one hand, this is likely to promote overall diversity in the stand; on the other hand, it may also provide economically damaging pest species.

Not all workers report such similarities between forest understorey and canopy. In one instance, the understorey was actually found to harbour more diversity than associated forest canopy. In Ecuador, butterfly species richness and individual abundance was lower in the canopy than in the understorey (DeVries *et al.*, 1999a). Willott (1999) studied moth populations in Sabah and detected a canopy fauna distinct from that in understorey. Certainly, the most specialized species might be expected to be adversely affected by tropical forest management and conversion to other types of forest. None of the sampling methods discussed in this chapter can be expected to collect representatives of the very specialized fauna of high canopies in tropical forests – these arthropods may be the most susceptible to habitat change resulting from forest management. Nonetheless, results from the present studies strongly suggest that the recruitment of new, non-natural forest species is highly likely to redress this balance at least in terms of overall diversity.

35

Conclusion: arthropods, canopies and interpretable patterns

Yves Basset, Vojtech Novotny, Scott E. Miller and Roger L. Kitching

ABSTRACT

This synthesis chapter examines patterns of vertical stratification, temporal distribution, resource use and host specificity of arthropods in tropical rainforest canopies with particular regard to previous contributions in this volume. We stress research themes that are likely to be promising in future entomological research in tropical canopies and highlight lacunae in the understanding of arthropod distribution within tropical canopies. We are particularly interested in two simple questions that may stimulate further research on arthropods within tropical canopies. First, are distribution patterns of and resource use by arthropods congruent between geographical locations and, especially, forest types, such as lowland wet, lowland dry and montane tropical forests? Second, are particular arthropod taxa distributed similarly along vertical profiles in the forest and horizontally in the canopy? We discuss these and related issues from both evolutionary and conservation perspectives.

INTRODUCTION

This concluding chapter of the volume attempts to summarize spatial and temporal patterns of arthropod distribution in tropical forest canopies, with particular reference to the data presented in the preceding chapters. In particular, we emphasize congruent and/or different patterns among geographical locations, forest types and arthropod taxa for each of the three major research themes covered by this volume (vertical distribution, temporal patterns and resource use). The last section of the chapter discusses future directions of research relevant to canopy arthropods in the tropics, adding to the items already touched upon in Ch. 2.

VERTICAL GRADIENTS IN TROPICAL FORESTS

Congruent patterns among geographical locations and forest types?

One congruent pattern among geographical locations is the existence of distinct arthropod faunas in the lower (understorey) and higher parts (mid- and upper canopy) of rainforests, illustrated by all the contributions in this volume that discussed the vertical distribution of arthropods (Chs. 3, 6–11, 22, 25–27). This pattern is confirmed by several other studies in tropical rainforests (see reviews in Ch. 3 and in Basset *et al.*, 2001a) and contrasts with studies performed in temperate forests (e.g. Fowler, 1985; Le Corff & Marquis, 1999; Lowman *et al.*, 1993b).

Not unexpectedly, this pattern is most obvious when comparing the litter and canopy faunas (Ch. 9), but it is also apparent between the understorey and the upper canopy, particularly for insect herbivores (Chs. 3, 22 and 29). However, the extent to which a strict stratification of arthropod faunas exists from the litter to the overstorey is highly debatable (Ch. 3). Ant and bark beetle assemblages, for example, are not well stratified within the crowns of oaks in Sabah (Ch. 6). The extent of arthropod stratification may depend both on the arthropod taxa concerned (see below) and the forest type. For the latter, in addition to other determinants discussed in Ch. 3, at least three factors may affect arthropod stratification in tropical forests: (i) the vertical continuity of habitats; (ii) the physiognomy of the forest; and (iii) edaphic patterns.

First, differences in the continuity (or discontinuity) of available habitats for arthropods from the litter to the upper canopy may explain the extent of arthropod stratification (Basset, 2001b). For example, in montane rainforests, the trunks of trees often bear a thick cover of mosses and epiphytes and the canopy is often lower than in lowland rainforests (Richards, 1996). Accordingly, the continuity between the soil/litter and canopy in montane forests may be greater than in lowland forests, and this could explain the higher occurrence in the former of taxa well represented in the soil/litter habitat, such as Collembola, Acari and Arachnida (e.g. Stork & Brendell, 1990). Arthropod stratification would not be obvious in this case. Nadkarni & Longino (1990) reported relatively few differences in the fauna of suspended soils compared with that of the litter in a cloud forest in Costa Rica. Similarly, Sørensen (Ch. 9) reports few differences in the spider faunas of the understorey and the higher canopy of a montane forest in Tanzania.

Collembola seem to be particularly sensitive to forest type (Basset, 2001b) and this may be related to the high accumulation of organic matter in the canopy or slower decomposition rates, as discussed by Kitching *et al.* (1997) and Palacios-Vargas *et al.* (1998). Palacios-Vargas and Castaño-Meneses (Ch. 15) show that springtails can use both bromeliads on the ground and epiphytes in the canopy, migrating upward when water stress is considerable. As convincingly discussed by Shaw and Walter (Ch. 26), discrete and cryptic habitats may promote frequent inbreeding in arthropod species, thus leading to conservative arthropod assemblages. Depending on the vertical distribution of these habitats, this may also reinforce arthropod stratification in tropical forests.

Second, the forest physiognomy (i.e. forest cover and height of canopy), which is also related in part to the continuity of available habitats, may have substantial effects upon arthropod stratification. The leaf area index, for example, usually varies significantly from forest floor to the upper canopy (e.g. Parker, 1995) and this variance is maximal in tall wet tropical forests (Smith, 1973). Koike & Nagamitsu (Ch. 8) emphasize the need to measure adequately canopy structure and its influence on the flight patterns of certain taxa. A rather heterogeneous vertical distribution of values of the leaf area index may promote different insect flight patterns and, ultimately, different patterns of oviposition and vertical distribution.

In comparison with tall wet tropical forests, dry tropical forests and savanna forests are often characterized by more homogeneous environmental gradients, including abiotic and biotic factors such as illumination, relative humidity, leaf area index and the presence of flowers and seeds (e.g. Richards, 1996). Accordingly, it is less likely that arthropod faunas will be well stratified in dry forests or savannas. For example, E. Charles (personal communication) found a higher similarity of chrysomelid assemblages between the understorey and upper canopy of a dry forest than between these strata in a wet forest in Panama.

Further, in dry (seasonal) rainforests, many tree species are deciduous or partly deciduous and nutrient cycling there may be quicker than in wet rainforests, where the foliage turnover of evergreen trees is slower. Canopy arthropods may possess a variety of adaptations to cope with gradual or sudden leaf exchange and increase in irradiance and water stress, or they may migrate to other locations, as is known to occur in a variety of moths (e.g. Janzen, 1988a). This suggests that arthropods may be pre-adapted to use habitats exhibiting varying leaf area indices. The data of Janzen (Ch. 33), for example, do not appear to indicate a stratification of saturniids in the dry forest of Guanacaste in Costa Rica.

It is also of interest to question whether the vertical distribution of habitats in mixed wet forests is similar to that in monodominant (or less speciose) wet forests, which may similarly promote arthropod stratification in both forest types. Again the physiognomy and leaf area index may appear superficially more homogeneous in monodominant than in mixed forests, and this may promote arthropod stratification in mixed forests. Schulze and Fielder (Ch. 7) report on the stratification of Pyraloidea in a dipterocarp forest, perhaps with more homogeneous environmental gradients than in highly diverse forests. They point out that differences are not extensive and that the moth assemblages of the understorey and upper canopy appear to be influenced by similar factors.

Last, edaphic factors may influence migrations from the forest floor to the canopy and, in consequence, arthropod stratification. Many soil taxa migrate upward in the canopy of Central Amazonian inundation forests in response to flooding (e.g. Adis, 1981, 1997a; Erwin & Adis, 1982). This results in distinct arthropod assemblages at different heights in the short term, but less so in the long term. In this case, the extent of arthropod stratification depends on the time frame being considered.

In sum, we predict that the extent of arthropod stratification will be relatively low in montane, dry and savanna forests, and perhaps also in subtropical forests, but higher in tall wet tropical forests. Disturbance and opening of the canopy in the last is likely to affect significantly populations of canopy arthropods there.

Congruent patterns among arthropod taxa?

This issue has already been discussed at length in Ch. 3. Following the same approach, we discuss whether patterns of vertical distribution seem to be congruent for representatives of different feeding guilds, in light of the various contributions of this volume.

With regard to scavengers, fungal-feeders and dead-wood eaters, Winchester and Behan-Pelletier (Ch. 10) show that the habitat represented by suspended soils is modified by height, and that these effects can have implications for the composition of arthropod assemblages between 32 and 42 m. The review compiled by Prinzing and Woas (Ch. 24) indicates that many factors, including ecophysiological conditions, immigration from population pools and interspecific interactions, can affect the redistribution of springtails and mites among different habitats. Most likely, these patterns of redistribution are responsible for the faunal differences observed between the forest floor and the canopy. Itioka *et al.* (Ch. 12) report that the Blattodea respond positively to humus accumulation in the canopy, following flowering in dipterocarp forests, possibly promoting vertical migrations of this taxon. Barrios (Ch. 25) indicates that the scavenging fauna is more abundant on seedlings of *Castilla elastica* in the understorey than on conspecific mature trees in the canopy. These observations stress that habitats for scavengers and fungal-feeders, such as dead wood and associated fungi, appear to be relatively discontinuous and discrete along a vertical transect of rainforest (Ch. 26). Consequently, we might expect rather different assemblages of scavengers and fungal-feeders at different heights in rainforests, perhaps with a specialized fauna able to cope with the harsh environmental conditions of the upper canopy. However, the abundance and diversity of this guild should be highest near ground level (see Ch. 3).

With regard to herbivores, particularly insect herbivores, De Dijn (Ch. 11) notes the highest abundance and diversity of Homoptera is in the canopy in Surinam. Itioka *et al.* (Ch. 12) stress that anthophilous insects respond best to general flowering in the canopy of dipterocarp forests in Malaysia. Amédégnato (Ch. 22) discusses the stratification of habitats available to grasshoppers in the Amazon and the different assemblages occurring at different heights. In Panama, Barrios (Ch. 25) shows that mature trees of *C. elastica* are foraged and fed upon by a more abundant and diverse fauna of herbivores than their conspecific seedlings and saplings in the understorey. In chapter 27, Basset *et al.* compare the assemblages of insect herbivores in the understorey and upper canopy in Gabon, showing that their abundance and diversity is higher in the latter. They emphasize the low faunal turnover across the two strata, both during day and night.

There are also exceptions to this pattern of higher abundance and diversity of insect herbivores in the higher part of the canopy compared with the understorey, and this parallels the available literature (see Ch. 3). Although Schulze and Fiedler (Ch. 7) observed distinct assemblages of Pyraloidea in the understorey and upper canopy of a dipterocarp forest in Malaysia, species richness was highest in the understorey. As the authors note, whether this results from the low diversity of Lepidoptera that feed on dipterocarps in general is not known. Scolytinae were rather indifferent to height when their fine distribution in the crown of oaks in Malaysia was considered (Ch. 6). The influence of illumination on different arthropod resources, such as dead or live wood and leaves, warrants further investigation.

In sum, since many food resources, such as leaves, flowers and fruits, are more abundant in the upper canopy than in the understorey of wet rainforests, the abundance and diversity of many herbivore taxa are expected to be higher in the former strata. Differences in foliage quality between the upper canopy and understorey may induce a clear stratification of herbivores, as reported in several studies, particularly when taxa have a narrow host range (see details in Chs. 3 and 5).

Within the predatory and parasitoidal guilds, generalist predators such as spiders appear to be more diverse at ground level and the similarity of their assemblages in the understorey and canopy is rather high in Tanzania (Ch. 9). Others, such as wasps, forage indifferently in both the understorey and canopy and do not respond well to flowering events in the canopy (Ch. 12). On *C. elastica*, both insect predators and parasitoids were more abundant on mature trees than on conspecific seedlings or saplings (Ch. 25). As already noted in Ch. 3, the extent of stratification of predators and parasitoids depends on whether or not they specialize on certain prey or host species and whether they can tolerate environmental differences across different forest levels.

For Malaysia, Simon *et al.* (Ch. 6) report that alate ants prefer the lower parts of the crown, where litter accumulates and they are more likely to establish a nest, but that ant workers forage indiscriminately over the whole crowns. Ants are also most abundant in suspended soils situated in the lower part of the canopy (Ch. 10). Both De Dijn (Ch. 11) in Surinam and Barrios (Ch. 9) in Panama indicate that ants are more abundant in the understorey than in the upper canopy. As discussed in Ch. 3, strict stratification of ant species in the canopy may sometimes occur, but the abundance and biomass of ants may not necessarily be higher in the canopy than near ground level, and their observed distributions may depend on nesting ecology.

In sum, many arthropod species are likely to forage at preferred levels within the rainforest canopy. Strict stratification in closed and wet tropical forests has been reported for certain scavengers and fungal-feeders, herbivores and ants, but it appears less likely for generalist predators and biting flies (see Ch. 3).

Stratification and speciation in the canopy

Beyond the effects of latitudinal gradients and ambient available environmental energy (e.g. Turner *et al.*, 1996; Kolasa *et al.*, 1998; Gaston, 2000), additional hypotheses have been suggested to explain high species richness in tropical rainforests (e.g. Simpson & Haffer, 1978; Endler, 1982; Erwin & Adis, 1982; Brown, 1999; Schneider *et al.*, 1999; Gascon *et al.*, 2000; Hubbell, 2001; Whittaker *et al.*, 2001). These are as follows.

1. The island hypothesis, in which the formation of numerous islands caused by high sea levels during the Tertiary period subsequently resulted in species differentiation.
2. The riverine barrier hypothesis, in which emerging rivers during the Quaternary period separated species and caused them to diverge.
3. The montane isolate hypothesis, in which periods of low rainfall stranded animal populations on different parts of rainforested mountains, where they subsequently diverged.
4. The Pleistocene refuge hypothesis, in which rainforests shrank during a cool and dry period. The biota left in the fragments diverged and then rejoined expanding rainforest tracks during subsequent warming and moistening of the climate.
5. The gradient hypothesis, in which speciation results from the action of diversifying selection across environmental gradients.
6. The peripheral divergence hypothesis, in which peripheral areas of forest receive gene flow and dispersing individuals from fewer directions than do central populations with forests on all sides. This allows more differentiation to occur in peripheral populations.
7. Sympatric speciation may result from ecological separation, such as in the case of fig wasps feeding on different plants.

Basically, hypotheses 1–4 emphasize geographic isolation as the driving force behind speciation, whereas 5–7 represent the outcome of ecology-driven speciation processes. Geographic speciation has long been held as the most important factor regulating speciation in tropical forests and it may indeed be notable, despite recent criticisms of the Pleistocene refuge hypothesis (e.g. Colinvaux, 1998).

Recent studies suggest that natural selection across habitat gradients, without significant geographic or genetic isolation, may also be an important process in animal speciation in tropical rainforests (e.g. Schneider *et al.*, 1999; Bridle *et al.*, 2001). In other words, isolation through vicariance in rainforests may in certain cases be insufficient to produce phenotypic divergence among populations and, therefore, may be less important in species formation than previously thought (Schneider *et al.*, 1999). For example, Schneider *et al.* (1999) compared mitochondrial and morphological divergence in ground-dwelling rainforest lizards in Australia and concluded that natural selection operating across ecological gradients can be more important than geographic isolation in similar habitats in generating phenotypic diversity. Their study contrasted lizards occurring in rainforests, savannas and on the forest edge. One could summarize their findings by stating that 'species are more likely to evolve if they live at the rainforest edge' (Brown, 1999).

Following this line of reasoning, we consider that an ecotone such as the upper canopy (or canopy surface) of closed wet tropical forests may represent a significant arena for natural selection promoting speciation, especially for small organisms such as arthropods. For these taxa, environmental gradients from the canopy to the overstorey may be similar to those occurring from the forest interior to the edge. The implications for the conservation of rainforest biota may be important. First, it may be imperative to preserve the processes that promote genetic variation in rainforest species (Erwin, 1991c; Brown, 1999; Schneider *et al.*, 1999; Levin & Levin, 2002). Second, a distinct arthropod fauna restricted to the upper canopy of tropical rainforest may represent one of the biotas that is most endangered by anthropogenic disturbance (Basset *et al.*, 2001a). In brief, there is enough inferential evidence to stimulate entomologists to pay more attention to the arthropods of the upper canopy and the gradients of environmental factors that they must face there.

TEMPORAL PATTERNS IN TROPICAL CANOPIES

Congruent patterns among geographical locations and forest types?

Temporal patterns of distribution in canopy arthropods may be compared at three different scales of increasing resolution: (i) among forest types, especially between wet and dry forests; (ii) among host or support plants; and (iii) between undisturbed and disturbed forest sites. With regard to the first scale of resolution, many temporal factors shape the structure of arthropod communities. Some, including both climatic and biotic factors (such as biotic resource tracking, resource competition and predation), are often predictable and act at the regional scale (Ch. 4). Others appear to be stochastic and may act at a much finer scale (Chs. 18, 19 and 23), although spatial scale and predictability may not be related necessarily.

Even when considering a particular host-plant species in differing geographical situations, historical factors may constrain the local diversity of its associated arthropods (e.g. Cornell & Lawton, 1992). Gruner and Polhemus (Ch. 13), for example, in their analysis of *Metrosideros* sp. in Hawaii over a 4-million-year period, conclude that ecological factors affecting arthropod densities are significant at the scale of the ecosystem, but are less significant at the level of the forest patch or individual tree. They also contend that if *Metrosideros* sp. is a recent arrival in Hawaii, then contemporary arthropod assemblages are a subset of the regional species pool that has shifted hosts within a short evolutionary time. Interestingly, the situation appears similar for *Ficus* spp. and its associated herbivores in New Guinea (Basset & Novotny, 1999).

It is of interest that certain arthropod taxa in various forests growing under extreme amplitudes of rainfall show similar temporal patterns, moving upward in the canopy during the most unfavourable periods. In dry forests in Mexico, springtails migrate upwards into bromeliads to escape water stress during the dry season, increasing the diversity in the canopy (Ch. 15). In inundated forests of the Amazon, Carabidae, Pseudoscorpiones, Chilopoda and many other taxa also migrate upwards in the canopy during flooding (e.g. Adis, 1981, 1997a; Erwin & Adis, 1982). There is a seasonal component in arthropod vertical distribution that may be obvious in tropical forests enduring large climatic amplitudes. Similarly, arthropod vertical migrations may occur in montane forests during the most unfavourable periods (perhaps moving down during the coolest periods).

Other effects may be more subtle, as indicated by the study of Wagner (Ch. 14) in Uganda. There, arthropod seasonality depends on forest type and local irrigation (i.e. swamp forest), even in forest plots located close to each other. Wagner stresses that in this case seasonal factors and habitat structure are more important than host specificity in explaining assemblage composition and structure. This is an excellent illustration of the contention that temporal variation may be as important as spatial variation in tropical canopies (See Ch. 4). In Uganda, many small beetles aggregate in trees during the dry season without any long-lasting association with the trees: yet another example of migration during an unfavourable period.

In Ch. 4, Didham and Springate explain convincingly the effects of host phylogeny on the patterns of temporal distribution of arthropods in tropical canopies (see also Ch. 16 for an example of such patterns). Host phylogeny includes factors acting at the second scale of resolution, as listed above. Didham and Springate refer to host traits such as tree phenology, growth rate and senescence pattern to predict cascading effects on arthropod assemblages. Host phenology and growth rates, in affecting the amount or quality of resources available to insect herbivores, are prime factors in this equation. Despite this, it is perhaps frustrating to realize that, to date, we have basically no data comparing the influence of phenology of evergreen versus deciduous hosts on the structure of associated arthropod assemblages.

Patterns of tree senescence are also likely to affect deeply the resources available to arthropods. Senescence may affect the balance between live and dead wood in the canopy and the amount of suspended soils, but it may also promote the growth of other plants, such as epiphytes. This, in turn, may affect the seasonality of arthropods – the working hypothesis of Stuntz *et al.* (Ch. 17). In particular, the presence of epiphytes in the crowns of host trees could act as an environmental buffer for arthropods during the dry season. The hypothesis is attractive but not supported by the data put forward in Ch. 17 for *Annona glabra*, its epiphytes and their arthropod inhabitants in Panama.

Didham and Springate (Ch. 4) also suggest that at the third scale of resolution, the disturbance of originally similar forest sites, host phylogeny most likely leads to changes in long-term temporal variation in the structure of arthropod assemblages. Indeed, Schowalter and Ganio (Ch. 28) report that annual patterns of arthropod abundance are often related to hurricane disturbance or drought in the neotropics. In their study, herbivores had variable but important effects on canopy structure (notably leaf loss), depending on tree species. Leaf loss may influence canopy porosity (i.e. penetration of light, water and air flow) and hence the turnover of water, carbon and other nutrients.

Congruent patterns among arthropod taxa?

To discuss congruence and differences of temporal patterns among arthropod taxa in the canopy, it is best to examine patterns at the following three scales: (i) diel activity patterns; (ii) seasonal patterns (i.e. within an annual cycle); and (iii) multiannual patterns. With regard to the first, several studies showed convincingly that different arthropod taxa forage either during day or during the night in the canopy (e.g. Hammond, 1990; Basset & Springate, 1992; Springate & Basset, 1996; Compton *et al.*, 2000; Basset *et al.*, 2001a; Ch. 27). Some subdominant ant species, for example, are active either during day or night, sharing territory with other ants (Ch. 30). Most of these studies also indicate that arthropod diel activity is overall higher during the day than at night in rainforests. Schowalter and Ganio (Ch. 28) report distinct diel patterns for different taxa but overall find little difference in arthropod activity during day and night. Their methods targeted rather inactive insects and did not specifically monitor arthropod activity.

Since scents are often better propagated when the relative humidity is high (i.e. at night), fungal-feeders or scavengers relying to some extent on scents for the location of their food resources may be particularly active during night. Conversely, herbivores that may rely to some extent on vision (guided by foliage reflectance etc.) may be particularly active during day. The studies cited above generally reflect these trends, although no generalization is possible at the species level. Each guild includes representative species that may be more active during either day or night, resulting in distinct faunas foraging during day or night.

Janzen (1983a) proposed that herbivores in rainforests might be more active during night than day, to escape day-active predators and parasitoids. Various studies using different methods (e.g. Springate & Basset, 1996; Novotny *et al.*, 1999a) showed that this was unlikely. Night was confirmed as a relatively enemy-free time, which, however, was not exploited to the fullest by herbivores.

With regard to seasonal patterns, several studies emphasized differences in the seasonal distribution of both arthropod guilds and species in the canopy during the annual cycle (e.g. Wolda, 1978b; Basset, 1991c; Ribeiro & Pimmenta, 1991; Janzen, 1993b; Kato *et al.*, 1995; DeVries *et al.*, 1997; Chs. 7, 15 and 16). In addition, the seasonal distribution of arthropods intimately associated with certain host trees may be affected by certain traits such as host phenology and growth, often dependent on host phylogeny, as well as on other abiotic and biotic factors (Ch. 4). Therefore, seasonal patterns are likely to be highly variable and difficult to predict, unless arthropods specialize on a particular resource. Indeed, two studies in New Guinea and Guyana indicated that specialist herbivores are likely to be more seasonal (and predictable) than generalist herbivores (Novotny & Basset, 1998; Basset, 2000).

Resource availability over time also greatly influences multiannual patterns of arthropod distribution (Chs. 4, 12 and 32). For example, arthropod guilds specializing on different resources respond differently to general flowering in dipterocarp forests (Ch. 12). Further, multiannual modification of the resource base (Price, 1992) available to arthropod assemblages through natural disturbance, such as El Niño events, may also affect them greatly and trigger cascading effects (Ch. 28).

Are temporal patterns similar in the understorey and upper canopy?

Most likely, the answer to this question will depend on the availability and predictability of resources in these two forest strata, and on other factors discussed by Didham and Springate (Ch. 4). The interpretation of patterns may also prove to be different depending on whether detailed information exists on arthropods and their resources. Schulze and Fiedler (Ch. 7) indicate that seasonal fluctuations of pyraloid moths appear to be similar in both strata, and that both faunas are probably influenced by similar factors in the understorey and upper canopy. This may be related to the relatively uniform floral composition of this forest, perhaps providing a more uniform temporal environment with synchronous leaf-flush, flowering and fruiting. Without information about the resources used by the moths collected, it is difficult to discuss this contention.

Itioka *et al.* (Ch. 12) show that generalist predators foraging both in the understorey and upper canopy do not respond strongly to general flowering; that is, their seasonal patterns are similar in both strata. Patterns may be different for more specialized species, as noted above. For example, the seasonal availability of the young foliage of mature trees of *Pourouma bicolor* in Panama is significantly less variable than that of conspecific saplings (Basset, 2001a). Since certain herbivore species specialize on *Pourouma* sp. saplings in the understorey, their seasonal distribution is likely to differ from that of counterparts specializing on mature foliage in the canopy. Alternatively, understorey specialists may be more generalist with regard to plant use, a hypothesis that is also plausible. In Guyana, insect herbivores foraging on seedlings included a majority of generalist species (Basset, 1999b). In any case, the high leaf turnover in the upper canopy of rainforests, compared with their understorey (e.g. van Schaik *et al.*, 1993; Coley & Kursar, 1996), is likely to have a significant influence on herbivore seasonality.

RESOURCE USE IN TROPICAL CANOPIES

Congruent patterns among geographical locations and forest types?

What are the most significant differences in terms of habitat and resource use among different types of tropical forest? In the tropics, most studies interested in altitudinal gradients have focussed on litter and understorey arthropods, often targeting specific taxa (e.g. Janzen *et al.*, 1976; Wolda, 1987; Hanski & Niemelä, 1990; Holloway *et al.*, 1990; Olson, 1994a). Altitudinal transects targeting canopy arthropods have been more rarely employed (e.g. Gagné, 1979; Stork & Brendell, 1990; Allison *et al.*, 1997) and specific studies on the fauna of canopies in montane forests are even scarcer (e.g. Chs. 6 and 9). Often, these studies show bell-shaped curves of diversity, with maxima at the transition zone of lowland and montane forests, diminishing thereafter with increasing and decreasing altitude. Mid-elevation diversity appears to be inflated by faunal overlap from both lower and higher elevation zones (e.g. Lees *et al.*, 1999).

In addition to the high continuity of habitats from the ground to the canopy, discussed above, the harshness of the climate in montane forests is also likely to affect the habitats available to arthropods there. For example, the ratio of ants to other taxa appears to decrease from lowland wet forests to montane forests (Stork & Brendell 1990; Basset, 2001b). Montane rainforest may be too cold to allow ants to forage efficiently or for their larvae to develop fast enough (Brown 1973), or the resource base may be too low to allow large populations of their homopteran associates to exist. In lowland forests, high ratios of ants to Homoptera are more common and this may be partly related to mutualism and the tending of homopterans in the canopy in favourable situations (Basset, 2001b).

Although information on the effects of altitudinal gradients on canopy arthropods is relatively scarce, that on the effects of rainfall gradients is more abundant, as evidenced by several contributions in this volume. Ribeiro (Ch. 31) reviews the major differences between tropical savannas and wet forests. The former experience more leaf damage and support higher abundance of more host-specific herbivores and more gall-forming species than the latter. These differences appear to reflect the concentrated plant resources in savannas (as opposed to diluted resources in mixed tropical forests – the 'trees as islands' of Janzen (1968, 1973c) – and evolutionary responses to sclerophylly. Ribeiro also suggests that the canopy surface in savannas is rather different from that in tall wet forests, and its properties need to be better studied.

In support of Ribeiro's views, Mody *et al.* (Ch. 19) stress that African savanna trees can support high numbers of arthropods and a species richness intermediate between that on temperate and tropical trees (see also Krüger & McGavin, 1998a). Further, beetles, but not so much ants, show a high fidelity to both tree species and individual trees among study years. The data clearly stress the individual trees as the source of variation for arthropod assemblages. Even if the structure of the arthropod assemblage varies little between years on these trees, it is difficult to predict from one conspecific tree to the other. In a dry (deciduous) forest in Costa Rica, many species of saturniid moths are rather specialized (Janzen, 1988a; Ch. 33). In a wetter, semideciduous forest in Panama, Ødegaard (2000c; Ch. 21) shows that many phytophagous beetles, dominated by Chrysomelidae and Curculionidae, appear relatively specialized, many dependent upon lianas.

In contrast, recent studies using large sample sizes in wet forests indicate relatively lower proportions of highly specialized herbivores (e.g. Basset, 1999b; Basset & Novotny, 1999; Novotny *et al.*, 1999b, 2002a,b; Ch. 29). It is possible that the proportion of insect herbivores that are specialized may decrease as one moves from savannas through dry to increasingly wetter tropical forests. Possible explanations may include the increasing dilution of hosts in mixed wet forests and the resulting constraints on host location (e.g. Basset, 1992a); and the increasing number of taxonomically related hosts in mixed wet forests, favouring host switches (e.g. Novotny *et al.*, 1999b, 2002a; Ch. 5).

Congruent patterns among arthropod taxa?

We discuss briefly here the ecological specialization (habitat and resource use) of a certain number of taxa, that is, their 'niche width', with particular emphasis on the studies reported in this volume. Care must be taken to distinguish habitat and resource specialists in certain cases (Ch. 18). In Panama, for example, the caterpillars of the nymphalid *Tigridia acesta* (L.) were only collected from the saplings of *P. bicolor*, not from conspecific mature trees (Basset, 2001a). They would feed on the young foliage of trees, but with apparent difficulty. This species probably occurs rarely if at all, on the foliage of mature *P. bicolor*. It may be restricted to saplings of *Pourouma* spp. in the understorey but feed also on relatively small and shaded *Cecropia* spp. in the understorey, as indicated by host records (Young, 1986; DeVries, 1987a). The adults prefer to fly in the understorey (DeVries & Walla, 2001). This could be an example of a habitat specialist (in the understorey) able to feed on different resources (various Cecropiaceae). A similar situation is observed in many species of herbivorous beetles that specialize on roots or wood as larvae but perform 'maturation feeding' on a range of hosts in the canopy as adults (Ch. 15; and see paragraph on grasshoppers, below).

Shaw and Walter (Ch. 26) show that primitive arthropods may be common in tree hollows in the canopy because of frequent inbreeding in these discrete habitats. In particular, mites are diverse in these habitats and also occur in nests of vertebrates, which provide many transitional forms from free-living to ectocommensal or parasitic. Therefore, sampling from cryptic canopy habitats is important to understand mite evolution. Tropical forest canopies are likely to include many such discrete habitats and associated specialized arthropod inhabitants. Other arthropods, such as free-feeding herbivores, are often good fliers that disperse readily and may not be so closely tied to particular habitats. However, it is this latter category of arthropods that are most likely to be collected in the canopy by fogging or trapping. Hence, our estimates of niche breadth may be biased against cryptic arthropods when relying solely on these collecting methods. Further, cryptic arthropods may also redistribute in subtle ways among habitats (Ch. 24).

The full life cycle of most canopy arthropods is unknown (but see Paarmann & Paarmann, 1997). One other notable exception is represented by the canopy grasshoppers of the Amazon, which have been well studied by Amédégnato (1997; Ch. 22). Their spatial distribution is extremely complex as they exploit different types of food from the litter to the overstorey, lay eggs on different substrates and represent different assemblages in the succession of the vegetation. They include both specialists and generalists and it is very possible that many species show rather narrow habitat specificity (oviposition substrate, stratification), but a wider use of plant resources.

Recent studies of beetle assemblages by fogging in wet forests (e.g. Basset *et al.*, 1996a; Chs. 14, 18 and 19) indicate that many representatives of this taxon may not be as host specific as early fogging studies claimed (e.g. Erwin & Scott, 1980). In particular, Kitching *et al.* (Ch. 29) show that as inter-tree taxonomic distance increases, there is a clear decline in beetle similarity for faunal assemblages. However, inconsistencies emerge when taxonomic sample size gets smaller, as different taxa do not respond similarly.

Using a different approach based on collecting *in situ*, Ødegaard (Ch. 33) records that each plant species studied in a small area of dry forest in Panama supported an average of 51 species of phytophagous beetles. Effective specialization was 13.9 beetle species per plant, so that host specificity in the whole forest was estimated to be 7–10%. Such a forest may harbour 1600–2000 species of phytophagous beetles, but probably only 40% of the species belongs to described taxa. This last figure scales the magnitude of the task awaiting taxonomists, even in a country where the arthropod fauna has been reasonably well studied, such as Panama. The estimates of species richness appear surprisingly low and probably result from the small area being studied (the crane perimeter) and low replicates in terms of individual trees. Here again, estimates of host specificity are lower than early guesses of 20% of beetle species being host specific (Erwin, 1982).

Beetles feeding on flowers appear not to be very host specific and they may form random assemblages not consistent in space and time (Ch. 23). Beetles may be flying from far away to reach the flower resources. Flowering events are significant since they represent a conspicuous, easy to access resource and further provide important meeting points ('concourses') for conspecific insects. Indeed, loose niches in plant–pollinator systems within rainforest canopies appear relatively common and often involve generalist plant–pollinator relationships (Ch. 32). Although this remains to be proven, flower visitors may be specialized with regard to habitat (flowers in the canopy, or understorey, perhaps certain features of flowers), but less discriminating with regard to the identity of the flowers.

Janzen (Ch. 33) provides a comprehensive overview of host use for 31 saturniid species on 370 woody dicot species, representing 66 families, in a dry forest in Costa Rica. Comparable datasets, even for other Lepidoptera, are currently not available. Janzen notes the difference between oviposition and feeding records, and focusses on the latter. Although many species are rather specialized, probably few can really be considered as being monophagous. A similar pattern exists in leaf-chewing insects in lowland wet forests of New Guinea (Novotny *et al.*, 2002a). Interestingly, many plant species or families locally available in the study area in Costa Rica are not fed upon, even by the most polyphagous species. Further, Janzen contends that saturniid–plant interaction at Guanacaste probably do not result from intricate pairwise coevolution, but rather from ecological 'fitting' (Janzen, 1985b) of immigrant species. He predicts that these patterns may apply to other diverse tropical canopy and large caterpillars.

In an ingenious experiment, Jaffé *et al.* (Ch. 20) show that flowers in a savanna display more features that repel ants than do canopy flowers in a wet forest. Canopy flowers appear to be adapted to tolerate ants. This sheds some new light on the intricate relationships between ants and plants. It also suggests that specialized ecological interactions or coevolutionary processes have occurred less commonly in the canopy of wet forests than in savannas.

The emerging picture from this compilation of studies is that tropical canopy arthropods may be less specialized than previously thought with regard to resource use (early arguments: Beaver, 1979; Basset, 1992a). At the same time, they may be more specialized than previously thought with regard to habitat use, particularly with regard to specific habitats situated within different strata in tall wet forests (see Ch. 3). The conservation implications are not comforting, implying that small levels of disturbance that alter arthropod habitats, breaking the canopy surface for example, may have significant negative consequences.

IMPROVING OUR KNOWLEDGE OF CANOPY ARTHROPODS IN THE TROPICS

Themes currently lacking in canopy entomology and future directions of research

In biogeographic terms, most of our knowledge originates from the neotropical region (12 contributions out of 28 in this volume and see Basset, 2001b). African canopies are poorly known, heavily threatened by anthropogenic disturbance and deserve a concerted rapid 'action plan' for their study. Lowland wet forests are also the most studied forest type (Basset, 2001b). Research should also be expanded to focus on montane forests, savannas and dry forests, the last representing one of the most endangered tropical biomes (Janzen, 1988c). Data on resource and habitat use, among other things, for particular higher taxa occurring in different types of tropical forest would be most useful. To date, these data are mostly restricted to butterflies (e.g. Spitzer *et al.*, 1993; Brown & Hutchings, 1997), which are neither particularly speciose nor particularly tied to forest canopies (Basset *et al.*, 1998). In addition, next to nothing is known about how insect herbivores, for example, use resources in monodominant and mixed wet lowland forests, respectively. Monodominant stands are relatively common in the tropics (e.g. Hart, 1990) and their comparative study would certainly be a most instructive example. Since monodominant species often grow slowly (e.g. Hart, 1990), they may be well defended chemically and relatively unpalatable to insects; consequently, they may support a relatively high proportion of generalist insect herbivores (Basset, 1999b).

It has already been emphasized that our current knowledge of canopy arthropods is strongly biased towards beetles and ants (Ch. 5). Taxonomical biasses against invertebrates are reviewed elsewhere (Basset, 2001b); see also Ch. 16 for a plea to study Dipteran assemblages in the canopy. In particular, insect predators and parasitoids (mostly Hymenoptera Parasitica) are two guilds for which the information relevant to tropical canopies is particularly scarce. Apart from studies of community structure focussing on predator–prey ratios (e.g. Gaston *et al.*, 1992; Basset, 1995, Krüger & McGavin, 2001) and predatory arboreal ants (review in Ch. 30), few studies have been concerned with these guilds and the relevant arthropod taxa. Hymenopteran parasitoids are particularly diverse in tropical canopies (e.g. Noyes, 1989b; Askew, 1990; Stork, 1991; Horstmann *et al.*, 1999), but their autoecology and synecology have been rarely studied (e.g. Memmott *et al.*, 1993; Godfray *et al.*, 1999).

With regard to other resources available and prominent in canopies, sap-sucking and fruit-feeding/seed-eating herbivores have been rarely studied (see Ch. 5). For the last group, although studies of post-dispersal seed attack by insects are plentiful, studies of predispersal attack in the canopies of tropical forests are few (e.g. Janzen, 1983b; Forget, 1994; Delobel *et al.*, 1995; Peredo Cervantes *et al.*, 1999). Predispersal seed predation is most likely to be limited by either predator satiation or environmental factors affecting adult oviposition and larval development, whereas postdispersal predation may reach 100% through an expansion of the insect population in highly disturbed forest habitats (e.g. Wright, 1990).

The significance of the upper canopy has been stressed several times in this volume. In particular, the data of Basset *et al.* (Ch. 27) suggest that insect herbivores of the upper canopy may be resident and well adapted to environmental conditions there. Whether the fauna collected in the upper canopy is very specialized and whether it is different from that foraging a few metres below in the canopy represents a key follow-up question for exploration. Since the upper canopy may well be distinguished from the lower canopy only in closed, relatively flat and undisturbed rainforests, the implications for the conservation of tropical rainforest arthropods may also be important. In this context, it is perhaps surprising that the biological significance of the 'inversion surface' (*surface d'inversion*) of Oldeman (1974) has never been assessed for the arthropod fauna and, particularly, for insect herbivores. Briefly, this is the zone where water becomes the limiting factor for tree growth and where photosynthesis is performed with a minimum of transpiration. It roughly corresponds to the first branching of dominant trees (Oldeman, 1974). In French Guiana, Sterck and Bongers (2001) recently showed that regressions of plant traits against tree height were linear with study trees below 25 m in height, and become nonlinear with taller trees. Light availability was not considered to be an important selection force acting on architectural changes with tree height. Thus, branching patterns in the upper canopy may be different from those below (e.g. Hallé *et al.*, 1978; Bell *et al.*, 1999; Sterck & Bongers, 2001), with likely consequences for the arthropod fauna.

The consequences of the high leaf turnover in the upper canopy (e.g. van Schaik *et al.*, 1993; Coley & Kursar, 1996) should not be underestimated for insect herbivores, which are often highly dependent on windows of availability of young foliage (e.g. Aide & Londoño, 1989; Basset, 1991d). High leaf turnover may select for extremely specialized herbivores feeding on young foliage, providing that the resource base is large enough (Basset, 1996a); for specialists feeding on mature foliage and able to cope with its toughness and low nutrient content; or for generalists feeding on young foliage. It would be particularly stimulating to compare the habitat and resource use of arthropods in the understorey and upper canopy of tall wet forests. In the same vein, a study of faunal similarity between the upper canopy and forest gaps in a wet forest would be of the utmost interest to understand what proportion of fauna may be lost when the upper canopy is disturbed by logging.

Recent advances in community models could help us to understand better the effects of fluctuating resources in the canopy on arthropod assemblages. In particular, nonequilibrium models explain diversity as a balance between immigration and extinction, with the species composition itself constantly changing (e.g. Bell, 2001; Hubbell, 2001; Chave *et al.*, 2002). Roubik *et al.* (Ch. 32) show that this approach is promising for studying flower resources and their associated visitors, but it may also be relevant to other resources and their associated arthropods, as well as to the study of arthropod vertical distribution, especially when daily or seasonal vertical migrations occur (see Ch. 3).

Our knowledge of canopy arthropods in the tropics is particularly thin when discussing food-webs (Godfray *et al.*, 1999). However, this situation is unlikely to improve for tall wet forests unless the important issues of canopy access and replication can be solved (Ch. 2). More humbly, we will have to initiate work in plantations (e.g. Memmott *et al.*, 1993) or forests with easily accessible canopies, such as savannas (Ch. 31). As our methods of study, databases and canopy science mature, we may be able to seek the answer to more ambitious questions and to perform more consequential studies.

How to study canopy arthropods inexpensively

Most tropical forests are situated in the Third World. As a consequence, local entomologists rarely have the opportunity to study tropical canopies with sufficient infrastructure and funding (Kitching & Clarke, 1989) and the present volume reflects this unfortunate state of affairs. Long-term arthropod studies are more likely to originate from international collaborative programmes based in the tropics, often well funded by developed countries (e.g. Guanacaste area: Janzen, 1998; La Selva: Longino & Colwell, 1997; Barro Colorado Island: Leigh, 1999; Manaus: Fonseca *et al.*, 1998; Lambir Hills: Yumoto *et al.*, 1996; Danum Valley: Marshall & Swaine, 1992). Unfortunately, most of these programmes are based in the neotropics.

We emphasize here that it is crucial to encourage the study of canopy arthropods by local entomologists. It can be achieved through the above-mentioned programmes, although concentration of resources and research may not always be desirable, particularly when studying highly heterogeneous environments. The other strategy is to implement inexpensive methods of collecting and studying arthropods in canopies, such as those presented by De Dijn, and Koike and Nagamitsu in Chs. 8 and 11. Arguably, there are methodological problems related to the use of these methods, but these are no more severe than for many other collecting methods. The key point is to be aware of the limitations of the method selected, to ensure that it is relevant to the question(s) asked, and that it will allow a straightforward interpretation of the data (e.g. Basset *et al.*, 1997b). A supplementary approach is to train local parataxonomists to help with the research (Ch. 2).

Conservation of tropical canopies and their inhabitants

This is an important topic of research (see Soulé & Orians, 2001; Ch. 2, 18 and 34). In Malaysia, loss of diversity in selectively logged forests is not drastic compared with the diversity of primary forests (Ch. 34). Alteration in guilds and loss of species is obvious only in exotic plantations. Still, these provide refuges for the fauna, providing that the understorey is well developed (unlike in oil palm plantations). It is not clear whether these considerations would still apply if the fauna of the upper canopy had been specifically targeted by the study (see Willott, 1999).

In contrast to these rather optimistic views, Floren & Linsenmair (Ch. 18) report, also from Malaysia, strong effects of anthropogenic disturbance. For example, 40 years after disturbance, the fauna of the disturbed forest is still different from that in the primary forest. There is a transition from deterministically structured communities to random ones through forest succession. In particular, assemblages of Coleoptera and Formicidae show deterministic patterns in disturbed forests, whereas nonequilibrium conditions mediate species coexistence in primary forests (random assemblages). This argument in itself is worthy of further investigation since, for example, ant mosaics appear to exist and are deterministically structured in primary rainforests (Ch. 30; counter-argument in Floren & Linsenmair, 2000a).

What are the effects on arboreal arthropods of the opening of the canopy after the creation of natural or anthropogenic gaps? Do the upper canopy and its fauna 'fall' to the ground? As far as insect herbivores are concerned, the short answer to this is most likely no, since forest gaps typically include a different set of plant species (pioneers) than are present in the mature canopy (shade-tolerant species), and many insect species are relatively specialized. In addition, herbivores foraging on mature trees in Guyana do not invade conspecific seedlings subsisting in logged and more illuminated patches (Basset *et al.*, 2001b). Taxa less tied to resources occurring specifically in the upper canopy, such as dung beetles, do not appear to suffer much from canopy loss and survive well in the understorey of disturbed forests (Davis & Sutton, 1998). This and related issues warrant further investigations.

And . . . how many species?

We cannot close this volume without commenting briefly on the fundamental question of how many species; this has been at the core of early studies in canopy entomology in the tropics (e.g. Erwin, 1982; Stork, 1988). It is becoming increasingly clear that global estimates of biodiversity cannot be based on a handful of canopy studies. There are various reasons for this but most significantly it is because of the tremendous variance in the distribution of arthropod diversity among biogeographical locations and forest types (e.g. May, 1990; Thomas, 1990a; Basset *et al.*, 1996a; Ødegaard, 2000c; Novotny *et al.*, 2002b). The research has now shifted towards understanding how biodiversity is distributed in tropical forests, how it is maintained and how it can be preserved from anthropogenic disturbance. We hope that this volume has brought some element of response to these questions, has stimulated confirmed canopy entomologists to seek even more ambitious investigations, preferably of a collaborative nature, and that it will stimulate students to pay attention to one of the most fascinating habitats on Earth and its little-known inhabitants.

References

Achtziger, R., Nigmann, U. & Zwölfer, H. (1992). Rarefaction-Methoden und ihre Einsatzmöglichkeiten bei der zooökologischen Zustandsanalyse und Bewertung von Biotopen. *Zeitschrift für Ökologie und Naturschutz*, **1**, 89–105.

Adams, E. S. (1994). Territory defense by the ant *Azteca trigona*: maintenance of an arboreal ant mosaic. *Oecologia*, **97**, 202–208.

Adis, J. (1979). Problems of interpreting arthropod sampling with pitfall traps. *Zoologische Anzeigen, Jena*, **202**, 177–184.

Adis, J. (1981). Comparative ecological studies of the terrestrial arthropod fauna in Central Amazonian inundation forests. *Amazonia*, **7**, 87–173.

Adis, J. (1982). Eco-entomological observations from the Amazon. II. Carabids are adapted to inundation-forests! *Coleopterists Bulletin*, **36**, 440–441.

Adis, J. (1984a). Seasonal igapó-forests of Central Amazonian black-water rivers and their terrestrial arthropod fauna. In *The Amazon. Limnology and Landscape Ecology of a Mighty Tropical River and its Basin*, ed. H. Sioli, pp. 245–268. Dordrecht: W. Junk.

Adis, J. (1984b). Vertical distribution of arthropods on trees in black water inundation forests (Central Amazonia, Brazil). In *Tropical Rain-Forest – The Leeds Symposium*, eds. A. C. Chadwick & S. L. Sutton, pp. 123–126. Leeds: Leeds Philosophical and Literary Society.

Adis, J. (1988). On the abundance and density of terrestrial arthropods in central Amazonian dryland forests. *Journal of Tropical Ecology*, **4**, 19–24.

Adis, J. (1990). Thirty million arthropod species – too many or too few? *Journal of Tropical Ecology*, **6**, 115–118.

Adis, J. (1997a). Terrestrial invertebrates: survival strategies, group spectrum, dominance and activity patterns. In *The Central Amazon Floodplain. Ecological Studies*, Vol. 126, ed. W. J. Junk, pp. 318–330. Berlin: Springer-Verlag.

Adis, J. (1997b). Estratégias de sobrevivência de invertebrados terrestres em florestas inundáveis da Amazônia Central: Uma reposta a inundação de longo periodo. *Acta Amazonica*, **27**, 43–54.

Adis, J., Basset, Y., Floren, A., Hammond, P. M. & Linsenmair, K. E. (1998b). Canopy fogging of an overstorey tree – recommendations for standardization. *Ecotropica*, **4**, 93–97.

Adis, J., Harada, A. Y., da Fonseca, C. R. V., Paarman, W. & Rafael, J. A. (1998a). Arthropods obtained from the Amazonian tree species 'Cupiuba' (*Goupia glabra*) by repeated canopy fogging with natural pyrethrum. *Acta Amazonica*, **28**, 273–283.

Adis, J., Lubin, Y. D. & Montgomery, G. G. (1984). Arthropods from the canopy of inundated and terra firme forest near Manaus, Brazil, with critical consideration on the pyrethrum-fogging technique. *Studies on Neotropical Fauna and Environment*, **19**, 223–236.

Adis, J., Paarmann, W., da Fonseca, C. R. V. & Rafael, J. A. (1997). Knockdown efficiency of natural pyrethrum and survival rate of living arthropods obtained by canopy fogging in Central Amazonia. In *Canopy Arthropods*, eds. N. E. Stork, J. Adis & R. K. Didham, pp. 67–81. London: Chapman & Hall.

Adis, J., Ribeiro, E. F., de Morais, J. W. & Cavalcante, E. T. S. (1989). Vertical stratification and abundance of arthropods from white sands soil of a Neotropical Campinarana forest during the dry season. *Studies in Neotropical Fauna and Environment*, **24**, 201–211.

Adis, J. & Schubart, H. O. R. (1984). Ecological research on arthropods in Central Amazonian forest ecosystems with recommendation for study procedures. In *Trends in Ecological Research for the 1980s*, eds. J. Cooley & F. B. Golley, pp. 111–144. New York: Plenum Press.

Aguiar, G. M. D., Schuback, P. D., Vilela, M. L. & Azevedo, A. C. R. D. (1985). Aspectos da ecologia dos flebotomos do Parque Nacional da Serra dos Orgaos, Rio de Janeiro 2. Distribuição vertical (Diptera, Psychodidae, Phlebotominae). *Memorias do Instituto Oswaldo Cruz*, **80**, 187–194.

Aguilera, N. (1962). Algunas notas sobre suelos de coníferas de México. Publicación Especial. In *Seminario y viaje de estudio del Instituto Nacional de Investigaciones Forestales*, ed. N. Aguilera, pp. 132–240. México: Instituto Nacional de Investigaciones Forestales.

Aide, T. M. (1988). Herbivory as a selective agent on the timing of leaf production in a tropical understory community. *Nature*, **336**, 574–575.

Aide, T. M. (1992). Dry season leaf production: an escape from herbivory. *Biotropica*, **24**, 532–537.

Aide, T. M. (1993). Patterns of leaf development and herbivory in a tropical understory community. *Ecology*, **74**, 455–466.

Aide, T. M. & Londoño, E. C. (1989). The effects of rapid leaf expansion on the growth and survivorship of a lepidopteran herbivore. *Oikos*, **55**, 66–70.

Aizen, M. A. & Feinsinger, P. (1994a). Habitat fragmentation, native insect pollinators, and feral honey bees in Argentine 'Chaco Serrano'. *Ecological Applications*, **4**, 378–392.

Aizen, M. A. & Feinsinger, P. (1994b). Forest fragmentation, pollination, and plant reproduction in Chaco dry forest, Argentina. *Ecology*, **75**, 330–351.

Alberti, G. (1991). Spermatology in the Acari: systematic and functional implications. In *The Acari: Reproduction, Development and Life-history Strategies*, eds. R. Schuster & P. W. Murphy, pp. 77–106. Cambridge: Chapman & Hall.

Albrectsen, B. R. (2000). Flowering phenology and seed predation by a tephritid fly: escape of seeds in time and space. *Ecoscience*, **7**, 433–438.

Alencar, J. C. (1986). Estudos silviculturais de uma população natural de *Copaifera multijuga* Hayne, Leguminosae, na Amazônia Central. I. Germinação. *Acta Amazonica*, **11**, 3–11.

Alencar, J. C., De Almeida, R. A. & Fernandes, N. P. (1979). Fenologia de espécies florestais em floresta tropical úmida de terra firme na Amazônia central. *Acta Amazonica*, **9**, 163–198.

Allee, W. C. (1926). Distribution of animals in a tropical rain forest with relation to environmental factors. *Ecology*, **7**, 445–468.

Allen, W. (2001). *Green Phoenix: Restoring the Tropical Forests of Guanacaste, Costa Rica*. New York: Oxford University Press.

Allison, A., Samuelson, G. A. & Miller, S. E. (1993). Patterns of beetle species diversity in New Guinea rain forest as revealed by canopy fogging: preliminary findings. *Selbyana*, **14**, 16–20.

Allison, A., Samuelson, G. A. & Miller, S. E. (1997). Patterns of beetles species diversity in *Castanopsis acuminatissima* (Fagaceae) trees studied with canopy fogging techniques in mid-montane New Guinea rain forest. In *Canopy Arthropods*, eds. N. E. Stork, J. Adis & R. K. Didham, pp. 224–236. London: Chapman & Hall.

Allwood, A. J., Chinajariyawong, A., Drew, R. A. I., Homacek, E. L., Hancock, D. L., Hengsawad, C., Jipanin, M., Kong Krong, C., Kritsaneepaiboon, S., Leong, C. T. S. & Vijaysegaran, S. (1999). Host plant records for fruit flies (Diptera: Tephritidae) in South East Asia. *Raffles Bulletin of Zoology*, Suppl. **7**, 1–92.

Alvarez, T., Frampton, G. K. & Goulson, D. (2000). The role of hedgerows in the recolonisation of arable fields by epigeal Collembola. *Pedobiologia*, **44**, 516–526.

Amédégnato, C. (1997). Diversity of an Amazonian canopy grasshopper community in relation to resource partitioning and phylogeny. In *Canopy Arthropods*, eds. N. E. Stork, J. Adis & R. K. Didham, pp. 281–319. London: Chapman & Hall.

Amédégnato, C. (1998). *Biodiversité des Acridiens Néotropicaux: Origine et Facteurs Environnementaux de ses Variations*. Paris: Ministère de l'Environnement, programme SOFT.

Amédégnato, C. & Descamps, M. (1980a). Evolution des populations d'orthoptères d'Amazonie du nord-ouest dans les cultures et les formations secondaires d'origine anthropique. *Acrida*, **9**, 1–33.

Amédégnato, C. & Descamps, M. (1980b). Etude comparative de quelques peuplements acridiens de la forêt néotropicale. *Acrida*, **9**, 171–216.

Amédégnato, C. & Descamps, M. (1982). Dispersal centers of the Amazonian acridids. *Acta Amazonica*, **12**, 155–165.

Amédégnato, C. & Poulain, S. (1987). Les acridiens néotropicaux I: Proctolabine Amazoniens (Orthoptera, Acrididae). *Annales de la Société Entomologique de France (NS)*, **23**, 399–434.

Andersen, A. N. & Reichel, H. (1994). The ant (Hymenoptera: Formicidae) fauna of Holmes Jungle, a rainforest patch in the seasonal tropics of Australia's Northern Territory. *Journal of the Australian Entomology Society*, **33**, 153–158.

Anderson, J. M. (1977). The organisation of soil animal communities. *Ecological Bulletin (Stockholm)*, **25**, 15–23.

Anderson, J. M. (1978a). Competition between two unrelated species of soil Cryptostigmata (Acari) in experimental microcosms. *Journal of Animal Ecology*, **47**, 787–803.

Anderson, J. M. (1978b). Inter- and intra-habitat relationships between woodland Cryptostigmata species diversity and the diversity of the soil and litter microhabitats. *Oecologia*, **32**, 341–348.

Anderson, M. C. (1964). Study of woodland light climate I. The photographic computation of light conditions. *Journal of Ecology*, **52**, 27–41.

Anderson, R. S. (1995). An evolutionary perspective on diversity in Curculionoidea. *Memoir of the Entomological Society of Washington*, **14**, 103–114.

André, H. M. (1975). Observations sur les Acariens corticoles de Belgique. *Fondation Universitaire Luxembourgeoise, Série 'notes de recherche'*, **4**, 5–31.

André, H. M. (1976). Introduction à l'étude écologique des communautés des microarthropodes corticoles soumises à la pollution atmosphérique. I. Les microhabitats corticoles. *Bulletin d'Écologie*, **7**, 431–444.

André, H. M. (1983). Notes on the ecology of corticolous epiphyte dwellers. 2. Collembola. *Pedobiologia*, **25**, 271–278.

André, H. M. (1984). Notes on the ecology of corticolous epiphyte dwellers. 3. Oribatida. *Acarologia*, **25**, 385–395.

André, H. M. (1985). Associations between corticolous microarthropod communities and epiphytic cover on bark. *Holarctic Ecology*, **8**, 113–119.

André, H. M., Lebrun, P. & Noti, M.-I. (1992). Biodiversity in Africa: a plea for more data. *Journal of African Zoology*, **106**, 3–15.

André, H. M., Noti, M.-I. & Lebrun, P. (1994). The soil fauna: the other last biotic frontier. *Biodiversity and Conservation*, **3**, 45–56.

Andres, L. A. & Bennet, F. D. (1975). Biological control of aquatic weeds. *Annual Review of Entomology*, **20**, 31–46.

Andrewatha, H. G. & Birch, L. C. (1954). *The Distribution and Abundance of Animals*. Chicago: University of Chicago Press.

Anhuf, D., Motzer, T., Rollenbeck, R., Schröder, B. & Scarzynski, J. (1999). Water budget of the Surumoni crane site (Venezuela). *Selbyana*, **20**, 179–185.

Aoki, J.-I. (1967). Microhabitats of oribatid mites on a forest floor. *Bulletin of the National Science Museum, Tokyo*, **10**, 133–138.

Aoki, J.-I. (1971). Soil mites (oribatids) climbing trees. In *Proceedings of the 3rd International Congress on Acarology*, eds. M. Daniels & B. Rosický, pp. 59–65. The Hague: Junk.

Appanah, S. (1985). General flowering in the climax rain forests of south-east Asia. *Journal of Tropical Ecology*, **1**, 225–240.

Appanah, S. (1993). Mass flowering of dipterocarp forests in the aseasonal tropics. *Journal of Bioscience*, **18**, 457–474.

Arias, J. R. & Freitas, R. A. D. (1982). The vectors of cutaneous leishmaniasis in the Central Amazon of Brazil. 3. Phlebotomine sand fly stratification in a forest. *Acta Amazonica*, **12**, 599–608.

Armbrecht, I., Jiménez, E., Alvarez, G., Uloa-Chacon, P. & Armbrecht, H. (2001). An ant mosaic in the Colombian rain forest of Chocó (Hymenoptera: Formicidae). *Sociobiology*, **37**, 491–509.

Armbruster, W. S. (1992). Phylogeny and the evolution of plant–animal interactions. *BioScience*, **42**, 12–20.

Ashton, P. S. (1988). Dipterocarp biology as a window to the understanding of tropical forest structure. *Annual Review of Ecology and Systematics*, **19**, 347–370.

Ashton, P. S. (1991). Toward a regional classification of the humid tropics of Asia. *Tropics*, **1**, 1–12.

Ashton, P. S., Appanah, S. & Chan, H. T. (1995). Tree prosthesis for crown access. *Selbyana*, **16**, 174–178.

Ashton, P. S., Givnish, T. J. & Appanah, S. (1988). Staggered flowering in the Dipterocarpaceae: new insights into floral induction and the evolution of mast fruiting in the aseasonal tropics. *American Naturalist*, **132**, 44–66.

Ashton, P. S. & Hall, P. (1992). Comparisons of structure among mixed dipterocarp forests of north-western Borneo. *Journal of Ecology*, **80**, 459–481.

Askew, R. R. (1980). The biology of gall-wasps. In *The Biology of Gall Insects*, ed. T. N. Ananthakrtshnan, pp. 223–271. London: Academic Press.

Askew, R. R. (1990). Species diversity of hymenopteran taxa in Sulawesi. In *Insects and the Rain Forests of South East Asia (Wallacea)*, eds. W. J. Knight & J. D. Holloway, pp. 255–260. London: The Royal Entomological Society of London.

Athias-Binche, F. (1982). A redescription of *Thinozercon michaeli* Halbert, 1915 (Uropodina: Thinozerconoidea) with notes on its systematic status. *Proceedings of the Royal Irish Academy, Section B: Biological Geological and Chemical Science*, **82**, 261–276.

Athias-Binche, F. (1993). Dispersal in varying environments: The case of phoretic uropodid mites. *Canadian Journal of Zoology*, **71**, 1793–1798.

Athias-Binche, F., Schwarz, H. H. & Meierhofer, I. (1993). Phoretic association of *Neoseius novus* (Ouds., 1902) (Acari: Uropodina) with *Necrophorus* spp. (Coleoptera: Silphidae): A case of sympatric speciation? *International Journal of Acarology*, **19**, 75–86.

Aubréville, A. (1950). *Flore forestière soudano-guinéenne: A.O.F. – Cameroun – A.E.F.* Paris: Société d'Editions Géographiques, Maritimes et Coloniales.

Augspurger, C. K. (1980). Mass-flowering of a tropical shrub (*Hybanthus prunifolius*): Influence on pollination attraction and movement. *Evolution*, **34**, 475–488.

Azarbayjani, F. F., Burgin, S. & Richardson, B. J. (1999). Arboreal arthropod biodiversity in woodlands. II. The pattern of recovery of diversity on *Melaleuca linariifolia* following defaunation. *Australian Journal of Ecology*, **24**, 655–661.

Azevedo, A. C. R., Luz, S. L. B., Vilela, M. L. & Rangel, E. F. (1993). Studies on the sandfly fauna of Samuel Ecological Station, Porto Velho municipality, Rondonia State, Brazil. *Memorias do Instituto Oswaldo Cruz*, **88**, 509–512.

Bach, C. E. & Carr, D. S. (1990). Aggregation behavior of a willow flea beetle *Altica subplicata* Coleoptera Chrysomelidae. *Great Lakes Entomologist*, **23**, 65–76.

Baker, H. G. & Baker, I. (1983). Floral nectar sugar constituents in relation to pollinator type. In *Handbook of Experimental Pollination Biology*, eds. C. E. Jones & R. J. Little, pp. 117–141. New York: Van Nostrand Reinhold.

Baker, H. G., Bawa, K. S., Frankie, G. W. & Opler, P. A. (1983). Reproductive biology of plants in tropical forests. In *Tropical Rain Forest Ecosystems: Structure and Function. Ecosystems of the World*, Vol. 14A, ed. F. B. Golley, pp. 183–215. Amsterdam: Elsevier Scientific.

Baker, H. H. & Baker, I. (1973). Studies of nectar-constitution and pollinator–plant coevolution. In *Coevolution of Animals and Plants*, eds. L. E. Gilbert & P. H. Raven, pp. 100–140. Austin: University of Texas Press.

Balslev, H. & Renner, S. S. (1989). Ecuadorean forests east of the Andes. In *Tropical Forests. Botanical Dynamics, Speciation and Diversity*, eds. L. B. Holm-Nielsen, I. C. Nielsen & H. Balslev, pp. 287–295. London: Academic Press.

Barbosa, P. & Wagner, M. R. (1989). *Introduction to Forest and Shade Tree Insects*. San Diego, CA: Academic Press.

Bardgett, R. D. & Chan, K. F. (1999). Experimental evidence that soil fauna enhances nutrient mineralization and plant nutrient uptake in montane grassland ecosystem. *Soil Biology and Biochemistry*, **31**, 1007–1014.

Barker, M. G. (1996). Vertical profiles in a Brunei rain forest: I. Microclimate associated with a canopy tree. *Journal of Tropical Forest Science*, **8**, 505–519.

Barker, M. G. (1997). An update on low-tech methods for forest canopy access and on sampling a forest canopy. *Selbyana*, **18**, 61–71.

Barker, M. G. & Pinard, M. A. (2001). Forest canopy research: sampling problems, and some solutions. *Plant Ecology*, **153**, 23–38.

Barker, M. G. & Sutton, S. L. (1997). Low-tech methods for forest canopy access. *Biotropica*, **29**, 243–247.

Barlow, H. S. & Woiwod, I. P. (1989). Moth diversity of a tropical forest in Peninsular Malaysia. *Journal of Tropical Ecology*, **5**, 37–50.

Barlow, H. S. & Woiwod, I. P. (1990). Seasonality and diversity of Macrolepidoptera in two lowland sites in the Dumoga-Bone National Park, Sulawesi Utara. In *Insects and the Rain Forests of South East Asia (Wallacea)*, eds. W. J. Knight & J. D. Holloway, pp. 167–172. London: The Royal Entomological Society of London.

Barnola, L. F., Hasegawa, M. & Cedeno, A. (1994). Mono- and sesquiterpene variation in *Pinus caribaea* needles and its relationship to *Atta laevigata* herbivory. *Biochemical Systematics and Ecology*, **22**, 437–445.

Barone, J. A. (1998). Host-specificity of folivorous insects in a moist tropical forest. *Journal of Animal Ecology*, **67**, 400–409.

Barone, J. A. (2000). Comparison of herbivores and herbivory in the canopy and understory for two tropical tree species. *Biotropica*, **32**, 307–317.

Basset, Y. (1988). A composite interception trap for sampling arthropods in tree canopies. *Journal of the Australian Entomological Society*, **27**, 213–219.

Basset, Y. (1990). The arboreal fauna of the rainforest tree *Argyrodendron actinophyllum* as sampled with restricted canopy fogging: composition of the fauna. *Entomologist*, **109**, 173–183.

Basset, Y. (1991a). The taxonomic composition of the arthropod fauna associated with an Australian rainforest tree. *Australian Journal of Zoology*, **39**, 171–190.

Basset, Y. (1991b). The spatial distribution of herbivory, mines and galls within an Australian rainforest tree. *Biotropica*, **23**, 271–281.

Basset, Y. (1991c). The seasonality of arboreal arthropods foraging within an Australian rainforest tree. *Ecological Entomology*, **16**, 265–278.

Basset, Y. (1991d). Leaf production of an overstorey rainforest tree and its effects on the temporal distribution of associated insect herbivores. *Oecologia*, **88**, 211–219.

Basset, Y. (1991e). Influence of leaf traits on the spatial distribution of insect herbivores associated with an overstorey rainforest tree. *Oecologia*, **87**, 388–393.

Basset, Y. (1992a). Host specificity of arboreal and free-living insect herbivores in rain forests. *Biological Journal of the Linnean Society*, **47**, 115–133.

Basset, Y. (1992b). Synecology and aggregation patterns of arboreal arthropods associated with an overstorey rainforest tree. *Journal of Tropical Ecology*, **8**, 317–327.

Basset, Y. (1992c). Influence of leaf traits on the spatial distribution of arboreal arthropods within an overstorey rainforest tree. *Ecological Entomology*, **17**, 8–16.

Basset, Y. (1993). Patterns in the organisation of the arthropod community associated with an Australian rainforest tree: how distinct from elsewhere? *Selbyana*, **14**, 13–15.

Basset, Y. (1994). Palatability of tree foliage to chewing insects: a comparison of a temperate and a tropical site. *Acta Oecologica*, **15**, 181–191.

Basset, Y. (1995). Arthropod predator–prey ratios on vegetation at Wau, Papua New Guinea. *Science in New Guinea*, **21**, 103–112.

Basset, Y. (1996). Local communities of arboreal herbivores in Papua New Guinea: predictors of insect variables. *Ecology*, **77**, 1909–1916.

Basset, Y. (1997). Species-abundance and body size relationships in insect herbivores associated with New Guinean forest trees, with particular reference to insect host-specificity. In *Canopy Arthropods*, eds. N. E. Stork, J. Adis & R. K. Didham, pp. 237–264. London: Chapman & Hall.

Basset, Y. (1999a). Diversity and abundance of insect herbivores collected on *Castanopsis acuminatissima* (Fagaceae) in New Guinea: relationships with leaf production and surrounding vegetation. *European Journal of Entomology*, **96**, 381–391.

Basset, Y. (1999b). Diversity and abundance of insect herbivores foraging on seedlings in a rain forest in Guyana. *Ecological Entomology*, **24**, 245–259.

Basset, Y. (2000). Insect herbivores foraging on seedlings in an unlogged rain forest in Guyana: spatial and temporal considerations. *Studies on Neotropical Fauna and Environment*, **35**, 115–129.

Basset, Y. (2001a). Communities of insect herbivores foraging on saplings versus mature trees of *Pourouma bicolor* (Cecropiaceae) in Panama. *Oecologia*, **129**, 253–260.

Basset, Y. (2001b). Invertebrates in the canopy of tropical rain forests: how much do we really know? *Plant Ecology*, **153**, 87–107.

Basset, Y., Aberlenc, H.-P., Barrios, H., Curletti, G., Béranger, J.-M., Vesco, J.-P., Causse, P., Haug, A., Hennion, A.-S., Lesobre, L., Marques, F. & O'Meara, R. (2001a). Stratification and diel activity of arthropods in a lowland rainforest in Gabon. *Biological Journal of the Linnean Society*, **72**, 585–607.

Basset, Y., Aberlenc, H.-P. & Delvare, G. (1992). Abundance and stratification of foliage arthropods in a lowland rain forest of Cameroon. *Ecological Entomology*, **17**, 310–318.

Basset, Y. & Burckhardt, D. (1992). Abundance, species richness, host utilization and host specificity of insect folivores from a woodland site, with particular reference to host architecture. *Revue Suisse de Zoologie*, **99**, 771–791.

Basset, Y., Charles, E., Hammond, D. S. & Brown, V. K. (2001b). Short-term effects of canopy openness on insect herbivores in a rain forest in Guyana. *Journal of Applied Ecology*, **38**, 1045–1058.

Basset, Y., Charles, E. & Novotny, V. (1999). Insect herbivores on parent trees and conspecific seedlings in a rain forest in Guyana. *Selbyana*, **20**, 146–158.

Basset, Y. & Höft, R. (1994). Can apparent leaf damage in tropical trees be predicted by herbivore load or host-related variables? A case study in Papua New Guinea. *Selbyana*, **15**, 3–13.

Basset, Y. & Kitching, R. L. (1991). Species number, species abundance and body length from arboreal arthropods associated with an Australian rainforest tree. *Ecological Entomology*, **16**, 391–402.

Basset, Y. & Novotny, V. (1999). Species richness of insect herbivores on *Ficus* in Papua New Guinea. *Biological Journal of the Linnean Society*, **67**, 477–499.

Basset, Y., Novotny, V., Miller, S. E. & Pyle, R. (2000). Quantifying biodiversity: experience with parataxonomists and digital photography in New Guinea and Guyana. *BioScience*, **50**, 899–908.

Basset, Y., Novotny, V., Miller, S. E. & Springate, N. D. (1998). Assessing the impact of forest disturbance on tropical invertebrates: some comments. *Journal of Applied Ecology*, **35**, 461–466.

Basset, Y., Novotny, V. & Weiblen, G. (1997a). *Ficus*: a resource for arthropods in the tropics, with particular reference to New Guinea. In *Forests and Insects: 18th Symposium of the Royal Entomological Society*, eds. A. Watt, N. E. Stork & M. Hunter, pp. 341–361. London: Chapman & Hall.

Basset, Y. & Samuelson, G. A. (1996). Ecological characteristics of an arboreal community of Chrysomelidae in Papua New Guinea. In *Chrysomelidae Biology*. Vol. 2. *Ecological Studies*, eds. P. H. A. Jolivet & M. L. Cox, pp. 243–262. Amsterdam: SPB Academic.

Basset, Y., Samuelson, G. A., Allison, A. & Miller, S. E. (1996a). How many species of host-specific insect feed on a species of tropical tree? *Biological Journal of the Linnean Society*, **59**, 201–216.

Basset, Y., Samuelson, G. A. & Miller, S. E. (1996b). Similarities and contrasts in the local insect faunas associated with ten forest tree species of New Guinea. *Pacific Science*, **50**, 157–183.

Basset, Y. & Springate, N. D. (1992). Diel activity of arboreal arthropods associated with a rainforest tree. *Journal of Natural History*, **26**, 947–952.

Basset, Y. & Springate, N. D. (1993). Les insectes ont besoin des arbres dont on besoin les hommes. Geneva, Switzerland: Le Nouveau Quotidient, p. 30.

Basset, Y., Springate, N. D., Aberlenc, H.-P. & Delvare, G. (1997b). A review of methods for collecting arthropods in tree canopies. In *Canopy Arthropods*, eds. N. E. Stork, J. Adis & R. K. Didham, pp. 27–52. London: Chapman & Hall.

Bates, H. W. (1884). Insecta, Carabidae, Cicindelidae. *Biologia Centrali-Americana, Supplement*, **1**, 257–299.

Bates, M. (1944). Observations on the distribution of diurnal mosquitos in a tropical forest. *Ecology*, **25**, 159–170.

Bauer, R. & Christian, E. (1993). Adaptations of three springtail species to granite boulder habitats (Collembola). *Pedobiologia*, **37**, 280–290.

Bauer, T. (1979). Die Feuchtigkeit als steuernder Faktor für das Kletterverhalten der Collembolen. *Pedobiologia*, **19**, 165–175.

Bauer, T. & Christian, E. (1987). Habitat dependent differences in the flight behaviour of Collembola. *Pedobiologia*, **30**, 233–239.

Bawa, K. S. (1990). Plant–pollinator interactions in tropical rainforests. *Annual Review of Ecology and Systematics*, **21**, 399–422.

Beaman, J. H. & Beaman, R. S. (1990). Diversity and distribution patterns in the flora of Mount Kinabalu. In *The Plant Diversity of Malaysia*, eds. P. Baas, K. Kalkman & R. P. Geesink, pp. 147–160. Amsterdam: Kluwer Academic.

Beattie, A. J., Turnbull, C., Knox, R. B. & Williams, E. G. (1984). Ant inhibition of pollen function: a possible reason why ant pollination is rare. *American Journal of Botany*, **71**, 421–426.

Beaver, R. A. (1979). Host specificity of temperate and tropical animals. *Nature*, **281**, 139–141.

Beccaloni, G. W. (1997). Vertical stratification of ithomiine butterfly (Nymphalidae: Ithomiinae) mimicry complexes: the relationship between adult flight height and larval host-plant height. *Biological Journal of the Linnean Society*, **62**, 313–341.

Becerra, J. X. (1994). Squirt-gun defense in *Bursera* and the chrysomelid counterploy. *Ecology*, **75**, 1991–1996.

Beck, J., Schulze, C. H., Linsenmair, K. E. & Fiedler, K. (2002). From forest to farmland: diversity of geometrid moths along two habitat gradients on Borneo. *Journal of Tropical Ecology*, **17**, 33–51.

Beck, L. (1963). Zur Ökologie und Taxonomie der neotropischen Bodentiere. I. Zur Oribatiden-Fauna Perus. *Zoologische Jahrbücher, Abteilung für Systematik, Ökologie und Geographie der Tiere*, **90**, 299–392.

Beck, L. (1968). Sôbre a biologia de alguns Aracnídeos na floresta tropical da Reserva Ducke (I.N.P.A., Manaus/Brasil). *Amazoniana*, **1**, 247–250.

Beck, L. (1971). Bodenzoologische Gliederung und Charakterisierung des amazonischen Regenwaldes. *Amazoniana*, **3**, 69–132.

Beck, L. (1993). Zur Bedeutung der Bodentiere für den Stoffkreislauf in Wäldern. *Biologie in unserer Zeit*, **23**, 286–294.

Beck, L., Dumpert, K., Franke, U., Römbke, J., Mittmann, H.-W. & Schönborn, W. (1988). Vergleichende ökologische Untersuchungen in einem Buchenwald nach Einwirkung von Umweltchemikalien. In *Auffindung von Indikatoren zur prospektiven Bewertung der Belastung von Ökosystemen. Band 9 – Endberichte der geförderten Vorhaben*, Vol. 1, eds. B. Scheele & H. Verfonderen, pp. 548–701. Jülich: KFA.

Beck, L., Höfer, H., Martius, C., Römbke, J. & Verhaagh, M. (1997). Bodenbiologie tropischer Regenwälder. *Geographische Rundschau*, **49**, 24–31.

Becker, P. (1992). Colonization of islands by carnivorous and herbivorous Heteroptera and Coleoptera: effects of island area, plant species richness, and 'extinction' rates. *Journal of Biogeography*, **19**, 163–171.

Begon, M., Harper, J. L. & Townsend, C. R. (1995). *Ecology: Individuals, Populations and Communities*. Oxford: Blackwell Science.

Behan-Pelletier, V. M., Paoletti, M. G., Bissett, B. & Stinner, B. R. (1993). Oribatid mites of forest habitats in northern Venezuela. *Tropical Zoology Special Issue*, **1**, 39–54.

Behan-Pelletier, V. M. & Walter, D. E. (2000). Biodiversity of oribatid mites (Acari: Oribatida) in tree-canopies and litter. In *Invertebrates as Webmasters*, eds. D. C. Coleman & P. Hendrix, pp. 187–202. Wallingford: CABI Publication.

Behan-Pelletier, V. M. & Winchester, N. N. (1998). Arboreal oribatid mite diversity: colonizing the canopy. *Applied Soil Ecology*, **9**, 45–51.

Bell, A. D., Bell, A. & Dines, T. D. (1999). Branch construction and bud defence status at the canopy surface of a West African rainforest. *Biological Journal of the Linnean Society*, **66**, 481–499.

Bell, G. (2001). Neutral macroecology. *Science*, **293**, 2413–2418.

Bell, S. S., McCoy, E. D. & Mushinsky, H. R. (1991). *Habitat Structure – The Physical Arrangement of Objects in Space*. London: Chapman & Hall.

Bellido, A. (1979). Ecologie de *Carabodes willmanni* Bernini, 1975 (Acari, Oribatei) dans les formations pionnières de la lande armoricaine. *Revue d'Ecologie et Biologie du Sol*, **16**, 195–218.

Bengtsson, G., Hedlund, K. & Rundgren, S. (1994). Food- and density-dependent dispersal: evidence from a soil collembolan. *Journal of Animal Ecology*, **63**, 513–520.

Benson, W. W., Brown, K. S., Jr & Gilbert, L. E. (1976). Coevolution of plants and herbivores: passion flower butterflies. *Evolution*, **29**, 659–680.

Benzing, D. H. (1983). Vascular epiphytes: a survey with special reference to their interactions with other organisms. In *Tropical Rain Forest: Ecology and Management*, eds. S. L. Sutton, T. C. Whitmore & A. C. Chadwick, pp. 11–24. Oxford: Blackwell.

Benzing, D. H. (1990). *Vascular Epiphytes*. Cambridge: Cambridge University Press.

Benzing, D. H. (2000). *Bromeliads – Profile of an Adaptive Radiation*. Cambridge: Cambridge University Press.

Berg, J. (1991). Distribution pattern and phenology of Phthiracaridae and Euphthiracaridae in a beech forest.

In *Modern Acarology*, Vol. 2, eds. F. Dusbabek & V. Bukva, pp. 521–527. Prague: Academia.

Berger, W. H. & Parker, F. L. (1970). Diversity of planktonic Foraminifera in deep-sea sediments. *Science*, **168**, 1345–1347.

Berkov, A. & Tavakilian, G. (1999). Host utilization of the Brazil nut family (Lecythidaceae) by sympatric wood-boring species of *Palame* (Coleoptera, Cerambycidae, Lamiinae, Acanthocinini). *Biological Journal of the Linnean Society*, **67**, 181–198.

Berlow, E. L., Navarrete, S. A., Briggs, C. J., Power, M. E. & Menge, B. A. (1999). Quantifying variation in the strengths of species interactions. *Ecology*, **80**, 2206–2224.

Bernays, E. A. & Chapman, R. F. (1994). *Host-plant Selection by Phytophagous Insects*. New York: Chapman & Hall.

Bernays, E. A., Driver, G. C. & Bilgener, M. (1989). Herbivores and plant tannins. *Advances in Ecological Research*, **19**, 263–302.

Bernays, E. A. & Janzen, D. H. (1988). Saturniid and sphingid caterpillars: two ways to eat leaves. *Ecology*, **69**, 1153–1160.

Berry, P. E., Holst, B. K. & Kierych, K. Y. (eds.) (1998). *Flora of the Venezuelan Guyana*, Vol. 4: *Caesalpiniaceae-Ericaceae*. St Louis: Missouri Botanical Garden Press.

Berryman, A. A. (1996). What causes population cycles of forest Lepidoptera? *Trends in Ecology and Evolution*, **11**, 28–32.

Beugnon, G. & Dejean, A. (1992). Adaptive properties of the chemical trail system of the African weaver ant. *Insectes Sociaux*, **39**, 341–346.

Bigger, M. (1976). Oscillations of tropical insect populations. *Nature*, **259**, 207–209.

Bigger, M. (1993). Ant–hemipteran interactions in a tropical ecosystem. Description of an experiment on cocoa in Ghana. *Bulletin of Entomological Research*, **83**, 475–505.

Billick, I. & Case, T. J. (1994). Higher order interactions in ecological communities: what are they and how can they be detected? *Ecology*, **75**, 1529–1543.

Birnbaum, P. (2001). Canopy surface topography in a French Guiana forest and the folded forest theory. *Plant Ecology*, **153**, 293–300.

Black, W. C., Klompen, J. S. H. & Keirans, J. E. (1997). Phylogenetic relationships among tick subfamilies (Ixodida: Ixodidae: Argasidae) based on the18S nuclear rDNA gene. *Molecular Phylogenetics and Evolution*, **7**, 129–144.

Blanc, P. (1990). Bioclimatologie comparée de la canopée et du sous-bois. In *Biologie d'une Canopée de Forêt équatoriale. Rapport de Mission Radeau des Cimes Octobre–Novembre 1989, Petit Saut-Guyane Française*, eds. F. Hallé & P. Blanc, pp. 42–43. Montpellier/Paris: Montpellier II et CNRS-Paris VI.

Blanche, K. R. & Westoby, M. (1995). Gall-forming insect diversity is linked to soil fertility via host plant taxon. *Ecology*, **76**, 2334–2337.

Blanton, C. M. (1990). Canopy arthropod sampling: a comparison of collapsible bag and fogging methods. *Journal of Agricultural Entomology*, **7**, 41–50.

Blüthgen, N., Verhaagh, M., Goitía, W. & Blüthgen, N. (2000a). Ant nests in tank bromeliads – an example of non-specific interaction. *Insectes Sociaux*, **47**, 313–316.

Blüthgen, N., Verhaagh, M., Goitía, W., Jaffé, K., Morawetz, W. & Barthlott, W. (2000b). How plants shape the ant community in the Amazonian rainforest canopy: the key role of extrafloral nectaries and homopteran honeydew. *Oecologia*, **125**, 229–240.

Bogacheva, I. A. (1984). Distribution of phyllophagous insects in the crown of drooping birch at the northern boundary of tree vegetation. *Soviet Journal of Ecology*, **15**, 153–159.

Boinski, S. & Fowler, N. L. (1989). Seasonal patterns in a tropical lowland forest. *Biotropica*, **21**, 223–233.

Boiteau, G., Bousquet, Y. & Osborn, W. (1999). Vertical and temporal distribution of Coccinellidae (Coleoptera) in flight over an agricultural landscape. *Canadian Entomologist*, **131**, 269–277.

Bolton, B. (1994). *Identification Guide to the Ant Genera of the World*. Cambridge, Massachusetts: Harvard University Press.

Bolton, B. (1996). A preliminary analysis of the ants (Formicidae) of Pasoh Forest Reserve. In *Conservation, Management and Development of Forest Resources. Proceedings of the Malaysia–United Kingdom Programme Workshop, 1996*, eds. L. S. See, D. Y. May, I. D. Gauld & J. Bishop, pp. 84–96. Kuala Lumpur: Malaysia–United Kingdom Programme.

Bongers, F. (2001). Methods to assess tropical rain forest canopy structure: an overview. *Plant Ecology*, **153**, 263–277.

Bonnet, L., Cassagnau, P. & Travé, J. (1975). L'écologie des arthropodes muscicoles à la lumière de l'analyse des correspondances: Collemboles et Oribates du Sidobre (Tarn, France). *Oecologia*, **21**, 359–373.

Borchert, R. (1998). Responses of tropical trees to rainfall seasonality and its long-term changes. *Climatic Change*, **39**, 381–393.

Borges, P. A. V. & Brown, V. K. (1999). Effect of island geological age on the arthropod species richness of Azorean pastures. *Biological Journal of the Linnean Society*, **66**, 373–410.

Bowden, J. (1976). Weather and phenology of some African Tabanidae. *Journal of the Entomological Society of South Africa*, **39**, 207–245.

Bowden, J. (1982). An analysis of the factors affecting catches of insects in light traps. *Bulletin of Entomological Research*, **72**, 535–556.

Bowden, J., Haines, I. H. & Mercer, D. (1976). Climbing Collembola. *Pedobiologia*, **16**, 298–312.

Bowles, I. A., Rice, R. E., Mittermeier, R. A. & da Fonseca, G. A. B. (1998). Logging and tropical forest conservation. *Science*, **280**, 1899–1900.

Braby, M. F. (2000). *Butterflies of Australia. Their Identification, Biology and Distribution*. Melbourne: CSIRO.

Bradshaw, W. E. (1974). Phenology and seasonal modeling in insects. In *Phenology and Seasonality Modeling*, ed. H. Lieth, pp. 127–137. New York: Springer-Verlag.

Braganca, M. A. L., Zanuncio, J. C., Picanco, M. & Laranjeiro, A. J. (1998). Effects of environmental heterogeneity on Lepidoptera and Hymenoptera populations in eucalyptus plantations in Brazil. *Forest Ecology and Management*, **103**, 287–292.

Bray, J. R. & Curtis, J. T. (1957). An ordination of the upland forest communities of southern Wisconsin. *Ecological Monographs*, **27**, 325–349.

Breed, M. D. & Harrison, J. (1989). Arboreal nesting in the giant tropical ant, *Paraponera clavata* (Hymenoptera: Formicidae). *Journal of the Kansas Entomological Society*, **62**, 133–135.

Brévignon, C. (1990). Quelques élevages guyanais: *Napeocles jucunda* Hübner (Lep. Nymphalidae), *Papilio torquatus* Cramer, *Papilio anchisiades* Esper, *Eurytides ariarathes* Esper (Lep. Papilionidae). *Sciences Naturelles*, **68**, 19–21.

Brévignon, C. (1992). Elevage de deux Riodininae guyanais, *Napaea beltiana* Bates et *Cremna thasus* Stoll. I. A propos de la myrmécophilie des chenilles. *Alexanor*, **17**, 403–413.

Bridle, J. R., Garn, A.-K., Monk, K. A. & Butlin, R. K. (2001). Speciation in *Chitaura* grasshoppers (Acrididae: Oxyinae) on the island of Sulawesi: colour patterns, morphology and contact zones. *Biological Journal of the Linnean Society*, **72**, 373–390.

Brown, E. S. & Taylor, L. R. (1971). Lunar cycles in the distribution and abundance of airborne insects in the equatorial highlands of East Africa. *Journal of Animal Ecology*, **40**, 767–779.

Brown, K. (1999). Life on the edge. *New Scientist*, **2213**, 46–49.

Brown, K. S., Jr (1997). Diversity, disturbance, and sustainable use of Neotropical forests: insects as indicators for conservation monitoring. *Journal of Insect Conservation*, **1**, 25–42.

Brown, K. S., Jr & Hutchings, R. W. (1997). Disturbance, fragmentation, and the dynamics of diversity in Amazonian forest butterflies. In *Tropical Forest Remnants. Ecology, Management and Conservation of Fragmented Communities*, eds. W. F. Laurance & R. O. Bierregaard Jr, pp. 91–110. Chicago: University of Chicago Press.

Brown, K. W. (1961). Entomological studies from a high tower in Mpanga Forest, Uganda. XI. Observations on Coleoptera. *Transactions of the Royal Entomological Society of London*, **113**, 353–355.

Brown, S. & Lugo, A. E. (1982). Storage and production of organic matter in tropical forests and their role in the global carbon cycle. *Biotropica*, **14**, 161–187.

Brown, S. & Lugo, A. E. (1990). Tropical secondary forests. *Journal of Tropical Ecology*, **6**, 1–31.

Brown, S., Lugo, A. E., Silander, S. & Liegel, L. (1983). *Research history and communities in the Luquillo Experimental Forest*. (General Technical Report SO-44.) Southern Forest Experiment Station, New Orleans, LA: US Department of Agriculture, Forest Service.

Brown, W. L., Jr (1973). A comparison of the Hylean and Congo–West African rain forest ant faunas. In *Tropical Forest Ecosystems in Africa and South America: A Comparative Review*, eds. B. J. Meggers, E. S. Ayensu & W. D. Duckworth, pp. 161–185. Washington: Smithsonian Institution Press.

Brühl, C. A., Gunsalam, G. & Linsenmair, K. E. (1998). Stratification of ants (Hymenoptera, Formicidae) in a primary rain forest in Sabah, Borneo. *Journal of Tropical Ecology*, **14**, 285–297.

Bryant, G. E. (1919). Coleoptera in Borneo. *Entomologist's Monthly Magazine*, **55**, 70–76.

Büchs, W. (1988). *Stamm- und Rindenzoozönosen verschiedener Baumarten des Hartholzauenwaldes und ihr Indikatorwert für die Früherkennung von Baumschäden*. Dissertation. Bonn: University of Bonn.

Bull, E. L., Holthausen, R. S. & Henjum, M. G. (1992). Roost trees used by pileated woodpeckers in northeastern Oregon. *Journal of Wildlife Management*, **56**, 786–793.

Bulla, L. (1994). An index of evenness and its associated diversity measure. *Oikos*, **70**, 167–171.

Bullock, S. H. (1986). Climate of Chamela, Jalisco, and trends in the south coastal region of Mexico. *Archives of Meteorology, Geophysic and Bioclimatology*, **36**, 297–316.

Bullock, S. H. (1988). Rasgos del ambiente físico y biológico de Chamela, Jalisco, México. *Folia Entomológica Mexicana*, **77**, 5–17.

Bullock, S. H. & Solís-Magallanes, J. A. (1990). Phenology of canopy trees of a tropical deciduous forest in Mexico. *Biotropica*, **22**, 22–35.

Bunge, J. & Fitzpatrick, M. (1993). Estimating the number of species – a review. *Journal of the American Statistical Association*, **88**, 364–373.

Burns, J. M. (1996). Genitalia and the proper genus: *Codatractus* gets *mysie* and *uvydixa*–in a compact *cyda* group–as well as a hysterectomy, while *Cephise* gets part of *Polythrix* (Hesperiidae: Pyrginae). *Journal of the Lepidopterist's Society*, **50**, 173–216.

Burns, J. M. & Janzen, D. H. (1999). *Drephalys*: division of this showy neotropical genus, plus a new species and the immatures and food plants of two species from Costa Rican dry forest (Hesperiidae: Pyrginae). *Journal of the Lepidopterist's Society*, **53**, 77–89.

Burns, J. M. & Janzen, D. H. (2001). Biodiversity of pyrrhopygine skipper butterflies (Hesperiidae) in the Area de Conservación Guanacaste, Costa Rica. *Journal of the Lepidopterist's Society*, **55**, 15–43.

Butcher, P. A., Doran, J. C. & Slee, M. U. (1994). Intraspecific variation in leaf oils of *Melaleuca alternifolia* (Myrtaceae). *Biochemical Systematics and Ecology*, **22**, 419–430.

Butler, L. & Kondo, V. (1991). Macrolepidopterous moths collected by blacklight trap at Cooper's Rock State Forest, West Virginia: a baseline study. *Bulletin of the Agricultural and Forestry Experiment Station, West Virginia University*, **705**, 1–25.

Butler, L., Kondo, V., Barrows, E. M. & Townsend, E. C. (1999). Effects of weather conditions and trap types on sampling for richness and abundance of forest Macrolepidoptera. *Environmental Entomology*, **28**, 795–811.

Byers, J. A., Anedebrant, O. & Lofqvist, J. (1989). Effective attraction radius: a method for comparing species attractants and determining densities of flying insects. *Journal of Chemical Ecology*, **15**, 749–766.

Caceres, C. E. (1997). Dormancy in invertebrates. *Invertebrate Biology*, **116**, 371–383.

Cachan, P. (1964). Analyse statistique des pullulations de Scolytoidea mycétophages en forêt sempervirente de Côte d'Ivoire. *Annales de la Faculté des Sciences, Université de Dakar*, **14**, 5–70.

Cachan, P. (1974). Importance écologique des variations verticales microclimatiques du sol à la canopée dans la forêt tropicale humide. In *Ecologie Forestière. La Forêt: son Climat, son Sol, ses Arbres, sa Faune*, ed. P. Pesson, pp. 21–42. Paris: Gauthier-Villars.

Caley, M. J. & Schluter, D. (1997). The relationship between local and regional diversity. *Ecology*, **78**, 70–80.

Camin, J. H. (1953a). Metagynellidae, a new family of uropodine mite, with the description of *Metagynella parvula*, a new species from tree holes. *Bulletin of the Chicago Academy of Sciences*, **9**, 391–409.

Camin, J. H. (1953b). A revision of the cohort Trachytina Trägårdh, 1938, with the description of *Dyscritaspis whartoni*, a new genus and species of polyaspid mite from tree holes. *Bulletin of the Chicago Academy of Sciences*, **9**, 335–385.

Camin, J. H. (1955). Uropodellidae, a new family of mesostigmatid mites based on *Uropodella lacinata* Berlese, 1988 (Acarina, Liroaspina). *Bulletin of the Chicago Academy of Sciences*, **10**, 65–81.

Capecki, Z. (1969). *Insects Damaging Wood of the Beech (*Fagus sylvatica *(L.)) on the Area of its Natural Occurrence in Poland* (in Polish). Warszawa: Prace Institut Badawczego Lesnuictwa.

Cappuccino, N. & Price, P. W. (eds.) (1995). *Population Dynamics: New Approaches and Synthesis*. New York: Academic Press.

Caraglio, Y., Nicolini, E. & Petronelli, P. (2001). Observations on the links between the architecture of a tree (*Dicorynia guianensis* Amshoff) and Cerambycidae activity in French Guiana. *Journal of Tropical Ecology*, **17**, 459–463.

Carbonell, C. S. (1964). Habitat, ecología y ontogenia de *Paulinia acuminata* (DG.) (Acridoidea, Pauliniidae) en el Uruguay. *Revista de la Sociedad Uruguaya de Entomología*, **6**, 39–48.

Carbonell, C. S. & Arrillaga, B. (1959). Sobre la relación anatómica de las ootecas de *Marellia remipes* Uvarov (Orthoptera, Acrid. Pauliniidae) con las hojas de su planta huesped, y su posible significación fisiológica. *Revista de la Sociedad Uruguaya de Entomología*, **3**, 45–56.

Carneiro, M. A. A., Ribeiro, S. P. & Fernandes, G. W. (1995). Artrópodos de um gradiente altitudinal na Serra do Cipó, Minas Gerais, Brasil. *Revista Brasileira de Entomologia*, **39**, 597–604.

Carroll, C. R. & Janzen, D. H. (1973). Ecology of foraging by ants. *Annual Review of Ecology and Systematics*, **4**, 231–257.

Carroll, G. C. (1980). Forest canopies: complex and independent subsystems. In *Forests: Fresh Perspectives From Ecosystem Analysis*, ed. R. H. Waring, pp. 87–108. Corvallis, Oregon: Oregon State University.

Carson, H. L. & Clague, D. A. (1995). Geology and biogeography of the Hawaiian Islands. In *Hawaiian Biogeography: Evolution on a Hotspot Archipelago*, eds. W. L. Wagner & V. A. Funk, pp. 14–29. Washington, DC: Smithsonian Institution Press.

Cassagnau, P. (1961). *Ecologie du Sol dans les Pyrénées centrales. Les Biocoénoses de Collemboles*. Paris: Hermann.

Casson, D. S. & Hodkinson, I. D. (1991). The Hemiptera (Insecta) communities of tropical rain forest in Sulawesi. *Zoological Journal of the Linnean Society*, **102**, 253–275.

Cates, R. G. & Orians, G. H. (1975). Successional status and the palatability of plants to generalized herbivores. *Ecology*, **56**, 410–418.

Cavalcanti, M. J. (2001). *Mantel for Windows, Test for Association between Two Symmetric Distance Matrices with Permutation Iterations*, Version 1.14. http://life.bio.sunysb.edu/morph/

Cedeño, A., Mérida, T. & Zegarra, J. (1999). Ant gardens of Surumoni, Venezuela. *Selbyana*, **20**, 125–132.

Cervantes, L. (1988). Intercepción de lluvia por el dosel en una comunidad tropical. *Ingeniería Hidráulica en México*, **3**, 30–41.

Chadwick, O. A., Derry, L. A., Vitousek, P. M., Huebert, B. J. & Hedin, L. O. (1999). Changing sources of nutrients during four million years of ecosystem development. *Nature*, **397**, 491–497.

Chai, P. & Srygley, R. B. (1990). Predation and the flight, morphology, and temperature of Neotropical rain-forest butterflies. *American Naturalist*, **135**, 748–765.

Chao, A. (1984). Non-parametric estimation of the number of classes in a population. *Scandinavian Journal of Statistics*, **11**, 265–270.

Chapman, C. A., Balcomb, S. R., Gillespie, T. R., Skorupa, J. P. & Struhsaker, T. T. (2000). Long-term effects of logging on African primate communities: a 28-year comparison from Kibale National Park, Uganda. *Conservation Biology*, **14**, 207–217.

Chapman, C. A. & Chapman, L. J. (1990). Density and growth rates of some tropical dry forest trees: comparisons between successional forest types. *Bulletin of the Torrey Botanical Club*, **117**, 226–231.

Chapman, C. A. & Chapman, L. J. (1996). Exotic tree plantations and the regeneration of natural forests in Kibale National Park, Uganda. *Biological Conservation*, **76**, 253–257.

Chappell, N. A., Bidin, K. & Tych, W. (2001). Modelling rainfall and canopy controls on a net-precipitation beneath selectively logged tropical forest. *Plant Ecology*, **153**, 215–229.

Chase, M. R., Moller, C., Kesseli, R. & Bawa, K. S. (1996). Distant gene flow in tropical trees. *Nature*, **383**, 398–399.

Chave, J., Muller-Landau, H. C. & Levin, S. A. (2002). Comparing classical community models: theoretical consequences for patterns of diversity. *American Naturalist*, **159**, 1–23.

Chazdon, R. & Fetcher, N. (1984). Photosythetic light environments in a lowland tropical rainforest in Costa Rica. *Journal of Ecology*, **72**, 553–564.

Chazdon, R. L. & Whitmore, T. C. (eds.) (2002). *Foundations of Tropical Forest Biology: Classic Papers with Commentaries*. Chicago: University of Chicago Press.

Chen, Y.-C., Hwang, W.-H., Chao, A. & Kuo, C.-Y. (1995). Estimating the number of common species. Analysis of the number of common bird species in Ke-Yar Stream and Chung-Kang Stream (in Chinese with English abstract). *Journal of the Chinese Statistical Association*, **33**, 373–393.

Cherrill, A. J. & Sanderson, R. A. (1994). Comparison of sweep-net and pitfall trap samples of moorland Hemiptera: evidence for vertical stratification within vegetation. *Entomologist*, **113**, 70–81.

Chey, V. K. (1994). Comparison of Biodiversity between Plantation and Natural Forests in Sabah using Moths as Indicators. PhD thesis. Oxford, UK: University of Oxford.

Chey, V. K., Holloway, J. D., Hambler, C. & Speight, M. R. (1998). Canopy knockdown of arthropods in exotic plantations and natural forest in Sabah, north-east Borneo, using insectidal mist-blowing. *Bulletin of Entomological Research*, **88**, 15–24.

Chey, V. K., Holloway, J. D. & Speight, M. R. (1997). Diversity of moths in forest plantations and natural forests in Sabah. *Bulletin of Entomological Research*, **87**, 371–385.

Chiesa, S., Confortini, F. & Madesani, R. (1998). Edición especial geología del ACG. *Rothschildia*, **5**, 1–33.

Christensen, O. (1980). Aspects of distribution patterns of *Liebstadia humerata* (Acari, Cryptostigmata) in a Danish oak forest. *Pedobiologia*, **20**, 24–30.

Chung, A. Y. C. (1999). The Diversity of Coleoptera Assemblages in Different Habitat Types in Sabah, Malaysia, with Special Reference to Herbivory. PhD thesis. Oxford, UK: University of Oxford.

Chung, A. Y. C., Chey, V. K., Eggleton, P., Hammond, P. M. & Speight, M. R. (2001). Variation in beetle (Coleoptera) diversity at different heights of tree canopy in a native forest and forest plantation in Sabah, Malaysia. *Journal of Tropical Forest Science*, **13**, 369–385.

Chung, A. Y. C., Eggleton, P., Speight, M. R., Hammond, P. M. & Chey, V. K. (2000). The diversity of beetle assemblages in different habitat types in Sabah, Malaysia. *Bulletin of Entomological Research*, **90**, 475–496.

Churchill, D. M. & Christensen, P. (1970). Observations on pollen harvesting by brush-tongued lorikeets. *Australian Journal of Zoology*, **18**, 427–437.

Clague, D. & Dalrymple, G. B. (1989). Tectonics, geochronology, and origin of the Hawaiian-Emperor volcanic chain. In *A Natural History of the Hawaiian Islands*, ed. E. A. Kay, pp. 5–40. Honolulu, Hawaii: University of Hawaii Press.

Claridge, M. F. & Wilson, M. R. (1981). Host plant associations, diversity and species-area relationships of mesophyll-feeding leafhoppers of trees and shrubs in Britain. *Ecological Entomology*, **6**, 217–238.

Clarke, J. F. G. (1983). A new species of *Eomichla* from Costa Rica (Oecophoridae). *Journal of the Lepidopterist's Society*, **37**, 155–159.

Clarke, K. R. (1993). Non-parametric multivariate analyses of changes in community structure. *Australian Journal of Ecology*, **18**, 117–143.

Cleyet-Marrel, D. (2000). La bulle des cimes. In *Biologie d'une Canopée de Forêt équatoriale – IV. Rapport de la Mission du Radeau des Cimes à La Makandé, Forêt des Abeilles, Gabon. Janvier–Mars 1999*, ed. F. Hallé, pp. 20–21. Paris/Lyon, France: Pro-Natura International & Opération Canopée.

Clinton, B. D., Boring, L. R. & Swank, W. T. (1993). Canopy gap characteristics and drought influences in oak forests of the Coweeta Basin. *Ecology*, **74**, 1551–1558.

Cloudsley-Thompson, J. L. (1980). *Microecologia. Temas de Biologia*, Vol. 2. São Paulo, Brasil: Editora Pedagógica e Universitária.

Cockburn, A. & Lazenby-Cohen, K. (1992). Use of nest trees by *Antechinus stuartii*, a semelparous lekking marsupial. *Journal of Zoology*, **226**, 657–680.

Cockburn, P. F. (1972). *Fagaceae. Tree Flora of Malaysia*, Vol. 1. Petaling Jaya: Longman Malaysia.

Coddington, J. A., Griswold, C. E., Dávila, D. S., Peñaranda, E. & Larcher, S. F. (1991). Designing and testing sampling protocols to estimate biodiversity in tropical ecosystems. In *The Unity of Evolutionary Biology: Proceedings of the Fourth International Congress of Systematic and Evolutionary Biology*, ed. E. C. Dudley, pp. 44–60. Portland: Dioscorides Press.

Coddington, J. A., Young, L. H. & Coyle, F. A. (1996). Estimating spider species richness in a southern Appalachian cove hardwood forest. *The Journal of Arachnology*, **24**, 111–124.

Coe, M. & Collins, N. M. (1986). *Kora: An Ecological Inventory of the Kora National Reserve, Kenya*. London: Royal Geographical Society.

Coley, P. D. (1980). Effects of leaf age and plant life history patterns on herbivory. *Nature*, **284**, 545–546.

Coley, P. D. (1983). Herbivory and defensive characteristics of tree species in a lowland tropical forest. *Ecological Monographs*, **53**, 209–233.

Coley, P. D. & Aide, T. M. (1991). Comparison of herbivory and plant defense in temperate and tropical broad-leaved forests. In *Plant–Animal Interactions: Evolutionary Ecology in Tropical and Temperate Regions*, eds. P. W. Price, T. M. Lewinsohn, G. W. Fernandes & W. W. Benson, pp. 25–49. New York: John Wiley & Sons.

Coley, P. D. & Barone, J. A. (1996). Herbivory and plant defenses in tropical forests. *Annual Review of Ecology and Systematics*, **27**, 305–335.

Coley, P. D. & Kursar, T. A. (1996). Anti-herbivore defenses of young tropical leaves: physiological constraints and ecological trade-offs. In *Tropical Forest Plant Ecophysiology*, eds. S. S. Mulkey, R. L. Chazdon & A. P. Smith, pp. 305–336. New York: Chapman & Hall.

Colinvaux, P. A. (1998). A new vicariance model for Amazonian endemics. *Global Ecology and Biogeography Letters*, **7**, 95–96.

Colwell, R. K. (1995). Effects of nectar consumption by the hummingbird flower mite *Procyolaelaps kirmsei* on nectar availability in *Hamelia patens*. *Biotropica*, **27**, 206–217.

Colwell, R. K. (1997a). *Biota: the Diversity Database Manager*. Sunderland, MA: Sinauer Associates.

Colwell, R. K. (1997b). *EstimateS: Statistical Estimation of Species Richness and Shared Species from Samples*. Version 5. User's guide and application published at: http://viceroy.eeb.uconn.edu/estimates

Colwell, R. K. & Coddington, J. A. (1994). Estimating terrestrial biodiversity through extrapolation. *Philosophical Transactions of the Royal Society of London, Series B*, **345**, 101–118.

Colwell, R. K. & Lees, D. C. (2000). The mid-domain effect: geometric constraints on the geography of species richness. *Trends in Ecology and Evolution*, **15**, 70–76.

Common, I. F. B. (1990). *Moths of Australia*. Melbourne: Melbourne University Press.

Compton, G. L. & Krantz, G. W. (1978). Mating behaviour and related morphological specialisation in the uropodine mite, *Caminella peraphora*. *Science*, **200**, 1300–1301.

Compton, S. G., Ellwood, M. D. F., Davis, A. J. & Welch, K. (2000). The flight heights of chalcid wasps (Hymenoptera, Chalcidoidea) in a lowland Bornean rain forest: fig wasps are the high fliers. *Biotropica*, **32**, 515–522.

Compton, S. G. & Hawkins, B. A. (1992). Determinants of species richness in southern African fig wasp assemblages. *Oecologia*, **91**, 68–74.

Compton, S. G., Lawton, J. H. & Rashbrook, V. K. (1989). Regional diversity, local community structure and vacant niches: the herbivorous arthropods of bracken in South Africa. *Ecological Entomology*, **14**, 365–373.

Condit, R. (1996). Defining and mapping vegetation types in mega-diverse tropical forests. *Trends in Ecology and Evolution*, **11**, 4–5.

Condit, R. (1997). *Tropical Forest Census Plots: Methods and Results from Barro Colorado Island, Panama and a Comparison with other Plots*. New York: Springer Verlag.

Condit, R., Hubbell, S. P., LaFrankie, J. V., Sukumar, R., Manokaran, N., Foster, R. & Ashton, P. S. (1996). Species–area and species–individual relationships for tropical trees: a comparison of three 50-hectare plots. *Journal of Ecology*, **84**, 549–562.

Connell, J. H. (1971). On the role of natural enemies in preventing competitive exclusion in some marine animals and in rain forest trees. In *Dynamics of Populations*, eds. P. J. den Boer & G. R. Gradwell, pp. 298–312. Wageningen, the Netherlands: PUDOC.

Connell, J. H., Lowman, M. D. & Noble, I. R. (1997). Subcanopy gaps in temperate and tropical forests. *Australian Journal of Ecology*, **22**, 163–168.

Connor, E. F., Faeth, S. H., Simberloff, D. & Opler, P. A. (1980). Taxonomic isolation and the accumulation of herbivorous insects: a comparison of introduced and native trees. *Ecological Entomology*, **5**, 205–211.

Corbet, P. S. (1961a). Entomological studies from a high tower in Mpanga Forest, Uganda. IV. Mosquito breeding at different levels in and above the forest. *Transactions of the Royal Entomological Society of London*, **113**, 275–283.

Corbet, P. S. (1961b). Entomological studies from a high tower in Mpanga Forest, Uganda. VI. Nocturnal flight activity of Culicidae and Tabanidae as indicated by light-traps. *Transactions of the Royal Entomological Society of London*, **113**, 301–314.

Corbet, P. S. (1966). The role of rhythms in insect behaviour. In *Insect Behaviour*, ed. P. T. Haskell, pp. 13–28. London: Royal Entomological Society.

Cordell, S., Goldstein, G., Meinzer, F. C. & Vitousek, P. M. (2001). Regulation of leaf life-span and nutrient-use efficiency of *Metrosideros polymorpha* trees at two extremes of a long chronosequence in Hawaii. *Oecologia*, **127**, 198–206.

Cordell, S. G., Goldstein, G., Mueller-Dombois, D., Webb, D. & Vitousek, P. M. (1998). Physiological and morphological variation in *Metrosideros polymorpha*, a dominant Hawaiian tree species, along an altitudinal gradient: the role of phenotypic plasticity. *Oecologia*, **113**, 188–196.

Cornell, H. V. (1985). Local and regional richness of cynipine wasps on California oaks. *Ecology*, **66**, 1247–1260.

Cornell, H. V. & Lawton, J. H. (1992). Species interactions, local and regional processes, and limits to the richness of ecological communities: a theoretical perspective. *Journal of Animal Ecology*, **61**, 1–12.

Corrales, J. F. & Epstein, M. E. (1997). Review of Costa Rican *Venacodicodia*, with descriptions of two new species and localities for *V. ruthea* (Lepidoptera: Limacodidae). *Revista de Biología Tropical*, **45**, 1093–1105.

Costa, J. T., Fitzgerald, T. D. & Janzen, D. H. (2002). Trail-following behavior and natural history of the social caterpillar of *Arsenura armida* in Costa Rica (Lepidoptera: Saturniidae: Arsenurinae). *Tropical Lepidoptera*, in press.

Cowie, R. H. (1995). Variation in species diversity and shell shape in Hawaiian land snails: in situ speciation and ecological relationships. *Evolution*, **49**, 1191–1202.

Cox, R. T. & Carlton, C. E. (1998). A commentary on prime numbers and life cycles of periodical cicadas. *American Naturalist*, **152**, 162–164.

Coxson, D. S. & Nadkarni, N. M. (1995). Ecological roles of epiphytes in nutrient cycles of forest ecosystems. In *Forest Canopies*, eds. M. D. Lowman & N. M. Nadkarni, pp. 495–543. San Diego, CA: Academic Press.

Cracraft, J. (2000). Charting the biosphere: building global capacity for systematics science. In *Nature and Human Society: The Quest for a Sustainable World. Proceedings of the 1997 Forum on Biodiversity*, eds. P. H. Raven & T. Williams, pp. 374–386. Washington, DC: National Academy Press.

Cranston, P. & Hillman, T. (1992). Rapid assessment of biodiversity using 'biological diversity technicians'. *Australian Biologist*, **5**, 144–154.

Crawley, M. J. & Akhteruzzaman, M. (1988). Individual variation in the phenology of oak trees and its consequence for herbivorous insects. *Functional Ecology*, **2**, 409–415.

Crews, T. E., Kitayama, K., Fownes, J. H., Riley, R. H., Herbert, D. A., Mueller-Dombois, D. & Vitousek, P. M. (1995). Changes in soil phosphorus fractions and ecosystem dynamics across a long chronosequence in Hawaii. *Ecology*, **76**, 1407–1424.

Croat, T. B. (1978). *Flora of Barro Colorado Island*. Stanford: Stanford California Press.

Crome, F. H. J. & Richards, G. C. (1988). Bats and gaps: Microchiropteran community structure in a Queensland rain forest. *Ecology*, **69**, 1960–1969.

Crouther, M. M. (1983). Fauna associated with white-rumped swiftlets at Doolamai Cave. *Australian Bird Watcher*, **10**, 32.

Curletti, G. (2000). Gli *Agrilus* della spedizione 'Radeau des cimes' – Gabon 1999 (Coleoptera, Buprestidae). *Lambillionea*, **100**, 459–470.

Curran, L. M., Caniago, I., Paoli, G. D., Astianti, D., Kusneti, M., Leighton, M., Nirarita, C. E. & Haeruman, H. (1999). Impact of El Niño and logging on canopy tree recruitment in Borneo. *Science*, **286**, 2184–2188.

Curran, L. M. & Leighton, M. (2000). Vertebrate responses to spatiotemporal variation in seed production of mast-fruiting Dipterocarpaceae. *Ecological Monographs*, **70**, 101–128.

Curran, L. M. & Webb, C. O. (2000). Experimental tests of the spatiotemporal scale of seed predation in mast-fruiting Dipterocarpaceae. *Ecological Monographs*, **70**, 129–148.

Dafni, A. (1992). *Pollination Ecology: A Practical Approach.* Oxford: Oxford University Press.

da Fonseca, C. R. V., Adis, J. & Martius, C. (1998). Mechanisms that maintain tropical diversity – a project of Teuton–Brazilian cooperation 1991–1996. *Acta Amazonica*, **28**, 205–215.

Daily, G. C. & Ehrlich, P. R. (1995). Preservation of biodiversity in small rainforest patches: rapid evaluations using butterfly trapping. *Biodiversity and Conservation*, **4**, 35–55.

Dajoz, R. (1980). *Ecologie des Insectes Forestiers*. Paris: Gauthier-Villars.

Dajoz, R. (1982). *Précis d'Ecologie*, 4th edn. Paris: Gauthier-Villars.

Dajoz, R. (2000). *Insects and Forests. The Role and Diversity of Insects in the Forest Environment*. London: Intercept.

Dangerfield, P. C., Whitfield, J. B., Sharkey, M. J., Janzen, D. H. & Mercado, I. (1996). *Hansonia*, a new genus of cardiochiline Braconidae (Hymenoptera) from Costa Rica, with notes on its biology. *Proceedings of the Entomological Society of Washington*, **98**, 592–596.

Davidson, D. W. (1988). Ecological studies of Neotropical ant gardens. *Ecology*, **69**, 1138–1152.

Davidson, D. W. (1997). The role of resource imbalances in the evolutionary ecology of tropical arboreal ants. *Biological Journal of the Linnean Society*, **61**, 153–181.

Davidson, D. W. (1998). Resource discovery versus resource domination in ants: a functional mechanism for breaking the trade-off. *Ecological Entomology*, **23**, 484–490.

Davidson, D. W. & Patrell-Kim, L. (1996). Tropical arboreal ants: why so abundant? In *Neotropical Biodiversity and Conservation*, ed. A. C. Gibson, pp. 127–140. Los Angeles: Mildred E. Mathias Botanical Garden, University of California.

Davies, J. G. (1999). Beetles (Coleoptera) of Mkomazi. In *Mkomazi: the Ecology, Biodiversity and Conservation of a Tanzanian Savanna*, eds. M. J. Coe, N. C. McWilliam, G. N. Stone & M. J. Packer, pp. 249–268. London: Royal Geographic Society and the Institute of British Geographers.

Davies, J. G., Stork, N. E., Brendell, M. J. D. & Hine, S. J. (1997). Beetle species diversity and faunal similarity in Venezuelan rainforest tree canopies. In *Canopy Arthropods*, eds. N. E. Stork, J. Adis & R. K. Didham, pp. 85–103. London: Chapman & Hall.

Davis, A. J. (1999a). Species packing in tropical forests: diel flight activity of rainforest dung-feeding beetles (Coleoptera: Aphodiidae, Scarabaeidae, Hybosoridae) in Borneo. *Raffles Bulletin of Zoology*, **47**, 473–486.

Davis, A. J. (1999b). Perching behaviour in Bornean dung beetles (Coleoptera: Scarabaeidae). *Coleopterists Bulletin*, **53**, 365–370.

Davis, A. J., Huijbregts, J. & Krikken, J. (2000). The role of local and regional process in shaping dung beetle communities in tropical forest plantations in Borneo. *Global Ecology and Biogeography*, **9**, 281–292.

Davis, A. J., Huijbregts, J., Kirk-Spriggs, A. H., Krikken, J. & Sutton, S. L. (1997). The ecology and behaviour of arboreal dung beetles in Borneo. In *Canopy Arthropods*, eds. N. E. Stork, J. Adis & R. K. Didham, pp. 417–432. London: Chapman & Hall.

Davis, A. J. & Sutton, S. L. (1998). The effects of rainforest canopy loss on arboreal dung beetles in Borneo: implications for the measurement of biodiversity in derived tropical ecosystems. *Diversity and Distributions*, **4**, 167–173.

Dawson, J. W. & Stemmerman, L. (1990). *Metrosideros* (Myrtaceae). In *Manual of the Flowering Plants of Hawaii*, eds. W. L. Wagner, D. R. Herbst & S. H. Sohmer, pp. 964–970. Honolulu, Hawaii: Bernice P. Bishop Museum Press.

De Abreu, R. L. S., da Fonseca, C. R. V., Hurtado Guerrero, J. C. & de Paula, E. V. C. M. (2001). Preferência de vôo de nove especies da familia Scolytidae (Insecta: Coleoptera) na Amazonia central. *Acta Amazonica*, **31**, 61–68.

De Graaf, N. R. (1986). *A Silvicultural System for Natural Regeneration of Tropical Rain Forest in Suriname*. Wageningen: Agricultural University Wageningen.

De Oliveira, M. L. & Campos, L. M. O. (1996). Preferência por estratos florestais e por substâncias odoríferas em abelhas euglossinae. *Revista Brasileira de Zoologia*, **13**, 1075–1085.

Deane, L. M., Ferreira Neto, J. A. & Lima, M. M. (1984). The vertical dispersion of *Anopheles* (*Kerteszia*) *cruzi* in a forest in southern Brazil suggests that human cases of malaria of simian origin might be expected. *Memorias do Instituto Oswaldo Cruz*, **79**, 461–463.

DeAngelis, D. L. (1980). Energy flow, nutrient cycling, and ecosystem resilience. *Ecology*, **61**, 764–771.

Dejean, A. (1990a). Prey capture strategy of the African weaver ant. In *Applied Myrmecology: A World Perspective*, eds. R. K. Vander Meer, K. Jaffe & A. Cedeno, pp. 472–481. Boulder, CO: Westview Press.

Dejean, A. (1990b). Circadian rhythm of *Oecophylla longinoda* in relation with territoriality and predatory behaviour. *Physiological Entomology*, **15**, 393–403.

Dejean, A. (1991). Adaptation d'*Oecophylla longinoda* aux variations spatio-temporelles de la densité en proies. *Entomophaga*, **36**, 29–54.

Dejean, A. (2000). Ant protection (Hymenoptera: Formicidae) of two pioneer plant species against the variegated locust. *Sociobiology*, **36**, 217–226.

Dejean, A., Akoa, A., Djiéto-Lordon, C. & Lenoir, A. (1994a). Mosaic ant territories in an African secondary forest (Hymenoptera: Formicidae). *Sociobiology*, **23**, 275–292.

Dejean, A., Belin, M. & McKey, D. (1992b). Les relations plantes-fourmis dans la Canopée. In *Biologie d'une Canopée de Forêt Équatoriale – II. Rapport de Mission: Radeau des Cimes Octobre-Novembre 1991, Réserve de Campo, Cameroun*, eds. F. Hallé & O. Pascal, pp. 76–80. Paris: Fondation Elf.

Dejean, A. & Corbara, B. (1990). Predatory behavior of a neotropical arboricolous ant: *Pachycondyla villosa* (Formicidae, Ponerinae). *Sociobiology*, **17**, 271–286.

Dejean, A., Corbara, B. & Orivel, J. (1999). The arboreal ant mosaic in two Atlantic forests. *Selbyana*, **20**, 133–145.

Dejean, A., Corbara, B., Orivel, J., Snelling, R. R., Delabie, J. H. C. & Belin-Depoux, M. (2000a). The importance of ant gardens in the pioneer vegetal formations of French Guiana. *Sociobiology*, **35**, 425–439.

Dejean, A., Djiéto-Lordon, C. & Durand, J. L. (1997). Ant mosaic in oil palm plantations of the south-west province of Cameroon: impact on a leaf miner chrysomelid beetle. *Journal of Economic Entomology*, **90**, 1092–1096.

Dejean, A., Durou, S., Corbara, B., Lewis, S., Bolton, B., Pascal, O. & Breteler, F. J. (2000b). La mosaïque des fourmis arboricoles dominantes dans la forêt de la Makandé. In *Biologie d'une Canopée de Forêt Équatoriale – IV. Rapport de la Mission du Radeau des Cimes à la Makandé, Forêt des Abeilles, Gabon. Janvier–Mars 1999*, ed. F. Hallé, pp. 59–66. Paris: Pro-Natura International & Opération Canopée.

Dejean, A. & Gibernau, M. (2000). A rainforest ant mosaic: the edge effect (Hymenoptera: Formicidae). *Sociobiology*, **35**, 385–401.

Dejean, A., Gibernau, M. & Bourgoin, T. (2000c). A new case of trophobiosis between ants and heteropterans. *Comptes Rendus de l'Académie des Sciences*, **323**, 447–454.

Dejean, A., Gibernau, M., Durand, J. L., Abehassera, D. & Orivel, J. (2000d). Pioneer plant protection against herbivory: impact of different ant species (Hymenoptera: Formicidae) on a proliferation of the variegated locust. *Sociobiology*, **36**, 227–236.

Dejean, A., Lenoir, A. & Godzinska, E. J. (1994b). The hunting behaviour of *Polyrhachis laboriosa*, a non-dominant ant of the African equatorial forest (Hymenoptera: Formicinae; Formicinae). *Sociobiology*, **23**, 293–313.

Dejean, A., McKey, D., Gibernau, M. & Belin-Depoux, M. (2000e). The arboreal ant mosaic in a Cameroonian rainforest (Hymenoptera: Formicidae). *Sociobiology*, **35**, 403–423.

Dejean, A., Nkongmeneck, B., Corbara, B. & Djiéto-Lordon, C. (1991). Impact des fourmis arboricoles sur une prolifération d'*Achaea catocaloides* (Lepidoptera, Noctuidae) dans une cacaoyère du Cameroun. *Acta Oecologica*, **12**, 471–488.

Dejean, A. & Olmsted, I. (1997). Ecological studies on *Aechmea bracteata* (Swartz) (Bromeliaceae). *Journal of Natural History*, **31**, 1313–1334.

Dejean, A., Olmsted, I. & Camal, J. F. (1992a). Interaction between *Atta cephalotes* and arboreal ants in the Biosphere Reserve Sian Ka'an (Quintana Roo, Mexico): efficient protection of the trees (Hymenoptera: Formicidae). *Sociobiology*, **20**, 57–76.

Dejean, A., Olmsted, I. & Snelling, R. R. (1995). Tree–epiphyte–ant relationships in the low inundated forest of Sian Ka'an Biosphere Reserve, Quintana Roo, Mexico. *Biotropica*, **27**, 57–70.

Dejean, A., Orivel, J., Corbara, B., Delabie, J. & Teillier, L. (1998). La mosaïque de fourmis arboricoles. In *Biologie d'une Canopée de Forêt Equatoriale – III. Rapport de la Mission d' Exploration Scientifique de la Canopée de Guyane, Octobre–Décembre 1996*, ed. F. Hallé, pp. 140–153. Paris: Pro-Natura International & Opération Canopée.

Del Claro, K. & Oliveira, P. S. (1999). Ant–homopteran interactions in a Neotropical savanna: the honeydew-producing treehopper, *Guayaquila xiphias* (Membracidae), and its associated ant fauna on *Didymopanax vinosum* (Araliaceae). *Biotropica*, **31**, 135–144.

Delamare-Debouteville, C. (1948). Etude quantitative du peuplement animal des sols suspendus et des épiphytes en forêt tropicale. *Comptes Rendus de l'Académie des Sciences*, **226**, 1544–1546.

Delamare-Debouteville, C. (1951). *Microfaune du Sol dans les Pays Tempérés et Tropicaux*. Paris: Hermann.

Delobel, A., Couturier, G., Kahn, F. & Nilsson, J. A. (1995). Trophic relationships between palms and bruchids (Coleoptera: Bruchidae: Pachymerini) in Peruvian Amazonia. *Amazoniana*, **13**, 209–219.

Delvare, G. & Aberlenc, H. P. (1989). *Les Insectes d' Afrique et d' Amérique Tropicale. Clés pour la Reconnaissance des Familles*. Montpellier, France: CIRAD/PRIFAS.

den Boer, P. J. (1990). The survival value of dispersal in terrestrial arthropods. *Biological Conservation*, **54**, 175–192.

Denlinger, D. L. (1980). Seasonal and annual variation of insect abundance in the Nairobi National Park, Kenya. *Biotropica*, **12**, 100–106.

Denlinger, D. L. (1986). Dormancy in tropical insects. *Annual Review of Entomology*, **31**, 239–264.

de Souza, A. L. T., Fernandes, G. W., Figueira, J. E. C. & Tanaka, M. O. (1998). Natural history of a gall-inducing weevil *Collabismus clitellae* (Coleoptera: Curculionidae) and some effects on its host plant *Solanum lycocarpum* (Solanaceae) in Southeastern Brazil. *Ecology and Population Biology*, **91**, 404–409.

Dettner, K. & Liepert, C. (1994). Chemical mimicry and camouflage. *Annual Review of Entomology*, **39**, 129–154.

DeVries, P. J. (1987a). *The Butterflies of Costa Rica and Their Natural History*. Vol. I, *Papilionidae, Pieridae, Nymphalidae*. Princeton, New Jersey: Princeton University Press.

DeVries, P. J. (1987b). Stratification of fruit-feeding nymphalid butterflies in a Costa Rican rainforest. *Journal of Research on the Lepidoptera*, **26**, 98–108.

DeVries, P. J., Lande, R. & Murray, D. (1999b). Associations of co-mimetic ithomiine butterflies on small spatial and temporal scales in a neotropical rainforest. *Biological Journal of the Linnean Society*, **67**, 73–85.

DeVries, P. J., Murray, D. & Lande, R. (1997). Species diversity in vertical, horizontal, and temporal dimensions of a fruit-feeding butterfly community in an Ecuadorian rainforest. *Biological Journal of the Linnean Society*, **62**, 343–364.

DeVries, P. J. & Walla, T. R. (2001). Species diversity and community structure in neotropical fruit-feeding butterflies. *Biological Journal of the Linnean Society*, **74**, 1–15.

DeVries, P. J., Walla, T. R. & Greeney, H. F. (1999a). Species diversity in spatial and temporal dimensions of fruit-feeding butterflies from two Ecuadorian rainforests. *Biological Journal of the Linnean Society*, **68**, 333–353.

Dial, R. & Roughgarden, J. (1995). Experimental removal of insectivores from rain forest canopy: direct and indirect effects. *Ecology*, **76**, 1821–1834.

Dial, R. & Tobin, S. C. (1994). Description of arborist methods for forest canopy access and movement. *Selbyana*, **15**, 24–37.

Dickinson, D. H. & Pugh, G. L. F. (eds.) (1974). *Biology of Plant Litter Decomposition,* Vols. 1 & 2. London: Academic Press.

Didham, R. K. (1997). Dipteran tree-crown assemblages in a diverse southern temperate rainforest. In *Canopy Arthropods*, eds. N. E. Stork, J. Adis & R. K. Didham, pp. 320–343. London: Chapman & Hall.

Didham, R. K., Ghazoul, J., Stork, N. E. & Davis, A. J. (1996). Insects in fragmented forests: a functional approach. *Trends in Ecology and Evolution*, **11**, 255–260.

Didham, R. K., Hammond, P. M., Lawton, J. H., Eggleton, P. & Stork, N. E. (1998). Beetle species richness responses to tropical forest fragmentation. *Ecological Monographs*, **68**, 295–323.

Diniz, I. R. & Morais, H. C. (1995). Larvas de lepidoptera e suas plantas hospedeiras em um cerrado de Brasilia, DF, Brasil. *Revista Brasileira de Entomologia*, **39**, 755–770.

Diniz, I. R. & Morais, H. C. (1997). Lepidopteran caterpillar fauna of cerrado host plants. *Biodiversity and Conservation*, **6**, 817–836.

Dippenaar-Schoeman, A. S. & Jocqué, R. (1997). *African Spiders. An Identification Manual. Plant Protection Research Institute Handbook No. 9*. Pretoria, South Africa: ARC-Plant Protection Research Institute.

Diserud, O. H. & Ødegaard, F. (2000). The beta binomial model for host specificity between organisms in trophic interactions. *Biometrics*, **56**, 855–861.

Dixon, A. F. G. (1976). Factors determining the distribution of sycamore aphids on sycamore leaves during summer. *Ecological Entomology*, **1**, 275–278.

Dixon, A. F. G., Kindlmann, P., Leps, J. & Holman, J. (1987). Why there are so few species of aphids, especially in the tropics. *American Naturalist*, **129**, 580–592.

Djiéto-Lordon, C. & Dejean, A. (1999). Tropical arboreal ant mosaic: innate attraction and imprinting determine nesting site selection in dominant ants. *Behavioral Ecology and Sociobiology*, **45**, 219–225.

Djiéto-Lordon, C., Nkonmeneck, B., Lowman, M. & Dejean, A. (2002). Fauna of the tangle roots of *Platycerium* (Polypodiaceae) in southern Cameroon. *Selbyana*, in press.

Djiéto-Lordon, C., Orivel, J. & Dejean, A. (2001a). Predatory behaviour of an African ponerine ant *Platythyrea modesta* (Hymenoptera: Formicidae). *Sociobiology*, **38**, 303–316.

Djiéto-Lordon, C., Orivel, J. & Dejean, A. (2001b). Consuming large prey on the spot: the case of the arboreal foraging ponerine ant *Platythyrea modesta* (Hymenoptera, Formicidae). *Insectes Sociaux*, **48**, 324–326.

Djiéto-Lordon, C., Richard, F. J., Owona, C., Orivel, J. & Dejean, A. (2001c). The predatory behavior of the dominant ant species *Tetramorium aculeatum* (Hymenoptera, Formicidae). *Sociobiology*, **38**, 765–775.

Dobler, S. & Farrell, B. D. (1999). Host use evolution in *Chrysochus* milkweed beetles: evidence from behaviour, population genetics and phylogeny. *Molecular Ecology*, 8, 1297–1307.

Docherty, M. & Leather, S. R. (1997). Structure and abundance of arachnid communities in Scots and lodgepole pine plantations. *Forest Ecology and Management*, **95**, 197–207.

Domrow, R. (1964). The *ulysses* species-group, genus *Haemolaelaps* (Acarina: Laelapidae). *Proceedings of the Linnean Society of New South Wales*, **84**, 155–162.

Domrow, R. (1979). Dermanyssine mites from Australian birds. *Records of the Western Australian Museum*, **7**, 403–413.

Domrow, R. (1980). A new species of the *ulysses* group, genus *Haemolaelaps* Berlese (Acari: Dermanyssidae). *Proceedings of the Linnean Society of New South Wales*, **104**, 222–227.

Domrow, R. (1981). Oriental Mesostigmata (Acari). 6. A Malesian member of the *mesopicos* group (*Haemolaelaps*). *Acarologia*, **22**, 115–119.

Domrow, R. (1988). Acari Mesostigmata parasitic on Australian vertebrates: an annotated checklist, keys and bibliography. *Invertebrate Taxonomy*, **1**, 817–948.

Doucet, J. L. (1996). Régénération Naturelle dans la Forêt des Abeilles (Gabon). U.E.R. Sylviculture. Gembloux, Belgium: Faculté Universitaire des Sciences Agronomiques de Gembloux.

Downum, K., Lee, D., Hallé, F., Quirke, M. & Towers, N. (2001). Plant secondary compounds in the canopy and understorey of a tropical rain forest in Gabon. *Journal of Tropical Ecology*, **17**, 477–481.

Drake, C. J. & Ruhoff, F. A. (1965). Lacebugs of the world: a catalog. *Smithsonian Institution, US National Museum Bulletin*, **243**, 1–634.

Drake, J. A. (1990). Communities as assembled structures: do rules govern pattern? *Trends in Ecology and Evolution*, **5**, 159–164.

Duan, J. J., Weber, D. C. & Dorn, S. (1996). Spring behavioral patterns of the apple blossom weevil. *Entomologia Experimentalis et Applicata*, **79**, 9–17.

Duffey, E. (1966). Spider ecology and habitat structure. *Senckenbergiana Biologica*, **47**, 45–49.

Dunger, W. (1983). *Tiere im Boden*. Wittenberg: Ziemsen.

Dupuy, J. M. & Chazdon, R. L. (1998). Long-term effects of forest regrowth and selective logging on the seed bank of tropical forests in NE Costa Rica. *Biotropica*, **30**, 223–237.

Dussourd, D. E. & Eisner, T. (1987). Vein-cutting behavior: insect counterploy to the latex defense of plants. *Science*, **237**, 898–901.

Duviard, D. & Pollet, A. (1973). Spatial and seasonal distribution of Diptera, Homoptera and Hymenoptera in a moist shrub savanna: ecological behavior of winged insect populations in the savannas of Ivory Coast. I. *Oikos*, **24**, 42–57.

Dyer, L. A. (1995). Tasty generalists and nasty specialists? Antipredator mechanisms in tropical lepidopteran larvae. *Ecology*, **76**, 1483–1496.

Ebersolt, G. (1990). Opération Radeau des cimes, Mission 1989 – Guyane. Bilan technique partie radeau. In *Biologie d'une Canopée de Forêt Équatoriale. Rapport de Mission. Radeau des Cimes Octobre-Novembre 1989, Guyane Française*, eds. F. Hallé & P. Blanc, pp. 15–26. Paris/Montpellier: Montpellier II et CNRS-Paris VI.

Ebersolt, G. (2000). Icos et radeau. In *Biologie d'une Canopée de Forêt Équatoriale – IV. Rapport de la Mission du Radeau des Cimes à La Makandé, Forêt des Abeilles, Gabon. Janvier–Mars 1999*, ed. F. Hallé, pp. 22–24. Paris/Lyon, France: Pro-Natura International & Opération Canopée.

Edsberg, E. & Hagvar, S. (1999). Vertical distribution, abundance, and biology of oribatid mites (Acari) developing inside decomposing spruce needles in a podsol soil profile. *Pedobiologia*, **43**, 413–421.

Edwards, P. B., Wanjura, W. J. & Brown, W. V. (1993). Selective herbivory by Christmas beetles in response to intraspecific variation in *Eucalyptus* terpenoids. *Oecologia*, **95**, 551–557.

Edwards, P. J. & Wratten, S. D. (1981). *Ecologia das Interações Entre Insetos e Plantas. Temas de Biologia*. Vol. 27. São Paulo, Brasil: Editora Pedagógica e Universitária.

Eggeling, W. J. (1947). Observations on the ecology of the Budongo rain forest, Uganda. *Journal of Ecology*, **34**, 20–87.

Eggleton, P. & Vane-Wright, R. I. (eds.) (1994). *Phylogenetics and Ecology*. London: Academic Press.

Ehrlich, P. R. & Raven, P. H. (1964). Butterflies and plants: a study in coevolution. *Evolution*, **18**, 586–608.

Eisenbeis, G. (1989). Allometric relationships of the body surface area to body mass in different life forms of soil arthropods: 2. Surface area and climate. In *Third International Seminar on Apterygota*, ed. R. Dallai, pp. 387–400. Siena: University of Siena.

Eisenbeis, G. & Wichard, W. (1987). *Atlas on the Biology of Soil Arthropods*. Berlin: Springer.

Eiten, G. (1972). The cerrado vegetation of Brazil. *Botanical Review*, **38**, 201–341.

Eiten, G. (1978). Delimitation of the cerrado concept. *Vegetatio*, **36**, 169–178.

Elliott, F. R. (1930). An ecological study of the spiders of the beech–maple forest. *Ohio Journal of Science*, **30**, 1–22.

Ellwood, M. D. F. & Foster, W. A. (2000). How well does fogging sample a high canopy epiphyte? In *XXI International Congress of Entomology Abstracts*, ed. D. L. Gazzoni, pp. 903. Londrina, Brazil: Embrapa Soja.

Elton, C. S. (1966). *The Pattern of Animal Communities*. London: Methuen.

Elton, C. S. (1973). The structure of invertebrate populations inside neotropical rain forest. *Journal of Animal Ecology*, **42**, 55–104.

Emmons, L. H. (1984). Geographic variation in densities and diversities of non flying Amazonian mammals. *Biotropica*, **16**, 210–222.

Enders, F. (1974). Vertical stratification in orb-web spiders (Araneidae, Araneae) and a consideration of other methods of coexistence. *Ecology*, **55**, 317–328.

Endler, J. A. (1982). Pleistocene forest refuges: fact or fancy? In *Biological Diversification in the Tropics*, ed. G. T. Prance, pp. 641–657. New York: Columbia University Press.

Endress, P. K. (1994). *Diversity and Evolutionary Biology of Tropical Flowers*. Cambridge: Cambridge University Press.

Engen, S. & Lande, R. (1996). Population dynamic models generating species abundance distributions of the gamma type. *Journal of Theoretical Biology*, **178**, 325–331.

Erwin, T. L. (1982). Tropical forests: their richness in Coleoptera and other arthropod species. *Coleopterists Bulletin*, **36**, 74–75.

Erwin, T. L. (1983a). Tropical forest canopies: the last biotic frontier. *Bulletin of the Entomological Society of America*, **29**, 14–19.

Erwin, T. L. (1983b). Beetles and other insects of tropical forest canopies at Manaus, Brazil, sampled by insecticidal fogging. In *Tropical Rain Forest: Ecology and Management*, eds. S. L. Sutton, T. C. Whitmore & A. C. Chadwick, pp. 59–76. Oxford: Blackwell.

Erwin, T. L. (1989). Canopy arthropod biodiversity: a chronology of sampling techniques and results. *Revista Peruana de Entomología*, **32**, 71–77.

Erwin, T. L. (1991a). How many species are there: revisited. *Conservation Biology*, **5**, 330–333.

Erwin, T. L. (1991b). Establishing a tropical species co-occurrence database. Part 1. A plan for developing consistent biotic inventories in temperate and tropical habitats. *Memoria del Museo de Historia Natural, Universidad Nacional Mayor de San Marcos, Peru*, **20**, 1–16.

Erwin, T. L. (1991c). An evolutionary basis for conservation strategies. *Science*, **253**, 750–752.

Erwin, T. L. (1995). Measuring arthropod biodiversity in the tropical forest canopy. In *Forest Canopies*, eds. M. D. Lowman & N. M. Nadkarni, pp. 109–127. San Diego, CA: Academic Press.

Erwin, T. L. & Adis, J. (1982). Amazonian inundation forests: their role as short-term refuges and generator of species richness and taxon pulses. In *Biological Diversification in the Tropics*, ed. G. T. Prance, pp. 358–371. New York: Columbia University Press.

Erwin, T. L. & Johnson, P. J. (2000). Naming species, a new paradigm for crisis management in taxonomy: rapid journal validation of scientific names enhanced with more complete descriptions on the Internet. *Coleopterists Bulletin*, **54**, 269–278.

Erwin, T. L. & Scott, J. C. (1980). Seasonal and size patterns, trophic structure and richness of Coleoptera in the tropical arboreal ecosystem: the fauna of the tree *Luehea seemannii* Triana and Planch in the Canal Zone of Panama. *Coleopterists Bulletin*, **34**, 305–322.

Espinosa, J. (1962). Vegetación de una corriente de lava de formación reciente localizada en el declive meridional de la Sierra del Chichinautzin. *Boletín de la Sociedad Botánica de México*, **27**, 67–114.

Espírito-Santo, M. M. & Fernandes, G. W. (1998). Pattern of abundance of *Neopalma baccharidis* Burck. (Homoptera: Psyllidae) galls on *Baccharis dracunculifolia* DC (Asteraceae). *Environmental Entomology*, **27**, 870–876.

Estrada, A. & Coates-Estrada, R. (1986). Use of leaf resources by howling monkeys (*Alouatta palliata*) and leaf-cutting ants (*Atta cephalotes*) in the tropical rain-forest of Los Tuxtlas, Mexico. *American Journal of Primatology*, **10**, 51–66.

Evans, G. O. (1992). *Principles of Acarology*. Wallingford, UK: C.A.B.

Evans, G. O. & Till, W. M. (1979). Mesostigmatic mites of Britain and Ireland (Chelicerata: Acari: Parasitioformes): an introduction to their external morphology and classification. *Transactions of the Zoological Society of London*, **35**, 139–270.

Evans, J. (1996). *Plantation Forestry in the Tropics*. Oxford, UK: Clarendon Press.

Faber, J. H. & Joosse, E. N. G. (1993). Vertical distribution of Collembola in a *Pinus nigra* organic soil. *Pedobiologia*, **37**, 336–350.

Faeth, S. H. (1980). Invertebrate predation of leaf-miners at low densities. *Ecological Entomology*, **5**, 111–114.

Faeth, S. H. (1986). Indirect interactions between temporally separated herbivores mediated by the host plant. *Ecology*, **67**, 479–494.

Faeth, S. H., Mopper, S. & Simberloff, D. (1981). Abundance and diversity of leaf-mining insects on three oak species: effects of host-plant phenology and nitrogen content of leaves. *Oikos*, **37**, 238–251.

Fagan, L. L. (1999). Arthropod Colonization of Needle Litter on the Ground and in the Canopy of Montane *Abies amabilis* Trees on Vancouver Island, British Columbia. MSc thesis. Victoria, British Columbia: University of Victoria.

Fain, A. & Galloway, T. D. (1993). Mites (Acari) from nests of sea birds in New Zealand II. Mesostigmata and Astigmata. *Bulletin de l'Institut Royal des Sciences Naturelles de Belgique*, **63**, 95–111.

Fain, A. & Philips, J. R. (1977). Astigmatic mites from nests of birds of prey in the USA. I. Description of four new species of Glycyphagidae. *International Journal of Acarology*, **3**, 105–114.

Falesi, I. & Rodrigues, B. N. S. (1969). *Os solos da área de Manaus-Itacoatiara. Série Estudados e Ensaios 1*. Belém, Pará, Brazil: Secretaria de Produção do Amazonas. IPEAN.

Faria, M. L. & Fernandes, G. W. (2001). Vigour of a dioecious shrub and attack by a galling herbivore. *Ecological Entomology*, **26**, 37–45.

Farrell, B. D. (1998). 'Inordinate Fondness' explained: why are there so many beetles? *Science*, **281**, 555–559.

Farrell, B. D., Dussourd, D. E. & Mitter, C. (1991). Escalation of plant defense: do latex and resin canals spur plant diversification? *American Naturalist*, **138**, 881–900.

Farrell, B. D. & Erwin, T. L. (1988). Leaf-beetle community structure in an amazonian rainforest canopy. In *Biology of Chrysomelidae*, eds. P. Jolivet, E. Petitpierre & T. H. Hsiao, pp. 73–90. Dordrecht, the Netherlands: Kluwer Academic.

Farrell, B. D. & Mitter, C. (1993). Phylogenetic determinants of insect/plant community diversity. In *Species Diversity in Ecological Communities: Historical and Geographical Perspectives*, eds. R. E. Ricklefs & D. Shluter, pp. 253–266. Chicago: University of Chicago Press.

Farrow, R. A. & Greenslade, P. (1992). A vertical migration of Collembola. *Entomologist*, **111**, 38–45.

Fauth, J. E., Bernardo, J., Camara, M., Resetarits, W. J., Jr, van Buskirk, J. & McCollum, S. A. (1996). Simplifying the jargon of community ecology: a conceptual approach. *American Naturalist*, **147**, 282–286.

Federle, W., Rohrseitz, K. & Hölldobler, B. (2000). Attachment forces of ants measured with a centrifuge: better 'wax-runners' have a poorer attachment to a smooth surface. *Journal of Experimental Biology*, **203**, 505–512.

Feduccia, A. (1999). *The Origin and Evolution of Birds*, 2nd edn. New Haven, CT: Yale University Press.

Feeny, P. P. (1970). Seasonal changes in oak leaf tannins and nutrients as a cause of spring feeding by winter moth caterpillars. *Ecology*, **51**, 565–581.

Feeny, P. P. (1976). Plant apparency and chemical defense. In *Recent Advances in Phytochemistry*, Vol. 10, ed. J. B. Harborne, pp. 163–206. London: Academic Press.

Feinsinger, P. (1983). Coevolution and pollination. In *Coevolution*, eds. D. J. Futuyma & M. Slatkin, pp. 283–310. Sunderland, MA: Sinauer.

Feinsinger, P. L. & Swarm, L. A. (1978). How common are ant-repellent nectars? *Biotropica*, **10**, 238–239.

Fernandes, G. W. (1998). Hypersensitivity as a phenotypic basis of plant induced resistance against a galling insect (Diptera: Cecidomyiidae). *Population Ecology*, **27**, 260–266.

Fernandes, G. W., Araújo, L. M., Carneiro, M. A. A., Cornelissen, T. G., Barcelos-Greco, M. C., Lara, A. C. F. & Ribeiro, S. P. (1997). Padrões de riqueza de insetos em gradientes altitudinais na Serra do Cipó, Minas Gerais. In *Contribuição ao conhecimento ecológico do cerrado – trabalhos selecionados do 3o Congresso de Ecologia do Brasil*, eds. L. L. Leite & C. H. Saito, pp. 191–195. Brasília: Depto Ecologia, Universidade de Brasília.

Fernandes, G. W., Carneiro, M. M. A., Lara, A. C. F., Allain, L. R., Andrade, G. I., Julião, G. R., Reis, T. R. & Silvia, I. M. (1996). Galling insects on neotropical species of *Baccharis* (Asteraceae). *Tropical Zoology*, **9**, 315–332.

Fernandes, G. W., Espírito-Santo, M. M. & Faria, M. L. (1999). Cynipid gall growth dynamics and enemy attack: effects of gall size, toughness and thickness. *Anais da Sociedade Entomológica do Brasil*, **28**, 211–218.

Fernandes, G. W. & Negreiros, D. (2001). The occurrence and effectiveness of hypersensitive reaction against galling herbivores across host taxa. *Ecological Entomology*, **26**, 46–55.

Fernandes, G. W. & Price, P. W. (1988). Biogeographical gradients in galling species richness: tests of hypotheses. *Oecologia*, **76**, 161–167.

Fernandes, G. W. & Price, P. W. (1991). Comparison of tropical and temperate galling species richness: the roles of environmental harshness and plant environmental status. In *Plant–Animal Interactions: Evolutionary Ecology in Tropical and Temperate Regions*, eds. P. W. Price, T. M. Lewinsohn, G. W. Fernandes & W. W. Benson, pp. 91–115. New York: John Wiley & Sons.

Fernandes, G. W. & Price, P. W. (1992). The adaptative significance of insect gall distribution: survivorship of species in xeric and mesic habitats. *Oecologia*, **90**, 14–20.

Fiedler, K. (1995). Lycaenid butterflies and plants: host-plant relationships, tropical versus temperate. *Ecotopica*, **1**, 51–58.

Fiedler, K. (1998). Diet breath and host plant diversity of tropical- vs. temperate-zone herbivores: South-East Asian and West Palaearctic butterflies as a case study. *Ecological Entomology*, **23**, 285–297.

Filip, V., Dirzo, R., Maass, J. M. & Sarukhán, J. (1995). Within- and among-year variation in the levels of herbivory on the foliage of trees from a Mexican tropical deciduous forest. *Biotropica*, **27**, 78–86.

Fisher, R. A., Corbet, A. S. & Williams, C. B. (1943). The relation between the number of species and the number of individuals in a random sample of an animal population. *Journal of Animal Ecology*, **12**, 42–58.

Fittkau, E. J. & Klinge, H. (1973). On biomass and trophic structure of the Central Amazonian rain forest ecosystem. *Biotropica*, **5**, 2–14.

Fjeldså, J. (1999). The impact of human, a forest disturbance, on the endemic avifauna of the Udzungwa Mountains, Tanzania. *Bird Conservation International*, **9**, 47–62.

Floren, A. (1995). *Diversität und Wiederbesiedelungsdynamik arborikoler Arthropodengemeinschaften in einem Tieflandregenwald auf Borneo, Sabah, Malaysia*. Berlin: W & T Verlag.

Floren, A., Freking, A., Biehl, M. & Linsenmair, K. E. (2001a). Anthropogenic disturbance changes the structure of arboreal tropical ant communities. *Ecography*, **24**, 547–554.

Floren, A. & Linsenmair, K. E. (1994). Zur Diversität und Wiederbesiedlungsdynamik von Arthropoden auf drei Baumarten in einem Regenwald in Sabah, Malaysia. *Andrias*, **13**, 23–28.

Floren, A. & Linsenmair, K. E. (1997a). Diversity and recolonisation dynamics of canopy arthropod communities in a lowland rain forest in Sabah, Malaysia. In *Tropical Biodiversity and Systematics. Proceedings of the International Symposium on Biodiversity and Systematics in Tropical Ecosystems*, Bonn, 1994, ed. H. Ulrich, pp. 245–249. Bonn: Zoologisches Forschunginstitut und Museum Alexander Koenig.

Floren, A. & Linsenmair, K. E. (1997b). Diversity and recolonization dynamics of selected arthropod groups on different trees species in a lowland rainforest in Sabah, Malaysia with special reference to Formicidae. In *Canopy Arthropods*, eds. N. E. Stork, J. Adis & R. K. Didham, pp. 344–381. London: Chapman & Hall.

Floren, A. & Linsenmair, K. E. (1998a). Non-equilibrium communities of Coleoptera in trees in a lowland rain forest of Borneo. *Ecotropica*, **4**, 55–67.

Floren, A. & Linsenmair, K. E. (1998b). Diversity and recolonization of arboreal Formicidae and Coleoptera in a lowland rain forest in Sabah, Malaysia. *Selbyana*, **19**, 155–161.

Floren, A. & Linsenmair, K. E. (1999). Changes in arboreal arthropod comunities along a disturbance gradient. *Selbyana*, **20**, 284–289.

Floren, A. & Linsenmair, K. E. (2000a). Do ant mosaics exist in pristine lowland rain forests? *Oecologia*, **123**, 129–137.

Floren, A. & Linsenmair, K. E. (2000b). Biodiveristät und Dschungel. In *Sieben Hügel – Bilder und Zeichen des 21. Jahrhunderts.* Band 2, pp. 33–37. Berlin: Henschel Verlag.

Floren, A. & Linsenmair, K. E. (2001). The influence of anthropogenic disturbances on the structure of arboreal arthropod communities. *Plant Ecology*, **153**, 153–167.

Floren, A., Linsenmair, K. E. & Biun, A. (1998). Structure and dynamics of arboreal arthropod communities. *Sabah Parks Nature Journal*, **1**, 69–82.

Floren, A., Riede, K. & Ingrisch, S. (2001b). Diversity of Orthoptera from Bornean lowland rain forest trees. *Ecotropica*, **7**, 33–42.

Floren, A. & Schmidl, J. (2000). Faunistisch-ökologische Ergebnisse eines Baumkronen-Benebelungsprojektes in einem Eichenhochwald des Steigerwaldes. *Beiträge zur bayerischen Entomofaunistik*, **3**, 179–195.

Flowers, R. W. & Janzen, D. H. (1997). Feeding records of Costa Rican leaf beetles (Coleoptera: Chrysomelidae). *Florida Entomologist*, **80**, 334–366.

Foggo, A., Ozanne, C. M. P., Speight, M. R. & Hambler, C. (2001). Edge effects and tropical forest canopy invertebrates. *Plant Ecology*, **153**, 347–359.

Forget, P.-M. (1994). Regeneration pattern of *Vouacapoua americana* (Caesalpiniaceae), a rodent-dispersed tree species in French Guiana. *Biotropica*, **26**, 420–426.

Foster, R. B. (1974). Seasonality of Fruit Production and Seed Fall in a Tropical Forest Ecosystem in Panama. PhD Thesis. Duke University, Durham, NC.

Foster, R. B. (1982). The seasonal rhythm of fruitfall on Barro Colorado Island. In *The Ecology of a Tropical Forest*, eds. E. G. Leigh Jr, A. S. Rand & D. M. Windsor, pp. 67–81. Washington, DC: Smithsonian Institution Press.

Fowler, S. V. (1985). Difference in insect species richness and faunal composition of birch seedlings, saplings and trees: the importance of plant architecture. *Ecological Entomology*, **10**, 159–169.

Fowler, S. V. & Lawton, J. H. (1982). The effects of host plant distribution and local abundance on the species richness of agromyzid flies attacking British umbellifers. *Ecological Entomology*, **7**, 257–265.

Fox, L. R. (1988). Diffuse coevolution within complex communities. *Ecology*, **69**, 906–907.

Fox, L. R. & Morrow, P. A. (1981). Specialization: species property or local phenomenon ? *Science*, **211**, 887–893.

Frangi, J. L. & Lugo, A. E. (1991). Hurricane damage to a flood plain forest in the Luquillo Mountains of Puerto Rico. *Biotropica*, **23**, 324–335.

Frank, J. H. (1999). Bromeliad-eating weevils. *Selbyana*, **20**, 40–48.

Frankie, G. W. (1975). Tropical forest phenology and pollinator plant coevolution. In *Coevolution of Animals and Plants*, eds. L. E. Gilbert & P. H. Raven, pp. 192–209. Austin: University of Texas Press.

Frankie, G. W., Baker, H. O. & Opler, P. A. (1974). Comparative phenological studies of trees in tropical wet and dry forests in the lowlands of Costa Rica. *Journal of Ecology*, **62**, 881–919.

Frankie, G. W., Haber, W. A., Opler, P. A. & Bawa, K. S. (1983). Characteristics and organization of the large bee pollination system in the Costa Rican dry forest. In *Handbook of Experimental Pollination Biology*, eds. C. E. Jones & R. J. Little, pp. 411–447. New York: Van Nostrand Reinhold.

Franklin, E., Adis, J. & Woas, S. (1997). The oribatid mites. In *The Central Amazon Floodplain*, ed. W. J. Junk, pp. 331–349. Berlin: Springer.

Franklin, E. N., Woas, S., Schubart, H. O. R. & Adis, J. A. (1998). Ácaros oribatídeos (Acari: Oribatida) arborícolas de duas florestas inundáveis da Amazônia Central. *Revista Brasilia de Biologia*, **58**, 317–335.

Freitas, R. O. (1951). Ensaio sôbre o relêvo tectônico do Brasil. *Revista Brasileira de Geografia*, **30**, 3–52.

Fretwell, S. D. (1972). *Populations in a Seasonal Environment*. New Jersey: Princeton University Press.

Fréty, T. & Dewynter, M. (1998). Amphibiens Anoures de la Forêt des Abeilles (Gabon). *Journal of African Zoology*, **112**, 171–184.

Freude, H., Harde, K. W. & Lohse, G. A. (1965–1983). *Die Käfer Mitteleuropas*. Krefeld: Goecke & Evers.

Frith, C. B. & Frith, D. W. (1985). Seasonality of insect abundance in an Australian upland tropical rainforest. *Australian Journal of Ecology*, **10**, 237–248.

Fritz, R. S. (1992). Community structure and species interactions of phytophagous insects on resistant and susceptible host plants. In *Plant Resistance to Herbivores and Pathogens. Ecology, Evolution and Genetics*, eds. R. S. Fritz & E. L. Simms, pp. 240–277. Chicago: University of Chicago Press.

Fujikawa, T. (1974). Comparison among oribatid faunas from different microhabitats in forest floors. *Applied Entomological Zoology*, **9**, 105–114.

Funke, W. (1979). Wälder, Objekte der Ökosystemforschung. Die Stammregion – Lebensraum und Durchgangszone von Arthropoden. *Jahresberichte des naturwissenschaftlichen Vereins zu Wuppertal*, **32**, 45–50.

Futuyma, D. J. & Gould, F. (1979). Associations of plants and insects in a deciduous forest. *Ecological Monographs*, **49**, 33–50.

Futuyma, D. J. & Mayer, G. C. (1980). Non-allopatric speciation in animals. *Systematic Zoology*, **29**, 254–271.

Futuyma, D. J. & Moreno, G. (1988). The evolution of ecological specialization. *Annual Review of Ecology and Systematics*, **19**, 207–233.

Gagné, W. C. (1979). Canopy-associated arthropods in *Acacia koa* and *Metrosideros* tree communities along an altitudinal transect on Hawaii Island. *Pacific Insects*, **21**, 56–82.

Gagné, W. C. (1981). Canopy-associated arthropods. In *Island Ecosystems: Biological Organization in Selected Hawaiian Communities*, eds. D. Mueller-Dombois, K. W. Bridges & H. L. Carson, pp. 118–127. Stroudsburg, PA: Hutchinson Ross.

Gagné, W. C. & Howarth, F. G. (1981). Arthropods associated with foliar crowns of structural dominants. In *Island Ecosystems: Biological Organization in Selected Hawaiian Communities*, eds. D. Mueller-Dombois, K. W. Bridges & H. L. Carson, pp. 273–288. Stroudsburg, PA: Hutchinson Ross.

Gagné, W. C. & Martin, J. L. (1968). The insect ecology of red pine plantations in Central Ontario. V. The Coccinellidae (Coleoptera). *Canadian Entomologist*, **100**, 835–846.

Galindo, P., Trapido, H., Carpenter, S. J. & Blanton, F. S. (1956). The abundance cycles of arboreal mosquitoes during six years at a sylvan yellow fever locality in Panama. *Annals of the Entomological Society of America*, **49**, 543–547.

García, E. (1988). *Modificaciones al sistema de clasificación climática de Köppen. Tercera edición*. México: Instituto de Geografía, Universidad Nacional Autónoma de México.

García-Oliva, F., Ezcurra, E. & Galicia, L. (1991). Pattern of rain distribution in the central Pacific coast of Mexico. *Geography Annals*, **73-A**, 179–186.

Gardette, E. (1996). The effect of timber logging on the diversity of woody climbers at Pasoh. In *Conservation, Management and Development of Forest Resources. Proceedings of the Malaysia–United Kingdom Programme Workshop, 1996*, eds. L. S. See, D. Y. May, I. D. Gauld & J. Bishop, pp. 115–126. Kuala Lumpur: Malaysia–United Kingdom Programme.

Gascon, C., Lovejoy, T. E., Bierregaard, R. O., Jr, Malcolm, J. R., Stouffer, P. C., Vasconcelos, H. L., Laurance, W. F., Zimmerman, B., Tocher, M. & Borges, S. (1999). Matrix habitat and species richness in tropical forest remnants. *Biological Conservation*, **91**, 223–230.

Gascon, C., Malcom, J. R., Patton, J. L., da Silva, M. N. F., Bogart, J. P., Lougheed, S. C., Peres, C. A., Neckel, S. & Boag, P. T. (2000). Riverine barriers and the geographic distribution of Amazonian species. *Proceedings of the National Academy of Sciences of the United States of America*, **97**, 13672–13677.

Gaston, K. J. (1991a). Estimates of the near-imponderable: a reply. *Conservation Biology*, **5**, 564–566.

Gaston, K. J. (1991b). The magnitude of global insect species richness. *Conservation Biology*, **5**, 283–296.

Gaston, K. J. (1992). Regional number of insects and plant species. *Functional Ecology*, **6**, 243–247.

Gaston, K. J. (1993). Herbivory at the limits. *Trends in Ecology and Evolution*, **8**, 193–194.

Gaston, K. J. (2000). Global patterns in biodiversity. *Nature*, **405**, 220–227.

Gaston, K. J. & Blackburn, T. M. (2000). *Pattern and Process in Macroecology*. Oxford: Blackwell.

Gaston, K. J., Warren, P. H. & Hammond, P. M. (1992). Predator–non-predator ratios in beetle assemblages. *Oecologia*, **90**, 417–421.

Gaston, K. J. & Williams, P. H. (1993). Mapping the world's species – the higher taxon approach. *Biodiversity Letters*, **1**, 2–8.

Gaud, J. & Atyeo, W. T. (1996). *Feather Mites of the World (Acarina: Astigmata): The Supraspecific Taxa*. Tervuren, Belgium: Musee Royal de l'Afrique Centrale.

Gauer, U. (1997a). The Collembola. In *The Central Amazon Floodplain*, ed. W. Junk, pp. 351–359. Berlin: Springer.

Gauer, U. (1997b). Collembola in Central Amazon inundation forests – strategies for surviving floods. *Pedobiologia*, **41**, 69–73.

Gauld, I. D. (1988). A survey of the Ophioninae (Hymenoptera: Ichneumonidae) of tropical Mesoamerica with special reference to the fauna of Costa Rica. *Bulletin of the British Museum (Natural History)*, **57**, 1–309.

Gauld, I. D. (1991). The Ichneumonidae of Costa Rica, 1. *Memoirs of the American Entomological Institute*, **47**, 1–589.

Gauld, I. D. (1997). The Ichneumonidae of Costa Rica, 2. *Memoirs of the American Entomological Institute*, **57**, 1–485.

Gauld, I. D. (2000). The Ichneumonidae of Costa Rica, 3. *Memoirs of the American Entomological Institute*, **63**, 1–400.

Gauld, I. D., Gaston, K. G. & Janzen, D. H. (1992). Plant allelochemicals, tritrophic interactions and the anomalous diversity of tropical parasitoids: the 'nasty' host hypothesis. *Oikos*, **65**, 353–357.

Gauld, I. D. & Janzen, D. H. (1994). The classification, evolution and biology of the Costa Rican species of *Cryptophion* (Hymenoptera: Ichneumonidae). *Zoological Journal of the Linnean Society*, **110**, 297–324.

Gauld, I. D. & Ward, S. (1987). The callajoppine parasitoids of sphingids in Central America (Hymenoptera: Ichneumonidae). *Systematic Entomology*, **12**, 503–508.

Gautier-Hion, A. & Michaloud, G. (1989). Are figs always keystone resources for tropical frugivorous vertebrates? A test in Gabon. *Ecology*, **70**, 1826–1833.

Geiser, R. (1994). Artenschutz für holzbewohnende Käfer. *Berichte der Akademie für Natur und Landschaft*, **18**, 89–114.

Gentili, P. & Solis, M. A. (1998). Checklist and key of New World species of *Omiodes* Guenée with descriptions of four new Costa Rican species (Lepidoptera: Crambidae). *Entomologica Scandinavica*, **28**, 471–492.

Gentry, A. H. (1974). Coevolutionary patterns in Central American Bignoniaceae. *Annals of the Missouri Botanical Garden*, **61**, 728–759.

Gentry, A. H. (1982). Neotropical floristic diversity: Phytogeographical connections between Central and South America, Pleistocene climatic fluctuations, or an accident of the Andean orogeny? *Annals of the Missouri Botanical Garden*, **69**, 557–593.

Gentry, A. H. (ed.) (1990). *Four Neotropical Rainforests*. New Haven, CA: Yale University Press.

Gentry, A. H. & Dobson, C. (1987). The contribution of non-trees to species richness of a tropical rain forest. *Biotropica*, **19**, 149–156.

Gentry, A. H. & Emmons, L. H. (1987). Geographical variation in fertility, phenology, and composition of the understorey of neotropical forests. *Biotropica*, **19**, 216–227.

Gerrish, G. (1989). Comparing crown growth and phenology of juvenile, early mature, and late mature *Metrosideros polymorpha* trees. *Pacific Science*, **43**, 211–223.

Gibbs, D. G., Pickett, A. D. & Leston, D. (1968). Seasonal population changes in cocoa capsids (Hemiptera, Miridae) in Ghana. *Bulletin of Entomological Research*, **58**, 279–293.

Gibernau, M. & Dejean, A. (2001). Ant protection of a Heteroptera trophobiont against a parasitoid wasp. *Oecologia*, **126**, 53–57.

Gibson, W. W. (1947). An ecological study of the spiders of a river terrace forest in western Tennessee. *Ohio Journal of Science*, **47**, 38–44.

Gilbert, L. E. (1983). Co-evolution and mimicry. In *Co-evolution*, eds. D. J. Futuyma & M. Slatkin, pp. 263–281. Sunderland, MA: Sinauer.

Gillespie, R. G., Croom, H. B. & Hasty, G. L. (1997). Phylogenetic relationships and adaptive shifts among major clades of *Tetragnatha* spiders (Araneae: Tetragnathidae) in Hawai'i. *Pacific Science*, **51**, 380–394.

Gilyarov, M. S. & Bregetova, N. G. (1977). *Description of Soil-Dwelling Acarines – Mesostigmata*. Leningrad: Nauka.

Gisin, H. (1943). Ökologie und Lebensgemeinschaften der Collembolen im schweizerischen Exkursionsgebiet Basels. *Revue Suisse de Zoologie*, **50**, 131–224.

Giulietti, A. M. & Pirani, J. R. (1988). Patterns of geographic distribution of some plant species from the Espinhaço range, Minas Gerais and Bahia, Brazil. In *Proceedings of a Workshop on Neotropical Distribution Patterns*, eds. P. E. Vanzolini & W. R. Heyer, pp. 39–69. Rio de Janeiro: Academia Brasileira de Ciência.

Gjelstrup, P. (1979). Epiphytic cryptostigmatic mites on some beech- and birch-trees in Denmark. *Pedobiologia*, **19**, 1–8.

Glantz, S. (1992). *A Primer of Biostatistics*. New York: McGraw-Hill.

Godfray, H. C. (1985). The absolute abundance of leaf miners on plants of different successional stages. *Oikos*, **45**, 17–25.

Godfray, H. C., Lewis, O. T. & Memmott, J. (1999). Studying insect diversity in the tropics. *Philosophical Transactions of the Royal Society, Biological Sciences*, **354**, 1811–1824.

Gómez-Pompa, A. (1971). Possible papel de la vegetación secundaria en la evolución de la flora tropical. *Biotropica*, **3**, 125–135.

Gómez-Pompa, A. & Dirzo, R. (1995). *Reservas de la biosfera y otras áreas naturales protegidas de México*. México: Instituto Nacional de Ecología y Consejo Nacional para el Conocimiento y Uso de la Biodiversidad.

Gonzalez, A., Lawton, J. H., Gilbert, F. S., Blackburn, T. M. & Evans-Freke, I. (1998). Metapopulation dynamics, abundance, and distribution in a microecosystem. *Science*, **281**, 2045–2047.

Goodland, R. (1971). A physiognomic analysis of the 'cerrado' vegetation of central Brasil. *Journal of Ecology*, **59**, 411–419.

Goodland, R. & Ferri, M. G. (1979). *Ecologia do Cerrado*. São Paulo: Editora da Universidade de São Paulo.

Goodland, R. & Pollard, R. (1973). The Brazilian cerrado vegetation: a fertility gradient. *Journal of Ecology*, **61**, 219–224.

Gottsberger, G. (1989a). Comments on flower evolution and beetle pollination in the genera *Annona* and *Rollinia* (Annonaceae). *Plant Systematics and Evolution*, **167**, 189–194.

Gottsberger, G. (1989b). Beetle pollination and flowering rhythm of *Annona* spp. (Annonaceae) in Brazil. *Plant Systematics and Evolution*, **167**, 165–187.

Gottsberger, G. (1990). Flowers and beetles in the South American tropics. *Acta Botanica*, **103**, 360–365.

Gottsberger, G. & Silberbauer-Gottsberger, I. (1991). Olfactory and visual attraction of *Erioscelis emarginata* (Cyclocephalini, Dynastinae) to the inflorescences of *Philodendron selloum* (Araceae). *Biotropica*, **23**, 23–28.

Gotwald, W. H., Jr (1996). Mites that live with army ants: a natural history of some myrmecophilous hitch-hikers, browsers, and parasites. *Journal of the Kansas Entomological Society*, **69**, 232–237.

Götzke, A. (1993). Ameisenzönosen ausgewählter tropischer Baumkronen: Struktur, Diversität und Ressourcennutzung der Gemeinschaft. Diploma thesis. Würzburg: University Würzburg.

Götzke, A. & Linsenmair, K. E. (1996). Does beta-diversity allow reliable discrimination between deterministically and stochastically assembled communities? *Ecotropica*, **2**, 79–82.

Gould, F. (1979). Rapid host range evolution in population of the phytophagous mite *Tetranychus urticae* Koch. *Evolution*, **33**, 791–802.

Grant, S. & Moran, V. C. (1986). The effects of foraging ants on arboreal insect herbivores in an undisturbed woodland savanna. *Ecological Entomology*, **11**, 83–93.

Grassle, J. F. & Smith, W. (1976). A similarity measure sensitive to the contribution of rare species and its use in investigation of variation in marine benthic communities. *Oecologia*, **25**, 13–22.

Greenslade, P. J. M. (1983). Adversity selection and the habitat templet. *American Naturalist*, **122**, 352–365.

Greenstone, M. H. (1982). Determinants of web species diversity: vegetation structural diversity vs. prey availability. *Oecologia*, **62**, 299–304.

Griffiths, D. (1997). Local and regional species richness in North American lacustrine fish. *Journal of Animal Ecology*, **66**, 49–56.

Gross, N. D., Torti, S. D., Feener, D. H. & Coley, P. D. (2000). Monodominance in an African rain forest: is reduced herbivory important? *Biotropica*, **32**, 430–439.

Gross, S. W. & Fritz, R. S. (1982). Differential stratification, movement and parasitism of sexes of the bagworm, *Thyridopteryx ephemeraeformis* on redcedar. *Ecological Entomology*, **7**, 149–154.

Grove, S. J. & Stork, N. E. (2000). An inordinate foundness for beetles. *Invertebrate Taxonomy*, **14**, 733–739.

Grover, J. P. (1997). *Resource Competition. Population and Community Biology Series, No.19*. London: Chapman & Hall.

Grüne, S. (1979). *Brief Illustrated Key to European Bark Beetles*. Hannover, Germany: Verlag & Schaper.

Guerrant, E. O., Jr & Fiedler, P. L. (1981). Flower defense against nectar-pilferage by ants. *Biotropica Suppl. Reproductive Botany*, **13**, 25–33.

Guilbert, E., Baylac, M. & Najt, J. (1995). Canopy arthropod diversity in a New Caledonian primary forest sampled by fogging. *Pan-Pacific Entomologist*, **71**, 3–12.

Gullan, P. J. & Cranston, P. S. (2000). *The Insects: An Outline of Entomology*, 2nd edn. Oxford: Blackwell Science.

Gunnarsson, B. (1990). Vegetation structure and the abundance and size distribution of spruce-living spiders. *Journal of Animal Ecology*, **59**, 743–752.

Haarlov, N. (1960). Microarthropods from Danish soils. Ecology, phenology. *Oikos Supplement*, **3**, 1–176.

Häckel, H. (1993). *Metereologie*. Stuttgart: UTB-Ulmer.

Haddow, A. J. (1961). Entomological studies from a high tower in Mpanga Forest, Uganda. VII. The biting behaviour of mosquitoes and tabanids. *Transactions of the Royal Entomological Society of London*, **113**, 315–335.

Haddow, A. J. & Corbet, P. S. (1961a). Entomological studies from a high tower in Mpanga Forest, Uganda. V. Swarming activity above the forest. *Transactions of the Royal Entomological Society of London*, **113**, 284–300.

Haddow, A. J. & Corbet, P. S. (1961b). Entomological studies from a high tower in Mpanga Forest, Uganda. II. Observations on certain environmental factors at different levels. *Transactions of the Royal Entomological Society of London*, **113**, 257–269.

Haddow, A. J., Corbet, P. S. & Gillett, J. D. (1961). Entomological studies from a high tower in Mpanga Forest, Uganda. I. Introduction. *Transactions of the Royal Entomological Society of London*, **113**, 249–256.

Haddow, A. J. & Ssenkubuge, Y. (1965). Entomological studies from a high steel tower in Zika Forest, Uganda. Part I. The biting activity of mosquitoes and tabanids as shown by twenty-four-hour catches. *Transactions of the Royal Entomological Society of London*, **117**, 215–243.

Hadwen, W. L., Kitching, R. L. & Olsen, M. F. (1998). Folivory levels of seedlings and canopy trees in tropical and subtropical rainforests in Australia. *Selbyana*, **19**, 162–171.

Halaj, J., Ross, D. W. & Moldenke, A. R. (1998). Habitat structure and prey availability as predictors of the abundance and community organization of spiders in western Oregon forest canopies. *Journal of Arachnology*, **26**, 203–220.

Halaj, J., Ross, D. W. & Moldenke, A. R. (2000). Importance of habitat structure to the arthropod food-web in Douglas-fir canopies. *Oikos*, **90**, 139–152.

Hallé, F. (1995). Canopy architecture in tropical trees: a pictorial approach. In *Forest Canopies*, eds. M. D. Lowman & N. M. Nadkarni, pp. 27–44. San Diego, CA: Academic Press.

Hallé, F. (1998). Distribution verticale des métabolites secondaires en forêt équatoriale – une hypothèse. In *Biologie d'une Canopée de Forêt Equatoriale – III. Rapport de la Mission d' Exploration Scientifique de la Canopée de Guyane, Octobre–Décembre 1996*, ed. F. Hallé, pp. 129–138. Paris: Pro-Natura International & Opération Canopée.

Hallé, F. (ed.) (2000). *Biologie d'une Canopée de Forêt Tropicale – IV. Rapport de la Mission du Radeau des Cimes à la Makandé, Forêt des Abeilles, Gabon, Janvier–Mars 1999*. Paris: Pro-Natura International & Opération Canopée.

Hallé, F. & Blanc, P. (eds.) (1990). *Biologie d'une Canopée de Forêt Equatoriale. Rapport de Mission. Radeau des Cimes Octobre-Novembre 1989, Guyane Française*. Montpellier/Paris: Montpellier II et CNRS-Paris VI.

Hallé, F., Cleyet-Marrell, D. & Ebersolt, G. (2000). *Le Radeau des Cimes. L' Exploration des Canopées Forestières*. Paris: Jean-Claude Lattès.

Hallé, F., Oldeman, R. A. A. & Tomlinson, P. B. (1978). *Tropical Trees and Forests, an Architectural Analysis*. Berlin: Springer-Verlag.

Halliday, R. B. (1998). *Mites of Australia: A Checklist and Bibliography*. Melbourne: CSIRO Publishing.

Hamilton, W. D. (1978). Evolution and diversity under bark. In *Diversity of Insect Faunas*, eds. L. A. Mound & N. Waloff, pp. 154–175. Oxford: Blackwell Scientific.

Hamilton, W. D. (1996). Funeral feasts: evolution and diversity under bark. In *Narrow Roads of Gene Land: Vol 1. Evolution of Social Behaviour*, ed. W. D. Hamilton, pp. 386–420. Oxford: Freeman.

Hammel, B. A. (1986). The vascular flora of La Selva Biological Station, Costa Rica. Cecropiaceae. *Selbyana*, **9**, 192–195.

Hammer, M. (1972). Microhabitats of oribatid mites on a Danish woodland floor. *Pedobiologia*, **12**, 412–423.

Hammond, P. M. (1990). Insect abundance and diversity in the Dumoga-Bone National Park, N. Sulawesi, with special reference to the beetle fauna of lowland rain forest in the Toraut region. In *Insects and the Rain Forests of South East Asia (Wallacea)*, eds. W. J. Knight & J. D. Holloway, pp. 197–254. London: The Royal Entomological Society of London.

Hammond, P. M. (1992). Species inventory. In *Global Biodiversity, Status of the Earth's Living Resources*, ed. B. Groombridge, pp. 17–39. London: Chapman & Hall.

Hammond, P. M. (1994). Practical approaches to the estimation of the extent of biodiversity in speciose groups. *Philosophical Transactions of the Royal Society of London, Series B*, **345**, 119–136.

Hammond, P. M. (1995). Magnitude and distribution of biodiversity. In *Global Biodiversity Assessment*, eds. V. T. Heywood & R. T. Watson, pp. 113–138. Cambridge: Cambridge University Press.

Hammond, P. M., Kitching, R. L. & Stork, N. E. (1996). The composition and richness of the tree-crown Coleoptera assemblage in an Australian subtropical forest. *Ecotropica*, **2**, 99–108.

Hammond, P. M., Stork, N. E. & Brendell, M. J. D. (1997). Tree-crown beetles in context: a comparison of canopy and other ecotone assemblages in a lowland tropical forest in Sulawesi. In *Canopy Arthropods*, eds. N. E. Stork, J. Adis & R. K. Didham, pp. 184–223. London: Chapman & Hall.

Hanski, I. & Gyllenberg, M. (1997). Uniting two general patterns in the distribution of species. *Science*, **275**, 397–400.

Hanski, I. & Hammond, P. M. (1995). Biodiversity in boreal forests. *Trends in Ecology and Evolution*, **10**, 5–6.

Hanski, I. & Niemelä, J. (1990). Elevational distributions of dung and carrion beetles in northern Sulawesi. In *Insects and the Rain Forests of South East Asia (Wallacea)*, eds. W. J. Knight & J. D. Holloway, pp. 145–152. London: The Royal Entomological Society of London.

Harada, A. Y. & Adis, J. (1997). The ant fauna of tree canopies in central Amazonia: a first assessment. In *Canopy Arthropods*, eds. N. E. Stork, J. Adis & R. K. Didham, pp. 382–400. London: Chapman & Hall.

Harada, A. Y. & Adis, J. (1998). Ants obtained from trees of a 'Jacareuba' (*Calophyllum brasiliense*) forest plantation in central Amazonia by canopy fogging: first results. *Acta Amazonica*, **28**, 309–318.

Harborne, J. B. (1980). Plant phenolics. In *Secondary Plant Products: Encyclopedia of Plant Physiology*. Vol. 8, eds. E. A. Bell & B. V. Charlwood, pp. 129–402. Berlin: Springer Verlag.

Hart, T. B. (1990). Monospecific dominance in tropical rain forests. *Trends in Ecology and Evolution*, **5**, 6–11.

Hartley, S. (1998). A positive relationship between local abundance and regional occupancy is almost inevitable (but not all positive relationships are the same). *Journal of Animal Ecology*, **67**, 992–994.

Harvey, P. H. & Pagel, M. D. (1991). *The Comparative Method in Evolutionary Biology*. Oxford: Oxford University Press.

Haslam, E. (1988). Plant polyphenols (syn. vegetable tannins) and chemical defence – a reappraisal. *Journal of Chemical Ecology*, **14**, 1789–1805.

Hassall, M., Visser, S. & Parkinson, D. (1986). Vertical migration of *Onychiurus subtenuis* (Collembola) in relation to rainfall and microbial activity. *Pedobiologia*, **29**, 175–182.

Hawksworth, D. L. & Ritchie, J. M. (1993). *Biodiversity and Biosystematic Priorities: Microorganisms and Invertebrates*. Wallingford, UK: CAB International.

Hawthorne, D. J. & Via, S. (2001). Genetic linkage of ecological specialization and reproductive isolation in pea aphids. *Nature*, **412**, 904–907.

Hayek, L.-A. C. & Buzas, M. A. (1997). *Surveying Natural Populations*. New York: Columbia University Press.

Heatwole, H., Lowman, M. D., Donovan, C. & McCoy, M. (1997). Phenology of leaf-flushing and macroarthropod abundances in canopies of *Eucalyptus* saplings. *Selbyana*, **18**, 200–214.

Heck, K. L. J., van Belle, G. & Simberloff, D. (1975). Explicit calculation of the rarefaction diversity measurement and the determination of sufficient sample size. *Ecology*, **56**, 1459–1461.

Hegarty, E. E. & Caballé, G. (1991). Distribution and abundance of vines in forest communities. In *The Biology of Vines*, eds. F. E. Putz & H. A. Mooney, pp. 313–335. Cambridge: Cambridge University Press.

Hemming, J. D. C. & Lindroth, R. L. (1995). Intraspecific variation in aspen phytochemistry: Effects on performance of gypsy moth and forest tent caterpillars. *Oecologia*, **103**, 79–88.

Henderson, P. A. & Seaby, R. M. H. (1998). *Species Diversity and Richness*, version 2.62. Pennington, UK: Pisces Conservation Ltd.

Herbert, D. A. & Fownes, J. H. (1999). Forest productivity and efficiency of resource use across a chronosequence of tropical montane soils. *Ecosystems*, **2**, 242–254.

Herms, D. A. & Mattson, W. J. (1992). The dilemma of plants: to grow or defend. *Quarterly Review in Biology*, **67**, 283–335.

Herrera, C. M., Herrera, J. & Espadaler, X. (1984). Nectar thievery by ants from southern Spanish insect-pollinated flowers. *Insectes Sociaux*, **31**, 142–154.

Herrera, C. M., Jordano, P., Guitian, J. & Traveset, A. (1998). Annual variability in seed production by woody plants and the masting concept: reassessment of the principles and relationship to pollination and seed dispersal. *American Naturalist*, **152**, 576–594.

Hertzberg, K. (1997). Migration of Collembola in a patchy environment. *Pedobiologia*, **41**, 494–505.

Hertzberg, K., Leinaas, H. P. & Ims, R. A. (1994). Patterns of abundance and demography: Collembola in a habitat patch gradient. *Ecography*, **17**, 349–359.

Hertzberg, K., Yoccoz, N. G., Ims, R. A. & Leinaas, H. F. (2000). The effects of spatial habitat configuration on recruitment, growth and population structure in Arctic Collembola. *Oecologia*, **124**, 381–390.

Hespenheide, H. A. (1991). Bionomics of leaf-mining insects. *Annual Review of Entomology*, **36**, 535–560.

Hespenheide, H. A. (1993). An overview of faunal studies. In *La Selva. Ecology and Natural History of a Neotropical Rain Forest*, eds. L. A. Mcdade, K. S. Bawa, H. A. Hespenheide & G. S. Hartshorn, pp. 238–243. Chicago: University of Chicago Press.

Hesse, R. (1924). *Tiergeographie auf ökologischer Grundlage*. Jena: Fischer.

Hijii, N. (1983). Arboreal arthropod fauna in a forest. I. Preliminary observation on seasonal fluctuations in density, biomass and faunal composition in a *Chamaecyparis obtusa* plantation. *Japanese Journal of Ecology*, **33**, 435–444.

Hijii, N. (1987). Seasonal changes in abundance and spatial distribution of the soil arthropods in a Japanese cedar (*Cryptomeria japonica* D. Don) plantation, with special reference to Collembola and Acarina. *Ecological Research*, **2**, 159–173.

Hill, C. J. & Cermak, M. (1997). A new design and some preliminary results for a flight intercept trap to sample forest canopy arthropods. *Australian Journal of Entomology*, **36**, 51–55.

Hill, J. K. (1999). Butterfly spatial distribution and habitat requirements in a tropical forest: impacts of selective logging. *Journal of Applied Ecology*, **36**, 364–372.

Hill, J. K., Hamer, K. C., Lace, L. A. & Banham, W. M. T. (1995). Effects of selective logging on tropical forest butterflies on Buru, Indonesia. *Journal of Applied Ecology*, **32**, 754–760.

Hill, K. D. (1998). 'Gymnosperms' – the paraphyletic stem of seed plants. *Flora of Australia*, **48**, 505–526.

Hilliard, J. R. (1982). Endophyitic oviposition by *Leptysma marginicollis marginicollis* and *Stenacris vitreipennis* (Orthoptera: Acrididae: Leptysminae) with life history notes. *Transactions of the American Entomological Society*, **108**, 153–180.

Hingston, R. W. G. (1930). The Oxford University Expedition to British Guiana. *Geographical Journal*, **76**, 1–24.

Hingston, R. W. G. (1932). *A Naturalist in the Guiana Forest*. London: Edward Arnold.

Hodges, R. W. (1985). A new species of *Dichomeris* from Costa Rica (Lepidoptera: Gelechiidae). *Proceedings of the Entomological Society of Washington*, **87**, 456–459.

Hodkinson, I. D. & Casson, D. (1991). A lesser predilection for bugs: Hemiptera (Insecta) diversity in tropical rain forests. *Biological Journal of the Linnean Society*, **43**, 101–109.

Hodkinson, I. D. & Casson, D. S. (2000). Patterns within patterns: abundance–size relationships within the Hemiptera of tropical rain forest or why phylogeny matters. *Oikos*, **88**, 509–514.

Höfer, H., Brescovit, A. D., Adis, J. & Paarman, W. (1994). The spider fauna of neotropical tree canopies in Central Amazonia: first results. *Studies on Neotropical Fauna and Environment*, **29**, 23–32.

Holdridge, L. R. (1978). *Ecología Basada en Zonas de Vida*. San José, Costa Rica: Instituto Interamericano de Cooperación para la Agricultura.

Holdridge, L. R., Grenke, W. H., Hatheway, W. H., Liang, T. & Tosi, J. J. A. (1971). *Forest Environments in Tropical Life Zones, a Pilot Study*. Oxford: Pergamon Press.

Hölldobler, B. (1979). Territories of the African weaver ant (*Oecophylla longinoda* Latreille): a field study. *Zeitschrift für Tierpsychologie*, **51**, 201–213.

Hölldobler, B. (1983). Territorial behavior of the green tree ant (*Oecophylla longinoda*). *Biotropica*, **15**, 241–250.

Hölldobler, B. & Lumdsen, C. J. (1980). Territorial strategies in ants. *Science*, **210**, 732–739.

Hölldobler, B. & Wilson, E. O. (1977). Colony-specific territorial pheromone in the African weaver ant *Oecophylla longinoda* (Latreille). *Proceedings of the National Academy of Sciences USA*, **74**, 2072–2075.

Hölldobler, B. & Wilson, E. O. (1978). The multiple recruitment systems of the African weaver ant *Oecophylla longinoda* (Latreille) (Hymenoptera: Formicidae). *Behavioral Ecology and Sociobiology*, **3**, 19–60.

Hölldobler, B. & Wilson, E. O. (1990a). *The Ants*. Cambridge, MA: Belknap Press of Harvard University.

Hölldobler, B. & Wilson, E. O. (1990b). Host tree selection by the neotropical ant *Paraponera clavata* (Hymenoptera: Formicidae). *Biotropica*, **22**, 213–214.

Hollier, J. A. & Belshaw, R. D. (1993). Stratification and phenology of a woodland Neuroptera assemblage. *Entomologist*, **112**, 169–175.

Holloway, J. D. (1984a). Moths as indicator organisms for categorizing rain-forest and monitoring changes and regulation processes. In *Tropical Rain-Forest: The Leeds Symposium*, eds. A. C. Chadwick & S. L. Sutton, pp. 235–242. Leeds: Leeds Philosophical and Literary Society.

Holloway, J. D. (1984b). The larger moths of the Gunung Mulu National Park; a preliminary assessment of their distribution, ecology, and potential as environmental indicators. *Sarawak Museum Journal*, **30** (Special Issue 2), 149–190.

Holloway, J. D. (1984c). Notes on the butterflies of the Gunung Mulu National Park. *Sarawak Museum Journal*, **30** (Special Issue 2), 89–131.

Holloway, J. D. (1998). The impact of traditional and modern cultivation practices, including forestry, on Lepidoptera diversity in Malaysia and Indonesia. In *Dynamics of Tropical Ecosystems*, eds. N. Brown, D. M. Newbery & H. Prins, pp. 567–597. Oxford: Blackwell Scientific.

Holloway, J. D., Robinson, G. S. & Tuck, K. R. (1990). Zonation in the Lepidoptera of northern Sulawesi. In *Insects and the Rain Forests of South East Asia*

(Wallacea), eds. W. J. Knight & J. D. Holloway, pp. 153–166. London: The Royal Entomological Society of London.

Honciuc, V. (1996). Laboratory studies of the behaviour and life cycle of *Archegozetes longisetosus* Aoki 1965 (Oribatida). In *Proceedings IX International Acarology Congress*, Columbus, Ohio, Vol. 1, eds. R. Mitchell, D. J. Horn, G. R. Needham & W. C. Welbourn, pp. 637–640. Columbus: Ohio Biological Survey.

Hood, W. G. & Tschinkel, W. R. (1990). Desiccation resistance in arboreal and terrestrial ants. *Physiological Entomology*, **15**, 23–35.

Hooper, D. U., Bignell, D. E., Brown, V. K., Brussaard, L., Dangerfield, J. M., Wall, D. H., Wardle, D. A., Coleman, D. C., Giller, K. E., Lavelle, P., van der putten, W. H., De Ruiter, P. C., Rusek, J., Silver, W. L., Tiedje, J. M. & Wolters, V. (2000). Interactions between aboveground and belowground biodiversity in terrestrial ecosystems: patterns, mechanisms, and feedbacks. *BioScience*, **50**, 1049–1061.

Hopkin, S. P. (1997). *Biology of Springtails (Insecta: Collembola)*. Oxford: Oxford University Press.

Hopkins, M. J. G. (1983). Unusual diversities of seed beetles (Coleoptera : Bruchidae) on *Parkia* (Leguminosae: Mimosoideae) in Brazil. *Biological Journal of the Linnean Society*, **19**, 329–338.

Horstmann, K., Floren, A. & Linsenmair, K. E. (1999). High species richness of Ichneumonidae (Hymenoptera) from the canopy of a Malaysian rain forest. *Ecotropica*, **5**, 1–12.

Hossaert-McKey, M., Orivel, J., Labeyrie, E., Pascal, L., Delabie, J. H. C. & Dejean, A. (2001). Differential associations with ants of three co-occurring extrafloral nectary-bearing plants. *Ecoscience*, **8**, 325–335.

Hotchkiss, S., Vitousek, P. M., Chadwick, O. A. & Price, J. (2000). Climate cycles, geomorphological change, and the interpretation of soil and ecosystem development. *Ecosystems*, **3**, 522–534.

Houck, M. A. & O'Connor, B. M. (1991). Ecological and evolutionary significance of phoresy in the Astigmata. *Annual Review of Entomology*, **36**, 611–636.

Hovestadt, T. (1997). *Fruchtmerkmale, endozoochore Samenausbreitung und ihre Bedeutung für die Zusammensetzung der Pflanzengemeinschaft*. Berlin: Wissenschaft & Technik Verlag.

Howard, P. C., Viskanic, P., Davenport, T. R. B., Kigenyi, F. W., Baltzer, M., Dickinson, C. J., Lwanga, J. S., Matthews, R. A. & Balmford, A. (1998). Complementarity and the use of indicator groups for reserve selection in Uganda. *Nature*, **394**, 472–475.

Howe, H. F. & Westley, L. C. (1986). Ecology of pollination and seed dispersal. In *Plant Ecology*, ed. M. J. Crawley, pp. 185–215. London: Blackwell Scientific.

Howe, H. F. & Westley, L. C. (1988). *Ecological Relationships of Plants and Animals*. Oxford: Oxford University Press.

Huang, H. T. & Yang, P. (1987). The ancient cultured citrus ant. A tropical ant is used to control insect pests in southern China. *BioScience*, **37**, 665–671.

Hubbell, S. P. (2001). *The Unified Neutral Theory of Biodiversity and Biogeography. Monographs in Population Biology*, No. 32. Princeton, NJ: Princeton University Press.

Hubbell, S. P., Foster, R. B., O'Brien, S. T., Harms, K. E., Condit, R., Wechsler, B., Wright, S. J. & Loo de Lao, S. (1999). Light-gap disturbances, recruitment limitation, and tree diversity in a Neotropical forest. *Science*, **283**, 554–557.

Huber, J. T. (1998). The importance of voucher specimens, with practical guidelines for preserving specimens of the major invertebrate phyla for identification. *Journal of Natural History*, **32**, 367–385.

Huber, O. (1986). The vegetation of the Caroni River basin. *Interciencia*, **11**, 301–310.

Huchon, D., Catzeflis, F. M. & Douzery, E. J. P. (2000). Variance of molecular datings, evolution of rodents and the phylogenetic affinities between Ctenodactylidae and Hystricognathi. *Proceedings of the Royal Society of London, Series B*, **267**, 393–402.

Hudson, P. J., Dobson, A. P. & Newborn, D. (1998). Prevention of population cycles by parasite removal. *Science*, **282**, 2256–2258.

Hunter, M. D. & Price, P. W. (1992). Playing shutes and ladders: heterogeneity and the relative roles of bottom-up and top-down forces in natural communities. *Ecology*, **73**, 724–732.

Hunter, M. D. & Price, P. W. (1998). Cycles in insect populations: delayed density dependence or exogenous driving variables? *Ecological Entomology*, **23**, 216–222.

Hunter, P. E. (1993). Mites associated with New World passalid beetles (Coleoptera: Passalidae). *Acta Zoologica Mexicana Nueva Serie*, **1**, 1–37.

Hurlbert, S. H. (1971). The non-concept of species diversity; a critique and alternative parameters. *Ecology*, **52**, 577–586.

Huston, M. A. (1994). *Biological Diversity. The Coexistence of Species on Changing Landscapes*. Cambridge: Cambridge University Press.

Hyland, K. E. (1979). Specificity and parallel host-parasite evolution in the Turbinoptidae, Cytoditdae and Ereynetidae living in the respiratory passages of birds. In *Recent Advances in Acarology*, Vol. 2, ed. J. G. Rodriguez, pp. 363–370. New York: Academic Press.

Illueca, J. B. (1985). Demografía histórica y ecológica del istmo de Panamá, 1500–1945. In *Loteria*, eds. S. Heckadon Moreno & J. Espinosa, pp. 43–56. Panamá: Instituto de Investigaciones Agropecuarias de Panamá and the Smithsonian Tropical Research Institute.

Inoue, T. & Hamid, A. A. (eds.) (1994). *Plant Reproductive Systems and Animal Seasonal Dynamics: Long-term Study of Dipterocarp Forests in Sarawak. Canopy Biology Program in Sarawak (CBPS) Series I*. Kyoto, Japan: Center of Ecological Research, Kyoto University.

Inoue, T., Yumoto, T., Hamid, A. A., Ogino, K. & Lee, H. S. (1994). Construction of a canopy observation system in the tropical rainforest of Sarawak. In *Plant Reproductive Systems and Animal Seasonal Dynamics: Long-term Study of Dipterocarp Forests in Sarawak* (Canopy Biology Program in Sarawak Series I), eds. T. Inoue & A. A. Hamid, pp. 4–18. Kyoto, Japan: Center of Ecological Research, Kyoto University.

Inoue, T., Yumoto, T., Hamid, A. A., Seng, L. H. & Ogino, K. (1995). Construction of a canopy observation system in a tropical rainforest of Sarawak. *Selbyana*, **16**, 24–35.

Intachat, J. (1995). Assessment of Moth Diversity in Natural and Managed Forests in Peninsular Malaysia. PhD thesis. Oxford, UK: University of Oxford.

Intachat, J. & Holloway, J. D. (2000). Is there stratification in diversity or preferred flight height of geometroid moths in Malaysian lowland tropical forest? *Biodiversity and Conservation*, **9**, 1417–1439.

Intachat, J., Holloway, J. D. & Speight, M. R. (1997). The effects of different forest management practices on geometroid moth populations and their diversity in Peninsular Malaysia. *Journal of Tropical Forest Science*, **9**, 411–430.

Intachat, J., Holloway, J. D. & Speight, M. R. (1999). The impact of logging on geometroid moth populations and their diversity in lowland forests of Peninsular Malaysia. *Journal of Tropical Forest Science*, **11**, 61–78.

Intachat, J., Holloway, J. D. & Staines, H. (2001). Effects of weather and phenology on the abundance and diversity of geometroid moths in a natural Malaysian tropical rain forest. *Journal of Tropical Ecology*, **17**, 411–429.

Intachat, J. & Woiwod, I. P. (1999). Trap design for monitoring moth biodiversity in tropical rainforests. *Bulletin of Entomological Research*, **89**, 153–163.

Isaac, R. A. & Johnson, W. C. (1976). Determination of total nitrogen in plant tissue, using a block digestor. *Journal of the Association of Official Analytical Chemists*, **59**, 98–100.

Ismail, G. & Din, L. B. (eds.) (1995). *A Scientifc Journey Through Borneo, Sayap-Kinabalu Park, Sabah*. Selangor Darul Ehsan, Malaysia: Faculty of Resource Science and Technology, Universiti Malaysia Sarawak & Pelanduk.

Itino, T. & Yamane, S. (1994). Vertical distribution of ants in the canopy of a lowland mixed dipterocarp forest of Sarawak. In *Plant Reproductive Systems and Animal Seasonal Dynamics: Long-term Study of Dipterocarp Forests in Sarawak* (Canopy Biology Program in Sarawak Series I), eds. T. Inoue & A. A. Hamid, pp. 227–230. Kyoto, Japan: Center for Ecological Research, Kyoto University.

Itino, T. & Yamane, S. (1995). The vertical distribution of ants on canopy trees in a Bornean lowland rain forest. *Tropics*, **4**, 277–281.

Itioka, T., Inoue, T., Kaliang, H., Kato, M., Nagamitsu, T., Momose, K., Sakai, S., Yumoto, T., Mohamad, S. U., Hamid, A. A. & Yamane, S. K. (2001). Six-year population fluctuation of the giant honey bee *Apis dorsata* F. (Hymenoptera: Apidae) in a tropical lowland dipterocarp forest in Sarawak. *Annals of the Entomological Society of America*, **94**, 545–549.

Jackson, D. A. (1984). Ant distribution patterns in a Cameroonian cocoa plantation: investigation of the ant-mosaic hypothesis. *Oecologia*, **62**, 318–324.

Jackson, R. V. (1996). A new technique for accessing tree canopies to measure insect herbivory. *Australian Journal of Entomology*, **35**, 93–95.

Jaenike, J. (1990). Host specialization in phytophagous insects. *Annual Review of Ecology and Systematics*, **21**, 243–273.

Jaffé, K., Pavis, C., Vansuyt, H. & Kermarrec, A. (1989). Extrafloral nectaries on flowers of the orchid *Spathoglottis plicata* Blume. *Biotropica*, **21**.

James, D. G., Stevens, M. M., O'Malley, K. J. & Faulder, R. J. (1999). Ant foraging reduces the abundance of beneficial and incidental arthropods in citrus canopies. *Biological Control*, **14**, 121–126.

Janzen, D. H. (1968). Host plants as islands in evolutionary and contemporary time. *American Naturalist*, **102**, 592–595.

Janzen, D. H. (1970). Herbivores and the number of tree species in tropical forests. *American Naturalist*, **104**, 501–528.

Janzen, D. H. (1971a). Seed predation by animals. *Annual Review of Ecology and Systematics*, **2**, 465–492.

Janzen, D. H. (1971b). Escape of *Cassia grandis* L. beans from predators in time and space. *Ecology*, **52**, 964–979.

Janzen, D. H. (1973a). Sweep samples of tropical foliage insects: effects of seasons, vegetation types, elevation, time of day and insularity. *Ecology*, **54**, 687–708.

Janzen, D. H. (1973b). Comments on host-specificity of tropical herbivores and its relevance to species richness. In *Taxonomy and Ecology*, ed. V. H. Heywood, pp. 201–211. London: Academic Press.

Janzen, D. H. (1973c). Host plants as islands. II. Competition in evolutionary and contemporary time. *American Naturalist*, **107**, 786–790.

Janzen, D. H. (1974). Tropical blackwater rivers, animals, and mast fruiting by the Dipterocarpaceae. *Biotropica*, **6**, 69–103.

Janzen, D. H. (1977). Why don't ants visit flowers? *Biotropica*, **9**, 14.

Janzen, D. H. (1980a). Specificity of seed-attacking beetles in a Costa Rican deciduous forest. *Journal of Ecology*, **68**, 929–952.

Janzen, D. H. (1980b). When is it coevolution? *Evolution*, **34**, 611–612.

Janzen, D. H. (1982). Guia para la identificación de mariposas nocturnas de la familia Saturniidae del Parque Nacional Santa Rosa, Guanacaste, Costa Rica. *Brenesia*, **19/20**, 255–299.

Janzen, D. H. (1983a). Food webs: who eats what, why, how, and with what effects in a tropical forest. In *Tropical Rain Forest Ecosystems. Structure and Function*, ed. F. B. Golley, pp. 167–182. Amsterdam: Elsevier.

Janzen, D. H. (1983b). Larval biology of *Ectomyelois muriscis*, (Pyralidae: Phyticinae), a Costa Rican fruit parasite of *Hymenaeae courbaril* (Leguminosae: Caesalpinioideae). *Brenesia*, **21**, 387–393.

Janzen, D. H. (ed.) (1983c). *Costa Rican Natural History*. Chicago, IL: University of Chicago Press.

Janzen, D. H. (1984a). Two ways to be a tropical big moth: Santa Rosa saturniids and sphingids. *Oxford Surveys in Evolutionary Biology*, **1**, 85–140.

Janzen, D. H. (1984b). Natural history of *Hylesia lineata* (Saturniidae: Hemileucinae) in Santa Rosa National Park, Costa Rica. *Journal of the Kansas Entomological Society*, **57**, 490–514.

Janzen, D. H. (1985a). A host plant is more than its chemistry. *Illinois Natural History Survey Bulletin*, **33**, 141–174.

Janzen, D. H. (1985b). On ecological fitting. *Oikos*, **45**, 308–310.

Janzen, D. H. (1986). Biogeography of an unexceptional place: what determines the saturnid and sphingid moth fauna of Santa Rosa National Park, Costa Rica, and what does it mean to conservation biology? *Brenesia*, **25/26**, 51–87.

Janzen, D. H. (1987a). How moths pass the dry season in a Costa Rican dry forest. *Insect Science and its Application*, **8**, 489–500.

Janzen, D. H. (1987b). Insect diversity of a Costa Rican dry forest: why keep it, and how? *Biological Journal of the Linnean Society*, **30**, 343–356.

Janzen, D. H. (1987c). When, and when not to leave. *Oikos*, **49**, 241–243.

Janzen, D. H. (1988a). Ecological characterization of a Costa Rican dry forest caterpillar fauna. *Biotropica*, **20**, 120–135.

Janzen, D. H. (1988b). The migrant moths of Guanacaste. *Orion Nature Quarterly*, **7**, 38–41.

Janzen, D. H. (1988c). Tropical dry forests. The most endangered major tropical ecosystem. In *Biodiversity*, ed. E. O. Wilson, pp. 130–137. Washington, DC: National Academy Press.

Janzen, D. H. (1993a). Taxonomy: universal and essential infrastructure for development and management of tropical wildland biodiversity. In *Proceedings of the Norway/UNEP Expert Conference on Biodiversity*, eds. O. T. Sandlund & P. J. Schei, pp. 100–113. Trondheim, Norway: Directorate for Nature Management and Norwegian Institute for Nature Research.

Janzen, D. H. (1993b). Caterpillar seasonality in a Costa Rican dry forest. In *Ecological and Evolutionary Constraints on Caterpillars*, eds. N. E. Stamp & T. M. Casey, pp. 448–477. London: Chapman & Hall.

Janzen, D. H. (1996). ATBI priorities set. *Association of Systematics Collections Newsletter*, **24**, 45–56.

Janzen, D. H. (1998). How to grow a wildland: the gardenification of nature. *Insect Science and Application*, **17**, 269–276.

Janzen, D. H. (2000). Costa Rica's Area de Conservación Guanacaste: a long march to survival through non-damaging biodevelopment. *Biodiversity*, **1**, 7–20.

Janzen, D. H. (2003). Ecology of dry forest wildland insects in the Area de Conservación Guanacaste, northwestern Costa Rica. In *Biodiversity Conservation in Costa Rica: Learning the Lessons in Seasonal Dry Forests*, eds. G. W. Frankie, A. Mata & S. B. Vinson, in press. Berkeley, CA: University of California Press.

Janzen, D. H., Ataroff, M., Fariñas, M., Reyes, S., Rincón, N., Soler, A., Soriano, P. & M., V. (1976). Changes in the arthropod community along an elevational gradient in the Venezuelan Andes. *Biotropica*, **8**, 193–203.

Janzen, D. H. & Gauld, I. D. (1997). Patterns of use of large moth caterpillars (Lepidoptera: Saturniidae and Sphingidae) by ichneumonid parasitoids (Hymenoptera) in Costa Rican dry forest. In *Forests and Insects: 18th Symposium of the Royal Entomological Society*, eds. A. D. Watt, N. E. Stork & M. D. Hunter, pp. 251–271. London: Chapman & Hall.

Janzen, D. H., Hallwachs, W., Jimenez, J. & Gamez, R. (1993). The role of the parataxonomists, inventory managers, and taxonomists in Costa Rica's national biodiversity inventory. In *Biodiversity Prospecting: Using Generic Resources for Sustainable Development*, eds. W. V. Reid, S. A. Laird, C. A. Meyer, R. Gamez, A. Sittenfeld, D. H. Janzen, M. A. Gollin & C. Juma, pp. 223–254. Washington, DC: World Resources Institute.

Janzen, D. H. & Liesner, R. (1980). Annotated check-list of plants of lowland Guanacaste Province, Costa Rica, exclusive of grasses and non-vascular cryptogams. *Brenesia*, **18**, 15–90.

Janzen, D. H. & Schoener, T. W. (1968). Difference in insect abundance and diversity between wetter and drier sites during a tropical dry season. *Ecology*, **49**, 96–110.

Janzen, D. H., Sharkey, M. J. & Burns, J. M. (1998). Parasitization biology of a new species of Braconidae (Hymenoptera) feeding on larvae of Costa Rican dry forest skippers (Lepidoptera: Hesperiidae: Pyrginae). *Tropical Lepidoptera*, **9**(Suppl.), 33–41.

Jeanne, R. L. (1979). A latitudinal gradient in rates of ant predation. *Ecology*, **60**, 1211–1224.

Jermy, T. (1976). Insect–host-plant relationship: coevolution or sequential evolution. In *The Host Plant in Relation to Insect Behavior and Development*, ed. T. Jermy, pp. 109–113. New York: Plenum Press.

Jermy, T. (1993). Evolution of insect–plant relationships – a devil's advocate approach. *Entomologia Experimentalis et Applicata*, **66**, 3–12.

Jessop, L. & Hammond, P. M. (1993). Quantitative sampling of Coleoptera in north-east woodlands using flight interception traps. *Transactions of the Natural History Society of Northumbria*, **56**, 41–60.

Joel, G., Aplet, G. & Vitousek, P. M. (1994). Leaf morphology along environmental gradients in Hawaiian *Metrosideros polymorpha*. *Biotropica*, **26**, 17–22.

Johns, R. J. (1992). The influence of deforestation and selective logging operations on plant diversity in Papua New Guinea. In *Tropical Deforestation and Species Extinction*, eds. T. C. Whitmore & J. A. Sayer, pp. 143–147. London: Chapman & Hall.

Johnson, M. P. & Mueller, A. J. (1990). Flight and diel activity of the three cornered alfalfa hopper (Homoptera: Membracidae). *Environmental Entomology*, **19**, 677–683.

Johnston, D. E. (1982). Acari. In *Synopsis and Classification of Living Organisms*, ed. S. P. Parker, pp. 111–117. New York: McGraw and Hill.

Jolivet, P. (1987). Remarques sur la biocénose des *Cecropia* (Cecropicaeae). Biologie des *Coelomera* Chevrolat avec la description d'une nouvelle espèce du Brésil (Coleoptera Chrysomelidae Galerucinae). *Bulletin mensuel de la Société Linnéenne de Lyon*, **56**, 255–276.

Jolivet, P. (1996a). *Ants and Plants. An example of Coevolution*. Leiden: Backhuys.

Jolivet, P. (1996b). *Les Fourmis et les Plantes. Un example de Coevolution*. Paris: Fondation Singer-Polinac, Editions Boubée.

Jolivet, P. & Hawkeswood, T. J. (1995). *Host-plants of Chrysomelidae of the World. An Essay about the Relationships between the Leaf-beetles and their Food-plants*. Leiden: Backhuys.

Jones, C. G. (1996). *Linking Species & Ecosystems*. New York: Chapman & Hall.

Jones, G. (1990). Prey selection by the greater horseshoe bat (*Rhinolophus ferrumequinum*): optimum foraging by echolocation? *Journal of Animal Ecology*, **59**, 587–602.

Joosse, E. N. G. & Veltkamp, E. (1970). Some aspects of growth, moulting and reproduction in five species of surface dwelling Collembola. *Netherlands Journal of Zoology*, **20**, 315–328.

Jordal, B. H. & Kirkendall, L. R. (1998). Ecological relationships of a guild of tropical beetles breeding in *Cecropia* petioles in Costa Rica. *Journal of Tropical Ecology*, **14**, 153–176.

Jordal, B. H., Normark, B. B. & Farrell, B. D. (2000). Evolutionary radiation of an inbreeding haplodiploid beetle lineage (Curculionidae, Scolytinae). *Biological Journal of the Linnean Society*, **71**, 483–499.

Jørum, P. (2000). Billefaunaen i Hald Egeskov (Coleoptera). *Entomologiske Meddelelser*, **68**, 1–46.

Juniper, B. E. & Jeffree, C. E. (1983). *Plant Surfaces*. London: Edward Arnold.

Kaczmarek, M. (1975). An analysis of Collembola communities in different pine forest environments. *Ekologia Polska*, **23**, 265–293.

Kahn, F. & de Granville, J.-J. (1992). *Palms in Forest Ecosystems of Amazonia*. Berlin: Springer-Verlag.

Kalko, E. K. V. (1998). Organisation and diversity of tropical bat communities through space and time. *Zoology*, **101**, 281–297.

Kalshoven, L. G. E. (1981). *Pests of Crops in Indonesia*. Translated by P. A. van der Laan & G. H. L. Rothschild. Jakarta: P. T. Ichtiar Baru & van Hoeve.

Kampichler, C. (1990). Community structure and vertical distribution of Collembola and Cryptostigmata in a dry-turf cushion plant. *Biology and Fertility of Soils*, **9**, 130–134.

Kampichler, C. (1999). Fractal concepts in studies of soil fauna. *Geoderma*, **88**, 283–300.

Kampichler, C., Dzeroski, S. & Wieland, R. (2000). The application of machine learning techniques to the analysis of soil ecological data bases: relationships between habitat features and Collembola community characteristics. *Soil Biology & Biochemistry*, **32**, 107–209.

Kampichler, C. & Hauser, M. (1993). Roughness of soil pore surface and its effect on available habitat space of microarthropods. *Geoderma*, **56**, 223–232.

Karban, R. (1997). Evolution of prolonged development: a life table analysis for periodical cicadas. *American Naturalist*, **150**, 446–461.

Karban, R. & Baldwin, I. T. (1997). *Induced Responses to Herbivory*. Chicago, IL: University of Chicago Press.

Kareiva, P. (1990). Population dynamics in spatially complex environments: theory and data. *Philosophical Transactions of the Royal Society of London, Series B*, **330**, 175–190.

Kareiva, P. (1994a). Higher order interactions as a foil to reductionist ecology. *Ecology*, **75**, 1527–1528.

Kareiva, P. (1994b). Space: the final frontier for ecological theory. *Ecology*, **75**, 1.

Karg, W. (1991). New species of the genus *Androlaelaps* Berlese (Mesostigmata: Laelapidae) from a cockroach in Madagascar. *International Journal of Acarology*, **17**, 165–168.

Kaspari, M., Pickering, J. & Windsor, D. (2001). The reproductive flight phenology of a neotropical ant assemblage. *Ecological Entomology*, **26**, 245–257.

Kaspari, M. & Yanoviak, S. P. (2001). Bait use in tropical litter and canopy ants – evidence of differences in nutrient limitation. *Biotropica*, **33**, 207–211.

Kato, M., Inoue, T., Hamid, A. A., Itino, T., Merdek, M. B., Nona, A. R., Nagamitsu, T., Yamane, S. & Yumoto, T. (1994). Seasonality and vertical structure of light-attracted insect communities in a tropical lowland dipterocarp forest in Sarawak. In *Plant Reproductive Systems and Animal Seasonal Dynamics: Long-term Study of Dipterocarp Forests in Sarawak* (Canopy Biology Program in Sarawak Series I), eds. T. Inoue & A. A. Hamid, pp. 199–221. Kyoto, Japan: Center of Ecological Research, Kyoto University.

Kato, M., Inoue, T., Hamid, A. A., Nagamitsu, T., Merdek, M. B., Nona, A. R., Itino, T., Yamane, S. & Yumoto, T. (1995). Seasonality and vertical structure of light-attracted insect communities in a Dipterocarp forest in Sarawak. *Research on Population Ecology*, **37**, 59–79.

Kato, M., Itioka, T., Sakai, S., Momose, K., Yamane, S., Hamid, A. A. & Inoue, T. (2000). Various population fluctuation patterns of light-attracted beetles in a tropical lowland dipterocarp forest in Sarawak. *Research on Population Ecology*, **42**, 97–104.

Keeley, J. E. & Bond, W. J. (1999). Mast flowering and semelparity in bamboos: the bamboo fire cycle hypothesis. *American Naturalist*, **154**, 383–391.

Keesing, F. (2000). Cryptic consumers and the ecology of an African savanna. *BioScience*, **50**, 205–215.

Kelly, C. K. & Southwood, T. R. E. (1999). Species richness and resource availability: a phylogenetic analysis of insects associated with trees. *Proceedings of the National Academy of Sciences USA*, **96**, 8013–8016.

Kenne, M. & Dejean, A. (1999). Spatial distribution, size density of nests of *Myrmicaria opaciventris* Emery (Formicidae: Myrmicinae). *Insectes Sociaux*, **46**, 179–185.

Kenne, M., Schatz, B. & Dejean, A. (2000). Hunting strategy of a generalist ant species proposed as a biological control agent against termites. *Entomologia Experimentalis et Applicata*, **94**, 31–40.

Kennedy, C. E. J. & Southwood, T. R. E. (1984). The number of species of insects associated with British trees: a re-analysis. *Journal of Animal Ecology*, **53**, 455–478.

Kenney, A. J. & Krebs, C. J. (2000). *Programs for Ecological Methodology*, version 5.2, published at http://www.zoology.ubc.ca/~krebs

Kethley, J. B. (1977). An unusual parantennuloid, *Philodana johnstoni* n.g., n. sp. (Acari: Parasitiformes: Philodanidae, n. fam.) associated with *Neatus tenebrioides* (Coleoptera: Tenebrionidae) in North America. *Annals of the Entomological Society of America*, **70**, 487–494.

Kevan, P. G. & Baker, H. G. (1983). Insect as flower visitors and pollinators. *Annual Review of Entomology*, **28**, 407–453.

Khoo, K. C., Ooi, P. A. C. & Ho, C. T. (1991). *Crop Pests and Their Management in Malaysia*. Kuala Lumpur: Tropical Press.

Kinn, D. N. (1971). The life cycle and behavior of *Cercoleipus coelonotus* (Acarina: Mesostigmata) including a survey of phoretic mite associates of California Scolytidae. *University of California Publications in Entomology*, **65**, 1–66.

Kinn, D. N. (1980). Mutualism between *Dendrolaelaps neodisetus* and *Dendrolaelaps frontalis*. *Environmental Entomology*, **9**, 756–758.

Kinn, D. N. (1984). Life cycle of *Dendrolaelaps neodisetus* (Mesostigmata: Digamasellidae), a nematophagous mite associated with pine bark beetles (Coleoptera: Scolytidae). *Environmental Entomology*, **13**, 1141–1144.

Kirch, P. V. (1982). The impact of the prehistoric Polynesians on the Hawaiian ecosystem. *Pacific Science*, **36**, 1–14.

Kirkendall, L. R. (1993). Ecology and Evolution of biased sex ratios in bark and ambrosia beetles. In *Evolution and Diversity of Sex Ratios in Insects and Mites*, eds. D. L. Wrensch & M. A. Ebbert, pp. 235–345. London: Chapman & Hall.

Kitayama, K. (1992). An altitudinal transect study of the vegetation on Mount Kinabalu, Borneo. *Vegetatio*, **102**, 149–171.

Kitayama, K., Lakim, M. & Wahab, M. Z. (1999). Climate profile of Mount Kinabalu during late 1995–early 1998 with special reference to the 1998 drought. *Sabah Parks Nature Journal*, **2**, 85–100.

Kitayama, K. & Mueller-Dombois, D. (1995). Vegetation changes along gradients of long-term soil development in the Hawaiian montane rainforest zone. *Vegetatio*, **120**, 1–20.

Kitching, R. L. (1987). Aspects of the natural history of the lycaenid butterfly *Allotinus major* in Sulawesi. *Journal of Natural History*, **21**, 535–544.

Kitching, R. L. (1993). Biodiversiy and taxonomy: impediment or opportunity? In *Conservation Biology in Australia and Oceania*, eds. C. Moritz & J. Kikkawa, pp. 253–268. Chipping Norton, Australia: Surrey Beatty.

Kitching, R. L. (2000). *Food Webs and Container Habitats: The Natural History and Ecology of Phytotelmata*. Cambridge: Cambridge University Press.

Kitching, R. L. & Arthur, M. (1993). The biodiversity of arthropods from Australian rain forest canopies: summary of projects and the impact of drought. *Selbyana*, **14**, 29–35.

Kitching, R. L., Bergelson, J. M., Lowman, M. D., McIntyre, S. & Carruthers, G. (1993). The biodiversity of arthropods from Australian rainforest canopies: general introduction, methods, sites and ordinal results. *Australian Journal of Ecology*, **18**, 181–191.

Kitching, R. L. & Clarke, C. (1989). Indigenous tropical ecology. *Trends in Ecology and Evolution*, **4**, 85–87.

Kitching, R. L., Floater, G. & Mitchell, H. (1994). The biodiversity of arthropods from Australian rainforest canopies: ecological questions and management challenges. In *Biodiversity: Its Complexity and Role*, eds. M. Yasuno & M. M. Watanabe, pp. 119–137. Tokyo: Global Environmental Forum.

Kitching, R. L., Li, D. Q. & Stork, N. E. (2001). Assessing biodiversity 'sampling packages': how similar are arthropod assemblages in different tropical rainforests? *Biodiversity and Conservation*, **10**, 793–813.

Kitching, R. L., Mitchell, H., Morse, G. & Thebaud, C. (1997). Determinants of species richness in assemblages of canopy arthropods in rainforests. In *Canopy Arthropods*, eds. N. E. Stork, J. Adis & R. K. Didham, pp. 131–150. London: Chapman & Hall.

Kitching, R. L., Vickerman, G., Laidlaw, M. & Hurley, K. (2000). *The Comparative Assessment of Arthropod and Tree Diversity on Old-World Rainforests: the Rainforest CRC/Earthwatch Manual. Technical Report*. Cairns: Cooperative Research Centre for Tropical Rainforest Ecology and Management.

Kitching, R. L. & Zalucki, J. (1996). The biodiversity of arthropods from Australian rain forest canopies: some results on the role of tree species. In *Tropical Rainforest Research – Current Issues. Proceedings of the Conference held in Bandar Seri Begawan, April 1993*, eds. D. S. Edwards, W. E. Booth & S. C. Choy, pp. 21–28. Amsterdam: Kluwer Academic.

Klima, J. (1954). Die Oribatiden und ihre Cönosen in der Umgebung von Innsbruck. Dissertation. Innsbruck: University of Innsbruck.

Knowles, O. H. & Parrotta, J. A. (1997). Phenological observations and tree seed characteristics in an equatorial moist forest at Trombetas, Para State, Brazil. In *Phenology in Seasonal Climates I*, eds. H. Lieth & M. D. Schwartz, pp. 67–84. Leiden, The Netherlands: Backhuys.

Kochmer, J. P. & Handel, S. N. (1986). Constraints and competition in the evolution of flowering phenology. *Ecological Monographs*, **56**, 303–325.

Köhler, F. (1997). Bestandserfassung xylobionter Käfer im Nationalpark Bayerischer Wald. *Beiträge zur bayerischen Entomofaunistik*, **118**, 73–118.

Köhler, W., Schachtel, G. & Voleske, P. (1995). *Biostatistik*. Berlin: Springer-Verlag.

Koike, F. (1985). Reconstruction of two-dimensional tree and forest canopy profiles using photographs. *Journal of Applied Ecology*, **22**, 921–929.

Koike, F. (1994). Structure and light environment in the forest canopy of Lambir Hills National Park, Sarawak. In *Plant Reproductive Systems and Animal Seasonal Dynamics: Long-term Study of Dipterocarp Forests in Sarawak* (Canopy Biology Program in Sarawak Series I), eds. T. Inoue & A. A. Hamid, pp. 40–42. Kyoto, Japan: Center of Ecological Research, Kyoto University.

Koike, F. & Hotta, M. (1996). Foliage canopy structure and height distribution of woody species in climax forests. *Journal of Plant Research*, **109**, 53–60.

Koike, F., Riswan, S., Partomihardjo, T., Suzuki, E. & Hotta, M. (1998). Canopy structure and insect community distribution in a tropical rain forest of West Kalimantan. *Selbyana*, **19**, 147–154.

Koike, F. & Syahbuddin (1993). Canopy structure of a tropical rain forest and the nature of an unstratified upper layer. *Functional Ecology*, **7**, 230–235.

Koike, F., Tabata, H. & Malla, S. B. (1990). Canopy structure and its effect on shoot growth and flowering in subalpine forests. *Vegetatio*, **86**, 101–113.

Kolasa, J., Hewitt, C. L. & Drake, J. A. (1998). Rapoport's rule: an explanation or a byproduct of the latitudinal gradient in species richness? *Biodiversity and Conservation*, **7**, 1447–1455.

König, B. & Linsenmair, K. E. (1996). Biologische Diversität – Ein Phänomen und seine Dimensionen. In *Biologische Vielfalt*, eds. B. König & K. E. Linsenmair, pp. 8–15. Heidelberg: Spektrum, Akademischer Verlag.

Koponen, S., Rinne, V. & Clayhills, T. (1997). Arthropods on oak branches in SW Finland, collected by a new trap type. *Entomologica Fennica*, **8**, 177–183.

Koptur, S. & Truong, N. (1998). Facultative ant–plant interactions: nectar sugar preferences of introduced pest ant species in south Florida. *Biotropica*, **30**, 179–189.

Krantz, G. W. (1978). *A Manual of Acarology*. Corvallis, OR: Oregon State University. (Emended 1986.)

Krebs, C. J. (1989). *Ecological Methodology*. New York: Harper & Row.

Krebs, C. J. (1999). *Ecological Methodology*, 2nd edn. Menlo Park, USA: Benjamin/Cummings.

Krebs, C. J., Boutin, S., Boonstra, R., Sinclair, A. R. E., Smith, J. N. M., Dale, M. R. T., Martin, K. & Turkington, R. (1995). Impact of food and predation on the Snowshoe Hare cycle. *Science*, **269**, 1112–1115.

Krüger, O. & McGavin, G. C. (1997). The insect fauna of *Acacia* species in Mkomazi Game Reserve, north-east Tanzania. *Ecological Entomology*, **22**, 440–444.

Krüger, O. & McGavin, G. C. (1998a). Insect diversity of *Acacia* canopies in Mkomazi reserve, north-east Tanzania. *Ecography*, **21**, 261–268.

Krüger, O. & McGavin, G. C. (1998b). The influence of ants on the guild structure of *Acacia* insect communities in Mkomazi Game Reserve, north-east Tanzania. *African Journal of Ecology*, **36**, 213–220.

Krüger, O. & McGavin, G. C. (2000). Macroecology of local insect communities. *Acta Oecologica*, **21**, 21–28.

Krüger, O. & McGavin, G. C. (2001). Predator–prey ratio and guild constancy in a tropical insect community. *Journal of Zoology, London*, **253**, 265–273.

Krylov, V. V. (1971). On the station–species curve. In *Statistical Ecology.* Vol. 3. *Many Species Populations, Ecosystems and System Analysis*, eds. G. P. Patil, E. C. Pielou & W. E. Waters, pp. 233–235. Pennsylvania, PA: Pennsylvania State University Press.

Kugler, H. (1955). *Einführung in die Blütenökologie*. Jena: Gustav Fischer Verlag.

Kursar, T. A. & Coley, P. D. (1992). Delayed greening in tropical leaves: an antiherbivore defense? *Biotropica*, **24**, 256–262.

Laidlaw, M., Olsen, M., Kitching, R. L. & Greenway, M. (2000). Tree floristic and structural characteristics of one hectare of subtropical rainforest in Lamington National Park, Queensland. *Proceedings of the Royal Society of Queensland*, **109**, 91–105.

Laidlaw, R. K. (1996). A comparison between populations of primates, squirrels, tree shrews, and other mammals inhabiting virgin, logged, fragmented and plantation forests in Malaysia. In *Conservation, Management and Development of Forest Resources. Proceedings of the Malaysia–United Kingdom Programme Workshop, 1996*, eds. L. S. See, D. Y. May, I. D. Gauld & J. Bishop, pp. 141–159. Kuala Lumpur: Malaysia–United Kingdom Programme.

Laman, T. G. (1995). Safety recommendations for climbing rain forest trees with 'single rope technique'. *Biotropica*, **27**, 406–409.

Lambert, F. R. & Marshall, A. G. (1991). Keystone characteristics of bird-dispersed *Ficus* in a Malaysian lowland rain forest. *Journal of Ecology*, **79**, 793–809.

Lamprey, H. F., Havely, G. & Makachas, S. (1974). Interactions between *Acacia*, bruchid seed beetles and large herbivores. *East African Wildlife Journal*, **12**, 81–85.

Lande, R. (1996). Statistics and partitioning of species diversity, and similarity among multiple communities. *Oikos*, **76**, 5–13.

Landsberg, J. & Gillieson, D. (1982). Repetitive sampling of the canopies of tall trees using a single rope technique. *Australian Forestry*, **45**, 59–61.

Langenheim, J. H. & Stubblebine, W. H. (1983). Variation in leaf resin between parent tree and progeny in *Hymenaea*: implications for herbivory in the humid tropics. *Biochemical and Systematic Ecology*, **11**, 97–106.

Lara, A. C. F. & Fernandes, G. W. (1996). The highest diversity of galling insects: Serra do Cipó, Brazil. *Biodiversity Letters*, **3**, 111–114.

Laskar, J., Joutel, F. & Boudin, F. (1993). Orbital, precessional and insolation quantities for the Earth from 20 Myr to 10 Myr. *Astronomy and Astrophysics*, **270**, 522–533.

Lattin, J. D. (1999). Bionomics of the Anthocoridae. *Annual Review of Entomology*, **44**, 207–231.

Laurance, W. F., Gascon, C. & Rankin-de Merona, J. M. (1999). Predicting effects of habitat destruction on plant communities: a test of a model using Amazonian trees. *Ecological Applications*, **9**, 548–554.

Laurance, W. F., Laurance, S. G., Ferreira, L. V., Rankin-de Merona, J. M., Gascon, C. & Lovejoy, T. E. (1997). Biomass collapse in Amazonian forest fragments. *Science*, **278**, 1117–1118.

Laurance, W. F., Pérez-Salicrup, D., Delamônica, P., Fearnside, P. M., D'Angelo, S., Jerozolinski, A., Pohl, L. & Lovejoy, T. E. (2001). Rain forest fragmentation and the structure of Amazonian liana communities. *Ecology*, **82**, 105–116.

Lavelle, P., Lattaud, C., Trigo, D. & Barois, I. (1995). Mutualism and biodiversity in soils. *Plant & Soil*, **170**, 23–33.

Lawrence, J. F. (1982). Coleoptera. In *Synopsis and Classification of Living Organisms.* Vol. 2, ed. S. P. Parker, pp. 482–553. New York: McGraw-Hill.

Lawrence, J. F. & Newton, A. F. (1995). Families and subfamilies of Coleoptera (with selected genera, notes, references and data on family-group names). In *Biology, Phylogeny, and Classification of Coleoptera: Papers Celebrating the 80th Birthday of Roy A. Crowson*, eds. J. Pakaluk & S. A. Slipinski, pp. 779–913. Warszawa: Museum i Instytut Zoologii PAN.

Lawrence, R. F. (1953). *The Biology of the Cryptic Fauna of Forests.* Cape Town, South Africa: A. A. Balkema.

Lawton, J. H. (1978). Host plant influences on insect diversity: the effect of space and time. In *Diversity of Insect Faunas*, eds. L. A. Mound & N. Waloff, pp. 105–125. Oxford: Blackwell.

Lawton, J. H. (1983). Plant architecture and the diversity of phytophagous insects. *Annual Review of Entomology*, **28**, 23–29.

Lawton, J. H. (1999). Are there general laws in ecology? *Oikos*, **84**, 177–192.

Lawton, J. H., Bignell, D. E., Bolton, B., Bloemers, G. F., Eggleton, P., Hammond, P. M., Hodda, M., Holt, R. D., Larsen, T. B., Mawdsley, N. A., Stork, N. E., Srivastava, D. S. & Watt, A. D. (1998). Biodiversity inventories, indicator taxa and effects of habitat modification in tropical forest. *Nature*, **391**, 72–76.

Lawton, J. H. & May, R. M. (eds.) (1995). *Extinction Rates.* Oxford: Oxford University Press.

Lawton, J. H. & Schröder, D. (1977). Effects of plant type, size of geographical range and taxonomic isolation on number of insect species associated with British plants. *Nature*, **265**, 137–140.

Lawton, J. H. & Strong, D. R. (1981). Community patterns and competition in folivorous insects. *American Naturalist*, **118**, 317–338.

Le Corff, J. & Marquis, R. J. (1999). Differences between understorey and canopy in herbivore community composition and leaf quality for two oak species in Missouri. *Ecological Entomology*, **24**, 46–58.

Le Moult, E. (1955). *Mes chasses aux papillons.* Paris: Editions Pierre Horey.

Lees, D. C., Kremen, C. & Andriamampianina, L. (1999). A null model for species richness gradients: bounded range overlap of butterflies and other rainforest endemics in Madagascar. *Biological Journal of the Linnean Society*, **67**, 529–584.

Legendre, P. & Legendre, L. (1998). *Numerical Ecology.* 2nd English Ed. Amsterdam: Elsevier.

Lehmacher, W., Wassmer, G. & Reitmeir, P. (1991). Procedures for two sample comparisons with multiple endpoints controlling the experimentwise error rate. *Biometrics*, **47**, 511–521.

Leigh, E. G., Jr (1994). Do insect pests promote mutualism among tropical trees? *Journal of Ecology*, **82**, 677–680.

Leigh, E. G., Jr (1999). *Tropical Forest Ecology. A View from Barro Colorado Island.* New York: Oxford University Press.

Leigh, E. G., Jr, Rand, A. S. & Windsor, D. M. (1982). *The Ecology of a Tropical Forest. Seasonal Rhythms and Long-term Changes.* Washington, DC: Smithsonian Institution Press.

Leigh, E. G., Jr & Smythe, N. (1978). Leaf production, leaf consumption, and the regulation of folivory on Barro Colorado Island. In *The Ecology of Arboreal Folivores*, ed. G. G. Montgomery, pp. 33–50. Washington, DC: Smithsonian Institution Press.

Leigh, E. G., Jr & Windsor, D. M. (1982). Forest production and regulation of primary consumers on Barro Colorado Island. In *The Ecology of a Tropical Forest: Seasonal Rhythms and Long-Term Changes*, eds. E. G. Leigh Jr, A. S. Rand & D. M. Windsor, pp. 111–122. Washington, DC: Smithsonian Institution Press.

Leinaas, H. P. & Fjellberg, A. (1985). Habitat structure and life history strategies of two partly sympatric and closely related, lichen feeding collembolan species. *Oikos*, **44**, 448–458.

Lemaire, C. (1978). *The Attacidae of America (=Saturniidae). Attacinae.* Neuilly-sur-Seine, France: C. Lemaire, 42, Boulevard Victor Hugo.

Lemaire, C. (1980). *The Attacidae of America (=Saturniidae). Arsenurinae.* Neuilly-sur-Seine, France: C. Lemaire, 42, Boulevard Victor Hugo.

Lemaire, C. (1988). *The Saturniidae of America. Ceratocampinae.* San José, Costa Rica: Museo Nacional de Costa Rica.

Lemaire, C. (2002). *The Saturniidae of America. Hemileucinae.* Germany: Eric Bauer.

Lepointe, J. (1956). Méthodes de capture dans l'écologie des arbres. *Vie et Milieu*, **7**, 233–241.

Leps, J., Novotny, V. & Basset, Y. (2001). Effect of habitat and successional optimum of host plants on the composition of their herbivorous communities: leaf-chewing insects on shrubs and trees in New Guinea. *Journal of Ecology*, **89**, 186–199.

Leston, D. (1973). The ant mosaic–tropical tree crops and the limiting of pests and diseases. *Pest Articles and News Summaries*, **19**, 311–341.

Leston, D. (1978). A neotropical ant mosaic. *Annals of the Entomological Society of America*, **71**, 649–653.

Levin, D. A. (1976). Alkaloid-bearing plants: an ecogeographical perspective. *American Naturalist*, **110**, 261–284.

Levin, P. S. & Levin, D. A. (2002). The real biodiversity crisis. *American Scientist*, **90**, 6–8.

Levin, S. A. (1992). The problem of pattern and scale in ecology. *Ecology*, **73**, 1943–1967.

Levings, S. C. & Windsor, D. M. (1985). Litter arthropod populations in a tropical deciduous forest: relationships between years and arthropod groups. *Journal of Animal Ecology*, **54**, 61–70.

Lewis, K. J. (1997). Growth reduction in spruce infected by *Inonotus tomentosus* in central British Columbia. *Canadian Journal of Forest Research*, **27**, 1669–1674.

Liebherr, J. K. & Zimmerman, E. C. (2000). *Hawaiian Carabidae (Coleoptera), Part 1: Introduction and Tribe Platynini, Insects of Hawaii*, Vol. 16. Honolulu, Hawaii: University of Hawaii Press.

Lindenmayer, D. B., Welsh, A., Donnelly, C. F. & Meggs, R. A. (1996). Use of nest trees by the mountain brushtail possum (*Trichosurus caninus*) (Phalangeridae: Marsupialia). I. Number of occupied trees and frequency of tree use. *Wildlife Research*, **23**, 343–361.

Lindquist, E. E. (1975). *Digamasellus* Berlese, 1905, and *Dendrolaelaps* Halbert, 1915, with descriptions of new taxa of Digamasellidae (Acarina: Mesostigmata). *Canadian Entomologist*, **107**, 1–43.

Lindquist, E. E. (1976). Transfer of the Tarsocheylidae to the Heterostigmata, and reassignment of the Tarsonemina and Heterostigmata to lower hierarchic status in the Prostigmata. *Canadian Entomologist*, **108**, 23–48.

Lindquist, E. E. (1986). The world genera of Tarsonemidae (Acari: Heterostigmata): a morphological, phylogenetic and systematic revision, with a reclassification of family-group taxa in the Heterostigmata. *Memoirs of the Entomological Society of Canada*, **136**, 1–517.

Link, W. A. & Nichols, J. D. (1994). On the importance of sampling variance to investigations of temporal variation in animal population size. *Oikos*, **69**, 539–544.

Linsenmair, K. E. (1990). Tropische Biodiversität: Befunde und offene Probleme. *Verhandlungen der Deutschen Zoologischen Gesellschaft*, **83**, 245–261.

Linsenmair, K. E., Davis, A. J., Fiala, B. & Speight, M. R. (eds.) (2001). *Tropical Forest Canopies: Ecology and Management*. Dordrecht, the Netherlands: Kluwer Academic.

Lockwood, J. A., Struttmann, J. M. & Miller, C. J. (1996). Temporal patterns in feeding of grasshoppers (Orthoptera: Acrididae): importance of nocturnal feeding. *Environmental Entomology*, **25**, 570–581.

Lockwood, J. P., Lipman, P. W., Peterson, L. D. & Warshauer, F. R. (1988). *Generalized Ages of Surface Flows of Mauna Loa Volcano, Hawaii*. Washington, DC: US Government Printing Office.

Lohr, B. (2000). Introducing COPAS, a new access tool for canopy research. *What's Up, The Newsletter of the International Canopy Network*, **6**, 3–4.

Longino, J. T. (2000). What to do with data. In *Ants. Standard Methods for Measuring and Monitoring Biodiversity*, eds. D. Agosti, J. D. Majer, L. E. Alonso & T. R. Schultz, pp. 186–203. Washington, DC: Smithsonian Institution Press.

Longino, J. T. & Colwell, R. C. (1997). Biodiversity assessment using structured inventory: capturing the ant fauna of a tropical rain forest. *Ecological Applications*, **7**, 1263–1277.

Longino, J. T. & Nadkarni, N. M. (1990). A comparison of ground and canopy leaf litter ants (Hymenoptera: Formicidae) in a neotropical montane forest. *Psyche*, **97**, 81–93.

Lopes, J., Arias, J. R. & Yood, J. D. C. (1983). Evidências preliminares de estratificação vertical de postura de ovos por alguns Culicidae (Diptera), em floresta no municipio de Manaus-Amazonas. *Acta Amazonica*, **13**, 431–439.

Loranger, G., Bandyopadhyaya, I., Razaka, B. & Ponge, J.-F. (2001). Does soil acidity explain altitudinal gradients in collembolan communities? *Soil Biology and Biochemistry*, **33**, 381–391.

Loranger, G., Ponge, J. F., Blanchart, E. & Lavelle, P. (1998). Impact of earthworms on the diversity of microarthropods in a vertisol (Martinique). *Biology and Fertility of Soils*, **27**, 21–26.

Loreau, M. (2000). Are communities saturated? On the relationship between alpha, beta and gamma diversity. *Ecology Letters*, **3**, 73–76.

Losos, J. B. (1996). Phylogenetic perspectives on community ecology. *Ecology*, **77**, 1344–1354.

Lott, E. J. (1985). *Listados florísticos de México. III. La Estación de Biología Chamela, Jalisco*. México: Instituto de Biología, Universidad Nacional Autónoma de México.

Lott, E. J., Bullock, S. H. & Solís-Magallanes, J. A. (1987). Floristic diversity and structure of upland and arroyo forests of coastal Jalisco. *Biotropica*, **19**, 228–235.

Loughrin, J. H., Potter, D. A. & Hamilton, K. T. R. (1995). Volatile compounds induced by herbivory act as aggregation kairomones for the Japanese beetle (*Popillia japonica* Newman). *Journal of Chemical Ecology*, **21**, 1457–1467.

Lounibos, L. P. (1981). Habitat segregation among African treehole mosquitoes. *Ecological Entomology*, **6**, 129–154.

Lovejoy, T. E. & Bierregaard, R. O., Jr (1990). Central Amazonian forests and the minimum critical size. In *Four Neotropical Rainforests*, ed. A. H. Gentry, pp. 60–71. New Haven, CA: Yale University Press.

Loveless, A. R. (1962). Further evidence to support a nutritional interpretation of sclerophylly. *Annals of Botany*, **26**, 551–561.

Lovett, J. C. & Wasser, S. (eds.) (1993). *Biogeography and Ecology of the Rainforest of Eastern Africa*. Cambridge: Cambridge University Press.

Lowman, M. D. (1982a). Effects of different rates and methods of leaf area removal on rain forest seedlings of Coachwood (*Ceratopetalum apetalum*). *Australian Journal of Botany*, **30**, 477–483.

Lowman, M. D. (1982b). Seasonal variation in insect abundance among three Australian rain forests, with particular reference to phytophagous types. *Australian Journal of Ecology*, **7**, 353–361.

Lowman, M. D. (1984). An assesment of techniques for measuring herbivory: is rainforest defoliation more intense than we thought? *Biotropica*, **16**, 264–268.

Lowman, M. D. (1985). Temporal and spatial variability in insect grazing of the canopies of five Australian rainforest tree species. *Australian Journal of Ecology*, **10**, 7–24.

Lowman, M. D. (1992). Leaf growth dynamics and herbivory in five species of Australian rain-forest canopy trees. *Journal of Ecology*, **80**, 433–447.

Lowman, M. D. (1995). Herbivory as a canopy process in rain forest trees. In *Forest Canopies*, eds. M. D. Lowman & N. M. Nadkarni, pp. 431–455. San Diego, CA: Academic Press.

Lowman, M. D. (1997). Herbivory in forests: from centimetres to megametres. In *Forests and Insects: 18th Symposium of the Royal Entomological Society*, eds. A. D. Watt, N. E. Stork & M. D. Hunter, pp. 135–149. London: Chapman & Hall.

Lowman, M. D. & Bouricius, B. (1993). Canopy walkways – techniques for their design and construction. *Selbyana*, **14**, 4.

Lowman, M. D. & Bouricius, B. (1995). The construction of platforms and bridges for forest canopy access. *Selbyana*, **16**, 179–184.

Lowman, M. D. & Box, J. D. (1983). Variation in leaf toughness and phenolic content among five species of Australian rain forest trees. *Australian Journal of Ecology*, **8**, 17–25.

Lowman, M. D., Brown, M., Desrosiers, A. & Randle, D. C. (1999). Temporal variation in herbivory of a Peruvian bromeliad. *Journal of the Bromeliad Society*, **49**, 81–84.

Lowman, M. D., Hallé, F., Bouricius, B., Coley, P., Nadkarni, N., Parker, G., Saterson, K. & Wright, S. J. (1995). What's up? Perspectives from the first international canopy conference at Sarasota, Florida, 1994. *Selbyana*, **16**, 1–11.

Lowman, M. D. & Heatwole, H. (1992). Spatial and temporal variability in defoliation of Australian *Eucalyptus*. *Ecology*, **73**, 129–142.

Lowman, M. D. & Moffett, M. (1993). The ecology of tropical rain forest canopies. *Trends in Ecology and Evolution*, **8**, 104–107.

Lowman, M. D., Moffett, M. & Rinker, H. B. (1993a). A new technique for taxonomic and ecological sampling in rain forest canopies. *Selbyana*, **14**, 75–79.

Lowman, M. D. & Nadkarni, N. M. (eds.) (1995). *Forest Canopies*. San Diego, CA: Academic Press.

Lowman, M. D., Taylor, P. & Block, N. (1993b). Vertical stratification of small mammals and insects in the canopy of a temperate deciduous forest: a reversal of tropical forest distribution? *Selbyana*, **14**, 25.

Lowman, M. D. & Wittman, P. K. (1996). Forest canopies: methods, hypotheses, and future directions. *Annual Review of Ecology and Systematics*, **27**, 55–81.

Lowman, M. D., Wittman, P. K. & Murray, D. (1996). Herbivory in a bromeliad of the Peruvian rain forest canopy. *Journal of the Bromeliad Society*, **46**, 52–55.

Luczak, J. (1966). The distribution of wandering spiders in different layers of the environment as a result of interspecies competition. *Ekologia Polska Series A*, **14**, 233–244.

Ludwig, J. A. & Reynolds, J. F. (1988). *Statistical Ecology, a Primer on the Methods and Computing*. New York: John Wiley & Sons.

Mabberley, D. J. (1997). *The Plant-Book. A portable Dictionary of the Vascular Plants*. Cambridge: Cambridge University Press.

Macedo, C. A. & Langenheim, J. H. (1989). Intra- and interplant leaf sesquiterpene variability in *Copaifera langsdorfii*: relation to microlepidopteran herbivory. *Biochemical Systematics and Ecology*, **17**, 551–557.

Macia, A. & Bradshaw, W. E. (2000). Seasonal availability of resources and habitat degradation for the western tree-hole mosquito, *Aedes sierrensis*. *Oecologia*, **125**, 55–65.

MacKay, W. P., Solange, S. & Whitford, W. G. (1987). Diurnal activity patterns and vertical migration in desert soil microarthropods. *Pedobiologia*, **30**, 65–71.

Madeira, J. A., Ribeiro, K. T. & Fernandes, G. W. (1998). Herbivory, tannins and sclerophylly in *Chamaecrista linearifolia* (Fabaceae) along an altitudinal gradient. *Brazilian Journal of Ecology*, **2**, 24–29.

Magurran, A. E. (1988). *Ecological Diversity and its Measurement*. London: Chapman & Hall.

Majer, J. D. (1972). The ant mosaic in Ghana cocoa farms. *Bulletin of Entomological Research*, **62**, 151–160.

Majer, J. D. (1976a). The maintenance of the ant mosaic in Ghana cocoa farms. *Journal of Applied Ecology*, **13**, 123–144.

Majer, J. D. (1976b). The ant mosaic in Ghana cocoa farms: further structural considerations. *Journal of Applied Ecology*, **13**, 145–155.

Majer, J. D. (1990). The abundance and diversity of arboreal ants in northern Australia. *Biotropica*, **22**, 191–199.

Majer, J. D. (1993). Comparison of the arboreal ant mosaic in Ghana, Brazil, Papua New Guinea and Australia – its structure and influence on arthropod diversity. In *Hymenoptera and Biodiversity*, eds. J. LaSalle & I. D. Gauld, pp. 115–141. Walligford, UK: CAB International.

Majer, J. D. (1996). Ant recolonisation of rehabilitated bauxite mines at Trombetas, Para, Brazil. *Journal of Tropical Ecology*, **12**, 257–273.

Majer, J. D. & Camer-Pesci, P. (1991). Ant species in tropical Australian tree crop and native ecosystems – is there a mosaic? *Biotropica*, **23**, 173–181.

Majer, J. D., Delabie, J. H. C. & Smith, M. R. B. (1994). Arboreal ant community patterns in Brazilian cocoa farms. *Biotropica*, **26**, 73–83.

Majer, J. D. & Recher, H. F. (1988). Invertebrate communities on western Australian eucalypts, a comparison of branch clipping and chemical knockdown procedures. *Australian Journal of Ecology*, **13**, 269–278.

Majer, J. D. & Recher, H. F. (1999). Are eucalypts Brazil's friend or foe? An entomological viewpoint. *Anais da Sociedade Entomológica do Brasil*, **28**, 185–200.

Majer, J. D., Recher, H. F. & Ganesh, S. (2000). Diversity patterns of eucalypt canopy arthropods in eastern and western Australia. *Ecological Entomology*, **25**, 295–306.

Malhi, Y. & Grace, J. (2000). Tropical forests and atmospheric carbon dioxide. *Trends in Ecology and Evolution*, **15**, 332–337.

Mallet, J. (1986). Dispersal and gene flow in a butterfly with home range behavior: *Heliconius erato* (Lepidoptera: Nymphalidae). *Oecologia*, **68**, 210–217.

Mallet, J. & Joron, M. (1999). Evolution of diversity in warning color and mimicry: polymorphisms, shifting balance, and speciation. *Annual Review of Ecology and Systematics*, **30**, 201–233.

Manokaran, N. (1996). Effect, 34 years later, of selective logging in the lowland Dipterocarp forest at Pasoh, Peninsular Malaysia, and implications on present-day logging in hill forests. In *Conservation, Management and Development of Forest Resources. Proceedings of the Malaysia–United Kingdom Programme Workshop, 1996*, eds. L. S. See, D. Y. May, I. D. Gauld & J. Bishop, pp. 41–61. Kuala Lumpur: Malaysia–United Kingdom Programme.

Mantel, N. (1967). The detection of disease clustering and a generalized regression approach. *Cancer Research*, **27**, 209–220.

Mapongmetsem, P. M., Duguma, B., Nkongmeneck, B. A. & Puig, H. (1998). Déterminisme de la défeuillaison chez quelques essences forestières tropicales du Cameroun. *Revue d' Ecologie (Terre et Vie)*, **53**, 193–210.

Maraun, M., Alphei, J., Bonkowski, M., Buryn, R., Migge, S., Peter, M., Schaefer, M. & Scheu, S. (1999). Middens of the earthworm *Lumbricus terrestris* (Lumbricidae): microhabitats for micro- and mesofauna in forest soil. *Pedobiologia*, **43**, 276–287.

Marques, A. R., Garcia, Q. S., Rezende, J. L. P. & Fernandes, G. W. (2000). Variations in leaf characteristics of two species of *Miconia* in the Brazilian cerrado under different light intensities. *Tropical Ecology*, **41**, 47–60.

Marquis, R. J. (1991). Hervivore fauna of *Piper* (Piperaceae) in a Costa Rican wet forest: diversity, specificity and impact. In *Plant–Animal Interactions: Evolutionary Ecology in Tropical and Temperate Regions*, eds. P. W. Price, T. M. Lewinsohn, G. W. Fernandes & W. W. Benson, pp. 179–208. New York: John Wiley & Sons.

Marquis, R. J. (1992). Selective impact of herbivores. In *Plant Resistance to Herbivores and Pathogens: Ecology, Evolution, and Genetics*, eds. R. S. Fritz & E. L. Simms, pp. 301–325. Chicago, IL: University of Chicago Press.

Marquis, R. J. & Braker, H. E. (1993). Plant–herbivore interactions: diversity, specificity, and impact. In *La Selva: Ecology and Natural History of a Neotropical Rain Forest*, eds. L. A. McDade, K. S. Bawa, H. A. Hespenheide & G. S. Hartshorn, pp. 261–281. Chicago, IL: University of Chicago Press.

Marshall, A. G. & Swaine, M. D. (eds.) (1992). *Tropical rain forest: disturbance and recovery. Philosophical Transactions of the Royal Society of London, Series B*, **335**, 323–462.

Martin, S. J. (1995). Hornets (Hymenoptera: Vespinae) of Malaysia. *Malayan Nature Journal*, **49**, 71–82.

Martínez-Yrizar, A., Maass, J. M., Pérez-Jiménez, A. & Sarukhán, J. (1996). Net primary productivity of a tropical deciduous forest ecosystem in western Mexico. *Journal of Tropical Ecology*, **12**, 169–175.

Martius, C. & Bandeira, A. G. (1998). Wood litter stock in tropical moist forests in central Amazonia. *Ecotropica*, **4**, 115–118.

Maschwitz, U. & Hänel, H. (1988). Biology of the Southeast Asian nocturnal wasp *Provespa anomala* (Hymenoptera: Vespidae). *Entomologia Generalis*, **14**, 47–52.

Mason, D. (1996). Responses of Venezuelan understorey birds to selective logging, enrichment strips, and vine cutting. *Biotropica*, **28**, 296–309.

Masters, G. J. (1995). The effect of herbivore density on host plant mediated interactions between two insects. *Ecological Research*, **10**, 125–133.

Masters, G. J. & Brown, V. K. (1992). Plant-mediated interactions between two spatially separated insects. *Functional Ecology*, **6**, 175–179.

MathSoft (1996). *S-Plus 4.5: Guide to Statistics*. Seattle, WA: MathSoft Incorporated.

Matson, P. A. & Hunter, M. D. (1992). The relative contributions of top-down and bottom-up forces in population and community ecology. *Ecology*, **73**, 723.

Mattingly, P. F. (1949). Studies on West African forest mosquitoes. Part I. The seasonal distribution, biting cycle and vertical distribution of four of the principal species. *Bulletin of Entomological Research*, **40**, 149–168.

Mattson, W. J. (1980). Herbivory in relation to plant nitrogen content. *Annual Review of Ecology and Systematics*, **11**, 119–161.

Mawdsley, N. A. & Stork, N. E. (1997). Host-specificity and the effective specialization of tropical canopy beetles. In *Canopy Arthropods*, eds. N. E. Stork, J. Adis & R. K. Didham, pp. 104–130. London: Chapman & Hall.

May, R. M. (1975). Patterns of species abundance and diversity. In *Ecology and Evolution of Communities*, eds. M. L. Cody & J. M. Diamond, pp. 81–120. Cambridge, MA: Harvard University Press.

May, R. M. (1986). How many species are there? *Nature*, **324**, 514–515.

May, R. M. (1988). How many species are there on Earth? *Science*, **241**, 1441–1449.

May, R. M. (1990). How many species? *Philosophical Transactions of the Royal Society of London, Series B*, **330**, 293–304.

May, R. M. (1992). How many species inhabit the Earth? *Scientific American*, **267**, 42–48.

May, R. M. (1994). Conceptual aspects of the quantification of the extent of biological diversity. *Philosophical Transactions of the Royal Society of London, Series B*, **345**, 13–20.

May, R. M. & MacArthur, R. H. (1972). Niche overlap as a function of environmental variability. *Proceedings of the National Academy of Science USA*, **69**, 1109–1113.

Maynard Smith, J. (1966). Sympatric speciation. *American Naturalist*, **100**, 637–650.

McArdle, B. H., Gaston, K. J. & Lawton, J. H. (1990). Variation in the size of animal populations: patterns, problems and artefacts. *Journal of Animal Ecology*, **59**, 439–454.

McClure, H. E. (1966). Flowering, fruiting and animals in the canopy of a tropical rain forest. *Malayan Forester*, **29**, 182–203.

McCune, B. & Mefford, M. J. (1999). *PC-ORD. Multivariate Analysis of Ecological Data*, Version 4. Gleneden Beach, CA: MjM Software Design.

McDonald, J. F. & Thomas, M. B. (1990). *Flora of Lamington National Park*. Brisbane: Queensland Department of Primary Industries.

McDowell, W. H. & Estrada-Pinto, A. (1988). *Rainfall at El Verde Field Station, 1964–1986*. (Technical Report CEER-T-228.) San Juan, Puerto Rico: US Department of Energy, Center for Energy and Environmental Research.

McGavin, G. C. (1999). Arthropod diversity and the tree flora of Mkomazi. In *Mkomazi: the Ecology, Biodiversity and Conservation of a Tanzanian Savanna*, eds. M. Coe, N. McWilliams, G. Stone & M. Packer, pp. 371–390. London: Royal Geographical Society and the Institute of British Geographers.

McGraw, J. R. & Farrier, M. H. (1969). Mites of the superfamily Parasitoidea (Acarina: Mesostigmata) associated with *Dendroctonus* and *Ips* (Coleoptera: Scolytidae). *North Carolina Agricultural Experiment Station. Technical Bulletin*, **192**, 1–162.

McQuillan, P. B. (1993). *Nothofagus* (Fagaceae) and its invertebrate fauna – an overview and preliminary synthesis. *Biological Journal of the Linnean Society*, **49**, 317–354.

Medeiros, M. A., Fowler, H. G. & Delabie, J. H. C. (1996). O mosaico de formigas (Hymenoptera; Formicidae) em cacauais do sul da Bahia. *Científica*, **23**, 291–300.

Medianero, E. (1999). Riqueza de insectos formadores de agallas en dos zonas ecológicas tropicales. MSc Thesis, University of Panamá.

Medina, M. C., Robbins, R. K. & Lamas, G. (1996). Vertical stratification of flight by Ithomiine butterflies (Lepidoptera: Nymphalidae) at Pakitza, Manu National Park, Perú. In *Manu: The Biodiversity of Southeastern Peru*, eds. D. E. Wilson & A. Sandoval, pp. 211–216. Lima, Peru: Smithsonian Institution & Editorial Horizonte.

Medway, L. (1972). Phenology of a tropical rain forest in Malaya. *Biological Journal of the Linnean Society*, **4**, 117–146.

Meffe, G. K. & Carrol, R. (1997). *Principles of Conservation Biology*. Sunderland, MA: Sinauer.

Melo, C. & López, J. (1994). Parque Nacional El Chico, marco geográfico-natural y propuesta de zonificación para su manejo operativo. *Investigaciones Geográficas Boletín*, **28**, 65–128.

Memmott, J., Godfray, H. C. J. & Bolton, B. (1993). Predation and parasitism in a tropical herbivore community. *Ecological Entomology*, **18**, 348–352.

Mendonça, M. S., Jr (2001). Gall insects diversity patterns: the resource synchronisation hypothesis. *Oikos*, **95**, 171–176.

Menken, S. B. J. (1996). Pattern and process in the evolution of insect-plant associations: *Yponomeuta* as an example. *Entomologia Experimentalis et Applicata*, **80**, 297–305.

Mercier, J. L. (1997). Les communications sociales chez la fourmi *Polyrhachis laboriosa* (Hymenoptera: Formicidae). PhD Thesis, University of Paris-Nord.

Mercier, J. L. & Dejean, A. (1996). Ritualized behaviour during competition for food between two Formicinae. *Insectes Sociaux*, **43**, 17–29.

Mercier, J. L., Dejean, A. & Lenoir, A. (1998). Limited aggressiveness among African arboreal ants sharing the same territories: the result of a co-evolutionary process. *Sociobiology*, **32**, 139–150.

Mercier, J. L., Lenoir, A. & Dejean, A. (1997). Ritualized versus aggressive behaviours displayed by *Polyrhachis laboriosa* (F. Smith) during intraspecific competition. *Behavioural Processes*, **41**, 39–50.

Michaud, J. P. (1990). Conditions for the evolution of polyphagy in herbivorous insects. *Oikos*, **57**, 278–279.

Miko, L. & Stanko, M. (1991). Small mammals as carriers of non-parasitic mites (Oribatida, Uropodina). In *Modern Acarology*, Vol. 1, eds. F. Dusbábek & V. Bukva, pp. 395–402. The Hague & Prague: SPB Academic & Academia.

Miles, D. B. & Dunham, A. E. (1993). Historical perspectives in ecology and evolutionary biology: the use of phylogenetic comparative analyses. *Annual Review of Ecology and Systematics*, **24**, 587–619.

Miller, J. S. (1987). Host-plant relationship in the Papilionidae (Lepidoptera): parallel cladogenesis or colonization? *Cladistics*, **3**, 105–120.

Miller, J. S., Janzen, D. H. & Franclemont, J. G. (1997). New species of *Euhapigioides*, new genus, and *Hapigiodes* in Hapigiini, new tribe, from Costa Rica, with notes on their life history and immatures (Lepidoptera: Notodontidae). *Tropical Lepidoptera*, **8**, 81–99.

Miller, M. F. (1996). Acacia seed predation by bruchids in an African savanna ecosystem. *Journal of Applied Ecology*, **33**, 1137–1144.

Miller, S. E. (1991). Biological diversity and the need to nurture systematic collections. *American Entomologist*, **37**, 76.

Miller, S. E. (1994). Development of world identification services: networking. In *Identification and Characterization of Pest Organisms*, ed. D. L. Hawksworth, pp. 69–80. Wallingford, UK: CAB International.

Miller, S. E. (2000). Taxonomy for understanding biodiversity – the strengths of the BioNet International model. In *Proceedings of the Second BioNet–International Global Workshop (BIGW2)*, eds. T. Jones & S. Gallagher, pp. 13–18. Egham, UK: BioNet International.

Miller, S. E. & Rogo, L. M. (2002). Challenges and opportunities in understanding and utilising African insect diversity. *Cimbebasia*, **17**, 197–218.

Millidge, A. F. & Russell-Smith, A. (1992). Linyphiidae from rain forests of Southeast Asia (Araneae). *Journal of Natural History*, **26**, 1367–1404.

Milne, A. (1957). The natural control of insect populations. *Canadian Entomologist*, **89**, 193–213.

Minchin, P. R. (1987). An evaluation of the relative robustness of techniques for ecological ordination. *Vegetatio*, **69**, 89–107.

Minckley, R. L., Cane, J. H., Kervin, L. & Roulston, T. H. (1999). Spatial predictability and resource specialization of bees at a superabundant, widespread resource. *Biological Journal of the Linnean Society*, **67**, 119–147.

Minitab, I. (1998). *MINITAB. Version 12.1.* State College, PA: Minitab, Inc.

Missa, O. (2000). Diversity and Spatial Heterogeneity of a Weevil Fauna Living in the Canopy of a Tropical Lowland Rainforest in Papua New Guinea. PhD thesis. Bruxelles: Université Libre de Bruxelles.

Mitchell, A. W. (1982). *Reaching the Rain Forest Roof.* Leeds: Leeds Philosophical and Literary Society.

Mitchell, A. W. (2001). Introduction – Canopy science: time to shape up. *Plant Ecology*, **153**, 5–11.

Mitter, C., Farrell, B. & Futuyma, D. J. (1991). Phylogenetic studies of insect–plant interactions: insights into the genesis of diversity. *Trends in Ecology and Evolution*, **6**, 290–293.

Moffett, M. W. (1993). *The High Frontier. Exploring the Tropical Rainforest Canopy*. Cambridge, MA: Harvard University Press.

Moffett, M. W. (2000). What's 'up'? A critical look at the basic terms in canopy biology. *Biotropica*, **32**, 569–596.

Moffett, M. W. & Lowman, M. D. (1995). Canopy access techniques. In *Forest Canopies*, eds. M. D. Lowman & N. M. Nadkarni, pp. 3–26. San Diego, CA: Academic Press.

Mole, S., Ross, J. A. M. & Waterman, P. G. (1988). Light-induced variation in phenolic levels in foliage of rain-forest plants. I. Chemical changes. *Journal of Chemical Ecology*, **14**, 1–22.

Momose, K., Nagamitsu, T. & Inoue, T. (1996). The reproductive ecology of an emergent dipterocarp in a lowland rain forest in Sarawak. *Plant Species Biology*, **11**, 189–198.

Momose, K., Nagamitsu, T., Sakai, S., Inoue, T. & Hamid, A. A. (1994). Climatic data in Lambir Hills National

Park and Miri Airport, Sarawak. In *Plant Reproductive Systems and Animal Seasonal Dynamics: Long-term Study of Dipterocarp Forests in Sarawak* (Canopy Biology Program in Sarawak Series I), eds. T. Inoue & A. A. Hamid, pp. 26–39. Kyoto, Japan: Center for Ecological Research, Kyoto University.

Momose, K., Yumoto, T., Nagamitsu, T., Kato, M., Nagamasu, H., Sakai, S., Harrison, R. D., Itioka, T., Hamid, A. A. & Inoue, T. (1998). Pollination biology of a lowland dipterocarp forest. I. Characteristics of the plant-pollinator community in a lowland dipterocarp forest. *American Journal of Botany*, **85**, 1477–1501.

Monteith, J. L. (1965). Light distribution and photosynthesis in field crops. *Annals of Botany*, **29**, 17–37.

Moore, L. V., Myers, J. H. & Eng, R. (1988). Western tent caterpillars prefer the sunny side of the tree, but why? *Oikos*, **51**, 321–326.

Morais, H. C., Diniz, I. R. & Baumgarten, L. (1995). Padroes de produção de folhas e sua utilização por larvas de Lepidoptera em um cerrado brasileiro. *Revista Brasileira de Botânica*, **18**, 163–170.

Moran, V. C., Hoffmann, J. H., Impson, F. A. C. & Jenkins, J. F. G. (1994). Herbivorous insect species in the tree canopy of a relict South African forest. *Ecological Entomology*, **19**, 147–154.

Moran, V. C. & Southwood, T. R. E. (1982). The guild composition of arthropod communities in trees. *Journal of Animal Ecology*, **51**, 289–306.

Morawetz, W. (1998). The Surumoni project: the botanical approach toward gaining an interdisciplinary understanding of the functions of the rain forest canopy. In *Biodiversity: A Challenge for Development Research and Policy*, eds. W. Barthlott & M. N. Winiger, pp. 71–80. Berlin: Springer-Verlag.

Morris, R. F. (1960). Sampling insect populations. *Annual Review of Entomology*, **5**, 243–264.

Morris, R. F. (1963). The dynamics of epidemic spruce budworm populations. *Memoirs of the Entomological Society of Canada*, **31**, 1–322.

Morris, W., Grevstad, F. & Herzig, A. (1996). Mechanisms and ecological functions of spatial aggregations in chrysomelid beetles. In *Chrysomelidae Biology*, Vol. 2, Ecological Studies, eds. P. H. A. Jolivet & M. L. Cox, pp. 303–322. Amsterdam: SPB Academic Publishing.

Morris, W. F., Wiser, S. D. & Klepetka, B. (1992). Causes and consequences of spatial aggregation in the phytophagous beetle *Altica tombacina*. *Journal of Animal Ecology*, **62**, 49–58.

Morse, D. R., Stork, N. E. & Lawton, J. H. (1988). Species number, species abundance and body length relationships of arboreal beetles in Bornean lowland rain forest trees. *Ecological Entomology*, **13**, 25–37.

Mrciak, M. & Sixl, W. (1979). Faunistic–Ecological investigations on mites living in the nests of tree cavities in Austria. In *Proceedings of the 4th International Congress of Acarology*, ed. E. Piffl, pp. 611–614. Budapest: Akadémiai Kiado.

Muirhead-Thomson, R. C. (1991). *Trap Responses of Flying Insects*. London: Academic Press.

Mulkey, S. S., Kitajima, K. & Wright, S. J. (1996). Plant physiological ecology of tropical forest canopies. *Trends in Ecology and Evolution*, **11**, 408–412.

Munroe, E. & Solis, M. A. (1999). The Pyraloidea. In *Lepidoptera, Moths and Butterflies.* Vol. 1, *Evolution, Systematics, and Biogeography. Handbuch der Zoologie*, Bd. IV *Arthropoda: Insecta*, part 35, ed. N. P. Kristensen, pp. 233–261. Berlin: Walter de Gruyter.

Murali, K. S. & Sukumar, R. (1993). Leaf flushing phenology and herbivory in a tropical dry deciduous forest, southern India. *Oecologia*, **94**, 114–119.

Murillo, B. C., Astaiza, V. R. & Fajardo, O. P. (1988). Biología de *Anopheles* (*Kerteszia*) *neivai* H., D. & K., 1913 (Diptera: Culicidae) en la costa Pacífica de Colombia. 1. Fluctuación de la población larval y características de sus criaderos. *Revista de Salud Pública*, **22**, 94–100.

Muul, I. (1999). Development of canopy walkways by Illar Muul and associates: a brief history. *Selbyana*, **20**, 186–190.

Muul, I. & Lim, B. L. (1970). Vertical zonation in a tropical rain forest in Malaysia: methods of study. *Science*, **169**, 788–789.

Nadkarni, N. M. (1981). Canopy roots: convergent evolution in rainforest nutrient cycles. *Science*, **213**, 1024–1025.

Nadkarni, N. M. (1988). Use of a portable platform for observation of tropical forest canopy animals. *Biotropica*, **20**, 350–351.

Nadkarni, N. M. (1994). Diversity of species and interactions in the upper tree canopy of forest ecosystems. *American Zoologist*, **34**, 70–78.

Nadkarni, N. M. (1995). Good-bye, Tarzan. The science of life in the treetops gets down to business. *The Sciences*, **Jan/Feb**, 28–33.

Nadkarni, N. M. & Longino, J. T. (1990). Invertebrates in canopy and ground organic matter in a Neotropical montane forest, Costa Rica. *Biotropica*, **22**, 286–289.

Nadkarni, N. M. & Parker, G. G. (1994). A profile of forest canopy science and scientists – Who we are, what we want to know, and obstacles we face: results of an international survey. *Selbyana*, **15**, 38–50.

Nadkarni, N. M., Parker, G. G., Ford, E. D., Cushing, J. B. & Stallman, C. (1996). The international canopy network: a pathway for interdisciplinary exchange of scientific information on forest canopies. *Northwest Science*, **70**, 104–108.

Nagamasu, H., Momose, K. & Nagamitsu, T. (1994). Flora of the canopy biology plot and surrounding areas in Lambir Hills National Park, Sarawak. In *Plant Reproductive Systems and Animal Seasonal Dynamics: Long-term Study of Dipterocarp Forests in Sarawak* (Canopy Biology Program in Sarawak Series I), eds. T. Inoue & A. A. Hamid, pp. 47–56. Kyoto, Japan: Center of Ecological Research, Kyoto University.

Nakagawa, M., Tanaka, K., Nakashizuka, T., Ohkubo, T., Kato, T., Maeda, T., Sato, K., Miguchi, H., Nagamasu, H., Ogino, K., Teo, S., Hamid, A. A. & Seng, L. H. (2000). Impact of severe drought associated with the 1997–1998 El Niño in a tropical forest in Sarawak. *Journal of Tropical Ecology*, **16**, 355–367.

Nakamura, K., Abbas, I. & Hasyim, A. (1990). Seasonal fluctuations of the lady beetle *Epilachna vigintioctopunctata* (Coccinellidae: Epilachninae) in Sumatra and comparisons to other tropical insect population cycles. In *Natural History of Social Wasps and Bees in Equatorial Sumatra*, eds. S. F. Sakagami, R. Ohgushi & D. W. Roubik, pp. 13–29. Sapporo, Japan: Hokkaido University Press.

Nannelli, R., Turchetti, T. & Maresi, G. (1998). Corticolous mites (Acari) as potential vectors of *Cryphonectria parasitica* (Murr.) Barr hypovirulent strains. *International Journal of Acarology*, **24**, 237–244.

Nascimento, M. T. & Proctor, J. (1994). Insect defoliation of a monodominant Amazonian rainforest. *Journal of Tropical Ecology*, **10**, 633–636.

Nascimento, M. T. & Proctor, J. (1997). Population dynamics of five tree species in a monodominant *Peltogyne* forest and two other forest types on Maracá island, Roraima, Brazil. *Forest Ecology and Management*, **94**, 115–128.

Nason, J. D., Herre, E. A. & Hamrick, J. L. (1996). Paternity analysis of the breeding structure of strangler fig populations: evidence for substantial long-distance wasp dispersal. *Journal of Biogeography*, **23**, 501–512.

Nepstad, D. C., Veríssimo, A., Alencar, A., Nobre, C., Lima, E., Lefebvre, P., Schlesinger, P., Potter, C., Moutinho, P., Mendoza, E., Cochrane, M. & Brooks, V. (1999). Large-scale impoverishment of Amazonian forests by logging and fire. *Nature*, **398**, 505–508.

Neter, J., Wasserman, W. & Kutner, M. H. (1990). *Applied Linear Statistical Models*, 3rd edn. Boston, MA: Richard D. Irwin.

Netuzhilin, I., Chacon, P., Cerda, H., López-Hernández, D., Torres, F., Paoletti, M. G. & Reddy, M. V. (1999). Assessing agricultural impact using ant morphospecies as bioindicators in the Amazonian savanna–forest ecotone (Puerto Ayacucho, Amazon State), Venezuela. In *Management of Tropical Agroecosystems and the Beneficial Soil Biota*, ed. M. V. Reddy, pp. 291–352. Enfield, NH: Science Publishers.

Ng, R. & Lee, S. S. (1980). Environmental factors affecting the vertical distribution of Diptera in a tropical primary lowland dipterocarp forest in Malaysia. In *Tropical Ecology and Development. Proceedings of the Vth International Symposium of Tropical Ecology, 16–21 April 1979, Kuala Lumpur, Malaysia*, ed. J. I. Furtado, pp. 123–129. Kuala Lumpur: The International Society of Tropical Ecology.

Ngnegueu, P. R. & Dejean, A. (1992). Les relations plantes–fourmis au niveau des troncs d'arbres. In *Biologie d'une Canopée de Forêt Equatoriale – II. Rapport de Mission: Radeau des Cimes Octobre–Novembre 1991, Réserve de Campo, Cameroun*, eds. F. Hallé & O. Pascal, pp. 81–82. Paris: Fondation Elf.

Ngnegueu, P. R. & Dejean, A. (1994). Prey capture by the carpenter ant *Atopomyrmex mocquerysi* André (Formicidae- Myrmicinae). In *Les Insectes Sociaux*, eds. A. Lenoir, G. Arnold & M. Lepage, pp. 480–480. Paris: Presses de l'Université Paris Nord.

Nichols, J. D., Agyeman, V. K., Agurgo, F. B., Wagner, M. R. & Cobbinah, J. R. (1999). Patterns of seedling survival in the tropical African tree, *Milicia excelsa*. *Journal of Tropical Ecology*, **15**, 451–461.

Nicolai, V. (1985). Die ökologische Bedeutung verschiedener Rindentypen bei Bäumen. Dissertation. Marburg: University of Marburg.

Nicolai, V. (1986). The bark of trees: thermal properties, microclimate and fauna. *Oecologia*, **69**, 148–160.

Nicolai, V. (1989). Thermal properties and fauna of the bark of trees in two different African ecosystems. *Oecologia*, **80**, 421–430.

Nieder, J., Engwald, S., Klawun, M. & Barthlott, W. (2000). Spatial distribution of vascular epiphytes in a lowland Amazonian rainforest in southern Venezuela (Surumoni crane project). *Biotropica*, **32**, 385–396.

Nieder, J., Prosperí, J. & Michaloud, G. (2001). Epiphytes and their contribution to canopy diversity. *Plant Ecology*, **153**, 51–63.

Nielsen, B. O. (1978). Food resource partition in the beech leaf-feeding guild. *Ecological Entomology*, **3**, 193–201.

Nielsen, B. O. (1987). Vertical distribution of insect populations in the free air space of beech woodland. *Entomologiske Meddelelser*, **54**, 169–178.

Nishida, G. M. (1997). Hawaiian Terrestrial Arthropod Checklist, 3rd edn. *Bishop Museum Technical Report*, **12**, iv, 263.

Nishida, T., Haramoto, F. H. & Nakahara, L. M. (1980). Altitudinal distribution of endemic psyllids (Homoptera: Psyllidae) in the *Metrosideros* ecosystem.

Proceedings of the Hawaiian Entomological Society, **23**, 255–262.

Normark, B. B., Jordal, B. H. & Farrell, B. D. (1999). Origin of a haplodiploid beetle lineage. *Proceedings of the Royal Society of London Series B*, **266**, 2253–2259.

Norton, R. A. (1980). Observations on phoresy by oribatid mites. *International Journal of Acarology*, **6**, 121–130.

Norton, R. A. (1998). Morphological evidence for the evolutionary origin of Astigmata (Acari: Acariformes). *Experimental and Applied Acarology*, **22**, 559–594.

Norton, R. A. (2001). Systematic relationships of Nothrolohmanniidae, and the evolutionary plasticity of body form in the Enarthronota (Acari: Oribatida). In *Acarology. Proceedings of the 10th International Congress*, eds. R. B. Halliday, D. E. Walter, H. C. Proctor, R. A. Norton & M. J. Colloff, pp. 58–75. Melbourne: CSIRO Publishing.

Norton, R. A., Bonamo, P. M., Grierson, J. D. & Shear, W. A. (1988). Oribatid mite fossils from a terrestrial Devonian deposit near Gilboa, New York. *Journal of Palaeontology*, **62**, 259–269.

Norton, R. A., Kethley, J. B., Johnston, D. E. & O'Connor, B. M. (1993). Phylogenetic perspectives on genetic systems and reproductive modes of mites. In *Evolution and Diversity of Sex Ratio in Insects and Mites*, eds. D. Wrensch & M. A. Ebbert, pp. 8–89. New York: Chapman & Hall.

Norton, R. A. & Palacios-Vargas, J. G. (1987). A new arboreal Scheloribatidae, with ecological notes on epiphytic oribatid mites of Popocatépetl, Mexico. *Acarologia*, **28**, 75–89.

Norton, R. A. & Palmer, S. C. (1991). The distribution, mechanisms and evolutionary significance of parthenogenesis in oribatid mites. In *The Acari: Reproduction, Development and Life History Strategies*, eds. R. Schuster & P. W. Murphy, pp. 107–136. London: Chapman & Hall.

Novotny, V. (1993). Spatial and temporal components of species diversity in Auchenorrhyncha (Insecta: Hemiptera) communities of Indochinese montane rain forest. *Journal of Tropical Ecology*, **9**, 93–100.

Novotny, V. (1994). Association of polyphagy in leafhoppers (Auchenorrhyncha, Hemiptera) with unpredictable environments. *Oikos*, **70**, 223–231.

Novotny, V. (2000). *Ol binatang kisim nem bilong Madang.* Wantok. Madang, Papua New Guinea: 13 April 2000.

Novotny, V. & Basset, Y. (1998). Seasonality of sap-sucking insects (Auchenorrhyncha, Hemiptera) feeding on *Ficus* (Moraceae) in a lowland rain forest in Papua New Guinea. *Oecologia*, **115**, 514–522.

Novotny, V. & Basset, Y. (1999). Body size and host plant specialisation: a relationship from a community of herbivorous insects from New Guinea. *Journal of Tropical Ecology*, **15**, 315–328.

Novotny, V. & Basset, Y. (2000). Rare species in communities of tropical insect herbivores: pondering the mystery of singletons. *Oikos*, **89**, 564–572.

Novotny, V., Basset, Y., Auga, J., Boen, W., Dal, C., Drozd, P., Kasbal, M., Isua, B., Kutil, R., Manumbor, M. & Molem, K. (1999a). Predation risk for herbivorous insects on tropical vegetation: a search for enemy-free space and time. *Australian Journal of Ecology*, **24**, 477–483.

Novotny, V., Basset, Y., Miller, S. E., Allison, A., Samuelson, G. A. & Orsak, L. J. (1997). The diversity of tropical insect herbivores: an approach to collaborative international research in Papua New Guinea. In *Proceedings of the International Conference on Taxonomy and Biodiversity Conservation in the East Asia*, eds. B. H. Lee, J. C. Choe & H. Y. Han, pp. 112–125. Chonju: Korean Institute for Biodiversity Research of Chonbuk National University.

Novotny, V., Basset, Y., Miller, S. E., Drozd, P. & Cizek, L. (2002a). Host specialisation of leaf-chewing insects in a New Guinean rain forest. *Journal of Animal Ecology*, **71**, 400–412.

Novotny, V., Basset, Y., Miller, S. E., Weiblen, G. D., Bremer, B., Cizek, L. & Drozd, P. (2002b). Low host specificity of herbivorous insects in a tropical forest. *Nature*, **416**, 841–844.

Novotny, V., Basset, Y., Samuelson, G. A. & Miller, S. E. (1999b). Host use by chrysomelid beetles feeding on *Ficus* (Moraceae) and Euphorbiaceae in New Guinea. In *Advances in Chrysomelidae Biology 1*, ed. M. Cox, pp. 343–360. Leyden: Backhuys.

Novotny, V. & Leps, J. (1997). Distribution of leafhoppers (Auchenorrhyncha, Hemiptera) on their host plant *Oxyspora paniculata* (Melastomataceae) in the understorey of a diverse rainforest. *Ecotropica*, **3**, 83–90.

Novotny, V., Tonner, M. & Spitzer, K. (1991). Distribution and flight behaviour of the junglequeen butterfly, *Stichophthalma louisa* (Lepidoptera: Nymphalidae), in an Indochinese montane rainforest. *Journal of Research on the Lepidoptera*, **30**, 279–288.

Novotny, V. & Wilson, M. R. (1997). Why are there no small species among xylem-sucking insects? *Evolutionary Ecology*, **11**, 419–437.

Nowicki, C. (1998). Diversität epiphytischer und terrestrischer Pflanzen eines ecuadorianischen Bergnebelwaldes im Vergleich. Diploma thesis. Botanisches Institut, University of Bonn.

Noyes, J. S. (1989a). A study of five methods of sampling Hymenoptera (Insecta) in a tropical rainforest, with special reference to the Parasitica. *Journal of Natural History*, **23**, 285–298.

Noyes, J. S. (1989b). The diversity of Hymenoptera in the tropics with special reference to Parasitica in Sulawesi. *Ecological Entomology*, **14**, 197–207.

Nummelin, M. & Fuersch, H. (1992). Coccinellids of the Kibale Forest, Western Uganda: a comparison between virgin and managed sites. *Tropical Zoology*, **5**, 155–166.

Nunes, A. L. & Adis, J. (1992). Observaciones sobre el comportamiento sexual y la oviposición de *Stenacris fissicauda fissicauda* (Bruner, 1908) (Orthoptera, Acrididae). *Etologia*, **2**, 59–63.

Oba, G., Stenseth, C. N. & Lusigi, W. J. (2000). New perspectives on sustainable grazing management in arid zones of Sub-Saharan Africa. *BioScience*, **50**, 35–51.

Oberprieler, R. (2003). Weevils – evolution's ultimate success in diversity? In *Gondwana Alive, Biodiversity and the Evolving Terrestrial Biosphere*, eds. J. M. Anderson, M. de Wit, F. Thackeray & B. van Wyk. Johannesburg: Witwatersrand University Press, in press.

O'Brien, C. W. & Wibmer, G. J. (1982). Annotated Checklist of the Weevils (Curculionidae *sensu lato*) of North America, Central America, and the West Indies. *Memoirs of the American Entomological Institute*, **34**, ix, 382.

O'Brien, L. B. & Wilson, S. W. (1985). Planthopper systematics and external morphology. In *The Leafhoppers and Planthoppers*, eds. L. R. Nault & J. G. Rodriguez, pp. 61–102. New York: John Wiley & Sons.

OConnor, B. M. (1982). Evolutionary ecology of astigmatid mites. *Annual Review of Entomology*, **27**, 385–409.

OConnor, B. M. (1984). Acarine–Fungal Relationships: the evolution of symbiotic associations. In *Fungus–Insect Relationships: Perspectives in Ecology and Evolution*, eds. Q. Wheeler & M. Blackwell, pp. 354–381. New York: Columbia University Press.

Ødegaard, F. (2000a). The relative importance of trees versus lianas as hosts for phytophagous beetles (Coleoptera) in tropical forests. *Journal of Biogeography*, **27**, 283–296.

Ødegaard, F. (2000b). *Færre dyrearter enn tidligere antatt.* (Fewer species of animals than earlier believed.) Schrødingers katt, 25 January 2000. Studiogjest/innslag *c.* 7 min [Norwegian TV-show].

Ødegaard, F. (2000c). How many species of arthropods? Erwin's estimate revised. *Biological Journal of the Linnean Society*, **71**, 583–597.

Ødegaard, F., Diserud, O. H., Engen, S. & Aagaard, K. (2000). The magnitude of local host specificity for phytophagous insects and its implications for estimates of global species richness. *Conservation Biology*, **14**, 1182–1186.

Ohsaki, N. (1995). Preferential predation of female butterflies and the evolution of Batesian mimicry. *Nature*, **378**, 173–175.

Økland, R. H. (1996). Are ordination and constrained ordination alternative or complementary strategies in general ecological studies? *Journal of Vegetation Science*, **7**, 289–292.

Oldeman, R. A. A. (1974). Ecotopes des arbres et gradients écologiques verticaux en forêt guyanaise. *La Terre et la Vie*, **28**, 487–520.

Oliveira, P. & Gibbs, P. (1994). Pollination biology and breeding systems of six *Vochysia* species (Vochysiaceae) in Central Brazil. *Journal of Tropical Ecology*, **10**, 509–522.

Oliver, C. D. & Larson, B. C. (1990). *Forest Stand Dynamics*. New York: McGraw-Hill.

Olson, D. M. (1992). Rates of predation by ants (Hymenoptera: Formicidae) in the canopy, understory, leaf litter, and edge habitats of a lowland rainforest in Southwestern Cameroon. In *Biologie d'une Canopée de Forêt Equatoriale – II. Rapport de Mission: Radeau des Cimes Octobre–Novembre 1991, Réserve de Campo, Cameroun*, eds. F. Hallé & O. Pascal, pp. 101–109. Paris: Fondation Elf.

Olson, D. M. (1994a). The distribution of leaf litter invertebrates along a Neotropical altitudinal gradient. *Journal of Tropical Ecology*, **10**, 129–150.

Olson, E. J. (1994b). Dietary Ecology of a Tropical Moth Caterpillar, *Rothschildia lebeau* (Lepidoptera: Saturniidae). PhD Thesis, University of Pennsylvania, Philadelphia.

Opler, P. A., Frankie, G. W. & Baker, H. G. (1976). Rainfall as a factor in the release, timing and synchronization of anthesis by tropical trees and shrubs. *Journal of Biogeography*, **3**, 231–236.

Orivel, J. & Dejean, A. (1999). L'adaptation à la vie arboricole chez les fourmis. *L'Année Biologique*, **38**, 131–148.

Orivel, J. & Dejean, A. (2000). Comparative effect of the venoms of ants of the genus *Pachycondyla* (Hymenoptera: Ponerinae). *Toxicon*, **39**, 195–201.

Orivel, J., Malherbe, M. C. & Dejean, A. (2001a). Relationships between pretarsus morphology and arboreal life in Ponerine ants of the genus *Pachycondyla* (Formicidae: Ponerinae). *Annals of the Entomological Society of America*, **94**, 449–456.

Orivel, J., Redeker, V., Le Caer, J. P., Krier, F., Revol-Junelles, A. M., Longeon, A., Chaffote, A., Dejean, A. & Rossier, J. (2001b). Ponericin, new antibacterial and insectidal peptides from the venom of the ant, *Pachycondyla goeldii*. *Journal of Biological Chemistry*, **276**, 17823–17829.

Orivel, J., Souchal, A., Cerdan, P. & Dejean, A. (2000). Prey capture behavior of the arboreal ant *Pachycondyla goeldii* (Hymenoptera: Formicidae). *Sociobiology*, **35**, 131–140.

Orr, A. G. & Häuser, C. L. (1996). Temporal and spatial patterns of butterfly diversity in a lowland tropical rainforest. In *Tropical Rainforest Research – Current Issues. Proceedings of the Conference held in Bandar Seri Begawan, April 1993*, eds. D. S. Edwards, W. E. Booth & S. C. Choy, pp. 125–138. Dordrecht: Kluwer Academic.

Osler, G. H. R. & Beattie, A. J. (2001). Contribution of oribatid and mesostigmatid soil mites in ecologically based estimates of global species richness. *Austral Ecology*, **26**, 70–79.

Otto, H.-J. (1994). *Waldökologie*. Stuttgart: Ulmer.

Paarmann, W. & Kerck, K. (1997). Advances in using the canopy fogging technique to collect living arthropods from tree-crowns. In *Canopy Arthropods*, eds. N. E. Stork, J. Adis & R. K. Didham, pp. 53–66. London: Chapman & Hall.

Paarmann, W. & Paarmann, D. (1997). Studies on the biology of a canopy-dwelling carabid beetle collected by canopy fogging in the rainforest of Sulawesi (Indonesia). In *Canopy Arthropods*, eds. N. E. Stork, J. Adis & R. K. Didham, pp. 433–441. London: Chapman & Hall.

Paarmann, W. & Stork, N. E. (1987a). Canopy fogging, a method of collecting living insect for investigations of life history strategies. *Journal of Natural History*, **21**, 563–566.

Paarmann, W. & Stork, N. E. (1987b). Seasonality of ground beetles (Coleoptera: Carabidae) in the rain forests of N. Sulawesi (Indonesia). *Insect Science and its Application*, **8**, 483–487.

Palacios-Vargas, J. G. (1981). Collembola asociados a *Tillandsia* (Bromeliaceae) en el derrame lávico del Chichinautzin, Morelos, México. *Southwestern Entomologist*, **6**, 87–98.

Palacios-Vargas, J. G., Castaño-Meneses, G. & Cutz-Pool, L. (2000). Collembola associated to *Tillandsia violacea* (Bromeliaceae) in Mexican Quercus-Abies forests. In *Abstracts X International Colloquium on Apterygota*, ed. J. Rusek, pp. 22. Ceske Budejovice, Czech Republic: ICARIS.

Palacios-Vargas, J. G., Castaño-Meneses, G. & Gómez-Anaya, J. A. (1998). Collembola from the canopy of a Mexican tropical decidous forest. *Pan-Pacific Entomologist*, **74**, 47–54.

Palacios-Vargas, J. G., Castaño-Meneses, G. & Pescador, A. (1999). Phenology of canopy arthropods of a tropical deciduous forest in Western Mexico. *Pan-Pacific Entomologist*, **75**, 200–211.

Palacios-Vargas, J. G. & Gómez-Anaya, J. A. (1993). Los Collembola (Hexapoda: Apterygota) de Chamela, Jalisco, México (Distribución ecológica y claves). *Folia Entomológica Mexicana*, **89**, 1–34.

Palmer, M. W. (1990). The estimation of species richness by extrapolation. *Ecology*, **71**, 1195–1198.

Palmer, M. W. (1993). Putting things in even better order: the advantages of canonical correspondence analysis. *Ecology*, **74**, 2215–2230.

Paoletti, M. G., Taylor, R. A. J., Stinner, B. R., Stinner, D. H. & Benzing, D. H. (1991). Diversity of soil fauna in the canopy and forest floor of a Venezuelan cloud forest. *Journal of Tropical Ecology*, **7**, 373–383.

Papageorgis, C. (1975). Mimicry in neotropical butterflies. *American Scientist*, **63**, 522–532.

Paradise, C. J. & Dunson, W. A. (1997). Insect species interactions and resource effects in treeholes: are helodid beetles bottom-up facilitators of midge populations? *Oecologia*, **109**, 303–312.

Parker, G. G. (1995). Structure and microclimate of forest canopies. In *Forest Canopies*, eds. M. D. Lowman & N. M. Nadkarni, pp. 431–455. San Diego, CA: Academic Press.

Parker, G. G. (1997). Canopy structure and light environment of an old-growth Douglas-fir western Hemlock forest. *Northwest Science*, **71**, 261–270.

Parker, G. G. & Brown, M. J. (2000). Forest canopy stratification – Is it useful? *American Naturalist*, **155**, 473–484.

Parker, G. G., Smith, A. P. & Hogan, K. P. (1992). Access to the upper forest canopy with a large tower crane. *BioScience*, **42**, 664–670.

Paulian, R. (1947). *Observations écologiques en forêt de Basse Côte d'Ivoire*. Paris: Lechevalier.

Paulson, G. S. & Akre, R. D. (1991). Behavioral interactions among formicid species in the ant mosaic of an organic pear orchard. *Pan-Pacific Entomologist*, **67**, 288–297.

Peakall, R. (1994). Interactions between orchids and ants. In *Orchid Biology: Review and Perspectives, VI*, ed. J. Arditti, pp. 103–134. New York: John Wiley and Sons.

Pearson, D. L. & Derr, J. A. (1986). Seasonal patterns of lowland forest floor arthropods in southeastern Peru. *Biotropica*, **18**, 244–256.

Peck, S. B., Wigfull, P. & Nishida, G. (1999). Physical correlates of insular species diversity: the insects of the Hawaiian Islands. *Annals of the Entomological Society of America*, **92**, 529–536.

Pemberton, R. W. (1998). The occurrence and abundance of plants with extrafloral nectaries, the basis for antiherbivore defensive mutualisms, along a latitudinal gradient in east Asia. *Journal of Biogeography*, **25**, 661–668.

Pence, D. B. (1979). Congruent inter-relationships of the Rhinonyssinae (Dermanyssidae) with their avian hosts.

In *Recent Advances in Acarology*, Vol. 2, ed. J. G. Rodriguez, pp. 371–377. New York: Academic Press.

Peng, R. K., Christian, K. & Gibb, K. (1997). The effect of the green ant, *Oecophylla smaragdina* (Hymenoptera: Formicidae), on insect pests of cashew trees in Australia. *Bulletin of Entomological Research*, **85**, 279–284.

Penny, N. D. & Arias, J. R. (1982). *Insects of an Amazon Rain Forest*. New York: Columbia University Press.

Peredo Cervantes, L., Lyal, C. H. C. & Brown, V. K. (1999). The stenomatine moth *Stenoma catenifer* Walsingham: a pre-dispersal seed predator of Greenheart (*Chlorocardium rodiei* (Schomb.) Rohwer, Richter & van der Werff) in Guyana. *Journal of Natural History*, **33**, 531–542.

Perevolotsky, A. (1994). Tannins in Mediterranean woodland species: lack of responses to browsing and thinning. *Oikos*, **71**, 333–340.

Perfect, T. J. & Cook, A. G. (1982). Diurnal periodicity of flight in some Delphacidae and Cicadellidae associated with rice. *Ecological Entomology*, **7**, 317–326.

Perry, D. R. (1978). A method of access into crowns of emergent and canopy trees. *Biotropica*, **10**, 155–157.

Perry, D. R. & Williams, J. (1981). The tropical rainforest canopy: a method of providing total access. *Biotropica*, **13**, 283–285.

Peterson, A. T. & Slade, N. A. (1998). Extrapolating inventory results into biodiversity estimates and the importance of stopping rules. *Diversity and Distributions*, **4**, 95–105.

Pfeiffer, M. & Linsenmair, K. E. (2000). Contributions to the life history of the Malaysian giant ant *Camponotus gigas* (Hymenoptera, Formicidae). *Insectes Sociaux*, **47**, 123–132.

Phillipson, J. & Thompson, D. J. (1983). Phenology and intensity of phyllophage attack on *Fagus sylvatica* in Wytham Woods, Oxford. *Ecological Entomology*, **8**, 315–330.

Pianka, E. R. (1966). Latitudinal gradients in species diversity: a review of concepts. *American Naturalist*, **100**, 33–46.

Pianka, E. R. (1988). *Evolutionary Ecology*. New York: Harper and Row.

Pickett, S. T. A. & Cadenasso, M. L. (1995). Landscape ecology: spatial heterogeneity in ecological systems. *Science*, **269**, 331–334.

Pik, A. J., Oliver, I. & Beattie, A. J. (1999). Taxonomic sufficiency in ecological studies of terrestrial invertebrates. *Australian Journal of Ecology*, **24**, 555–562.

Pitman, N. C. A., Terborgh, J., Silman, M. R. & Nunez, V. P. (1999). Tree species distribution in an upper Amazonian forest. *Ecology*, **80**, 2551–2661.

Pittendrigh, C. S. (1948). The Bromeliad–*Anopheles*–Malaria complex in Trinidad. I. The Bromeliad flora. *Evolution*, **2**, 60–89.

Plotkin, J. B., Potts, M. D., Yu, D. W., Bunyavejchewin, S., Condit, R., Foster, R., Hubbell, S., LaFrankie, J., Manokaran, N., Lee, H.-S., Sukumar, R., Nowak, M. A. & Ashton, P. S. (2000). Predicting species diversity in tropical forests. *Proceedings of the National Academy of Sciences USA*, **97**, 10850–10854.

Plumptre, A. J. (1996). Changes following 60 years of selective timber harvesting in the Budongo Forest Reserve, Uganda. *Forest Ecology and Mangement*, **89**, 101–113.

Plumptre, A. J. & Reynolds, V. (1994). The effect of selective logging on the primate populations in the Budongo Forest Reserve, Uganda. *Journal of Applied Ecology*, **31**, 631–641.

Pogue, M. G. & Janzen, D. H. (1998). A revision of the genus *Obrima* Walker (Lepidoptera: Noctuidae), with a description of a new species from Costa Rica and notes on its natural history. *Proceedings of the Entomological Society of Washington*, **100**, 566–575.

Poilecot, P., Bonfou, K., Dosso, H., Lauginie, F., N'Dri, K., Nicole, M. & Sangare, Y. (1991). *Un écosystème de savane soudanienne: le Parc National de la Comoé*. Paris: UNESCO.

Poinsot-Balaguer, N. (1976). Dynamique des communautés de Collemboles en milieu xérique méditerranéen. *Pedobiologia*, **16**, 1–17.

Polis, G. A., Anderson, W. B. & Holt, R. D. (1997). Toward an integration of landscape and food web ecology: the dynamics of spatially subsidized food webs. *Annual Reviews in Ecology and Systematics*, **28**, 289–316.

Polis, G. A. & Strong, D. R. (1996). Food web complexity and community dynamics. *American Naturalist*, **147**, 813–846.

Pollard, D. G. (1968). Stylet penetration and feeding damage of *Eupteryx melissae* Curtis (Hemiptera, Cicadellidae) on sage. *Bulletin of Entomological Research*, **58**, 55–71.

Ponge, J.-F. (1991). Food resources and diets of soil animals in a small area of Scots pine litter. *Geoderma*, **49**, 33–62.

Ponge, J.-F. (1993). Biocoenoses of Collembola in Atlantic temperate grass-woodland ecosystems. *Pedobiologia*, **37**, 223–244.

Ponge, J.-F. (2000). Vertical distribution of Collembola (Hexapoda) and their food resources in organic horizons of beech forests. *Biology and Fertility of Soils*, **32**, 508–522.

Pontin, A. J. (1982). *Competition and Coexistence of Species*. Boston, MA: Pitman.

Popma, J., Bongers, F. & Meave del Castillo, J. (1988). Patterns in the vertical structure of the tropical lowland rain forest of Los Tuxtlas, Mexico. *Vegetatio*, **74**, 81–91.

Porembski, S. (1991). Beiträge zur Pflanzenwelt des Comoé-Nationalparks (Elfenbeinküste). *Natur und Museum*, **121**, 61–83.

Porter, J. R. (1972). The Growth and Phenology of *Metrosideros* in Hawaii. Dissertation. Honolulu, Hawaii: University of Hawaii.

Potter, D. A. & Kimmerer, T. W. (1986). Seasonal allocation of defense investment in *Ilex opaca* Aiton and constraints on a specialist leafminer. *Oecologia*, **69**, 217–224.

Price, P. W. (1987). The role of natural enemies in insect populations. In *Insect Outbreaks*, eds. P. Barbosa & J. C. Schultz, pp. 287–312. New York: Academic Press.

Price, P. W. (1991). The plant vigor hypothesis and herbivore attack. *Oikos*, **62**, 244–251.

Price, P. W. (1992). The resource-based organization of communities. *Biotropica*, **24**, 273–282.

Price, P. W. (1997). *Insect Ecology*. New York: John Wiley & Sons.

Price, P. W., Bouton, C. E., Gross, P., McPheron, B. A., Thompson, J. N. & Weis, A. E. (1980). Interactions among three trophic levels : influence of plants on interactions between insect herbivores and natural enemies. *Annual Review of Ecology and Systematics*, **11**, 41–65.

Price, P. W., Diniz, I. R., Morais, H. C. & Marques, E. S. A. (1995). The abundance of insect herbivore species in the tropics: the high local richness of rare species. *Biotropica*, **27**, 468–478.

Price, P. W., Fernandes, G. W., Lara, A. C. F., Brawn, J., Barrios, H., Wright, M., Ribeiro, S. P. & Rothcliff, N. (1998). Global patterns in local number of insect galling species. *Journal of Biogeography*, **25**, 581–591.

Prinzing, A. (1996). Arthropods on solitary tree trunks: microclimatic opportunities, use of climatic gradients and of microhabitats and the effects on a preferably grazed lichen. In *Checklist of the Collembola*, ed. F. Jannsens, published at http://www.geocities.com/~fransjanssens/publicat/prinzing.htm

Prinzing, A. (1997). Spatial and temporal use of microhabitats as a key strategy for the colonization of tree bark by *Entomobrya nivalis* L. (Collembola Entomobryidae). In *Canopy Arthropods*, eds. N. Stork, J. Adis & R. K. Didham, pp. 453–476. London: Chapman & Hall.

Prinzing, A. (1999). Wind-acclimated thallus morphogenesis in a lichen *(Evernia prunastri*, Parmeliaceae) probably favored by grazing disturbances. *American Journal of Botany*, **86**, 173–183.

Prinzing, A. (2001). Use of shifting microclimatic mosaic by arthropods on exposed tree trunks. *Annals of the Entomological Society of America*, **94**, 210–218.

Prinzing, A., Kretzler, S. & Beck, L. (2000). Resistance to disturbance is a diverse phenomenon and does not increase with abundance: the case of oribatid mites. *Ecoscience*, **7**, 452–460.

Prinzing, A. & Wirtz, H. P. (1997). The epiphytic lichen, *Evernia prunastri* L., as a habitat for arthropods: shelter from desiccation, food limitation and indirect mutualism. In *Canopy Arthropods*, eds. N. E. Stork, J. Adis & R. K. Didham, pp. 477–494. London: Chapman & Hall.

Proctor, M., Yeo, P. & Lack, A. (1996). *The Natural History of Pollination*. Portland, OR: Timber Press.

Pschorn-Walcher, H. & Gunhold, P. (1957). Zur Kenntnis der Tiergemeinschaften in Moos– und Flechtenrasen an Park– und Waldbäumen. *Zeitschrift für Morphologie und Ökologie der Tiere*, **46**, 342–354.

Pung, O. J., Carlile, L. D., Whitlock, J., Vives, S. P., Durden, L. A. & Spadgenske, E. (2000). Survey and host fitness effects of red-cockaded woodpecker blood parasites and nest cavity arthropods. *Journal of Parasitology*, **86**, 506–510.

Purvis, A. & Hector, A. (2000). Getting the measure of biodiversity. *Nature*, **405**, 212–219.

Putz, F. E. (1979). Aseasonality in Malaysian tree phenology. *Malaysia Forester*, **42**, 1–24.

Radovsky, F. J. (1969). Adaptive radiation in the parasitic Mesostigmata. *Acarologia*, **11**, 450–478.

Radovsky, F. J. (1985). Evolution of mammalian mesostigmate mites. In *Coevolution of Parasitic Arthropods and Mammals*, ed. K. C. Kim, pp. 441–504. New York: John Wiley & Sons.

Radovsky, F. J. (1994). The evolution of parasitism and the distribution of some dermanyssoid mites (Mesostigmata) on vertebrate hosts. In *Mites: Ecological and Evolutionary Analyses of Life History Patterns*, ed. M. A. Houck, pp. 186–217. London: Chapman & Hall.

Radtkey, R. R. & Singer, M. C. (1995). Repeated reversals of host-preference evolution in a specialist insect herbivore. *Evolution*, **49**, 351–359.

Rahman, M. A., Abidin, Z. Z., Nor, B. M. & Abdullah, M. T. (1995). A brief study of bird fauna at Sayap-Kinabalu Park, Sabah. In *A Scientifc Journey Through Borneo, Sayap-Kinabalu Park, Sabah*, eds. G. Ismail, L. B. Din, S. D. Ehsan, pp. 1–7. Malaysia: The Faculty of Resource Science and Technology, University of Malaysia Sarawak & Pelanduk Publications.

Ranius, T. & Hedin, J. (2001). The dispersal rate of a beetle, *Osmoderma eremita*, living in tree hollows. *Oecologia*, **126**, 363–370.

Ranta, E., Kaitala, V. & Lundberg, P. (1997). The spatial dimension in population fluctuations. *Science*, **278**, 1621–1623.

Raupach, M. R. (1989). Turbulent transfer in plant canopies. In *Plant Canopies: Their Growth, Form and Function*, eds. G. Russell, B. Marshall & P. G. Jarvis, pp. 41–61. Cambridge: Cambridge University Press.

Raupp, M. J. (1985). Effects of leaf toughness on mandibular wear of the leaf beetle, *Plagioclera versicolora*. *Ecological Entomology*, **10**, 72–79.

Reader, P. M. & Southwood, T. R. E. (1981). The relationships between palatability to invertebrates and the successional status of a plant. *Oecologia*, **51**, 271–275.

Real, L. (1983). *Pollination Biology*. New York: Academic Press.

Recher, H. F., Majer, J. D. & Ganesh, S. (1996a). Seasonality of canopy invertebrate communities in eucalypt forests of eastern and western Australia. *Australian Journal of Ecology*, **21**, 64–80.

Recher, H. F., Majer, J. D. & Ganesh, S. (1996b). Eucalypts, arthropods and birds: on the relation between foliar nutrients and species richness. *Forest Ecology and Management*, **85**, 177–196.

Redborg, K. E. & Redborg, A. H. (2000). Resource partitioning of spider hosts (Arachnida, Araneae) by two mantispid species (Neuroptera, Mantispidae) in an Illinois woodland. *Journal of Arachnology*, **28**, 70–78.

Rees, C. J. C. (1983). Microclimate and the flying Hemiptera fauna of a primary lowland rain forest in Sulawesi. In *Tropical Rain Forest: Ecology and Management*, eds. S. L. Sutton, T. C. Whitmore & A. C. Chadwick, pp. 121–136. Oxford: Blackwell Scientific.

Reich, P. B. (1995). Phenology of tropical forests: patterns, causes, and consequences. *Canadian Journal of Botany*, **73**, 164–174.

Reid, C. A. M. (1995). A cladistic analysis of subfamilial relationships in the Chrysomelidae *sensu lato* (Chrysomelidae). In *Biology, Phylogeny, and Classification of Coleoptera: Papers Celebrating the 80th Birthday of Roy A. Crowson*, eds. J. Pakaluk & S. A. Slipinski, pp. 559–632. Warszawa: Museum i Instytut Zoologii PAN.

Ribeiro, J. E. L. S., Hopkins, M. J. G., Vicentini, A., Sothers, C. A., Costa, M. A. S., Brito, J. M., Souza, M. A., Martins, L. H. P., Lohmann, L. G., Assunção, P. A. C. L., Pereira, E. C., Silva, C. F., Mesquita, M. R. & Procópio, L. C. (1999b). *Flora da Reserva Ducke: Guia de Identificação das plantas vasculares de uma florestas de terra-firme na Amazônia Central*. Manaus, Amazonas, Brazil: Projeto Flora da Reserva Ducke (INPA-DFID).

Ribeiro, J. E. L. S., Nelson, B. W., Da Silva, M. F., Martins, L. S. S. & Hopkins, M. (1994b). Reserva Florestal Ducke: Diversidade e composição da flora vascular. *Acta Amazonica*, **23**, 19–30.

Ribeiro, M. N. G. & Adis, J. (1984). Local rainfall variability a potential bias for bioecological studies in the Central Amazon. *Acta Amazonica*, **14**, 159–174.

Ribeiro, S. P. (1999). The Role of Herbivory in *Tabebuia* spp. Life History and Evolution. PhD thesis. London: Imperial College at Silwood Park.

Ribeiro, S. P., Braga, A. O., Silva, C. H. L. & Fernandes, G. W. (1999a). Leaf polyphenols in Brazilian Melastomataceae: sclerophylly, habitats, and insect herbivores. *Ecotropica*, **5**, 137–146.

Ribeiro, S. P. & Brown, V. K. (1999). Insect herbivory in tree crowns of *Tabebuia aurea* and *T. ochracea* (Bignoniaceae) in Brazil: contrasting the Cerrado with the 'Pantanal Matogrossense'. *Selbyana*, **20**, 159–170.

Ribeiro, S. P., Carneiro, M. A. A. & Fernandes, G. W. (1998). Free-feeding insect herbivores along environmental gradients in Serra do Cipó: basis for a management plan. *Journal of Insect Conservation*, **2**, 107–118.

Ribeiro, S. P. & Pimenta, H. R. (1991). Padrões de abundância e distribuição temporal de herbívoros de vida livre em *Tabebuia ochracea* (Bignoniaceae). *Anais da Sociedade Entomologica do Brasil*, **20**, 428–448.

Ribeiro, S. P., Pimenta, H. R. & Fernandes, G. W. (1994a). Herbivory by chewing and sucking insects on *Tabebuia ochracea*. *Biotropica*, **26**, 302–307.

Richard, F. J., Fabre, A. & Dejean, A. (2001). Predatory behavior in dominant arboreal ant species: the case of *Crematogaster* sp. (Hymenoptera: Formicidae). *Journal of Insect Behavior*, **14**, 271–282.

Richards, P. W. (1952). *The Tropical Rain Forest*. Cambridge, UK: Cambridge University Press.

Richards, P. W. (1996). *The Tropical Rain Forest. An Ecological Study*, 2nd edn. Cambridge: Cambridge University Press.

Richardson, B. A. (1999). The bromeliad microcosm and the assessment of faunal diversity in a neotropical forest. *Biotropica*, **31**, 321–336.

Ricklefs, R. E. (1975). Seasonal occurrence of night-flying insects on Barro Colorado Island, Panama Canal Zone. *Journal of the New York Entomological Society*, **83**, 19–32.

Ricklefs, R. E. (1987). Community diversity: relative roles of local and regional processes. *Science*, **235**, 167–171.

Ricklefs, R. E. (2000). Rarity and diversity in Amazonian forest trees. *Trends in Ecology and Evolution*, **15**, 83–84.

Ricklefs, R. E. & Schluter, D. (eds.) (1993). *Species Diversity in Ecological Communities: Historical and Geographical Perspectives*. Chicago, IL: University of Chicago Press.

Riha, G. (1951). Zur Ökologie der Oribatiden in Kalksteinböden. *Zoologische Jahrbücher, Abteilung für Systematik, Ökologie und Geographie der Tiere*, **80**, 407–450.

Rincon, R. M., Roubik, D. W., Finegan, B., Delgado, D. & Zamora, N. (1999). Regeneration in tropical rain forest managed for timber production: understory bees and their floral resources in a logged and silviculturally treated Costa Rican forest. *Journal of the Kansas Entomological Society*, **72**, 379–393.

Rizzini, C. T. (1997). *Tratado de Fitogeografia do Brasil. Aspectos Ecológicos, Sociológicos e Florísticos*, 2nd edn. Rio de Janeiro: Âmbito Cultural Edições.

Roberts, H. R. (1973). Arboreal Orthoptera in the rain forests of Costa Rica collected with insecticide: a report on the grasshoppers (Acrididae), including new species. *Proceedings of the Academy of Natural Sciences of Philadelphia*, **125**, 46–66.

Roberts, H. R. (1993). The tribe Diapodini (Coleoptera: Platypodidae) of Papua New Guinea. *Bishop Museum Occasional Papers*, **35**, 1–39.

Robinson, G. S. (1998). Bugs, hollow curves and species-diversity indexes. *Stats Magazine for the Students of Statistics*, **21**, 8–13.

Robinson, G. S. & Tuck, K. R. (1993). Diversity and faunistics of small moths (Microlepidoptera) in Bornean rainforest. *Ecological Entomology*, **18**, 385–393.

Robinson, G. S. & Tuck, K. R. (1996). Describing and comparing high invertebrate diversity in tropical forest – a case study of small moths in Borneo. In *Tropical Rainforest Research – Current Issues. Proceedings of the Conference held in Bandar Seri Begawan, April 1993*, eds. D. S. Edwards, W. E. Booth & S. C. Choy, pp. 29–42. Amsterdam: Kluwer Academic.

Robinson, G. S., Tuck, K. R. & Intachat, J. (1995). Faunal composition and diversity of smaller moths (Microlepidoptera and Pyraloidea) in lowland tropical rainforest at Temengor, Hulu Perak, Malaysia. *Malayan Nature Journal*, **48**, 307–317.

Robinson, G. S., Tuck, K. R. & Shaffer, M. (1994). *A Field Guide to the Smaller Moths of South-East Asia*. Kuala Lumpur: Malaysian Nature Society.

Rockwood, L. L. (1974). Seasonal changes in the suceptibility of *Cresentia alata* leaves to the flea beetle, *Ocdionychus* sp. *Ecology*, **55**, 142–148.

Rödel, M. O. (2000). *Herpetofauna of West Africa*. Vol. I, *Amphibians of the West African Savanna*. Frankfurt/Main: Edition Chimaira.

Roderick, G. K. & Gillespie, R. G. (1998). Speciation and phylogeography of Hawaiian terrestrial arthropods. *Molecular Ecology*, **7**, 519–531.

Rodgers, D. J. & Kitching, R. L. (1998). Vertical stratification of rainforest collembolan (Collembola: Insecta) assemblages: description of ecological patterns and hypotheses concerning their generation. *Ecography*, **21**, 392–400.

Roland, J. (1993). Large-scale forest fragmentation increases the duration of tent caterpillar outbreak. *Oecologia*, **93**, 25–30.

Ronquist, F. (1999). Phylogeny of the Hymenoptera (Insecta): the state of the art. *Zoologica Scripta*, **28**, 3–11.

Room, P. M. (1971). The relative distribution of ant species in Ghana's cocoa farms. *Journal of Animal Ecology*, **40**, 735–751.

Room, P. M. (1975). Relative distributions of ant species in cocoa plantations in Papua New Guinea. *Journal of Applied Ecology*, **12**, 47–61.

Root, R. B. (1973). Organization of a plant–arthropod association in simple and diverse habitats: the fauna of collards (*Brassica oleracea*). *Ecological Monographs*, **43**, 95–124.

Roques, A. (1988). The larch cone fly in the french alps. In *Dynamics of Forest Insect Populations: Patterns, Causes, Implications*, ed. A. A. Berrymann, pp. 1–28. New York: Plenum Press.

Rosenzweig, M. L. (1987). Habitat selection as a source of biological diversity. *Evolutionary Ecology*, **1**, 315–330.

Rosenzweig, M. L. (1995). *Species Diversity in Space and Time*. Cambridge: Cambridge University Press.

Ross, J. (1981). *The Radiation Regime and Architecture of Plant Stands*. The Hague: W. Junk.

Roth, I. (1987). *Stratification of a Tropical Forest as Seen in Dispersal Types*. Dordrecht: W. Junk.

Roubik, D. W. (1989). *Ecology and Natural History of Tropical Bees*. Cambridge: Cambridge University Press.

Roubik, D. W. (1991). Aspects of Africanized honey bee ecology in tropical America. In *The African Honey Bee*, eds. M. Spivak, M. D. Breed & D. J. C. Fletcher, pp. 147–158. Boulder, CO: Westview Press.

Roubik, D. W. (1992). Loose niches in tropical communities: why are there so many trees and so few bees? In *Resource Distribution and Animal–Plant Interactions*, eds. M. D. Hunter, T. Ohgushi & P. W. Price, pp. 327–354. San Diego, CA: Academic Press.

Roubik, D. W. (1993). Tropical pollinators in the canopy and understorey: field data and theory for stratum 'preferences'. *Journal of Insect Behavior*, **6**, 659–673.

Roubik, D. W. (1996a). Diversity in the real world: tropical forests as pollinator reserves. In *Tropical Bees and the Environment*, eds. M. Mardan, A. Sipat, Kamaruddin, M. Yusoff, R. Kiew & M. M. Abdullah, pp. 111–122. Selangor, Malaysia: UPM Press.

Roubik, D. W. (1996b). African honey bees as exotic pollinators in French Guiana. In *The Conservation of Bees*, eds. A. Matheson, S. L. Buchmann, C. O'Toole, P. Westrich & I. H. Williams, pp. 173–182. London: Academic Press.

Roubik, D. W. (1996c). Measuring the meaning of honey bees. In *The Conservation of Bees*, eds. A. Matheson, S. L. Buchmann, C. O'Toole, P. Westrich & I. H. Williams, pp. 163–172. London: Academic Press.

Roubik, D. W. (2000). Pollination system stability in tropical America. *Conservation Biology*, **14**, 1235–1236.

Roubik, D. W. (2002). Tropical bee colonies, pollen dispersal and reproductive gene flow in forest trees. In *Symposium on Modelling and Experimental Research on Genetic Processes in Tropical and Temperate Forests*, online at http://kourou.cirad.fr/genetique/Symposium_e.html. Belem and Kourou: EMBRAPA and Silvolab.

Roubik, D. W., Inoue, T. & Hamid, A. A. (1995). Canopy foraging by two tropical honeybees: bee height fidelity and tree genetic neighborhoods. *Tropics*, **5**, 81–93.

Roubik, D. W. & Wolda, H. (2001). Do competing honey bees matter? Dynamics and abundance of native bees before and after honey bee invasion. *Population Ecology*, **43**, 10–20.

Rowe, W. J. & Potter, D. A. (1996). Vertical stratification of feeding by Japanese beetles within linden tree canopies: selective foraging or height per se? *Oecologia*, **108**, 459–466.

Rowell, C. H. F. (1978). Food plant specificity in neotropical rain-forest acridids. *Entomologia Experimentalis et Applicata*, **24**, 651–662.

Rowell, C. H. F., Rowell-Rahier, M., Braker, H. E., Cooper-Driver, G. & Gomez, L. D. (1983). The palatability of ferns and the ecology of two tropical forest grasshoppers. *Biotropica*, **15**, 207–216.

Ruedas, L. A., Salazar-Bravo, J., Dragoo, J. W. & Yates, T. L. (2000). The importance of being earnest: what, if anything, constitutes a 'specimen examined?'. *Molecular Phylogenetics and Evolution*, **17**, 129–132.

Russell-Smith, A. & Stork, N. E. (1994). Abundance and diversity of spiders from the canopy of tropical rainforests with particular reference to Sulawesi, Indonesia. *Journal of Tropical Ecology*, **10**, 545–558.

Russell-Smith, A. & Stork, N. E. (1995). Composition of spider communities in the canopies of rainforest trees in Borneo. *Journal of Tropical Ecology*, **11**, 223–235.

Rydell, J. (1995). Echolocating bats and hearing moths: who are the winners? *Oikos*, **73**, 419–424.

Rydell, J. & Lancaster, W. C. (2000). Flight and thermoregulation in moths were shaped by predation from bats. *Oikos*, **88**, 13–18.

Rzedowski, J. (1988). *Vegetación de México*. México: Limusa.

Sachs, L. (1997). *Angewandte Statistik. Anwendung statistischer Methoden*. Berlin: Springer Verlag.

Safranyik, L., Linton, D. A. & Shore, T. L. (2000). Temporal and vertical distribution of bark-beetles (Coleoptera: Scolytidae) captured in barrier traps at baited and unbaited lodgepole pines the year following attack by the mountain pine beetle. *Canadian Entomologist*, **132**, 799–810.

Sakai, S., Momose, K., Yumoto, T., Kato, M. & Inoue, T. (1999a). Beetle pollination of *Shorea parvifolia* (section *Mutica*, Dipterocarpaceae) in a general flowering period in Sarawak, Malaysia. *American Journal of Botany*, **86**, 62–69.

Sakai, S., Momose, K., Yumoto, T., Nagamitsu, T., Nagamasu, H., Hamid, A. A., Nakashizuka, T. & Inoue, T. (1999b). Plant reproductive phenology over four years including an episode of general flowering in a lowland dipterocarp forest, Sarawak, Malaysia. *American Journal of Botany*, **86**, 1414–1436.

Salati, E. (1987). The forest and the hydrologic cycle. In *The Geophysiology of Amazonia: Vegetation and Climate Interactions*, ed. R. E. Dickinson, pp. 273–296. New York: John Wiley & Sons.

Salatino, A. (1993). Chemical ecology and the theory of oligotrophic scleromorphism. *Anais da Academia Brasileira de Ciência*, **65**, 1–13.

Salmon, S. & Ponge, J.-F. (1999). Distribution of *Heteromurus nitidus* (Hexapoda, Collembola) according to soil acidity: interactions with earthworms and predator pressure. *Soil Biology and Biochemistry*, **31**, 1161–1170.

Samways, M. J. (1990). Ant assemblage structure and ecological management in citrus and subtropical fruit orchards in southern Africa. In *Applied Myrmecology: A World Perspective*, eds. R. K. Van der Meer, A. Jaffe & A. Cedeno, pp. 570–587. Boulder, CO: Westview Press.

Samways, M. J., Nel, M. & Prins, A. J. (1982). Ants (Hymenoptera: Formicidae) foraging in citrus trees and attending honeydew-producing Homoptera. *Phytophylactica*, **14**, 155–157.

Sarmiento, G. (1983). The savannas of tropical America. In *Ecosystems of the World*. Vol. 13. *Tropical Savannas*, ed. F. Bourlière, pp. 245–288. Amsterdam: Elsevier Scientific.

SAS (1982). *SAS/STAT User's Guide*, Version 6, 4th edn. Cary, NC: SAS Institute.

SAS (1995). *SAS/STAT User's Guide*, Version 6. Cary, NC: SAS Institute.

Schaefer, M. (1974). Auswirkungen natürlicher und experimenteller Störungen in Grenzzonen von Ökosystemen, untersucht am Beispiel der epigäischen Arthropodenfauna. *Pedobiologia*, **14**, 51–60.

Schal, C. (1982). Intraspecific vertical stratification as a mate-finding mechanism in tropical cockroaches. *Science*, **215**, 1405–1407.

Schal, C. & Bell, W. J. (1986). Vertical community structure and resource utilization in neotropical forest cockroaches. *Ecological Entomology*, **11**, 411–423.

Schatz, G. E. (1990). Some aspects of pollination biology in Central American forests. In *Reproductive Ecology of Tropical Forest Plants*, eds. K. S. Bawa & M. Hadley, pp. 69–84. Paris: UNESCO and Parthenon Publishing.

Schauff, M. E. & Janzen, D. H. (2001). Taxonomy and ecology of Costa Rican *Euplectus* (Hymenoptera: Eulophidae), parasitoids of caterpillars (Lepidoptera). *Journal of Hymenoptera Research*, **10**, 181–230.

Schedl, K. E. (1981). Scolytidae. In *Die Käfer Mitteleuropas*. Vol. 10, eds. H. Freude, K. W. Harde & G. A. Lohse, pp. 34–99. Krefeld, Germany: Goecke & Evers.

Scheidler, M. (1990). Influence of habitat structure and vegetation architecture on spiders. *Zoologische Anzeiger*, **225**, 333–340.

Schenker, R. (1984). Spatial and seasonal distribution patterns of oribatid mites (Acari, Oribatei) in a forest soil ecosystem. *Pedobiologia*, **27**, 133–149.

Schenker, R. & Block, W. (1986). Micro-arthropod activity in three contrasting terrestrial habitats on Signy Island, maritime Antarctic. *British Antarctic Survey Bulletin*, **71**, 31–43.

Scheu, S. & Falca, M. (2000). The soil food web of two beech forests (*Fagus sylvatica*) of constrasting humus type: stable isotope analysis of a macro- and a mesofauna-dominanted community. *Oecologia*, **123**, 285–296.

Schimper, A. F. W. (1888). Die Wechselbeziehungen zwischen Pflanzen und Ameisen im tropischen Amerika. *Botanische Mitteilungen Tropen*, **1**, 1–98.

Schmidt, G. & Zotz, G. (2000). Herbivory in the epiphyte, *Vriesea sanguinolenta* Cogn. & Marchal (Bromeliaceae). *Journal of Tropical Ecology*, **16**, 829–839.

Schmidt, G. & Zotz, G. (2001). Ecophysiological consequences of differences in plant size – *in situ* carbon gain and water relations of the epiphytic bromeliad, *Vriesea sanguinolenta* Cogn. & Marchal. *Plant, Cell and Environment*, **24**, 101–112.

Schmidt, T. (1999). *Die Arthropodengemeinschaften in der Baumkrone von Jungbirken (* Betula pendula*Roth): Welche Faktoren beeinflussen die Struktur dieser Lebensgemeinschaft?* Berlin: Wissenschaft und Technik Verlag.

Schneider, C. J., Smith, T. B., Larison, B. & Moritz, C. (1999). A test of alternative models of diversification in tropical rainforests: ecological gradients vs. rainforest refugia. *Proceedings of the National Academy of Sciences USA*, **96**, 13869–13873.

Schoeller, M. (1996). *Ökologie mitteleuropäischer Blattkäfer, Samenkäfer und Breitrüssler. Die Blatt- und Samenkäfer von Vorarlberg und Lichtenstein*. Bürs: Selbstverlag.

Schoener, T. W. (1974). Resource partitioning in ecological communities. *Science*, **185**, 27–39.

Scholtz, C. H. & Holm, E. (1989). *Insects of Southern Africa*. Durban: Butterworths.

Schoonhoven, L. M., Jermy, T. & Van Loon, J. J. A. (eds.) (1998). *Insect – Plant Biology from Physiology to Evolution*. London: Chapman & Hall.

Schowalter, T. D. (1995). Canopy arthropod communities in relation to forest age and alternative harvest practices in western Oregon. *Forest Ecology and Management*, **78**, 115–125.

Schowalter, T. D. (2000). *Insect Ecology: An Ecosystem Approach*. San Diego, CA: Academic Press.

Schowalter, T. D. & Ganio, L. M. (1998). Vertical and seasonal variation in canopy arthropod communities in an old-growth conifer forest in southwestern Washington, USA. *Bulletin of Entomological Research*, **88**, 633–640.

Schowalter, T. D. & Ganio, L. M. (1999). Invertebrate communities in a tropical rain forest canopy in Puerto Rico following Hurricane Hugo. *Ecological Entomology*, **24**, 191–201.

Schowalter, T. D., Hargrove, W. W. & Crossley, D. A., Jr (1986). Herbivory in forested ecosystems. *Annual Review of Entomology*, **31**, 177–196.

Schowalter, T. D. & Lowman, M. D. (1999). Insect herbivory in forests. In *Ecosystems of the World: Ecosystems of Disturbed Ground*, ed. L. R. Walker, pp. 253–269. Amsterdam: Elsevier.

Schowalter, T. D., Webb, J. W. & Crossley, D. A., Jr (1981). Community structure and nutrient content of canopy arthropods in clearcut and uncut ecosystems. *Ecology*, **62**, 1010–1019.

Schubert, H. (1998). *Untersuchungen zur Arthropodenfauna in Baumkronen - Ein vergleich von Natur- und Wirtschaftswäldern*. Berlin: Wissenschaft & Technik Verlag.

Schultz, J. P. (1960). *The Vegetation of Suriname*, Vol. II, *Ecological Studies on the Rain Forest in Northern Suriname*. Amsterdam: van Eedenfonds.

Schulze, C. H. (2000). Auswirkungen anthropogener Störungen auf die Diversität von Herbivoren – Analyse von Nachtfalterzönosen entlang von Habitatgradienten in Ost-Malaysia. PhD thesis. Bayreuth, Germany: University of Bayreuth.

Schulze, C. H. & Fiedler, K. (1998). Habitat preference and flight activity of Morphinae butterflies in a Bornean rainforest, with a note on sound production by adult *Zeuxidia* (Lepidoptera: Nymphalidae). *Malayan Nature Journal*, **52**, 163–176.

Schulze, C. H., Linsenmair, K. E. & Fiedler, K. (2001). Understorey versus canopy – patterns of vertical stratification and diversity among Lepidoptera in a Bornean rainforest. *Plant Ecology*, **153**, 133–152.

Schuman, G. E., Stanley, M. A. & Knudsen, D. (1973). Automated total nitrogen analysis of soil and plant samples. *Soil Science Society America Proceedings*, **37**, 480–481.

Schütz, S., Weissbecker, B., Hummel, H., Apel, K.-H., Schmitz, H. & Bleckmann, H. (1999). Insect antenna as a smoke detector. *Nature*, **398**, 298–299.

Scott, J. A. (1975). Flight patterns among eleven species of diurnal Lepidoptera. *Ecology*, **56**, 1367–1377.

Scriber, J. M. & Slansky, F. (1981). The nutritional ecology of immature insects. *Annual Review of Entomology*, **26**, 183–212.

Seeman, O. D. (2001). Myriad Mesostigmata associated with log-inhabiting arthropods. In *Acarology X: Proceedings of the 10th International Congress of Acarology*, eds. R. B. Halliday, R. A. Norton, H. C. Proctor & M. J. Coloff, pp. 272–276. Melbourne: CSIRO Publishing.

Seeman, O. D. & Walter, D. E. (1997). A new species of Triplogyniidae (Mesostigmata: Celaenopsoidea) from Australian rainforests. *International Journal of Acarology*, **23**, 49–59.

Seniczak, S. & Plichta, W. (1978). Structural dependence of moss mite populations (Acari, Oribatei) in a forest soil ecosystem. *Pedobiologia*, **17**, 305–319.

Senn, J., Hanhimaki, S. & Haukioja, E. (1992). Among-tree variation in leaf phenology and morphology and its correlation with insect performance in the mountain birch. *Oikos*, **63**, 215–222.

Seyd, E. L. & Seaward, M. R. D. (1984). The association of oribatid mites with lichens. *Journal of the Linnean Society*, **80**, 369–420.

Sharkey, M. J. & Janzen, D. H. (1995). Review of the world species of *Sigalaphus* (Hymenoptera: Braconidae: Sigalaphinae) and biology of *Sigalaphus romeroi*, new species. *Journal of Hymenoptera Research*, **4**, 99–109.

Sharrocks, B., Marsters, J., Ward, I. & Evenett, P. J. (1991). The fractal dimension of lichens and the distribution of arthropod body lengths. *Functional Ecology*, **5**, 457–460.

Shelly, T. E. (1985). Ecological comparisons of robber fly species (Diptera : Asilidae) coexisting in a neotropical forest. *Oecologia*, **67**, 57–70.

Shelly, T. E. (1988). Relative abundance of day-flying insects in treefall gaps vs shaded understorey in a Neotropical forest. *Biotropica*, **20**, 114–119.

Shmida, A. & Wilson, M. V. (1985). Biological determinants of species diversity. *Journal of Biogeography*, **12**, 1–20.

Siepel, H. (1994). Life-history tactics of soil microarthropods. *Biology and Fertility of Soils*, **18**, 263–278.

Silva, D. (1996). Species composition and community structure of Peruvian rainforest spiders: A case study from seasonally inundated forest along the Samiria river. (*Proceedings of the XIIIth International Congress of Arachnology*, Geneva, 3–8. IX.1995.) *Revue Suisse de Zoologie*, **vol. hors série**, 597–610.

Silveira-Guido, A. & Perkins, B. D. (1974). Biology and host specificity of *Cornops aquaticum* (Bruner) (Orthoptera: Acrididae), a potential biological control agent for water hyacinth. *Environmental Entomology*, **4**, 400–404.

Sime, K. R. & Brower, A. V. Z. (1998). Explaining the latitudinal gradient anomaly in ichneumonid species richness: evidence from butterflies. *Journal of Animal Ecology*, **67**, 387–399.

Simon, U. (1995). *Untersuchung der Stratozönosen von Spinnen und Weberknechten (Arachn.: Araneae, Opilionida) an der Waldkiefer (* Pinus sylvestris*L.*). Berlin: Wissenschaft & Technik Verlag.

Simon, U. & Linsenmair, K. E. (2001). Arthropods in tropical oaks: differences in their spatial distributions within tree crowns. *Plant Ecology*, **153**, 179–191.

Simpson, B. B. & Haffer, J. (1978). Speciation patterns in the Amazonian forest biota. *Annual Review of Ecology and Systematics*, **9**, 497–518.

Singer, R. B. & Cocucci, A. A. (1999). Pollination mechanisms in four sympatric southern Brazilian epidendroideae orchids. *Lindleyana*, **14**, 47–56.

Sittenfeld, A., Uribe-Lorio, L., Mora, M., Nielsen, V., Arrieta, G. & Janzen, D. H. (2002). Does a polyphagous caterpillar have the same gut microbiota when feeding on different species of food plants? *Revista de Biología Tropical*, **50**, 547–560.

Skutch, A. F. (1945). The most hospitable tree (*Cecropia*). *Scientific Monthly (New York)*, **60**, 1–17.

Slotow, R. & Hamer, M. (2000). Biodiversity research in South Africa: comments on current trends and methods. *South African Journal of Science*, **96**, 222–224.

Smith, A. P. (1973). Stratification of temperate and tropical forests. *American Naturalist*, **107**, 671–683.

Smith, B. (1986). *Evaluation of Different Similarity Indices Applied to Data from the Rothamsted Insect Survey*. York: University of York.

Smith, D. R. (1992). A synopsis of the sawflies (Hymenoptera: Symphyta) of America south of the United States: Argidae. *Memoirs of the American Entomological Society*, **39**, 1–201.

Smrz, J. & Kocourkova, J. (1999). Mite communities of two epiphytic lichen species (*Hypogymnia physodes* and *Parmelia sulcata*) in the Czech Republic. *Pedobiologia*, **43**, 385–390.

Smythe, N. (1982). The seasonal abundance of night-flying insects in a Neotropical forest. In *The Ecology of a Tropical Forest: Seasonal Rhythms and Long-Term Changes*, eds. E. G. Leigh, Jr, A. S. Rand & D. M. Windsor, pp. 309–318. Washington, DC: Smithsonian Institution Press.

Smythe, N. (1996). The seasonal abundance of night-flying insects in a Neotropical forest. In *The Ecology of a Tropical Forest: Seasonal Rhythms and Long-Term Changes*, 2nd edn, eds. E. G. Leigh, Jr, A. S. Rand & D. M. Windsor, pp. 309–318. Washington, DC: Smithsonian Institution Press.

Soberón, J. M. & Llorente, J. B. (1993). The use of species accumulation functions for the prediction of species richness. *Conservation Biology*, **7**, 480–488.

Sokal, R. R. & Braumann, C. A. (1980). Significance tests for coefficients of variation and variability profiles. *Systematic Zoology*, **29**, 50–66.

Sokal, R. R. & Rohlf, F. J. (1995). *Biometry*, 3rd edn. New York: Freemann.

Solís, E. (1993b). Características físico-químicas del suelo en un ecosistema tropical caducifolio de Chamela, Jalisco. Tesis Profesional. México: Fac. de Ciencias, Universidad Nacional Autónoma de México.

Solis, M. A. (1992). Checklist of the Old World Epipaschiinae and the related New World genera *Macalla* and *Epipaschia* (Pyralidae). *Journal of the Lepidopterists' Society*, **46**, 280–297.

Solis, M. A. (1993a). A phylogenetic analysis and reclassification of the genera of the *Pococera* complex (Lepidoptera: Pyralidae: Epipaschiinae). *Journal of the New York Entomological Society*, **101**, 1–83.

Solomon, M. (1980). *Dinâmica de populações. Temas de Biologia*. Vol. 3. São Paulo: Editora Pedagógica e Universitária.

Solow, A. R. (1993). A simple test for change in community structure. *Journal of Animal Ecology*, **62**, 191–193.

Soltis, D. E., Soltis, P. S., Chase, M. W., Mort, M. K., Albach, D. C., Zanis, M., Savolainen, V., Hahn, W. H., Hoot, S. B., Fay, M. F., Axtell, M., Swensen, S. M., Prince, L. M., Kress, W. J., Nixon, K. C. & Ferris, J. S. (2000). Angiosperm phylogeny inferred from 18S rDNA, *rbcL*, and *atpB* sequences. *Botanical Journal of the Linnean Society*, **133**, 381–461.

Sorensen, J. T., Campbell, B. C., Gill, R. J. & Steffen-Campbell, J. D. (1995). Non-monophyly of Auchenorrhyncha ('Homoptera'), based upon 18S rDNA phylogeny: eco-evolutionary and cladistic implications within pre-heteropterodea Hemiptera (s.l.) and a proposal for new monophyletic suborders. *Pan-Pacific Entomologist*, **71**, 31–60.

Sørensen, L. L. (2000). Biodiversity in Space – a Three-Dimensional Comparison of Spider Faunas in East African Forests. Ph.D. dissertation. Copenhagen, Denmark: University of Copenhagen.

Soulé, M. E. & Orians, G. H. (2001). *Conservation Biology: Research Priorities for the Next Decade*. Washington, DC: Island Press.

Southwood, T. R. E. (1960). The abundance of the Hawaiian trees and the number of their associated insect species. *Proceedings of the Hawaiian Entomological Society*, **17**, 299–303.

Southwood, T. R. E. (ed.) (1968). *Insect Abundance. Symposium of the Royal Entomological Society of London*, No. 4. Oxford: Blackwell Scientific.

Southwood, T. R. E. (1978). *Ecological Methods*. London: Chapman & Hall.

Southwood, T. R. E. (1988). Tactics, strategies and templets. *Oikos*, **52**, 3–18.

Southwood, T. R. E. & Henderson, P. A. (2000). *Ecological Methods*. Oxford: Blackwell Science.

Southwood, T. R. E. & Kennedy, C. E. J. (1983). Trees as islands. *Oikos*, **41**, 359–371.

Southwood, T. R. E., Moran, V. C. & Kennedy, C. E. J. (1982a). The assessment of arboreal insect fauna: comparisons of knockdown sampling and faunal lists. *Ecological Entomology*, **7**, 331–340.

Southwood, T. R. E., Moran, V. C. & Kennedy, C. E. J. (1982b). The richness, abundance and biomass of the arthropod communities on trees. *Journal of Animal Ecology*, **51**, 635–649.

Spain, A. V. & Harrison, R. A. (1968). Some aspects of the ecology of arboreal cryptostigmata (Acari) in New Zealand with special reference to the species associated with *Olearia colensoi* Hook.f. *New Zealand Journal of Science*, **11**, 452–458.

Speidel, W. (1998). Studies on the phylogeny of the Acentropinae (Lepidoptera, Crambidae). *Mémoires de la Société Royale Belge d'Entomologie*, **38**, 25–30.

Speight, M. R. & Wylie, F. R. (2001). *Insect Pests in Tropical Forestry*. Wallingford: CABI.

Spencer, J. L., Gewax, L. J., Keller, J. E. & Miller, J. R. (1997). Chemiluminiscent tags for tracking insect movement in darkness: application to moth photo-orientation. *Great Lakes Entomologist*, **30**, 33–43.

Spitzer, K., Novotny, V., Tonner, M. & Leps, J. (1993). Habitat preferences, distribution and seasonality of the butterflies (Lepidoptera, Papilionoidea) in a montane tropical rain forest, Vietnam. *Journal of Biogeography*, **20**, 109–121.

Springate, N. D. & Basset, Y. (1996). Diel activity of arboreal arthropods associated with Papua New Guinean trees. *Journal of Natural History*, **30**, 101–112.

SPSS (1994). *SPSS Advanced Statistics*, version 6.1. Chicago, IL: SPSS.

Srivastava, D. S. (1999). Using local–regional richness plots to test for species saturation: pitfalls and potentials. *Journal of Animal Ecology*, **68**, 1–16.

Srygley, R. B. & Chai, P. (1990). Flight morphology of Neotropical butterflies: palatability and distribution of mass to the thorax and abdomen. *Oecologia*, **84**, 491–499.

StatSoft (1995). *Statistica User Guide. The Complete Statistical System*. Tulsa, OK: StatSoft.

StatSoft (1997). *STATISTICA for Windows*. Tulsa, OK: StatSoft.

StatSoft (1999). *STATISTICA for Windows, 5.5*. Tulsa, OK: StatSoft.

Staudt, M., Mandl, N., Joffre, R. & Rambal, S. (2001). Intraspecific variability of monoterpene composition emitted by *Quercus ilex* leaves. *Canadian Journal of Forest Research*, **31**, 174–180.

Stebayeva, S. K. (1975). Phytogenic microstructure of Collembola associations in steppes and forests of Siberia. In *Progress in Soil Zoology*, ed. J. Vanek, pp. 77–84. The Hague: W. Junk.

Steel, R. G. D. & Torrie, J. H. (1980). *Principles and Procedures of Statistics*, 2nd edn. New York: McGraw-Hill.

Sterck, F. & Bongers, F. (2001). Crown development in tropical rain forest trees: patterns with tree height and light availability. *Journal of Ecology*, **89**, 1–13.

Sterck, F., van der Meer, P. & Bongers, F. (1992). Herbivory in two rain forest canopies in French Guyana. *Biotropica*, **24**, 97–99.

Sterck, F. J., Bongers, F. & Newbery, D. M. (2001). Tree architecture in a Bornean lowland rain forest: intraspecific and interspecific patterns. *Plant Ecology*, **153**, 279–292.

Stevens, G. C. (1990). The latitudinal gradient in geographical range: How so many species coexist in the tropics. *American Naturalist*, **133**, 240–256.

Stevenson, B. G. & Dindal, D. L. (1982). Effect of leaf shape in forest litter spiders: community organization and microhabitat selection of immature *Enoplognatha ovata* (Clerck) (Theridiidae). *Journal of Arachnology*, **10**, 165–178.

Stewart-Oaten, A. (1995). Rules and judgments in statistics: three examples. *Ecology*, **76**, 2001–2009.

Stewart-Oaten, A., Murdoch, W. W. & Walde, S. J. (1995). Estimation of temporal variability in populations. *American Naturalist*, **146**, 519–535.

Stokes, A. E., Schultz, B. B., Degraaf, R. M. & Griffin, C. R. (2000). Setting mist net from platforms in the forest canopy. *Journal of Field Ornithology*, **71**, 57–65.

Stoops, C. A., Adler, P. H. & McCreadie, J. W. (1998). Ecology of aquatic Lepidoptera (Crambidae: Nymphulinae) in South Carolina, USA. *Hydrobiologia*, **379**, 33–40.

Stork, N. E. (1987a). Arthropod faunal similarity of Bornean rain forest trees. *Ecological Entomology*, **12**, 219–226.

Stork, N. E. (1987b). Guild structure of arthropods from Bornean rain forest trees. *Ecological Entomology*, **12**, 69–80.

Stork, N. E. (1988). Insect diversity: facts, fiction and speculation. *Biological Journal of the Linnean Society*, **35**, 321–337.

Stork, N. E. (1991). The composition of the arthropod fauna of Bornean lowland rain forest trees. *Journal of Tropical Ecology*, **7**, 161–180.

Stork, N. E. (1993). How many species are there? *Biodiversity and Conservation*, **2**, 215–232.

Stork, N. E. (1994). Inventories of biodiversity: more than a question of numbers. In *Systematics and Conservation Evaluation*, eds. P. I. Forey, C. J. Humphries & R. I. Vane-Wright, pp. 81–100. Oxford: Clarendon Press.

Stork, N. E. (1997). Measuring global biodiversity and its decline. In *Biodiversity II*, eds. M. L. Reaka-Kudla, D. E. Wilson & E. O. Wilson, pp. 41–68. Washington, DC: Joseph Henry Press.

Stork, N. E. (1999). The magnitude of biodiversity and its Decline. In *The Living Planet in Crisis: Biodiversity Science and Policy*, eds. J. Cracraft & F. T. Grifo, pp. 3–32. New York: Columbia University Press.

Stork, N. E., Adis, J. A. & Didham, R. K. (eds.) (1997a). *Canopy Arthropods*. London: Chapman & Hall.

Stork, N. E. & Best, V. (1994). European Science Foundation – Results of a survey of European canopy research in the tropics. *Selbyana*, **15**, 51–62.

Stork, N. E. & Brendell, M. J. D. (1990). Variation in the insect fauna of Sulawesi trees with season, altitude and forest type. In *Insects and the Rain Forests of South East Asia* (*Wallacea*), eds. W. J. Knight & J. D. Holloway, pp. 173–190. London: The Royal Entomological Society of London.

Stork, N. E. & Brendell, M. J. D. (1993). Arthropod abundance in lowland rain forest of Seram. In *Natural History of Seram, Maluku, Indonesia*, eds. I. D. Edwards, A. A. MacDonald & J. Proctor, pp. 115–130. Andover, MA: Intercept.

Stork, N. E., Didham, R. K. & Adis, J. (1997b). Canopy arthropod studies for the future. In *Canopy Arthropods*, eds. N. E. Stork, J. Adis & R. K. Didham, pp. 551–561. London: Chapman & Hall.

Stork, N. E. & Gaston, K. J. (1990). Counting species one by one. *New Scientist*, **1729**, 43–47.

Stork, N. E. & Hammond, P. M. (1997). Sampling arthropods from tree-crowns by fogging with knockdown insecticides: lessons from studies of oak tree beetle assemblages in Richmond Park (UK). In *Canopy Arthropods*, eds. N. E. Stork, J. Adis & R. K. Didham, pp. 3–26. London: Chapman & Hall.

Stork, N. E., Hammond, P. M., Russell, B. L. & Hadwen, W. L. (2001). The spatial distribution of beetles within the canopies of oak trees in Richmond Park, U.K. *Ecological Entomology*, **26**, 302–311.

Stork, N. E., Wright, S. J. & Mulkey, S. S. (1997c). Craning for a better view: the Canopy Crane Network. *Trends in Ecology and Evolution*, **12**, 415–420.

Stoutjesdijk, P. H. & Barkman, J. J. (1992). *Microclimate, Vegetation and Fauna*. Knivsta: Opulus Press.

Strandtmann, R. W. & Wharton, G. W. (1958). A Manual of Mesostigmatid Mites Parasitic on Vertebrates. *Contributions of the Institute of Acarology, University of Maryland*, **4**, 1–330.

Strauss, S. Y. (1991). Indirect effects in community ecology: their definition, study and importance. *Trends in Ecology and Evolution*, **6**, 206–210.

Strenzke, K. (1952). Untersuchungen über die Tiergemeinschaften des Bodens. Die Oribatiden und ihre Synusien in den Böden Norddeutschlands. *Zoologica*, **37**, 1–104.

Strong, D. R., Lawton, J. H. & Southwood, T. R. E. (1984). *Insects on Plants. Community Patterns and Mechanisms*. Oxford: Blackwell.

Stubbs, C. S. (1989). Patterns of distribution and abundance of corticolous lichens and their invertebrate associates on *Quercus rubra* in Maine. *Bryologist*, **92**, 453–460.

Stuntz, S. (2001). The Influence of Epiphytes on Arthropods in the Tropical Forest Canopy. PhD thesis. Würzburg, Germany: Universität Würzburg.

Stuntz, S., Simon, U. & Zotz, G. (1999). Assessing the potential influence of vascular epiphytes on arthropod diversity in tropical tree crowns: hypotheses, approaches, and preliminary data. *Selbyana*, **20**, 276–283.

Stuntz, S., Simon, U. & Zotz, G. (2002a). Rainforest airconditioning: the moderating influence of epiphytes on the microclimate in tropical tree crowns. *International Journal of Biometeorology*, **46**, 53–59.

Stuntz, S., Ziegler, C., Simon, U. & Zotz, G. (2002b). Structure and diversity of the arthropod fauna within three canopy epiphyte species in Central Panama. *Journal of Tropical Ecology*, **18**, 161–176.

Sugden, A. M. (1985). Aerial walkways in rain forest canopies, construction and use in Panama, Papua New Guinea, and Sulawesi. In *The Botany and Natural History of Panama*, eds. W. G. d'Arcy & M. D. A. Correa, pp. 207–219. St Louis, MO: Missouri Botanical Garden.

Süßenbach, D., Brehm, G., Häuser, C. L. & Fiedler, K. (2001). Assessment of β-diversity with rich, incompletely sampled animal communities. *Zoology*, **104**, (Suppl. IV), 22.

Sutton, S. L. (1979). A portable light trap for studying insects of the upper canopy. *Brunei Museum Journal*, **4**, 156–160.

Sutton, S. L. (1983). The spatial distribution of flying insects in tropical rain forests. In *Tropical Rain Forest: Ecology and Management*, eds. S. L. Sutton, T. C. Whitmore & A. C. Chadwick, pp. 77–92. Oxford: Blackwell.

Sutton, S. L. (1989). The spatial distribution of flying insects. In *Tropical Rain Forest Ecosystems. Biogeographical and Ecological Studies*, eds. H. Lieth & M. J. A. Werger, pp. 427–436. Amsterdam: Elsevier.

Sutton, S. L. (2001). Alice grows up: canopy science in transition from Wonderland to reality. *Plant Ecology*, **153**, 13–21.

Sutton, S. L., Ash, C. P. & Grundy, A. (1983a). The vertical distribution of flying insects in the lowland rain forest of Panama, Papua New Guinea and Brunei. *Zoological Journal of the Linnean Society*, **78**, 287–297.

Sutton, S. L. & Collins, N. M. (1991). Insects and tropical forest conservation. In *The Conservation of Insects and Their Habitats*, eds. N. M. Collins & J. A. Thomas, pp. 405–424. New York: Academic Press.

Sutton, S. L. & Hudson, P. J. (1980). The vertical distribution of small flying insects in the lowland rain forest of Zaire. *Zoological Journal of the Linnean Society*, **68**, 111–123.

Sutton, S. L., Whitmore, T. C. & Chadwick, A. C. (eds.) (1983b). *Tropical Rain Forest: Ecology and Management*. Oxford: Blackwell.

Swezey, O. H. (1954). *Forest Entomology in Hawaii. Special Publication 76*. Honolulu, Hawaii: Bishop Museum.

Synnott, T. J. (1985). *A Checklist of the Flora of Budongo Forest Reserve, Uganda, with Notes on Ecology and Phenology. Commonwealth Forestry Institute Occasional Papers, 27*. Oxford, UK: Commonwealth Forestry Institute.

Szarzynski, J. & Anhuf, D. (2001). Micrometerorological conditions and canopy energy exchanges of a Neotropical rainforest (Surumoni-crane project, Venezuela). *Plant Ecology*, **153**, 231–239.

Szujecki, A. (1987). *Ecology of Forest Insects*. Dordrecht, the Netherlands: W. Junk and Kluwer Academic.

Takeda, H. (1987). Dynamics and maintenance of collembolan community structure in a forest soil system. *Research in Population Ecology*, **29**, 291–346.

Tauber, M. J. & Tauber, C. A. (1976). Insect seasonality: diapause maintenance, termination and postdiapause development. *Annual Review of Entomology*, **21**, 81–107.

Tauber, M. J., Tauber, C. A. & Masaki, S. (1986). *Seasonal Adaptations of Insects*. New York: Oxford University Press.

Tavakilian, G., Berkov, A., Meurer-Grimes, B. & Mori, S. (1997). Neotropical tree species and their faunas of xylophagous longicorns (Coleoptera: Cerambycidae) in French Guiana. *Botanical Review*, **63**, 303–355.

Taylor, B. (1977). The ant mosaic on cocoa and other tree crops in western Nigeria. *Ecological Entomology*, **2**, 245–255.

Taylor, L. R. (1974). Insect migration, flight periodicity and the boundary layer. *Journal of Animal Ecology*, **43**, 225–238.

Taylor, L. R. & Carter, C. I. (1961). The analysis of numbers and distribution in an aerial population of Macrolepidoptera. *Transactions of the Royal Entomological Society London*, **113**, 369–386.

Taylor, V. A. (1978). A winged elite in a subcortical beetle as a model for a prototermite. *Nature*, **276**, 73–74.

Taylor, V. A. (1981). The adaptive and evolutionary significance of wing polymorphism and parthenogenesis in *Ptinella* Motschulsky (Coleoptera: Ptiliidae). *Ecological Entomology*, **6**, 89–98.

Terborgh, J. (1985). The vertical component of plant species diversity in temperate and tropical forests. *American Naturalist*, **126**, 760–776.

Terborgh, J. (1986). Keystone plant resources in the tropical forest. In *Conservation Biology. The Science of Scarcity and Diversity*, ed. M. E. Soulé, pp. 330–344. Sunderland, MA: Sinauer.

Terborgh, J. W. & Faaborg, J. (1980). Saturation of bird communities in the West Indies. *American Naturalist*, **116**, 178–195.

Theenhaus, A., Scheu, S. & Schaefer, M. (1999). Contramensal interactions between two collembolan species: effects on population development and on soil processes. *Functional Ecology*, **13**, 238–246.

Théry, M. (2001). Forest light and its influence on habitat selection. *Plant Ecology*, **153**, 251–261.

Thomas, C. D. (1990a). Fewer species. *Nature*, **347**, 237.

Thomas, C. D. (1990b). Herbivore diets, herbivore colonization, and the escape hypothesis. *Ecology*, **71**, 610–615.

Thompson, J. D. (1981). Spatial and temporal components of resource assessment by flower-feeding animals. *Journal of Animal Ecology*, **50**, 49–59.

Thompson, J. N. (1994). *The Coevolutionary Process*. Chicago, IL: University of Chicago Press.

Thomson, J. D., Herre, E. A., Hamrick, J. L. & Stone, J. L. (1991). Genetic mosaics in strangler fig trees: implications for tropical conservation. *Science*, **254**, 1214–1216.

Thornton, I. W. B. (1984). Pscocoptera of the Hawaiian Islands. Part III. The endemic *Ptycta* complex (Psocidae): systematics, distribution, and evolution. *International Journal of Entomology*, **26**, 1–128.

Till, W. M. (1963). Ethiopian mites of the genus *Androlaelaps* Berlese *s.lat.* (Acari: Mesostigmata). *Bulletin of the British Museum (Natural History) Zoology*, **10**, 1–104.

Tipton, V. J. (1960). The genus *Laelaps* with a review of the Laelaptinae and a new subfamily Alphalaelaptinae (Acarina : Laelaptidae). *University of California Publications in Entomology*, **16**, 233–356.

Tobin, J. E. (1991). A neotropical rainforest canopy, ant community: some ecological considerations. In *Ant–Plant Interactions*, eds. C. R. Huxley & D. F. Cutler, pp. 536–538. Oxford: University Press.

Tobin, J. E. (1994). Ants as primary consumers: diet and abundance in the Formicidae. In *Nourishment and Evolution in Insect Societies*, eds. J. H. Hunt & C. A. Napela, pp. 279–307. Boulder, CO: Westview Press.

Tobin, J. E. (1995). Ecology and diversity of tropical forest canopy ants. In *Forest Canopies*, eds. M. D. Lowman & N. M. Nadkarni, pp. 129–147. San Diego, CA: Academic Press.

Toda, M. J. (1977). Vertical microdistribution of Drosophilidae (Diptera) within various forests in Hokkaido. I. Natural broad-leaved forest. *Japanese Journal of Ecology*, **27**, 207–214.

Toda, M. J. (1992). Three-dimensional dispersion of drosophilid flies in a cool temperate forest in northern Japan. *Ecological Research*, **7**, 283–295.

Tonn, W. M., Magnuson, J. J., Rask, M. & Toivonen, J. (1990). Intercontinental comparison of small-lake fish assemblages: the balance between local and regional processes. *American Naturalist*, **136**, 345–375.

Tovar, E. (1999). Estructura de las comunidades de artrópodos epífitos asociados a los encinos (*Quercus* spp.) del Valle de México. Tesis Maestría. México: Fac. de Ciencias, Universidad Nacional Autónoma de México.

Toy, R. J. (1991). Interspecific flowering patterns in the Dipterocarpaceae in west Malaysia: implications for predator satiation. *Journal of Tropical Ecology*, **7**, 49–57.

Toy, R. J., Marshall, A. G. & Tho, Y. P. (1992). Fruiting phenology and the survival of insect fruit predators: a case study from the South-east Asian Dipterocarpaceae. *Philosophical Transactions of the Royal Society of London, Series B*, **335**, 417–423.

Trägårdh, I. (1942). Microgyniina, a new group of Mesostigmata. *Entomologisk Tidskrift*, **63**, 120–133.

Trägårdh, I. (1950). Studies on the Celaenopsidae, Diplogyniidae and Schizogyniidae (Acarina). *Arkiv för Zoologi*, **1**, 361–451.

Travé, J. (1963). Ecologie et biologie des Oribates (Acariens) saxicoles et arboricoles. *Vie et Milieu, Supplément*, **14**, 1–267.

Trivers, R. (2000). William Donald Hamilton (1936–2000). *Nature*, **404**, 828.

Turchin, P. (1987). The role of aggregation in the response of Mexican bean Beetles to host-plant density. *Oecologia*, **71**, 577–582.

Turchin, P., Taylor, A. D. & Reeve, J. D. (1999). Dynamical role of predators in population cycles of a forest insect: an experimental test. *Science*, **285**, 1068–1070.

Turk, S. Z. (1985). Acridios del NOA. VI. Ciclo de vida de *Cornops frenatus cannae* Roberts & Carbonell (Acrididae, Leptysminae) con especial referencia a su oviposición endofitica. *Revista de la Sociedad Entomológica Argentina*, **43**, 91–102.

Turk, S. Z. & Barrera, M. (1976). Acridios del NOA. I. Estudios biológicos, morfométricos y aspectos ecológicos de *Chromacris speciosa* (Thunberg) (Acrididae, Romaleinae). *Acta Zoologica Lilloana, Tucuman*, **32**, 121–145.

Turnbull, A. L. (1960). The spider population of a stand of oak (*Quercus robur* L.) in Wytham Wood, Berks., England. *Canadian Entomologist*, **92**, 110–124.

Turnbull, A. L. (1973). The ecology of the true spiders (Aranemorphae). *Annual Review of Entomology*, **18**, 305–348.

Turner, I. M. (1994). A quantitative analysis of leaf form in woody plants from the world's major broad-leaved forest types. *Journal of Biogeography*, **21**, 413–419.

Turner, J. R. G., Lennon, J. J. & Greenwood, J. J. D. (1996). Does climate cause the global biodiversity gradient? In *Aspects of the Genesis and Maintenance of Biological Diversity*, eds. M. E. Hochberg, J. Clobert & R. Barbault, pp. 199–220. Oxford: Oxford University Press.

Turner, R. G., Gatehouse, C. M. & Corey, C. A. (1987). Does solar energy control organic diversity? Butterflies, moths and the British climate. *Oikos*, **48**, 195–205.

Tutin, C. E. G. & White, L. J. T. (1998). Primates, phenology and frugivory: present, past and future patterns in the Lopé Reserve, Gabon. In *Dynamics of Tropical Communities. 37th Symposium of the British Ecological Society*, eds. D. M. Newbery, H. H. T. Prins & N. D. Brown, pp. 309–337. Oxford: Blackwell Science.

Uetz, G. W. (1979). The influence of variation in litter habits on spider communities. *Oecologia*, **40**, 29–42.

Uetz, G. W. (1991). Habitat structure and spider foraging. In *Habitat Structure. The Physical Arrangement of Objects in Space*, eds. S. S. Bell, E. D. McCoy & H. R. Muhinsky, pp. 325–348. London: Chapman & Hall.

Ungar, P. S., Teaford, M. F., Glander, K. E. & Pastor, R. F. (1995). Dust accumulation in the canopy: a potential cause of dental microwear in primates. *American Journal of Physical Anthropology*, **97**, 93–99.

University of Suriname (1996). *Wane Hills Baseline Study, Final Report*. Paramaribo: University of Suriname.

Usua, E. J. (1973). Induction of diapause in the maize stemborer, *Busseola fusca*. *Entomologia Experimentalis et Applicata*, **16**, 322–328.

Valderrama, A. C. (1999). Diversidad de insectos minadores en un bosque tropical. Tesis de Maestría. Panamá: Programa de Maestría en Entomología, Universidad de Panamá.

Van der Pijl, L. (1954). *Xylocopa* and flowers in the tropics. *Proceedings of the Nederlandse Koninklijke Akademie van Wetenschappen, series C*, **57**, 413–423.

van Heezik, Y. & Seddon, P. J. (1999). Effects of season and habitat on bird abundance and diversity in a steppe desert, northern Saudi Arabia. *Journal of Arid Environments*, **43**, 301–317.

Van Klinken, R. D. & Walter, G. H. (2001). Subtropical drosophilids in Australia can be characterized by adult distribution across vegetation type and by height above forest floor. *Journal of Tropical Ecology*, **17**, 705–718.

van Mele, P. & Cuc, N. T. T. (2000). Evolution and status of *Oecophylla smaragdina* (Fabricius) as a pest control agent in citrus in the Mekong Delta, Vietnam. *International Journal of Pest Management*, **46**, 295–301.

van Schaik, C. P., Terborgh, J. W. & Wright, J. S. (1993). The phenology of tropical forests: adaptative significance and consequences for primary consumers. *Annual Review of Ecology and Systematics*, **24**, 353–377.

Vannier, G. (1970). Réactions des microarthropodes aux variations de l'état hydrique du sol. In *Recherche coopérative sur programme du CNRS, No. 40*, pp. 23–258. Paris: Editions du Centre National de la Recherche Scientifique.

Vareschi, V. (1980). *Vegetationsökologie der Tropen*. Stuttgart: Ulmer.

Vargas Márquez, F. (1984). *Parques Nacionales de México y Reservas Equivalentes. Pasado, Presente y Futuro*. México: Instituto de Investigaciones Económicas, Universidad Nacional Autónoma de México.

Vasconcelos, H. L. (1999). Levels of leaf herbivory in Amazonian trees from different stages in forest regeneration. *Acta Amazonica*, **29**, 615–623.

Vasconcelos, H. L., Vilhena, J. M. S. & Caliri, G. J. A. (2000). Responses of ants to selective logging of a central

Amazonian forest. *Journal of Applied Ecology*, **37**, 508–514.

Vegter, J. J. (1983). Food and habitat specialization in coexisting springtails (Collembola, Entomobryidae). *Pedobiologia*, **25**, 253–262.

Vegter, J. J. (1987). Phenology and seasonal partitioning in forest floor Collembola. *Oikos*, **48**, 175–185.

Vegter, J. J., De Bie, P. & Dop, H. (1988). Distributional ecology of forest floor Collembola (Entomobryidae) in the Netherlands. *Pedobiologia*, **31**, 65–73.

Vepsälänen, K. (1982). Assembly of island ant communities. *Annales Zoologici Fennici*, **19**, 327–335.

Veras, R. S. & Castellon, E. G. (1998). *Culicoides* Latreille (Diptera, Ceratopogonidae) in Brazilian Amazon. V. Efficiency of traps and baits and vertical stratification in the Forest Reserve Adolpho Ducke. *Revista Brasileira de Zoologia*, **15**, 145–152.

Verhoef, H. A. (1996). The role of soil microcosms in the study of ecosystem processes. *Ecology*, **77**, 685–690.

Verhoef, H. A. & Witteven, J. (1980). Water balance in Collembola and its relation to habitat selection, cuticular water loss and water uptake. *Journal of Insect Physiology*, **26**, 201–208.

Vitousek, P. M. (1995). The Hawaiian Islands as a model system for ecosystem studies. *Pacific Science*, **49**, 2–16.

Vitousek, P. M., Loope, L. L. & Stone, C. P. (1987). Introduced species in Hawaii: biological effects and opportunities for ecological research. *Trends in Ecology and Evolution*, **2**, 224–227.

Vitousek, P. M., Mooney, H. A., Lubchenco, J. & Melillo, J. M. (1997). Human domination of Earth's ecosystems. *Science*, **277**, 494–499.

Vitousek, P. M., Turner, D. R. & Kitayama, K. (1995). Foliar nutrients during long-term soil development in Hawaiian montane rain forest. *Ecology*, **76**, 712–720.

von Allmen, H. & Zettel, J. (1982). Populationsbiologische Untersuchungen zur Art *Entomobrya nivalis* (Collembola). *Revue Suisse de Zoologie*, **89**, 919–926.

von Martius, C. F. P. (1897). *Flora Brasiliensis. Bd. XII.* München/Leipzig: R. Oldenbourg.

Wagner, T. (1996a). Zussammensetzung der baumbewohnenden Arthropodenfauna in Wäldern Zentalafrikas; mit Anmerkungen zur Nebelmethode und zum Morphotypen-Verfahren. *Mitteilungen des Internationalen Entomologischen Vereins*, **21**, 25–42.

Wagner, T. (1996b). Artenmannigfaltigkeit baumkronenbewohnender Arthropoden in zentralafrikanischen Wäldern, unter besonderer Berücksichtigung der Käfer. Dissertation. Rheinische Friedrich-Wilhelm University, Bonn.

Wagner, T. (1997). The beetle fauna of different tree species in forests of Rwanda and East-Zaire. In *Canopy Arthropods*, eds. N. E. Stork, J. Adis & R. K. Didham, pp. 169–183. London: Chapman & Hall.

Wagner, T. (1998). Influence of tree species and forest type on the chrysomelid community in the canopy of an Ugandan tropical forest. In *Proceedings of the IV. International Symposium on the Chrysomelidae*, eds. M. Biondi, M. Daccordi & D. G. Furth, pp. 253–269. Torino: Museo Regionale di Scienze Naturale di Torino.

Wagner, T. (1999). Arboreal chrysomelid community structure and faunal overlap between different types of forests in Central Africa. In *Advances in Chrysomelidae Biology 1*, ed. M. L. Cox, pp. 247–270. Leiden: Backhuys.

Wagner, T. (2000). Influence of forest type and tree species on canopy-dwelling beetles in Budongo forest, Uganda. *Biotropica*, **32**, 502–514.

Wagner, T. (2001). Seasonal changes in the canopy arthropod fauna in *Rinorea beniensis* in Budongo Forest, Uganda. *Plant Ecology*, **153**, 169–178.

Wallwork, J. A. (1976). *The Distribution and Diversity of Soil Fauna*. London: Academic Press.

Wallwork, J. A. (1983). Oribatids in forest ecosystems. *Annual Review of Entomology*, **28**, 109–130.

Wallwork, J. A., Kamill, B. W. & Whitford, W. G. (1984). Life styles of desert litter-dwelling microarthropods: a reappraisal on the reproductive behaviour of cryptostigmatic mites. *South African Journal of Science*, **80**, 163–169.

Walter, D. E. (1985). The effects of litter type and elevation on colonization of mixed coniferous litterbags by oribatid mites. *Pedobiologia*, **28**, 383–387.

Walter, D. E. (1995). Dancing on the head of a pin: mites in the rainforest canopy. *Records of the Western Australian Museum, Supplement*, **52**, 49–53.

Walter, D. E. (2000). A jumping mesostigmatan mite, *Saltiseius hunteri* n.g., n. sp. (Acari: Mesostigmata: Trigynaspida: Saltiseidae, n. fam.) from Australia. *International Journal of Acarology*, **26**, 25–31.

Walter, D. E. (2001). Endemism and cryptogenesis in 'segmented' mites: a review of Australian Alicorhagiidae, Terpnacaridae, Oehserchestidae and Grandjeanicidae (Acari, Sarcoptiformes). *Australian Journal of Entomology*, **40**, 207–218.

Walter, D. E. & Behan-Pelletier, V. M. (1993). Systematics and ecology of *Adhaesozetes polyphyllos*, sp. nov. (Acari: Oribatida: Licneremaeoidea) leaf-inhabiting mites from Australian rainforests. *Canadian Journal of Zoology*, **71**, 1024–1040.

Walter, D. E. & Behan-Pelletier, V. M. (1999). Mites in forest canopies: filling the size shortfall? *Annual Review of Entomology*, **44**, 1–19.

Walter, D. E. & Krantz, G. W. (1999). New early derivative Mesostigmatans from Australia: *Nothogynus* n.g., Nothogynidae n. fam. *International Journal of Acarology*, **25**, 67–76.

Walter, D. E. & Lindquist, E. E. (1989). Life history and behaviour of mites in the genus *Lasioseius* (Acari: Mesostigmata: Ascidae) from grassland soils in Colorado, with taxonomic notes and description of a new species. *Canadian Journal of Zoology*, **67**, 2797–2813.

Walter, D. E. & Norton, R. A. (1984). Body size distribution in sympatric oribatid mites (Acari: Sarcoptiformes) from California pine litter. *Pedobiologia*, **27**, 99–106.

Walter, D. E. & O'Dowd, D. J. (1995). Life on the forest phylloplane: hairs, little houses, and myriad mites. In *Forest Canopies*, eds. M. D. Lowman & N. M. Nadkarni, pp. 325–351. San Diego, CA: Academic Press.

Walter, D. E., O'Dowd, D. J. & Barnes, V. (1994). The forgotten arthropods: foliar mites in the forest canopy. *Memoirs of the Queensland Museum*, **36**, 221–226.

Walter, D. E. & Proctor, H. C. (1998). Feeding behaviour and phylogeny: observations on early derivative Acari. *Experimental & Applied Acarology*, **22**, 39–50.

Walter, D. E. & Proctor, H. C. (1999). *Mites: Ecology, Evolution and Behaviour*. Sydney: UNSW Press.

Walter, D. E., Seeman, O., Rodgers, D. & Kitching, R. L. (1998). Mites in the mist: how unique is a rainforest canopy-knockdown fauna? *Australian Journal of Ecology*, **23**, 501–508.

Walter, H. & Breckle, S.-W. (1991). *Ökologie der Erde, Band 2 Spezielle Ökologie der Tropen und Subtropen*. Stuttgart: Fischer.

Walter, P. (1983). Contribution à la connaissance des Scarabéides coprophages du Gabon (Col.). 2. Présence de populations dans la canopée de la forêt Gabonaise. *Bulletin de la Société Entomologique de France*, **88**, 514–521.

Warchalowski, A. (1985). *Chrysomelidae – Leaf beetles (Insecta: Coleoptera) I. Fauna Polski 10* (in Polish). Warszawa: PAN.

Ward, L. K. & Spalding, D. F. (1993). Phytophagous British insects and mites and their food-plant families: total numbers and polyphagy. *Biological Journal of the Linnean Society*, **49**, 257–276.

Waring, G. L. & Cobb, N. S. (1992). The impact of plant stress on herbivore population dynamics. In *Plant–Insect Interactions*, Vol. 4, ed. E. A. Bernays, pp. 167–226. Boca Raton, FL: CRC Press.

Wasserman, S. S. & Futuyma, D. J. (1981). Evolution of host plant utilization in laboratory populations of the southern cowpea weevil, *Callosobruchus maculatus* (Coleoptera: Bruchidae). *Evolution*, **35**, 605–617.

Watanabe, H. (1997). Estimation of arboreal and terrestrial arthropod densities in the forest canopy as measured by insecticide smoking. In *Canopy Arthropods*, eds. N. E. Stork, J. Adis & R. K. Didham, pp. 401–414. London: Chapman & Hall.

Watanabe, H. & Ruaysoongnern, S. (1989). Estimation of arboreal arthropod density in a dry evergreen forest in Northeastern Thailand. *Journal of Tropical Ecology*, **5**, 151–158.

Waterhouse, D. F. (ed.) (1991). *The Insects of Australia*. Carlton, Victoria: Melbourne University Press.

Watling, R., See, L. S. & Turnbull, E. (1996). Putative ectomycorrhizal fungi of Pasoh Forest Reserve, Negri Sembilan, Malaysia. In *Conservation, Management and Development of Forest Resources. Proceedings of the Malaysia–United Kingdom Programme Workshop, 1996*, eds. L. S. See, D. Y. May, I. D. Gauld & J. Bishop, pp. 105–114. Kuala Lumpur: Malaysia–United Kingdom Programme.

Watt, A. D., Stork, N. E. & Bolton, B. (2002). The diversity and abundance of ants in relation to forest disturbance and plantation establishment in southern Cameroon. *Journal of Applied Ecology*, **39**, 18–30.

Watt, A. D., Stork, N. E. & Hunter, M. D. (eds.) (1997a). *Forests and Insects: 18th Symposium of the Royal Entomological Society*. London: Chapman & Hall.

Watt, A. D., Stork, N. E., McBeath, C. & Lawson, G. L. (1997b). Impact of forest management on insect abundance and damage in a lowland tropical forest in southern Cameroon. *Journal of Applied Ecology*, **34**, 985–998.

Wauthy, G., Noti, M.-I. & Dufrene, M. (1989). Geographic ecology of soil oribatid mites. *Pedobiologia*, **33**, 399–416.

Way, M. J. (1953). The relationship between certain ant species with particular reference to biological control of the coreid, *Theraptus* sp. *Bulletin of Entomological Research*, **44**, 669–671.

Way, M. J. (1963). Mutualism between ants and honeydew-producing Homoptera. *Annual Review of Entomology*, **8**, 307–344.

Way, M. J. & Bolton, B. (1997). Competition between ants for coconut palm nesting sites. *Journal of Natural History*, **31**, 439–455.

Way, M. J. & Khoo, C. (1992). Role of ants in pest management. *Annual Review of Entomology*, **37**, 479–503.

Webb, C. O. (2000). Exploring the phylogenetic structure of ecological communities: an example for rain forest trees. *American Naturalist*, **156**, 145–155.

Webb, L. J., Tracey, J. G. & Williams, W. T. (1984). A floristic framework of Australian rainforests. *Australian Journal of Ecology*, **9**, 169–198.

Weber, E. (1986). *Grundriss der biologischen Statistik*. Jena: VEB Gustav Fischer Verlag.

Weiher, E. & Keddy, P. A. (1995). Assembly rules, null models, and trait dispersion: new questions from old patterns. *Oikos*, **74**, 159–164.

Weintraub, P. G. & Horowitz, A. R. (1996). Spatial and diel activity of the pea leafminer (Diptera: Agromyzidae) in potatoes, *Solanum tuberosum*. *Environmental Entomology*, **25**, 722–726.

Weiss, I. (1995). Spinnen und Weberknechte auf Baumstämmen im Nationalpark Bayrischer Wald. In *Proceedings of the 15th European Colloquium of Arachnology*, ed. V. Ruzicka, pp. 184–192. Ceske Budejovice, Czech Republic: Czech Academy of Sciences.

West, C. (1986). Insect communities in tree canopies. In *Kora: An Ecological Inventory of the Kora National Reserve, Kenya*, eds. M. Coe & N. M. Collins, pp. 209–222. London: Royal Geographical Society.

Wheeler, W. M. (1942). Studies of neotropical ant-plants and their ants. *Bulletin of the Museum of Comparative Zoology, Harvard University*, **90**, 1–263.

Whitacre, D. F. (1981). Additional techniques and safety hints for climbing tall trees and some equipment and information sources. *Biotropica*, **13**, 286–291.

White, K. & Abernethy, K. (1997). *A Guide to the Vegetation of the Lopé Reserve Gabon*. New York: Wildlife Conservation Society.

White, T. C. R. (1978). The importance of a relative shortage of food in animal ecology. *Oecologia*, **33**, 71–86.

White, T. C. R. (1984). The abundance of invertebrate herbivores in relation to the availability of nitrogen in stressed food plants. *Oecologia*, **63**, 90–105.

Whitmore, T. C. (1975). *Tropical Rain Forests of the Far East*. Oxford, UK: Clarendon Press.

Whitmore, T. C. (1979). Gaps in the forest canopy. In *Tropical Trees as Living Systems*, eds. P. B. Tomlinson & M. H. Zimmerman, pp. 639–655. New York: Cambridge University Press.

Whitmore, T. C. (1984). *Tropical Rain Forests of the Far East*. Oxford, UK: Clarendon Press.

Whitmore, T. C. (1989). Canopy gaps and the two major groups of forest trees. *Ecology*, **70**, 536–538.

Whitmore, T. C. (1990). *An Introduction to Tropical Rain Forests*. Oxford, UK: Oxford University Press.

Whittaker, R. H. (1975). *Communities and Ecosystems*. 2nd edn. New York: MacMillan.

Whittaker, R. J., Willis, K. J. & Field, R. (2001). Scale and species richness: towards a general, hierarchical theory of species diversity. *Journal of Biogeography*, **28**, 453–470.

Wibmer, G. J. & O'Brien, C. W. (1986). Annotated checklist of the weevils (Curculionidae *sensu lato*) of South America. *Memoirs of the American Entomological Institute*, **39**, xvi, 563.

Wich, S. A. & van Schaik, C. P. (2000). The impact of El Niño on mast fruiting in Sumatra and elsewhere in Malesia. *Journal of Tropical Ecology*, **16**, 563–577.

Wiens, J. A. (1984). On understanding a non-equilibrium world: myth and reality in community patterns and processes. In *Ecological Communities. Conceptual Issues and the Evidence*, eds. D. R. Strong, D. Simberloff, L. G. Abele & A. B. Thistle, pp. 439–457. Princeton, NJ: Princeton University Press.

Wiens, J. A. (1989). *The Ecology of Bird Communities*. Vols. I, II. Cambridge: Cambridge University Press.

Wigglesworth, V. B. (1984). *Insect Physiology*, 8th edn. London: Chapman & Hall.

Williams, R. J., Myers, B. A., Muller, W. J., Duff, G. A. & Eamus, D. (1997). Leaf phenology of woody species in a North Australian tropical savanna. *Ecology*, **78**, 2542–2558.

Williamson, M. H. & Lawton, J. H. (1991). Fractal geometry of ecological habitats. In *Habitat structure – The Physical Arrangement of Objects in Space*, eds. S. S. Bell, E. D. McCoy & H. R. Mushinsky, pp. 69–86. London: Chapman & Hall.

Willis, K. J., Kleckowski, A. & Crowhurst, S. J. (1999). 124,000-year periodicity in terrestrial vegetation change during the late Pliocene epoch. *Nature*, **397**, 685–688.

Willmer, P. G. (1982). Microclimate and the environmental physiology of insects. *Advances in Insect Physiology*, **16**, 1–57.

Willmott, K. R., Constantino, L. M. & Hall, J. P. W. (2001). A review of *Colobura* (Lepidoptera: Nymphalidae) with comments on larval and adult ecology and description of a sibling species. *Annals of the Entomological Society of America*, **94**, 185–196.

Willott, S. J. (1999). The effects of selective logging on the distribution of moths in a Bornean rainforest. *Philosophical Transactions of the Royal Society, Biological Sciences*, **354**, 1783–1790.

Wilms, W., Imperatriz-Fonseca, V. L. & Engels, W. (1996). Resource partitioning between highly eusocial bees and possible impact of the introduced Africanized honey bee on native stingless bees in the Brazilian Atlantic rainforest. *Studies on Neotropical Fauna and Environment*, **31**, 137–151.

Wilson, D. E. & Sandoval, A. (eds.) (1997). *Manu. The Biodiversity of Southeastern Peru*. Lima: Smithsonian Institution & Editorial Horizonte.

Wilson, E. O. (1958). Patchy distributions of ant species in New Guinea rain forest. *Psyche*, **65**, 26–38.

Wilson, E. O. (1959). Some ecological characteristics of ants in New Guinea rain forests. *Ecology*, **40**, 437–447.

Wilson, E. O. (1985). Time to revive systematics. *Science*, **230**, 1227.

Wilson, E. O. (1987). The arboreal ant fauna of Peruvian Amazon forest: a first assessment. *Biotropica*, **19**, 245–251.

Wilson, E. O. (ed.) (1988a). *Biodiversity*. Washington, DC: National Academy Press.

Wilson, E. O. (1988b). The current state of biological diversity. In *Biodiversity*, ed. E. O. Wilson, pp. 3–18. Washington, DC: National Academy Press.

Wilson, E. O. (1992). *The Diversity of Life*. Cambridge, Massachusetts: Harvard University Press.

Winchester, N. N. (1997a). Canopy arthropods of coastal Sitka spruce trees on Vancouver Island, British Columbia, Canada. In *Canopy Arthropods*, eds. N. E. Stork, J. Adis & R. K. Didham, pp. 151–168. London: Chapman & Hall.

Winchester, N. N. (1997b). Conservation of Biodiversity: Guilds, Microhabitat Use and Dispersal of Canopy Arthropods in the Ancient Sitka Spruce Forests of the Carmanah Valley, Vancouver Island, British Columbia. Ph.D. thesis. Victoria, BC: University of Victoria.

Winchester, N. N., Behan-Pelletier, V. & Ring, R. A. (1999). Arboreal specificity, diversity and abundance of canopy-dwelling oribatid mites (Acari: Oribatida). *Pedobiologia*, **43**, 391–400.

Windsor, D. M. (1990). *Climate and Moisture Variability in a Tropical Forest: Long-term Records from Barro Colorado Island, Panama*. Washington, DC: Smithsonian Institution Press.

Wint, G. R. W. (1983). Leaf damage in tropical rain forest canopies. In *Tropical Rain Forest: Ecology and Management*, eds. S. L. Sutton, T. C. Whitmore & A. C. Chadwick, pp. 229–240. Oxford: Blackwell.

Winter, C. (1963). Zur Ökologie und Taxonomie der neotropischen Bodentiere. II. Zur Collembolen-Fauna Perus. *Zoologische Jahrbücher, Abteilung für Systematik, Ökologie und Geographie der Tiere*, **90**, 393–520.

Wirth, V. (1980). *Flechtenflora*. Stuttgart: UTB Ulmer.

Wojtusiak, J., Godzinska, E. J. & Dejean, A. (1995). Capture and retrieval of very large prey by workers of the African weaver ant, *Oecophylla longinoda* (Latreille 1802). *Tropical Zoology*, **8**, 309–318.

Wolda, H. (1978a). Fluctuations in abundance of tropical insects. *American Naturalist*, **112**, 1017–1045.

Wolda, H. (1978b). Seasonal fluctuations in rainfall, food and abundance of tropical insects. *Journal of Animal Ecology*, **47**, 369–381.

Wolda, H. (1979). Abundance and diversity of Homoptera in the canopy of a tropical forest. *Ecological Entomology*, **4**, 181–190.

Wolda, H. (1981). Similarity indices, sample size and diversity. *Oecologia*, **50**, 296–302.

Wolda, H. (1982). Seasonality of leafhoppers (Homoptera) on Barro Colorado Island, Panama. In *Ecology of a Tropical Forest: Seasonal Rhythms and Long-Term Changes*, eds. E. G. Leigh, Jr, A. S. Rand & D. M. Windsor, pp. 319–330. Washington, DC: Smithsonian Institution Press.

Wolda, H. (1987). Altitude, habitat and tropical insect diversity. *Biological Journal of the Linnean Society*, **30**, 313–323.

Wolda, H. (1988). Insect seasonality: why? *Annual Review of Ecology and Systematics*, **19**, 1–18.

Wolda, H. (1989). Seasonal cues in tropical organisms. Rainfall? Not necessarily! *Oecologia*, **80**, 437–442.

Wolda, H. (1992). Trends in abundance of tropical forest insects. *Oecologia*, **89**, 47–52.

Wolda, H. (1996). Between-site similarity in species composition of a number of Panamanian insect groups. *Miscellània Zoològica*, **19**, 39–50.

Wolda, H. & Flowers, R. H. (1985). Seasonality and diversity of mayfly adults (Ephemeroptera) in a 'non-seasonal' environment. *Biotropica*, **17**, 330–335.

Wolda, H., Marek, J., Spitzer, K. & Novák, I. (1994). Diversity and variability of Lepidoptera populations in urban Brno, Czech Republic. *European Journal of Entomology*, **91**, 213–226.

Wolda, H., O'Brien, C. W. & Stockwell, H. P. (1998). Weevil diversity and seasonality in tropical Panama as deduced from light-trap catches (Coleoptera: Curculionoidea). *Smithsonian Contributions to Zoology*, **590**, 1–79.

Wolda, H. & Wong, M. (1988). Tropical insect diversity and seasonality. Sweep samples vs. light traps. *Proceedings of the Nederlandse Koninklijke Akademie van Wetenschappen, series C*, **91**, 203–216.

Wolfe, E. W. & Morris, J. (1996). *Geologic Map of the Island of Hawaii*. Washington, DC: US Department of the Interior, US Geological Survey.

Wolfe, K. L. (1993). The *Copaxa* of Mexico and their immature stages (Lepidoptera: Saturniidae). *Tropical Lepidoptera*, **4**(Suppl. 1), 1–26.

Woltemade, H. (1982). Zur Ökologie baumrindenbewohnender Hornmilben (Avari, Oribatei). *Sitzungsberichte der Gesellschaft Naturforschender Freunde zu Berlin*, **22**, 118–139.

Wolters, V. (1985). Untersuchungen zur Habitatbindung und Nahrungsbiologie der Springschwänze (Collembola) eines Laubwaldes unter besonderer Berücksichtigung ihrer Funktion in der Zersetzerkette. Dissertation. Göttingen: University of Göttingen.

Woodley, N. E. & Janzen, D. H. (1995). A new species of *Melanagromyza* (Diptera: Agromyzidae) mining leaves of *Bromelia pinguin* (Bromeliaceae) in a dry forest in Costa Rica. *Journal of Natural History*, **29**, 1329–1337.

Wootton, A. (1993). *Insects of the World*. Poole, UK: Blandford Press.

Wootton, J. T. (1994). Predicting direct and indirect effect: an integrated approach using experiments and path analysis. *Ecology*, **75**, 151–165.

Wright, S. D., Yong, C. G., Dawson, J. W., Whittaker, D. J. & Gardner, R. C. (2000). Riding the ice age El Nino? Pacific biogeography and evolution of *Metrosideros* subg. *Metrosideros* (Myrtaceae) inferred from nuclear ribosomal DNA. *Proceedings of the National Academy of Science USA*, **97**, 4118–4123.

Wright, S. J. (1983). The dispersion of eggs by bruchid beetle among *Scheelea* palm seeds and the effects of distance to the parent palm. *Ecology*, **64**, 1016–1021.

Wright, S. J. (1990). Cumulative satiation of a seed predator over the fruiting season of its host. *Oikos*, **58**, 1016–1021.

Wright, S. J. (1992). Seasonal drought, soil fertility and the species density of tropical forest plant communities. *Trends in Ecology and Evolution*, **7**, 260–263.

Wright, S. J. (1995). The canopy crane. In *Forest Canopies*, eds. M. D. Lowman & N. M. Nadkarni, p. 15. San Diego, CA: Academic Press.

Wright, S. J. & Colley, M. (1994). *Accessing the Canopy. Assesment of Biological Diversity and Microclimate of the Tropical Forest Canopy: Phase I*. Nairobi, Kenya: United Nation Environmental Program.

Wright, S. J. & Colley, M. (1996). *Tropical Forest Canopy Programme*. Nairobi, Kenya: UNEP/STRI.

Wright, S. J. & Cornejo, F. H. (1990). Seasonal drought and the timing of flowering and leaf fall in a Neotropical forest. In *Reproductive Ecology of Tropical Forest Plants*, eds. K. S. Bawa & M. Hadley, pp. 49–61. Paris: UNESCO and Parthenon Press.

Wunderle, I. (1992a). Die Baum- und Bodenbewohnenden Oribatiden (Acari) im Tieflandregenwald von Panguana, Peru. *Amazoniana*, **12**, 119–142.

Wunderle, I. (1992b). Die Oribatiden-Gemeinschaften (Acari) der verschiedenen Habitate eines Buchenwaldes. *Carolinea*, **50**, 79–144.

Yanoviak, S. P. (1999). Community structure in water-filled tree holes of Panama: effects of hole height and size. *Selbyana*, **20**, 106–115.

Yanoviak, S. P. (2001). Predation, resource availability, and community structure in Neotropical water-filled tree holes. *Oecologia*, **126**, 125–133.

Yanoviak, S. P. & Kaspari, M. (2000). Community structure and the habitat templet: ants in the tropical forest canopy and litter. *Oikos*, **89**, 259–266.

Yap, S. K. & Chan, H. T. (1990). Phenological behaviour of some *Shorea* species in peninsula Malaysia. In *Reproductive Ecology of Tropical Forest Plants*, eds. K. S. Bawa & M. Hadley, pp. 21–35. Paris: UNESCO and Parthenon Press.

Yela, J. L. & Holyoak, M. (1997). Effects of moonlight and meteorological factors on light and bait trap catches of noctuid moths. (Lepidoptera: Noctuidae). *Environmental Entomology*, **26**, 1283–1290.

Young, A. M. (1983). Patterns of distribution and abundance in small samples of litter inhabiting Orthoptera in some Costa-Rican cacao plantations. *Journal of the New York Entomological Society*, **91**, 312–327.

Young, A. M. (1986). *Tigridia asesta* (Linnaeus) (Nymphalidae) is not associated with *Theobroma cacao* L. (Sterculiaceae). *Journal of the Lepidopterists' Society*, **39**, 146–150.

Young, A. M. & Hermann, H. R. (1980). Notes on foraging of the giant tropical ant *Paraponera clavata* (Hymenoptera: Formicidae: Ponerinae). *Journal of the Kansas Entomological Society*, **53**, 35–55.

Young, O. P. (1984). Perching of Neotropical dung beetles on leaf surfaces: an example of behavioral thermoregulation. *Biotropica*, **16**, 324–327.

Yukawa, J. (2000). Synchronization of gallers with host plant phenology. *Population Ecology*, **42**, 105–113.

Yumoto, T. (1987). Pollination systems in a warm temperate evergreen broad-leaved forest on Yaku Island. *Ecological Research*, **2**, 133–145.

Yumoto, T., Inoue, T. & Hamid, A. A. (1996). Monitoring and inventoring system in canopy biology program in Sarawak, Malaysia. In *DIWPA Series* Vol. 1. *Biodiversity and the Dynamics of Ecosystems*, eds. I. M. Turner, C. H. Diong, S. S. L. Lim & P. K. L. Ng, pp. 203–215. Singapore: National University of Singapore.

Zar, J. H. (1974). *Biostatistical Analysis*. Englewood Cliffs, NJ: Prentice Hall.

Zar, J. H. (1994). *Biostatistical Analysis*, 3rd edn. Englewood Cliffs, NJ: Prentice Hall.

Zar, J. H. (1999). *Biostatistical Analysis*, 4th edn. Englewood Cliffs, NJ: Prentice Hall.

Zavortink, T. J. (1973). Mosquito studies (Diptera, Culicidae) XXIX. A review of the subgenus *Kerteszia* of *Anopheles*. *Contributions of the American Entomological Institute*, **9** (part 3), 1–54.

Zitani, N. M., Shaw, S. R. & Janzen, D. H. (1997). Description and biology of a new species of *Meteorus* Haliday (Hymenoptera: Braconidae, Meteorinae) from Costa Rica, parasitizing larvae of *Papilio* and *Parides* (Lepidoptera: Papilionidae). *Journal of Hymenoptera Research*, **6**, 178–185.

Zitani, N. M., Shaw, S. R. & Janzen, D. H. (1998). Systematics of Costa Rican *Meteorus* (Hymenoptera: Braconidae: Meteorinae) species lacking a dorsope. *Journal of Hymenoptera Research*, **7**, 182–208.

Zotz, G. (1995). How fast does an epiphyte grow? *Selbyana*, **16**, 150–154.

Zotz, G. (1998). Demography of the epiphytic orchid, *Dimerandra emarginata*. *Journal of Tropical Ecology*, **14**, 725–741.

Zotz, G. & Andrade, J.-L. (1998). Water relations of two co-occurring epiphytic bromeliads. *Journal of Plant Physiology*, **152**, 545–554.

Zotz, G., Dietz, G. & Bermejo, P. (1999). The epiphyte community of *Annona glabra* on Barro Colorado Island, Panama. *Journal of Biogeography*, **26**, 761–776.

Zotz, G. & Tyree, M. T. (1996). Water stress in the epiphytic orchid, *Dimerandra emarginata* (G. Meyer) Hoehne. *Oecologia*, **107**, 151–159.

Index

Abies religiosa, 161
Acacia collinsii, 369
Acacia mangium, 388
Acanthaceae, 160
Acari, 102–9, 271–81, 291–303
Achilixiidae, 46
Acradenia euodiiformis, 333
Acrididae, 237–55
Acronychia pubescens, 330
Adeloneivaia isara, 374, 378
Adeloneivaia jason, 373
adherence
  to substrate, 344
Agaonidae, 23, 91, 111, 122, 304
Agauria salicifolia, 93
Agave horrida, 161
aggregation, 23, 42, 146, 154, 198–212
Albizia gummifera, 93
algae, 271
allomones, 343
altitudinal gradients, 400
Amazon, 170–5, 237–55
Anacardiaceae, 317, 365, 376
Anacardium excelsium, 283
Anapidae, 96
Androlaelaps casalis, 299
Annona glabra, 176–85
Annona purpurea, 375
Annona spraguei, 283
Annonaceae, 317, 375, 378
annual events, 315–28
Anogeissus leiocarpa, 198–212
ant
  dominant, 341–7
  garden, 343
  mosaic, 341–7
  nest, 342
  see also Formicidae
Anthicidae, 192, 205, 209
Anthribidae, 46, 263
Apeiba membranacea, 367
Aphididae, 346
Aphloia theifomis, 93
Apidae, 33, 308, 310, 313, 360–8
Apionidae, 146, 150, 154
Apis dorsata, 127
Apis mellifera, 361, 366, 367
Apocynaceae, 51, 365
Apodidae, 88, 91
Aporusa lagenocarpa, 190–7
Araliaceae, 357
Araneae, 92–101
Araucaria cunninghamii, 330, 333
Araucariaceae, 330
Arctiidae, 83, 376
Arecaceae, 316
Argasidae, 299
Argyrodendron actinophyllum, 21, 68, 333, 339, 352
Arsenura armida, 374, 378
Arsenura batesii, 371
Ascidae, 298
Asclepiadaceae, 51
assemblage
  random, 190–7
  structure, 28–39, 198–212, 315–28
Asteraceae, 38, 160, 355
Astraptes fulgerator, 376
Astronium graveolens, 283
Atopomyrmex mocquerisi, 344, 345
Attelabidae, 46
Aucoumea klaineana, 345
Australia, 329–40, 344
Automeris io, 371, 378, 379
Automeris pallidior, 371
Automeris tridens, 374, 378
Automeris zugana, 369, 371, 373, 375, 378
Automeris zurobara, 373
Aves, 86–91

Baccharis dracunculifolia, 355
bacteria, 108
Bactrocera papayae, 50
bark, 244, 272, 277, 278, 280, 291, 292
beating, 94, 284, 306
behaviour, 17–27
  foraging, 67
  predatory, 344
Berlese funnel, 277
Bignoniaceae, 39, 348–59, 365
biodiversity templet, 338
biological control agent, 346
birds, see Aves
Blattaria, 126–34
Blattidae, 315, 319
body size, 45, 84, 146, 154, 273, 276
Bombacaceae, 378
Bombacopsis quinatum, 369
Borneo, 59–91, 126–34, 190–7, 344, 380–93
Brachychiton acerifolius, 333
branch clipping, 317
Brazil, 170–5, 237–55, 348–59
Brentidae, 46, 222
Bromeliaceae, 24, 159–69, 176–85
Brosimum alicastrum, 160
Brosimum utile, 317
Bruchinae, 149
Buprestidae, 25, 30, 220–36
Burkea africana, 198–212
Bursera cuneata, 161
Burseraceae, 110–22, 316
Busseola fusca, 29

Caeculidae, 296
Caio championi, 369
Callitrichia sellafrontis, 96
Calycophyllum candidissimum, 375
Cameroon, 342, 343, 345
Camponotus acvapimensis, 345
Camponotus brutus, 343
Camponotus femoratus, 343
Camponotus gigas, 393
canopy
  access, 7–16
  crane, 12, 214, 222, 256, 257, 283, 316, 317, 361
  definition of, 4
  opening, 320

canopy (*cont.*)
raft, 12, 103, 306
roots, 323
roughness, 348, 359
sledge, 12, 103, 306
surface, 4, 88, 357, 358
tomography, 86–91
tower, 11, 87, 91, 127
upper, 358, 360, 394–405
volume, 148
walkway, 11, 70, 87
cantharophily, 256
Carabidae, 25, 135, 138, 141–3, 150, 209, 398
Carabodidae, 108
carbon flux, 316
carrying capacity, 44
Casearia arborea, 316
Casearia corymbosa, 375, 376
Castanopsis acuminatissima, 352
Castilla elastica, 282–90, 396
caterpillar, 45, 84, 290, 352, 369–79, 401
Cecidomyiidae, 174, 175
Cecropia schreberiana, 316, 317
Cecropiaceae, 42, 49, 113, 316, 401
Cedrela odorata, 378
Celastraceae, 66
Cephalotes atratus, 218
Cerambycidae, 46, 133, 220–36, 256, 260, 263
Ceratophysella gibbosa, 166, 169
Ceratopogonidae, 24, 174, 310
Cercopidae, 286
Ceroplastes rubens, 315, 318
cerrado, 348–59
Chamaecrista desvauxii, 214
Chamaecrista linearifolia, 355
Chironomidae, 174
Chronosequence, 135–45
Chrysobalanaceae, 317
Chrysomelidae, 43, 44, 46, 49, 133, 148, 150, 152–154, 190–7, 205, 220–36, 256–65, 282, 286, 287, 308, 312, 313, 346, 350, 353, 388, 401
Chrysophyllum brenesii, 373
Cicadellidae, 45, 138, 141, 282, 286, 287, 312, 350
Cicadidae, 37
Citheronia lobesis, 374
citrus plantation, 345, 346
cladogenesis, 292, 300
Cleridae, 34, 192, 263
Coccidae, 343, 346
Coccinellidae, 209, 285, 392
Coccoidea, 315, 318–19
Coccus acutissimus, 315, 318
cocoa plantation, 342, 343, 345, 346
coevolution, 218, 377, 402
diffuse, 218
Cojoba rufescens, 283
Collembola, 173, 175, 271–81, 395, 396, 398
colonization, 136, 143
recolonization, 68, 196
Combretaceae, 198–212
compartmentalization, 237
competition, 23, 28–39, 276, 277, 360–8
complementarity, 94, 96
conditioning, 345
conservation, 39, 190–7, 404
Convolvulaceae, 160, 365
Copaxa moinieri, 369, 374, 377
Copaxa rufinans, 377
Copepoda, 104
Copiopteryx semiramis, 369, 373
Cordia alliodora, 283, 378
Corylophidae, 150, 153, 154, 204, 209, 211, 333
Corythophora alta, 170–5
Costa Rica, 369–79
Crematogaster depressa, 345
Crematogaster gabonensis, 346
Crematogaster limata parabiotica, 343
Crossopteryx febrifuga, 198–212
Culicidae, 18, 21, 24
Curculionidae, 33, 43, 44, 46, 146, 148, 150, 154, 190, 205, 220–36, 256, 260, 261, 263, 264, 286, 287, 290, 308, 313, 333, 388, 401
Cyatholipidae, 96
Cymbaeremaeidae, 108
Cynometra alexandri, 146–58
Cyrtoxipha gundlachi, 318

Dacryodes excelsa, 316
decomposition, 102
Dendroctonus frontalis, 34
Dendrolaelaps neodisetus, 296
density of insect populations, 352
Derbidae, 46, 312
Dermanyssidae, 300
Dermestidae, 263
desiccation, 32, 219, 274
deterministic patterns, 190–7
Deuterosminthurus maassius, 164, 166
diapause, 30, 32, 43
Dictyopharidae, 286
Dicyrtomidae, 169
Didymopanax vinosum, 357
diel activity, 36, 256, 261, 264, 304–28, 343
Digamasellidae, 296, 303
Dimerandra emarginata, 176–85
Dipterocarpaceae, 32, 33, 37, 38, 43, 87
Dirphia avia, 374, 375, 378
dispersal, 23, 24, 32, 46, 229, 241, 274, 275, 293, 297, 345, 367, 368
disturbance, 39, 113, 190–7, 315–28, 404
diversity
alpha, 69–85
beta, 20, 67, 69–85
local, 136, 144, 220–36
regional, 135, 143, 144
dominance, 190–212, 341–7, 366, 367
dormancy, 154
Doryphora sassafras, 333
Drosophilidae, 86
drought, 315–28
dry forest, 25, 159–69, 220–36, 369–79, 402

Eacles imperialis, 373, 379
earthworm middens, 273, 276
Ecclinusa guianensis, 170–5
ecological drift, 361
ecological fitting, 369, 377, 402
ecosystem
function, 316
structure, 315–28
ecotone, 397
Ecpantheria icasia, 376
Ecpantheria suffusa, 376
Ecuador, 237–55
El Niño, 37
Elaeis guineensis, 388
Elaeocarpaceae, 316
Elateridae, 209, 260, 261, 263, 312, 388
Encyrtidae, 26
endemicity, 247
Endomychidae, 285
enemy-free space, 23, 29, 50, 163, 211, 278, 290
Entandrophragma utile, 147
Enterolobium ciclocarpum, 283
Entomobryidae, 285
epiphyte, 42, 67, 103, 106, 159, 176–85, 237–55, 277, 329, 339, 341, 343, 399
Ericaceae, 38
Erythroxylaceae, 350
Eschweilera atropetiolata, 170–5
Eschweilera rodriguesiana, 170–5
Eschweilera romeu-cardosoi, 170–5

Eschweilera wachenheimii, 170–5
Eschweilera pseudodecolorans, 170–5
Eucalyptus deglupta, 386
Euglycyphagidae, 300
Eulophidae, 285
Eumastacidae, 237–55
Euphorbiaceae, 160, 190–7
Eupodidae, 296
Eupomatia laurina, 330
Eupomatiaceae, 330
Exostema mexicanum, 376

Fabaceae, 350, 373, 375, 378
Fagaceae, 373, 378
Fedrizziidae, 296
feeding trials, 284
filariosis, 24
fire, 30, 353, 358
Flacourtiaceae, 316, 375, 376
Flatidae, 286
flight, 86–91
flowers, 32, 33, 38, 87, 126–34, 179, 213–19, 256–65, 290, 339, 360–8, 402
 general flowering, 126–34
 mass flowering, 43, 126–34, 263
foliage
 density, 86–91
 structure, 86–91
forest
 edge, 243
 gap, 243, 315–28, 404
 inundation, 395
 management, 380–93
 monodominant, 348–59, 395, 403
 montane, 59–68, 92–101, 247, 402
 physiognomy, 17–27, 395
 plantation, 380–93
 regeneration, 190–7, 243, 290
 swamp, 146–58
Formicidae, 26, 45, 59–68, 93, 102, 107, 114, 121, 148, 152, 170, 173–5, 183, 185, 190, 195, 197, 198–219, 264, 278, 280, 281, 285, 289, 308, 341–7, 350, 357, 358, 381, 386, 392, 394, 397, 402–4
fractal dimension, 273
fragmentation, 275
Frankliniella gardeniae, 365
Frankliniella parvula, 360, 365, 367
French Guiana, 237–55, 343
fruiting, 179
fungi, 108, 274, 330, 381

Gabon, 102–9, 304–14, 345
gall-forming insects, 348–59
gene flow, 143, 144
geographic distribution, 249, 371
geological age, 135–45
Geometridae, 25, 82–4, 121, 286, 376, 391
global estimates of species richness, 220, 228, 405
Goupia glabra, 66
gregarism, 240
Gryllacrididae, 318
Gryllidae, 318
guano, 296
Guarea excelsa, 378
Guatteria dumetorum, 317
guava plantation, 345
Guayaquila xiphias, 357
guild, 70, 237, 247, 284, 304, 315, 388
 structure, 190–7
Gustavia superba, 360, 362, 365, 367
Guyana, 289

habitat
 availability of, 394
 colonization of, 109
 continuity of, 106, 400
 fragmentation, 381
 structure, 146, 154
Hahnidae, 96
Halictidae, 367
haplodiploidy, 295, 298
haploidy, 292
Hawaii, 135–45
Hechtia podantha, 161
Heliconia latispatha, 283
herbivory, 11, 22, 177, 213, 283, 290, 315–317, 319, 320, 323, 340, 348–59, 393
Hesperiidae, 376
Heterocheylidae, 291, 296
Hirtella racemosa, 283
homozygosity, 295
honeydew, 219, 343, 344, 346
host shift, 145
host specificity, 5, 14, 40–53, 146, 154, 196, 220–36, 240, 256, 262, 264, 329–40, 369–79, 394
humidity, 31, 280
hurricane, 315–28, 381
Hydroptilidae, 179, 182, 185
Hylesia dalina, 378
Hylesia hamata, 371
Hylesia lineata, 373, 375, 378
Hymenaea altissima, 378
Hymenaea courbaril, 369, 378
Ichneumonidae, 197
immigration, 271, 275, 278, 377
inbreeding, 291–303
indicator group, 381
Inga vera, 373
insecticide knockdown, 6, 8, 69, 93, 99, 101, 111, 135–45, 148, 159, 161, 169, 170–5, 190–212, 304, 314, 329–40, 341, 380–93
insecticide treatment, 347
International Biodiversity Observation Year, 109
intraindividual tree characteristics, 198–212
invasion, 361, 367
inventory, 100, 101
island biogeography, 135–45, 361
Isotomidae, 277, 285
Issidae, 286
Ivory Coast, 198–212

Jadera aeola, 30
Jadera obscura, 30
Janzen–Connell model, 25, 282, 289

kairomones, 23
Khaya anthotheca, 147
Klinckowstroemiidae, 296

Ladella stali, 315, 318
Laelapidae, 299
Lampyridae, 318
latex, 48
Latridiidae, 146, 149, 150, 153, 205
Lauraceae, 377
leaf
 damage, 282, 283, 290, 315–28
 flush, 30, 34, 40–53, 138, 153, 184, 288, 289
 nutrients, 323
 palatability, 42, 49
 production, 32, 142, 316
 turnover, 395, 400, 403
leaf-mining insects, 30, 353
Lecythidaceae, 39, 170, 360, 362, 365
Leguminosae, 87, 160, 198–212
Lepidocyrtus finensis, 169
Lepidocyrtus finus, 169
liana, 22, 42, 220–36, 237, 281, 339, 360, 401
Licania hypoleuca, 317
lichen, 67, 103, 271, 273, 274, 280, 340
Licneremaeidae, 108
Linyphiidae, 96, 100, 101
litter, 92–101, 244, 272, 277, 280
 accumulation of, 107
 fauna, 17–27

logging, 27, 39, 190–7, 404
clear-felling, 380–93
selective, 147, 154, 359, 380–93
log-normal distribution, 71, 360–8
log-series distribution, 71, 382
Lohmanniidae, 297
Lonchocarpus minimiflorus, 375
Luehea seemannii, 288
Lycaenidae, 23, 88, 376
Lygaeidae, 142
Lythraceae, 365

Malachiidae, 192
malaria, 24
Malaysia, 59–91, 126–35, 190–7, 380–93
Malpighiaceae, 214, 352
Malvaceae, 160, 378
mango plantation, 345
Manilkara bidentata, 316
Manilkara chicle, 369, 373
Manilkara zapota, 288, 317
mast-seeding, 32, 34
Matayba guianensis, 256–65
meeting points, 256, 265
Megachilidae, 367
Megisthanidae, 296
Melanophila acuminata, 30
Melastomataceae, 355, 356
Meliaceae, 378
Melochia umbellata, 190–7
Membracidae, 43, 282, 286, 312, 313, 343, 357
Metagynellidae, 293
metapopulation dynamics, 275, 367
Metrosideros polymorpha, 135–45
Mexico, 159–69
microclimate, 67, 108, 175, 184, 282, 305
microhabitat, 237–55, 271–81
specialization, 103
Micropholis guyanensis, 170–5
migration, 25, 256, 305, 313
Milankovitch cycles, 37
mimicry, 23
Miridae, 285, 315, 318, 346, 350
Mochlozetidae, 108
monophagy, 50, 221, 374, 378, 402
moonlight, 81
Moraceae, 51, 317
Mordellidae, 133, 256, 260, 263
moss, 67, 103
Mymaridae, 26
Myristicaceae, 317
Myrmicaria opaciventris, 344–6

nectar, 213, 219, 367
robbing, 219
nectaries, extrafloral, 214, 342, 344
Nematoda, 102, 104, 108
Nemonychidae, 51
Neolitsea australiensis, 336
Neopelma baccharidis, 355
NESS index, 72
nest of vertebrates, 291–303
neutral theory of biodiversity and biogeography, 360–8
niche, 241, 243, 360–8
overlap, 276
Nitidulidae, 133, 260, 263
nitrogen, 142
foliar, 135, 142
Noctuidae, 23, 29, 286, 376
nonequilibrium
conditions, 195
models, 403
Notodontidae, 286, 287, 370, 376
Nymphalidae, 23, 25, 83, 88, 91, 286, 376

Ocotea veraguensis, 369, 377
Oecophylla longinoda, 342–6
Oecophylla smaragdina, 345, 346
Olacaceae, 103
oligophagy, 221, 329
Ongokea gore, 102–9
Oonopidae, 96
Orchidaceae, 176–85, 213
Oribatulidae, 108
Oripodidae, 108
Othorene verana, 373, 374, 379
Ourapteryx fulvinervis, 383
outer canopy, *see* canopy surface
overstorey, 4, 17–27, 86–91, 237, 245, 313
oviposition, 84, 240, 242, 243, 245, 358, 369, 374, 375, 378, 379
Oxyethira circaverna, 179, 182
Oxyethira maya, 179, 183

Pachycondyla goeldii, 344
palm, 237, 246
Panama, 176–85, 220–36, 282–90, 315–28, 360–8
Pantanal Matogrossense, 348–59
Papilionidae, 23, 88
Paraponera clavata, 344
Paraserianthes falcataria, 386
Parasitidae, 296
Parastasia bimaculata, 127
parataxonomist, 13, 15, 52, 93, 114
Paratrechina longicornis, 345

Parhypochthonius aphidinus, 297
Parinari excelsa, 93
Paronellidae, 159, 164
particulate feeding, 298
Passalidae, 293, 295–8
path analysis, 144
Peltogyne gracilipes, 350
Pentatomidae, 286
periodicity, 36
Periphoba arcaei, 374, 375
Persea americana, 378
Peru, 237–55
pesticide, 274
Phalacridae, 192
Phasmoptera, 313
phenology, 43, 82
of host tree, 28–39, 176–85
of trees, 263
phenols, 22, 354–6, 371
pheromones, 276, 342, 344
Philodana johnstoni, 295
Philodanidae, 295
Pholcidae, 96, 101
phoresy, 281, 291, 293, 296–8
Phoridae, 110–22, 312
phosphorus, 142
photoperiod, 30, 32
phylogeny, 45, 52, 296, 299, 303
of host plant, 23, 28–39, 52, 329–40, 398
Phytotelmata, 24
Pieridae, 23, 88
Pinus caribaea, 386
Piper reticulatum, 283
plant
architecture, 288
defences, 48
protection, 346
plantation, 341–7
citrus, 345, 346
cocoa, 342, 343, 345
guava, 345
mango, 345
Pleistocene refuge hypothesis, 397
pollen, 33, 37, 213
pollination, 213, 264, 360–8
syndromes, 90
polyphagy, 221, 329, 338
Polyrhachis laboriosa, 343, 344, 346
Polyrhachis weissi, 343
Popillia japonica, 23
Pourouma bicolor, 25, 42, 49, 282, 283, 288–90, 400, 401
Pouteria campechiana, 373
Pouteria glomerata, 170–5

predation, 34, 35, 84, 133, 174, 355, 357
  pressure, 91
predator satiation, 34
Prestoea montana, 316
prey, 101
productivity, 135, 136, 142, 392
Proscopiidae, 237–55
Protopulvinaria pyriformis, 315, 318
Provespa anomala, 126–34
Provespa nocturna, 126–34
Pseudisotoma sensibilis, 159, 164–6
Pseudococcidae, 315, 318, 343, 346
pseudoreplication, 12, 14
Psocidae, 143
Psocoptera, 170, 173, 175, 206, 207, 211
Psylloidea, 26, 43, 145, 286, 312, 313, 392
Ptiliidae, 333
Puerto Rico, 315–28
Pygmephoridae, 296
Pyralidae, 71
Pyraloidea, 69–85, 376
Pyrgomorphidae, 237–55
Pyroglyphidae, 300

Quercus castanea, 161
Quercus crassifolia, 161
Quercus crassipes, 161
Quercus greegii, 161
Quercus laeta, 161
Quercus laurina, 161
Quercus mexicana, 161
Quercus oleoides, 373, 378
Quercus rugosa, 161
Quercus subcericea, 59–68

rainfall, 29, 30, 32, 152, 316, 398
Raphia farinifera, 147
rarefaction, 224, 259, 263, 264
rarity, 47, 72, 78, 100, 196, 222, 368
redistribution, 271–81
regional factors, 28–39
resource
  availability, 17–27, 30, 39
  base, 142
  patchiness, 40–53
  tracking, 28–39
  use, 40–53, 394–405
Rhapignathidae, 296
Rhobdopterus fulvipes, 286
Rhopalidae, 30, 286
Rinorea beniensis, 146–58
Riodinidae, 376
Rollinia membranacea, 378
Romaleidae, 237–55
rope system, 86–91
Rothschildia forbesi, 379
Rothschildia lebeau, 375, 379
Rothschildia triloba, 371
Rubiaceae, 32, 87, 160, 198, 375, 376

Sabah, 59–85, 190–7, 380–93
Salina banksi, 159, 161, 164, 166, 168, 169
Salticidae, 94
sampling
  completeness, 230
  limitations, 7–16
  protocol, 68
sap, 40, 43, 44, 133
Sapindaceae, 30
sapling, 22, 49, 282–90
Sapotaceae, 87, 170–5, 316, 373
Sarawak, 86–91, 126–34
saturation deficit, 281
Saturniidae, 369–79, 401, 402
savanna, 198–219, 348–59, 400, 402, 404
Scarabaeidae, 21, 23, 127, 133, 209, 263, 304, 318, 392, 405
Scelionidae, 26, 312, 313
Schausiella denhezorum, 377, 378
Schausiella santarosensis, 369, 377, 378
Scheloribatidae, 108
Schizolobium parahybum, 160
Schoettella distincta, 166
Sciaridae, 174
sclerophylly, 348–59, 400
Scolytinae, 21, 27, 34, 46, 59–68, 222, 295, 298, 310, 313, 394, 396
seasonal cues, 28–39
seasonal distribution, 394–405
seasonal drivers, 28–39
seasonal fluctuations, 112, 118, 146, 152, 154
seasonality, 28–39, 126–34, 146–85, 286, 315–28, 361
  supra-annual, 37, 38, 360
seed, 30, 32, 34, 48, 403
seedling, 282–90, 357
Seira bipunctata, 168
Seira dubia, 168
Seira purpurea, 168
senescence, 399
similarity, 256, 262, 263, 286
  faunal, 92, 96, 100, 146, 150, 152, 165, 277, 278, 282, 283, 305, 313, 329–40
single-rope technique, 11, 61, 103
singleton, 62, 64, 68, 96, 100, 136, 149, 193, 256, 262, 264, 279
Sloanea berteriana, 317
Smarididae, 297
Sminthurinus quadrimaculatus, 159, 164, 166
soil
  fauna, 17–27
  suspended, 102–9, 159–69, 177, 277, 280
Solanaceae, 365
Spathoglottis plicata, 213
specialization
  ecological, 393, 401
  effective, 220, 221, 223, 402
speciation, 250, 397
species
  accumulation, 94, 98, 223, 224, 229, 286
  core, 361
  keystone, 44
  loss, 5, 381, 391
  packing, 276
  radiation, 145, 296, 338, 349, 358
  richness, 281, 349
  transient, 100, 175, 212, 220, 338, 352
Spelaeorhynchidae, 299
Sphingidae, 83, 91, 370, 371, 376
Spondias mombin, 376
stability, 69
Staphylinidae, 148, 150, 152–4, 195, 285, 333
stem borer, 30
Sterculiaceae, 190–7, 378
Stictococcidae, 346
Stigmaeidae, 296
stochastic events, 136, 190–212, 265
stratification, 6, 15, 17–27, 43, 59–122, 126, 153, 237–55, 271–90, 304–14, 348–59, 376, 394–405
stress, hygrothermal, 305
structural heterogeneity, 176
succession, 109, 190–7, 346, 380
Surinam, 110–22
swarming, 174
sweeping, 94, 98
Swietenia macrophylla, 160, 378
Syssphinx mexicana, 369

Tabanidae, 29
Tabebuia aurea, 350, 353, 357
Tabebuia ochracea, 352, 353, 357
Tachigali guianensis, 256–65
Tachigalia guianensis, 256–65
tannins, 22, 354, 355
Tanzania, 92–101
Tapirira guianensis, 317
Tarsocheylidae, 291, 296
taxonomic distance, 329–40
Tectocepheus velatus, 102, 106, 108, 109
temperature, 30, 31

temporal distribution, 394–405
temporal variation, 28–39, 69–85
Tenebrionidae, 209, 259, 295, 298, 388
Tephritidae, 50
Terminalia amazonia, 160
Terminalia ivorensis, 342
territoriality, 342
territory, defence of, 343
Tetragastris altissima, 110–22
Tetragnathidae, 96
Tetramorium aculeatum, 344–346
Tetramorium africanum, 345
Tetraponera anthracina, 346
Tettigometridae, 343
Thanasimus dubius, 34
Thomisidae, 94
Thripidae, 33, 286, 287, 289, 290, 360, 365
Thysanoptera, 42, 44, 46, 102, 104, 108, 170, 173, 175, 206, 207, 312, 313, 360–8
Thysanura, 206, 207, 211
Tigridia acesta, 401
Tiliaceae, 378
Tillandsia fasciculata, 176–85
Tingidae, 282, 286
Titaea tamerlani, 374
Tortricidae, 45
Trachyuropodidae, 293
Translocation, 30
trap
  flight interception, 60, 178, 256, 258, 263, 306, 380–93
  light, 4, 7, 8, 69–85, 126–34, 228, 304, 371, 380–93
  Malaise, 304
  pitfall, 61
  sticky, 307
  yellow pan, 110–22, 178
tree
  architecture, 17–27, 107, 109, 198–212, 339
  crown, 59–68
  emergent, 4, 86, 89
  felling, 240
  hollows, 291–303
  trunk, 108, 109, 272–275
treetop bubble, 12, 306
trichomes, 356
Trichosurus vulpecula, 295
Triplogyniidae, 295
Tropiduchidae, 315, 318
Turbinoptidae, 300
Uganda, 146–58
understorey, 17–27, 69–85, 92–101, 330, 380, 388, 393
  definition of, 4

vegetation, ruderal, 243
Venezuela, 213–19, 256–65
venom, 344
Verbenaceae, 87, 190–7
vertical gradients, 17–27
Vinsonia stellata, 315
Vinsonia stellifera, 318
Virola nobilis, 317
Virola sebifera, 317
Vitex pinnata, 190–7
Vochysia ferruginea, 367
Vriesea sanguinolenta, 176–85

water, availability of, 146, 154
Wilkiea austroqueenslandica, 333
wind, 21, 32
wing polymorphism, 293
wood, 291–303
  biomass of, 42

Xenylla humicola, 166, 169